# Lecture Notes in Intelligent Transportation and Infrastructure

**Series Editor**

Janusz Kacprzyk, Systems Research Institute, Polish Academy of Sciences, Warsaw, Poland

The series "Lecture Notes in Intelligent Transportation and Infrastructure" (LNITI) publishes new developments and advances in the various areas of intelligent transportation and infrastructure. The intent is to cover the theory, applications, and perspectives on the state-of-the-art and future developments relevant to topics such as intelligent transportation systems, smart mobility, urban logistics, smart grids, critical infrastructure, smart architecture, smart citizens, intelligent governance, smart architecture and construction design, as well as green and sustainable urban structures. The series contains monographs, conference proceedings, edited volumes, lecture notes and textbooks. Of particular value to both the contributors and the readership are the short publication timeframe and the world-wide distribution, which enable wide and rapid dissemination of high-quality research output.

Yi Lu Murphey · Ilya Kolmanovsky · Paul Watta
Editors

# AI-enabled Technologies for Autonomous and Connected Vehicles

Springer

*Editors*
Yi Lu Murphey
College of Engineering and Computer Science
University of Michigan–Dearborn
Dearborn, MI, USA

Ilya Kolmanovsky
Aerospace Engineering
University of Michigan
Ann Arbor, MI, USA

Paul Watta
College of Engineering and Computer Science
University of Michigan–Dearborn
Dearborn, MI, USA

ISSN 2523-3440 ISSN 2523-3459 (electronic)
Lecture Notes in Intelligent Transportation and Infrastructure
ISBN 978-3-031-06779-2 ISBN 978-3-031-06780-8 (eBook)
https://doi.org/10.1007/978-3-031-06780-8



This Springer imprint is published by the registered company Springer Nature Switzerland AG
The registered company address is: Gewerbestrasse 11, 6330 Cham, Switzerland

# Preface

Throughout the past decade, we have witnessed tremendous advances in automobile and mobility technologies as a result of the rapid proliferation of electric vehicles, the introduction of automated/autonomous vehicles (self-driving cars), and increasing vehicle connectivity enabled by vehicle-to-vehicle (V2V) and vehicle-to-everything (V2X) wireless communications. A major technology enabling these revolutionary changes in the automotive industry is artificial intelligence (AI), which is being applied to a rapidly growing and diverse number of automotive and ground vehicle-related applications. While AI-enabled autonomous vehicles have received the most attention from the public, this represents only one of many uses of AI in the automotive industry. This book aims to provide a comprehensive exposure to the state-of-the-art technologies and ongoing research in intelligent vehicle systems (IVS), i.e., AI-enabled systems developed to address a broad range of important automotive engineering challenges in the twenty-first century. The book includes chapters addressing numerous topics in this domain, including automated driving systems, connected vehicles, driver state detection and monitoring, optimal vehicle control through advanced machine learning, advanced driver assistance systems (ADAS), mobility solutions, in-vehicle infotainment, vehicle energy optimization to improve energy efficiency and reduce harmful emissions, and others. The book is beneficial to graduate students and researchers in the areas of AI, control, optimization, data analytics, and automotive engineering. The book also serves as a resource to practitioners in the automotive industry and policymakers in related government agencies.

Dearborn, USA — Yi Lu Murphey
Ann Arbor, USA — Ilya Kolmanovsky
Dearborn, USA — Paul Watta

# Contents

# Advances, Opportunities and Challenges in AI-enabled Technologies for Autonomous and Connected Vehicles

**Yi Lu Murphey, Ilya Kolmanovsky, and Paul Watta**

**Abstract** Like many industries, the automotive industry is experiencing a revolutionary change driven by the convergence of connectivity, electrification and changing customer needs. This book explores the state-of-the-art of such transformative technologies, including artificial intelligence-based systems for the sensing and control of autonomous vehicles, vehicle-to-vehicle (V2V) and vehicle-to-infrastructure (V2I) communications, and Internet of Things (IOT) and cloud-based services relevant to the automotive industry. By integrating vehicle autonomy with connectivity, significant improvements in safety, performance, environmental impact, and comfort/convenience can be achieved. This chapter addresses these advanced technologies and the nexus among them, and gives a brief introduction of the chapters in the book.

## 1 Introduction

The automotive industry is experiencing a revolutionary change in technologies, products, and perceived values stemming from innovation in hardware, software, and electronics. The following four areas are on the verge of disruptive technology innovation, as they will change the way people behave daily and the way cities and towns are organized:

- Autonomous driving
- Shared mobility

Y. L. Murphey (✉) · P. Watta
University of Michigan—Dearborn, Dearborn, MI, USA
e-mail: yilu@umich.edu

P. Watta
e-mail: watta@umich.edu

I. Kolmanovsky
University of Michigan, Ann Arbor, MI, USA
e-mail: ilya@umich.edu

 
Y. L. Murphey et al. (eds.), *AI-enabled Technologies for Autonomous and Connected Vehicles*, Lecture Notes in Intelligent Transportation and Infrastructure,
https://doi.org/10.1007/978-3-031-06780-8_1

- Connectivity
- Electrification.

These four trends are mutually reinforcing and show signs of following the so-called Kurzweil's law of accelerating returns, characterized by simultaneous exponential growth and price reduction. The development of autonomous vehicles is also a significant enabler of future intelligent transportation system design. Autonomous vehicles provide the following distinctive benefits:

- **Increased safety**: Autonomous vehicles offer the promise of increased safety for human drivers, passengers, and pedestrians. An important technology in modern vehicles is Advanced Driver Assistance Systems (ADAS) that assists drivers in avoiding accidents. ADAS contributes not only to increased safety but also to improved traffic management.
- **Increased lane capacity**: Autonomous vehicles can operate at higher speeds and shorter intra-vehicle spacing distances leading to greater lane capacity/throughput. By affecting traffic patterns, they also are able to mitigate traffic congestion.
- **Real-time path planning and route optimization**: Vehicles connect with other nearby vehicles to exchange information so that efficient and safe paths can be planned and executed. In addition, vehicles can interact with traffic management infrastructure to incorporate real-time information, such as road conditions, traffic levels, and weather events. Such information can be used to optimize route selection to reduce travel time and maintain safety.
- **Reduced energy consumption**: By being able to plan and follow a route and a vehicle speed trajectory (eco-driving) and through intelligent thermal management, autonomous vehicles can reduce fuel/energy consumption as well as pollutant emissions. For instance, ARPA-E's program—*Next-Generation Energy Technologies for Connected and Automated On-Road Vehicles* (NEXTCAR) [1] has successfully demonstrated that the use of automated and connected vehicle technologies has the potential to reduce fuel/energy consumption by 20% or more. Notably, the reduction of energy consumption due to a high level of onboard computing/processing power in autonomous vehicles will be important in providing these energy efficiency benefits. At high levels of proliferation of autonomous driving technologies, vehicle collisions could become fully preventable; since vehicle mass is driven by crash safety requirements, this opens up a possibility of making vehicles lighter and further reducing fuel/energy consumption.
- **Shared vehicles**: In the last few years, a number of level-4 autonomous vehicles have been developed specifically for e-hailing services—that is, cars designed for high utilization, robustness, extended lifetime and mileage, and passenger comfort [2]. If the cost of these services can be made sufficiently low, people would not need to own a vehicle; rather, they can just dial a ride up when needed. As noted in [3], in the private ownership of vehicles model, the car sits idle up to 95% of the time, representing a large inefficiency. As this example illustrates, autonomous vehicles can enable significant improvements in mobility and transportation.

In addition to technological breakthroughs, the design of future vehicles will be influenced by changing consumer values and preferences, societal trends, and tightening government regulation, leading to changes of products. Based on recent reports, in 2030, the share of electrified vehicles could range from 40 to 50% of new-vehicle sales [4–6], up to 15% of new cars sold could be fully autonomous, and 10% of the new cars sold may likely be a shared vehicle [2].

The design and development of a safe and cost-effective autonomous vehicle is an enormous technological challenge, and AI technologies are being used to meet these challenges. The book [7] by Alex Davies gives a colorful account of the challenges faced by the teams participating in DARPAs autonomous vehicle challenges, and the quest to develop driverless cars by tech companies such as Google, Uber and automotive OEM companies such as Ford, GM and many others. Over the last decade, artificial intelligence has firmly established itself as one of the enabling technologies in the automotive engineering field. AI has been used in new and innovative ways for autonomous-vehicle testing and development, vehicle control, and energy optimization. Modern vehicles are composed of complex networks of subsystems capable of sensing, wide-area connectivity, inference, and actuation actions. Such a vehicle system typically consists of up to 70 electronic control units (ECUs) capturing 2500 signals from the chassis, powertrain, user interfaces, and safety networks [8]. Many experts consider deep learning to be the most significant technology of AI behind autonomous-driving. Deep learning, which loosely mimics human neuron activity, has been shown to offer state-of-the-art performance for a surprisingly wide range of problems, such as voice and speech recognition and search, object recognition and scene understanding/processing, motion detection and prediction, and data analysis, as well as natural language processing. In the context of autonomous vehicle design, deep learning has been used for processing images captured by the on-board cameras, vehicle and pedestrian detection and trajectory prediction, and path and route planning, all in real time. Working together, these functions help vehicles recognize pedestrian traffic, other vehicles on the road, and traffic signals throughout previously mapped routes. These and complementary advances in optimization and control theory support the rapidly growing field of vehicle autonomy and intelligence.

This book focuses on autonomous driving, connectivity, and mobility. Research of the state-of-the-art technologies in these three areas are presented. In particular, the chapters that follow discuss the need for various types of vehicle data and applications thereof, enabling technologies, challenges, and identified opportunities. Connectivity and autonomous technologies will increasingly allow the car to become a platform for drivers and passengers to use their time in transit to consume various forms of media and services or to dedicate the freed-up time to other personal activities.

# 2 Autonomous Vehicles: Current Technologies and Challenges

This section presents an overview of the state-of-the-art technologies used in autonomous vehicle design, as well as the near-term challenges that remain.

## 2.1 *Levels of Autonomy*

In the terminology of the Society of Automotive Engineers (SAE), the task of operating a vehicle on the road is called the dynamic driving task. The SAE has defined 6 levels of driving automation ranging from 0 (fully manual) to 5 (fully autonomous) [9]. These levels have been adopted by the U.S. Department of Transportation, the automotive industry, and the research community [10].

**Level 0: No Driving Automation**. Most vehicles on the road today are Level 0, which require the human driver to control the vehicle manually. Although the human completes the dynamic driving task, there may be systems in place to help the driver. An example would be an emergency braking system—since it technically does not drive the vehicle, it does not qualify as automation.

**Level 1: Driver Assistance**. This level is considered as the lowest level of automation. Vehicles at this level have some form of automated driver assistance functions. An example here would be a vehicle with adaptive cruise control, which automatically controls the vehicle and maintains a safe distance behind another vehicle. The human driver monitors other aspects of driving, such as steering and braking.

**Level 2: Partial Driving Automation**. Vehicles at Level 2 involve automatic controls for both steering and accelerating/decelerating when ADAS is activated. Here the automation falls short of self-driving because the human driver must be able to take control of the car at any time. Tesla Autopilot and Cadillac (General Motors) Super Cruise systems, and ARGO AI (a Ford startup company startup) all qualify as Level 2.

**Level 3: Conditional Driving Automation**. Vehicles at this level have features of conditional driving automation, which provide environmental detection and have the capabilities of making competent driving task decisions, such as passing a slow-moving vehicle. However, autonomous systems at this level still require human override, and the human driver must, at all times, be ready to resume control of the vehicle when the system is unable to carry out the driving task. In terms of the underlying technology required, the jump from Level 2 to Level 3 is substantial. However, from the driver's point of view, there is not much change at all since their focused attention is still required. Some authors [11] are critical of this level of autonomy, as the attention of human drivers naturally tends to drift when not in full control of the vehicle; that is, humans are not able to uphold their end of the responsibilities. Honda [12], Mercedes [13], and Audi [14] all have introduced level 3 vehicles in recent years.

**Level 4: High Driving Automation**. Vehicles at this level can operate in self-driving mode, and in most cases do not require human interaction. However, a human still has the option to override the automated system and take full control of the vehicle. One example where the driver is required to assume full control is in areas where autonomous driving is restricted by law. For example, a municipality may forbid autonomous control on high-speed urban roads, a concept known as *geofencing*. Most Level 4 vehicles currently are geared toward ridesharing. The French company Navya has built and sold Level 4 shuttles and cabs that run fully on electric power and can reach a top speed of 55 mph [15]. Navya vehicles have been designed for *last mile* travel, i.e., the final leg of a trip. This is typically between a transport hub and the final destination; for examples, between a train station and a rider's home or between an airport and an airport hotel. One service already up and running is between the Charles de Gaulle Airport and Réseau Express Régional train station in Paris, France [15].

Waymo, which shares a parent company with Google, has built self-driving taxi cars at Level 4. Waymo has 600 fully driverless cars operating with geofencing in the suburbs outside Phoenix, Arizona, and ordinary people (non-experts) can ride in these Waymo vehicles. In October 2021, Waymo received permission to launch its taxi service (with a human monitor behind the wheel) in a second city—San Francisco—where hills, weather, and traffic complicate the task [16]. The company plans to eventually launch in other cities and license its automated driver technology to car manufacturers.

We expect to soon see more vehicles with Level 4 capabilities. Volvo has announced that it will equip its XC90 model with an intelligent system called *Highway Pilot*, which is a technology capable of controlling the vehicle on highways without the need for driver monitoring. However, it will not work on all highways. The feature will be activated only when it is verified to be safe for designated highways and conditions [10]. GM's Cruise is progressing with its efforts to launch robotaxis and has already received permits to offer robotaxi rides to public passengers in California [17]. Argo AI, Lyft, and Ford Motor Company are working together to commercialize autonomous ride hailing at scale. Ford self-driving cars, with safety drivers, on the Lyft network, and with passenger rides have been operating in Miami, Florida since late 2021 and plan to expand to other areas. This initial deployment phase will lay the groundwork for scaling operations aiming to deploy at least 1000 autonomous vehicles on the Lyft network across multiple markets over the next five years [18]. The Canadian automotive supplier Magna has taken a different approach. They are working with Lyft to supply high-tech kits (MAX4) that turn conventional vehicles into self-driving cars with Level 4 capabilities driving in both urban and highway environments [19].

**Level 5: Full Driving Automation**. Vehicles at this level do not require human attention and the dynamic driving task is completely automated. Level 5 vehicles will not even need a steering wheel or acceleration/braking pedals. They will perform at such a high level of safety and efficiency that geofencing will not be needed. Ford, GM, Tesla, Google Inc., BMW AG, and many other car makers are all racing to develop fully driverless cars. Fully autonomous cars are undergoing testing in several pockets of the world, but none are yet available to the general public.

Fully automated cars and trucks that drive us, instead of us driving them, will become a reality, though there is debate about how soon this might happen. These self-driving vehicles ultimately will integrate onto U.S. roadways by progressing through these six levels of driver assistance technology advancements in the coming years [19, 20].

## 2.2 AI Technologies in Autonomous Vehicles

As illustrated in Fig. 1, a computational framework of autonomous vehicles (also see Chapter "Robust AI Driving Strategy for Autonomous Vehicles") can be broadly categorized into the following core technical components:

- Sensing
- Perception
- Motion control
- Path planning
- Operation.

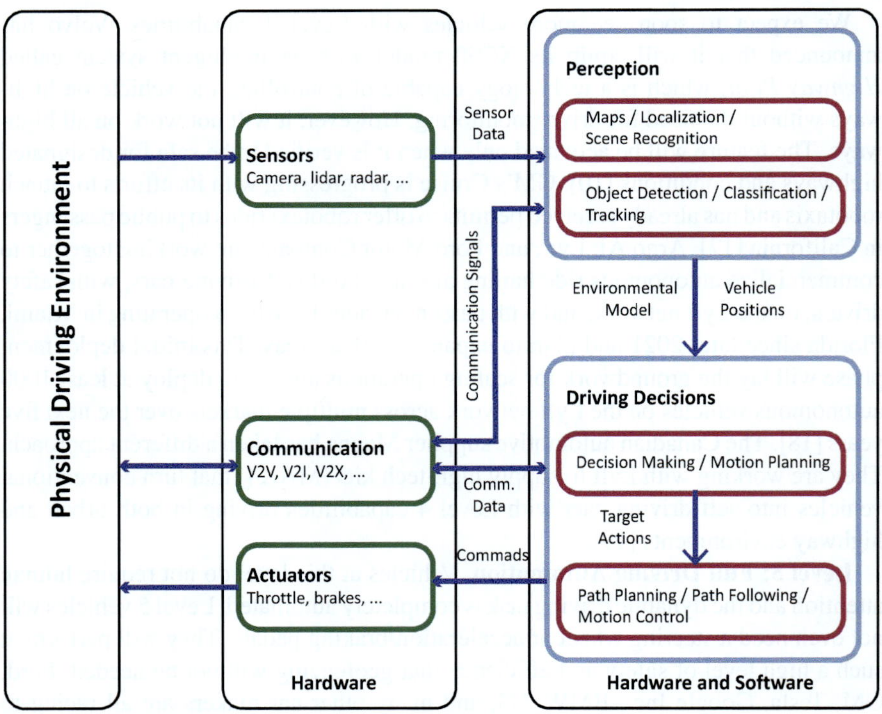


**Fig. 1** Overview of a typical autonomous vehicle system

In the first stage of processing, autonomous vehicles rely on sensors to provide information about the environment surrounding the vehicle. Commonly used sensors include cameras, radar, lidar, and ultrasound. The information from individual sensors can be leveraged and combined by a process called sensor fusion. The perception component analyzes the sensory data to achieve situational awareness; that is, an accurate understanding of both the state of the vehicle and its place and context in the surrounding environment. Here, localization techniques are used to generate a map of the environment, as well as to recognize scenes, objects, such as traffic vehicles, pedestrians and other road users. The planning component provides a navigation and motion plan for the vehicle, and involves a decision-making process that incorporates information about the route, the state of the AV, the current traffic, and the surrounding environment into an operation and motion plan, including control of vehicle speed, lane changing, path planning and following, merging, aborting the current maneuver, etc. These decisions are then carried out by vehicle-level control systems, which actuate vehicle propulsion, braking, and steering. AI is the major enabling technology used in the perception, planning, and decision-making processes in vehicle path and motion planning.

The following challenges and topics are addressed by the chapters in this book.

### 2.2.1 Sensors and Environmental Perception

Sensor systems, including radar, lidar, and cameras are rapidly evolving to meet the stringent demands of autonomous vehicles operations.

An example of a low-cost sensor system that does not rely on cameras or lidar is given in Chapter "Semi-autonomous Truck Platooning with a Lean Sensor Package". The authors propose a low-cost sensor package for the application to semi-autonomous truck platooning, whereby one lead truck is controlled by a human driver and one or more trucks follow the leader autonomously in convoy fashion. The proposed system is called *cooperative-adaptive cruise control* and utilizes radar, GPS, and dedicated short-range communication (DSRC)-based radios for vehicle-to-vehicle communication. The system was tested on a convoy of four class-8 semi-trucks on a 1.7-mile test track in Opelika, Alabama. After demonstrating that the convoy can be successfully maintained with the proposed sensor package, the authors study how robust the system is to various types of environmental factors, such as occlusions, antenna position, RF and GPS interference, and road curvature and road grade.

Environmental perception for autonomous vehicles generally focuses on the awareness and understanding of the driving environment based on various vehicle sensor data. Machine learning (ML) algorithms, especially deep neural networks (DNN), with applications to environmental perception have exploded in popularity during the last decade. In 2012, Krizhevsky et al. [21] proposed a DNN architecture, AlexNet, that achieved state-of-the-art accuracy higher than that of the second best architecture by more than 10% on the ImageNet dataset. It was a significant milestone in the field of deep learning and started a convolutional neural network

(CNN) renaissance. In the following years, while other neural networks similar to AlexNet offered improved accuracy, Simonyan et al. [22] proposed a different CNN configuration, known as Visual Geometry Group (VGG) Net, with $3 \times 3$ receptive fields throughout the network. They demonstrated that state-of-the-art performance on the ImageNet data could be achieved using the conventional ConvNet architecture with substantially increased network depth. That same year, Christian Szegedy et al. [23] presented GoogLeNet, which introduced a new level of organization in the form of inception. This book contains three chapters that discuss deep learning technologies used in environmental perception in autonomous vehicles.

The Chapter "Environmental Perception for Intelligent Vehicles" presents a detailed overview of core technologies used in the perception system in an autonomous vehicle. The chapter begins with an introduction to state-of-the-art sensor technologies used to capture multimodal environmental information, such as cameras, lidars, and radars. The chapter contains a thorough analysis of the core techniques used for interpreting the data obtained from these sensors, including sensor data restoration and denoising, semantic segmentation of vehicle environmental scenes, 2D and 3D object detection and tracking, vehicle localization and mapping, and multi-sensor fusion. The chapter also reviews deep learning algorithms that are effective for both qualitative and quantitative analyses of the sensor data.

Object detection and classification are core technologies in a vehicle environmental perception system because they can be used to provide precise range and size information of objects. The Chapter "3D Object Detection for Autonomous Driving" provides a broad overview of the recent advances in 3D object detection using 3D point cloud-based and camera-based information. In particular, the authors present a deep learning algorithm they developed for multi-class 3D object detection using the binocular image as input and a comparative study of its performance versus other well-known 3D object detection algorithms.

The recent development of deep neural learning achieved a remarkable breakthrough in object classification and detection with applications in advanced vehicle systems. Deep learning has the capability of learning features automatically from data using general-purpose learning procedures. However, because deep neural networks require large amounts of data to train the parameters in the network, it is challenging to develop object classification or detection systems with a relatively small dataset. Transfer learning is an important machine learning technique that learns representative low-level features in the lower layers of the model from large datasets that share similar object features. Those learned features can be transferred to a new system with a different application domain with a smaller dataset. Research has shown that a good transfer learning algorithm can offer an efficient training process and achieve improved system performance. The Chapter "Comparative Study on Transfer Learning for Object Classification and Detection" presents a comprehensive review of the state-of-art CNN models and extensive discussion on their architectures and learning algorithms, as well as a comparative study of transfer learning using six state-of-the-art CNN models.

The Chapter "Future Technology and Research Trends in Automotive Sensing" presents a comprehensive review of the progress that has been made in autonomous vehicles over the last two decades and articulates the importance of sensing technology in enabling intelligent autonomous vehicles. The focus is on radar- and lidar-based technologies, especially those that go beyond 2D and mechanical scanning. The importance of AI in improving sensor performance at marginal added cost is emphasized, and trends in optical computing, with their promise of substantial reduction of energy consumption while enhancing edge computing, are highlighted.

### 2.2.2 Learning to Drive Autonomous Vehicles as a Human Expert Driver

The interaction between an autonomous vehicle and the environment can be modeled as a stochastic Markov decision process (MDP), where an expert human driver is used as the target to be learned. A number of demonstrations from an expert driver can be collected, and reinforcement learning can be used to learn optimal driving strategies based on the collected training data. The unknown reward function of the expert driver can be approximated using a deep neural-network (DNN) [24]. The Chapter "Robust AI Driving Strategy for Automated Vehicles" provides an overview of reinforcement learning technologies used in decision making and control processes, and their own research in developing robust reinforcement learning solutions to support the design of real-world driving strategies for autonomous vehicles. The focus is on autonomous highway driving and the integration of reinforcement learning, vehicle motion control, and the control barrier function, leading to a robust AI driving strategy that can learn and adapt safely.

The Chapter "Artificially Intelligent Active Safety Systems" provides a comprehensive review of the state-of-the-art opportunities and challenges for applications of AI technologies to active safety systems in vehicles. An example of reinforcement learning applied to automatic emergency braking is elaborated and demonstrates that reinforcement learning agents can learn policies that are effective in avoiding collisions. This chapter concludes that AI technologies will play a critical role in the future of automotive safety systems.

### 2.2.3 Motion Planning and Control in AVs

In order to fully deploy highly automated driving technologies, vehicles need to be able not only to reliably sense their surrounding environment, but also to safely interact with it. Several chapters in this book address this issue.

The Chapter "Model Predictive Control for Safe Autonomous Driving Applications" is dedicated to model predictive control (MPC) and its application to autonomous vehicles. MPC has recently been introduced into automotive mass production [25–27], and the authors of this Chapter address issues that are critical in successfully employing MPC in such applications; in particular, for the issues of closed-loop stability in the case of user-provided arbitrary and possibly infeasible reference commands; and provide approaches to satisfying collision-avoidance requirements in the presence of pop-up or moving obstacles, such as human-driven vehicles, cyclists, pedestrians, animals, etc.

The Chapter "Energy-Efficient Autonomous Driving Using Cognitive Driver Behavioral Models and Reinforcement Learning" develops an eco-driving approach that accounts for the interactive nature of vehicles and human behavior in traffic. The contribution of this chapter is in connecting eco-driving (which historically has been focused primarily on longitudinal driving) with the contemporary advances in autonomous driving that enable autonomous intelligent vehicles to operate safely in traffic consisting of both automated and human-operated vehicles. The proposed approach is based on exploiting and fusing ideas in hierarchical game theory, reinforcement learning, and electric vehicle powertrain modeling and control. The results suggest significant opportunities for energy efficiency improvements with such approaches.

The Chapter "Self-learning Decision and Control for Highly Automated Vehicles" presents a comprehensive analysis of reinforcement learning (RL) and self-learning more broadly as a principled framework to generate decision and control policies for autonomous driving through interaction with the environment. RL can be successfully employed in autonomous driving system design if critical issues are addressed, including scalability, performance, interpretability, mixed modeling (model-free and model-based) and emergency handling. Many examples and illustrations are given, and opportunities and challenges in the application of learning to autonomous driving are discussed.

### 2.2.4 ADAS Systems

Through the last three decades, ADAS systems have been developed in the automotive industry to deliver improved safety and automated driving systems (ADS) that one day may handle the entire dynamic driving task. Many of the basic ADAS-system building blocks, such as automatic cruise control, automatic emergency braking, and lane-departure warning, are directly applicable to driverless cars. Some auto companies take the approach of incremental development toward full autonomy, where these ADAS systems are used as building blocks for a more capable system. A central control unit has the responsibility for managing the subsystems and ultimately driving the vehicle. The gradual introduction of autonomous driving will come about through the tempered deployment of self-driving capabilities. Many companies take the approach of deploying increasing levels of capabilities in progressive stages, starting from driver assistance to eventually fully autonomous, as the markets warm up to autonomous capabilities, the price points drop, and the technologies mature [27].

Modern ADAS systems follow complex rules governing when and where a given ADAS feature is activated. In cases where the path planning stage is found to have low confidence, for example from noisy sensor data or when the vehicle is operating in a region known to be problematic for the given ADAS feature, the feature will disengage and revert back to expecting human control of the vehicle. In the course of a drive, then, there is continuous human–computer interaction between the driver and the ADAS, as there may be several times when the ADAS feature is available and several other times when it is not. Like all areas of automotive design, ADAS designers are keenly interested in assessing the customer experience and satisfaction with using the technology. In the past, the typical approach to assessing customer experience was by surveying the customer. However, utilizing various vehicle and map data, this process can be automated, yielding a more accurate assessment of the customer experience. The Chapter "MAGMA: Mobility Analytics Generated from Metrics on ADAS" provides an analysis of how to automatically measure customer experience with various ADAS features. The chapter introduced several quality metrics that characterize and quantify customer experience, using data collected from both real trips and simulated trips. These measures can be used to evaluate ADAS system design and help optimize the experience of the customer.

The Chapter "Driver Assistance Systems and Safety—Assessment and Challenges" addresses challenges and opportunities in the testing of automated vehicles. As argued in this chapter, on-road testing alone is infeasible for validation of automated driving systems, due to the excessive time and effort required, and needs to be complemented by virtual testing, which is performed by simulation. However, virtual testing presents its own set of complexity and scalability challenges. To address these challenges, methods of choosing scenarios for testing and approaches to modeling to represent these scenarios are described. In particular, the scenarios for virtual testing can be generated or learned from measurements of the real-world environment. For the latter, the chapter provides an outlook on the application of importance sampling- and design of experiments-based approaches. Methods to represent the surrounding traffic and to determine the risk functions are also described as needed for the design and validation of safety-oriented systems.

Although the expectation is that technology will advance to a state where autonomous vehicles will be readily available to drive us safely and efficiently to whatever destination we choose, that eventuality will not happen in the near future, and even when such technology is available, the transition to level 5 autonomy will not happen overnight. In the meantime, we have to deal with human drivers in the loop, subject to making human types of driving mistakes.

One approach to ADAS development involves monitoring the driver as they drive the vehicle in order to understand and assess the driver's behavior. Real-time assessment and understanding of driver behavior is a challenging task as there are multiple factors that can influence the driver, such as inattention, fatigue, psychological state, ergonomic posture, traffic and weather conditions, and the behavior of the vehicle itself. Under certain conditions, each of these factors, or a combination thereof, can affect the driver to such an extent that the result is a traffic accident, or worse, a fatality. The Chapter "Human Factors Influencing Driver Behavior and Advances in Monitoring Methods" provides a comprehensive analysis of the factors affecting driver behavior and performance, as well as a survey of effective tools and methods for monitoring the driver in real time. The chapter also presents an unsupervised deep learning neural network for classifying driving style (aggressive, moderate, passive) using vehicle parameters as input.

## 3 Connectivity and Mobility

In order to coordinate the various systems involved in an autonomous vehicle, information must be shared among the subsystems, especially between the subsystems and the overall control unit. One way to share the data is using network communication. Rapid and consistent connectivity between autonomous vehicles and outside sources such as the cloud infrastructure ensures that signals get to and from the vehicles more quickly. The emergence of 5G wireless technology, which promises high-speed connections and data downloads, is expected to improve connectivity to these vehicles. V2X is an emerging vehicle communication technology that encompasses Vehicle-to-Vehicle (V2V) connectivity, Vehicle-to-Infrastructure (V2I), Vehicle-to-Pedestrians (V2P) and Vehicle-to-Network (V2N). The information can be used to improve fuel economy and prevent collisions. V2X is a very effective way to provide drivers or AVs with information about on-road hazards which they otherwise would be unable to see, allows for data exchange with the surrounding infrastructure to operate within the bounds of speed limits, traffic lights, and signage. V2X communication permits safe operation within traffic situations and is effective in preventing collisions or even near misses. By warning about vehicles or pedestrians, which are out of sight (for example obstructed by other vehicles, or just around the corner), V2X can considerably help drivers or vehicles avoid accidents.

When AVs are equipped with communication technologies, they are referred to as connected and automated vehicles, or CAVs, which are important enabling technologies in mobility related research. In order to ensure that CAVs operate safely at all times, advanced security technologies are needed to prevent malicious actors from eavesdropping or corrupting (hacking) the electronic components or networks used in the self-driving vehicle.

### *3.1 Mobility Research*

CAVs can be used to build mobility systems that solve common problems associated with urban transportation, such as congestion and traffic jams, by providing more accessible, safe, and efficient transportation. CAVs are mobile and situationally aware and can adapt to and communicate with their environment. CAVs will transform today's urban transportation system and revolutionize mobility [28].

The Chapter "Towards Learning-based Control of Connected and Automated Vehicles" presents a study on the challenges and perspectives of modeling and optimization-based control techniques for the safe coordination of multiple connected agents in various traffic scenarios, such as intersections, lane-changing and lane-merging maneuvers. The chapter is predominantly devoted to optimization-based control schemes, particularly to MPC, which explicitly accommodates constraints and exploits future trajectories of connected agents. The Chapter describes a general design of a controller that was developed to serve multiple use cases simultaneously with machine learning-based techniques to handle model uncertainties and mixed-traffic scenarios.

The Chapter "Virtual Rings on Highways: Traffic Control by Connected Automated Vehicles" is dedicated to traffic-flow dynamics and control in the case of CAVs. The benefits of vehicle connectivity in avoiding traffic congestion/jams, ensuring traffic smoothness, and reducing travel time and energy consumption are highlighted through modeling, theoretical derivations of traffic-flow control schemes, stability analysis, and numerical simulations. Dynamics and control of traffic flow involving rings and virtual rings are addressed.

Indeed, most research in CAV has principally been in engineering areas, with contributions from computer science, AI, biology, neuroscience, and psychology. However, it is also important to study the social and economic impacts of emerging mobility technologies [29]. In Chapter "Socioeconomic Impact of Emerging Mobility Markets and Implementation Strategies", the authors propose a mobility market that models the economic interactions of travelers in a smart city network with roads and public transit infrastructure. They show that the proposed mobility market is both incentive compatible and individually rational, the two properties that ensure that all selfish travelers are truthful in their communication with the social planner and voluntarily participate in the mobility market. The authors also show that the proposed market is economically sustainable, i.e., it generates revenue from each traveler and ensures that the operating costs of each mobility service are covered. It is through the appropriate design of monetary incentives that they successfully incentivize all travelers to truthfully report their travel preferences and voluntarily participate in the market. Thus, it is guaranteed a socially efficient mobility solution. The proposed mobility market also provides an incentive to central authorities to implement it, since the market ensures that there are minimum acceptable payments to cover the operating costs of the mobility services.

## *3.2 Prediction Model for V2X Communications*

The prospect of V2X communication will involve transmitting and receiving messages among many vehicles and many infrastructure devices, and all at the same time. At the hardware level, V2X communication systems require antennas and protocols that maximize bandwidth (how much and how fast information is transmitted) and maximize range (distance over which vehicles can reliably transmit and receive data), while simultaneously minimizing transmitted power and interference. For the past several years, DSRC (dedicated short-range communication) has been the only V2X technology available. Cellular vehicle-to-everything (C-V2X) is a more recent technology that has the same purpose of direct communication link between vehicles as DSRC. Initial tests of C-V2X show that it may have 20–30% more range than DSRC as well as significant improvement in performance with obstructions [30]. The Chapter "A Real-Time Seq2Seq Beamforming Prediction Model for C-V2X Links" presents a concise introduction of C-V2X technology, and discusses the challenges in C-V2X, in particular, the critically needed capability of long-haul communication without sacrificing congestion factor. To address this problem, the authors propose a deep learning-based real-time sequence-to-sequence (Seq2Seq) beamforming prediction model to forecast optimum beams within each beacon interval (BI).

### *3.3 Big Data Research in Transportation*

Advances in sensor and computing technologies used in modern vehicles and road networks have enabled an explosion of big data sources in transportation and mobility research. These new data sources have opened up whole new areas of research in mobility and enabled new insights. The recent developments in machine learning and artificial intelligence technologies have made prediction models substantially more accurate, robust, and flexible. The combination of the two has a great potential to enable researchers to find solutions to many challenging problems related to autonomous and connected vehicles, transportation, and mobility. The Chapter "Big Data in Road Transport and Mobility Research" introduces the reader to a vast scope of useful big data sources for autonomous vehicles and transportation, as well as to the key technologies of ML/AI effective for learning from big data for transportation research. The chapter presents rich sources of data collected through in-vehicle sensors and road infrastructure-based data acquisition devices and describes the contents of these datasets. In order to familiarize the reader with the applications of these cutting-edge datasets, the chapter includes detailed discussions of the ML/AI methods used in analyzing big data, and examples of research topics focused on the applications of these data sets in transportation research.

### *3.4 Automotive Cybersecurity*

In order for AVs to share information and operate cooperatively and efficiently, they are typically networked via various in-vehicle networks (IVNs). These connected autonomous vehicles can be viewed as a cyber-physical system that contains a large number of minicomputers called electronic control units (ECUs) networked through the IVNs. These IVNs are used to connect safety–critical components of the vehicle, including brakes, airbags, engine control, active safety devices, the electronic stability program and adaptive cruise control. Just like the internet, malicious actors can exploit vulnerabilities in these IVNs. Driving is a high-stakes real-time experience, and any potential data corruption and/or component malfunction pose a serious security threat. In order to ensure CAVs operate safely at all times, advanced security technologies are needed to prevent malicious actors from eavesdropping or corrupting (hacking) the electronic components of CAVs through IVNs. The Chapter "Machine Learning for Automotive Cybersecurity: Challenges, Opportunities and Future Directions" first gives an overview of the CAVs, an in-depth discussion of the types of cyberattacks that are possible in both in-vehicle networks and V2V networks, and an extensive discussion and security analysis of automotive networking protocols. Then the authors give a comprehensive presentation of a machine learning (ML) framework designed to defend against existing and emerging cyberattacks on IVNs. Significant challenges of using machine learning in IVN research are identified, and future directions of using the proposed machine learning approach are outlined in order to protect the next generation of vehicles from cyberattacks.

## References


1. Green Car Congress, https://www.greencarcongress.com/2021/03/20210312-nextcarii.html. Last visited 20 Feb 2022
2. Gao P, Kaas H-W, Mohr D, Wee D (2016) Technology-driven trends will revolutionize how industry players respond to changing consumer behavior, develop partnerships, and drive transformational change. Report, McKinsey & Company
3. Eastwood B, Why we aren't at "peak car" yet—and what can reverse the trend. https://mitsloan.mit.edu/ideas-made-to-matter/why-we-arent-peak-car-yet-and-what-can-reverse-trend. Last visited 20 Feb 2022
4. Marr B, The 5 biggest connected and autonomous vehicle trends in 2022. https://www.forbes.com/sites/bernardmarr/2021/12/20/the-5-biggest-connected-and-autonomous-vehicle-trends-in-2022/?sh=154dae71525f. Last visited 21 Feb 2022
5. Shepardson D, U.S. automakers to say they aspire to up to 50% of EV sales by 2030—sources. https://www.reuters.com/business/autos-transportation/us-automakers-say-they-aspire-up-50-ev-sales-by-2030-sources-2021-08-04/. Last visited on 18 Feb 2022
6. Global Electric Vehicle Market, Market Research Future, MRFR/AM/1261-CR, Oct 2020
7. Davies A (2021) Driven: the race to create the autonomous car. Simon & Schuster
8. Siegel JE, Erb DC, Sarma SE (2018) A survey of the connected vehicle landscape—architectures, enabling technologies, applications, and development areas. IEEE Trans Intell Transp Syst 19(8):2391–2406. https://doi.org/10.1109/TITS.2017.2749459
9. https://www.sae.org/standards/content/j3016_201806/. Last visited 19 Feb 2022
10. Vijayenthiran V, Volvo Highway Pilot promises unsupervised driving on highways. https://www.motorauthority.com/news/1117344_volvo-highway-pilot-promises-unsupervised-driving-on-highways. Last visited 21 Feb 2022
11. Torchinsky J (2019) Robot, take the wheel: the road to autonomous cars and the lost art of driving. Apollo Publishers
12. Davies C (2021) Honda very carefully sets loose its level 3 autonomous car, 4 March 2021. https://www.slashgear.com/honda-very-carefully-sets-loose-its-level-3-autonomous-car-04662291. Last visited 21 Feb 2022
13. Reyes A (2021) Mercedes-Benz wins world's first approval for level 3 autonomous cars: what's that mean? 13 Dec 2021. https://www.slashgear.com/mercedes-benz-wins-worlds-first-approval-for-level-3-autonomous-cars-whats-that-mean-13702207. Last visited 21 Feb 2022

14. Edelstein S (2020) Audi gives up on Level 3 autonomous driver-assist system in A8, 28 April 2020. https://www.motorauthority.com/news/1127984_audi-gives-up-on-level-3-autonomous-driver-assist-system-in-a8. Last visited on 17 Dec 2021
15. Vijayenthiran V (2018) Navya already sells fully self-driving cars, including in US, 17 Sept 2018. https://www.motorauthority.com/news/1118809_navya-already-sells-fully-self-driving-cars-including-in-us. Visited on 17 Dec 2021
16. Crowe S (2021) Cruise, Waymo OK'd for public robotaxi rides in California. https://www.therobotreport.com/cruise-waymo-dmv-approval-robotaxi-rides-california/. Last visited 11 Oct 2021
17. Ramey J (2022) Here's what a cruise Robotaxi ride looks like, 8 Nov 2021. https://www.autoweek.com/news/technology/a38189132/gm-cruise-autonomous-robotaxi-san-francisco/. Last visited 21 Feb 2022
18. Ford Motor Company, Argo AI and Ford to launch self-driving vehicles on Lyft Network by end of 2021. https://media.ford.com/content/fordmedia/fna/us/en/news/2021/07/21/argo-ai-ford-lyft-network.html. Last visited 21 Feb 2022
19. Magna (2022) Magna's formula for winning the self-driving car race. https://www.magna.com/innovation/driven-people-driving-change/article/max4-magna-s-formula-for-winning-the-self-driving-car-race. Last visited 21 Feb 2022
20. The Road to Full Automation, https://www.nhtsa.gov/technology-innovation/automated-vehicles-safety. Last visited 21 Feb 2022
21. Krizhevsky A, Sutskever I, Hinton G (2012) ImageNet classification with deep convolutional neural networks. In: Proceedings of the 25th international conference on neural information processing systems, vol 1, pp 1097–1105
22. Simonyan K, Zisserman A (2015) A very deep Convolutional Network for large-scale image recognition. In: Proceedings of the ICLR. https://arxiv.org/abs/1409.1556
23. Szegedy C, Liu W, Jia Y, Sermanet P, Reed S, Anguelov D, Erhan D, Vanhoucke V, Rabinovich A (2014) Going deeper with convolutions. In: Proceedings of the IEEE conference on computer vision and pattern recognition, pp 1–9
24. You C, Lu J, Filev D, Tsiotras P (2019) Advanced planning for autonomous vehicles using reinforcement learning and deep inverse reinforcement learning. Robot Auton Syst 114:1–18
25. Hrovat D, Di Cairano S, Tseng E, Kolmanovsky IV (2012) The development of model predictive control in automotive industry: a survey. In: Proceedings of 2012 IEEE international conference on control applications. IEEE, pp 295–302
26. Bemporad A, Bernardini D, Long R, Verdejo J (2018) Model predictive control of turbocharged gasoline engines for mass production. In: Proceedings of the SAE Paper 2018-01-0875
27. Garrett motion launches predictive control software with Hyundai Motor Company, https://investors.garrettmotion.com/news-releases/news-release-details/garrett-motion-launches-predictive-control-software-hyundai. Last visited 21 Feb 2022
28. van Schijndel-de Nooij M, Krosse B, van den Broek T, Maas S, van Nunen E, Zwijnenberg H (2010) Definition of necessary vehicle and infrastructure systems for Automated Driving. Study report, SMART 2010/0064, European Commission
29. Bissell D, Birtchnell T, Elliott A, Hsu EL (2020) Autonomous automobilities: the social impacts of driverless vehicles. Curr Sociol 68(1):116–134
30. Gettman, DSRC and C-V2X: similarities, differences, and the future of connected vehicle. https://www.kimley-horn.com/dsrc-cv2x-comparison-future-connected-vehicles/. Last visited 19 Feb 2022

# Sensors and Perception

# Semi-autonomous Truck Platooning with a Lean Sensor Package

**Sridhar Lakshmanan, Cristian Adam, Timothy Kleinow, Paul Richardson, Jacob Ward, Evan Stegner, David Bevly, and Mark Hoffman**

**Abstract** This chapter describes one method of approaching fuel-efficient truck platooning using a system called Cooperative Adaptive Cruise Control (CACC). The principal innovation in the system is its lean sensor package, including factory-ready standard ACC system utilizing a dual-beam radar, precision Global Positioning System (GPS), and a Vehicle-to-Vehicle (V2V) communication system. In other words, no imaging sensors such as camera or lidar, and no associated high-performance computing hardware such as Graphics Processing Units (GPU). Extensive test track and public road testing on class-8 semi-trucks, including edge case testing, reveals the efficacy and robustness of this system despite its leanness. Quantitative results are included in this chapter that trace cause and effect through the CACC system.

## 1 Overview

Semi-trucks, specifically class-8 (heavy duty, commercial) trucks, have recently become a platform of interest for autonomy systems. Platooning involves multiple trucks following each other in close proximity, with only the lead truck being manually driven and the rest being controlled autonomously. This approach to semi-truck autonomy is easily integrated on existing platforms, reduces delivery times, and reduces greenhouse gas emissions via energy efficiency benefits, with the associated reduction in fuel costs. Level 1 SAE fuel studies were performed on class-8 trucks operating with the Auburn Cooperative Adaptive Cruise Control (CACC) system, and fuel savings up to 10–12% were seen. Enabling platooning autonomy required the use of radar, global positioning systems (GPS), and wireless vehicle-to-vehicle (V2V) communication. Poor measurements and state estimates can lead to incorrect

S. Lakshmanan (✉) · C. Adam · T. Kleinow · P. Richardson
University of Michigan-Dearborn, Dearborn, MI, USA
e-mail: lakshman@umich.edu

J. Ward · E. Stegner · D. Bevly · M. Hoffman
Auburn University, Auburn, AL, USA

 
Y. L. Murphey et al. (eds.), *AI-enabled Technologies for Autonomous and Connected Vehicles*, Lecture Notes in Intelligent Transportation and Infrastructure,
https://doi.org/10.1007/978-3-031-06780-8_2

or missing positioning data, which can lead to unnecessary maneuvers and finally wasted fuel. This is especially an issue if deceleration (braking) is applied in response to a bad measurement. In this study, a faulty radar was shown to cause a greater than 5% increase in fuel consumption. The mechanism of this fuel consumption increase is investigated and applied to other types of sensor failures to indicate their potential effects on fuel economy. This analysis indicates that poor GPS signals over short time can be largely filtered out, with no real gain or loss of fuel economy. V2V communications were intentionally limited by causing interference, which resulted in dropped communication packets over a small physical area, but this did not lead to an appreciable impact on fuel economy.

This chapter is organized as follows. We begin with a review, including some rationale for the implementation of a CACC system, a broad overview of the technology, and a collection of relevant publications in the area of CACC. Next, the Auburn platooning system is introduced, including descriptions of the major sensor systems and their combined functionality in the platooning system. Subsequently, the main parameters for the test vehicle fleet and test track are provided. The impacts of sub-optimal sensor performance on the vehicle control, platoon dynamics, and fuel economy are then presented in the results section. Finally, the main takeaways are compiled in the conclusion.

## 2 Literature Survey

Class-8 vehicles typically travel up to six times as far as the average passenger car every year [1]. Furthermore, these class-8 vehicles account for approximately 71% by value of all freight shipped domestically in the United States [2]. Due to the fact that class-8 trucks account for such a large portion of the freight shipped domestically, small improvements in the efficiency of these vehicles result in substantial monetary savings for freight companies. Energy efficiency of these trucks is even more impactful because fuel-cost, up until 2015, was the highest cost-per-hour to operate these vehicles, even over driver-wages [3].

While some companies are pursuing standalone autonomy with full computer perception and deep-learning-based decision making, many other companies are pursuing a lower-level SAE autonomy, namely platooning. In this concept, autonomous vehicles follow a manually driven lead truck in close enough proximity to obtain fuel economy benefits from reduced aerodynamic drag, with far less upfront cost of system development and testing, deployment cost of vision sensors and high-performance computing, and operating cost in terms of power to said high-performance computing platforms.

Platooning systems critically rely on information from several sensors: radar, radio communication, and positioning either from GPS coordinates or base station infrastructure. The performance of these sensors and communications can impact measurements and coordination and as such are critical for platooning, and as a result, impact the platoon fuel economy. While a Kalman Filter is introduced to help mitigate poor measurements from sensors, not all bad measurements can be adequately filtered out. This chapter analyzes the platoon fuel economy in scenarios where the sensors perform sub-optimally. There is active research on identifying sensor faults in real-time utilizing complex algorithms such as in [4].

Prior work in this area is quite well established. A collection of selected works in this research area are provided here with respect to their interesting contributions:

- A rear mounted antenna appears to offer more reliable V2V communications [5].
- The usage of a Model Predictive Control (MPC) controller and message buffering appear to improve the reliability of a CACC system [6].
- V2V communications jamming attacks can be detected in real-time using an algorithm with a 90% chance of detection and no significant false-positive rate [7].
- LTE-V may not be able to support the strict specifications of ultra-reliable low-latency V2V communications [8].
- A mathematical framework may be constructed for optimizing the operations of a platoon based on the network delay and the stability of the vehicle control system [9].
- General conclusions may be made as to the effects of vehicle speed, antenna position, and track on packet error rate and therefore V2V performance in a truck platoon [10].
- A distributed consensus system can be used in a CACC which takes into account the characteristics associated with individual vehicles such as antenna placement and braking distance for improved safety and performance [11].
- CACC systems can be configured to explicitly take Information Flow Topology (IFT) dynamics into account for enhanced performance [12].
- A review of collected works on CACC architectures, controllers, and applications can be found in [13].
- An adaptive control scheme using leader information offers better performance than semi-autonomy and can be improved through the use of redundant wireless communication channels [14].
- Leader-based V2V-communications message scheduling can help to increase the Packet Reception Ratio (PRR) leading to increased stability and tighter platoons [15].
- Market penetration of CACC equipped vehicles has a significant effect on the performance of these systems, particularly after exceeding a critical point around 40% of vehicles equipped. The best strategy for performing CACC is dependent on the market penetration and will need to change over time [16].
- Dedicated Short Range Communication (DSRC) standards operating at 5.9 GHz and a 10 Hz update rate is sufficient in practice to maintain platooning. Dual antennas are needed to maintain direct line-of-sight between trucks even around curves or with intermediate vehicles between them [17].
- The use of CACC as an alternative to traditional ACC offers better performance but may incur a less comfortable driving experience for passengers due to increased jerk [18].

- CACC offers significantly increased fuel economy across a comprehensive sample of realistic scenarios including differing truck models, loads, road surfaces and curvature, speeds, following distances, and non-platoon vehicle interruptions [19].
- Evaluation of driver assisted platooning technology led to no significant concerns with regard to safety or traffic interactions, leading the research team to recommend a permissive stance on platooning to Florida state agencies [20].
- Two-truck platooning offers an alternative to higher levels of autonomy with still-significant fuel savings and reduced driver stresses [21].
- Dynamic frequency band allocation can be used to improve the wireless V2V link leading to improved platoon performance [22].
- 5G NR-V2X communication can support the requirements of truck platooning even with an unoptimized prototype implementation [23].
- C-V2X can be enhanced for operation with platoons and offers benefits over other V2V communication technologies [24].

These results represent some of the most interesting developments in this field and show a lot of promise for this emerging technology both in terms of present applicability and future developments and optimizations.

## 3 Auburn Platooning System

Since 2015, Auburn University has been developing a platooning system that is referred to as Cooperative-Adaptive Cruise Control (CACC). This CACC system has been validated up to SAE Level 2 autonomy, which means lateral and longitudinal control is operational, but human supervision is required [25].

The Auburn test platforms are fairly unique in the way their autonomy is achieved. Rather than have an actuator physically command an acceleration through a throttle angle, the Auburn system acts as the vehicles automated cruise control (ACC) system. As a result, as long as the Auburn platooning system has access to the vehicles Controller Area Network (CAN), it can control the vehicle.

The Auburn platooning system also has a fallback method in the case of unexpected behavior. All the CAN traffic being commanded by the Auburn system must go through the CAN gateway, which has an internal physical disconnect switch that will stop all CAN traffic from the Auburn computer. The Auburn hardware setup is shown in Fig. 1. The CACC system relies on several sensors: a Delphi Electronic Scanning Radar (ESR), Novatel ProPak GPS receivers, Memsense (3020) IMUs and Cohda MK5 OBU Wireless radios for Vehicle-to-Vehicle (V2V) communication. The MK5 radio, Delphi RADAR and Novatel GPS receiver will all be touched on further in the following sections.

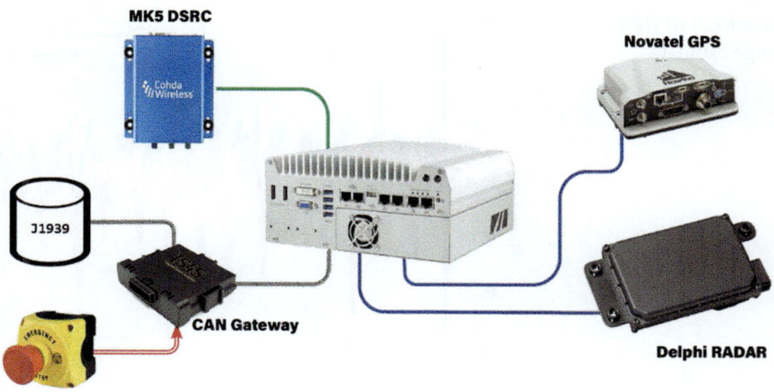


**Fig. 1** Hardware overview of the Auburn platooning system

## 3.1 Dynamic Base Real-Time Kinematic Positioning (DRTK)

GPS is the core measurement used in the Auburn platooning system. GPS allows for radar measurement initialization, provides centimeter to sub-centimeter accuracy range measurements, and allows for time-differenced carrier phase measurements to be used for accurate odometry measurements. While GPS alone does not provide an inter-vehicle range measurement, the use of V2V communication allows for raw GPS observables to be combined into a range measurement.

While GPS is a highly reliable measurement, it can suffer from atmospheric errors, receiver clock biases, multipath errors, and more. To help reduce these errors, there are static GPS receivers on earth, called base stations, which broadcast GPS corrections to atmospheric errors. These corrections are referred to as real-time kinematic (RTK) positioning.

DRTK operates on the same principle as RTK. A GPS pseudo-range measurement for vehicle '*a*', denoted $\rho_a$, can be formulated as

$$\rho_a = \|r_a\| + c(\delta t_a - \delta t) + \lambda(T + I) + M_\rho + \varepsilon_\rho$$

where $\|r_a\|$ is the true range from vehicle *a*'s receiver to the satellite, *c* is the speed of light, $\delta t$ and $\delta t_a$ are the satellite and receiver clock errors, *T* and *I* are the tropospheric and ionospheric effects, respectively, $\lambda$ is the carrier frequency, $M_\rho$ is multipath error, and $\varepsilon_\rho$ is measurement error. An important note is that atmospheric effects are considered constant within a certain region, usually within a few kilometers.

The details of the DRTK algorithm are beyond the scope of this chapter, but a concise summary is included herein for thoroughness. If two vehicles, *a* and *b*, are platooning with V2V communication, and the pseudo-ranges are passed from leader to follower, the following vehicle can take the difference of the two pseudo-range values, $\rho_a$ and $\rho_b$, at which point the atmospheric errors will be cancelled, and a relative position vector (RPV) between vehicles *a* and *b*, denoted $\vec{\rho}_{a/b}$, is returned.

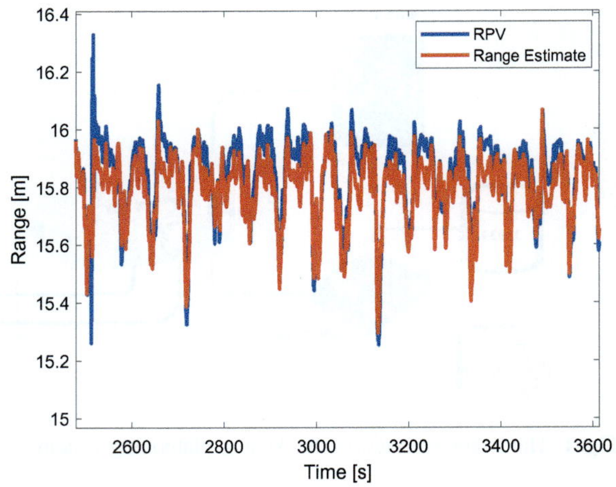


**Fig. 2** Accuracy of RPV versus range estimate

This RPV solution can be anywhere between decimeter to centimeter level accurate. This means the DRTK solution has an extremely low variance and can be used as a "truth" measurement for inter-vehicle distances. Figure 2 presents the DRTK solution overlaid on the best estimate of the inter-vehicle distance, or headway.

The GPS measurement is extrinsic and not inertial, so it is unaffected by engine, powertrain, and road vibrations. Therefore, static data from Auburn's National Center for Asphalt Technology (NCAT) test track allowed for a statistical categorization of the measurement. The noise characteristics for the GPS measurement are $\sigma = 0.125$ m and $\mu = 0$ m.

## *3.2 Delphi Electronically Scanning Radar (ESR)*

The Delphi radar is another critical sensor for the CACC system. Delphi ESRs provide a measurement of range, range rate, and an estimated bearing between the radar and target object. Delphi ESRs have 64 channels, which means 64 separate radar data points are retuned. Each point has a unique radar track ID between 1 and 64, with each track having an associated range, range-rate and bearing. Selecting the correct radar track is done by using the DRTK measurements to create a "bounding box" on the trailer of the lead vehicle. Each radar track is then sequentially checked to see if it falls within the bounding box.

Figure 3a depicts range returns that were used by the platooning system at the NCAT test track, as well as a subplot of the track ID that was used for the range update. While the measurement may seem "noisy", this is really just a result of a long timescale and the frequently changing radar track ID selected for the range measurement. Figure 3b provides a look at the data at a smaller timescale and at a time where the track ID only changes between two tracks (track 15 and 18, in

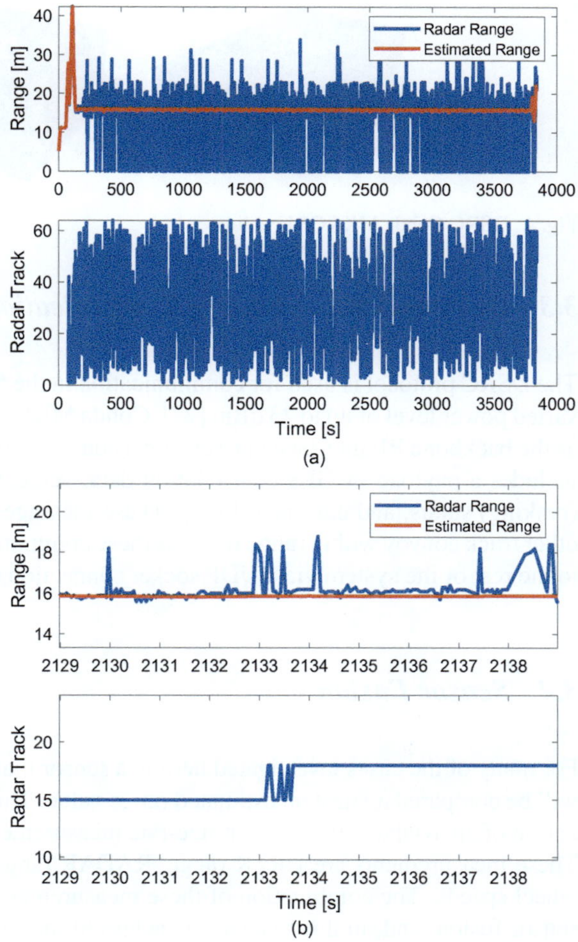


**Fig. 3** **a** High-level radar range versus time; **b** Selected radar range versus time

this example) to show that the range measurement variance falls within the expected range.

While vibrations can cause reductions in the Probability of Detection (PD) for a target due to shifts in the radar echo phase [26], the actual measurement is relatively unaffected. This allows for a mean and standard deviation value to be established for the sensor. According to the Delphi datasheet, the noise characteristics on the range measurement are $\sigma = 0.25\,\text{m}$ and $|\mu| < 0.0125\,\text{m}$.

**Fig. 4** DSRC Cohda MK5 OBU radios

## 3.3 Dedicated Short Range Communications

The DSRC protocol is used for communication in the 5.9 GHz frequency range, at a varied power level of up to 23 dBm [27]. Cohda MK5 OBU (on-board units) are used as the backbone RF interface via a custom (non-standard) wave service message that includes a payload of GPS constellation data, vehicle acceleration or deceleration (braking) status, and current velocity. These messages are broadcast such that any other truck convoy within range receives these communications, which are interfaced to the rest of the system via a UDP socket connection (Fig. 4).

## 3.4 Sensor Fusion

For many of the cases investigated herein, a sensor that is performing sub-optimally will be compared against an estimated range value. This estimated range is a combination of all available range and range-rate measurements on the platooning vehicle. These measurements are DRTK range, RADAR range and range-rate, and relative wheel speeds. The combination of these measurements is commonly referred to as sensor fusion, and, in this instance, is achieved through the use of a Kalman Filter (KF).

A Kalman Filter is a probabilistic filter that uses measurement noise characteristics as well as the process model uncertainty to optimally estimate the system states. The system states estimated in the Auburn CACC are $\hat{x} = \begin{bmatrix} r & \dot{r} & \beta \end{bmatrix}^T$ which are the inter-vehicle range, range rate, and bearing, respectively. These states are outlined by Fig. 5.

While the full derivation of the Kalman Filter is beyond the scope of this chapter, it is important to note two design decisions. First, in the case that the Auburn platooning system loses radar line-of-sight and also loses radio communication, the Kalman Filter will assume a constant range-rate and continue to predict the states. The second design decision was the sensor selection, which is critically important because the Kalman Filter weighs sensor noise variance versus the process uncertainty and the covariance of the estimated states. Because the platooning system has slow dynamics and a largely linear driveline model, the process uncertainty remains small. However,

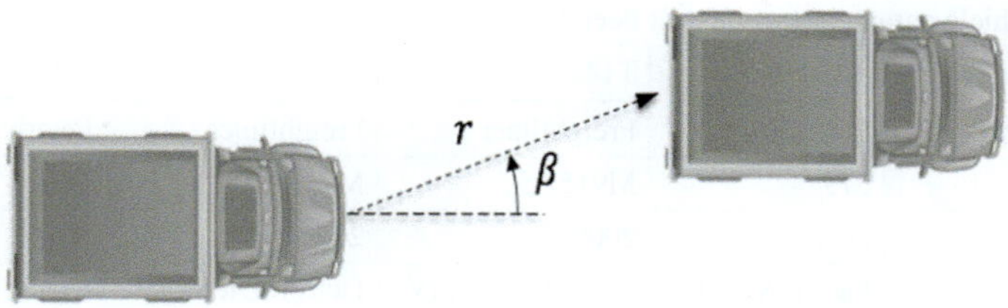


**Fig. 5** Estimated vehicle states

the Delphi RADAR and Novatel GPS receiver were specifically chosen because of their measurement stability and accuracy, which allows for accurate centimeter level ranging.

## 4 Testing Campaign

The current study utilizes four Class-8 semi-trucks operating with the Auburn CACC system, as shown in Fig. 6. Table 1 summarizes their basic parameters.

The results of this study were collected at the National Center for Asphalt Technology (NCAT), a 1.7-mile oval test track in Opelika, Alabama, which is shown in Fig. 7. All tests were performed with the four-truck platoon operating at 45 mph and circling the track in a counterclockwise direction, resulting in a lap time of approximately 136 s. When possible, multiple runs were performed to reduce variance from results. All tests were conducted after a one hour warm up period, mitigating the influence of varying tire pressures, engine coolant temperatures, driveline losses, etc.

**Fig. 6** The vehicle fleet during platoon operation at the American Center for Mobility. The vehicle order displayed is A1, T14, T13, and then A2

**Table 1** Vehicle parameters of the test fleet

| Truck | A1 | T14 | T13 | A2 |
|---|---|---|---|---|
| Manufacturer | Peterbilt | Freightliner | Freightliner | Peterbilt |
| Model | 579 | M915A5 | M915A5 | 579 |
| Model year | 2015 | 2009 | 2009 | 2015 |
| Engine | Paccar MX-13 | Detroit Diesel IV S60 | Detroit Diesel IV S60 | Cummins ISX15-415ST2 |
| Peak torque (ft lbs @ rpm) | 1750 @ 1000 | 1650 @ 1200 | 1650 @ 1200 | 1650 @ 1000 |
| Rated horsepower (bhp) | 430 | 500 | 500 | 415 |
| Truck and trailer gross weight (lbs.) | 35,660 | 37,996 | 46,947 | 38,020 |

**Fig. 7** National Center for Asphalt Technology (NCAT) test track

# 5 Results of Sensor Impairment

The system so far specified works quite well. It approaches a 12% savings of fuel with very little overhead in terms of expensive-to-integrate sensors and computing power. This section covers what happens when elements of this lean system drop out or fail to function correctly. We will look at failures of each of the three major components of the system and what effect the failures had on the control system, vehicle dynamics, and fuel economy.

## 5.1 *Effect of a Faulty Radar on Platooning*

As discussed in the Delphi radar section, the radar is tracking many different points during operation. If the radar fails to track the correct points, then the range measurement may be affected. To investigate the effect of such a scenario, a radar was installed backwards on one of the platooning trucks in the four-truck platoon, T13 from Table 1. This caused the radar to track incorrectly during the curves at the NCAT test track, as shown Fig. 8. Only several laps are shown, rather than the full span of 26 laps (an hour of operation). Filled data points represent time steps where the radar received a tracking update.

Figure 9 shows what the lap position in the subsequent figures translates to on the NCAT test track.

### 5.1.1 Control Effects

The lack of radar updates creates an erroneous headway estimate, as shown in Fig. 10a. The headway is the estimated range of the system (distance to the truck ahead), an output of the Kalman filter. As the radar updates sporadically, the range estimate chatters, leading to poor control and overshooting. The presence of a hill likely exacerbates the issue.

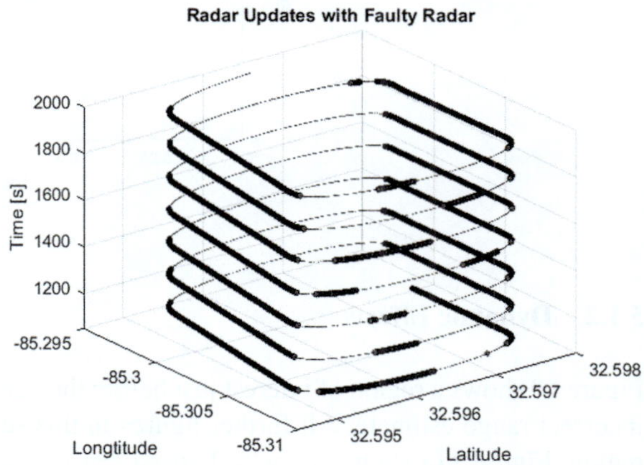


**Fig. 8** Radar updates with an incorrectly mounted radar. Filled points represent when radar updates were occurring

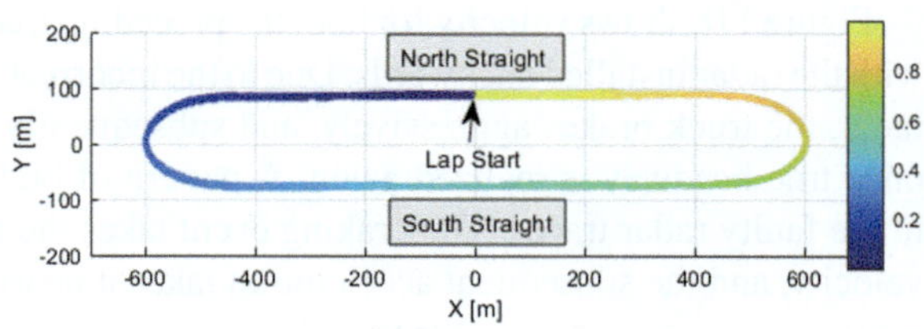


**Fig. 9** Demonstrating what lap position corresponds to on the NCAT test track

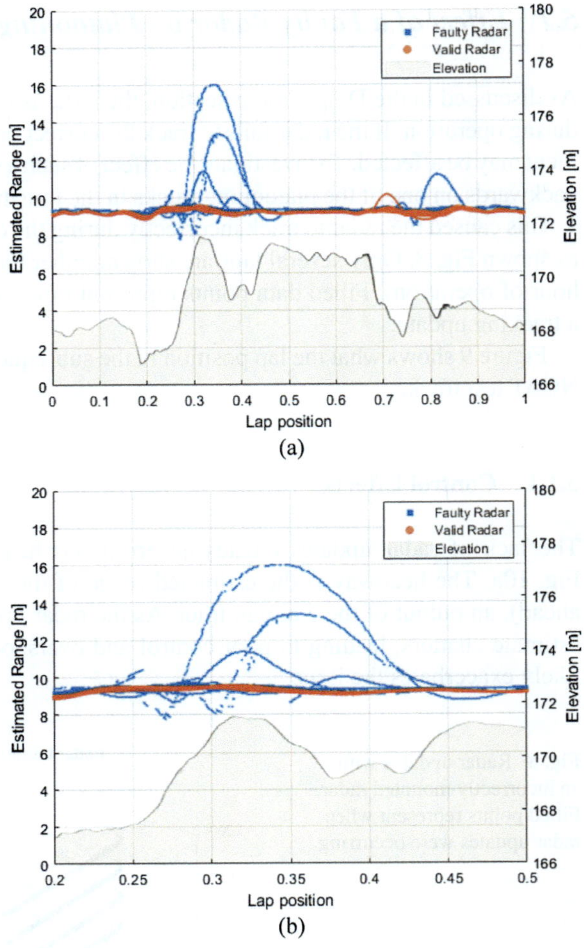


**Fig. 10** **a** Headway perceived by the CACC system for valid and faulty radar; **b** Range estimate in region of interest for truck T13 in a 4 T 100′ platoon with and without radar faults

### 5.1.2 Dynamic Effects

Figure 10 shows a region of interest just before the curve, highlighting a particularly incorrect range estimate. All further figures in this section will be focused on that region. Figure 11a shows the wheel-speed sensor data for T13 for both valid and faulty radar operation.

Figure 11b shows velocity for both the preceding truck and T13 during a single lap with the radar installed backwards. Due to the incorrect headway estimate from faulty radar, the truck brakes aggressively, and subsequently must accelerate aggressively once true headway is realized again. A pattern of braking and accelerating is clear in the faulty radar traces. The braking event takes the truck nearly 9% under the set velocity, and the subsequent acceleration takes it nearly 10% over the set speed.

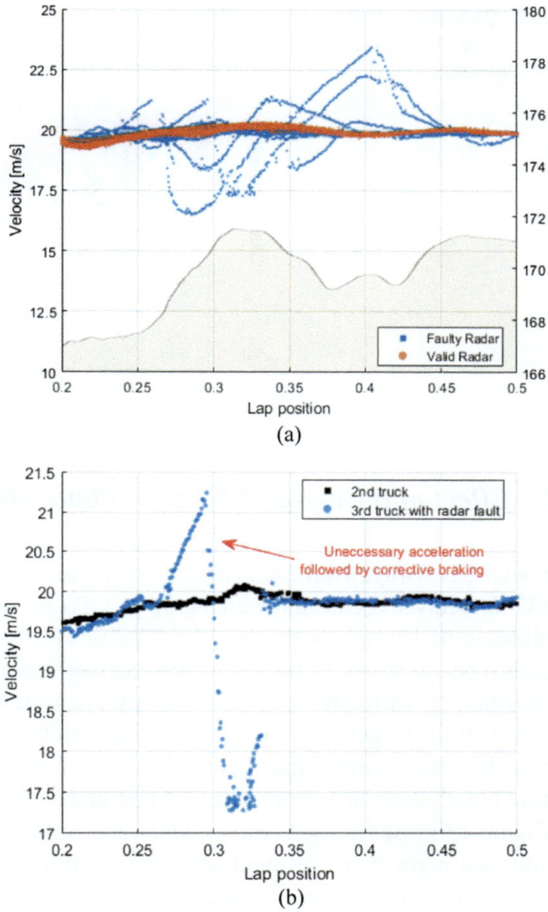


**Fig. 11** **a** Velocity trace for truck T13 in a 4 T 100′ platoon with and without radar faults; **b** Velocity profile for truck T13 in 4 T 100′ platoon and its immediate leader during one lap, showing aggressive correction

### 5.1.3 Fuel Effects

Aggressive acceleration events induced by the reversed radar and subsequently erroneous headway measurements force the truck to waste energy on:

1. Over-acceleration, especially in the event of a transmission downshift.
2. The braking event itself. Any time a truck in platoon actively decelerates, energy is wasted. Braking is herein generalized to mean both the retarder and the air brakes

Both effects can be seen in the CAN fuel rate data for the truck with faulty radar measurements in Fig. 12.

When the fuel rate data shown in Fig. 12 is integrated, fuel consumption over the course of faulty radar operation was 5.68% greater. The standard deviation over the section shown in Fig. 12 is much higher for the faulty radar, at $\sigma_{faulty} = 21.2$ L/h versus $\sigma_{valid} = 12.6$ L/h.

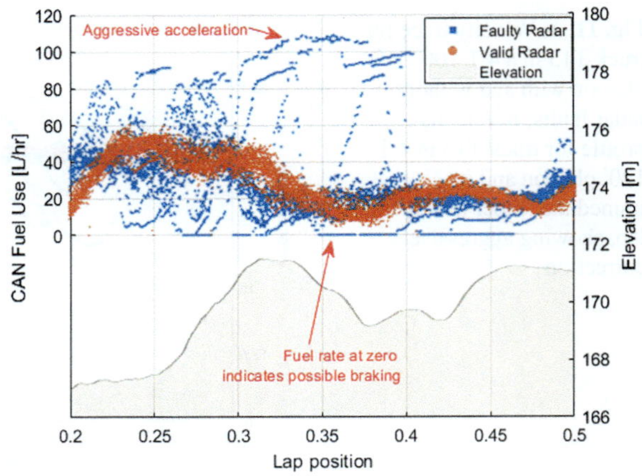


**Fig. 12** CAN Fuel rate for truck T13 with both valid and faulty radar

## *5.2 Effect of a Degraded GPS on Platooning*

While GPS is always available in this testing regimen, the number of satellites visible to a GPS receiver is constantly changing. A GPS receiver needs a minimum of four satellites to fix its position because the XYZ receiver position and the receiver clock bias must be estimated. While some GPS receivers have clocks stable enough to allow for positioning with three satellites, the Auburn system requires four. Additionally, visibility of four satellites does not guarantee quality estimates. Several factors can impact the measurement quality such as signal to noise ratio and elevation angle in addition to all the other error sources stated in the DRTK section.

Figure 13 presents a scenario in which the Auburn vehicles were platooning on an overcast night in a tree-lined area. While DRTK is able to maintain a position solution when five satellites are visible, the accuracy greatly drops when the number of visible satellites reduces to four. While this chatter remains relatively small, some GPS solutions can degrade to the point of returning RPVs of up to a mile.

Figure 13a provides a larger time scale for context and is generated from the same data as Fig. 13b. A scenario in which the same truck is platooning on the same terrain but has better GPS data is also shown for comparison.

### 5.2.1 Control Effects

While bad radar measurements were impactful enough to cause serious deviations in vehicle headway and CAN fuel rate, the sub-optimal GPS performance created almost no impact. This is primarily due to three key factors:

1. The headway controller on the vehicle penalizes range-rate errors 5X more than it penalizes range errors.
2. Radar is the dominant measurement in sensor fusion algorithm for range-rate updates.

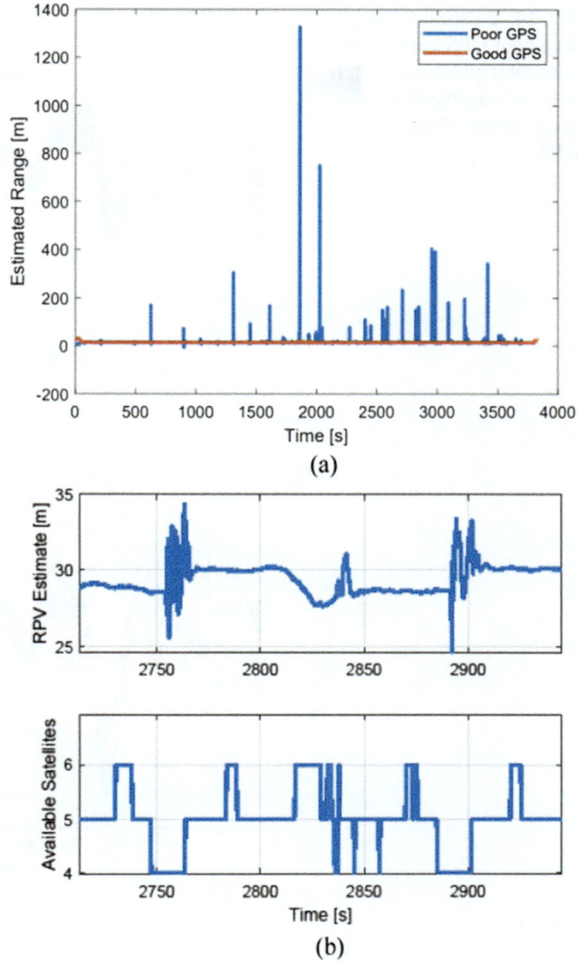


**Fig. 13** **a** RPVs in poor versus good GPS conditions; **b** RPV solution versus available satellites

3. DRTK measurements employ fault rejection.

While the reduced satellite GPS does create very poor range estimation on its own, the CACC headway controller successfully mitigated these disturbances. Figure 14a provides the range estimate vs normalized track position for both the "good" GPS versus "poor" GPS scenarios. Each of these runs contains 26 laps of NCAT test track operation. The scenario shows no clear signs of sub-optimal control performance except for one deviation in the 0.55–0.65 lap position range of the good GPS scenario. This deviation was due to the lead truck driver tapping the brakes at the end of the run and is thus not a control issue. The rejection of noise from the GPS signal is primarily accomplished through two methods. The first is the fusion of all available measurements as described in the sensor fusion section. The second is due

**Fig. 14** **a** Estimated range with poor GPS versus good GPS (all laps are shown); **b** Vehicle velocity results for good versus sub-optimal GPS

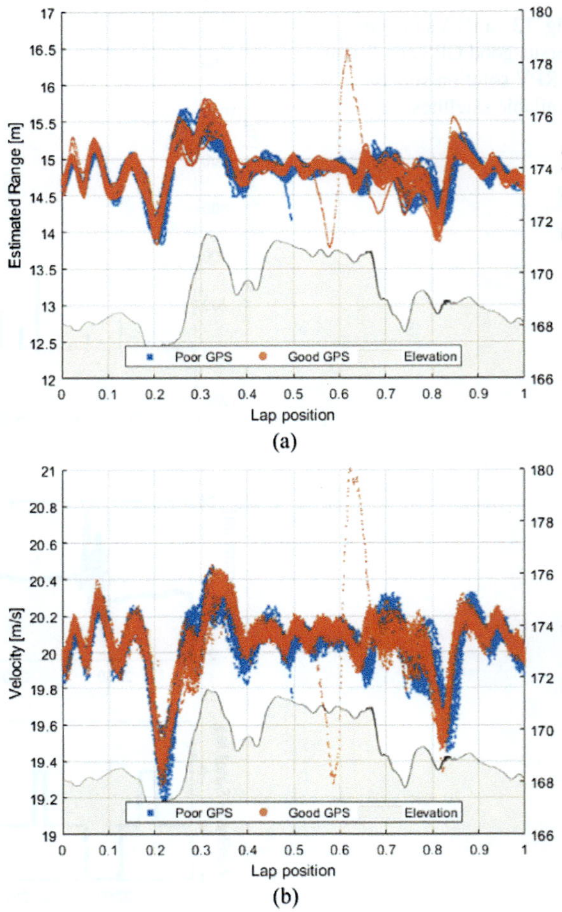


to performing a 3-sigma test. Namely, if the range measurement exceeds $3\sigma$ of the expected range, the measurement is rejected by the headway controller.

### 5.2.2 Dynamic Effects

Because the range estimate was largely unaffected, it was suspected that the velocity profile for the vehicle should also be largely unaffected. Figure 14b displays the velocity profile between the same good GPS and poor GPS runs. With the exception of one velocity perturbation on the good GPS run due to human error, the overall trends are almost identical between the two scenarios.

### 5.2.3 Fuel Effects

As a result of the sensor fusion algorithm being able to reject the sub-optimal GPS points, the fuel rate data shows almost no changes on a lap-to-lap basis through the run with only one exception. The good-GPS run consumed 0.255% more fuel over an hour of operation. While the human error does slightly confuse the issue, the amount of additional fuel wasted is not enough to substantively affect the overall fuel consumption. There is no clear negative platooning effects caused by the GPS as the tiny difference in fuel consumption is within the realm of random variation, see Fig. 15.

## *5.3 Effect of Radio Interference*

In order to test the effects of radio frequency (RF) interference, a 5.8 GHz video transmitter system was obtained. An RF channel close to the 5.9 GHz on which DSRC operates was selected and the transmitter was placed near the southeast testing loop corner, as shown in Fig. 16. The 5.8 GHz frequency is an industry, scientific, and medical (ISM) band as defined by the Federal Communications Commission (FCC), allowing for unregulated power levels up to 1 W (30 dBm), which is commonly used for Wi-Fi and other applications [28], making such a device a realistic enough threat to real-world applications. This RF interference source was set to transmit at 27.5 dBm, and using the standard COTS antenna, the sideband radiation was enough to generate noticeable interference in the 5.9 GHz range.

For clarity, Fig. 16a depicts a single lap of operation with interference. As a comparison, a lap of normal operation without interference is shown in Fig. 16b. In both tests, the data displayed are the packets received by truck T13 from the preceding vehicle, T14, while platooning at a 100′ headway gap distance and a platoon order of

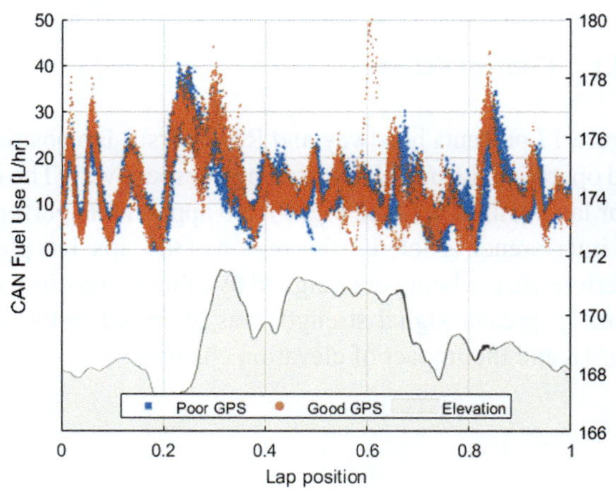


**Fig. 15** Fuel results for good versus sub-optimal GPS

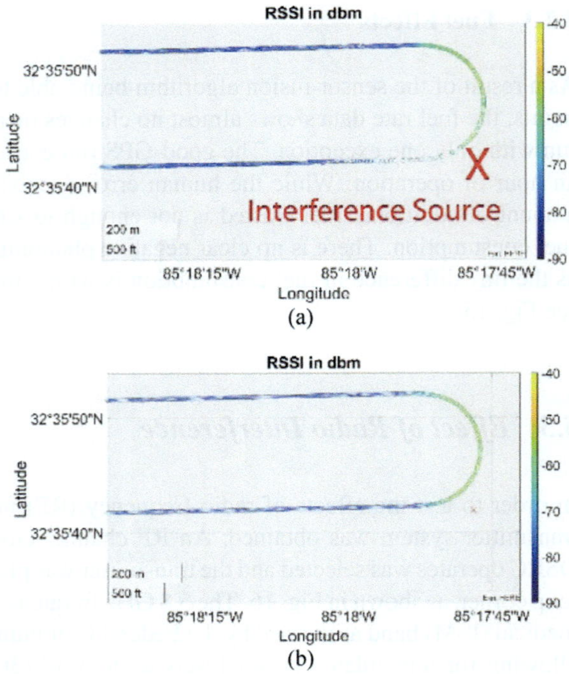


**Fig. 16** **a** Received packet signal strength in RSSI with the interference source active; **b** Received packet signal strength in RSSI by location during nominal operation

A1, T14, T13, and A2. Truck T13 was chosen so the impacts of the dropped packets caused by the RF interference on vehicles both ahead and behind the selected truck can be deciphered.

Because the interference device effectively increased the noise floor relative to the receiver, the effect on received signal strength indicator (RSSI) is small and does not reduce the signal power to a level that approaches the receiver's minimum sensitivity level, but does degrade the signal to noise ratio, leading to dropped packets.

### 5.3.1 Control Effect

Figure 17 presents headway and RSSI versus lap position for both normal operation and operation during RF interference, respectively. The interference device is located near lap position 0.7 (see Fig. 9). No appreciable performance degradation from the RF interference is noted in this region of the laps. It is plausible that the impacts of RF interference are being outweighed by other factors impacting the following distance. Namely, greater signal strength was observed in the curved area of the track, see Fig. 18 and the impact of elevation changes.

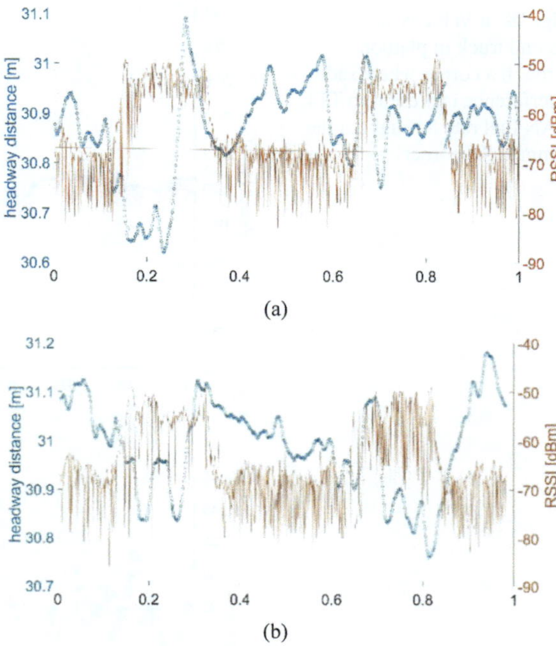


**Fig. 17** **a** Range versus RSSI versus lap position with no RF interference; **b** Range versus RSSI versus lap position with RF interference

### 5.3.2 Dynamic Effect

Because the range estimate was essentially unaffected, the velocity trace of a truck through a corner with radio interference was also unaffected relative to its velocity without radio interference. As an example, Fig. 18a shows the velocity of the second truck in platoon from the four-truck platoon, T14, during operation with and without radio interference at 100′ headway distance.

### 5.3.3 Fuel Effect

Because the velocity trace was unaffected by the interference, it follows that the fuel consumption was also unaffected. Figure 18b confirms that this interference strategy had no discernable impact on fuel consumption. The small differences between the with-interference and without-interference fuel rates could certainly be due to the influence of some other factor, such as wind, but no further effort was made to isolate this.

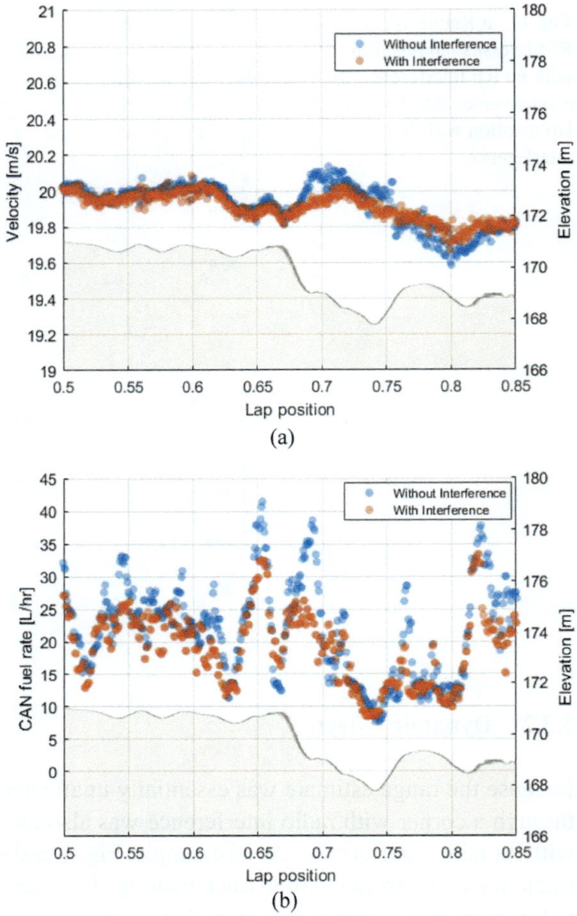


**Fig. 18** **a** Velocity of second truck in platoon through a corner where radio interference is present; **b** T14 CAN fuel rate throughout the jammed SE corner

## *5.4 Summary*

The CACC system outlined herein critically depends on GPS, radar, and V2V communications. As such, this study investigated three potential mechanisms in which sensor performance degradation could lead to degraded convoy platooning performance. The study was conducted in a controlled environment on a closed test track with minimal grade changes.

Installing a radar in reverse orientation led to missing radar updates and allowed erroneous range estimation in the short term, resulting in unnecessary vehicle transience. Due to aggressive acceleration events, operation with faulty radar led to a 5.68% increase in fuel consumption. However, the missing radar updates occurred during a relatively tight curve, an unlikely scenario for a real-world platooning application.

Additional scenarios where a lack of satellites caused poor GPS performance and where RF interference was intentionally created did not produce fuel consumption impacts. During these tests, the CACC range estimate was sufficiently accurate to maintain intended platooning performance. However, the RF interference should be expanded over longer durations and to stronger intensities before broader conclusions are reached.

It is worth noting that the results herein are specific to the control system and sensor suite of these trucks, and that other commercial or research-grade platooning systems may experience different results. Still, the mechanism of disturbance remains the same in all cases: sensor degradation leading to poor dynamic performance, leading to increased fuel consumption. Additionally, harsher environment such as rain or snow could pose unresolved challenges for the current sensors and merit further investigations.

The sensors used were standard and well-documented for this application, and there is a research opportunity in applying more novel sensors such as ultra-wide-band radar, lidar, and camera to platooning. These additional sensors could address some of the challenges identified here, particularly if paired with more advanced computer vision and deep-learned decision making. While the introduction of these features represents additional cost as previously noted, it is certainly possible for these costs to be offset by better optimization of fuel usage and increased safety and reliability through redundant data sources.

## 6 Conditional Effects

The performance and reliability of the existing V2V radio network have been investigated further under a variety of conditions, some of which are adversarial. The radio network performance was surprisingly robust and only degraded gradually. Data recorded on precision-instrumented trucks at both ACM and NCAT test tracks is used to provide an understanding of various effects on V2V network performance, including:

- Occlusions—non-line-of-sight (NLOS) between the Tx and Rx antenna may cause network signal loss.
- Rain—water droplets in the air may cause network signal degradation.
- Antenna position—antennas at higher elevation may have less ground clutter to deal with.
- RF interference—interference may cause network packet loss.
- GPS outage—outages caused by tree cover, tunnels, etc. may result in degraded performance.
- Road curvature—curves may affect antenna diversity.
- Road grade—antenna may have limited vertical coverage.

The data presentation and analysis build on those reported by others [29–37] and could be of interest to researchers, practitioners, and end-users in the broader autonomous-connected vehicle community. The work herein centers on conclusions drawn from processing sensor data acquired over several days of extensive testing. Antennas and sensors were mounted on our four class-8 trucks operating over a pair of test tracks. The first test track is a roughly circular loop at the American Center for Mobility (ACM) in Ypsilanti, Michigan. The track includes an overpass and tunnel and is about 2.3 miles in length. A satellite view is shown in Fig. 19.

The NCAT test track is capsule shaped and is about 1.7 miles in length with no special considerations such as bridges or tunnels. Figure 7 gives the aerial view of this test track.

The trucks operated as a linear convoy and maintained the same following order throughout each test. As test conditions were varied, external/environmental factors were monitored, providing a wide variety of interesting samples for processing. For example, the gap distance between each pair of subsequent trucks was set to 50′, 100′, and 150′ gap distances, as shown in Fig. 20, and data was collected during dry and wet operation. These variations enabled the elucidation of some key network performance indicators, allowing direct conclusions to be drawn about the effects of some of these scenarios on received signal integrity.

**Fig. 19** ACM test track (Ref: Google Maps). Semi-truck platoon testing was performed on the 2.3-mile outer (highway) loop

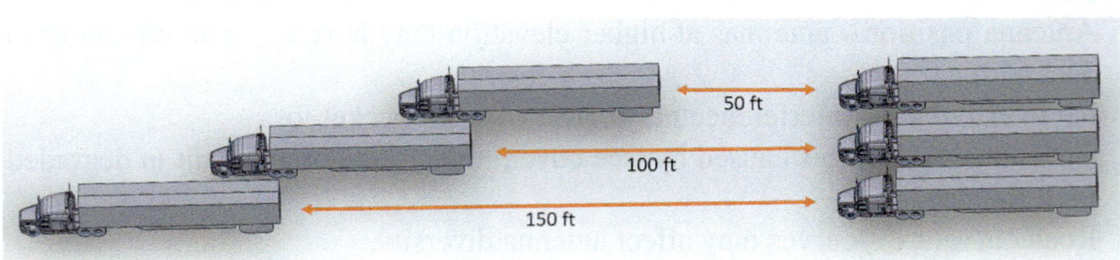


**Fig. 20** Class-8 trucks at varying platooning/following distances (to-scale)

**Table 2** Convoy details

| Truck | Leader (L) | Follower 1 | Follower 2 | Follower 3 |
|---|---|---|---|---|
| Manufacturer | Peterbilt | Freightliner | Freightliner | Peterbilt |
| Model | 579 | M915A5 | M915A5 | 579 |
| Year | 2015 | 2009 | 2009 | 2015 |

All testing used two Peterbilts from Auburn University and two Freightliners from the US Army GVSC, each hauling a 53′ trailer. Following order was consistent across for all tests, allowing convoy vehicles to be referenced descriptively as L (Leader), F1 (Follower immediately behind L), F2 (behind F1) and F3 (behind F2). Table 2 provides more specific details on the individual trucks.

Unfortunately, F2 did not capture the logs as required for some tests. While less than optimal, more than enough data was successfully gathered to reach useful conclusions in all test cases.

Cohda Wireless MK5 OBUs were utilized for the DSRC network. Each radio was connected to a pair of antennas mounted left and right on the truck cab for diversity. The antennas used were the ECOM6-5900 s from MobileMark, 5.9 GHz dipole antennas with 6 dBi of gain and the antenna pattern shown in Fig. 21.

While dipole antennas were utilized for availability reasons, it is worth noting that there are several interesting attempts at designing DSRC-specific antennas such as [38–40].

**Fig. 21** ECOM6-5900 antenna pattern over elevation angle. 0° points up from the antenna pole. Azimuthal pattern is isotropic

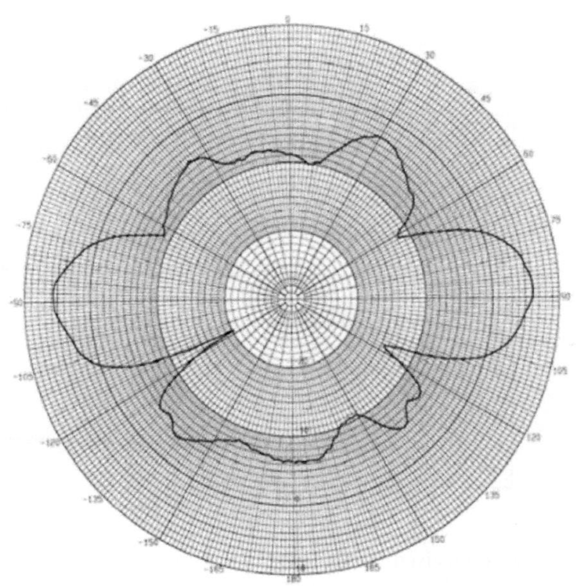

## 6.1 Occlusions

In order to examine the effects of an occluding obstacle in the line-of-sight (LOS) between a pair of communicating vehicles, the baseline communication effectiveness between L and F1 was documented at NCAT with a following distance of 150′ and truck speed of 45 mph, as shown in Fig. 22a. The baseline RSSI was −52.8 dBm along the curves and −67.7 along the straightaways.

This baseline performance was compared with the communication effectiveness between L and F2 with a 50′ following distance at the same speed, which results in close to the same total gap distance, but with the addition of F1 driving in between them, acting as a large metal obstacle and hindering a direct line-of-sight. These results are shown in Fig. 22b. RSSI was −52.7 dBm along the curves and −69.1 dBm along the straightaways.

These results indicate that the presence of an additional obstructing truck between the two communicating trucks adds ~1.4 dB in additional path loss in the straight portions of the track. However, this loss becomes negligible (~0.13 dB) on the track's sharp curves because there is a direct LOS between L and F2 that is not occluded by F1 around the curves.

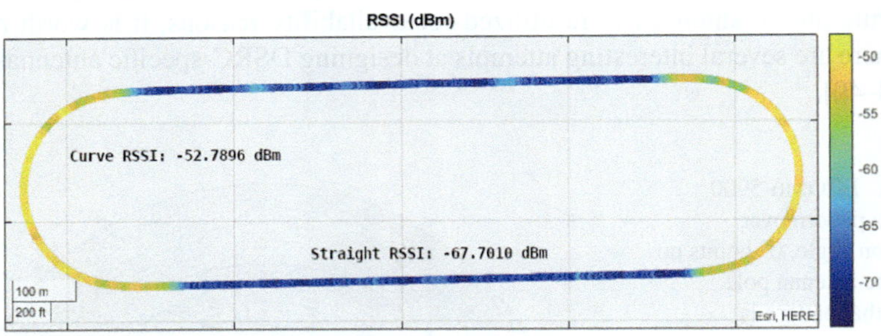


(a)

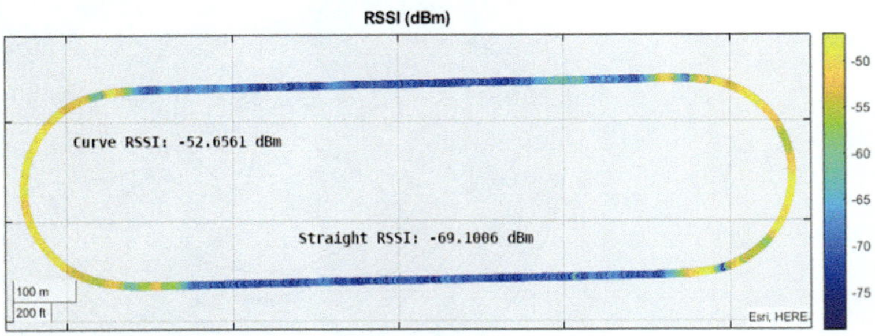


(b)

**Fig. 22** **a** Received Signal Strength Indicator (RSSI) of packets received by F1 from L at 150′ separation with no obstruction of the LOS between them; **b** RSSI of packets received by F2 from L at 150′ separation with F1 obstructing the LOS between L and F2

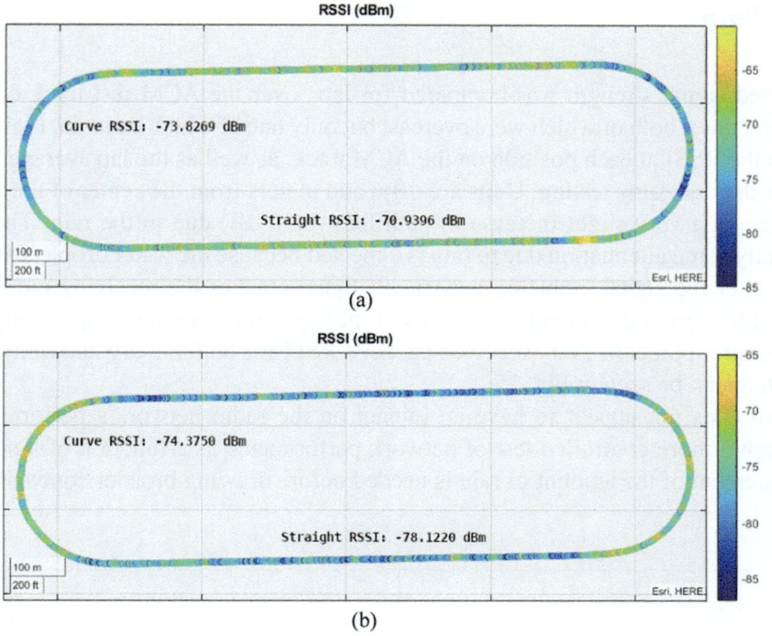


**Fig. 23** **a** RSSI of packets received by F3 from F2 at 150′ separation with no obstructions; **b** RSSI of packets received by F3 from F1 at 150′ separation with F2 obstructing the LOS between F1 and F3

The results were cross-verified by performing similar analysis on other vehicle pairs from the available convoy. The conclusions are analogous. For example, Fig. 23a shows the communication effectiveness between F3 and F2 at a following distance of 150′ and 45 mph producing RSSI of −73.8 dBm in the curves and −70.9 dBm on the straights.

Figure 23b shows the performance between F3 and F1 at the following distance of 50′, where F2 sits between them as the occluder. The resulting RSSI values are −74.4 dBm in the curves and -78.1 dBm on the straights.

The path loss due to occlusion is much more pronounced on F3—about ~7.2 dB on the straightaways, and ~0.6 dB on the curves—possibly due to the fact that occluding vehicle F2 is an armored truck and blocks the propagation of RF waves more than a normal truck such as F1.

The effects of an occluding truck hindering LOS are considered in [41, 42], and [43]. The latter of which showed a little over 5 dB loss caused by a vehicular obstruction in a 50 m gap.

## 6.2 Rain

Received signal strength was compared for laps over the ACM test track on two different days, both of which were overcast but only one of which had rain. Figure 24 shows the RSSI at each position on the ACM track, as well as the lap average RSSI for the dry and rainy testing. Units are dBm and meters from the center of the track.

There is a very slight increase in path loss (0.16 dB) due to the rain. There is generally some attenuation due to rain is expected because the water droplets absorb and scatter the signal, reducing how much signal reaches the receiving radio [44, 45]. Table 3 gives the statistical analysis of the dry and rainy days, respectively.

Histogram plots of every received packet's RSSI for both the dry and rainy days of testing can be seen in Fig. 25.

Rain does not appear to have an impact on the radio network's performance, although a more controlled test of network performance as a function of a precise measurement of the amount of rain is needed before drawing broader conclusions.

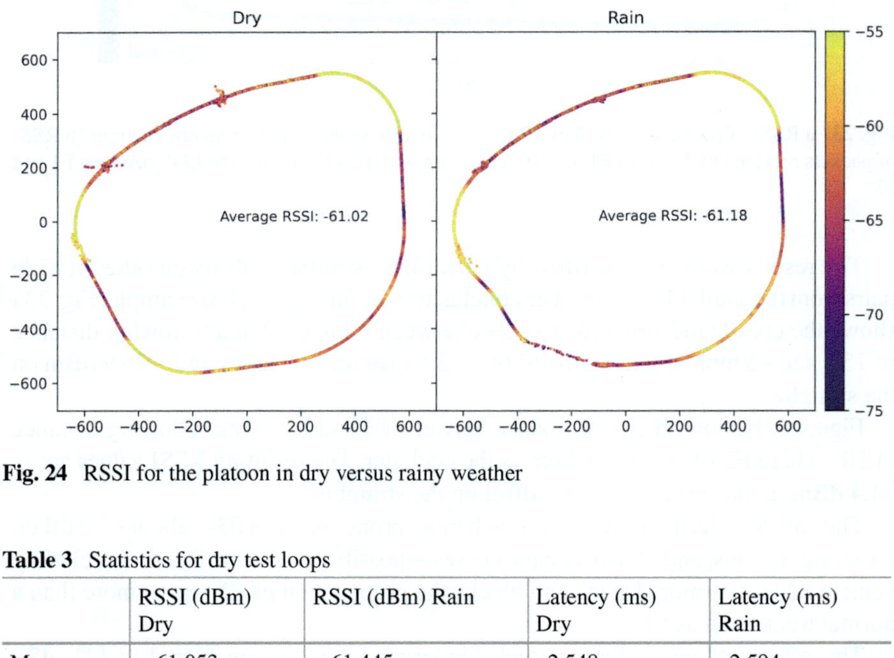


**Fig. 24** RSSI for the platoon in dry versus rainy weather

**Table 3** Statistics for dry test loops

| | RSSI (dBm) Dry | RSSI (dBm) Rain | Latency (ms) Dry | Latency (ms) Rain |
|---|---|---|---|---|
| Mean | −61.053 | −61.445 | 2.548 | 2.594 |
| St. dev. | 4.422 | 4.605 | 0.828 | 0.936 |
| Min | −76.000 | −77.000 | 1.713 | 1.666 |
| 25% | −64.000 | −65.000 | 2.156 | 2.174 |
| 50% | −61.000 | −62.000 | 2.264 | 2.276 |
| 75% | −57.000 | −57.000 | 2.496 | 2.496 |
| Max | −48.000 | −50.000 | 16.772 | 15.227 |

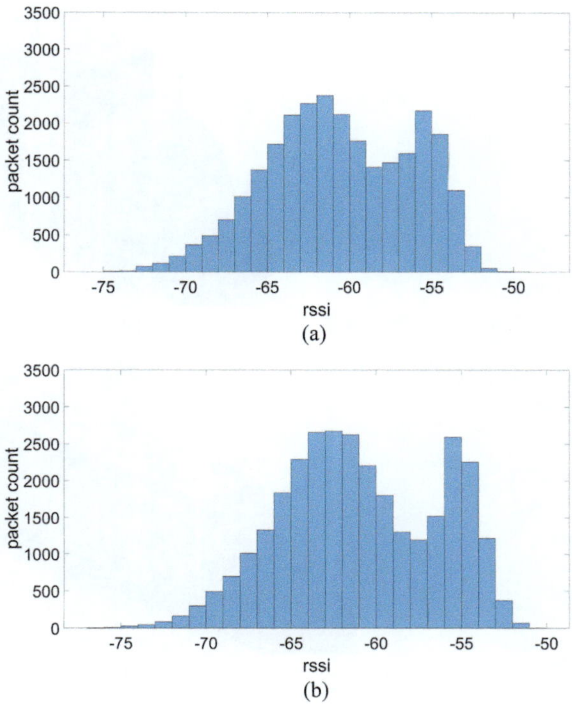


**Fig. 25** **a** RSSI histogram for received packets in dry weather; **b** RSSI histogram for received packets in rainy weather

## *6.3 Antenna Position*

In order to test the effects of the vertical antenna positioning on received signal integrity, antennas were mounted to the left and right sides of the cab at 8′ above ground level. The vehicles operated on the ACM track with the convoy continuously transmitting and receiving packets at 100′ gap distance. The GPS location of the trucks as well as the RSSI of all incoming packets for all vehicles were monitored. The test was repeated with the antennas raised to 13′ above the ground. Figure 26a illustrates the antenna mounting position for both of the Peterbilt trucks.

Figure 26b puts into perspective the entire length of the truck shown on a Peterbilt with attached trailer. Figure 27 shows the mounting on the top of the other two trucks used, which were Freightliner M915A5s. They used the same standard dry-van trailer as the Peterbilts and had the low (8′) antenna position mounted on top of the side view mirrors. As shown in the figure, the high (13′) position was instead mounted on a brace across the roof of the cab.

The overall RSSI averages for each truck with the antennas in both the low and high positions was considered. The higher antenna offered significant improvements of about 3 dB in RSSI for L and F1, with a very slight 0.2 dB loss on F3 at the very rear of the convoy. This is shown graphically in Fig. 28a, where the GPS loops have been colored according to the RSSI in dBm, as indicated by the color bar. The other axes once again give meters from the geometric center of the ACM highway loop.

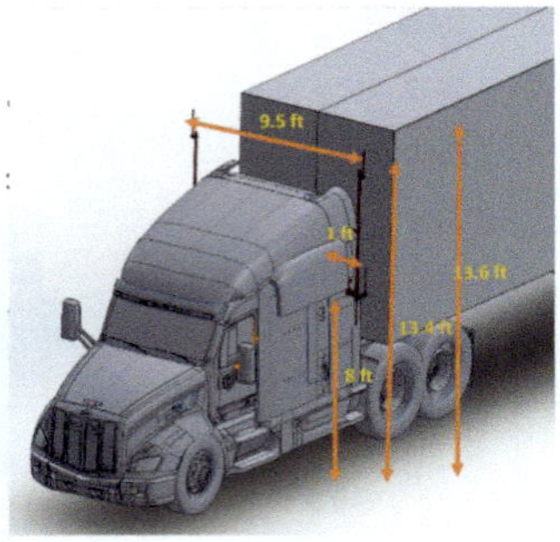


(a)

(b)

**Fig. 26** **a** Antenna mounting positions on the cab of a Peterbilt truck; **b** Overall Peterbilt truck cab with attached trailer dimensions

**Fig. 27** Antenna mounting on the Freightliner trucks

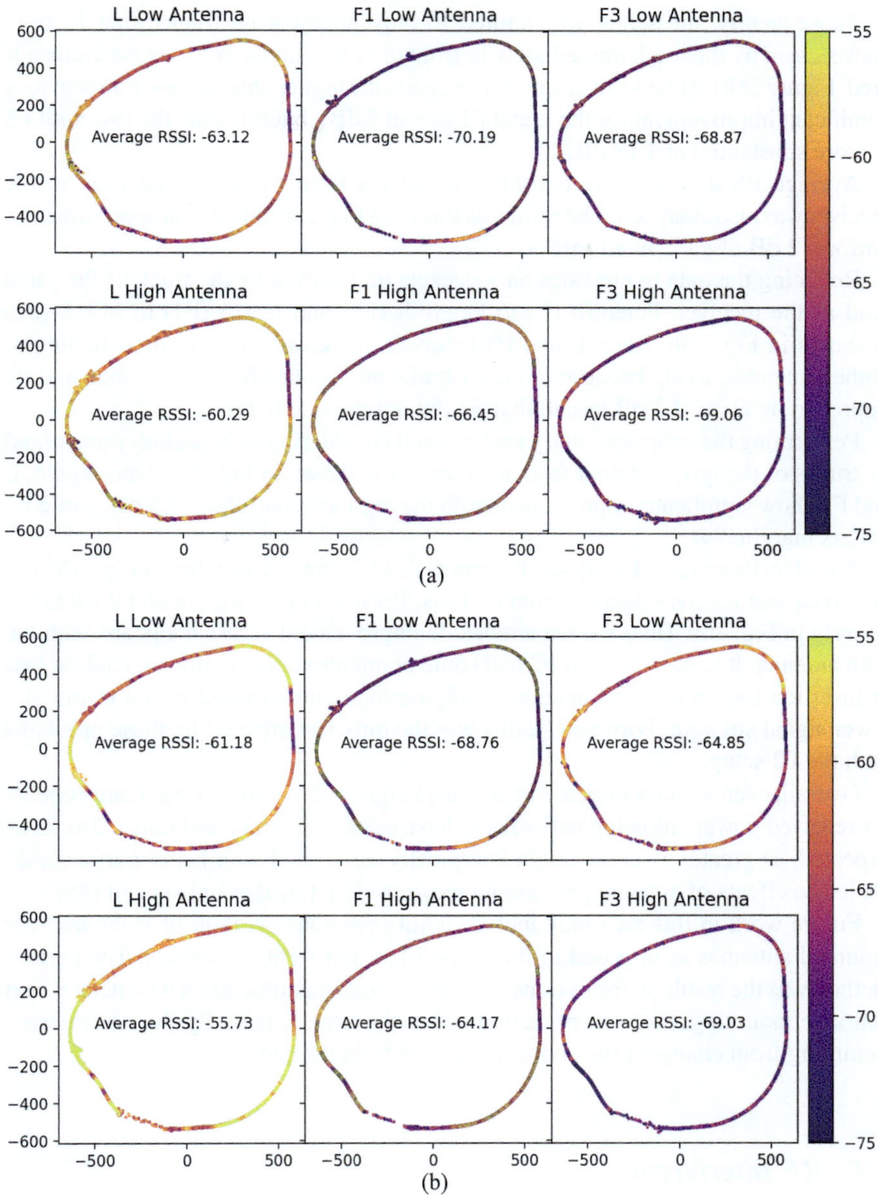


**Fig. 28** **a** Overall average RSSI per truck for the 8′ (low) and 13′ or 13′6″ (high) antenna elevations; **b** Average RSSI between neighboring trucks in the convoy

Subsequently, the scope was limited to only neighboring trucks, that is, only packets sent by the truck immediately in front of or behind the receiver were considered. Figure 28b displays the results. Once again, the higher antenna position provides significant improvement for the L and F1 (about 5 dB), but this time the loss with F3 is more substantial at 4.18 dB.

Average RSSI when only considering packets from trucks not adjacent to the receiver was also analyzed. The results as shown in Fig. 29a indicate an approximately uniform 2 dB of gain on all trucks.

Reducing the data to consider only packets transmitted by the truck of the same kind as the receiver, Peterbilt (L) to Peterbilt (F3) and M915 (F1) to M915 (F2) is shown in Fig. 29b. Here, L and F3 received the least benefit from switching to higher antennas, likely because of their significant distance from each other, and F1 improved by about 4.5 dB (recall that F2 did not provide logs).

Performing the complementary analysis and considering only packets transmitted by trucks of the type different from the receiver is shown in Fig. 30a. Once again, L and F1 show significant improvement with the high antenna, about 3.5 dB, while F3 suffers marginally.

Based on these varied analyses, it appears that F2 brings down the average RSSI of any truck that receives packets from it. Thus, RSSI averages were with F2 excluded entirely in Fig. 30b. Here we see noticeable improvement across the board with the high antenna. It is possible that F2 had some unintentionally significant cabling loss or insertion loss in the high antenna configuration, which caused it to transmit at a lower signal strength. Fortunately, this was the only test affected by these problems with the F2 setup.

Overall, even a minor increase in antenna height of 3–5′ offers significant benefits for received power, allowing networks to have better reliability and range. This is as expected, as greater antenna height is typically associated with better performance [46]. The effects of antenna positioning are considered further in [47] and [48].

Future work in this area may include additional consideration of front and rear mounted antennas as opposed to the current left and right mountings. The current method was the result of discussions with automotive manufacturers who determined that side mountings were more desirable for consumers, partially due to concerns stemming from changing the connected trailer from the cab.

## 6.4 RF Interference

In order to test the effects of RF interference, a variable frequency noise transmitter was placed near the testing loop while operating on an RF channel close to the 5.9 GHz of the vehicles' DSRC communication channel (Fig. 31).

This interference device was transmitting constant noise at 28 dBm and was placed at the southeast corner of the NCAT test loop. Figure 32 shows the RSSI and packet reception density for loops with the interference in place.

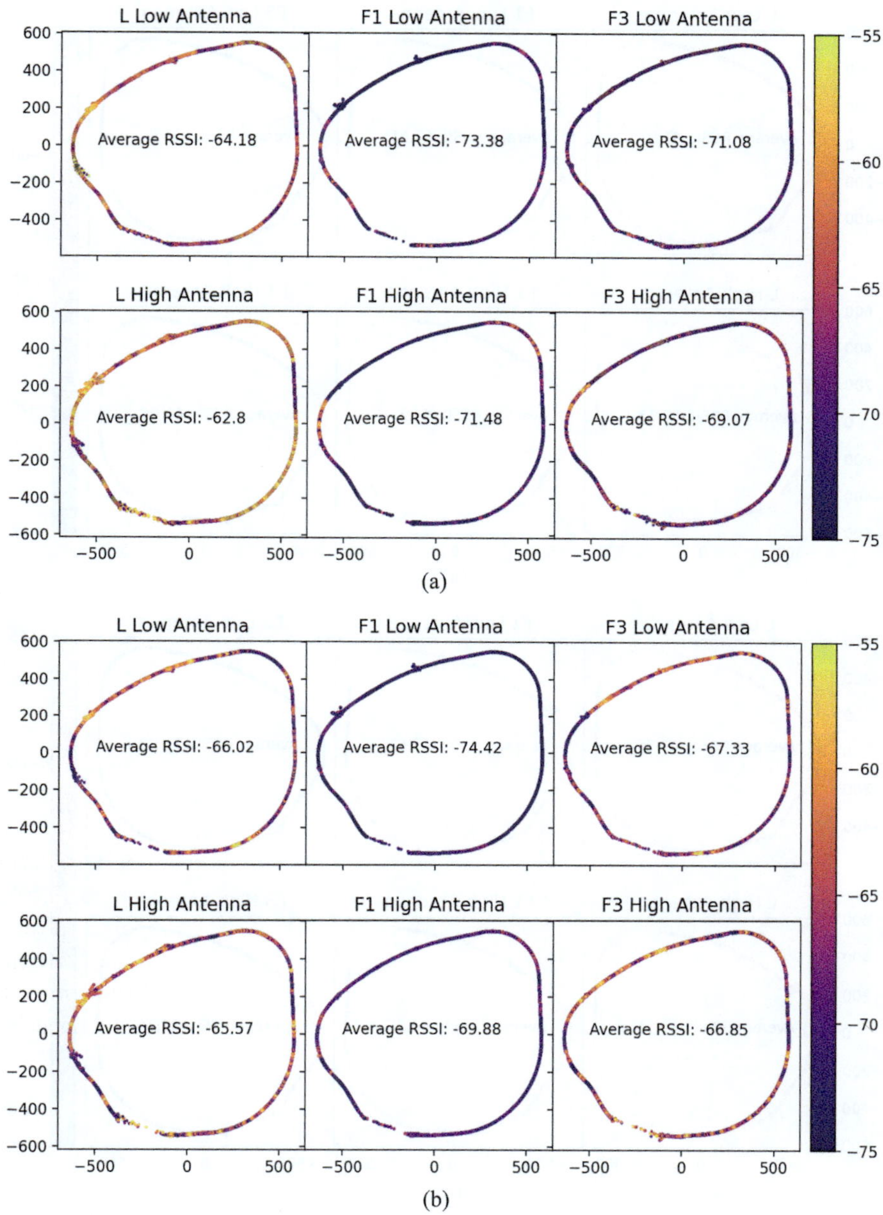


**Fig. 29** **a** Average RSSI between non-neighboring trucks in the convoy; **b** Average RSSI between trucks of the same type

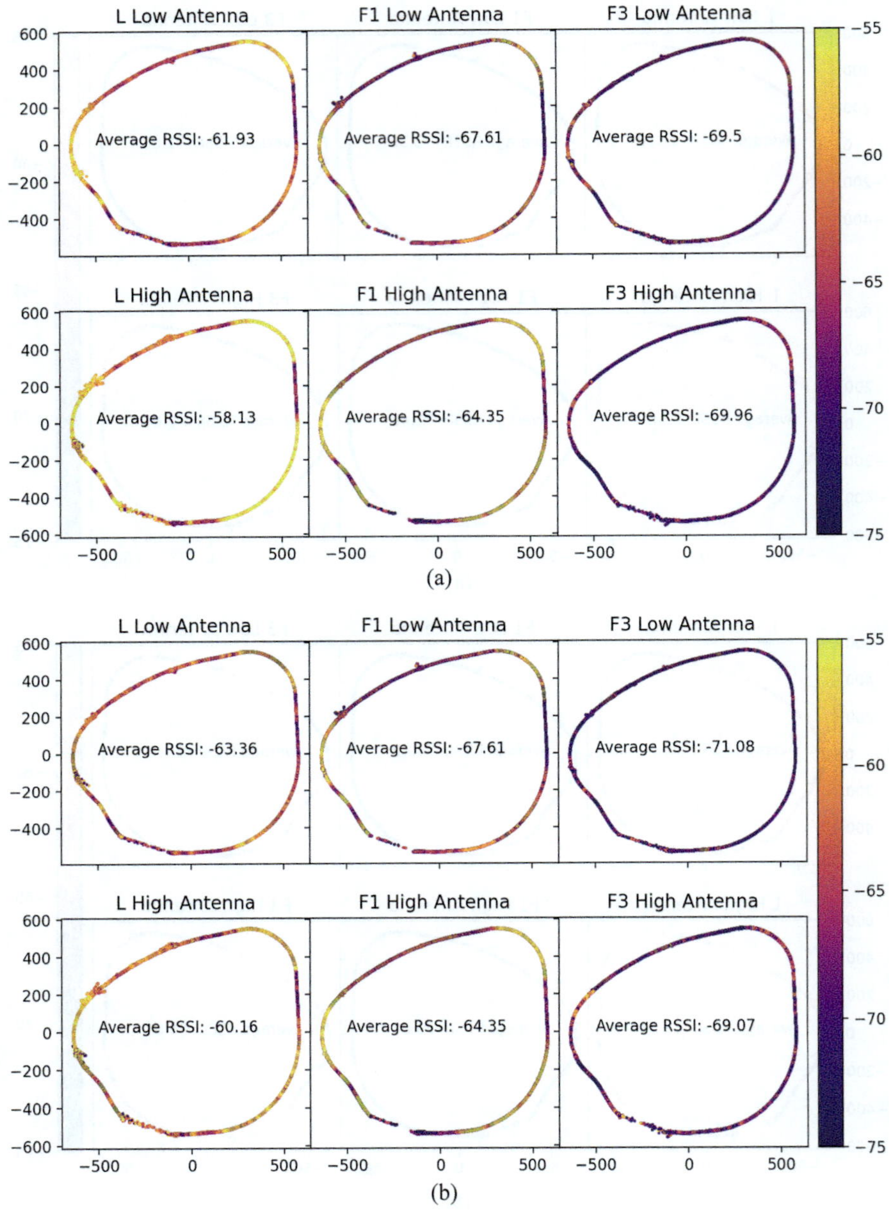


**Fig. 30** **a** Average RSSI between trucks of different types; **b** Average RSSI between trucks not including F2

**Fig. 31** Interference source (28 dBm) placed adjacent to the test track at NCAT

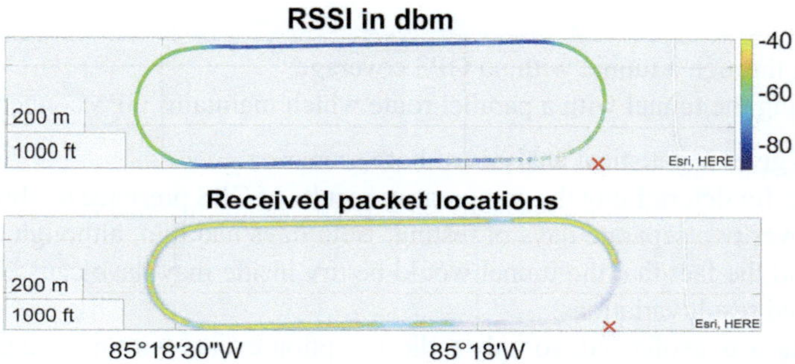


**Fig. 32** RSSI and received packet reception density graphs. Interference source marked with a red cross

Figure 33 shows the baseline loops, where no interference was present and far fewer packets were lost in the southeast corner. Average RSSI does not appear to be affected, which makes sense as the interference device only adds additional noise, not any additional path loss—meaning it affects the signal-to-noise ratio (SNR) but not the received signal strength.

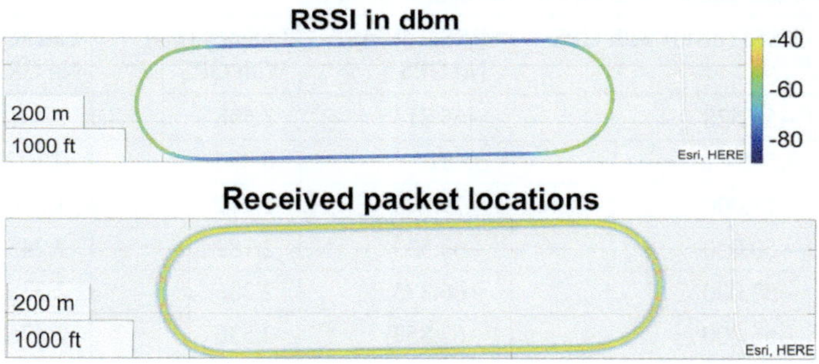


**Fig. 33** RSSI and received packets—Baseline (no interference)

The effects of interference on DSRC communications are explored more fully in [49] and [50]. Without either the interference device or the DSRC radio network turned on, the noise floor was measured at approximately −90 dBm using a spectrum analyzer. Given this, there were very few dropped packets while transmitting at 23 dBm and receiving at upwards of −80 dBm. However, with a 28 dBm interference device located southeast corner of the track, there is complete denial of communications in the surrounding portion of the track.

## 6.5 GPS Outage

To analyze the effects of GPS outage, two similar paths were compared on the ACM test track:

- Passing through a tunnel with no GPS coverage.
- Bypassing the tunnel with a parallel route which maintains GPS connection.

Table 4 gives a statistical analysis of both scenarios.

The data for determining the comparative results of GPS presence or absence was recorded over two separate days of testing. Both days had rain, although the rain's severity and the fact that the tunnel would be dry inside may have caused some of the observed result variations.

The data was geofenced, so only radio reception events within an area slightly larger than the length of the tunnel were extracted and processed. Figure 34 shows the RSSI along the path through and around the 700′ tunnel.

One thing to note is that the packets received inside the tunnel cannot be geo-located and all such packets were assigned to the last known geo-location of the receiver, which is at the entrance of the tunnel. The data inside the tunnel is otherwise perfectly valid and packets were sent and received without any problems. The tunnel's inside walls are made of metal, creating a GPS denied environment and causing internal reflections and wave-guiding effects.

**Table 4** RSSI and latency statistics for the path with GPS

| | RSSI (dBm) with GPS | RSSI (dBm) No GPS | Latency (ms) With GPS | Latency (ms) No GPS |
|---|---|---|---|---|
| Mean | −58.528 | −65.813 | 2.598 | 2.401 |
| St. dev. | 3.546 | 3.681 | 0.896 | 0.307 |
| Min | −73.000 | −79.000 | 1.717 | 1.816 |
| 25% | −60.000 | −68.333 | 2.182 | 2.245 |
| 50% | −57.000 | −66.375 | 2.286 | 2.335 |
| 75% | −56.000 | −62.857 | 2.530 | 2.476 |
| Max | −52.000 | −56.000 | 14.187 | 18.153 |

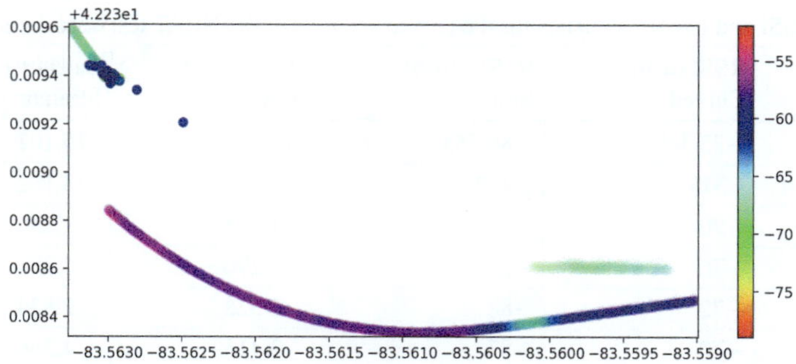


**Fig. 34** Visual depiction of the tunnel and bypass RSSI data. The lower path is the bypass, and the GPS-interrupted top path is through the tunnel

The RSSI data is also given as a histogram, shown in Fig. 35. The packet RSSI for the tunnel runs is bimodal, likely because of the difference between the tunnel and open-air conditions. From the open-air portion, it is clear that the larger mode in the tunnel shown in Fig. 35a is the RSSI state inside the tunnel. The entire tunnel graph is shifted slightly lower in RSSI, but this might be because of road curvature effects resulting from the shape of the path inside and outside the tunnel.

**Fig. 35** **a** Tunnel RSSI data. It is bimodal, likely representing the change in conditions between the tunnel (where the most time is spent) and outside; **b** Tunnel Bypass RSSI data. Here, the data has one mode, representing the signal strength outside of the tunnel

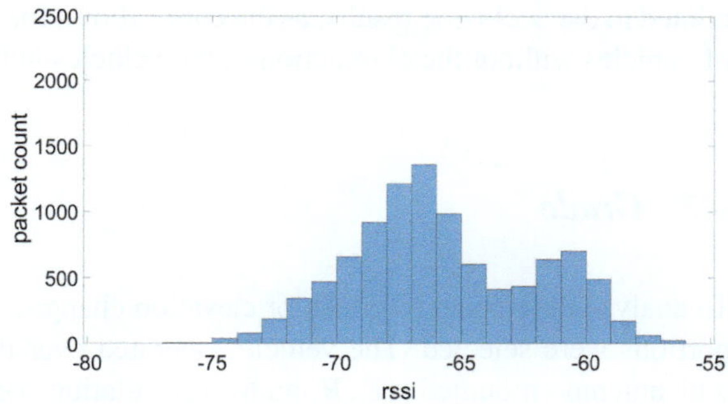


(a)

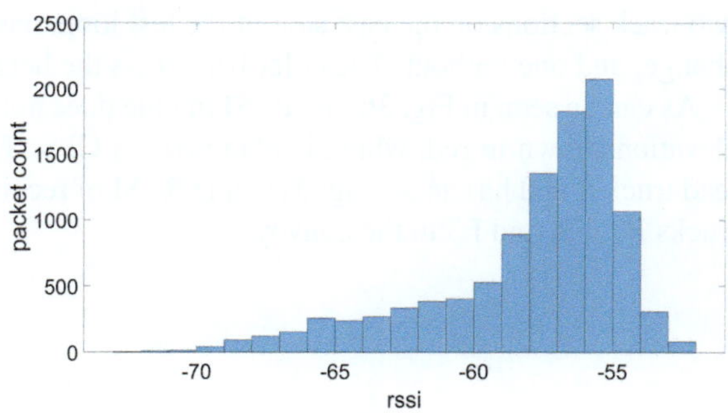


(b)

Table 5 RSSI and latency statistics for the curved sections of the NCAT test track

| | RSSI (dBm) Curved | RSSI (dBm) Straight | Latency (ms) Curved | Latency (ms) Straight |
|---|---|---|---|---|
| Mean | −73.349 | −88.553 | 18.283 | 15.107 |
| St. dev. | 4.318 | 2.472 | 93.654 | 73.182 |
| Min | −90 | −98 | 1.675 | 1.736 |
| 25% | −76 | −89 | 2.296 | 2.386 |
| 50% | −72 | −88 | 2.528 | 2.834 |
| 75% | −70 | −87 | 3.415 | 4.5665 |
| Max | −68 | −73 | 1299.465 | 992.392 |

## 6.6 Road Curvature

To analyze for road curvature, the NCAT data was filtered by splitting off the curved and straight sections of the track by GPS and analyzing both sections separately. Table 5 shows the results for both the straight and the curved sections of the NCAT loop.

A significant RSSI improvement exists when the trucks operate over a bend in the track. This is apparent when viewing, for example, Fig. 22. This is most likely related to the occlusion results, as the curve allows for line-of-sight between any pair of vehicles without the obstruction of the vehicles in between, or their own trailers.

## 6.7 Grade

To analyze the impact of grade, or elevation changes, on RSSI a couple ACM track portions were selected. The vehicles operated over these sections at 45 mph tests with antennas mounted at 8′. Running a correlation coefficient matrix on the selected stretches produced no significant correlation (< 0.2). Figure 36 is a sample of the two track sections on the east side of the test loop, one with a relatively large grade change, and one without. The color bar shows the height of the track in meters.

As can be seen in Fig. 36, the RSSI in blue does not appear to be correlated to the elevation shown in red, which is obtained via GPS. This data was collected on the lead truck L and has an averaged sample RSSI of received packets from all the other trucks F1, F2, and F3 in the convoy.

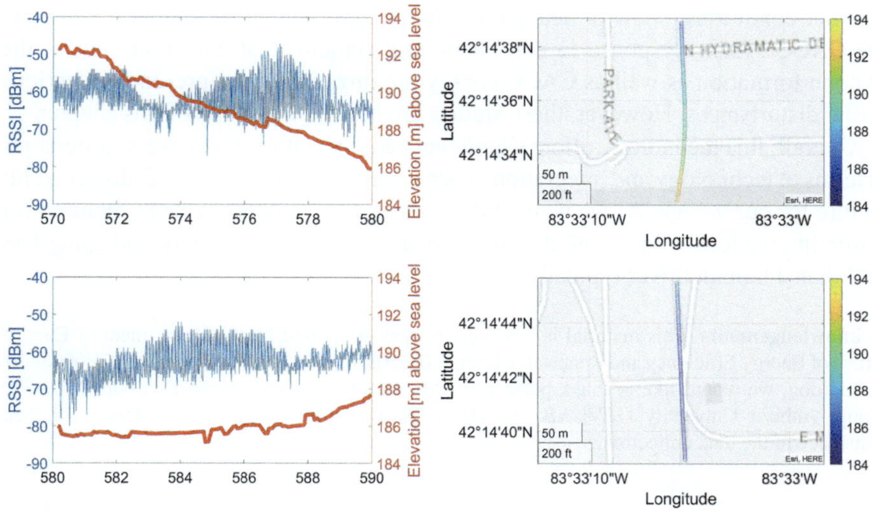


**Fig. 36** RSSI versus elevation in slope (above) and flat (below), 8′ antenna height

# 7 Conclusion

In this chapter we looked at a method for performing semi-autonomous leader–follower platooning using CACC and DSRC V2V communications and how it can be applied to commercial truck transport for the purpose of fuel savings and reduced driver workload. We gave a directory of some of the most important advances in the field for more of an in-depth consideration of the finer details of the technology. We gave a broad, high-level explanation of the major components of the system and their purposes, and then followed that with a discussion of tests showing just what happens when the sensor modules aren't working properly and how the system as a whole can be resilient to component failures. Finally, we discussed the impact of selected conditions that we believed could make a difference to the operation of the system including occluded line-of-sight between antennas, rain, antenna mounting height, additive RF interference, GPS denial, track curvature, and road slope. These tests gave us measurements ranging from not noticeable to significant differences in radio network performance; however, the physical real-world environment frequently includes multiple confounding factors and the more impactful considerations such as road curvature can overwhelm smaller subtleties when not isolated even on a closed test track.

The lean sensor suite and accompanying control architecture outlined herein has successfully provided energy efficiency benefits during on-road testing over stretches of I-85 and US 280 in Alabama, I-69 in Michigan, and logging roads in Canada. On-road implementation exposes the platoon to a host of exogenous disturbances not present in the track environment: interactions with non-platoon vehicles, traffic conditions forcing deviations from the desired speed set point, and a variety of weather conditions (snow or heavy rain). Ongoing research has sought to optimize the CACC system response to non-platooning vehicles that cut in and out of the platoon formation as well as CACC energy optimization in the presence of various traffic disturbances. However, these studies are beyond the scope of this chapter.

Overall, this technology offers significant benefits with very sparse requirements in terms of technology and integration expenses. By utilizing a leader–follower architecture for our vehicle AI we were able to perform the autonomy tasks required for traffic interactions with a high degree of trust in our system's safety and using less fuel than a human driver typically would.

**Acknowledgements** This material is based upon work supported by the Department of Energy, Office of Energy Efficiency and Renewable Energy (EERE), under Award Number DE- EE0008470. In addition, we would like to thank personnel from University of Michigan-Dearborn MDAS.ai group, Auburn University GAVLAB, NCAT, ACM, US Army GVSC, and Intertek for their assistance in the data collection.

## Definitions/Abbreviations

| | |
|---|---|
| **ACM** | American Center for Mobility |
| **CACC** | Connected Adaptive Cruise Control |
| **DSRC** | Dedicated Short-Range Communication |
| **GAVLAB** | GPS and Vehicle Dynamics Laboratory |
| **GPS** | Global Positioning System |
| **GVSC** | Ground Vehicle Systems Center |
| **NCAT** | National Center for Asphalt Technology |
| **NLOS** | Non-Line-of-Sight |
| **RF** | Radio Frequency |
| **RSSI** | Received Signal Strength Indicator |
| **SNR** | Signal to Noise Ratio |
| **V2V** | Vehicle to Vehicle |

## References

1. Federal Highway Administration, afdc.energy.gov [Online]. Available: https://afdc.energy.gov/data/10309. Accessed 10 Nov 2020
2. American Transportation Research Institute (2019). [Online]. Available: https://truckingresearch.org/wp-content/uploads/2019/11/ATRI-Operational-Costs-of-Trucking-2019-1.pdf. Accessed 10 Nov 2020
3. Bureau of Transportation Statistic, Freight shipments by mode [Online]. Available: https://www.bts.gov/topics/freight-transportation/freight-shipments-mode. Accessed 11 Oct 2020
4. Dadam SR, Jentz R, Ienzen T, Meissner H (2020) Diagnostic evaluation of exhaust gas recirculation (EGR) system on gasoline electric hybrid vehicle. SAE

5. Bergenhem C, Hedin E, Skarin D (2012) Vehicle-to-vehicle communication for a platooning system. Elsevier
6. van Nunen E, Verhaegh J, Silvas E, Semsar-Kazerooni E, Wouw N (2017) Robust model predictive cooperative adaptive cruise control subject to V2V impairments. In: IEEE 20th international conference on intelligent transportation systems (ITSC)
7. Lyamin N (2016) Performance evaluation of C-ACC/platooning under ITS-G5 communications. LICENTIATE THESIS I Halmstad University Dissertations no. 26
8. Yu T, Zhang S, Cao S, Xu S (2018) Performance evaluation for LTE-V based vehicle-to-vehicle platooning communication
9. Zeng T, Semiari O, Saad W, Bennis M (2019) Joint communication and control for wireless autonomous vehicle platoon systems. In: IEEE ICC, 2018
10. Bergenhem C, Johansson R, Coelingh E (2014) Measurements on V2V communication quality in a vehicle platooning application. In: MACOM 2014: multiple access communications, pp 35–48
11. Wang Z, Wu G, Barth MJ (2017) Developing a distributed consensus-based cooperative adaptive cruise control system for heterogeneous vehicles with predeessor following topology. J Adv Transp 2017(1023654)
12. Wang C, Gong S, Zhou A, Li T, Peeta S, Cooperative adaptive cruise control for connected autonomous vehicles by factoring communication-related constraints. In: Elsevier 23rd international symposium on transportation and traffic
13. Wang Z, Wu G, Barth MJ, A review on cooperative adaptive cruise control (CACC) systems: architectures, controls, and applications. In: IEEE 21st international conference on intelligent transportation systems (ITSC)
14. Gonçalves TR, Varma V, Elayoubi S (2020) Vehicle platooning schemes considering V2V communications: a joint communication/control approach. IEEE Wireless Commum, 1–6
15. Zhang C, Zang Y, Calvo JAL, Mathar R (2017) A novel V2V assisted platooning system: control scheme and MAC layer designs. In: 2017 IEEE 28th annual international symposium on personal, indoor, and mobile radio communications (PIMRC), 13 Oct 2017. IEEE
16. Shladover SE, Nowakowski C, Lu X-Y, Hoogendoorn R (2014) Using cooperative adaptive cruise control (CACC) to form high-performance vehicle streams. UC Berkeley Research Reports, 1 June 2014
17. Lu X-Y, Shladover SE, Automated truck platoon control and field test
18. Eilbert AC, Chouinard A-M, Tiernan TA, Smith SB (2020) Performance comparisons of cooperative and adaptive cruise control testing. In: A&WMA's 113th annual conference & exhibition
19. McAuliffe B, Lammert M, Lu X-Y, Shladover S, Surcel M-D, Kailas A, Influences on energy savings of heavy trucks using cooperative adaptive cruise control. SAE
20. Crane C, Bridge J, Bishop R (2018) Driver assistive truck platooning: considerations for florida state agencies
21. Roberts J, Mihelic R, Roeth M, Rondini D (2016) Confidence report: two-truck platooning. In: 2016 North American Council for Freight Efficiency
22. Sybis M, Kryszkiewicz P, Sroka P (2018) On the context-aware, dynamic spectrum access for robust intraplatoon communications. Mobile Inf Syst
23. Serizawa K, Mikami M, Moto K, Yoshino H (2019) Field trial activities on 5G NR V2V direct communication towards application to truck platooning. In: IEEE 90th vehicular technology conference
24. Nardini G, Virdis A, Campolo C, Molinaro A (2018) Cellular-V2X communications for platooning: design and evaluation. Sensors
25. Ward J, Smith P, Pierce D, Bevly D, Richardson P, Lakshmanan S, Argyris A, Smyth B, Adam C, Heim S (2019) Cooperative adaptive cruise control (CACC) in controlled and real-world environments: testing and results. In: 2019 NDIA ground vehicle systems engineering and technology symposium
26. Longman O (2019) Mitigation of vehicle vibration effect on automotive radar. IEEE Xplore, 16 Sept 2019

27. SAE V2X Core Technical Committee, Dedicated short range communications (DSRC) message set dictionary: a March 2016 update. Society of Automotive Engineers
28. Federal Communications Commission [Online]. Available: https://transition.fcc.gov/oet/ea/presentations/files/oct07/Oct_07-Basics_of_Unlicensed_Trans-JD.pdf. Accessed 12 Nov 2020
29. Francisco N, Vicente M (2019) Mixing V2V and Non-V2V equipped vehicles in car following. Transp Res Part C Emerg Technol 108:167–181
30. Amoozadeh M, Deng H, Chuah C-N, Zhang MH, Ghosal D (205) Platoon management with cooperative adaptive cruise control enabled by VANET. Vehic Commun 2:110–123
31. Sugimachi T, Fukao T, Suzuki Y, Kawashima H (2013) Development of autonomous platooning system for heavy-duty trucks. In: 7th IFAC symposium on advances in automotive control, vol 46, no 21, pp 52–57
32. SAE International (2016) Surface vehicle standard dedicated short range communication message set dictionary
33. Ma X, Chen X, Refai H (2009) Performance and reliability of DSRC vehicular safety communication: a formal analysis. EURASIP J Wirel Commun Netw 2009:1–13
34. Liu S, Xiang W, Punithan MX (2018) An empirical study on performance of DSRC and LTE-4G for vehicular communications. In: 2018 IEEE 88th vehicular technology conference (VTC-Fall)
35. 5GAA Automotive Association (2018) V2X functional and performance test procedures—selected assessment of device to device communication aspects
36. Bergenhem C, Johansson R, Coelingh E (2014) Measurements on V2V communication quality. In: MACOM, pp 35–48
37. Shladover SE, Nowakowski C, Lu X-Y, Using cooperative adaptive cruise control (CACC)
38. Liou C-Y, Mao S-G (2017) Miniaturized shark-fin rooftop antenna with integrated DSRC communication module for connected vehicles. In: 2017 XXXIInd General Assembly and scientific symposium of the international union of radio science (URSI GASS)
39. Ekiz L, Posselt A, Klemp O, Mecklenbrauker CF (2014) Assessment of design methodologies for vehicular 802.11p antenna systems. In: 2014 international conference on connected vehicles and expo (ICCVE)
40. Onishi H, Watanabe F, Mlinarsky F, Velasquez C (2013) DSRC performance assessment for crash warning applications. In: 2013 international conference on connected vehicles and expo (ICCVE)
41. Harri J, Tchouankem H, Klemp O, Demchenko O (2013) Impact of vehicular integration effects on the performance of DSRC communications. In: 2013 IEEE wireless communications and networking conference (WCNC)
42. CAMP LLC (2019) C-V2X performance assessment project
43. Meireles R, Boban M, Steenkiste P, Tonguz O, Barros J (2010) Experimental study on the impact of vehicular obstructions in VANETs. In: 2010 IEEE vehicular networking conference
44. Cheek E, Alghodhaifi H, Adam C, Andres R, Lakshmanan S (2020) Dedicated short range communications used as failsafe in autonomous navigation. In: Unmanned systems technology XXII
45. Kestwal MC, Joshi S, Garia LS (2014) Prediction of rain attenuation and impact of rain in wave propagation at microwave frequency for tropical region (Uttarakhand, India). Int J Microwave Sci Technol 2014
46. Kaul S, Ramachandran K, Shankar P, Oh S, Gruteser M, Seskar I, Nadeem T (2007) Effect of antenna placement and diversity on vehicular network communications. In: 4th annual IEEE communications society conference on sensor, mesh and ad hoc communications and networks, pp 112–121
47. Gao S, Lim A, Bevly D (2016) An empirical study of DSRC V2V performance in truck platooning scenarios. Digit Commun Netw 2(4):233–244

48. Bogard SE, Bao S, LeBlanc D, Jun L, Qiu S, Lin B (2017) Performance of DSRC during safety pilot model deployment. SAE Int J Passenger Cars-Electron Electr Syst 10:165–172
49. Cheng B, Lu H, Rostami A, Gruteser M, Kenney JB (2017) Impact of 5.9 GHz spectrum sharing on DSRC performance. In: 2017 IEEE vehicular networking conference (VNC)
50. Bainwol M, Ujkashevic D, Butler D, Kwant JFKJZ (2020) Comments of the Ford Motor Company before the Federal Communications Commission in the matter of use of the 5.850–5.925 GHz band

# Environmental Perception for Intelligent Vehicles

**Xiaoliang Tang, Yuanxiang Li, and Xian Wei**

**Abstract** Environmental Perception for Intelligent Vehicles (EPIV) generally focuses on the awareness and understanding of the driving environment around intelligent vehicles by various vehicle sensors. In recent years, a lot of excellent research has been conducted on developing novel methods and technologies for EPIV. This chapter overviews some of the main research topics in the field of EPIV. First, this chapter reviews various types of vehicle sensors which capture multimodal environmental information around intelligent vehicles and form the foundation for environmental perception. Second, this chapter focuses on data restoration and denoising technologies on camera and LiDAR sensors, which guarantees the quality of the data captured by the vehicle. Third, this chapter deals with methods on semantic segmentation, object detection and tracking with camera and LiDAR data, which play a central role in environmental understanding. Fourth, this chapter introduces technologies on location and mapping with multimodal sensor data, which is essential for the local path planning of intelligent vehicles. Finally, this chapter discusses the research technologies on fusing multimodal environmental data, which represents the frontier of EPIV development.

X. Tang
School of Information and Electronic Engineering, Zhejiang University of Science and Technology, 318 Liuhe Road, Xihu District, Hangzhou City, Zhejiang Province, China

Y. Li (✉)
School of Aeronautics and Astronautics, Shanghai Jiao Tong University, Shanghai 200240, China
e-mail: yuanxli@sjtu.edu.cn

X. Wei (✉)
Software Engineering Institute (SEI), East China Normal University, 3663 N. Zhongshan Rd., Shanghai 200062, China
e-mail: xwei@sei.ecnu.edu.cn

 
Y. L. Murphey et al. (eds.), *AI-enabled Technologies for Autonomous and Connected Vehicles*, Lecture Notes in Intelligent Transportation and Infrastructure,
https://doi.org/10.1007/978-3-031-06780-8_3

# 1 Sensors

Sensors make the vehicle capable of perceiving objects on the street and accumulating information for safe driving. Furthermore, the information captured by sensors is processed and analyzed to build the path from the start point to the end point and to send the appropriate instructions to the control system of the vehicle, such as accelerating, steering, and braking. In addition, the information captured by autonomous vehicle sensors, such as traffic jams and various kinds of obstacles on the road, can be shared with other vehicles if these vehicles are connecting through vehicle-to-everything equipments. This chapter mainly investigates three kinds of key sensors in intelligent autonomous vehicles: cameras, LiDARs, and radars.

## *1.1 Development of Sensors in Intelligent Autonomous Vehicles*

Autonomous vehicles require high-performance vehicle sensors with rigorous specifications for sensitivity, reliability, and data richness. In order to measure the autopilot level, the Society of Automotive Engineers (SAE) has defined 6 levels of autonomous driving, as shown in Fig. 1, from L0 (level 0, fully manual) to L5 (level 5, fully automatic). Higher levels of autopilot require not just more sensors and better sensors. For instance, the number of L2 autonomous driving sensors is about 6, that of L3 autonomous driving is about 13, and the number of L5 autonomous driving sensors will exceed 30 in the future. In the early stages of self-driving development, vehicles rely merely on cameras to achieve autonomous driving on structured roads. Similar to the human eye, the camera receives surrounding light information, yet its perception of the environment is highly vulnerable to bad conditions, such as glaring light, rain, fog, and snow.

With the development of LiDAR technology, LiDAR and millimeter-wave radar have gradually transferred from military to civilian use. Due to the characteristics of light travelling in a straight line, LiDAR has been used for ranging for a long

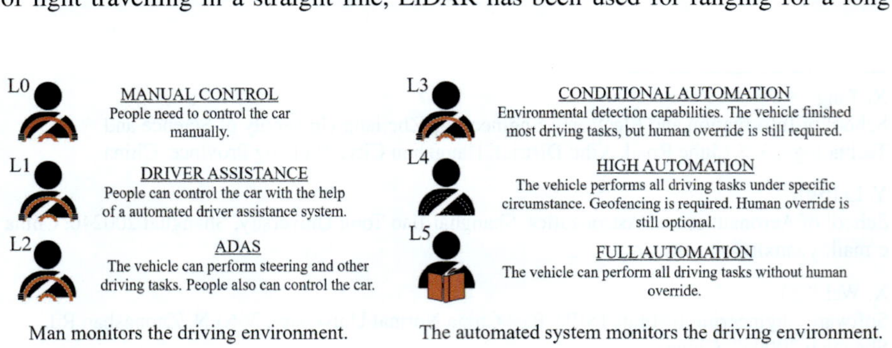


**Fig. 1** Levels of driving automation

time, such as for the guidance of aerial bombs. LiDAR began to be widely used in many civilian markets because of the emergence of vacuum cleaning robots. In 2010, Neato introduced a LiDAR on the sweeping robot and launched the Neato XV-11. This revolutionary breakthrough raised the popularization of the utilization of sweeping robots. Since then, LiDAR applications have become more widespread in commercial applications. Radar sensors followed a similar transformational path from military to civilian uses. Cameras, LiDAR, and radar are now poised to bring the arrival of the age of autonomous driving. More details about these three types of sensors are given in the sections below.

## *1.2 Camera*

Cameras are widely used in self-driving vehicles because they are well understood and relatively cheap. The mechanism of a camera can be formulated as follows: first, the camera captures the optimal image generated by the object and projects it onto the image sensor. Second, an electronic circuit transforms the optical information into an electrical signal; Third, Analog-to-Digital (A/D) conversion is used to transform the electrical signal into a digital one, amenable to computer processing. Finally, a Digital Signal Processing chip (DSP) processes the signal into a specific format image and can be displayed on a screen, as shown in Fig. 2. In the context of autonomous vehicles, cameras are commonly used for three types of applications: driving assistance, parking assistance, and in-vehicle driver monitoring. Each type of application may require a different type of camera.

Typically, monocular cameras and binocular cameras are employed for the detection of traffic lane lines, dynamic objects, traffic signs, etc. Compared with monocular cameras, binocular cameras provide better estimation for the distance of the target. Cameras with a fished-eyed lens are used in intelligent vehicles because they offer a broader field of vision, and hence helpful in monitoring dynamic targets in the driver's blind spot areas.

The workflow of the classification of the monocular camera involves 5 steps: image input, preprocessing, feature extraction, classification and matching, and output results. The range finding problem is to estimate the distance from the camera to an object according to the size in the image. The ranging principle of a monocular camera is shown in Fig. 3. The distance of the object to the image plane can be calculated according to

$$D = (W \times F)/P \tag{1}$$

**Fig. 2** The working principle of the camera

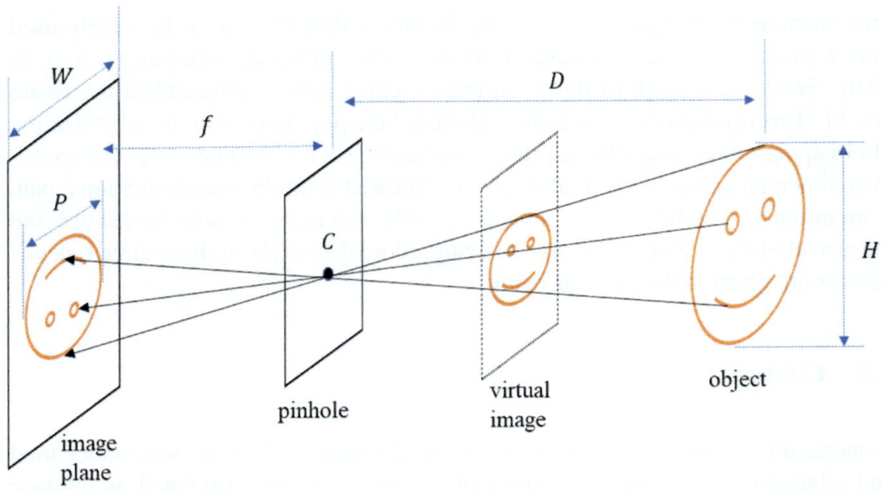


**Fig. 3** The principle of ranging for a monocular camera. In the figure, $F$ is the focal length of the camera, $C$ is the optical center of the lens, $D$ is the distance from the target to the camera, $H$ is the object height, $W$ is the image size, and $P$ is the size of the object in the image

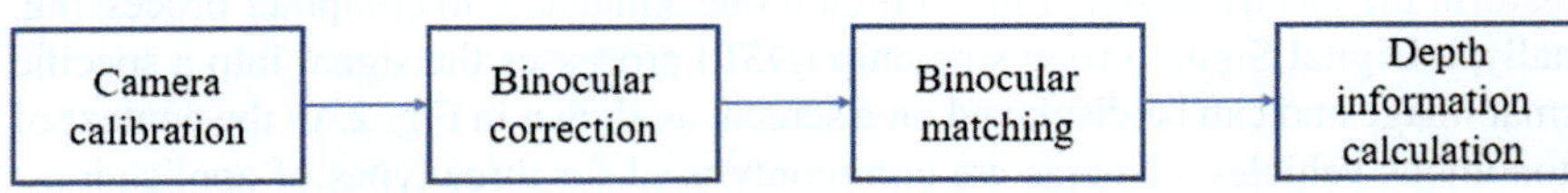


**Fig. 4** Ranging steps of a binocular camera

However, a monocular camera has many disadvantages: (1) monocular camera strictly requires object sizes (large objects may not be captured completely); (2) monocular camera usually leads to image distortion.

The ranging principle of the binocular camera is similar to the mechanism in human eyes. A system for computing object distance and depth using binocular cameras is given in Fig. 4. The first step is camera calibration, which calibrates each camera's internal parameters and measures the relative position between the two cameras. Next, the binocular correction step eliminates the distortion and the line alignment between the right and the left camera views. Then, corresponding pixels of the same scene on the left and the right camera views are matched in the Binocular matching phase. Lastly, the depth information is calculated at the end of the process.

The principle of binocular ranging is shown in Fig. 5, which illustrates the similarity relationship between triangles $\Delta P P_L P_R$ and $\Delta P O_L O_R$. The ratio of the distance between the left and the right camera imaging points $P_L$ and $P_R$ to the distance between the camera center points $O_L$ and $O_R$ equals the ratio of the distance $Z$ from the object to the baseline of the camera to the distance $Z - f$ from the object to the camera lens. The ratio relationship is given by:

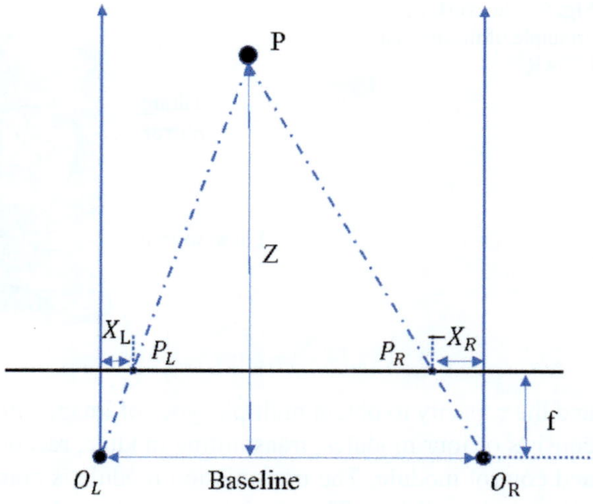


**Fig. 5** The principle of binocular camera ranging. $P$ is the object, which forms the target points on the left and right cameras respectively, as $P_L$ and $P_R$; $O_R$ and $O_L$ are the optical centers of the two cameras; $f$ is the camera's focal length, $B$ is the center distance between the two cameras, $Z$ denotes the depth information

$$\frac{B - (X_R - X_T)}{B} = \frac{Z - f}{Z} \tag{2}$$

Except for the variable $Z$, all other parameters are constants after the camera is calibrated. Solving for $Z$, then gives:

$$Z = \frac{f \cdot B}{X_R - X_T} \tag{3}$$

Car cameras have more stringent performance requirements than consumer cameras. For example, car cameras should function properly over the temperature range of $-40 \sim 85\,°\mathrm{C}$ and be robust to drastic changes of temperature. Vehicles produce strong vibrations when driving on uneven roads, so the car camera must resist vibrations of various intensities. When the vehicle starts, it produces an extremely high electromagnetic pulse, so the car camera needs to satisfy a highly antimagnetic performance. Moreover, because of the high speed of the vehicle and the frequently changing light environment, the camera electronics need to possess a high dynamic characteristic.

## 1.3 LiDAR

LiDAR uses an electromagnetic wave system (laser beam) to detect objects. Depending on the reflected or scattered signal parameter, such as arrival time and intensity, the LiDAR can determine the distance, azimuth, motion state, and surface optical characteristics of objects in its field of view. Compared with cameras, LiDAR has a higher angular resolution, higher range resolution, stronger anti-interference ability,

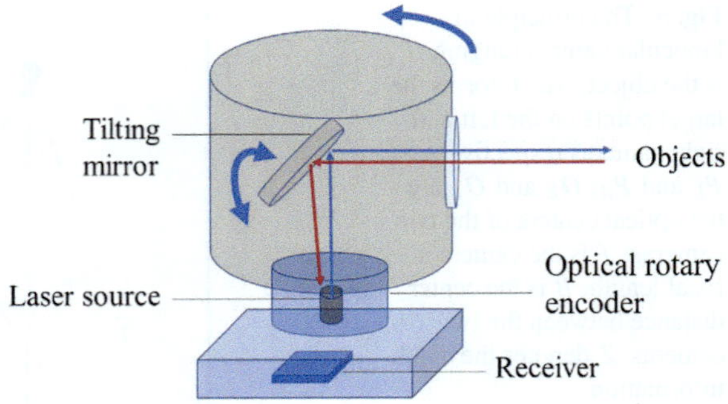


**Fig. 6** The working principle of mechanical LiDAR

and the capacity to obtain multiple types of image information of the target. LiDAR consists of four modules: transmitting module, receiving module, scanning module and control module. The transmission module is composed of a laser and an optical emission system. The receiving module includes an optical receiving system, optical filter device and photodetector. The scanning module consists of a motor, a miniature resonator mirror, and a phased array. The function of the scanning module is to change the spatial projection direction of the laser beam. The control module completes the control of the laser emission module, the receiving module, and the scanning module. Furthermore, the control module also processes the LiDAR data with the help of an external system.

There are 3 main types of LiDAR differentiated by the structure of its rotating parts: mechanical LiDAR, hybrid LiDAR and solid-state LiDAR. A schematic diagram of a mechanical LiDAR is shown in Fig. 6. The laser emitting components are arranged in a line array of laser light sources in the vertical direction and can generate laser beams of different directions in the vertical plane through the lens. Driven by the motor, the line array of the laser light source is rotating. As a result, the laser beam transforms from "line" into "plane" in the horizontal plane. Multiple laser "planes" are formed by rotating the laser light source around the roll axis. Therefore, producing a 3D scan of the environment.

For example, Velodyne LiDAR is a standard mechanical LiDAR. The performance parameters of Velodyne HDL-32E LiDAR are shown in Table 1. The number of laser beams of this LiDAR is 32, the scanning frequency is 20 Hz, the detection accuracy can reach centimeter-level, and the detection range is 0–70 m. Although mechanical LiDAR has a 360° viewing and high measurement accuracy, it has a large size, expensive cost, and low reliability of rotating parts.

Hybrid LiDAR utilizes a stepping motor rather than rotating parts. The hybrid LiDAR combines the Micro-Electro-Mechanical System (MEMS) with a vibrator to form a MEMS vibrator. The vibrator rotates to complete the laser scanning. Figure 7 shows the structure of the hybrid LiDAR. The control circuit drives the laser to generate laser pulses while driving the MEMS galvanometer to rotate, and the laser is reflected by the rotating galvanometer to realize scanning. Compared with the mechanical LiDAR, the hybrid LiDAR technology is more matured, and has a lower

**Table 1** Some important parameters of Velodyne HDL-32E LiDAR[a]

| Parameter | Value |
|---|---|
| Wavelength of electromagnetic waves | 905 nm |
| Detection range | 1m 70 m |
| Horizontal field of view | 360° |
| Vertical field of view | +10.67° ~ −30.67° |
| Distance accuracy | Vertical ~1.33°, horizontal ~0.16° |
| Scanning frequency | 20 Hz |

[a]Refer to https://velodynelidar.com/products/hdl-32e/

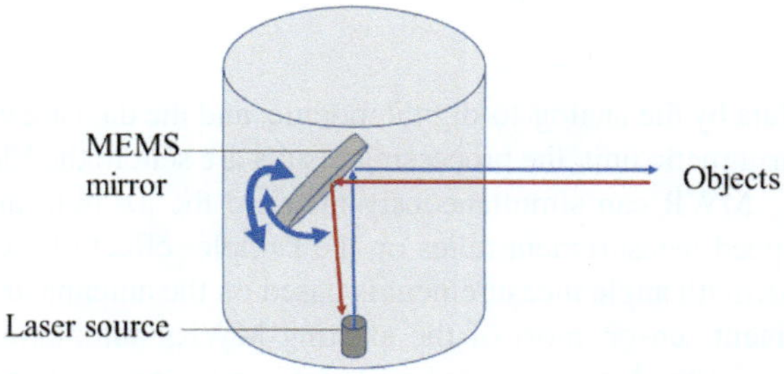


**Fig. 7** The working principle of hybrid LiDAR

cost. However, the receiving end of a hybrid laser radar has a complicated optical path, and a small scanning range.

The solid-state LiDAR completely eliminates the mechanical scanning structure, and achieves scanning rotation in both the vertical and horizontal directions. Unlike the introduced hybrid LiDAR and mechanical LiDAR, the solid-state LiDAR only retains the micro-movement, which reduces the loss caused by the mechanical movement like other kinds of LiDAR and which gives a longer working life for the solid-state LiDAR.

## 1.4 Radar

Millimeter-wave radar (MWR) refers to a long-wave radar that emits electromagnetic waves with a wavelength range of 1–10 mm and a frequency range of 30–300 GHz. Vehicle-mounted MWR is widely used in many key tasks, such as adaptive cruise control, lane change assistance, and collision warning systems. MWR has the advantages of small size, easy integration, and high spatial resolution. Compared with LiDAR, MWR has a lower cost and a stronger anti-interference ability.

The working principle of MWR is shown in Fig. 8. Firstly, MWR uses a voltage-controlled oscillator to generate electromagnetic waves with a specific modulation frequency. Secondly, the received signal is sent through the power amplifier module and the voltage-controlled oscillator module. Finally, the signal is transformed into

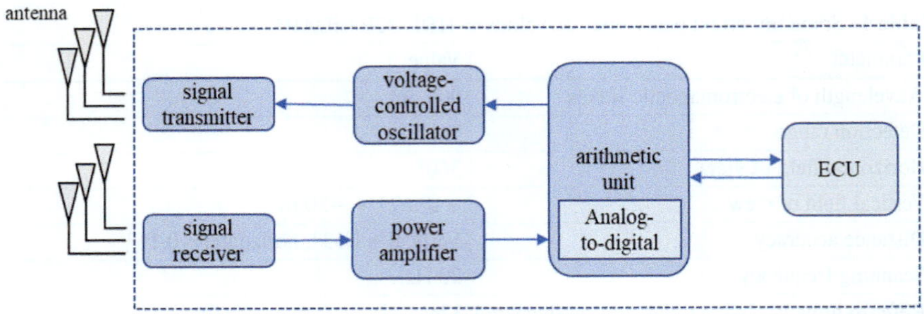


**Fig. 8** Schematic diagram of the basic working principle of MWR

data by the analog-to-digital module, and the data are collected and calculated in the arithmetic unit, the processing results are sent to the Electronic Control Unit (ECU).

MWR can simultaneously measure the azimuth angle of multiple targets. The speed measurement relies on the Doppler effect of electromagnetic waves, and the azimuth angle measurement is based on the antenna array of the MWR. As an intelligent sensor, most of the existing MWRs automatically process electromagnetic waves to obtain clustered millimeter-wave points. For example, the ARS408 MWR produced by Continental is a widely used intelligent MWR. The MWR data after intelligent processing is shown in Fig. 9. Each point represents a possible location of the target.

The detection range of ARS408 MWR is shown in Fig. 10. The long-range detection of ARS408 MWR can achieve 170 m, and the short-range detection can achieve 70m.

Each point output by the ARS MWR contains 18-dimensional information. As shown in Table 2, in addition to the basic three-dimensional space coordinates, the output includes radar point dynamic characteristics, such as target relative velocity, cluster false alarm probability, the Doppler state solution, point validity and other information.

## 1.5 Future of Sensors in Intelligent Vehicles

In order to identify and detect objects more accurately, there are several future research opportunities involving vehicle cameras and LiDAR. The development of vehicle camera technology has the following trends:

1. Higher dynamic range (HDR). The camera can successfully function in difficult lighting environments, such as the high contrast of bright and dark details in and approaching a tunnel under strong light. There are many such instances of high contrast in the environment. Therefore, the improvement of HDR is a critical factor in increasing the level of autonomous driving.

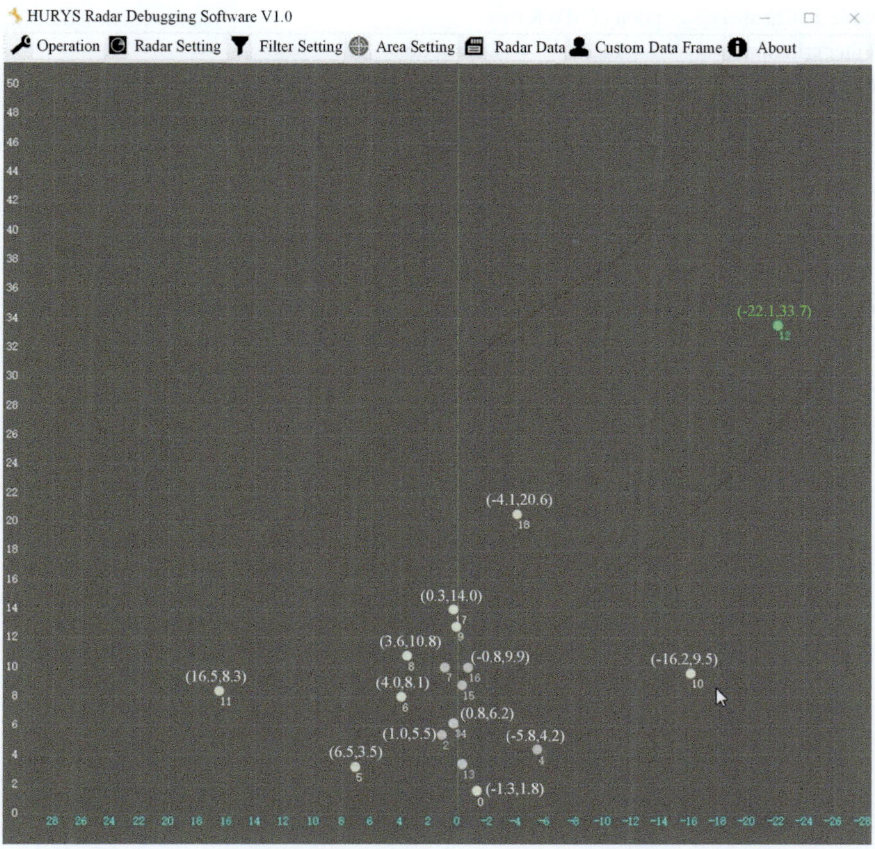


Fig. 9 The target point after ARS408 radar clustering

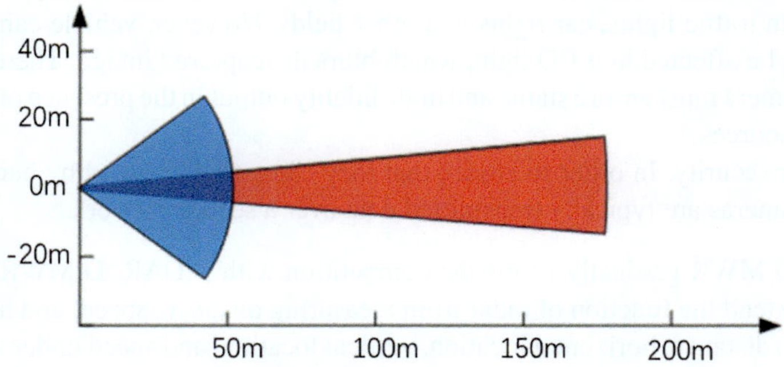


Fig. 10 ARS408 MWR detection range. The blue area is the short-range beam detection range. The red area is the long-range beam detection range

**Table 2** Channel description of MWR data[a]

| Filter criterion | Index | Description |
|---|---|---|
| Coordinate | 0,1,2 | Point space coordinates $(x, y, z)$ |
| Dyn_prop | 3 | 0: Moving, 1: Stationary, 2: Oncoming, 3: Stationary candidate, 4: Unknown, 5: Crossing stationary, 6: Crossing moving, 7: Stopped |
| ID | 4 | 125 Radar point ID |
| RCS | 5 | Radar reflection cross-sectional area |
| Velocities | 6, 7 | Speed in the x direction and y direction (m/s) |
| Compensation velocities | 8, 9 | The velocities in m/s compensated by the ego motion |
| Ambig_state | 11 | 0: Invalid, 1: Ambiguous, 2: Staggered ramp, 3: Unambiguous, 4: Stationary candidates |
| $x\_rms$, $y\_rms$ | 12, 13 | x-direction and y-direction position variance |
| Invalid_state | 14 | 0: Valid, 1: Invalid due to low RCS, 2: Invalid due to near-field artefact, 3: Invalid far range cluster because not confirmed in near range, 4: Valid cluster with low RCS, 5: Reserved, 6: Invalid cluster due to high mirror probability, 7: Invalid cluster because outside sensor field of view, 8: Valid cluster with azimuth correction due to elevation, 9: Valid cluster with high child probability, 10: Valid cluster with high multi-target probability, 11: Valid cluster with suspicious angle, 12: Valid cluster with high probability of being a 50 deg artefact, 13: Valid cluster but no local maximum,14: Valid cluster with high artefact probability, 15: Reserved, 16: Invalid cluster because it is a harmonics, 17: Valid cluster with above 95 m in near range |

[a]Refer to https://velodynelidar.com/products/hdl-32e/

2. Eliminate LED flicker (LEF). Due to the popularity of LEDs, they are widely used in traffic lights, car lights and other fields. However, vehicle cameras can easily be affected by LED light, which blurs the captured image. Therefore, the car camera must ensure stable and high-fidelity output in the presence of multiple LEF sources.
3. Cyber security. In order to ensure that they cannot be attacked by hackers, the car cameras are typically transmitted data over a secure network.

The 4D MWR gradually forms the competition with LiDAR. D MWR technology can extend the function of radar from measuring distance, speed, and horizontal azimuth to distance, horizontal location, vertical location, and speed under any lighting or weather conditions. Therefore, 4D millimeter-wave radar has the ability to combine the advantages of LiDAR and MWR to achieve accurate all-weather detection.

Therefore, 4D millimeter-wave radar has the ability to combine the advantages of LiDAR and MWR to achieve accurate all-weather detection.

## 2 Data Restoration

Data collected by sensors in self-driving systems suffer from a variety of noise sources, increasing the vulnerability of downstream tasks. This section explains the mechanism of deep neural networks for mitigating the effect of noise from both cameras and LiDARs.

### *2.1 RGB Image Restoration*

Images captured by AV vehicle cameras typically contain pedestrians, traffic signs, other vehicles, etc. Important image processing tasks include lane detection, and object tracking and segmentation. However, harsh weather such as rain, fog and snow can distort the images, making the above image processing tasks much more difficult. In addition, the process of image capture itself introduces noise which can also negatively affect the tasks. Image restoration, being an important step of image preprocessing, improves the performance of computer vision tasks for vehicles. This section introduces current rain and fog removal methods for RGB images applicable to autonomous driving.

Based on the deep learning algorithms, the system could restore the noise-affected image by recovering real-world noises. However, the challenge remains to be the quantitative demand for training image pairs. The utilization of synthesized rainy or foggy datasets is a part of data enhancement for training artificial models. Hence we can adopt artificially synthesized rainy or foggy datasets for training while the trained model would be tested on real-world noise. In detail, we randomly select 1000 images from the large traffic dataset Bdd100K [67] to artificially synthesize the rainy traffic scenarios, dubbed Bdd1000, and some examples are shown in Fig. 11.

Traditional image restoration methods tend to use physical priors, and dictionary learning is proper for representing priors effectively. These algorithms pay more attention to rain streaks themselves rather than probing into the formation of rain. A general formula for the physical prior which could be easily optimized [66] is given by:

$$\hat{B} = \arg\min_{B} \|I - B - S\|_2^2 + \gamma(B) + \psi(S) + \Omega(B, S) \tag{4}$$

**Fig. 11** Image samples of Bdd1000 with rain

where $B$ denotes the background layer, $S$ denotes the rain streak layer, $I$ denotes the identity matrix, $\gamma(B)$ represents priors on the non-rain layer, $\psi(S)$ denotes priors on the rain layer, and $\Omega(B, S)$ indicates a certain relationship between non-rain layer and rain layer.

Most of the traditional methods applied to rain removal, such as support vector machines [22], are computationally expensive and time-consuming, which are unsuitable for autonomous driving with high real-time requirements. Recently, owing to the powerful feature extraction capability, DNN has been widely used in computer vision tasks, such as image denoising, image de-raining, and image de-hazing. Especially, data-driven rain removal methods have been rapidly developing since 2017, which generally outperform traditional approaches. These methods usually use deep networks to automatically extract the features of images, and learn how to recover the original (undistorted) image. Deep learning-based models have more difficulties compared to traditional ones since the regularizers in Eq. (4) are not easy to fine-tune in huge parameter space. For deep learning methods, the most frequently used model is a relatively simpler one, a composite model:

$$O = B + S \tag{5}$$

where $B$ denotes background image as a non-rain layer, $S$ denotes rain streaks as rain layer, and $O$ is an original rainy image. The model is built upon the assumption that the rain image can be assumed as the superposition of background image and the rain streaks. Specifically, the background image is on the bottom, and the rain streaks are on the top and linearly superimposed on the background (as shown in Fig. 12). Thus, Eq. (5) can be rewritten as:

$$B = O - S \tag{6}$$

There are several popular deep learning-based rain removal methods, among which, using CNN to extract features occupies the mainstream and there are a great many variations. For instance, Fig. 13 shows the structure of PRENet [47], which consists of several residual modules, recurrent layers and progressive structures. The simple network structure results in less computational cost and guarantees real-time performance, even when they are combined with auxiliary tasks such as segmentation and classification.

**Fig. 12** Superimposed composite model

The computational efficiency of PRENet makes it competitive for self-driving systems. Some results on image removal and segmentation tasks provided by PRENet are shown in Fig. 14, where the restored images yield better segmentation results than rainy images.

There exist models jointly removing rain and fog within a simple-structured network [18] where PRENet is used for data preparation [18], YOLO v3 for object detection and PSPNet for image segmentation [55]. Figures 15 and 16 show the results of these two combinations respectively. Clearly, the images restored by PRENet demonstrate significant improvement for object detection and image segmentation tasks.

As for fog removal tasks, the haze formation model could be given by color attenuation prior:

$$I_f(x) = R_d(x) \times t_r(x) + A(1 - t_r(x)) \tag{7}$$

where $I_f(x)$, $R_d(x)$, and $A$ denote the intensity of the foggy image, the scene radiance of the recovered image (clear image), and the depth of the scene respectively, $t_r(x) =$

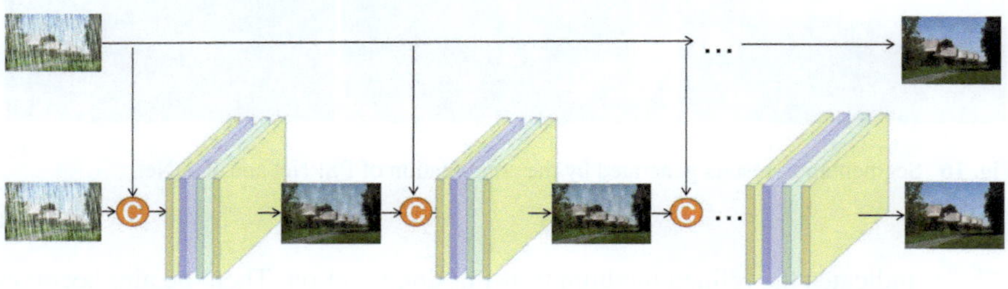

**Fig. 13** The architecture of PRENet. The circle $C$ means concatenation operator for tensors

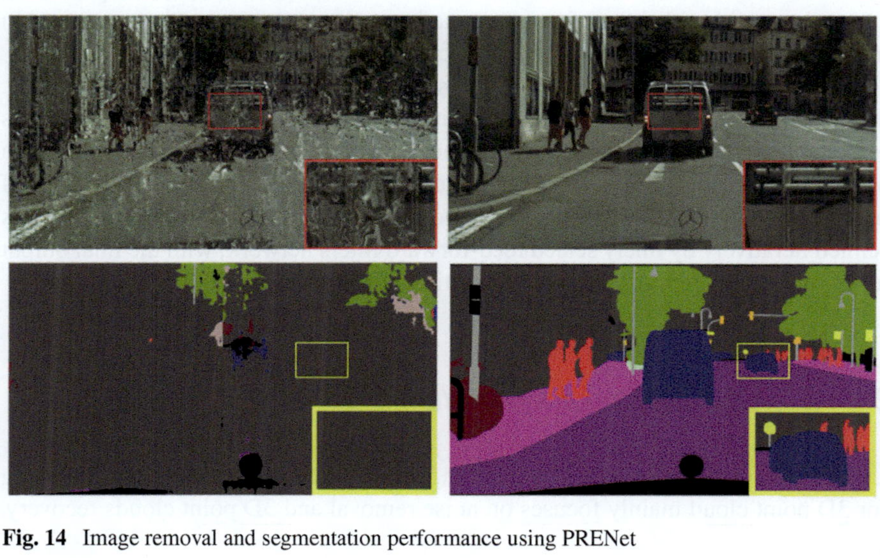

**Fig. 14** Image removal and segmentation performance using PRENet

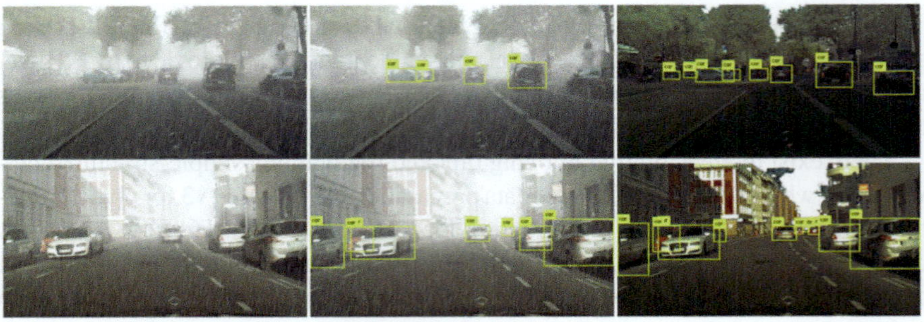

**Fig. 15** Object detection results generated by the combination of Yolov3 and JDDNet

**Fig. 16** Segmentation results generated by the combination of PSPNet and JDDNet

$e^{-\beta \cdot d(x)}$ indicates the refined medium transmission function. Then the aim becomes to restore scene radiance by recovering the unknown parameters $A$ and $t_r(x)$, which can be estimated with certain assumptions in traditional models [73]. Deep learning comes to an aid with better feature extraction power to estimate the transmission map, which could be refined with multi-scale architecture [1, 49].

For autonomous vehicles, the response speed is the most important for emergency. Therefore, the primary requirement of autonomous vehicles is efficiency, and it is reasonable to recover the unknown parameters with more robust assumptions, such as the HSI cue [7, 56, 57, 74, 76]. After converting the input hazy image into HSI mode, through the Patch extraction and nonlinear mapping, and the feature map is refined iteratively by finely scaled feed-forward filters network, with the final output given by local extremums and reconstruction of feature maps.

## 2.2 LiDAR Point Cloud Restoration

LiDAR also plays an important role among autonomous driving sensors. Denoising for 3D point cloud mainly focuses on noise removal and 3D point clouds recovery,

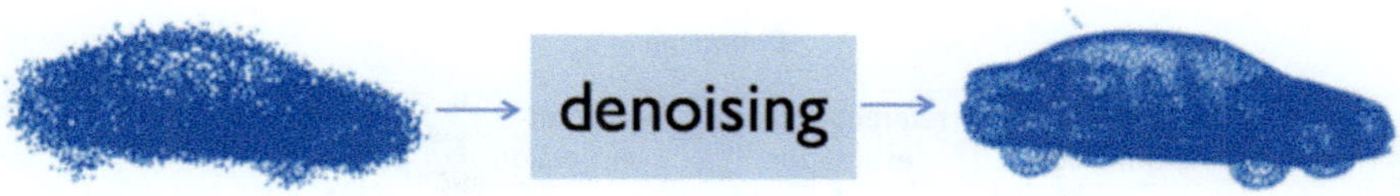


**Fig. 17** The denoising process on 3D point cloud

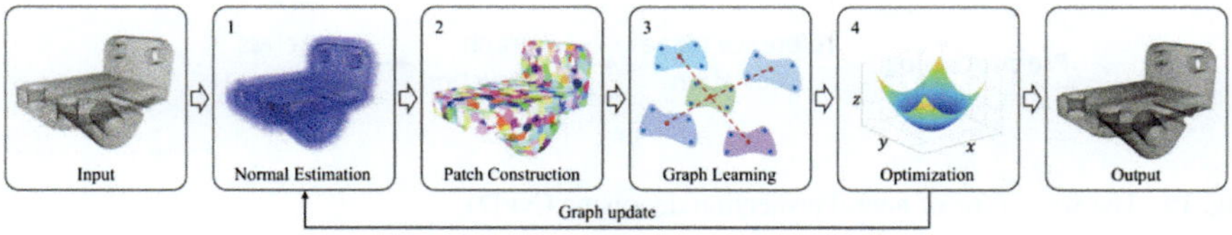


**Fig. 18** 3D point cloud denoising process using graph Laplacian regularizer [21]

which is essential for automated vehicle systems. Specifically, the denoising process on 3D point cloud is shown in Fig. 17 (refer to [9]).

Generally, point cloud restoration methods are built upon smoothness prior implemented by the regularizer $J(\hat{X})$:

$$\min_{\hat{X}} \left\| \hat{X} - X_\epsilon \right\|_F^2 + \lambda J(\hat{X}) \tag{8}$$

where $\hat{X}$ indicates the recovered point cloud and $X_\epsilon$ the point cloud with noise. There are several choices for $J(\hat{X})$, such as graph Laplacian regularization [69] which has been widely used to regularize the vertex and non-vertex terms of Mahalanobis distance matrix. A common pipeline of the method [21] on 3D point clouds is demonstrated in Fig. 18.

The DNN based models for denoising 3D point cloud data have gradually become the mainstream, such as neural projection denoising (NPD [13]). After projecting noise points onto an estimated rough reference plane, multi-projecting is implemented due to inevitable noise in the estimated rough reference plane NPD utilizes a neural network to estimate a reference plane so that to denoise 3D point cloud, which is shown in Fig. 19.

## 3 Semantic Segmentation

Semantic segmentation is one of the most important tasks in the computer vision, because it provides complex scene understanding. Semantic segmentation is the task of clustering parts of an image together which belong to the same object class [33]. It is a form of pixel-level prediction because each pixel in an image is classified according to a category. For example, the left of Fig. 20 is an original photograph,

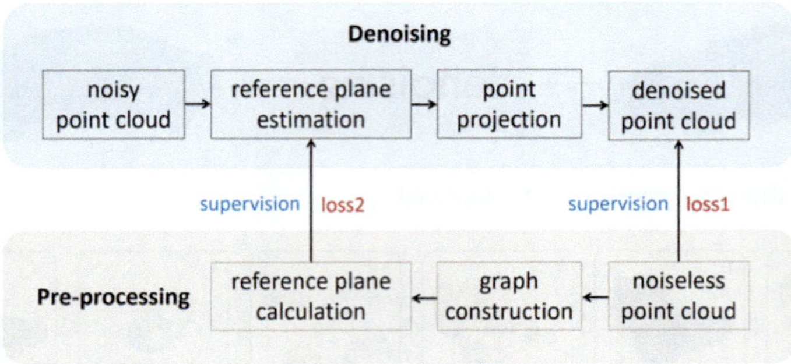


**Fig. 19** The work flow of neural projection denoising (NPD)

**Fig. 20** An instance of semantic segmentation

and after semantic segmentation, an image such as the one on the right is produced, where each color represents the class of a corresponding object. Nowadays, semantic segmentation is widely used in various fields, such as intelligent vehicle driving, geographic information systems, medical image analysis, and robotics.

Many types of methods have been proposed for semantic segmentation. Traditional algorithms divide an image into different parts by some region and edge detection techniques, and then gather similar pixels with clustering methods. Traditional algorithms of semantic segmentation are usually specific to the problem, and the universal algorithm of segmentation for all images does not exist. But deep learning approaches can solve this problem. With the advancement of computing power and the development of deep learning theory, convolutional neural networks are introduced to the field of semantic segmentation [68]. Different semantic segmentation methods for different data types are described in the following sections, respectively.

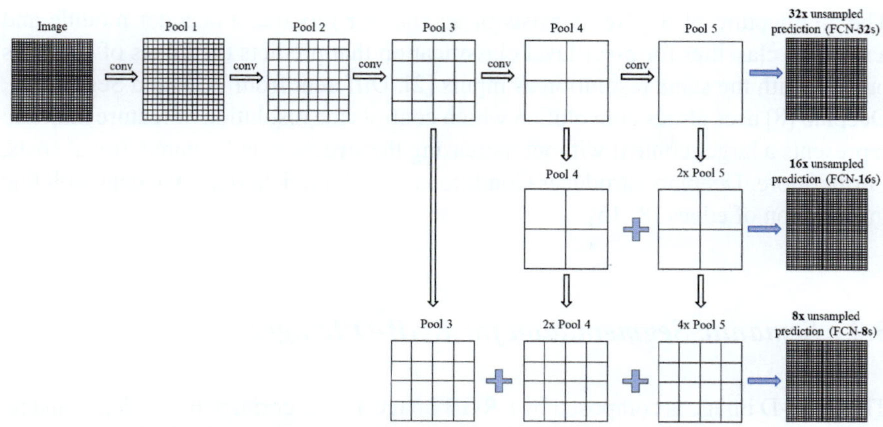


**Fig. 21** A skip architecture that combines outputs from low layers and high layers

## 3.1 Semantic Segmentation for RGB Images

An RGB image is a color pixel group of $M \times N \times 3$ size, where each color pixel corresponds to a three-value group, and the three values correspond to the red, green, and blue components of the RGB image at a specific spatial position, respectively.

There are a lot of RGB image datasets for training and testing semantic segmentation models. Pascal Visual Object Classes (VOC) 2012 dataset [14] consists of common objects in daily life, such as animals, vehicles, and electrical appliances. The Microsoft Common Objects in Context[1] (COCO) [30] dataset provides images of common objects in complex daily scenes for the broader scene understanding via semantic segmentation. The Cityscapes dataset [11] is suitable for training models for the control of autonomous vehicles, and it includes thousands of urban street scenes with fine annotations and coarse annotations.[2]

For RGB images, a classical model is Fully Convolutional Networks (FCN) [32] which is an end-to-end semantic segmentation method and achieves satisfied segmentation performance in PASCAL VOC. FCN is not sensitive to the size of input image, it can restore the images with various sizes to the same size as the input image by upsampling that is a learnable deconvolution. In addition, FCN adds a skip architecture (shown in Fig. 21) that upsamples the outputs of the high layer to make the output size keep the same as those of lower layers. Depending upon these special architectures, FCN performs more outstanding than contemporary deep networks in semantic segmentation.

Based on FCN, SegNet [2] introduces pooling indices that indicate the relative locations after pooling to reduce the computational cost and increase the accuracy.

---

[1] https://www.microsoft.com/en-us/research/publication/microsoft-coco-common-objects-in-context/.

[2] https://www.cityscapes-dataset.com/.

The architecture of SegNet consists of an encoder module, a decoder module and a softmax classifier for pixel-level classification that predicts the labels of pixels as outputs with the same resolution as inputs [2]. Different from FCN and SegNet [2], Deeplab [8] uses atrous convolution which controls the resolution of feature map and represents a larger context without increasing the structure and computational costs. Furthermore, Deeplab introduces Conditional Random Field (CRF) to deal with fine information of edges [8, 15].

## 3.2 *Semantic Segmentation for RGB-D Images*

The RGB-D image is composed by a RGB image and its corresponding depth image. A depth image is an image channel in which each pixel relates to a distance between the image plane and the corresponding object in the RGB image. RGB-D images contain not only 2D plane information but also spatial distance and geometric relationships.

NYU-Depth V2 dataset,[3] captured by the Microsoft Kinect, contains plenty of indoor scenes caught from video sequences [52]. SUN RGB-D dataset[4] is large-scale, similar to Pascal VOC in size, and contains tens of thousands of RGB-D images captured by four different sensors, which benefits training algorithms for scene understanding [53].

Images with distance information cannot directly input the models that only deal with RGB images. Instead, the depth data of RGB-D images are encoded as three channels on each pixel just like RGB images. In this way, the RGB-D images are transformed into RGB images, which is suitable for models designed for RGB data. Long Short-Term Memorized Context Fusion (LSTM-CF) model [28] takes advantage of some convolutional layers and a memory layer that encodes the short-range and long-range spatial relationships in the image to get the context in the depth channel. Besides, Zeng et al. [69] used the fully CNN to segment images and label every pixel from all viewpoints of RGB-D data, which utilizes multi-view data to improve the performance of current single-view approaches. Ma et al. [34] proposed a novel DNN model to perform semantic segmentation in RGB-D images. The deep neural network [34] can predict consistent semantic information from multiple perspectives and perform more consistently in semantic prediction than that trained by single-view images. Unlike those applying 2D semantic segmentation methods to RGB-D images directly, several works try to use 3D neural networks to treat data with color and geometric information. 3D Graph Neural Network (3DGNN), proposed by Qi et al. [43], leverages 3D point cloud data to generate k-nearest neighbors and trains the model by the time back-propagation algorithm so that each node finally outputs the predicted semantic labels of the corresponding pixel.

---

[3] https://cs.nyu.edu/~silberman/datasets/nyu_depth_v2.html.

[4] https://rgbd.cs.princeton.edu/.

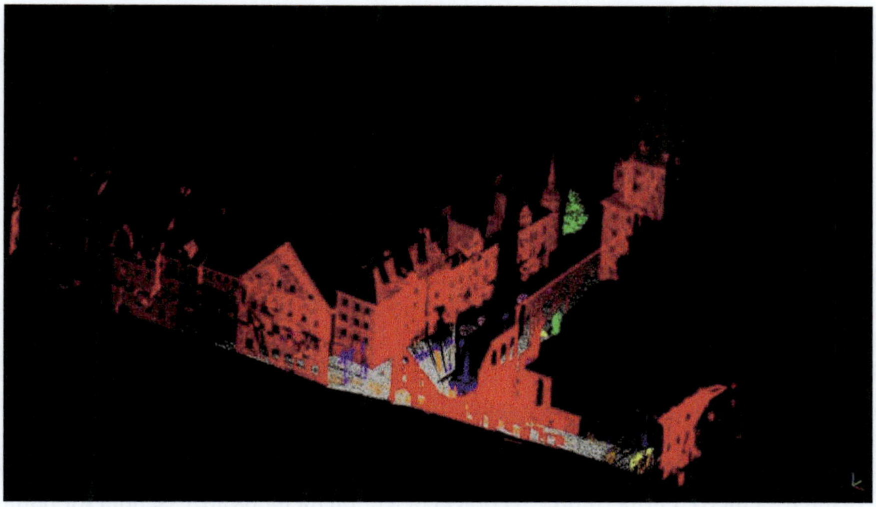

**Fig. 22** Sample of LiDAR point cloud

## *3.3 Semantic Segmentation for LiDAR Point Cloud*

Light Detection and Ranging (LiDAR) point cloud data are captured and analyzed by LiDAR sensor that detects and produces precise $x$, $y$, $z$ coordinates of target objects by laser light. Figure 22 demonstrates the sample of LiDAR point cloud data.

There are many point cloud datasets. Stanford 2D-3D-S[5] is an indoor spatial dataset with multiple modals includes more than 70,000 images with global XYZ measures and point clouds with semantic annotations. Sydney Urban Objects Dataset[6] contains various common urban road objects in Sydney, Australia and includes different categories of vehicles, pedestrians, signs, and trees independently scanned multiple times under non-ideal sensing conditions that represent actual urban sensing situation. Large-Scale Point Cloud Classification Benchmark[7] provides hand-labeled 3D point clouds of natural and urban scenes with the scale of about four billion labelled points.

3DGNN [43], introduced in Sect. 3.2 for processing RGB-D images, can also handle LiDAR point cloud. 3DGNN selects and analyzes the point cloud through a dense three-dimensional grid, and then employs a three-dimensional CNN to generate labels corresponding to each voxel. Finally, 3DGNN maps the labels back to the point cloud [43]. PointNet [40] deals with LiDAR point cloud data directly without transforming it to regular voxel grids, which reduces computational consumption. PointNet can solve the problem of spatial disorder of point cloud data through the

[5] http://buildingparser.stanford.edu/dataset.html.

[6] http://www.acfr.usyd.edu.au/papers/SydneyUrbanObjectsDataset.shtml.

[7] http://www.semantic3d.net/.

maximum pooling layer. PointNet learns discrete key points in the point cloud data that exactly represent the outlines of the object. PointNet has performed good robustness against missing data or small perturbations.

# 4 Object Detection

Object detection is the process of finding object instances of a certain class, such as bikes, persons, traffic lights, in videos or images. Different from classification tasks, object detection is able to detect multiple objects and their locations in the image. Generally speaking, the methods of 2D object detection can be divided into two main types. First, to generate the fixed number predictions on the grid (i.e., one-stage method); second, to make use of a proposal network to detect objects and utilize a second network to fine-tune the proposals and output the final predictions (i.e., two-stage method). Furthermore, 3D object detection methods are essential for point cloud data of intelligent vehicles. This chapter will introduce the application of object detection methods in intelligent vehicles, and then describe two types of 2D object detection methods; finally, the methods for 3D object detection are summarized.

## *4.1 2D Object Detection*

Object detection aims at recognizing instances of a predefined object class set (e.g. motorcycle, dog) and describing the locations of detected objects in the image with a bounding box. One example is shown in Fig. 23.

2D object detection methods can be divided into two categories, i.e., one-stage object detection methods and two-stage ones [33]. One-stage object detection methods usually employ a straightforward fully convolutional architecture whose outputs are classification probabilities and box offsets (w.r.t. pre-defined anchor box) at each spatial position. On the other hand, the workflow of two-stage object detection methods is more complicated than one-stage ones. Two-stage object detection methods first adopt the region proposal network (RPN) to filter out the regions with the high probability of containing an object from the entire image. Then the selected regions (named proposals) are fed into the region convolutional network (R-CNN) to get their classification scores and spatial offsets. One-stage detection methods are more efficient and elegant in design, while the two-stage detection methods usually achieve better accuracy than one-stage ones [77]. The discussions on these two kinds of object detection methods are provided as follows.

**Fig. 23** The example of object detection

### 4.1.1 One-Stage Object Detection

One-stage object detection methods usually involve a simpler and faster model architecture than two-stage ones. YOLO and its variants [44–46] compose the representative model family for one-stage object detection. YOLO v3 is the well-known representative one-stage object detection method. YOLO v3 utilizes a variant model of Darknet, which is composed by a 53-layer network that is trained via ImageNet dataset. For detecting objects, 53 more layers are stacked onto it, which composites a 106-layer fully convolutional underlying architecture for YOLO v3. Figure 24 shows YOLO v3's architecture.

The architecture of YOLO v3 is characterized by residual skip connections and upsampling operations. The most salient feature of YOLO v3 is the multi-scale detection that the detections are achieved at three different scales. YOLO is composed by a fully convolutional network whose output is made by a $1 \times 1$ kernel on the feature maps. In YOLO v3, the detection is implemented by multiple $1 \times 1$ detection kernels on feature maps of three different sizes at three different positions in the network.

The detection kernel is of size $1 \times 1 \times B \times (5 + C)$, where $B$ denotes the number of bounding boxes a cell on the feature map can predict, '5' represents the 4 bounding box attributes and one object confidence, and $C$ denotes the number of classes. The feature map generated by the kernel has identical height and width with the previous feature map, and involves detection attributes along with the depth. The cell (on the input image) that contains the center of the ground truth box for an object is chosen to be the one responsible for predicting the object. The working flow of the detection

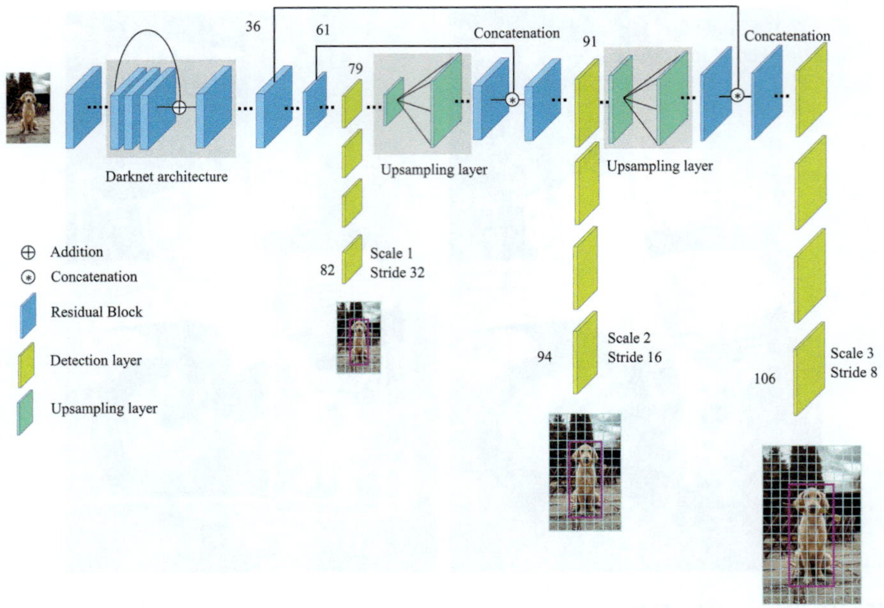


**Fig. 24** The architecture of YOLO v3 network

kernel in YOLO v3 is shown in Fig. 25 where the ground truth box is marked yellow whose center is contained in the cell marked green.

The predictions generated by YOLO v3 are at three scales, which are precisely produced by downsampling the input image dimensions 32, 16 and 8, respectively. The first detection is made by the 82nd layer (as shown in Fig. 24). For the first 81 layers, the image is down-sampled by the network, which guarantees that the 81st layer has a stride of 32. For example, given an image of $416 \times 416$, the resultant feature map is of size $13 \times 13$. The $1 \times 1$ detection kernel implements a detection operation and generates a detection feature map of size $13 \times 13 \times 255$, where the channel of the $1 \times 1$ kernel is 255. Then, the feature map from layer 79 is subjected to a few convolutional layers before being up sampled by $2\times$ (two times up-sampling) to dimensions of $26 \times 26$. The feature map is then concatenated with the feature map from layer 61. The combined feature maps are again subjected a few $1 \times 1$ convolutional layers to fuse the features from the earlier layer (layer 61). Then, the layer 94 makes the second detection and generates a detection feature map of $26 \times 26 \times 255$. A similar procedure is followed again, where the feature map from layer 91 is subjected to some convolutional layers before being depth concatenated with the feature map from layer 36.

YOLO v3 can achieve satisfied performance on small object detection tasks which is benefited from the multiscale detections (different scale detections generated by different layers). The concatenation of different scale feature maps helps preserve the fine-grained features and improve the accuracy of detecting small objects.

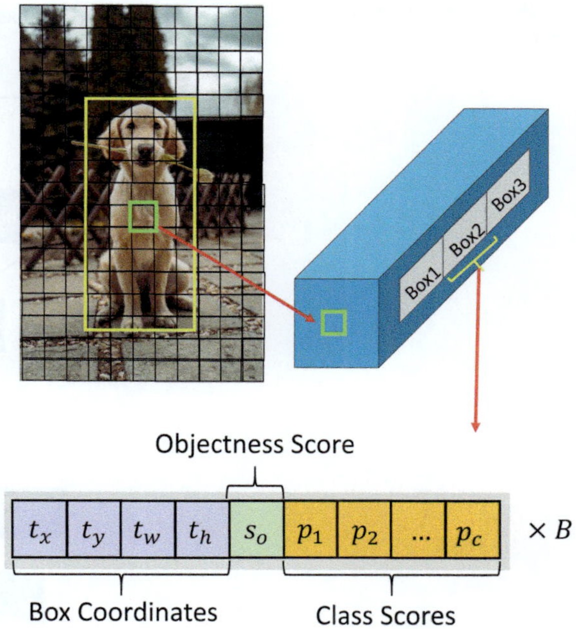


**Fig. 25** The working flow of the detection kernel in YOLO v3

YOLO v3 uses 9 anchor boxes. Three for each scale. Usually, the k-means algorithm needs to produce 9 anchors from customized data sets for training the YOLO v3 model in each scale. Then the anchors are arranged following the descending order of one dimension. The three biggest anchors of each scale are retained (three anchors are retained for each scale).

### 4.1.2 Two-Stage Object Detection

Compared with one-stage object detection methods, the two-stage methods have several superiorities:

- Two-stage object detection methods can filter out most negative proposals via sparse sampling of region proposals. On the contrary, one-stage object detection methods handle all the regions on the image and suffer from the risk of class imbalance.
- Depending upon the ROIAlign [19] operation, two-stage object detection methods can extract more representative features compared with one-stage ones.
- The object location can be regressed twice by two-stage detectors (once on each stage), the refinement of bounding boxes is better than that in one-stage methods.

Faster R-CNN [48] is a well-known two-stage object detection method that significantly outperforms its predecessor Fast R-CNN [17]. Fast R-CNN [17] uses a selective search algorithm to propose the regions where an object could be found.

**Fig. 26** The workflow of faster R-CNN

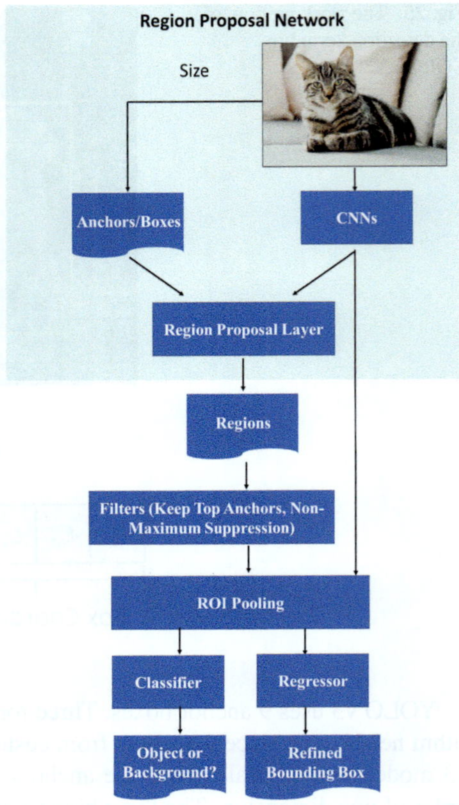


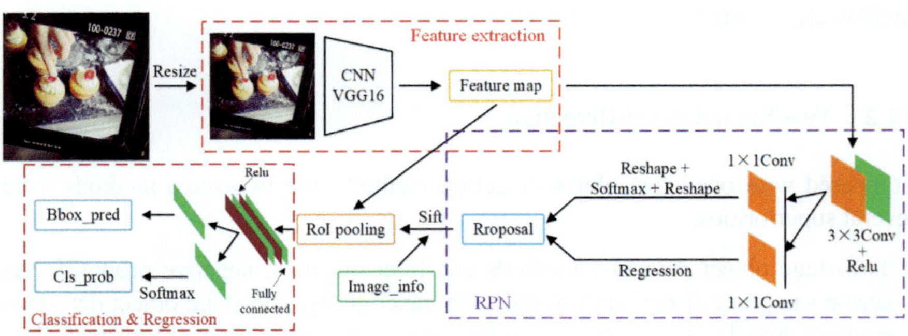


**Fig. 27** The architecture of faster R-CNN

However, the proposals are generated as part of the convolution operation in Faster R-CNN, which improves both efficiency and speed. The workflow and the architecture of Faster R-CNN are shown in Figs. 26 and 27, respectively.

Faster R-CNN is characterized by the module of Region Proposal Network (RPN). The RPN works by taking the output of a pre-trained deep CNN, such as VGG-16, and passing a small network over the feature map and generating multiple region proposals and a class prediction for each. Region proposals are bounding boxes that are based on so-called anchor boxes or pre-defined shapes. The class prediction is binary which indicates the presence of an object in the proposal region. A procedure of alternating training is used where both sub-networks are trained at the same time. This allows the parameters in the feature detector deep CNN to be fine-tuned for both tasks at the same time. The outputs of RPN (a bunch of boxes/proposals) are sent to a classifier and a regressor to check the occurrence of objects. RPN predicts the possibility of an anchor being background or foreground, and refines the anchor.

## 4.2 2D Object Detection of Fisheye Camera

Among all the visible light image sensors, fisheye cameras are popular for self-driving systems owning to the wide range of views it captures at one time. More specifically, a camera equipped with a canonical fisheye lens provides a rectangular projection up to 180° field of view but the objects around the lens are seriously distorted, rendering the detection task more difficult [38].

Object Detection of fisheye camera also needs a detection framework, yet fisheye images require further processing on coordinate conversion, and during the burdensome and repeating calculation, great importance should be attached to computation reduction. The detection framework could adopt object detection models taking a 2D perspective image $I_p$ as mentioned in Sect. 4.1, and we denote it as $N_p$. Then the problem could be formulated as fine-tuning $N_p$ to approximate $N_f$ to carry out object detection on fisheye dataset.

For coordinate conversion, a spherical image $I_s$ is defined in a spherical coordinate system $(\theta, \phi)$, and image $I_f \in I^{W_f \times H_f \times 3}$ captured by a fisheye camera on the image coordinate $(u, v)$ can be generated from $I_s$ by equirectangular projection $\rho_e : (u, v) \rightarrow (\theta, \phi)$, as shown in Fig. 28. While for redundant computation reduction, images are partitioned into multiple non-overlapping domains and equivariance is assumed across domains thus the problem can be further simplified:

$$N_f(\rho_e(I_s, L_{i,j}(\theta_i, \varphi_i))) \approx L_{i,j} N_f \rho_e(I_s, (\theta_i, \varphi_i)), i \neq j \tag{9}$$

For instance, Ren et al. [44] adopt YOLO architecture as $N_p$. Supposing the input image to be devided into $S \times S$ grid, each cell would correspond if the object center falls into the cell and then the cells become the basic units to generate distortions by comparing distortion levels defined as:

$$r_d = \frac{1}{\theta} \tan^{-1}(2r_u) \tan \frac{\theta}{2} \tag{10}$$

where $r_u$, $r_d$ and $\theta$ are demonstrated in Fig. 28.

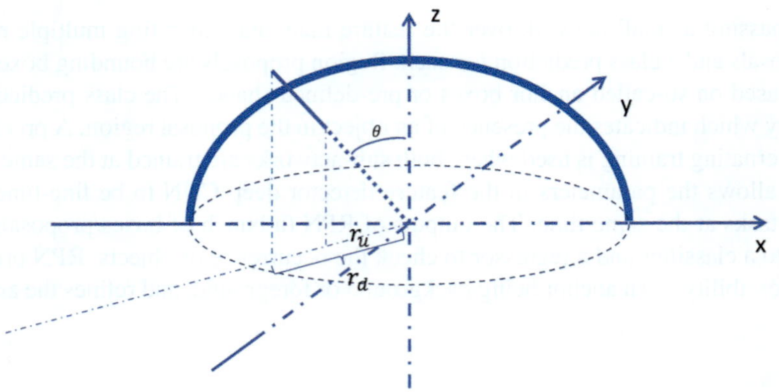


**Fig. 28** Relationship between image coordinates of fisheye image and the hemisphere coordinates of camera

Another difficulty to process on fisheye images comes from the imbalance of objects among data samples caused by severe distortion, as shown in Fig. 29. Catering to the imbalance in distortions, which directly results in the imbalance of object samples across a fisheye image, ∝-balanced cross-entropy (CE) loss rigorously categorizes samples positive and negative for each cell, for ground truth $\hat{p} \in -1, 1$, a piecewise function for any sample takes value as:

$$p_t = \begin{cases} p, & \text{if } \hat{p} = 1 \\ 1 - p, & \text{otherwise} \end{cases} \tag{11}$$

Then, focal loss (FL) differs all the samples by calling them easy or hard:

$$FL_{p_t} = -(1 - p_t)^{\gamma} \log(p_t) \tag{12}$$

where $\gamma \geq 0$ is a tunable parameter and CE becomes a particular case of FL. Ren et al. [44] formulates the problem with ideas in transfer learning, where domains are defined by the likelihood of image distortion.

To extract the domain-invariant features, the easy and hard cases are reformulated as two pseudo-domains and the goal becomes to better balance the easy and hard cases with information as much as possible. Therefore, the domains in one fisheye image are defined along the distance from the image center, and the spatial focal loss replaces per-sample weights with domain-based weights:

$$SFL(p_t^{i,c}) = -\log(p_t^{i,c}) \frac{1}{n_i} \sum_{j=1}^{n_i} (1 - p_t^{i,j})^{\gamma} \tag{13}$$

Applying SFL to YOLO v2, the loss function becomes:

**Fig. 29** Distortion imbalance among pedestrians of fisheye dataset

$$Loss = L_{location} + L_{SF} + L_{objclass} \quad (14)$$

where $L_{location}$ regularizes the bounding box for location estimation and $L_{objclass}$ penalizes classification result, where the bounding boxes are shown in red in Fig. 30.

## 4.3 3D Object Detection

3D object detection is essential for intelligent vehicles. The faithful representation of the 3D space around intelligent vehicles are usually required by various tasks of autonomous vehicles, such as scene understanding, planning, motion control. Recently, researchers have been leveraging the high precision LiDAR point cloud for accurate 3D object detection.

Monocular 3D object detection [35] predicts 3D bounding boxes with a single monocular, typically RGB images. 3D object detection from RGB images is fundamentally ill-posed because the critical information of depth is lacking in RGB images. Usually, intelligent vehicles are rigid bodies with known fixed shapes and sizes. It is possible to leverage the strong priors for intelligent vehicles to infer 3D bounding boxes based on 2D object detection.

2D object detection methods yield four degrees of freedom (DoF) axis-aligned bounding boxes with center $(x, y)$ and 2D size $(w, h)$. In contrast, the 3D bounding boxes in autonomous driving context generally have 7 DoF: 3D physical size $(w, h, l)$, 3D center location $(x, y, z)$ and yaw. The key problem is to recover the 7-DoF object from the 4-DoF one.

Deep3DBox [35] is proposed to regress the observation angle and 3D object size $(w, h, l)$ from the image patch enclosed by the 2D bounding box. The 7-DoF can be recovered by inferring the 3D location with three unknown center locations

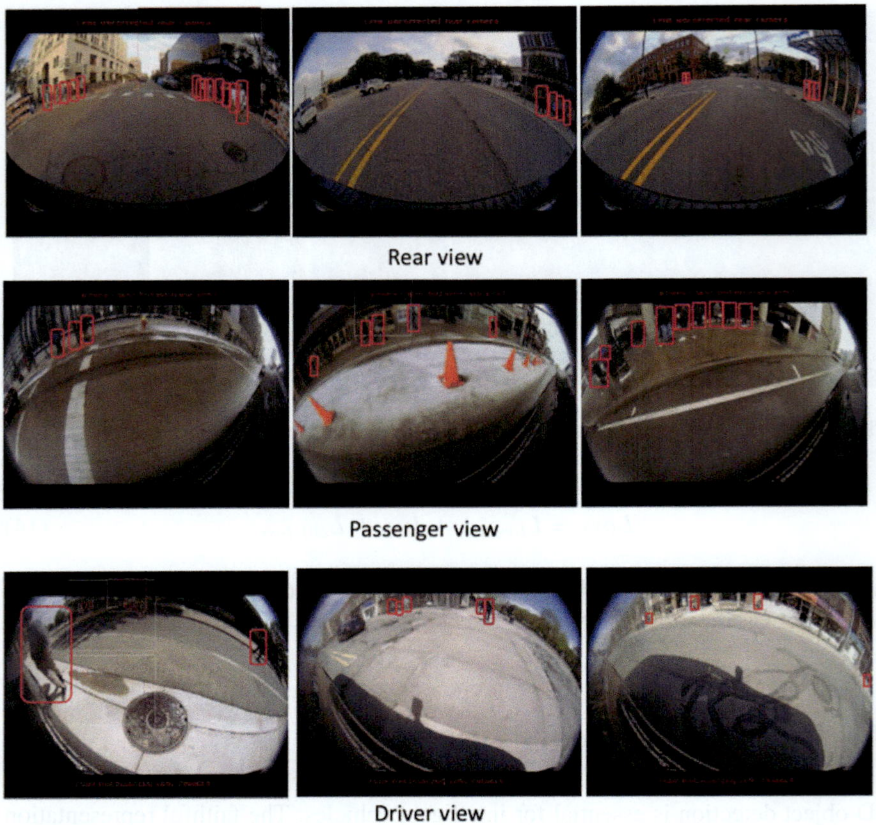


**Fig. 30** Object detection results on fisheye pedestrian image (FPI) dataset

$(x, y, z)$. The workflow that infers the 3D position based on 2D bound boxes is shown in Fig. 31.

Many recently proposed methods [31, 37, 75] on Deep3DBox introduce different forms of second-stage for fine-tuning the generated 3D cuboid. However, there exist two drawbacks for this kind of 3D object detection method. First, it depends upon the detection accuracy of 2D bounding box. If there exist moderate errors in 2D bounding box, the accuracy could be significantly hurt for 3D bounding box estimation. Second, the optimization relies on the size and position of bounding boxes but not the image appearance cue. Hence, it can hardly benefit from the large number of labeled training data.

The intelligent vehicles usually involve multiple sensors such as cameras, LiDAR. It is an attractive issue to fuse RGB images and LiDAR point cloud data for improving 3D detection performance. VoxelNet [75] is an end-to-end network that combines feature extraction and bounding box prediction, it is designed for object detection on the 3D point cloud. Frustum PointNets [39] significantly reduces the search space

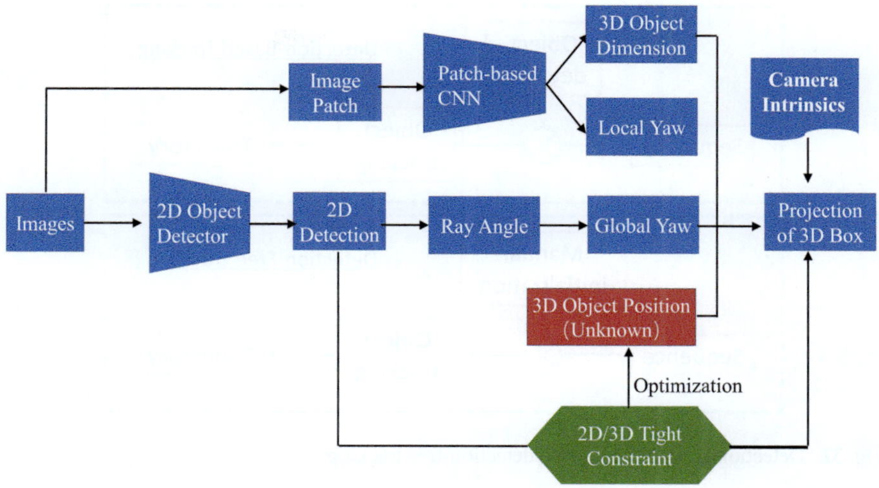


**Fig. 31** The pipeline inferring 3D position from 2D bounding box

via mature 2D object detection methods; it extracts the 3D bounding frustum of an object by extruding 2D bounding boxes from image detectors and then performs 3D object instance segmentation in the 3D space trimmed by the 3D frustums.

# 5 Object Tracking

Object tracking can be defined as the problem of estimating the trajectory of an object in the image plane as it moves around the scene [62]. Object detection performed by intelligent vehicles is a crucial operation that comes ahead of various autonomous driving tasks [59], such as object tracking, trajectories estimation, and collision avoidance. Object tracking methods vary with respect to different data types, e.g., RGB images and point cloud data. This chapter will introduce object tracking from two aspects, i.e., object tracking for RGB images and that for point cloud, respectively.

## *5.1 Object Tracking for RGB Images*

Object tracking aims to estimate the specified object position in subsequent frames based on its position in the first frame of the video. Similar to object detection, object tracking enhances the accuracy of subsequent object detection and grasping the object's subsequent position. In order to classify the methods of object tracking, there are several ways to categorize them, including the object acquisition and inference method.

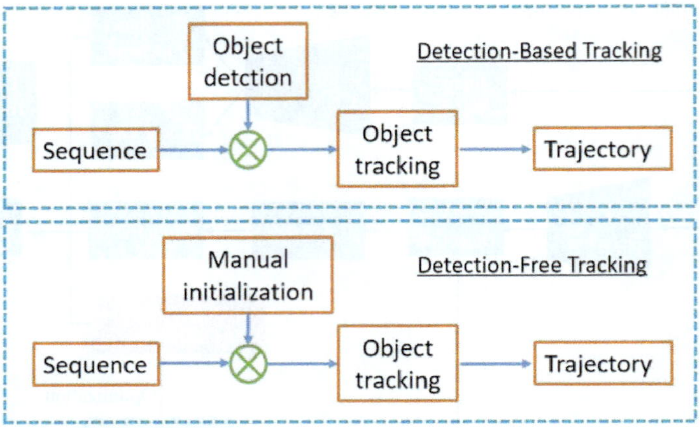


**Fig. 32** Detection-based tracking and detection-free tracking

According to the type of object acquisition, Object tracking methods can be divided into two categories. One relies on object detection results, called Detection-Based Tracking (DBT). The other is independent of object detection results, named Detection-Free Tracking (DFT). Figure 32 shows the difference between Detection-Based Tracking and Detection-Free Tracking.

In DBT, the object detector is pre-trained, which gives the DBT a powerful ability to capture objects, such as pedestrians and vehicles. Thus, the performance of DBT relies on the object detector to a large degree. Different from DBT, DFT needs to manually initialize the object position instead of the assistance of the target detector. Due to the capacity of automatically determining the object location, DBT is the mainstream method right now. For example, Wang et al. [63] proposed the DBT approach employed the RGB and Thermal (RGBT) image. Similarly, RGB and Depth(RGBD) images are usually used in the DBT tracking method owing their it valuable depth information [60, 61].

Toward different data processing approaches, the inference methods are different. The inference methods are divided into two categories: probabilistic inference and deterministic optimization. In detail, most of online tracking leverages probabilistic inference, and offline tracking tends to apply deterministic optimization. About the probabilistic inference, object tracking methods adopt several probabilistic inference models, which estimate object location in the future frames based on present observations. Hence, Markov property is assumed for estimation. Specifically, Markov property can be expressed as:

$$P(S_t \mid S_{1:t-1}) = P(S_t \mid S_{t-1}) \tag{15}$$

where $S_t$ denotes state of the t-th frame. $S_{1:t-1} = S_1, S_2, \cdots, S_{t-1}$ represents the state from the first frame to the t − 1 th frame. The state of the object and its observations are interrelated. The observations of the object can be described as:

$$P(O_{1:t} \mid S_{1:t}) = \prod_{i=1}^{t} P(O_t \mid S_t) \tag{16}$$

where $O_t$ denotes observations of the t-th frame. $O_{1:t} = O_1, O_2, \cdots, O_t$ represents observations from the first frame to the t-th frame. The algorithm establishes state models and observation models based on the above equations, which are employed to estimate and update the state. For instance, Kalman filter and particle filter-based method are prevalent probabilistic inference models.

Different from probabilistic inference-based methods, deterministic optimization-based methods tend to adopt maximum a posteriori to optimize the state of the object. Deterministic optimization is popular for offline tracking. There are several methods based on deterministic optimization, such as bipartite graph matching, conditional random field, and dynamic programming.

## 5.2 *Object Tracking for Point Cloud*

Compared to 2D tracking, 3D multi-object tracking is a more challenging job and more applicable for autonomous driving. The tracking objects in 3D multi-object tracking are three-dimensional and have the depth information which is not provided in RGB images. The drawback of 2D tracking algorithms is the limitation imposed by the camera. Because of the sensitivity to light and weather conditions, the camera affects the performance of the algorithm. In contrast, point cloud generated by LiDAR, which typically not be affected by light and weather conditions. In addition, point cloud data have more accurate information on the distance and the shape of objects, which is helpful in real time tracking. Thus, point cloud-based 3D object tracking algorithm attracted many experts and companies invest in this field.

Generally, 3D multi-object tracking can be regarded as a deeper task of 3D object detection. In a 3D multi-object tracking task, it needs to detect important objects in first frame as much as possible, and track them in the following frames with their unique IDs attained at the detection stage. There are mainly three components in 3D tracking, including 3D object detection, 3D Kalman filter and point cloud re-identification, as shown in Fig. 33.

Normally, it can be described as follow. Firstly, with the help of an existing 3D object detector, objects in the point cloud will be enveloped by 3D bounding boxes. The detection result $D$ often owns the following format: $D = (x, y, z, w, l, h, \theta, score)$ where $(x, y, z)$ represents the $x$, $y$, $z$ coordinates of the object respectively, $(w, l, h)$ represents the width, length and height of the 3D bounding box respectively, $\theta$ means the orientation angle of the object and $score$ is the detection score. Normally, many point cloud-based tracking algorithms adapt existing 3D object detection architecture, such as the Sparsely Embedding Convolutional Detection (SECOND). Secondly, the system needs to predict the state of the object in next frame, and shows its moving state with 3D Kalman filter. The current frame position is predicted based on the information of historical frames. In this stage, it needs to satisfy two equa-

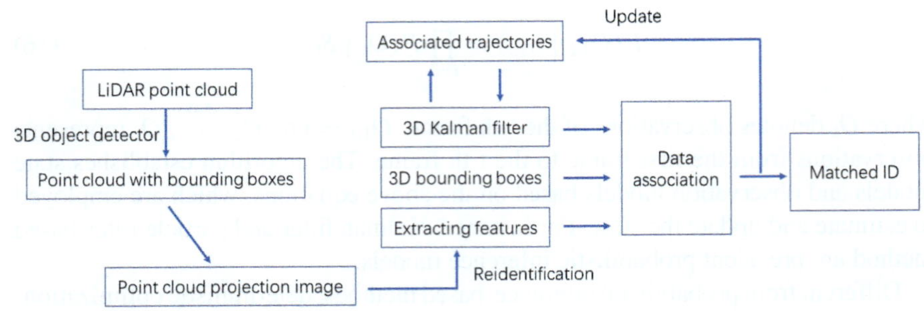


**Fig. 33** The workflow of object tracking for point cloud

tions: state equation and observation equation. At last, it is necessary to match the IDs of objects in the current frame and historical frames according to the similarity of features in different frames. In some cases, a certain object would be occluded by obstacles in some frames, which may cause the ID of this object to jump to the other objects. Therefore, the target of re-identification is to find the original trajectory by comparing the features of the tracked object.

However, there are several difficulties in point cloud-based tracking listed below:

1. Object occlusion. Like that in 3D object detection, the point cloud data of the occluded object is truncated, which will change the feature of object and hinder the algorithm from tracking the object. In tracking, the occlusion problem will cause the un-matching of features about the object between the current frame and historical frames.
2. Scale variance of shape. Owing to the property of LiDAR, objects near LiDAR have more points, while that far away from LiDAR have fewer points, which will make the same object have a different shape feature in different frames even though no occlusion. This problem will affect the performance of the tracking algorithm greatly.
3. Background disturbing. When a pedestrian, a dynamic object, is near a tree which is a static object, the points of pedestrians sometimes will be clustered together with the points of tree. And after the pedestrian walks away from the trees, the features of pedestrian will greatly change.
4. Computation burden. Due to the property of point cloud, like sparsity, the image's tracking algorithms can't be used directly. So, researchers tried to project 3D point cloud into different views. After that, fusion schemes can also be in 3D object recognition, e.g., multi-view convolutional Neural Network for 3D shape recognition [54]. However, projecting point cloud into different views will greatly increase the computation because each frame of point clouds will be projected. Reducing the number of point cloud frames need projected can help to lower the cost, but it will cause the lack of information, which will decrease the accuracy of the tracking algorithm.

Despite existing several difficulties in point cloud-based tracking, it still has many successful approaches proposed by researchers. P2B proposed in [42] is a point-to-box method for 3D object tracking. This method based on PointNet++ [41], which is a famous PointNet-based 3D object detection method. This method divides the tracking task into two parts. One is target-specific feature augmentation with a PointNet++ backbone. The second part is the target proposal and verification.

Because of the great success of Siamese networks in the 2D image object tracking task, some experts want to extend it into 3D tracking [16]. Giancola et al. [16] proposed the first 3D Siamese tracer. In this method, object candidates are generated by a Kalmen filter firstly, and then they will be passed into an encoding model for a representation. And then a cosine measurement function is used to measure the similarity with detected objects.

## 6 Simultaneous Localization and Mapping

Simultaneous localization and mapping (SLAM) is the standard technique for autonomous navigation of intelligent vehicles in an unknown environment. SLAM consists of localization and mapping. SLAM is usually applied in the tasks of outdoor and indoor navigation in environments that have sufficient landmarks, navigable terrain and distinct features. This chapter will introduce the concept of SLAM and describe 2D and 3D visual location and mapping.

### *6.1 SLAM Overview*

According to different positioning realization technologies, the location system of autonomous vehicles is usually summarized into three kinds: (1) Signal positioning based system. It utilizes the Global navigation satellite system (GNSS) to locate the position of vehicles. (2) Inertial measurement unit (IMU) based system. The location system employs an internal measurement unit to estimate the vehicle position and orientation. (3) Environmental characteristics based system. Leveraging the feature captured by the LiDAR or camera, the location system could obtain the current position and attitude of the vehicle by matching the feature stored in the database.

The current mainstream solutions for high-precision positioning of autonomous vehicles generally adopt a fusion method. It can be roughly divided into three categories: (1) GPS and IMU sensors fusion method, which integrates navigation and position function. (2) LiDAR and high-precision environment map fusion method, which matches the point could feature captured by LiDAR with the stored environmental characteristic feature. (3) GNSS, IMU, and environmental feature fusion method, which combines all the location approach mentioned above. The global positioning system belongs to absolute positioning. However, the GPS single-point positioning method is affected by the ionosphere, troposphere, signal propagation error, satellite clock error, orbit error, and signal propagation error. The positioning error

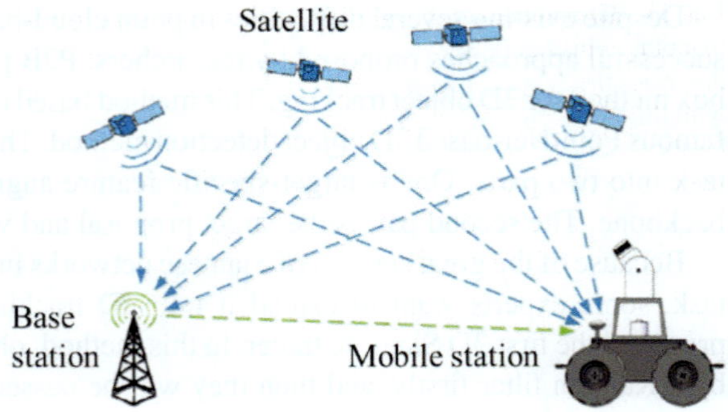


**Fig. 34** Schematic diagram of RTK principle

is at the meter level, which cannot meet the use of smart cars demand [3]. The RTK carrier phase differential positioning principle is shown in Fig. 34, which consists of satellites, reference stations, mobile stations and communication equipment. The base station and the mobile station receive satellite information at the same time, and then the base station sends the received information to the mobile station in real-time. The mobile station compares the satellite information received by itself with the satellite information from the base station to find the difference and solve the coordinates.

However, the RTK carrier phase differential positioning technology in practical applications is restricted by the use environment. For example, when a smart car is driving, it is inevitable to pass overpasses, high-rise buildings, continuous shades of trees, and other signal obstructions. Due to those obstructions, the location system generally loses the signal and is unable to output positioning information. Therefore, there are two positioning methods adopted. Firstly, an inertial navigation system based on the fusion of GPS and IMU is applied. IMU can output heading, attitude, position, and other information. When GPS signals are lost for a short time, the Kalman filter algorithm is used to fuse the IMU data set to obtain stable positioning information [27].

Another type of positioning method is to use high-precision maps. High-precision maps gradually change from simple to complex. SLAM can be divided into two categories: In order to meet the needs of autonomous vehicles, we need SLAM technology. SLAM is based on 2D vision or 3D vision. SLAM can use the image sensor to calculate the image information around the autonomous driving system, draw the map in real-time and give the vehicle's location at the same time. As shown in Fig. 35, visual SLAM is a SLAM method that uses a camera as a sensor [36]. According to the different camera sensors, it can be divided into monocular cameras, binocular cameras, and RGB-D cameras. 3D SLAM mainly refers to the location solution using LiDAR as the sensor. According to the difference of the LiDAR beam, it can be divided into 2D LiDAR SLAM based on single-line LiDAR and 3D LiDAR SLAM based on multi-line LiDAR.

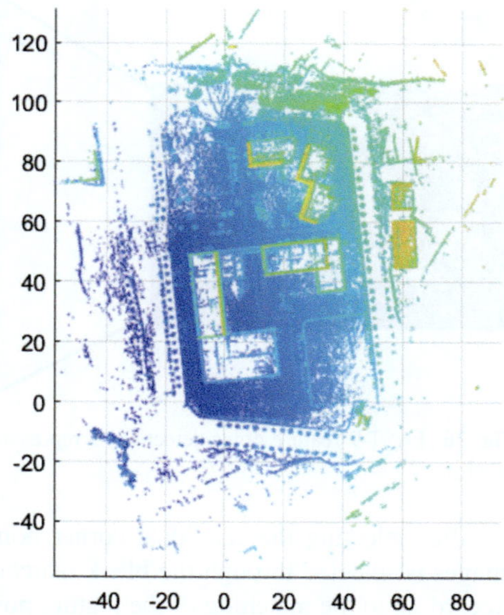


**Fig. 35** 3D point cloud mapping. This figure describes the 3D cloud data with different colors in a bird's view and denotes points of different elevations

## *6.2 2D Visual Location and Mapping*

Similar to the human eyes, the camera could obtain wealthy scene information. Due to the same reason, visual SLAM has become the hottest research direction in the SLAM field. In 2007, MonoSLAM [12] was the world's first monocular vision SLAM solution that meets real-time requirements. MonoSLAM is a SLAM method based on estimation theory to extend Kalman to track very sparse feature points. MonoSLAM solves the problem that visual SLAM could only rely on the camera to collect data in advance and then locate and build maps offline.

In 2007, Klein proposed the PTAM (Parallel Tracking and Mapping) method [23], which realized the parallel optimization of the tracking and mapping process. This method effectively reduces the amount of calculation by increasing the key frame mechanism and selecting images that meet the requirements for processing. Tracking process is mainly divided into 5 steps: (1) FAST [58] feature extraction. (2) Map initialization. (3) Tracking and location. (4) Select key frame. (5) Relocation. Mapping process is mainly divided into 4 steps: (1) Local bundle adjustment. (2) Global bundle adjustment. (3) Extract key frame to map. (4) Polar search and add points to the map.

The key frame-based monocular vision SLAM algorithm needs to extract image features at first. The FAST algorithm compares a pixel with enough pixels in its surrounding neighborhood to determine whether it is a corner point. Then, the corner points are filtered by non-maximum suppression. As shown in Fig. 36, the FAST algorithm selects a total of 16 pixels for pixel difference calculation. The probability that the center pixel is a corner point is determined by the magnitude of the difference.

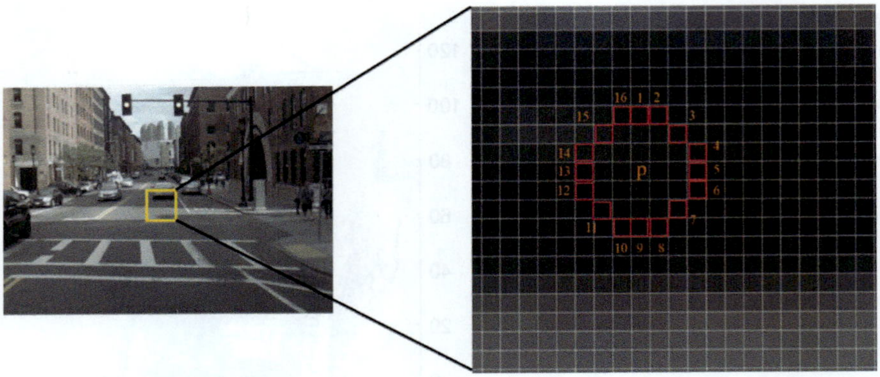


**Fig. 36** FAST-9 feature point detection, 9 represents a circle with a radius of 3 centered on the pixel *p*

After selecting the candidate corner points of the image, the second frame of images is selected through the block matching of the Single Shot Detector (SSD). According to the matching of the feature points between the two frames of images, the homography matrix between the two frames of images is calculated. Then, the corresponding rotation matrix and translation matrix are decomposed as the initial pose of the camera. According to the obtained rotation matrix, translation matrix and feature point pixel coordinates, the linear triangulation depth estimation method is used to estimate the world 3D coordinate system in the first frame of coordinate system. Finally, the world coordinate system of the feature points and the initial pose of the camera are optimized by the Bundle Adjustment method [64].

According to the camera pose of the previous frame, the camera pose of the current frame is predicted by using the motion model and the visual tracking algorithm named ESM. The PTAM method does not need to process each image finely, but string together several key images, and then optimize its trajectory and map. However, the positioning method of the camera's mapping is not suitable for large scenes, and the tracking is easier to lose.

## *6.3 3D Visual Location and Mapping*

Compared with camera sensors, LiDAR attracts much more attention. It is because of its high measurement accuracy, long-range ranging, and independency on external light.

Early research on 3D laser SLAM often remained on the 3D reconstruction technology of static objects. In 1992, Besl et al. [4] proposed the Iterative Closest Point (ICP), which is a high-level registration method based on free-form surfaces. This method uses the distance between the two point sets as the error function, and iteratively calculates the rotation and translation of the point sets to optimize the correspondence between the points in the two point sets. And it minimizes the error

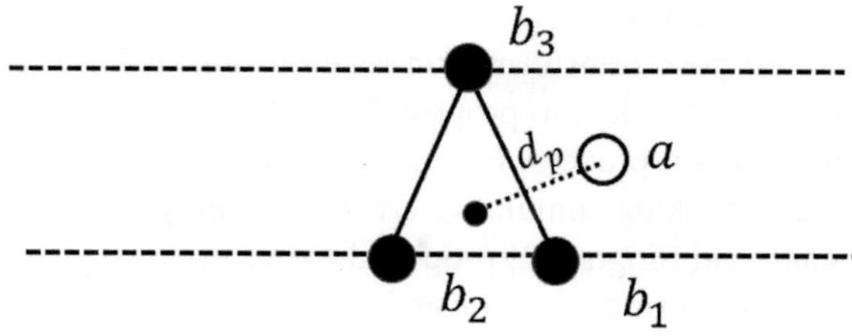


**Fig. 37** Plane point matching

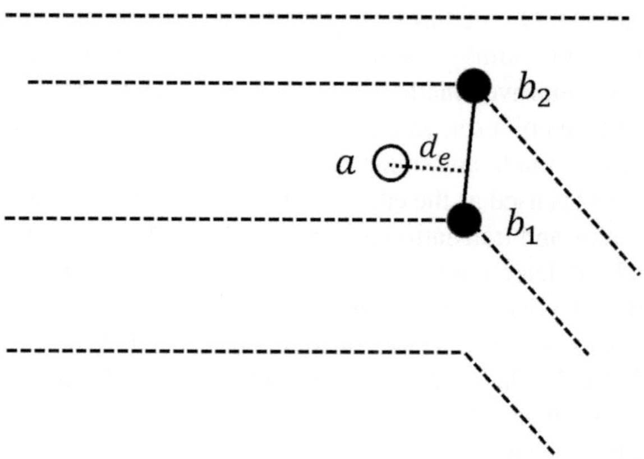


**Fig. 38** Edge point matching

function at last. Currently, the ICP method is the most widely used method in 3D point cloud matching.

Then in 2003, Biber et al. [5] proposed a method based on Normal Distributions Transform (NDT). Based on the assumption that the point cloud satisfies the normal distribution, this method achieves the inter-frame matching by calculating each grid's probability density function on the cloud network grid. The registration point cloud is transformed to the reference point cloud. Then, the probability of each point falling in the corresponding grid is calculated based on the normal distribution parameters. Finally, the sum of probabilities is taken as the objective function, and the transformation parameters are searched through iteration to achieve the minimum sum of probabilities.

In 2014, Zhang and Singh [70] proposed a 3D laser SLAM method suitable for large environments-LOAM. LOAM obtains radar odometer information through high-frequency motion estimation and low-frequency environment mapping to obtain point cloud map information. Firstly, an inertial measurement unit (IMU) is used to correct the point cloud distortion. Then, point-to-line and point-to-surface ICP registration is performed by extracting the line feature and surface feature of the adjacent point cloud as shown in Figs. 37 and 38 so as to obtain the odometer information.

Finally, the local map is established by selecting key frames, and the point-to-line and point-to-surface ICP algorithm is also adopted to realize the registration of key frames and local maps. In this way, the pose transformation relationship between the key frames is obtained, and a global map is established based on the key frames and the pose transformation relationship. In 2015, Zhang and Singh [71] merged vision sensor data to propose V-LOAM, and discussed the non-convergence of the optimization process. However, the V-LOAM method lacks loop closure detection module making cumulative errors inevitably appear under long-term work. In 2018, Zhang and Singh [72] proposed a data processing scheme LVIO, which is suitable for self-motion estimation and mapping. This scheme performs modules sequentially and recurrently for self-emotion estimation with 3D LiDAR, camera and IMU.

In 2018, Shan and Englot [50] improved the LOAM method and proposed LeGO-LOAM. On the basis of LOAM, light weight and ground optimization are realized. First, the point-to-surface ICP algorithm is improved by fusing the radar odometer. This improved method extracts ground points from each frame of the point cloud as a plane point set, and then based on the freedom of the plane, completes the solution of roll, pitch, and translation along the z-axis. In the same way, the remaining point cloud is used as the edge point to complete the solution of yaw, translation along the x-axis and translation along the y-axis. The LeGO-LOAM method also implements a loop detection module based on odometer information, and realizes loop detection by establishing a pose graph, which solves the LOAM's disadvantage about lacking back-end optimization. In 2020, Shan et al. [51] proposed LIO-SAM based on LeGO-LOAM, which adds IMU pre-integration, GPS information, and removes the inter-frame matching module. The flow chart of the LIO-SAM algorithm is shown in Fig. 39. The inputs are IMU, point cloud and GPS. The LIO-SAM method obtains the IMU pre-integration factor through IMU measurement between adjacent key frames. Then, the laser odometer factor is obtained by matching the key frame with the local map. When a new pose node is inserted into the factor graph, the GPS factor is added to the pose node. Finally, the loop factor is obtained by matching the key frame and the candidate loop key frame, and the overall map is optimized and constructed based on the optimization of the factor graph.

# 7 Multi-sensor Fusion

Human being makes decisions by combining multiple sensory data, such as auditory sense, visual sense. Similarly, the multi-sensor fusion is essentially important for autonomous vehicle. In addition, the multi-sensor fusion is widely applied in the fields of robotic systems, defense systems and transportation systems.

## *7.1 Multi-sensor Fusion Overview*

RGB image captured by camera has rich texture information and semantic information which are helpful to classification. But it lacks depth information. Point cloud from LiDAR contains detailed and accurate localization information, which is good

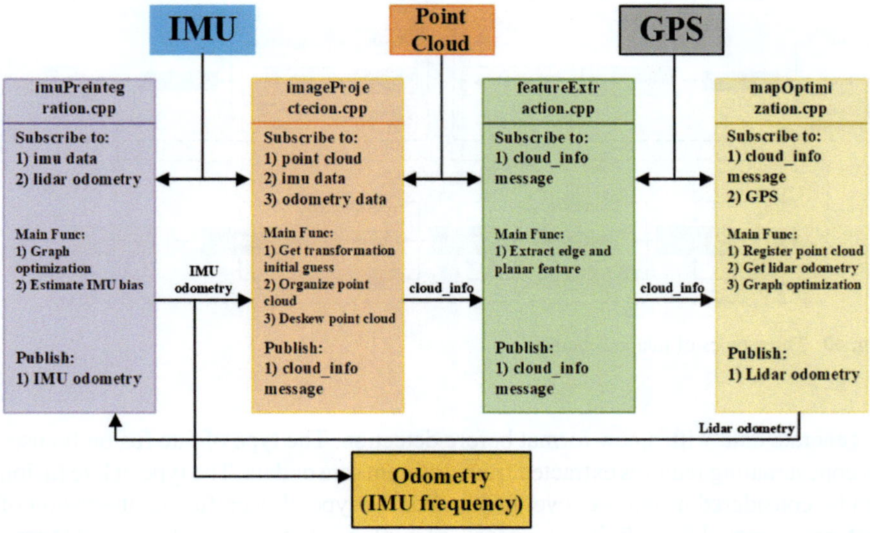


**Fig. 39** LIO-SAM framework

for estimating distance and position. While, it doesn't offer semantic information. Radar has advantages in velocity measurement. In a word, combining multiple sensors can supplement information needed in 3D object detection and improve detection accuracy and robustness. As shown in Table 3, a single type of sensor brings disadvantages if used independently. Thereby, to increase the overall performance, more and more studies attempt to adopt various sensors with different strategies and fusion schemes.

In general, the solution to multi-sensor fusion can be divided according to their pipeline into three types, i.e., early fusion, late fusion and deep fusion, as shown in Fig. 40. For the type of early fusion and the fusion operation is just concatenating data obtained by different sensors. The concatenated data are then fed into the detection network. The type of early fusion can be considered as data-level fusion which tends

**Table 3** Advantages and disadvantages of the three sensors

| Sensor | Range (m) | Advantage | Disadvantage |
|---|---|---|---|
| LiDAR | 100 | High accuracy and resolution, strong anti-interference ability and wide detection range | Expensive and easily affected by severe weather |
| Radar | 250 | 24 h-work, stable performance, precise velocity measurement and high resolution | Sensitive in some environment, cannot detect object's attribute and low accuracy |
| Camera | 50 | Low cost, rich image information, can capture object's attribute | Severely sensitive to lighting and weather condition, hard to detect distance |

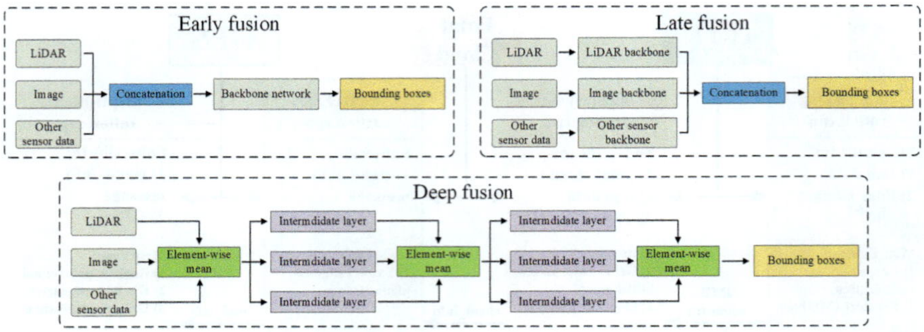


**Fig. 40** Three types of fusion schemes

to generate data with a new format before detection. The type of late fusion focuses on concatenating features extracted from different sensor data. The type of late fusion can be considered as feature-level fusion. For the type of deep fusion, the fusion of features extracted from different sensors are implemented by several different layers, the fusion operation can explore the interaction relationships among features. The type of deep fusion can be considered as the hierarchical fusion for features.

## 7.2 LiDAR and Camera Fusion

The fusion of LiDAR and Camera is widely used in the field of autonomous vehicles, because it not only utilizes the rich semantic information obtained from camera but also explores the location information from LiDAR point cloud.

MV3D [10] is one of the most famous methods in this category. MV3D [10] predicts 3D object based on bird-view image and front view image projected from point cloud and front-facing image from camera. Figure 41 shows its framework. Three images will be fed into VGG-based CNNs for feature extraction individually. After feature extraction, region candidates, generated based on the feature map extracted from the bird-view image, are projected to each view's feature map. A RoI pooling layer extracts the feature corresponding to each view's branch. At last, three feature maps fuse together.

Compared to MV3D [10], AVOD [25] accepts RGB image and bird-view image from the point cloud as input. Similarly, individual feature maps are extracted through extraction network. After cropping and resizing, feature maps are fused together for generating candidate boxes. Candidate boxes with high scores after Non-maximum Suppression (NMS) are projected back into the corresponding views' feature maps.

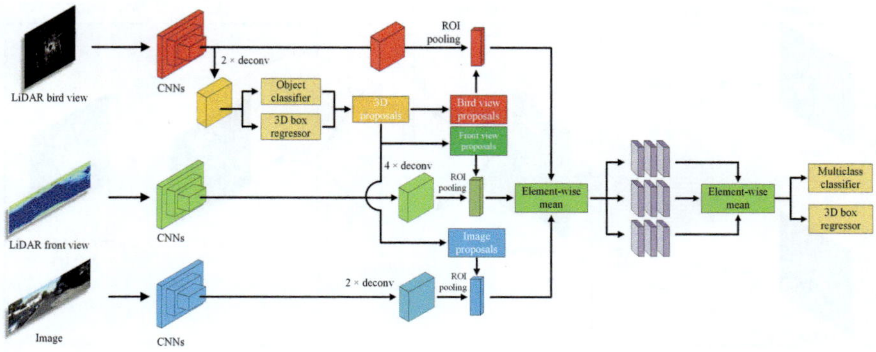


**Fig. 41** The framework of MV3D

## *7.3 LiDAR and Radar Fusion*

Because images have rich texture information, more works focus on LiDAR and camera fusion. However, neither LiDAR nor camera is friendly to detect occluded objects and measure velocity. Hence, a lot of works try to fuse LiDAR and Radar for detecting the occluded object. For example, Hollinger et al. [20] found LiDAR and Radar fusion is helpful for detecting object occluded by vegetation through exploiting the different properties of frequency spectra between LiDAR and Radar. Kwon et al. [26] proposed a detection scheme for partially occluded pedestrians based on LiDAR and Radar fusion. The authors introduce the new concept of occluded depth and occlusion RoI in order to determine whether an occluded object exists or not.

Radar has a great advantage in velocity measurement and is not sensitive to weather conditions. RadarNet [65] utilizes geometrical data and dynamic data from radar and adopts attention scheme in late fusion, which gains a good performance in nuScenes dataset [6]. Its overview is shown in Fig. 42. RadarNet is divided into three parts: voxel-based early fusion, detection network and attention-based late fusion. In voxel-based early fusion, data from LiDAR and Radar will be presented to the bird-view image and fused together after voxelization. In attention-based late fusion, outputs from the detection network and data from radar are fused together for precise velocity estimation based on the attention mechanism. RadarNet shows robustness for perceiving dynamic objects.

## *7.4 Radar and Camera Fusion*

The research works on Radar and camera fusion are relatively rare. Lim et al. [29] proposed FusionNet for Radar and camera fusion. A generalized architecture of FusionNet is shown in Fig. 43. Experiments have demonstrated that this structure can improve detection accuracy and model robustness. And the complementary nature of

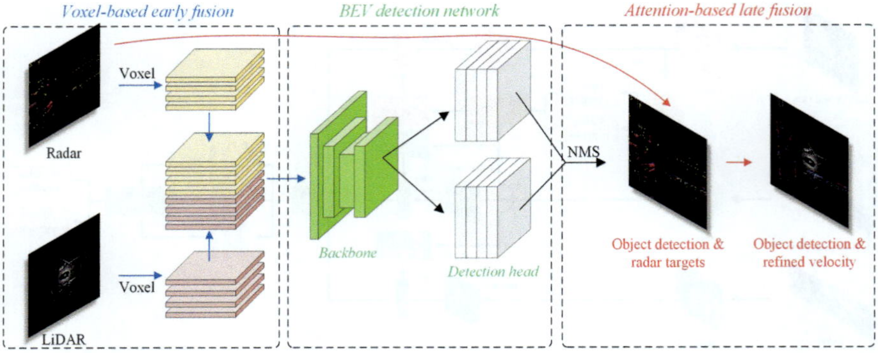


**Fig. 42** The overview of RadarNet

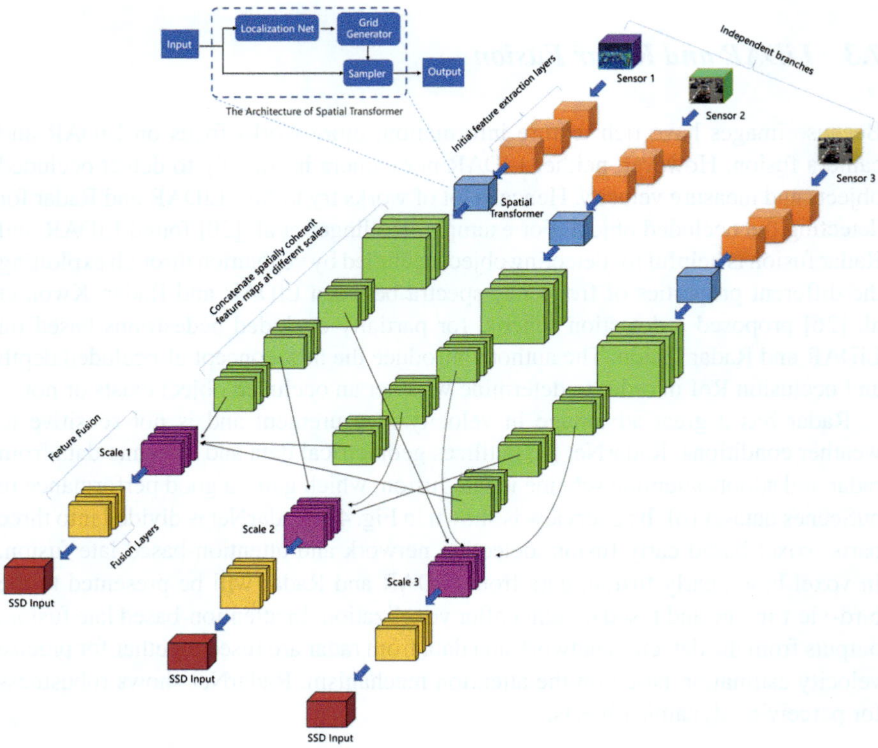


**Fig. 43** A generalized architecture of fusion network for multi-sensor fusion

radar and camera can be utilized to reduce the lateral error. In this structure, data from each sensor will be fed into a feature extraction branch individually. In the fusion stage, feature maps from different sensors will be concatenated a unified feature map with a double number of channels.

Recently, a new method named YOdar [24] was proposed, which is an uncertainty-based camera and Radar sensors fusion detection method. In this work, image and radar data will be fed into a YOLOv3 network and a FCN-8-network [32] for object detection respectively. Outputs from each branch will be aggregated and passed to a post-processing classifier for final vehicle detection.

In summary, for object detection, information of various sorts got from surroundings is helpful to improve the detection performance. Especially, the value of surrounding information become more important in the complicated task, such as 3D scenario. Single modality is not only unable to offer a variety of information, but also own its disadvantages. Thus, it is undoubted that combining multiple sensors is the wise way to achieve better performance. Besides, applying multiple sensors facilitates the accuracy and robustness of the model, which is critical to the stability of the intelligent application. It is reasonable for us to imagine that the fusion-based method will achieve better performance with the development of sensors.

## References

1. Ancuti CO, Ancuti C (2013) Single image dehazing by multi-scale fusion. IEEE Trans Image Process 22(8):3271–3282
2. Badrinarayanan V, Kendall A, Cipolla R (2017) SegNet: a deep convolutional encoder-decoder architecture for image segmentation. IEEE Trans Pattern Anal Mach Intell 39(12):2481–2495
3. Beltrán J, Guindel C, Moreno FM, Cruzado D, Garcia F, De La Escalera A (2018) BirdNet: a 3d object detection framework from lidar information. In: 2018 21st International conference on intelligent transportation systems (ITSC). IEEE, pp 3517–3523
4. Besl PJ, McKay ND (1992) Method for registration of 3-d shapes. In: Sensor fusion IV: control paradigms and data structures, vol 1611. International Society for Optics and Photonics, pp 586–606
5. Biber P. Straßer W (2003) The normal distributions transform: a new approach to laser scan matching. In: Proceedings 2003 IEEE/RSJ international conference on intelligent robots and systems (IROS 2003)(Cat. No. 03CH37453), vol 3. IEEE, pp 2743–2748
6. Caesar H, Bankiti V, Lang, AH, Vora S, Liong VE, Xu Q, Krishnan A, Pan Y, Baldan G, Beijbom O (2020) nuScenes: a multimodal dataset for autonomous driving. In: Proceedings of the IEEE/CVF conference on computer vision and pattern recognition, pp 11621–11631
7. Cai B, Xu X, Jia K, Qing C, Tao D (2016) DehazeNet: An end-to-end system for single image haze removal. IEEE Trans Image Process 25(11):5187–5198
8. Chen LC, Papandreou G, Kokkinos I, Murphy K, Yuille AL (2017) DeepLab: Semantic image segmentation with deep convolutional nets, atrous convolution, and fully connected CRFs. IEEE Trans Pattern Anal Mach Intell 40(4):834–848
9. Chen S, Liu B, Feng C, Vallespi-Gonzalez C, Wellington C (2020) 3d point cloud processing and learning for autonomous driving. arXiv:2003.00601
10. Chen X, Ma, H, Wan J, Li B, Xia T (2017) Multi-view 3d object detection network for autonomous driving. In: Proceedings of the IEEE conference on computer vision and pattern recognition, pp 1907–1915 (2017)
11. Cordts M, Omran M, Ramos S, Rehfeld T, Enzweiler M. Benenson R, Franke U, Roth S, Schiele B (2016) The cityscapes dataset for semantic urban scene understanding. In: Proceedings of the IEEE conference on computer vision and pattern recognition, pp 3213–3223
12. Davison AJ, Reid ID, Molton ND, Stasse O (2007) MonoSLAM: real-time single camera slam. IEEE Trans Pattern Anal Mach Intell 29(6):1052–1067
13. Duan C, Chen S, Kovacevic J (2019) 3d point cloud denoising via deep neural network based local surface estimation. In: ICASSP 2019—2019 IEEE international conference on acoustics, speech and signal processing (ICASSP). IEEE, pp 8553–8557

14. Everingham M, Eslami SA, Van Gool L, Williams CK, Winn J, Zisserman A (2015) The pascal visual object classes challenge: a retrospective. Int J Comput Vis 111(1):98–136
15. Garcia-Garcia A, Orts-Escolano S, Oprea S, Villena-Martinez V, Garcia-Rodriguez J (2017) A review on deep learning techniques applied to semantic segmentation. arXiv:1704.06857
16. Giancola S, Zarzar J, Ghanem B (2019) Leveraging shape completion for 3d Siamese tracking. In: Proceedings of the IEEE/CVF conference on computer vision and pattern recognition, pp 1359–1368
17. Girshick R (2015) Fast R-CNN. In: Proceedings of the IEEE international conference on computer vision, pp 1440–1448
18. Guo M, Chen M, Ma C, Li Y, Li X, Xie X (2020) High-level task-driven single image deraining: Segmentation in rainy days. In: International conference on neural information processing. Springer, pp 350–362
19. He K, Gkioxari G, Dollár P, Girshick R (2017) Mask R-CNN. In: Proceedings of the IEEE international conference on computer vision, pp 2961–2969
20. Hollinger J, Kutscher B, Close R (2015) Fusion of lidar and radar for detection of partially obscured objects. In: Unmanned systems technology XVII, vol 9468. International Society for Optics and Photonics, p 946806
21. Hu W, Gao X, Cheung G, Guo Z (2020) Feature graph learning for 3d point cloud denoising. IEEE Trans Signal Process 68:2841–2856
22. Velmurugan DK, Mathumitha B, Merylen Jenow B, Thamizh Oviyam R (2019) Automated vehicle: autonomous driving using SVM algorithm in supervised learning. Int J Eng Res Technol (IJERT) RTICCT
23. Klein G, Murray D (2007) Parallel tracking and mapping for small AR workspaces. In: 2007 6th IEEE and ACM international symposium on mixed and augmented reality. IEEE, pp 225–234
24. Kowol K, Rottmann M, Bracke S, Gottschalk H (2021) YOdar: uncertainty-based sensor fusion for vehicle detection with camera and radar sensors. In: Proceedings of the 13th international conference on agents and artificial intelligence - volume 2: ICAART, SciTePress, INSTICC, pp 177–186. ISBN 978–989–758–484–8, ISSN 2184–433X
25. Ku J, Mozifian M, Lee J, Harakeh A, Waslander SL (2018) Joint 3d proposal generation and object detection from view aggregation. In: 2018 IEEE/RSJ International conference on intelligent robots and systems (IROS). IEEE, pp 1–8
26. Kwon SK, Hyun E, Lee JH, Lee J, Son SH (2017) Detection scheme for a partially occluded pedestrian based on occluded depth in lidar-radar sensor fusion. Opt Eng 56(11):113112
27. Li D, Pan Z, Deng H, Peng T (2019) Trajectory tracking control law of multi-joint snake-like robot based on improved snake-like curve in flow field. Int J Adv Robot Syst 16(2):1729881419844665
28. Li Z, Gan Y, Liang X, Yu Y, Cheng H, Lin L (2016) LSTM-CF: unifying context modeling and fusion with LSTMs for RGB-D scene labeling. In: European conference on computer vision. Springer, pp 541–557
29. Lim TY, Ansari A, Major B, Fontijne D, Hamilton M, Gowaikar R, Subramanian S (2019) Radar and camera early fusion for vehicle detection in advanced driver assistance systems. In: Machine learning for autonomous driving workshop at the 33rd conference on neural information processing systems, vol 2
30. Lin TY, Maire M, Belongie S, Hays J, Perona P, Ramanan D, Dollár P, Zitnick CL (2014) Microsoft COCO: common objects in context. In: European conference on computer vision. Springer, pp 740–755
31. Liu L, Lu J, Xu C, Tian Q, Zhou J (2019) Deep fitting degree scoring network for monocular 3d object detection. In: Proceedings of the IEEE/CVF conference on computer vision and pattern recognition, pp 1057–1066
32. Long J, Shelhamer E, Darrell T (2015) Fully convolutional networks for semantic segmentation. In: Proceedings of the IEEE conference on computer vision and pattern recognition, pp 3431–3440
33. Lu X, Li Q, Li B, Yan J (2020) MimicDet: Bridging the gap between one-stage and two-stage object detection. In: Computer vision–ECCV 2020: 16th European conference, Glasgow, UK, 23–28 Aug 2020, Proceedings, Part XIV 16. Springer, pp 541–557

34. Ma L, Stückler J, Kerl C, Cremers D (2017) Multi-view deep learning for consistent semantic mapping with RGB-D cameras. In: 2017 IEEE/RSJ international conference on intelligent robots and systems (IROS). IEEE, pp 598–605
35. Mousavian A, Anguelov D, Flynn J, Kosecka J (2017) 3d bounding box estimation using deep learning and geometry. In: Proceedings of the IEEE conference on computer vision and pattern recognition, pp 7074–7082
36. Mur-Artal R, Montiel JMM, Tardos JD (2015) ORB-SLAM: a versatile and accurate monocular slam system. IEEE Trans Robot 31(5):1147–1163
37. Naiden A, Paunescu V, Kim G, Jeon B, Leordeanu M (2019) Shift R-CNN: deep monocular 3d object detection with closed-form geometric constraints. In: 2019 IEEE international conference on image processing (ICIP). IEEE, pp 61–65
38. Peng X, Murphey Y, Stent S, Li Y, Zhao Z (2019) Spatial focal loss for pedestrian detection in fisheye imagery. In: 2019 IEEE winter conference on applications of computer vision (WACV). IEEE, pp 561–569
39. Qi CR, Liu W, Wu C, Su H, Guibas LJ (2018) Frustum PointNets for 3d object detection from RGB-D data. In: Proceedings of the IEEE conference on computer vision and pattern recognition, pp 918–927
40. Qi CR, Su H. Mo K, Guibas LJ (2017) PointNet: deep learning on point sets for 3d classification and segmentation. In: Proceedings of the IEEE conference on computer vision and pattern recognition, pp 652–660
41. Qi CR, Yi L, Su H, Guibas LJ (2017) PointNet++: deep hierarchical feature learning on point sets in a metric space. Adv Neural Inf Process Syst 30
42. Qi H, Feng C, Cao Z, Zhao F, Xiao Y (2020) P2B: point-to-box network for 3d object tracking in point clouds. In: Proceedings of the IEEE/CVF conference on computer vision and pattern recognition, pp 6329–6338
43. Qi X, Liao R, Jia J, Fidler S, Urtasun R (2017) 3d graph neural networks for RGBD semantic segmentation. In: Proceedings of the IEEE international conference on computer vision, pp 5199–5208
44. Redmon J, Divvala S, Girshick R, Farhadi A (2016) You only look once: unified, real-time object detection. In: Proceedings of the IEEE conference on computer vision and pattern recognition, pp 779–788
45. Redmon J, Farhadi A (2017) Yolo9000: better, faster, stronger. In: Proceedings of the IEEE conference on computer vision and pattern recognition, pp 7263–7271
46. Redmon J, Farhadi A (2018) YOLOv3: an incremental improvement. arXiv:1804.02767
47. Ren D, Zuo W, Hu Q, Zhu P, Meng D (2019) Progressive image deraining networks: a better and simpler baseline. In: Proceedings of the IEEE/CVF conference on computer vision and pattern recognition, pp 3937–3946
48. Ren S, He K, Girshick R, Sun J (2015) Faster R-CNN: towards real-time object detection with region proposal networks. In: Advances in neural information processing systems, vol 28, pp 91–99
49. Ren W, Liu S, Zhang H, Pan J, Cao X, Yang MH (2016) Single image dehazing via multi-scale convolutional neural networks. In: European conference on computer vision. Springer, pp 154–169
50. Shan T, Englot B (2018) LeGO-LOAM: Lightweight and ground-optimized lidar odometry and mapping on variable terrain. In: 2018 IEEE/RSJ International conference on intelligent robots and systems (IROS). IEEE, pp 4758–4765
51. Shan T, Englot B, Meyers D, Wang W, Ratti C, Rus D (2020) LIO-SAM: Tightly-coupled lidar inertial odometry via smoothing and mapping. In: 2020 IEEE/RSJ international conference on intelligent robots and systems (IROS). IEEE, pp 5135–5142
52. Silberman N, Hoiem D, Kohli P, Fergus R (2012) Indoor segmentation and support inference from RGBD images. In: European conference on computer vision. Springer, pp 746–760
53. Song S, Lichtenberg SP, Xiao J (2015) Sun RGB-D: A RGB-D scene understanding benchmark suite. In: Proceedings of the IEEE conference on computer vision and pattern recognition, pp 567–576

54. Su H, Maji S, Kalogerakis E, Learned-Miller E (2015) Multi-view convolutional neural networks for 3d shape recognition. In: Proceedings of the IEEE international conference on computer vision, pp 945–953
55. Sun H, Ang MH, Rus D (2019) A convolutional network for joint deraining and dehazing from a single image for autonomous driving in rain. In: 2019 IEEE/RSJ international conference on intelligent robots and systems (IROS). IEEE, pp 962–969
56. Tan RT (2008) Visibility in bad weather from a single image. In: 2008 IEEE conference on computer vision and pattern recognition. IEEE, pp 1–8
57. Tang K, Yang J, Wang J (2014) Investigating haze-relevant features in a learning framework for image dehazing. In: Proceedings of the IEEE conference on computer vision and pattern recognition, pp 2995–3000
58. Viswanathan DG (2009) Features from accelerated segment test (fast). In: Proceedings of the 10th workshop on image analysis for multimedia interactive services, London, UK, pp 6–8
59. Wang Y, Luo X, Ding L, Fu S, Wei X (2019) Detection based visual tracking with convolutional neural network. Knowl-Based Syst 175:62–71
60. Wang Y, Wei X, Luo L, Wen W, Wang Y (2020) Robust RGB-D tracking via compact CNN features. Eng Appl Artif Intell 96:103974
61. Wang Y, Wei X, Shen H, Ding L, Wan J (2020) Robust fusion for RGB-D tracking using CNN features. Appl Soft Comput 92:106302
62. Wang Y, Wei X, Tang X, Shen H, Ding L (2020) CNN tracking based on data augmentation. Knowl-Based Syst 194:105594
63. Wang Y, Wei X, Tang X, Shen H, Zhang H (2021) Adaptive fusion CNN features for RGBT object tracking. IEEE Trans Intell Transport Syst
64. Wong KH, Ming M, Chang Y (2004) 3d model reconstruction by constrained bundle adjustment. In: Proceedings of the pattern recognition, 17th international conference on (ICPR'04), vol 3
65. Yang B, Guo R, Liang M, Casas S, Urtasun R (2020) RadarNet: exploiting radar for robust perception of dynamic objects. In: European conference on computer vision, pp 496–512. Springer
66. Yang W, Tan RT, Wang S, Fang Y, Liu J (2020) Single image deraining: from model-based to data-driven and beyond. IEEE Trans Pattern Anal Mach Intell 43(11):4059–4077
67. Yu F, Xian W, Chen Y, Liu F, Liao M, Madhavan V, Darrell T (2018) BDD100K: a diverse driving video database with scalable annotation tooling. 2(5):6. arXiv:1805.04687
68. Yuheng S, Hao Y (2017) Image segmentation algorithms overview. arXiv:1707.02051
69. Zeng J, Cheung G, Ng M, Pang J, Yang C (2019) 3d point cloud denoising using graph Laplacian regularization of a low dimensional manifold model. IEEE Trans Image Process 29:3474–3489
70. Zhang J, Singh S (2014) LOAM: Lidar odometry and mapping in real-time. In: Robotics: science and systems, vol 2
71. Zhang J, Singh S (2015) Visual-lidar odometry and mapping: Low-drift, robust, and fast. In: 2015 IEEE international conference on robotics and automation (ICRA). IEEE, pp 2174–2181
72. Zhang J, Singh S (2018) Laser-visual-inertial odometry and mapping with high robustness and low drift. J Field Robot 35(8):1242–1264
73. Zhao H, Shi J, Qi X, Wang X, Jia J (2017) Pyramid scene parsing network. In: Proceedings of the IEEE conference on computer vision and pattern recognition, pp 2881–2890
74. Zhou J, Zhou F (2013) Single image dehazing motivated by Retinex theory. In: 2013 2nd International symposium on instrumentation and measurement, sensor network and automation (IMSNA). IEEE, pp 243–247
75. Zhou Y, Tuzel O (2018) VoxelNet: end-to-end learning for point cloud based 3d object detection. In: Proceedings of the IEEE conference on computer vision and pattern recognition, pp 4490–4499
76. Zhu Q, Mai J, Shao L (2015) A fast single image haze removal algorithm using color attenuation prior. IEEE Trans Image Process 24(11):3522–3533
77. Zou Z, Shi Z, Guo Y, Ye J (2019) Object detection in 20 years: a survey. arXiv:1905.05055

# 3D Object Detection for Autonomous Driving

**Yihua Tan, Siwei Chen, and Pei Yan**

**Abstract** As an important application of Artificial Intelligence (AI), autonomous driving has developed rapidly in recent years. 3D object detection in autonomous driving has attracted more and more attention because it provides precise range and size information of an object. Some of the detection algorithms rely on point cloud data acquired from LiDAR. Compared with LiDAR, the optical image solution for autonomous driving is cheaper, so that image-based detection approaches are also popular. The chapter overview the recent advance on 3D object detection using 3D point cloud or image directly. For the computing simplicity, the approaches which detect 3D object from the input images without 3D point reconstruction benefit the vehicle computational platform. However, the object location accuracy of such approaches is limited. Therefore, a proposed method named as Descriptor Enhanced Stereo R-CNN (DESR-CNN) is specified in detail. Several 3D object detection algorithms are tested on KITTI dataset. The experimental results demonstrate that DESR-CNN outperforms most of the existing 3D object detection methods based on binocular images.

## 1 Introduction

In recent years, object detection has received more and more attention in computer vision because it is the key to environmental perception for autonomous driving [1]. It is essential to locate and identify cars, pedestrians, and other targets in autonomous driving tasks. Since 3D object gives more indicative information for intelligent driving, 3D box surrounding an object is the appropriate output from perception model, which can better assist automatic obstacle avoidance and emergency braking operation.

Y. Tan (✉) · S. Chen · P. Yan
School of Artificial Intelligence and Automation, Huazhong University of Science and Technology, Wuhan 430074, China
e-mail: yhtan@hust.edu.cn

 
Y. L. Murphey et al. (eds.), *AI-enabled Technologies for Autonomous and Connected Vehicles*, Lecture Notes in Intelligent Transportation and Infrastructure,
https://doi.org/10.1007/978-3-031-06780-8_4

The recent 3D object recognition methods are mainly divided into two categories: cloud point-based method [2–7] and image-based method [8–14]. Cloud point-based method can provide accurate 3D object location because cloud point can provide dense 3D samples of the environment. However, cloud point acquired from LiDAR has the disadvantage of expensive cost and complex calculation. On the other hand, image-based method can fully exploit the enriched texture information at low cost. Current image-based methods suffer from the inaccuracy of 3D object location. Meanwhile, most of image-based methods only focus on one type of target, which cannot meet the requirement of multiple kinds of target detection on complex road conditions.

In this chapter, after introducing some classical point cloud or image based approaches, we pay attention to overcome the limitations of the image-based method because of its application potential. More specifically, we aims to design a multi-class 3D object detection method with only the binocular image as input. Current methods achieve coarse location by recovering the 3D object bounding box only from the 2D boxes in left and right images. Li et al. designed 3D box alignment approach to improve the accuracy of object detection, which rectifies the coarse depth of objects by aligning the pixel RGB values between the left and right images [14]. Nevertheless, the representation of a pixel is not characterized enough only with RGB. Furthermore, not all the pixels in the object area are beneficial to 3D object alignment since some pixels possibly cause disturbance.

Focusing on the above problems, we also design a 3D Object Detection model named propose Descriptor Enhanced Stereo R-CNN (DESR-CNN). This model integrates an unsupervised local descriptor into the process of 3D box alignment to improve the accuracy of depth estimation. The idea of DESR-CNN is to use the description vectors rather than only the RGB values of pixels to compute the distance metric in the process of 3D box alignment. The benefit is twofold. First, the description vector is more robust as a distance metric compared with the pixel RGB value. Second, the responses of interest points corresponding to the descriptor can be used to determine the contributions of different pixels, which is more flexible than the original configuration that weights different pixels equally. Finally, the compared classical approaches and the proposed DESR-CNN are tested on KITTI dataset.

## 2 Recent Advance in 3D Object Detection

The task of 3D object detection is to find all Region of Interests (ROI) in the image and determine their categories and bounding box. Due to the object occlusion, truncation, as well as robustness of the surrounding dynamic environment, 3D object detection has continuously been a challenging problem in the field of computer vision.

We briefly review recent works of 3D object detection based on cloud point and image respectively.

## 2.1 Cloud Point-Based Method

With the rapid development of 3D acquisition technologies, there are more and more 3D sensors, including various types of 3D scanners, LiDAR, and RGB-D cameras. The 3D data acquired by these sensors can provide a wealth of geometric, shape and scale information. 3D data and 2D images complement each other, providing an opportunity to better understand the environment for the machine. 3D data can usually be represented with different formats, including depth images, point clouds, meshes, and volumetric grids. As a commonly used format, point cloud representation preserves the original geometric information in 3D space. Therefore, it is the preferred representation for autonomous driving [7].

Cloud point-based methods directly take raw point clouds as inputs to construct deep neural network. Single stage object detector (SSD) is extended to 3DSSD [2]. In the algorithm, all upsampling layers and refinement are abandoned to reduce the computation cost, and an anchor-free regression head with a 3D centerness assignment strategy is adopted. Yang et al. first generate an accurate proposal by seeding new spherical anchors for each point, then implement a parallel intersection and merge branch to improve position accuracy [3]. Shi et al. combine part location information and foreground segmentation information to generate proposals, then utilize the proposed ROI-aware pooling to refine the proposal [4]. Wang et al. propose a novel method termed Frustum ConvNet (F-ConvNet) for amodal 3D object detection from point clouds, in which F-ConvNet aggregates point-wise features as frustum level feature vectors to be passed to fully convolutional network [15]. Additionally, several other classical algorithms [1, 5, 16] are introduced in detail, which are tested in experiment in Sect. 4.

**Multi-view 3D Object Detection Network [1]** Shown as Fig. 1, the inputs to the network are the bird's eye view, front view of LiDAR point cloud and image. It first generates 3D object proposals from bird's eye view map and project them to three views. A deep fusion network is used to combine region-wise features obtained via

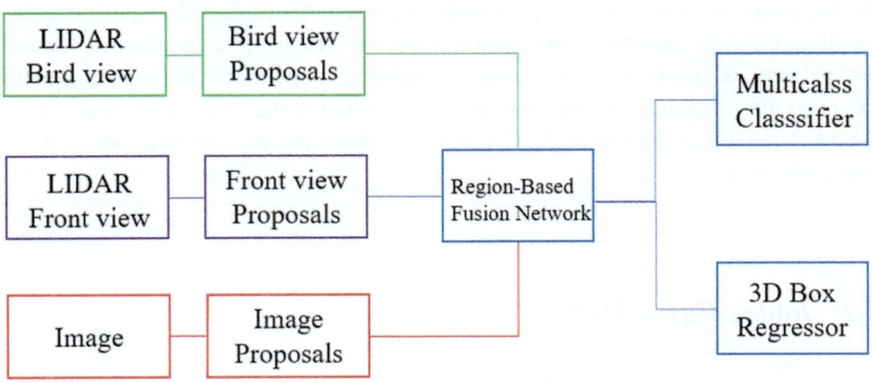


**Fig. 1** The overview of Multi-View 3D object detection network (MV3D)

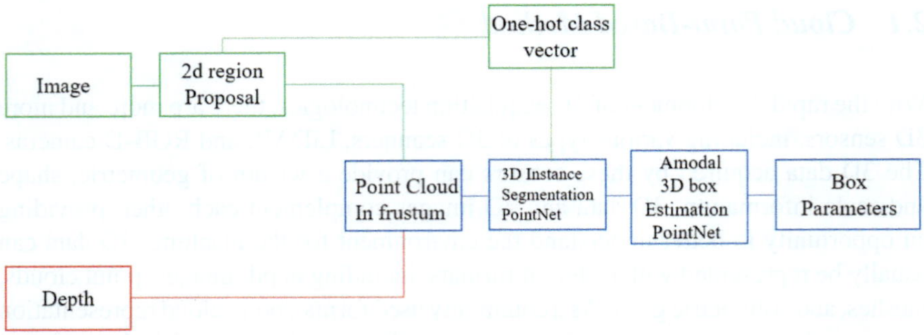


**Fig. 2** The network structure of Frustum PointNets (F-PointNet)

ROI pooling for each view. The fused features are then exploited to jointly predict object class and conduct 3D box regression.

**Frustum PointNet [5]** According to the network structure is shown in Fig. 2, it mainly consists of three modules: frustum proposal, 3D instance segmentation and amodal 3D box estimation. In frustum proposal module, 2D CNN object detector to propose and classify 2D regions, which are combined with point cloud to produce frustum. 3D instance segmentation module is a point net to group the instance point together, where the object instance is segmented by binary classification of each point based on intensity etc. Then, the segmented object point cloud is fed into a light-weight regression PointNet which tries to align points by translation such that their centroid is close to amodal box center. At last the box estimation net estimates the amodal 3D bounding box for the object.

**Structure Aware Single-Stage Detector (SA-SSD) [16]** The single-stage 3D object detector is constructed as Fig. 3. The network contains three sub-networks, a backbone network to extract the multi-stage features from point cloud, a back-end detection network to predict 3D bounding box and an auxiliary network to exploit point-wise supervisions. Specifically, the auxiliary network first converts the features from the backbone network back to point-wise representations, and then performs two auxiliary tasks: foreground segmentation and point-wise center estimation. The auxiliary network is jointly optimized with the backbone network in training stage and is removed after training, introducing no extra computational cost at inference stage. SA-SSD ranked at the top of KITTI 3D/BEV detection leaderboards when it was first published.

## 2.2 *Image-Based Method*

The image-based methods can fully exploit the shape and texture information, which generally reduce equipment and time costs. On the other hand, it also suffer the

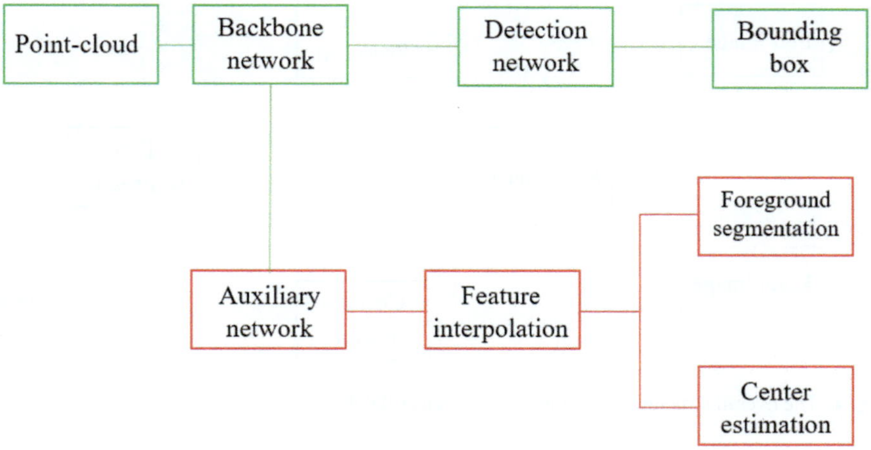


**Fig. 3** The framework of Structure Aware Single-Stage Detector (SA-SSD)

problem of depth information loss. Therefore, the location accuracy of these methods is generally lower than that of point cloud-based methods. Recently, the image-based methods also improve the location accuracy by keep better detection performance.

Some algorithms are monocular-based. Mousavian et al. first use a deep convolutional neural network to regress relatively stable 3D object attributes, and then combine these estimates with the geometric constraints provided by the 2D bounding box to generate a complete 3D bounding box [10]. Chen et al. utilize CNN to extract high-quality 3D object detection frames, then use several features to describe the object frame and make a preliminary score, finally regress the highest scored candidate [13]. Zia et al. solve the problem of scene understanding from the perspective of 3D shape modeling, and design a 3D scene representation of multi-object 3D shapes [11]. Xu et al. propose an end-to-end multi-level fusion approach to detect 3D object [12]. However, monocular-based methods rely on the input of prior information and each method is only suitable for some special situations.

Using binocular images for 3D object detection is also popular. Li et al. propose an easy-to-label 2D detection and discrete viewpoint classification to obtain coarse 3D measurements of objects through a lightweight semantic inference method [9]. Combined with the proposed BA approach, the temporal sparse feature correspondences and semantic 3D measurement model are merged into a unified optimization framework. Shi et al. [17] encode the points efficiently in a fixed radius near-neighbors graph and design a graph neural network to predict the category and shape of the object.

In the following paragaphs, some classical image-methods [8, 12, 14, 18] are introduced in detailed. All of them are also tested on KITTI dataset for comparison.

**3D Object Proposals [8]** The network structure is shown in Fig. 4. The conv layers make use of a stereo image pair to extract 3D information which is further passed to next two branches. One branch estimates the 3D box proposals by placing ground

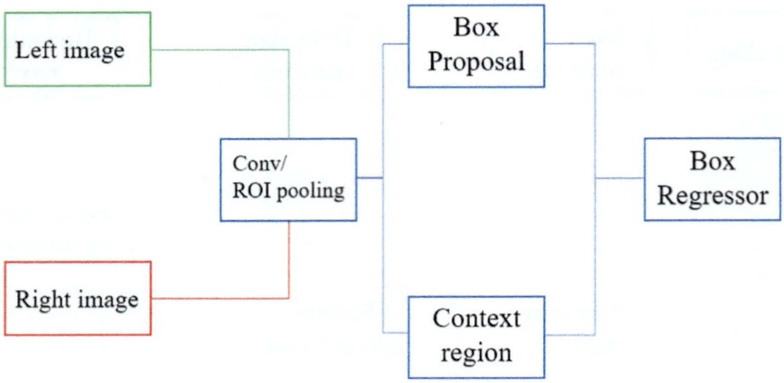


**Fig. 4** The overall structure of 3D Object Proposals (3DOP)

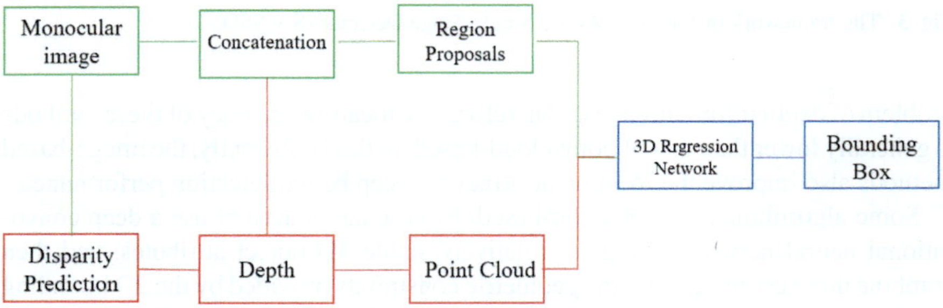


**Fig. 5** The overview of mulfusion

plane constraint. The other branch extract the contextual information for proposals. 3D object proposals and contextual information are exploited to predict accurate 3D bounding boxes and object orientation by defining multi-task loss.

**Multi-level Fusion (Mulfusion) [12]** As can be seen in Fig. 5, Xu et al. present an end-to-end multi-level fusion based framework for 3D object detection from a single monocular image. The whole network is composed of two parts: one for 2D region proposal generation and the other for simultaneously predictions of objects' 2D locations, orientations, dimensions, and 3D locations. With the help of a stand-alone module to estimate the disparity and compute the 3D point cloud, they introduce the multi-level fusion scheme. First, they encode the disparity information with a front view feature representation and fuse it with the RGB image to enhance the input. Second, features extracted from the original input and the point cloud are combined to boost the object detection. For 3D localization, the method introduces an extra stream to predict the location information from point cloud directly which is added to the aforementioned location prediction.

**Guidance and Surface Feature 3D Object Detection (GS3D) [18]** The overview of the proposed 3D object detection paradigm is shown in Fig. 6. The algorithm

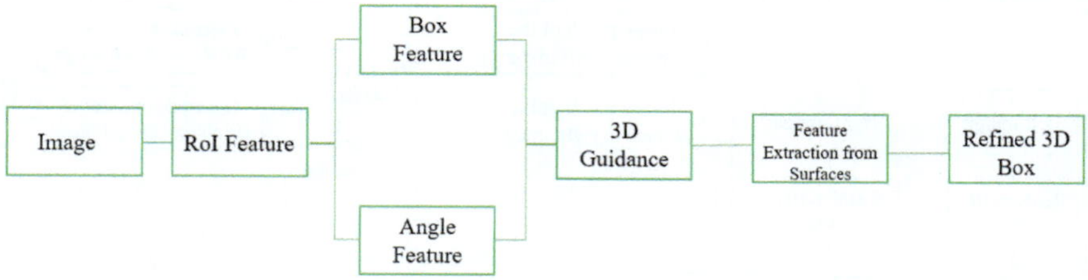


**Fig. 6** The overview of Guidance and Surface Feature 3D Object Detection (GS3D)

first utilizes a CNN based model (2D+ Orientation subnet) to obtain a 2D bounding box and observation orientation of the object. The guidance is then generated using the obtained 2D box and orientation with the projection matrix. Finally, the features extracted from visible surfaces as well as the 2D bounding box of the projected guidance are fed into the 3D box refinement model. Evaluated on the KITTI benchmark, the approach outperforms the most methods for single RGB image-based 3D object detection.

**Stereo RCNN (SRCNN) [14]** The network structure is shown in Fig. 7. Li et al. extend Faster R-CNN for stereo inputs to simultaneously detect and associate object in left and right images. They add extra branches after stereo Region Proposal Network (RPN) to predict sparse keypoints, viewpoints, and object dimensions, which are combined with 2D left–right boxes to calculate a coarse 3D object bounding box. They then recover the accurate 3D bounding box by a region-based photometric alignment using left and right RoIs (Fig. 7).

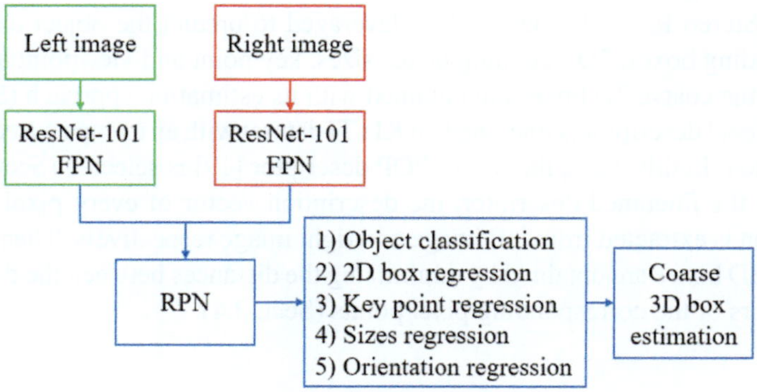


**Fig. 7** The overall flowchart of Stereo RCNN (SRCNN)

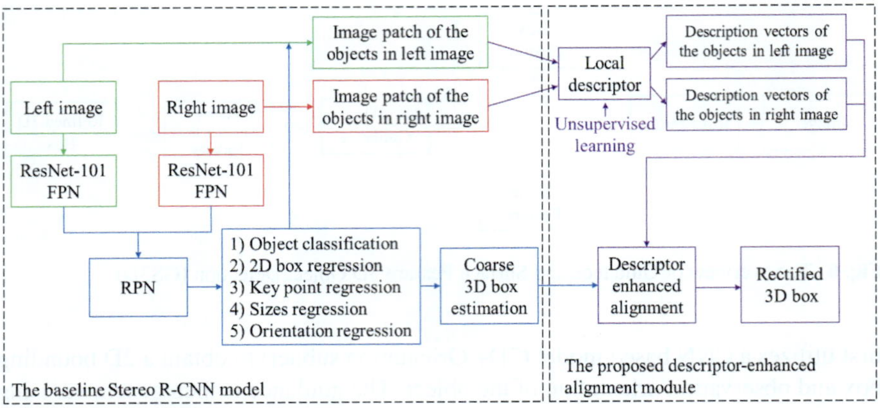


**Fig. 8** The overview of the proposed DESR-CNN. The coarse 3D boxes are first estimated with the baseline Stereo R-CNN model, and then the rectified 3D boxes are obtained by the proposed descriptor-enhanced alignment module

# 3 Descriptor Enhanced Stereo R-CNN

## 3.1 Overview

Figure 8 shows an overview of the proposed Descriptor Enhanced Stereo R-CNN (DESR-CNN). Taking Stereo R-CNN [14] as the baseline model, DESR-CNN designs the descriptor-enhanced alignment module to further improve the accuracy of 3D box estimation. This framework takes two RGB images as input and consists of the following steps:

1. The Stereo R-CNN model [14] is leveraged to predict the object classes, 2D bounding boxes, 3D bounding boxes sizes, key point and viewpoint angle, and then the coarse 3D boxes are obtained with its estimation approach (Sect. 3.2).
2. The local descriptor is finetuned on KITTI dataset with an unsupervised learning method. In this work, the recent POP descriptor [19] is selected (Sect. 3.3).
3. With the finetuned descriptor, the description vector of every pixel in object region is extracted from left image and right image respectively. Then the rectified 3D boxes are obtained by minimizing the distances between the description vectors of the corresponding pixel points (Sect. 3.4).

## 3.2 Coarse 3D Box Estimation

The 3D coordinate system is adopted from KITTI data set. The origin of the coordinate is on the camera center; x axis points to right on the 2D image plane; y axis points down and z axis points to the inner direction orthogonal to the image plane and stands for depth. 3D bounding box is represented as $B = (w, h, l, x, y, z, \theta, \varphi, \psi)$.

Here $w, h, l$ are the sizes of the box (width, height, and length respectively) and $x, y, z$ are the coordinates of the bottom center, following the KITTI annotation. The size and center coordinate are measured in meter. $\theta, \varphi, \psi$ are the rotation around $y$ axis, $x$ axis and $z$ axis respectively. Since the concerned objects are all on the ground, we only consider the $\theta$ rotation like the existing work [14] does.

For 2D detection, we modify the Faster-RCNN [20] framework by simultaneously detect and associate 2D bounding boxes for left and right images. The details are illustrated in Fig. 9. We use weight-share ResNet-101 [21] and FPN [22] as our backbone network to extract consistent features on left and right images. After feature extraction, we use Region Proposal Network (RPN) [20] to classify objects and regress box offsets for each input location. The boxes in the left and right images are detected and associated simultaneously with the approach in [14]. Then the objective function of multi-class classification is implemented as the cross-entropy loss.

After the above process of RPN, we have corresponding left–right proposal pairs. Then we use three branches of size, orientation and key point prediction.

For size prediction, we simply regress the offset between the ground-truth dimension with a pre-set dimension prior. As is standard, for each size we estimate the residual relative to the mean parameter value computed over the training dataset.

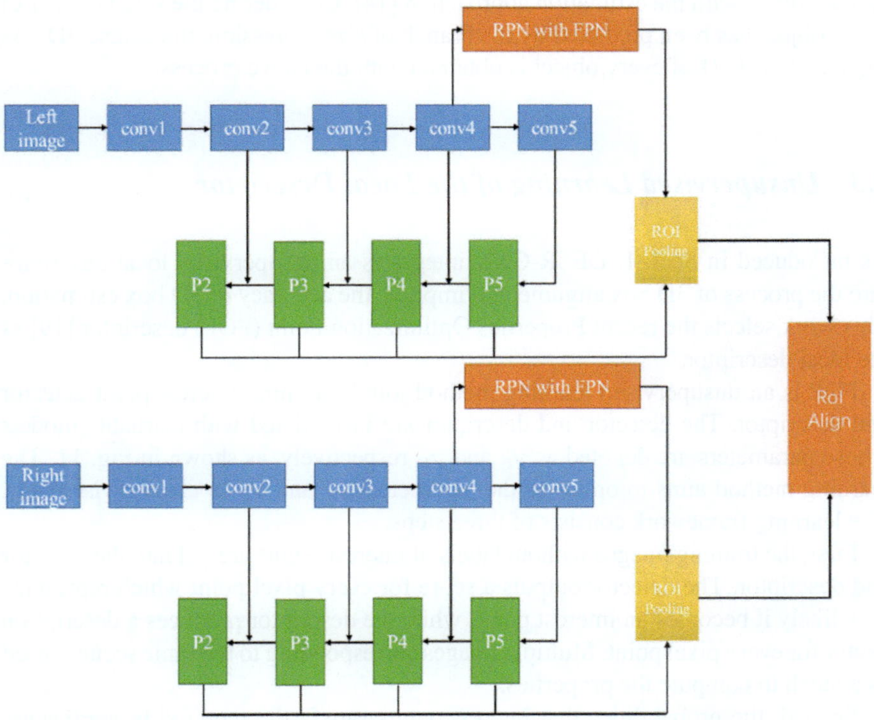


**Fig. 9** Details of the stereo RPN with FPN

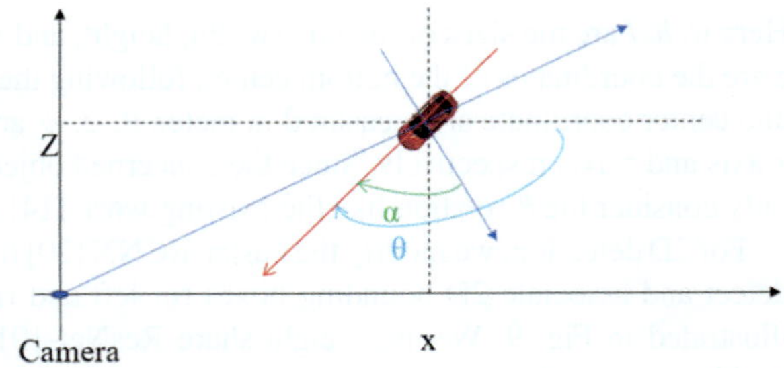


**Fig. 10** Top view of observation angle $\alpha$ and global rotation angle $\theta$. The blue arrows represent the observation axes and the red arrow indicates the heading of the car. Since it is a right-handed coordinate system, the positive direction of rotation is clockwise

The orientation estimated in the subnet is the observation angle of the object which is directly related to the appearance of the object. We denote the observation angle as $\alpha$ in order to distinguish it from the global rotation, $\theta$. Both $\alpha$ and $\theta$ are annotated in the KITTI data set and their geometry relationship is shown in Fig. 10.

Through the estimation from the network, we obtain the left–right 2D boxes, key points, and regressed sizes. The above variables are represented as $z = \{u_l, v_t, u_r, v_b, u_l^{'}, u_r^{'}, u_p\}$, which indicate the left, top, right, bottom edges of the left 2D box, left, right edges of the right 2D box, and the $u$ coordinate of the key point. Then the 3D coordinate $\{x, y, z\}$ and the global rotation angle θ of every object can be solved with the estimation approach in [14]. Considering the size $\{w, h, l\}$ of every object has been predicted by the branch of size regression, the coarse 3D box $\{x, y, z, \theta, w, h, l\}$ of every object is obtained with the above process.

### *3.3 Unsupervised Learning of the Local Descriptor*

As introduced in Sect. 1, DESR-CNN integrates an unsupervised local descriptor into the process of 3D box alignment to improve the accuracy of 3D box estimation. This work selects the recent Properties Optimization Point (POP) descriptor [19] as the local descriptor.

POP is an unsupervised learning method jointly training interest point detector and descriptor. The detector and descriptor are instantiated with learnable models whose parameters are denoted as $\gamma_F$ and $\gamma_D$ respectively, as shown in Fig. 11. The learning method aims to optimize the parameters $\gamma_F$ and $\gamma_D$ in unsupervised way. The learning framework consists of three steps.

First, the training images without labels of interest points are fed into the detector and descriptor. The detector outputs a score for every pixel point which represents how likely it becomes an interest point, while the descriptor produces a description vector for every pixel point. Multiple images corresponding to the same scene is used as a batch to compute the properties.

Second, the probabilities that interest points satisfy the required properties are computed. By introducing a latent variable to indicate interest point set, the probability for every property can be computed with the score and description of interest

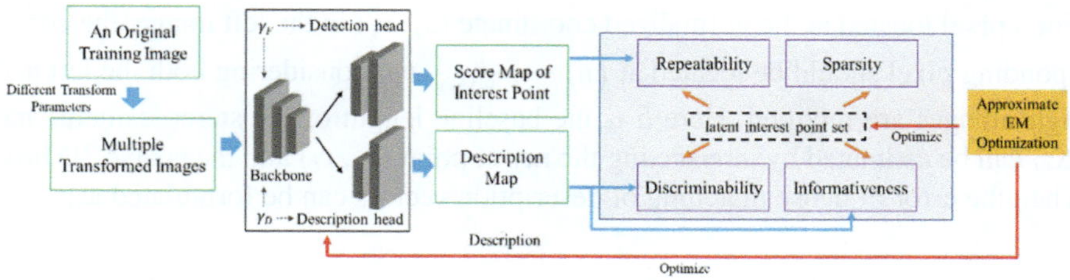


**Fig. 11** The overview of the Properties Optimization Point (POP) model [19]. $\gamma_F$ represents the learnable parameter of the detector containing the backbone and detection head, and $\gamma_D$ represents that of the descriptor containing the backbone and description head. The training process consists of three steps: (1) detector and descriptor are fed with multiple images corresponding to the same scene, and produce score and description of interest point, (2) joint probability of properties are computed with latent interest point set, (3) joint probability are optimized with approximate EM algorithm

point. In this paper, the required properties are concretized as sparsity, repeatability, invariability, discriminability and informativeness.

Third, the joint probability of required properties is maximized by our approximate Expectation Maximization (EM) algorithm, in which $\gamma_F$ and $\gamma_D$ are optimized jointly.

With the optimization algorithm detailed in [19], the POP detector and descriptor are first pre-trained on the large-scale MS COCO dataset, and then finetuned on the KITTI dataset. The entire optimization process is unsupervised without any human annotation. With the finetuned POP detector and descriptor, the response score map $S$ and description map $F$ can be obtained for the input image patch of the object. The $S(u_i, v_i)$ indicates the response score of the $i$ th pixel whose normalized coordinate is $(u_i, v_i)$, while the $F(u_i, v_i)$ represents the description vector of the $i$th pixel. The $S$ and $F$ will be applied in the descriptor-enhanced 3D box alignment module, which is introduced in Sect. 3.4.

### *3.4 Descriptor-Enhanced 3D Box Alignment*

As discussed in the existing work [14], the 3D box regressed by the above process is relatively coarse because it mainly relies on the low-resolution RoI maps. The existing work [14] proposes a 3D box alignment module to improve the 3D box precision by aligning the RGB values between the left and right images. Although this module is verified to be effective, it faces the problem that the RGB values are likely to be confused across different positions, which leads to the bottleneck of the alignment precision. To mitigate this problem, the descriptor-enhanced 3D box alignment module is designed in this work, which can be considered as a generalization of the existing 3D box alignment module in [14].

The idea of the descriptor-enhanced 3D box alignment is that the depth $z$ should be able to derive the correct pixel correspondences between the left and right images.

For a pixel located at the normalized coordinate $(u_i, v_i)$ in the left image, the corresponding pixel should be located at $\left(u_i - \frac{b}{(z+\Delta z_i)}, v_i\right)$ considering both the left and right images are rectified. Here $b$ is the baseline length of the stereo camera, and $\Delta z_i$ can be estimated by intersecting the ray of point $(u_i, v_i)$ and the coarse 3D box. Then the error of dense matching of description vectors can be formulated as:

$$E = \sum_{i=0}^{N}\left(S_l(u_i, v_i) \cdot \left\| F_l(u_i, v_i) - F_r\left(u_i - \frac{b}{z + \Delta z_i}, v_i\right)\right\|\right). \tag{1}$$

In Eq. (1), $F_l$ and $F_r$ denote the dense description map of the left and right images respectively, and $F_l(u_i, v_i)$ represents the description vector at the normalized coordinate $(u_i, v_i)$ in the left image. $S_l(u_i, v_i)$ indicates the response score of every pixel which is outputted by the interest point detector. In this work, the dense description map is obtained by combining the RGB vector and robust visual description extracted by the finetuned POP descriptor in Sect. 3.3. The RGB vector is the normalized RGB pixel value which is a 3-channels vector. And the robust visual description is a 64-channels vector. Therefore, the final description is the concatenation of the above two vectors which has 67 channels. Here only $z$ is the variable required to be optimized in Eq. (1), which can be solved with the searching-based algorithm introduced in [14]. This descriptor-enhanced alignment module is effective to improve the precision of 3D box, which is verified by the experiments in Sect. 4.4.

## 4 Experiments

We evaluate our framework on KITTI object detection benchmark [23]. The dataset contains 7481 training samples and 7518 testing samples. We further divide the training data into a training set with 3712 samples and a validation set with 3769 samples following the common protocol. We conduct experiments on the most commonly used car category and use average precision (AP) with an (IoU) threshold 0.5 and 0.7 as evaluation metric. However, in order to further verify the effect of multi-classification, we follow the same method to extend [10] to multi-classification. Only the image is also taken as the input in [10]. Following [16], the benchmark considers three levels of difficulties: easy, moderate, and hard based on the object size, occlusion state, and truncation level.

### 4.1 Implementation Details

We flip and exchange the left and right image, meanwhile mirror the viewpoint angle and key point respectively to form a new stereo imagery. The origin dataset is thereby doubled with different training targets. During training, we keep 1 stereo pair and

512 sampled RoIs in each mini-batch. We train the network using SGD with a weight decay of 0.0005 and a momentum of 0.9. The learning rate is initially set to 0.001 and reduced by 0.1 for every 5 epochs. We train 12 epochs in total.

## *4.2 Comparison with State-of-the-Art*

We compare our method with other state-of-the-art approaches in the car category by submitting the detection results to the KITTI server for evaluation. As shown in Table 1, our approach achieves the best performance among the approaches which use images as input only, and the performance among these approaches is very low with IoU threshold 0.7. Then we compare these approaches mentioned before deliberately with IoU threshold 0.5. Details are in Table 2. It can be clearly seen that our method is

**Table 1** Performance comparison with previous methods on KITTI test server

| Method | Input | Bev | | | 3D | | |
|---|---|---|---|---|---|---|---|
| | | Easy | Moderate | Hard | Easy | Moderate | Hard |
| MV3D [1] | LiDAR + RGB | 86.49 | 78.98 | 72.23 | 74.97 | 63.63 | 54.01 |
| F-PointNet [5] | LiDAR + RGB | 91.17 | 84.67 | 74.77 | 82.19 | 69.79 | 60.59 |
| AVOD [6] | LiDAR + RGB | 89.75 | 84.95 | 78.32 | 76.39 | 66.47 | 60.23 |
| F-ConvNet [15] | LiDAR + RGB | 91.51 | 85.84 | 76.11 | 87.36 | 76.39 | 66.69 |
| ContFuse [24] | LiDAR + RGB | 94.07 | 85.35 | 75.88 | 83.68 | 68.78 | 61.67 |
| MulFusion [12] | RGB | – | 19.54 | – | – | 9.8 | – |
| GSD [18] | RGB | – | – | – | 2.75 | 1.99 | 1.86 |
| 3DOP [8] | RGB | 12.63 | 9.49 | 7.59 | 6.55 | 5.07 | 4.1 |
| SROD [14] | RGB | 68.50 | 48.30 | 41.47 | 54.11 | 36.69 | 31.07 |
| DESR-CNN | RGB | 69.36 | 50.80 | 43.61 | 56.26 | 39.62 | 33.54 |

BEV and 3D object detection metric are used, reported by the Average Precision (AP) with IoU threshold 0.7

**Table 2** Performance comparison with previous methods on KITTI test server

| Method | Input | Bev | | | 3D | | |
|---|---|---|---|---|---|---|---|
| | | Easy | Moderate | Hard | Easy | Moderate | Hard |
| MulFusion[12] | RGB | – | 53.56 | – | – | 47.42 | – |
| 3DOP [8] | RGB | 55.04 | 41.25 | 34.55 | 46.04 | 34.63 | 30.39 |
| SROD [14] | RGB | 87.13 | 74.11 | 58.93 | 83.82 | 66.28 | 57.24 |
| GSD [18] | RGB | – | – | – | 21.66 | 15.47 | 14.75 |
| DESR-CNN | RGB | 87.16 | 76.44 | 66.95 | 86.49 | 74.29 | 59.14 |

BEV and 3D object detection metric are used, reported by the Average Precision (AP) with IoU threshold 0.5

**Table 3** Performance comparison with 3DBBX, reported by the Average Precision (AP) and mean Average Precision with IoU threshold 0.5

| Method | Car | Van | Truck | Pedestrian | Cyclist | mAP |
|---|---|---|---|---|---|---|
| 3DBBX | 45.32 | 43.12 | 45.36 | 10.91 | 15.39 | 32.02 |
| Ours | 86.49 | 82.13 | 83.26 | 27.56 | 19.12 | 59.71 |

better than other methods that only use image input, whether it is with IoU threshold 0.5 or IoU threshold 0.7.

As Table 1 shows, there is still a gap between our method with methods adding LiDAR. The reason should be that LiDAR can provide more accurate 3D information. In the other hand, our method's AP outperform other methods relying on only the image input. It shows that our method can improve accuracy without introducing expensive LiDAR equipment. The increase in accuracy is mainly due to the addition of Descriptor-enhanced 3D Box Alignment module. We make full use of the point correspondences between two input images. The results with IoU threshold 0.5 are shown in Table 2, where our method also outperforms state-of-the-art monocular-based methods [8, 12, 18] and stereo-method [14].

### *4.3 Multi-class Comparison with 3DBBX*

We expand [10] to multi-class object detection. There are 7 types of objects in KITTI: Car, Van, Truck, Pedestrian, Person_sitting, Cyclist and Tram. The class of Person_sitting in KITTI is less in quantity and the class is similar to pedestrian or cyclist, so we ignore this category.

As shown in Table 3, our approach performs better than 3DBBX [10] in all the classes with IoU threshold 0.5.

The result of Pedestrian, Cyclist, and Tram (hard class) is significantly worse than vehicle and there are three reasons for it. First of all, the perspective key point of the hard classes will be lost or not in the middle of the 2D box. Second, the sizes of hard classes are quite different so that size has no property of low-variance and unimodal. Finally, the orientation of hard class is hard to estimate.

### *4.4 Ablation Study*

In this work, the descriptor-enhanced 3D box alignment is designed to improve the precision of 3D box estimation. Therefore, we conduct experiments with two models and evaluate the performance to validate the contribution of the descriptor-enhanced alignment module. As shown in Table 4, the model with descriptor-enhanced 3D box alignment achieves significant improvement, which verifies its effectiveness.

**Table 4** Improvements of using descriptor-enhanced alignment, reported by the Average Precision (AP)

| Config | Bev | | | 3D | | |
|---|---|---|---|---|---|---|
| | Easy | Moderate | Hard | Easy | Moderate | Hard |
| No alignment with 0.5 IoU | 45.59 | 33.82 | 28.96 | 41.88 | 27.99 | 22.80 |
| RGB alignment with 0.5 IoU [14] | 87.13 | 74.11 | 58.93 | 85.84 | 66.28 | 57.24 |
| Descriptor-enhanced alignment with 0.5 IoU | 87.16 | 76.44 | 66.95 | 86.49 | 74.29 | 59.14 |
| No alignment with 0.7 IoU | 16.87 | 10.40 | 10.03 | 11.37 | 7.75 | 5.74 |
| RGB alignment with 0.7 IoU [14] | 68.50 | 48.30 | 41.47 | 54.11 | 33.39 | 31.07 |
| Descriptor-enhanced alignment with 0.7 IoU | 69.36 | 50.80 | 43.61 | 56.26 | 39.62 | 33.54 |

The method without alignment performs worst because the method mainly uses the high-level features learned by the network to predict. The complete model utilizes the more refined information of the matching relationship between the points of the same pixel in the binocular image to correct the bounding box. There is a similar module in SROD [14]. However, SROD [14] only uses gray difference as a measure of point matching with the same pixel. This method is affected by the confusion of similar regions, which leads to poor reliability. The local feature description used in our model is a more reliable metric for point matching, so more accurate three-dimensional point information can be estimated, which effectively improves the 3D positioning accuracy of the bounding box.

### *4.5 Runtime*

We evaluate the runtime of DESR-CNN on the KITTI dataset with a single Intel i9-9820X CPU and GeForce RTX 2080 Ti GPU. The total computational time to finish the 3D box detection on an image is 221 ms, which is suitable for some real-time applications. Compared with the baseline Stereo R-CNN, the extra computational time of DESR-CNN is 11 ms per image, which demonstrates that our descriptor-enhanced approach is efficient.

## 5 Conclusion and Future Work

In this chapter, we discuss 3D object detection for autonomous driving. The existing methods can be divided into cloud point-based and image-based types according to different input information. Both of the two kind of methods take convolution neural network to extract characteristic information. Even though cloud point based method can provide accurate object location, image-based method is still adopted in many practical solution in autonomous driving because it achieves better object classification performance with low cost. Recently, image-based methods are focusing on the more accurate 3D bounding box. Following the direction we present a new method named Descriptor Enhanced Stereo R-CNN (DESR-CNN) to improve 3D bounding box. The comparison experiments on KITTI dataset show that DESR-CNN outperforms most of the existing 3D object detection methods based on binocular images. For image-based 3D object detection, the challenges still exist, such as occlusion, small target, and computing efficiency.

## References


1. Chen X, Ma H, Wan J et al (2017) Multi-view 3d object detection network for autonomous driving. In: Proceedings of the IEEE conference on computer vision and pattern recognition, CVPR, pp 1907–1915
2. Yang Z, Sun Y, Liu S, et al (2020) 3dssd: point-based 3d single stage object detector. In: Proceedings of the IEEE conference on computer vision and pattern recognition, CVPR, pp 13–19
3. Yang Z, Sun Y, Liu S et al () Std: Sparse-to-dense 3d object detector for point cloud. In: Proceedings of the IEEE, CVF International Conference on Computer Vision, ICCV. 2019: 1951–1960
4. Shi S, Wang Z, Wang X et al (2019) Part-A 2 net: 3d part-aware and aggregation neural network for object detection from point cloud. 2(3). arXiv:1907.03670
5. Qi CR, Liu W, Wu C et al (2018) Frustum PointNets for 3d object detection from RGB-D data. In: Proceedings of the IEEE conference on computer vision and pattern recognition, CVPR, pp 918–927
6. Ku J, Mozifian M, Lee J et al (2018) Joint 3d proposal generation and object detection from view aggregation. In: Proceedings of the international conference on intelligent robots and systems, IROS, pp 1–8
7. Guo Y, Wang H, Hu Q et al (2020) Deep learning for 3d point clouds: a survey. IEEE Trans Pattern Anal Mach Intell 4338–4364
8. Chen X, Kundu K, Zhu Y et al (2017) 3d object proposals using stereo imagery for accurate object class detection. IEEE Trans Pattern Anal Mach Intell 40(5):1259–1272
9. Li P Qin T (2018) Stereo vision-based semantic 3d object and ego-motion tracking for autonomous driving. In: Proceedings of the European conference on computer vision, ECCV, pp 646–661
10. Mousavian A, Anguelov D, Flynn J, et al (2017) 3d bounding box estimation using deep learning and geometry. In: Proceedings of the IEEE conference on computer vision and pattern recognition, CVPR, pp 7074–7082
11. Zeeshan Zia M, Stark M, Schindler K (2014) Are cars just 3d boxes-jointly estimating the 3d shape of multiple objects. In: Proceedings of the IEEE conference on computer vision and pattern recognition, CVPR, pp 3678–3685
12. Xu B, Chen Z (2018) Multi-level fusion based 3d object detection from monocular images. In: Proceedings of the IEEE conference on computer vision and pattern recognition, CVPR, pp 2345–2353
13. Chen X, Kundu K, Zhang Z et al (2016) Monocular 3d object detection for autonomous driving. In: Proceedings of the IEEE conference on computer vision and pattern recognition, CVPR, pp 2147–2156

14. Li P, Chen X, Shen S (2019) Stereo R-CNN based 3d object detection for autonomous driving. In: Proceedings of the IEEE conference on computer vision and pattern recognition, CVPR, pp 7644–7652
15. Wang Z, Jia K (2019) Frustum ConvNet: sliding frustums to aggregate local point-wise features for amodal 3d object detection. In: RSJ international conference on intelligent robots and systems, IROS. IEEE, pp 1742–1749
16. He C, Zeng H, Huang J et al (2020) Structure aware single-stage 3d object detection from point cloud. In: Proceedings of the IEEE conference on computer vision and pattern recognition, CVPR, pp 11873–11882
17. Shi W, Rajkumar R (2020) Point-GNN: graph neural network for 3d object detection in a point cloud. In: Proceedings of the IEEE conference on computer vision and pattern recognition, CVPR, pp 1711–1719
18. Li B, Ouyang W, Sheng L et al (2019) Gs3d: an efficient 3d object detection framework for autonomous driving. In: Proceedings of the IEEE conference on computer vision and pattern recognition, CVPR, pp 1019–1028
19. Yan P, Tan Y, Tai Y et al (2021) Unsupervised learning framework for interest point detection and description via properties optimization. Pattern Recogn 112:107808
20. Ren S, He K, Girshick R, et al (2015) Faster R-CNN: towards real-time object detection with region proposal networks. In: Proceedings of the advances in neural information processing systems, NIPS, vol 28, pp 91–99
21. He K, Zhang X, Ren S et al (2016) Deep residual learning for image recognition. In: Proceedings of the IEEE conference on computer vision and pattern recognition, CVPR, pp 770–778
22. Lin TY, Dollár P, Girshick R et al (2017) Feature pyramid networks for object detection. In: Proceedings of the IEEE conference on computer vision and pattern recognition, CVPR, pp 2117–2125
23. Geiger A, Lenz P, Urtasun R (2012) Are we ready for autonomous driving the KITTI vision benchmark suite. In: Proceedings of the IEEE conference on computer vision and pattern recognition, CVPR, pp 3354–3361
24. Liang M, Yang B, Wang S et al (2018) Deep continuous fusion for multi-sensor 3d object detection. In: Proceedings of the European conference on computer vision, ECCV, pp 641–656

# Comparative Study on Transfer Learning for Object Classification and Detection

**Jungme Park, Wenchang Yu, Pawan Aryal, and Viktor Ciroski**

**Abstract** The recent development of deep neural learning achieved remarkable breakthroughs in object classification and detection. Deep learning has the capability of learning features automatically from data using general-purpose learning procedures. However, because deep neural networks require large amounts of data to train the parameters in the network, it is challenging to develop any object classification or detection system with a relatively small dataset. Transfer learning is an important machine learning technique that transfers the learned features in a pre-trained Convolution Neural Network (CNN) model into a new system. In this study, current state-of-the-art CNN models are reviewed in their architectures and characteristics. For the comparative study of transfer learning, the object classification and the detection systems are implemented using transfer learning with six state-of-the-art CNN models. The object classification model has achieved an accuracy of 97.01% for the three-class classification task using transfer learning. Furthermore, six different Faster R-CNN architectures are implemented for object detection. The performances of the different transferred models are compared in terms of the accuracy and the deploying speed of the new model. Experiments show that transfer learning saves training time and achieves accurate performance by fine-tuning the pre-existing deep learning model.

## 1 Introduction

The tasks for object classification and detection are core parts of environmental perception systems. The environmental perception system is an essential module for many automated intelligent systems such as Autonomous Driving (AD), Advanced Driver Assist System (ADAS), robotic vision, and surveillance camera, etc. Machine Learning (ML) is a common way to implement those environmental perception systems. ML finds the patterns existing in the data through the learning procedure

J. Park (✉) · W. Yu · P. Aryal · V. Ciroski
Kettering University, Flint, MI, USA
e-mail: jpark@kettering.edu

 
Y. L. Murphey et al. (eds.), *AI-enabled Technologies for Autonomous and Connected Vehicles*, Lecture Notes in Intelligent Transportation and Infrastructure,
https://doi.org/10.1007/978-3-031-06780-8_5

called 'training'. During the training step in ML, a huge amount of the data in the field is feed into the ML algorithm, and the ML algorithm is iterated until the system performance is satisfied.

Since LeCun et al. [1] presented a Convolutional Neural Network (CNN), called LeNet-5, for handwritten digit recognition in 1998, deep convolution neural learning has become a popular Deep Neural Network (DNN) architecture, and it has achieved remarkable breakthroughs in object classification and detection for several decades. In the traditional ML methods for object classification and detection, the ML system needs the input features carefully designed by human developers [2–4]. On the other hand, deep CNN methods contain the automated feature extraction layers, called 'Convolution Layer'. These convolution layers automatically extract input features from the huge data samples used in training. There are many reasons that deep-learning based object detection methods achieved state-of-the-art performances. First, massive sets of labeled data such as PASCAL-VOC [5, 6], ImageNet [7], MS-COCO [8], and Caltech [9, 10] are publically available for training on many different types of objects. In addition, the computing power has been increased with high-performance Graphic Processing Unit (GPU) computers.

In this book chapter, the authors researched transfer learning using various pre-existing deep CNN models with different datasets. The chapter is organized as follows. Section 2 presents a detailed literature review in deep CNNs and transfer learning. In Sect. 3, detailed architectures and characteristics of the state-of-the-art CNN models are discussed. In Sect. 4, transfer learning methodologies and experimental results are presented. Finally, the chapter is concluded in Sect. 5.

## 2 State-of-the-Art Review in DNNs and Transfer Learning

ML algorithms and applications have exploded in popularity during the last decade, especially in deep learning. In 2012, Krizhevsky et al. [11] proposed a DNN architecture 'AlexNet' that achieved the state-of-the-art accuracy higher than the second best architecture by more than 10% on the ImageNet dataset. It was a significant milestone in the field of deep learning and started a CNN renaissance. Following years, while other neural networks similar to AlexNet like ZFNet improved the accuracy, Simonyan et al. [12] proposed a different CNN configuration also known as 'Visual Geometry Group (VGG) Net' with $3 \times 3$ receptive fields throughout the network in 2014. They demonstrated that the state-of-the-art performance on the ImageNet data could be achieved using the conventional ConvNet architecture with substantially increased depth. Same year Szegedy et al. [13] presented 'GoogLeNet' that introduced a new level of organization in the form of inception.

After observing the success of AlexNet, VGG Net, and GoogLeNet, it seems that as the network goes deeper, the performance grows better. However, unexpectedly, when the network became deeper, the performance plateaued. In other words, it is also called the "vanishing gradient" problem. Batch normalization [14] is a technique for solving this problem. It calculates the mean, and standard deviation of a mini-batch then performs the normalization process. Batch normalization could stable the learning process and make the model converge faster. Residual Network (ResNet) [15] is another remarkable solution to the "vanishing gradient" problem. By introducing the residual block concept, the model's accuracy will not be degraded as the dimension goes deeper.

While development in deep learning frameworks was gaining attraction, object detection performance had plateaued during 2012–14 [16]. However, Grisch et al. [16] broke that deadlock by proposing Regions with CNN features (R-CNN) network by applying high capacity CNNs to the bottom-up region proposal to localize and segment objects. Later, Faster R-CNN was proposed by Ren et al. [17] that was the first end-to-end and the near real-time deep learning detector. Since then, several other significant frameworks have been proposed, including Feature Pyramid Networks [18] and CNN based one-stage detectors like You Only Look Once (YOLO) [19], Single Shot multi-box Detector (SSD) [20], and RetinaNet [21].

The computational resources and time required to train deep learning models from scratch are vast. Recent successful deep learning architectures for object detection [17, 19–22] have already been trained using the large dataset. It would save plenty of resources if the knowledge from the pre-trained DNN models can be reutilized to develop a new related task. This can be achieved using the ML technique called 'Transfer Learning' that enhances the learning of a new task by channeling the knowledge through a related task that has formerly been learned [23]. Early in 1995, the research on transfer learning started attracting a lot of attention. In [24], transfer learning was named with several different names such as "knowledge transfer", "multi-task learning", or "knowledge consolidation". Later, Pan and Yang [25] presented knowledge transfer to improve learning performance and reduce the effort to recollect the data. Moreover, Mengying [26] utilized transfer learning for the image classification system on a minimal dataset. The author showed that transfer learning saved training time and achieved accurate performance by fine-tuning the pre-exist deep learning model and data augmentation.

In [27], the authors conducted an intensive survey on transfer learning. Transfer learning was classified into four categories, which are instance-based deep transfer learning, mapping-based deep transfer learning, network-based deep transfer learning, and adversarial-based deep transfer learning. Xiao et al. [28] used a network-based deep transfer learning strategy and compared ultrasonic breast mass discrimination performances with traditional neural networks. Kim et al. [29] proposed an approach for a deep learning road lanes tracking system using transfer learning. While the transfer learning approach can be computationally efficient, it can also suffer from the negative transfer, which happens when the pre-training data contributes to negative learning on the target application [30].

## 3 Architecture and Characteristics of CNN Models

When image data samples are fed into a CNN model, there are two major steps for object classification. The first step is feature extraction. The feature extraction process identifies features that generally represent each class and discriminate each class from other classes. The second step is to conduct the classification task. The classification task predicts the object class among all given classes. In the CNN models, feature extraction is conducted by several convolutional layers, combined with activation functions and pooling layers. After the feature extraction step, the last convolutional layer outputs are fed to fully connected layers as input. Then the final fully connected layer generates the probabilities that belong to the corresponding classes.

Based on different tasks and objectives, the existing CNN models can be grouped into object classification and object detection. Classification refers to predicting the identity for a set of categories the observation belongs to. Object detection attempts to break up an image into smaller regions and identify the locations and the types of objects. The popular CNN models for object classification are, Alexnet [11], VGG-16 [12], VGG-19 [12], GoogLeNet [13], and ResNet [15]. The CNN Models for object detection have one additional procedure for localization to find the object's location in the image. The CNN models designed for object detection can be further summarized into two classes. One is implementing the two-stage method, like R-CNN [16], Fast R-CNN [31], and Faster R-CNN [17]. The other is utilizing the one-stage method like YOLO [19]. The two-stage method generates region proposals based on the input image, then based on the region to determine the object using the CNN's object classification feature. The one-stage method skips the region proposal step. It obtains the object class and location directly from the input image.

### 3.1 *CNN Models for Object Classification*

Due to advancements in GPU computing power, a bigger and deeper network can be designed and trained. Because of the groundbreaking work and success of AlexNet [11], it gathers much attention from AI researchers. AlexNet has 8 layers with 5 convolutional layers and 3 fully connected layers, as shown in Fig. 1. The visualized features in Fig. 2a–c presented the partial features learned on convolution layers in AlexNet. Edges and color filters are mainly learned in the first convolution layer, as presented in Fig. 2a, where the first 64 features are displayed. In Fig. 2b, c, more

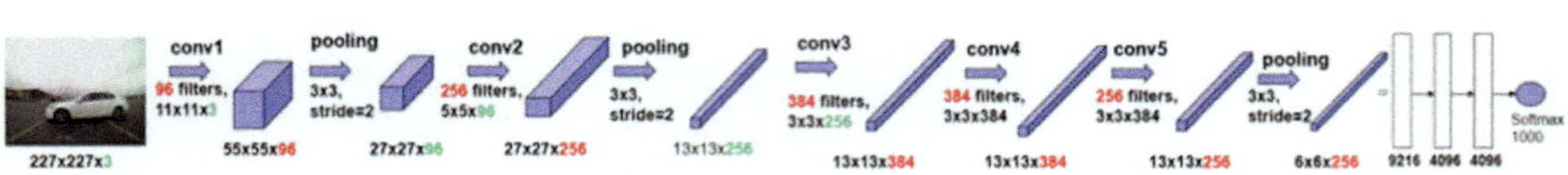


**Fig. 1** Architecture of AlexNet [11]

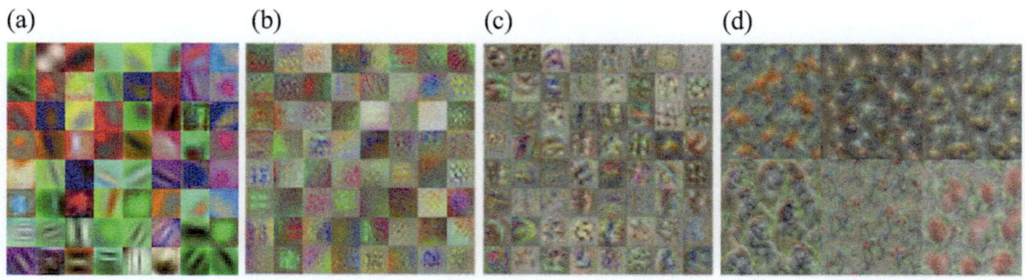


**Fig. 2** Features on convolution layers in AlexNet: **a** the first 64 features learned in the 1st convolution layer, **b**, **c**, patterns extracted in the 2nd and 4th convolution layers respectively, **d** patterns at the last fully connected layer

complicated patterns are extracted in the 2nd and 4th convolution layers, respectively. The deeper the layer, the more complicated patterns and textures are extracted. The feature images shown in Fig. 2d are generated from the last fully connected layer. Starting from the top to bottom and left to right, each feature image corresponds to the class for goldfish, minivan, cab, gorilla, tricycle, and strawberry, respectively.

Before AlexNet, the Rectified Linear Units (ReLU) activation function was not widely used. In the training stage, the amount of computation for the ReLU function is much lesser than the Sigmoid or the Tanh function so that the model can converge much faster. Further, the overlap pooling method is introduced in AlexNet. Compared to non-overlap pooling, overlap pooling can reduce errors and prevent the model from overfitting. In addition, the dropout technique is utilized in AlexNet. The dropout method disconnects certain neurons between two convolutional layers based on a predefined probability while maintaining the same number of neurons in the input and output layers.

Learning from the successful experience of AlexNet, the VGG network model [12] continues the effort to make the net deeper. The VGG architecture has two variations, VGG16 and VGG19. The VGG16 model has 13 convolutional layers and 3 fully connected layers, as shown in Fig. 3. On the other hand, VGG19 has three more convolutional layers than the VGG16 model. Although VGG has similarities with AlexNet, both use the convolutional layer structure followed by the ReLU activation function, the VGG model uses a small receptive field with a 3 × 3 kernel size compared to AlexNet. Further, a simple principle is used to design the VGG architecture, and the entire network uses the same 3 × 3 kernel size and the same 2 × 2 max pool size, making the model very concise. Moreover, the small kernel size brings more non-linearity to the model, and the amount of computation is controlled

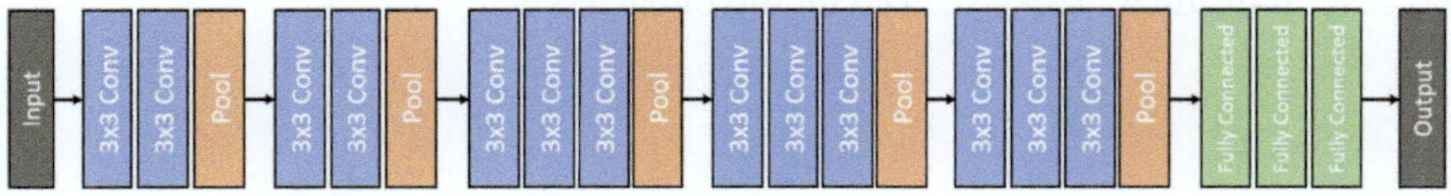


**Fig. 3** VGG16 architecture [12]

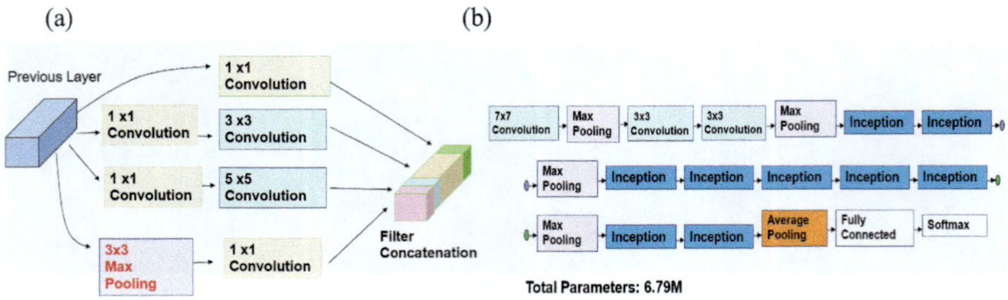


**Fig. 4** **a** Inception in GoogLeNet, **b** the architecture of GoogLeNet

when making the model deeper and wider. Because of its innovations, VGG is one of the most commonly used feature extractors.

GoogLeNet [13] implements 9 inception modules in its architecture. The inception module is presented in Fig. 4a. Combined with 3 other convolutional layers and 1 fully connected layer, the network model has a depth of 22, as presented in Fig. 4b. Although GoogLeNet has more layers than VGG, it contributes to strengthening the convolutional module. Adopted the methodology from [32], the inception modules are piled together to increase the network's width and depth while bringing limited performance punishment. The naïve version of the inception module combines 3 convolutional layers and 1 pooling layer, but the number of parameters is not optimized. For dimension reduction, another inception module is proposed by adding 3 additional $1 \times 1$ convolutional layers to the naïve version inception module, around half of the parameters can be reduced. The dimension reduction allows for a gain of computational efficiency and ensures a deeper and wider network capability. The inception module in GoogLeNet reduces the number of parameters to be learned up to 6.79M, which is dramatically reduced compared to 144M parameters in VGG19.

The ResNet model [15] has 152 layers, containing 151 convolutional layers and 1 fully connected layer. When the CNN model becomes deeper and deeper, the model's performance does not increase linearly with the number of layers. To solve this problem, the original gated shortcut connection method from Highway Networks [33] has been modified, and the residual block has been proposed along with the shortcut connection method. In traditional CNN, a layer N can only learn from the output of layer N − 1. In traditional CNN, a layer N can only learn from the output of layer N − 1. With shortcut connections, the Nth layer can learn from the output from the N − 2 layer or the N − 3 layer by adding the result from the previous layers. The goal is to learn the residue between previous layers and the Nth layer. These layers with a shortcut connection form a residual block. For ResNet, its architecture has several residual blocks stacked on top of each other.

## 3.2 NN Models for Object Detection

Object detection is considered one of the most important challenges in computer vision and the backbone of many deep learning applications. In general, object detection algorithms attempt to segment an image into smaller, more manageable regions to locate and classify areas of interest. The two most common approaches for localization are sliding windows and anchor boxes.

Sliding windows are one way to segment and localize objects within an image. The sliding window approach defines a rectangular region (a window) that "slides" across an image. Each area this window overlaps with is applied to an image classifier to localize the object. While sliding windows may be easy to implement, the most significant disadvantage is the computational cost. However, this sliding window becomes the backbone of the convolutional layer in many CNN architectures. The computational cost in CNNs can be reduced by applying a convolutional operation by sliding the filters over an image because the features are shared with neighboring pixels.

Anchor boxes can be considered simply a set of predefined bounding boxes, representing the scale, $s \in (0,1]$, and aspect ratio, $r > 0$, of the training dataset. Multiple anchor boxes are defined using all ground truth labels in the training data. Unlike the sliding window method, anchor boxes avoid repeated feature calculations by utilizing the feature map of the convolutional network. Thus, anchor boxes improve both the model's speed and accuracy. However, only several representative anchor boxes are selected due to the computational cost of including all anchor boxes in the training set [34]. As a result, it is possible to make errors on the unusual ratio bounding boxes. In addition, the detection performance is affected by selecting the size, aspect ratio, and the number of anchor boxes. Thus, tuning the parameters related to anchors is vital for overall performance. There are alternative approaches to anchor boxes, called anchor-free detectors [35–38]. Anchor-free detectors will attempt to find an object without the presents of anchors either using center-based methods [35, 36] or key point-based methods [37, 38]. Unlike anchor boxes, anchor-free detectors do not require hyper-parameters allowing them to be more generalized; however, this degrades the accuracy in comparison.

The Faster R-CNN [17] and YOLO [19] architectures utilize anchor boxes in different ways. The Faster R-CNN architecture [17] presented in Fig. 5a, evolved from R-CNN and Fast R-CNN, can be divided into four major parts: Convolutional layers as feature networks, Region Proposal Network (RPN), Region of Interest (ROI) pooling, and Classification. The major contribution of Faster R-CNN is to propose the RPN method. The goal of the RPN is to predict the best location for each anchor box. The RPN itself is a CNN that takes the feature map generated by the backbone CNN, such as VGG, as input. The RPN uses sliding windows over the feature map to generate region proposals, as shown in Fig. 5b. Several anchors are generated for each pixel on the feature map.

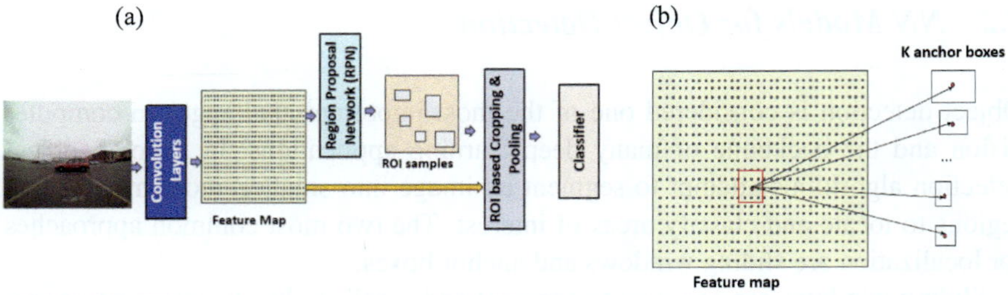


**Fig. 5** The faster R-CNN architecture: **a** faster R-CNN architecture, **b** anchor boxes

By default, the Faster R-CNN architecture defines 3 scales and 3 aspect ratios resulting in 9 total anchors. Then utilizing the softmax function, each anchor is determined to either foreground or background. To generate the final object detections, anchor boxes that belong to the background class are removed. For the remaining ones, their confidence scores are generated. The ROI pooling gathers the feature map and region proposals to extract the proposal feature maps and sending to the classifier. The classifier finishes the final step, determining the object class.

Unlike the Faster R-CNN network, the YOLO [19] architecture does not depend on a secondary network to determine anchor boxes. Instead, YOLO has done the pioneering work of a one-stage detection method. In [19], the input image is divided into S × S grids, as shown in Fig. 6. Each grid is responsible for detecting the object whose center is falling in that grid. Further, each grid predicts B bounding boxes, and each bounding box requires information about the center of the bounding box (x, y), the width and height of the bounding box (w, h), and the confidence score. The confidence score of the box, *C(object)*, represents the probability that the box contains the object. YOLO generates K-class probability values for each grid. As presented in Fig. 6, the cuboidal output has the size of $S \times S \times (5B + K)$. In [19], the grid size, $S = 7$, number of the bounding box in each grid, $B = 2$, and probability values of class within the grid, $K = 20$, are used.

YOLO uses a single bounding box regression to predict the height, width, center coordinate, and class of objects. The YOLO model uses sum-squared error loss function, *SQE*, which contains the *localization loss* for bounding box prediction,

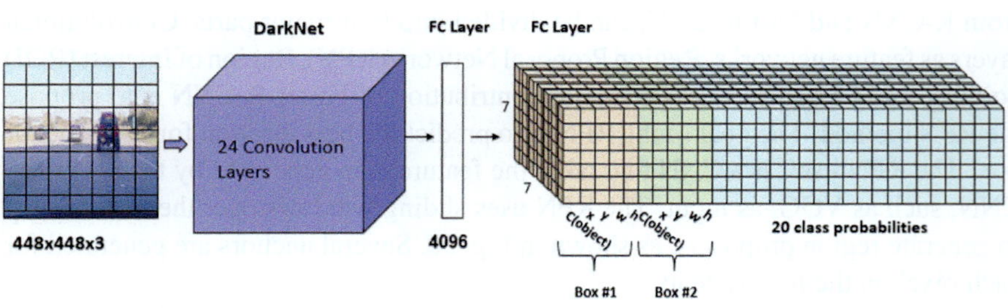


**Fig. 6** The YOLO architecture

*loss from the box confidence score*, and *the classification loss* for conditional class probabilities in Eq. 1–4 [19].

$$SQE = SQE_{localization} + SQE_{confidence} + SQE_{classification}, \tag{1}$$

$$\begin{aligned} SQE_{localization} &= \lambda_1 \sum_{i=0}^{S^2} \sum_{j=0}^{B} I_{ij}^{obj} \left[ \left(x_i - \hat{x}_i\right)^2 + \left(y_i - \hat{y}_i\right)^2 \right] \\ &+ \lambda_1 \sum_{i=0}^{S^2} \sum_{j=0}^{B} I_{ij}^{obj} \left[ \left(\sqrt{w_i} - \sqrt{\hat{w}_i}\right)^2 + \left(\sqrt{h_i} - \sqrt{\hat{h}_i}\right)^2 \right], \end{aligned} \tag{2}$$

$$SQE_{confidence} = \sum_{i=0}^{S^2} \sum_{j=0}^{B} \mathrm{I}_{ij}^{obj} \left(C_i - \hat{C}_i\right)^2 + \lambda_2 \sum_{i=0}^{S^2} \sum_{j=0}^{B} I_{ij}^{noobj} \left(C_i - \hat{C}_i\right)^2, \tag{3}$$

$$SQE_{classification} = \sum_{i=0}^{S^2} I_i^{obj} \sum_{k \in classes} \left(p_i(k) - \hat{p}_i(k)\right)^2, \tag{4}$$

where $I_{ij}^{obj} = 1$ if the jth bounding box in grid cl $i$ is responsible for detecting the object, otherwise 0; $I_{ij}^{noobj}$ is the complement of $I_{ij}^{obj}$ ; $\lambda_1$ and $\lambda_2$ are the parameters to control the penalties from the localization and confidence loss, respectively; $I_i^{obj}$ = 1 if object appears in grid cell *i*, otherwise 0; $\hat{p}_i(k)$ is the conditional probability for class k in the grid cell *i*. The final inferences of the given input are generated by applying the non-maximal suppression algorithm. YOLO is fast at the test time because it uses only a single CNN architecture to detect objects in the input image. However, YOLO [19] is not good at detecting small objects and its biggest drawback is not accurate compared to other object detection systems at that time.

## 4 Transfer Learning for Object Classification and Detection

In the state-of-the-art CNN architectures, representative low-level features are extracted in the lower layers of the model as presented in Fig. 2a, b. By extracting these lower-level features that are shared across similar datasets provides many benefits. Transfer learning is one of the most exciting fields in artificial intelligence. Transfer learning takes knowledge gained from one setting and exploits it to improve the performance of another task. If the existing trained model is defined as the source domain and the new task that needs to be trained as a target domain, mathematically, transfer learning can be defined as follows. The domain space, D, contains the feature space, X, and the probability distribution of the feature space, P(X): D = {X, P(X)}. The learning task, $\mathcal{L}$, is defined by the label, Y, and the predictive function to be learned, f(.): $\mathcal{L}$ = {Y, f(.)}. Two domain spaces, $D_s$ and $D_t$ are considered to be different if the feature spaces $X_t$ and $X_s$ are different or if the probability distributions $P(X_t)$ does not equal $P(X_s)$. Two learning tasks, $\mathcal{L}_s$ and $\mathcal{L}_t$ are different if the label space are not equal or if the conditional probability distributions are not equal, $P(Y_t|X_t) \neq P(Y_s|X_s)$. The goal of transfer learning for the target domain, $D_t$, is to learn the conditional probability for $P(Y_t|X_t)$ using the features in $D_s = \{X_s, P(X_s)\}$ and $\mathcal{L}_s = \{Y_s, f_s(.)\}$ [39].

When applying transfer learning, it is important to understand the initial data it was trained on. These shared lower-level features can affect the performance of the new task. For example, a CNN model learns the image's shapes, edges, and lighting with visual image data in its convolution layers. Because these features are generalized across most types of images, utilizing those learned features from big data in the existing CNN model to a new CNN model with relatively small data samples provides better accuracy than training the new model from scratch.

It is challenging to develop an object detection system with a small dataset. However, by utilizing the CNN models trained with a large amount of data, those learned features can be transferred to a new system with a smaller dataset. Transfer learning comes with a variety of benefits, other than just helping improve the performance of a small dataset. It also saves time during training. Because less data is required and low-level features are already learned, only a few weights need to be updated during the training process. In this study, two different transfer learning settings are implemented for object classification and detection.

## 4.1 Transfer Learning for Object Classification

In many sensor fused ADASs, it is possible to find the ROI of a potential object location in an image using automotive sensors such as radar or LiDAR. Then the identified ROI is fed into the CNN model to classify the object type, whether it is a car, bike, or pedestrian, etc. The object type classification is an important task to generate appropriate control signals. The Rear Cross-Traffic (RCT) detection system is an application of ADAS. Using automotive radar and camera sensors, the RCT detection system detects obstacles at the rear-end while the car is backing [40]. The object classification system is a sub-module inside the RCT detection system that classifies three different object types, 'car', 'bike', and 'pedestrian'. For the comparative study of transfer learning on the image classification task, six state-of-the-art CNN models are selected: AlexNet, VGG-16, VGG-19, Darknet19, Resnet-50, and GoogLeNet. Those selected CNN models were originally trained with 1.2M ImageNet data to classify 1000 classes.

To develop the object classification module in the RCT detection system, cropped image samples in the public data sets [41–44] are used to train the classification system. Total 8044 training samples are selected, including 2651 image patches for the bike class, 3381 for the car class, and 2012 samples for the pedestrian class. The sample image patches used in the training process are presented in Fig. 7. For

**Fig. 7** Sample image patches [41–44] used to train the object classification task in the RCT detection system

training of transfer learning, 90% of the collected data is used to retrain the CNN model, and 10% of the data is used for validation during training.

Figure 8 presents how the learned features in the pre-trained CNN models are transferred to the new model. First, three figures in Fig. 8a–c are the original features learned by GoogLeNet. In Fig. 8a, the first 64 low-level features such as edges, lines, and colors learned at the network's beginning are displayed. Figure 8b is the first 16 features learned in the first inception module of the network. Figure 8c shows the first three-channel outputs by the fully connected layer at the network's end for the class, 'tench', 'goldfish', and 'great white shark'.

On the other hand, the features learned by transfer learning are presented in Fig. 8d–f. Figure 8d shows that the features transferred in the low level of the network are very close to Fig. 8a. It means most of the original features are reused in the transferred system. Figure 8e is the first inception module features in the transferred system. By comparing Fig. 8e with Fig. 8b, the features are similar, but they have

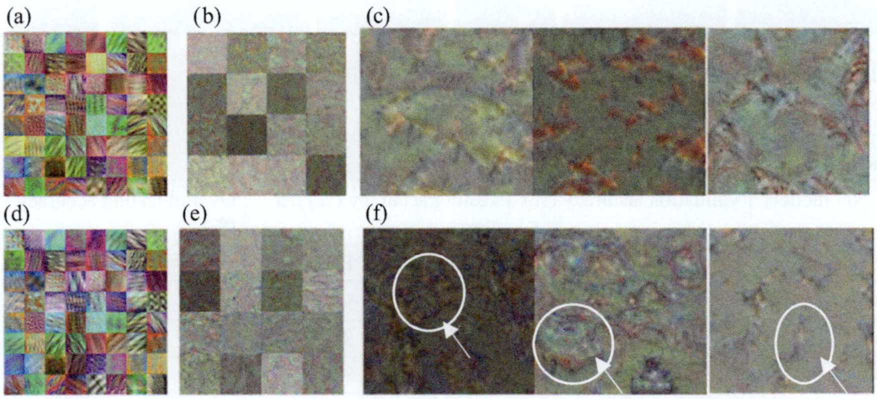


**Fig. 8** Features on convolution layers in GoogLeNet: **a–c**, features learned in the original GoogLeNet to classify 1000 classes, **d–f**, features learned through transfer learning to classify three classes

slightly changed. In Fig. 8f, the three-channel outputs by the fully connected layer at the end of the transferred network are displayed. The channel output images in Fig. 8f represent the selected classes such that the channel image for the 'bike' class contains distinct wheels of the bike, the channel image for the 'car' class contains the shape of vehicles, and the pedestrian shape is represented in the channel output for the pedestrian class.

The six transferred CNN models are evaluated with the new dataset recorded for the RCT detection system. The new dataset contains a total of 12,807 image samples, including 4796 samples for the 'bike' class, 3332 samples for the 'car' class, and 4679 for the 'pedestrian' class. Figure 9 displays the sample images for the testing of the transferred CNN models. In Fig. 9, the classification results on the testing images are depicted with the yellow bounding boxes and the classified class name. The overall classification performances are summarized in Table 1. According to the validation results during the training, VGG19, GoogLeNet, and VGG 16 have the top three accuracies of 97.01%, 96.89%, and 96.52%, respectively with the learning rate, $\alpha = 0.0001$. For the testing results on the dataset for the RCT detection system, the average accuracies of the top three models, VGG-19, GoogLeNet, and VGG-16,

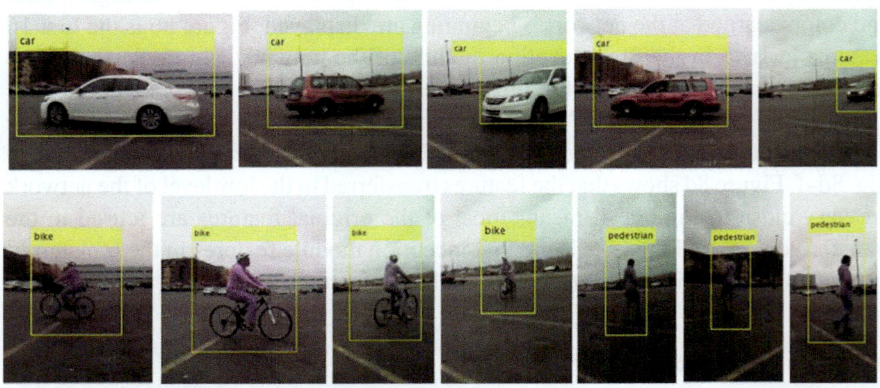


**Fig. 9** Object classification results by the transferred VGG-19 CNN model

**Table 1** Transfer learning results for object classification

| CNN models | Validation accuracy (%) | Testing accuracy (%) per class | | | Overall testing accuracy (%) |
|---|---|---|---|---|---|
| | | Bike | Car | Pedestrian | |
| AlexNet | 93.28 | 92.91 | 87.48 | 99.17 | 93.78 |
| VGG-16 | 96.52 | 92.47 | 97.72 | 95.77 | 95.04 |
| VGG-19 | 97.01 | 96.96 | 94.60 | 97.16 | 96.42 |
| Darknet-19 | 87.81 | 88.05 | 78.72 | 90.55 | 86.54 |
| Resnet-50 | 93.28 | 95.23 | 93.91 | 78.31 | 88.70 |
| GoogLeNet | 96.89 | 94.83 | 95.77 | 97.82 | 96.17 |

are 96.42%, 96.17%, and 95.04%, respectively. The validation results and the testing results are very similar. The experimental results prove that transfer learning for the object classification system in the RCT detection system is conducted successfully.

## 4.2 *Transfer Learning for Object Detection*

The recent CNN models such as Faster R-CNN or YOLO can conduct the localization and classification tasks in one system. To evaluate transfer learning for the object detection task with various CNN models, a vehicle detection system is implemented by transferring the knowledge in the pre-trained CNN models. As presented in Fig. 5a, the Faster R-CNN architecture contains the convolutional layers as the feature extraction network. Six different Faster R-CNN architectures are implemented by changing the feature extraction network inside the Faster R-CNN model. The selected CNN modules as the feature extraction network are Alexnet, VGG-16, VGG-19, GoogLeNet, Resnet50, and MobilNet. The transferred knowledge is mainly the features learned by those CNN models. In addition, the Faster R-CNN contains RPN to find the object's location using anchor boxes. Those anchor boxes that locate the objects in the image used by RPN are redefined using the new training data set.

The vehicle detection system's training data is the Udacity vehicle dataset [45] that contains 8738 images with the labeled data. The sample images are displayed in Fig. 10. The training of each architecture was conducted using the Dell Desktop PC with NVIDIA GPU GeForce RTX™ 2080. Figure 11 presents the vehicle detection

**Fig. 10** Udacity vehicle data set [45] to train the vehicle detection system

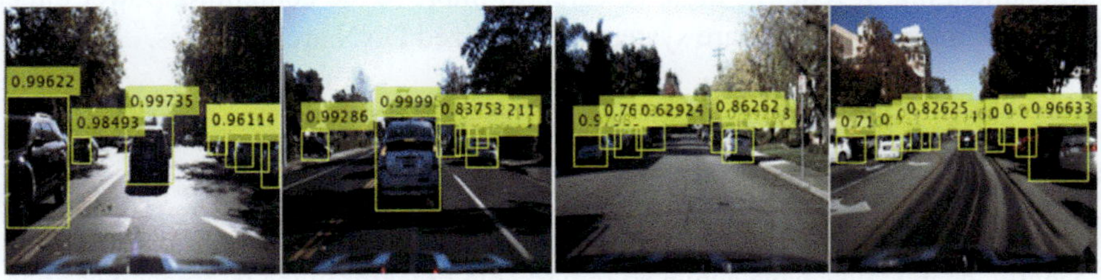

**Fig. 11** The validation results of the vehicle detection system on the Udacity vehicle data set

**Table 2** Transfer learning results for the vehicle detection system

| CNN models | Training time | Testing | | |
|---|---|---|---|---|
| | | Time/frame (s) | Precision *100 | Recall *100 |
| AlexNet-faster R-CNN | 6 h 39 m | 0.03 | 99.83 | 69.28 |
| VGG16-faster R-CNN | 9 h 26 m | 0.09 | 98.34 | 86.39 |
| VGG19-faster R-CNN | 9 h 47 m | 0.12 | 96.28 | 84.35 |
| MobilNet-faster R-CNN | 31 h 50 m | 0.31 | 80.34 | 90.58 |
| ResNet50-faster R-CNN | 49 h 10 m | 0.63 | 98.00 | 79.81 |
| GoogLeNet-faster R-CNN | 44 h 7 m | 0.42 | 96.23 | 83.19 |

system's validation results by the transferred CNN model, VGG16-Faster R-CNN, on the Udacity datasets. The training time for each architecture is summarized in Table 2. The training time for transfer learning depends on the number of layers in the network: The deeper the architecture, the longer the training time. The two architectures, AlexNet-Faster R-CNN and VGG16-Faster R-CNN have a quick training time of about 6 h and 9 h, respectively.

The evaluation of the transfer learning for the vehicle detection system is conducted on the new dataset collected in the RCT detection system. The testing image dataset contains 6233 image frames, and the example images are presented in Fig. 12. The testing is conducted using the laptop computer, Dell G7 with the processor, 8th Gen Intel® Core™ i7. The processing time per one image frame by each architecture is presented in Table 2. The testing performances for the vehicle detection system are measured with two metrics, precision and recall. The two metrics, precision and recall, are defined as follows: precision = TP/(TP + FP), recall = TP/(TP + FN), where TP = True Positive, FP = False Positive, FN = False Negative.

By considering precision and recall, the top three Faster-R-CNN architectures are VGG16-Faster R-CNN, VGG19-Faster R-CNN, and GoogLeNet-Faster R-CNN with the precision 98.34%, 96.28%, and 96.23% respectively. In addition, those models have the recall 86.39%, 84.35%, and 83.19%, respectively. Most of the missed detection cases have happened when either object size is too small or the whole shape of the object is not shown in the image, as shown in Fig. 13. Those missed cases are not critical in the RCT detection system because the missed objects are in the distance. By utilizing transfer learning, the training of the vehicle detection is completed within few days. However, deploying the trained DNN models for real-time application is still challenging. It requires more research to improve the computing power either in the AI inferencing hardware system or optimized DNN architectures.

## 5 Conclusion

Recently, deep CNNs have achieved remarkable breakthroughs in object classification and detection tasks. The state-of-the-art CNN models were trained with more

**Fig. 12** Vehicle detection results with VGG16-faster R-CNN

**Fig. 13** Missed detection cases: the cyan rectangular boxes are the labeled data

than a million data samples and learned rich features generalized across most types of images. Because DNNs require large amounts of data to train the parameters in the network, it is challenging to develop any object classification or detection systems with a relatively small dataset. Transfer learning is an important machine learning technique that transfers the learned features in a pre-existing deep CNN model into a new system with a relatively smaller dataset. In this study, transfer learning settings for object classification and detection are implemented.

The object classification system is developed for the RCT detection system using transfer learning to classify three different object types, 'car', 'bike', and 'pedestrian'. The dataset used to develop the new classification system is a relatively small dataset, and a total of 12,807 image samples are used to train the system. Six different CNN models are used for transfer learning. The top three models are VGG19, GoogLeNet, and VGG 16 and have an accuracy of 97.01%, 96.89%, and 96.52%, respectively. Using transfer learning, the classification system with a relatively small dataset generates good performances around 97–96%.

The vehicle detection system is implemented on the Udacity vehicle dataset using transfer learning for the object detection system. Inside the Faster R-CNN architecture, six pre-trained CNN modules trained on 1.2M ImageNet data to classify 1000 classes are inserted as the feature generation network. Six different Faster R-CNN architectures are retrained with the Udacity data to detect vehicles in the images. The top three Faster R-CNN architectures are VGG16-Faster R-CNN, VGG19-Faster R-CNN, and GoogLeNet-Faster R-CNN with the precision 98.34%, 96.28%, and 96.23%, respectively.

In transfer learning, the low-level features learned in the pre-existing CNN models did not change much, and only a few weights need to be updated during training. Because of that reason, the overall training time is saved. Experiments show that transfer learning saves training time and achieves accurate performance by fine-tuning the pre-existing deep learning model. However, when deploying the trained object detection system on real-time applications, the processing time per image must be improved. The computational cost is still high due to the object localization and a huge number of parameters inside the DNN architecture. Future research must improve the computing power either in the AI inferencing hardware system or optimized DNN architectures.

## References


1. LeCun Y, Bottou L, Bengio Y, Haffner P (1998) Gradient-based learning applied to document recognition. Proc IEEE. https://doi.org/10.1109/5.726791
2. Viola P, Jones M (2001) Rapid object detection using a boosted cascade of simple features. In: Proceedings of the 2001 computer vision and pattern recognition. IEEE computer society conference. https://doi.org/10.1109/CVPR.2001.990517
3. Viola P, Jones M (2004) Robust real-time face detection. Int J Comput Vis 57–2:137–154
4. Dalal N, Triggs B (2005) Histograms of oriented gradients for human detection. In: Computer vision and pattern recognition. IEEE computer society conference, vol 1, pp 886–893
5. Everingham M, Van Gool L, Williams C, Zisserman A (2010) The PASCAL visual object classes (VOC) challenge. Int J Comput Vis 88:303–338. https://doi.org/10.1007/s11263-009-0275-4
6. Everingham M, Eslami S, VanGool L, Williams C, Winn J, Zisserman A (2015) The Pascal visual object classes challenge: a retrospective. Int J Comput Vis 111:98–136. https://doi.org/10.1007/s11263-014-0733-5
7. Russakovsky O, Deng J, Su H, Krause J, Satheesh S, Ma S, Huang Z et al (2015) Imagenet large scale visual recognition challenge. Int J Comput Vis 115 (3):211–252. https://doi.org/10.1007/s11263-015-0816-y

8. Lin T, Maire M, Belongie S, Hays J, Perona P, Ramanan D, Dollár P, Zitnick C (2014) Microsoft coco: common objects in context. In: European conference on computer vision, pp 740–755. https://doi.org/10.1007/978-3-319-10602-1_48
9. Dollar P, Wojek C, Schiele B, Perona P (2009) Pedestrian detection: a benchmark. In: IEEE conference on computer vision and pattern recognition, pp 304–311
10. Dollar P, Wojek C, Schiele B, Perona P (2011) Pedestrian detection: an evaluation of the state of the art. IEEE Trans Pattern Anal Mach Intell 34(4). https://doi.org/10.1109/TPAMI.2011.155
11. Krizhevsky A, Sutskever I, Hinton G (2012) ImageNet classification with deep convolutional neural networks. In: Proceedings of the 25th international conference on neural information processing systems, vol 1, 1097–1105
12. Simonyan K, Zisserman A (2014) A very deep convolutional network for large-scale image recognition. https://arxiv.org/abs/1409.1556
13. Szegedy C, Liu W, Jia Y, Sermanet P, Reed S, Anguelov D, Erhan D, Vanhoucke V, Rabinovich A (2014) Going deeper with convolutions. In: Proceedings of the IEEE conference on computer vision and pattern recognition, pp 1–9
14. Ioffe S, Szegedy C (2015) Batch normalization: accelerating deep network training by reducing internal covariate shift. https://arxiv.org/abs/1502.03167
15. He K, Zhang S, Ren S, Sun J (2016) Deep residual learning for image recognition. Paper presented at the IEEE conference on computer vision and pattern recognition
16. Girshick R, Donahue J, Darrel T, Malik J (2014) Rich feature hierarchies for accurate object detection and semantic segmentation. In: Proceedings of the IEEE conference on computer vision and pattern recognition, pp 580–587
17. Ren S, He K, Girshick R, Sun J (2015) Faster R-CNN: towards real-time object detection with region proposal networks. Adv Neural Inf Process Syst 28:91–99
18. Lin T, Dollar P, Girshick R, He K, Hariharan B, Belongie S (2017) Feature pyramid networks for object detection. In: Proceedings of the IEEE conference on computer vision and pattern recognition, pp 2117–2125
19. Redmon J, Divvala S, Girshick R, Farhadi A (2016) You only look once: unified, real-time object detection. In: Proceedings of the IEEE conference on computer vision and pattern recognition, pp 779–788
20. Liu W, Anguelov D, Erhan D, Szegedy C, Reed S, Fu C, Berg A (2016) SSD: single shot multibox detector. In: European conference on computer vision, pp 21–37. https://arxiv.org/abs/1512.02325
21. Lin T, Goyal P, Girshick R, He K, Dollár P (2017) Focal loss for dense object detection. In: Proceedings of the IEEE international conference on computer vision, pp 2980–2988
22. Zou Z, Shi Z, Guo Y, Ye J (2019) Object detection in 20 years: a survey. https://arxiv.org/abs/1905.05055
23. Kulkarni R, Dhavalikar S, Bangar S (2018) Traffic light detection and recognition for self-driving cars using deep learning. In: Fourth international conference on computing communication control and automation, pp 1–4
24. Thrun S, Pratt L (1998) Learning to learn. Kluwer Academic Publishers, Norwell
25. Pan S, Yang Q (2010) A survey on transfer learning. IEEE Trans Knowl Data Eng 22:1345–1359
26. Mengying S (2019) Deep learning for image classification on very small datasets using transfer learning. Creative Compon 345
27. Tan C, Sun F, Kong T, Zhang W, Yang C, Liu C (2018) a survey on deep transfer learning. In: 27th international conference on artificial neural networks
28. Xiao T, Liu L, Li K, Qin W, Yu S, Li Z (2018) Comparison of transferred deep neural networks in ultrasonic breast masses discrimination. Biomed Res Int. https://doi.org/10.1155/2018/4605191
29. Kim J, Park C (2017) End-To-end ego lane estimation based on sequential transfer learning for self-driving cars. Paper presented at the IEEE conference on computer vision and pattern recognition workshops
30. Augiar A, Santos F, Sousa A, Oliveira P, Santos L (2020) Visual trunk detection using transfer learning and a depp learning-based coprocessor. IEEE Access 8:77308–77320

31. Girshick R (2015) Fast R-CNN. In: IEEE international conference on computer vision (ICCV), pp 1440–1448
32. Lin M, Chen Q, Yan S (2013) Network in network, arXiv:1312.4400.https://arxiv.org/pdf/1312.4400.pdf
33. Srivastava R, Greff K, Schmidhuber J (2015) Highway networks, arXiv:1505.00387. https://arxiv.org/pdf/1505.00387.pdf
34. Dive into deep learning: https://d2l.djl.ai/chapter_computer-vision/anchor.html
35. Tian Z, Shen C, Chen H, He T (2019) FCOS: fully convolutional one-stage object detection. In: Proceedings of the IEEE ICCV, pp 9627–9636
36. Kong T, Sun F, Liu H, Jiang Y, Shi J (2019) Foveabox: beyond anchor-based object detector, arXiv:1904.03797. https://arxiv.org/abs/1904.03797
37. Law H, Deng J (2018) Cornernet: detecting objects as paired keypoints. In: Proceedings of the ECCV, pp 734–750
38. Zhou X, Zhuo J, Krahenb P (2019) Extreme and bottom-up object detection by grouping center points. In: CVPR. https://arxiv.org/abs/1901.08043
39. Tsiakmaki M, Kostopoulo G, Kotsiantis S, Ragos O (2020) Transfer learning from deep neural networks for predicting student performance. Appl Sci 10:2145. https://doi.org/10.3390/app10062145
40. Park J, Yu W (2021) A sensor fused rear cross traffic detection system using transfer learning. Sensors 21:6055. https://doi.org/10.3390/s21186055
41. Krause J, Stark M, Deng J, Li F (2013) 3D object representations for fine-grained categorization. In: 4th IEEE workshop on 3D representation and recognition. In: ICCV 2013, Sydney, Australia, 8 Dec
42. Wang L, Shi J, Song G, Shen I (2007) Object detection combining recognition and segmentation. In: ACCV
43. Dalal N (2006) Finding people in images and videos. Dissertation, Institute National Polytechnique de Grenoble-INPG
44. Li X, Flohr F, Yang Y, Xiong H, Braun M, Pan S, Li K, Gavrila D (2016) A new benchmark for vision-based cyclist detection. Paper presented at the proceedings of the IEEE intelligent vehicles symposium (IV), Gothenburg, Sweden
45. Udacity-annotated driving data set. https://github.com/udacity/self-driving-car/tree/master/annotations

# Future Technology and Research Trends in Automotive Sensing

**Paul Schmalenberg, Jae S. Lee, Sean P. Rodrigues, and Danil Prokhorov**

**Abstract** We discuss the importance of sensing technology in enabling intelligence of future automotive vehicles. We briefly overview efforts of leading technology companies such as Waymo and Tesla which resulted in impressive progress toward highest levels of driving automation. We then describe our efforts in the areas of future radars and lidars, specifically, those which go beyond 2D and mechanical scanning emphasizing importance of AI in improving sensor performance at marginal added cost. We then discuss trends in optical computing with its promise of substantially reducing energy consumption while enhancing edge computing.

## 1 Introduction

Intelligent vehicles are making serious in-roads in our daily lives. Of primary importance are tech that promises to make the vehicles safer and more convenient. Twenty years ago the state of art in autonomous vehicles was illustrated by the results of the DARPA Grand Challenge 2004: none of the many vehicles competing in a desert race were able to come close at completing the mission of driving over 100 miles in an off-road setting (the top-scoring vehicle traveled only 7.5 miles). Clearly, technology was not ready [1].

Just three short years later the DARPA Urban Challenge showed that a specially designed and built vehicle could be driving autonomously in a relatively simple setting (mock-up town with no vulnerable road users present). Several vehicles completed the mission, and they did so without incidents [2, 3].

Emboldened by DUC 2007 success, Google decided to develop their own autonomous vehicles. Google hired talented researchers from different Universities, such as Stanford and CMU, under the leadership of Prof. Thrun and challenged them to produce a robot capable of driving autonomously on public roads in traffic [4]. Google's example is illustrative of a popular approach undertaken by many research

P. Schmalenberg · J. S. Lee · S. P. Rodrigues · D. Prokhorov (✉)
Toyota Motor North America R & D Center, 1555 Woodridge Ave., Ann Arbor, MI 48105, USA
e-mail: danil.prokhorov@toyota.com

 
Y. L. Murphey et al. (eds.), *AI-enabled Technologies for Autonomous and Connected Vehicles*, Lecture Notes in Intelligent Transportation and Infrastructure,
https://doi.org/10.1007/978-3-031-06780-8_6

groups throughout the world: if you want an autonomous vehicle, then you need a vehicle with a variety of sensors, each of which complements the others to allow the vehicle to build a more accurate picture of the world around it so that its motion can be both swift and accurate while navigating in changing and often uncertain driving environment.

Another popular feature of the original Google approach later improved and enhanced by its spin-off Waymo is the use of high-definition maps to ease the vehicle localization and navigation [5]. Such maps have not only detailed elements of the road structure, e.g., the location of curbs and lanes, but also road infrastructure elements, e.g., traffic lights, signs and adjacent buildings. The availability of these elements in the maps recorded as point clouds allows the vehicle to simplify significantly the task of driving, as anything which is not on the map could confidently be called obstacles, static or dynamic objects, even if only a few points of lidar beam returns have been registered for some of them.

Researchers at Tesla are pursuing arguably a more challenging path toward autonomous vehicles: relying on cameras as the primary perception sensor, and sometimes radar but absolutely no lidar! [6] Moreover, Tesla does not use pre-recorded high-fidelity maps to navigate its vehicles. They argue that human drivers use their eyes and may not need maps to drive safely in all kinds of driving conditions, including during the day and at night [7, 8].

Whether one belongs to the Google camp or the Tesla camp, the following question looms large: why can't we yet do what humans do so readily? Humans not just have two eyes (no camera can yet match capabilities of an eye, and that is why equating an eye with a camera should only be done metaphorically). They have a natural computer tuned by millions of years of evolution which computes according to yet-to-be-understood algorithms [9], and moreover can adapt existing algorithms to solving new tasks, learning new skills, often with little effort and remarkable ease.

In autonomous driving, a significant problem for decision making is a great variety of driving scenarios. Driving scenarios are classes of situations which may happen in the real world. Urban, suburban and rural settings are numerous and can come in very different shapes and forms. However, classes of such settings are limited and manageable by a suitably designed decision making system. Indeed, staying on the road in whatever lane is chosen, performing primitive maneuvers like moving left or right—and doing so without collisions, is basically all it takes to be a safe driver! Of course, the complexity of driving and associated expansion of driving situations may quickly grow once we have to include other objects on the road, signs, and traffic lights, but the basic primitives of driving remain the same. Thus, it may be argued that safe driving is a relatively easy human skill to acquire compared to, e.g., playing chess well, and this fact must have something to do not only with human excellent ability to recognize all kinds of contexts and generalize to driving accordingly but also with the simple nature of driving as a sequential perception-action problem; if it were not the case, then few people would have been able to start driving so quickly, after just a few hours of practicing [10].

Perception problems continue to be the greatest challenge of driving automation; see, e.g., [11–13]. Populating the obstacle map and maintaining reliable and easily computer-interpretable picture of the environment in the vicinity of the vehicle are the key features of a perception system of current and future intelligent vehicles including partially autonomous, active driver assistance systems such as, e.g., Toyota Safety Sense [14] or Toyota Guarding [15]. We believe that, in a foreseeable future, the intelligent vehicles will need a variety of sensors in order to be able to gradually approach the competency of an attentive human driver, whether the vehicle still relies on a pre-recorded high-fidelity map of the driving environment or not. A rich sensor suite becomes the necessary condition of sorts, while a sufficient condition is still expected to come from the relentless pace of advances of the autonomous driving algorithms for perception and decision making in terms of better processing of sensor information and making effective driving decisions. The algorithm analysis is beyond the scope of this chapter since we focus on sensing, specifically on the latest developments in radar, lidar and optical computing to enable smart sensing.

This chapter is arranged as follows. Section 2 discusses the advancements in radars and lidars. Section 3 describes our take on edge computing and photonic information process, followed by Conclusion.

## 2 Advancements in Radar and Lidar Sensing

In this section, characteristics of three typical sensors and their future trends are discussed. A sensor, by definition, detects the surrounding environment and translates it into different forms of information such as electric and mechanical signals. In the vehicle's perspective, sensors are subsystems detecting other objects such as vehicles, pedestrians or other vulnerable road users, elements of road infrastructure, and any other obstacles around the vehicle.

Camera is the essential among sensors because it is capable of identifying two-dimensional information, color, and texture of targets easily. For understanding traffic signs and signals, lane markings, roadside furniture, which are imperative information for autonomous driving, a camera can offer the most cost-efficient solution. Moreover, it is widely available across many industries. However, drawbacks are its inability of direct range estimation, susceptibility of dynamic range to illumination and weather impact, and the need of processing excessive amounts of data. The future of camera in a broad area of mobility depends on how to manage the combination of sensors and their data with the support of AI and machine learning [6]. Smart sensing emerges as an attractive proposition since much of the data is expected to be preprocessed at an individual sensor end before the central computation unit for increased overall efficiency of computation and energy. As discussed in Sect. 3, metasurface based special image processing technology is also drawing attention from the same perspective [16].

Ever since the commercial debut in heavy-duty trucks in the late 1990s by Eaton Corporation, automotive radar technology has undergone several generations of evolution in parallel with its commercial proliferation. In the early era of automotive radar, the quality and quantity of speed and range information was extremely limited so that its application was feasible for vehicles in well-defined highway condition. For many years radar was the enabling sensor for features such as Active Cruise Control (ACC). For this application absolute positioning of detections is not necessary, assigning the detection to a lane and computing the time-to-collision is sufficient for the system to actuate the accelerator and the brake so that a safety buffer to vehicles in front is maintained. As long as some simplifying assumptions are made such as other vehicles being on the road surface, simple motion models, and fixed widths of the lanes on a highway, the radar sensor only needs to determine the location of detections in range and azimuthal bearing. Therefore, the ability to scan in elevation is removed entirely which simplifies the system considerably by requiring only a single linear antenna array.

As more automated control features are developed for consumer vehicles, these radars originally designed for ACC applications are pressed into wider service. Radar's superior performance in all-weather conditions and detection of metal objects hundreds of meters away meant that they could not just be excluded from advanced driving sensor suites. However, the highly filtered output of the radar sensors left something lacking for the teams developing the algorithms operating on the raw sensor data. In fact, modern radars are demanded to provide multitudes of advanced features to accommodate other applications: from parking assistance to fully autonomous driving. Preferentially higher detection resolution in both azimuth and range, wider field of view, larger number of targets to be tracked and additional detection dimension in elevation have become requirements. Provisions have been offered by adding more channels of active or virtual elements in conjunction with widening usable bandwidth. Because of the strict regulation against spatially combined electromagnetic energy density, the number of channels or power per channel cannot be increased arbitrarily.

Locating detections in 3D space and determination of object shape would allow the radar to contribute useful data to advanced autonomous driving systems. A single scan direction is insufficient to accomplish 3D localization, so automotive radars will need to be expanded to two scan directions plus ranging. To determine object shape, improvement in resolution is necessary which corresponds to adding more elements to the antenna arrays. Measuring object shape especially when partially obscured through shallow angle multi-path and waveform processing techniques would fill a gap in current sensor suites. Figure 1 demonstrates an early example of target behind target detection [17]. Essentially, a high-resolution 3D imaging radar is required for future driving functions. To implement imaging function, antenna array should be scaled from one dimension to two dimensions; a simple and intuitive method is to extend the 'N array' of antenna into 'N x M array'. Alternatively, in most used antenna architectures a series of N gain elements is connected per channel and could be stacked up resulting in 'multi-N x M array'. In this case, the end-fire architecture may be preferred because of its feasible feed line connectivity [18].

Steering of high-resolution beams in 3D space at RF wavelengths have been commercialized for 5G applications. Automotive radar systems could adopt these technologies in a low-cost manner. One is multiple-input multiple-output (MIMO) 2D antenna technologies for base station communication in 5G networks [19]. MIMO

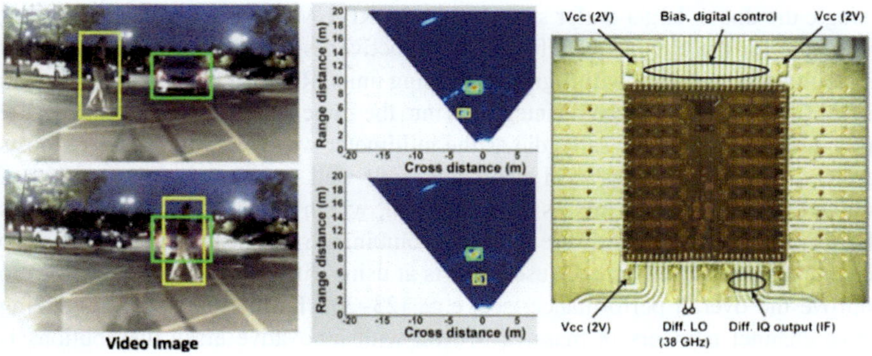


**Fig. 1** One of the authors walking in front of a vehicle (on the left) does not disturb the return from the vehicle, shown in a bird's eye view in center, in a phased array automotive radar prototype. This example is driven by a 16-channel RF phase shifter chip produced by one of our collaborators, on the right

technique can use non-uniform array spacing plus orthogonal waveforms to create virtual channels and reap the benefit of a fully populated array while minimizing physical channels [20], thus reducing cost. This is one way to achieve high resolution radar without the cost of a fully populated phased array. MIMO antennas have even been miniaturized for the use in cell phones, demonstrating that the concept is not limited to a fixed base station use [21]. On the receiver side, MIMO boosts the signal of several cellphone users simultaneously by localizing them and increasing the apparent gain through the MIMO technique. For radar applications, we simply consider the returns from multiple target reflections as the cell phone users when implementing MIMO.

Expanding the capability of the radar system comes with some drawbacks, mainly cost. A 2D MIMO array requires far more down-conversion mixers than current radars and computing the MIMO-related algorithms in an extra-dimension requires more on-board processing power. Both of these points clash with the expectation that radar sensors for vehicular applications should be less than $100 per unit. However, advancements in the area of RF-CMOS may allow us to overcome this challenge.

From the onset of automotive radar transition from radar's original military purpose, high frequency semiconductor industry played an important role by miniaturizing microwave component in integrated circuits. In the early 2000s, the RF semiconductor used for vehicle radar, especially for millimeter wave radar, was Gallium Arsenide (GaAs). At that time, the RF circuitry was made up of several discrete GaAs chips. With multiple chips and a high-priced fabrication process, the radar cost was exorbitant and such sensors could only be found in luxury vehicles. Heavy investment in cheaper RF-SiGe (Silicon–Germanium) technology pushed the maximum operating frequency beyond 100 GHz, enabling use for millimeter wave radar systems [22]. From the economic perspective, the number of operating channels per chip should be reduced as much as possible, in tune with integrating other functions. The advantage of SiGe technology is that many RF components can be integrated onto a single die due to larger wafer sizes. Recently, RF-CMOS has emerged competing with SiGe for the obvious benefit of cost attractiveness in mass production and its capability of integration with signal processing units. RF-CMOS is compatible with digital CMOS, and both can be integrated into the same die which means that digital processing will be on the same die as the millimeter wave circuitry, reducing costs further. This type of integration also lends itself well to increasing the number of unique down converting mixers in the system, which enables techniques such as 2D MIMO or a fusion hardware platform combining radar with camera. This is to be contrasted with software-focused efforts at using different sensing modalities to improve the overall performance; see, e.g., [23–26]. Furthermore, the optimizing active channel numbers in massive MIMO with innovative array distributions is expected to be challenging but attractive solutions [27].

Lidar is a useful sensor for automated driving applications as it can generate a high definition 3D point cloud of the surrounding area. Within the point cloud is information about all surrounding obstacles, and the features of fixed geometry such as ground and buildings can be matched to map data to improve self-localization. Short wavelengths of lasers are used to create focused beams which enable the high-resolution point clouds.

Off the back of the DARPA Challenges mentioned in Introduction prominent scanning lidars became a hallmark of vehicles fitted with automated driving systems [28]. These lidars contained an array of discrete lasers and detectors. Pulsed time-of-flight (ToF) methods determined the range to targets, mechanical scanning covered the azimuth scan plane and the array of lasers and detectors behind lenses covered the elevation scan plane. These bulky, heavy, and costly sensors meant that they were confined to research grade vehicles. Mechanical scanning of the whole sensor head with extreme precision meant that a robust brushless AC motor—a substantial cost, was a necessity. ToF methods accurately determine the range to a target but can be subject to interference from the sun or other light sources. These methods can only return a single target within a beam, which causes problems when aerosols or other small particulates are suspended in air, such as rain, snow, or dust. Recently, large strides have been made in silicon photonic integrated circuits (PICs). PICs allow processing of laser light signals in silicon chips. Investment in PICs has been bolstered by its importance in optical data communication, and one massive national collaborative effort is the American Institute for Manufacturing Integrated Photonics (AIM Photonics). AIM Photonics is a US National initiative targeting deployment of silicon photonic mass manufacturing methods throughout industry with funding from federal agencies, such as DARPA, state level, and private interests [29]. With silicon photonics technology an all-solid-state lidar has been demonstrated, and complex waveform encoding can be applied to the lidar signal.

Solid-state techniques utilizing silicon photonics have been demonstrated in two ways. One way is electronic beam steering by optical phased arrays (OPAs). Due to short wavelengths of the lasers used in near-infrared lidar, thousands of array elements can be packed into a single chip to enable high resolution, thin beam formation. One OPA [30] demonstrates a sub 9 $\text{cm}^2$ chip with more than 8000 array elements producing a half-power beamwidth of $0.01° \times 0.04°$ and scannable in two directions, with a scan angle of 100° by phase shifters and a scan angle of 17° in the second direction by wavelength modulation. Co-packaged CMOS dies demonstrate amazing miniaturization over classic mechanical beam scanning systems. One challenge of this type of beam steering due to the high resolution beam is covering all scan points in a reasonable amount of time. Wavelength division multiplex (WDM) techniques can naturally and easily be applied in silicon photonics to allow wavelength unique signals to co-exist in the same PIC. Lidar with 100+ comb generated signals being transformed into simultaneously scanned laser beams have been demonstrated [31], which would overcome the time crunch of scanning thin pencil beams.

Another way to realize solid-state lidar is utilizing integrated focal plane arrays, similar to digital cameras. These so-called flash lidars—due to the scene filling transmission of a laser pulse similar to a camera flash, are among the first solid-state lidars to be developed. Large scale flash lidar have been demonstrated [32] and are already on the market for OEM vehicle use [33]. In a flash lidar, the detector array is exposed to free space similar to mechanical lidar which meant they were limited to ToF ranging techniques. Recently, per-pixel integrated heterodyne circuitry though a hybrid CMOS-silicon photonics process offered by GlobalFoundries has been demonstrated [34]. This allows for the simplicity of flash type lidar while opening the door for complex encoding on the laser signal. Still, flash lidar suffers from a tradeoff between resolution and wide-angle field of view that does not exist in beam steering lidar, as a lens must be chosen to transform the focal plane array to angular detection.

One of the benefits in solid-state lidar using silicon photonics is the introduction of advanced encoding of the transmitted signal. Modulating the signal is natural in silicon photonics through ring resonators, Mach Zehnder interferometers or other similar active structures. Frequency Modulated Continuous Waveform (FMCW) method found often in vehicular radar sensors, is a simple modulation which generates a continuous ramp signal in frequency. This method allows determination of both range and velocity instantaneously (instead of estimating velocity by piecing together several detections). Moreover, multiple targets can be detected within a single beam, which is especially useful in poor weather conditions. Rain and snow clutter can be filtered out [35]. If the system has a sufficient number of unique heterodyne detectors, more advanced encoding could be imagined such as CDMA [36] where scene scanning is essentially done in the digital domain.

Before fully solid-state-scan lidar is introduced, Micro-Electro-Mechanical Systems (MEMS) based mirror architecture will be popular in the near future. The priority for solid-state lidar's success in market penetration is cost competitive miniaturization of essential components in IC. As learned from the precedent growth of radar market, it is important for semiconductor IC providers, tier-1 suppliers and OEMs to form a virtuous ecosystem.

With the betterment of all types of sensors mentioned above, i.e., higher detection resolution in all directions with more precise Doppler signature, it is now possible to predict even the rotation of a moving target. Not only the increased number of voxels but also innovative ways of their association can significantly improve overall data processing. For example, optimized data size and advanced waveforms can enhance refresh time while maintaining the detection accuracy. With advancement in sensor data processing by AI and machine learning algorithms, sensors can provide much more distinctive characteristics of targets, e.g., [37–39]. Algorithmic advances focused on embedded software will continue to drive sensor information processing because of its marginal added cost.

## 3 Toward Energy Efficient Edge Computing via Optical Advances

In the coming years the amount of data acquired from added sensors of intelligent vehicles is expected to increase significantly, and so will the amount of processing required to utilize this data. This leaves two options in terms of data processing: either to operate on edge computing systems in the vehicle or to partially rely on mobile networks in order to make timely transportation decisions.

Figure 2 provides a breakdown of connected computing technologies and where they process data. The role of this figure serves to elucidate technologies in the connected infrastructure of the automotive sector, rather than the broader IoT domain. Sensors lie at the base of the figure. These are broken into passive, active and smart

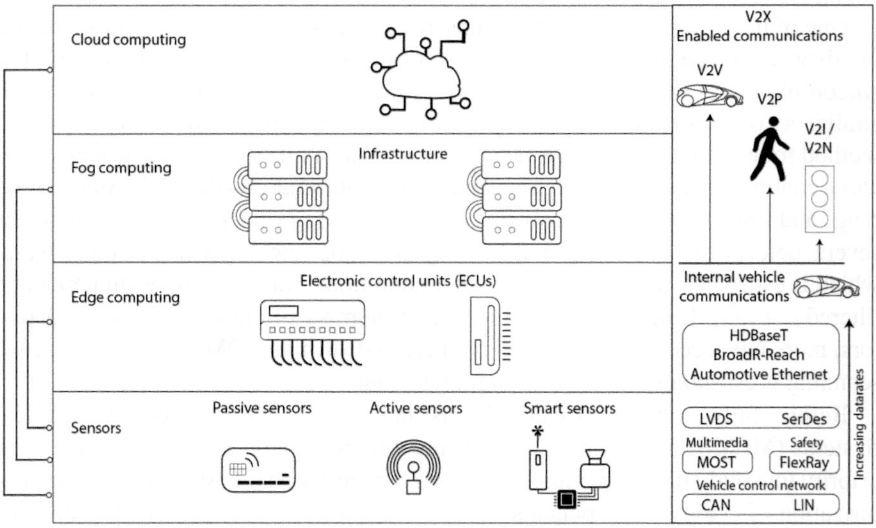


**Fig. 2** A schematic of layers at which processing may occur for the automotive sector: the cloud, fog, edge, or sensor level computing. The left of the figure designates the type of connections that might be activated between these compute layers. The right-hand side of the figure displays the type of vehicle communications that might occur, based on the computing structure on the left. For instance, V2X communications require fog or cloud computing, meanwhile internal communications within the vehicle rely on a slew of cabled connections

sensors. Passive sensors include items like RFID tags where no energy needs to be supplied to the sensor (more on passive optical sensors below). Active sensors are those that require power to sense. Smart sensors are those that can provide processed data using only the sensor signal; an example might be how a radar sensor can not only provide position information but also velocity information without the need to convert the signal to another domain. Additionally, we define edge computing as any system that can process data at the point of collection, as opposed to shuttling data via a network. While sensors and edge computing are separated in Fig. 2, to clarify their differences, any sensor that can actively process data can be included under the broader umbrella of edge computing; it is not required that the sensor have a microprocessor attached in order to be considered edge computing. Instead, edge computing is the existence of processing without the need to communicate information to a server or cloud network. Traditionally, edge computing included processing of sensor data on a programable controller, however this description has expanded to include systems with graphic processing units, tensor processing units and other more advanced computing infrastructure that was previously only found at data centers. For this reason, we define edge computing as computation through any system that does not rely on the access to fog or cloud computing services. Fog computing relies on the local area network architecture. Cloud computing relies on an internet access point that allows information to travel from the sensor to a globally accessible server.

Figure 2 also describes how these different processing locations impact specific vehicle communication functions. For instance, fog computing allows for vehicle-to-infrastructure processing; this may include data being transmitted wirelessly via Bluetooth or WiFi to local access points that are allowed to make decisions. For instance, one could imagine an intersection with traffic lights replaced via a local network server. The vehicles that enter the intersection would be sent wireless commands from the server that are based on immediate information gathered at the intersection.

With these definitions in mind, we can better understand how the modern and future vehicle may depend on each of these computing levels. For example, relying on the cloud has the potential for high latency due to large volumes of data and transmission rates, network congestion, and frequently occurring deadzones or urban canyons. However, traditional computer architectures remain energy intensive when running neural network algorithms in any intelligent vehicle—whether a car, a drone, or a remote sensor in IoT. The vehicles of today are even being recognized as cloud accessible hubs to be leased for bitcoin mining or algorithm training, as such datacenters-on-wheels sit idly in their garages.

As vehicles become more intelligent, the on-board power requirements of the vehicle must not only take on the most vital roles like safety and convenience but also play an increasing part in attending to diverse needs of vehicle's occupants. With the growth of autonomous functions, there will be a higher expectation for the vehicle to provide more in-vehicle services (entertainment, shopping, etc.). While power has always been a priority in the design of the vehicle, it will become an even more critical issue as electric vehicles become more autonomous. In our opinion, new hardware systems are required to achieve increasingly intensive computations at various edge interfaces without straining the power source of the vehicle [40].

While the decision to operate partially in the cloud, over mobile networks, or to operate solely on a mobile computing system of the vehicle are very much the subject of a transportation system design, any opportunity to unburden the power system from sensor processing and related computational costs is a universally welcomed proposition. For this reason, a new trend in intelligent vehicles is to off-load some of the needs of edge computing onto passive hardware. In this section, we review some opportunities to utilize passive optical systems for pre-processing prior to reaching the optical-electrical transduction interface [41].

One of the most immediate methods to relieve computational burden is to reformat the data of the sensors on the vehicle. For example, consider the image shown in Fig. 3a. This image can be described by a camera using a set of values from 1 to 256, however, the image conveys a range of information features such as depth cues, color, and lighting. In order to computationally process this image, feature extraction is applied across the data matrix.

The most common image kernel that is first applied to an image for feature extraction or image segmentation is a differentiator. Here we consider how adding a passive optical filter could be used to achieve arbitrary algorithmic computations of a scene.

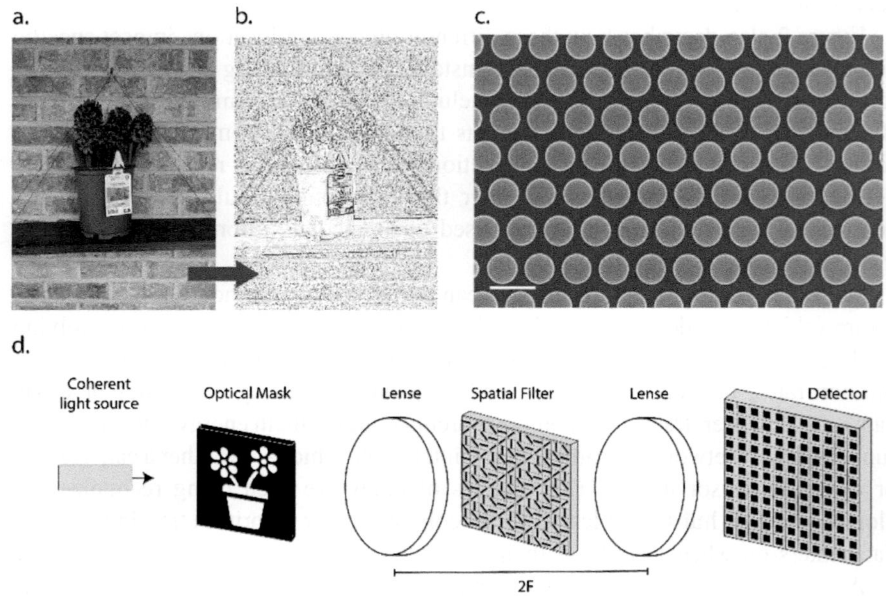


**Fig. 3** Methods for passive optical image differentiation. **a** Grayscale image of a flower. **b** Image shown in **a** passed through a numerically applied Laplacian image kernel. **c** Metasurface composed of an array of silicon pillars for image differentiation. The white horizontal bar represents 1 micron. **d** Optical train needed for coherent optical differentiation utilizing a spatial filter

One of the most obvious implementations of a passive optical filter for vehicle applications is image differentiation. While the technology is still at its early stage, researchers have demonstrated the ability to differentiate coherent light passed through an optical metasurface [42]. In this reference, a metasurface is composed of two sets of arrays of nanobeams that compute first- or second-order spatial differentiation in the x-direction. Rather than rely on digital software for processing, the image is passed via a passive, analog, optical filter. Figure 3d demonstrates how a metasurface like the one described in [42] can be utilized in a 4F optical system in order to implement image differentiation at the Fourier plane.

In fact, the concept of optical differentiation has been around since the 1960s, with the idea of utilizing spatial filters as a tool to alter the structure of a beam of light. The concept of spatial filtering plays off the unique Fourier optical transformation after an image passes through a lens. The unique direction of the light due to diffraction causes the sources of light to focus on different portions of the focal plane. By applying a spatial filter at the focal plane, a variety of analogue operations can be applied to an image. The example shown in Fig. 3d, utilizes this approach to spatial filtering, however, advances in the field have enabled researchers to utilize bound states in a continuum in order to achieve differentiation at any location along the propagation direction of the image [43, 44].

The advantage to utilize metasurfaces for analogue processing has gained significant momentum. Not only are the custom-made optical elements just several hundred nanometers thick, but they can also be applied at any location in the optical train. It should also be noted that metamaterials can often be designed to have high transparency. However, there still remains a serious disadvantage before the systems can be perfected for in-vehicle optical analogue processing. Most metasurface systems have an optical response that is limited to a narrow band of wavelengths, and the existing technologies are not capable of implementing this differentiation on incoherent light. This implies that a standard optical image cannot be differentiated optically. Incoherent optical processing is a hot topic of research and while some progress has been made [45], there still remain several hurdles toward implementing this technology in vehicles.

The ability of an optical system to apply image differentiation or edge determination is not only a useful technique in image processing, but it is typically one of the layers in convolutional neural networks. Thus, applying passive optical filters to sensing could enable reduced convolutional processing needed on these large matrix transformations for edge computing, thereby reducing one of the steps in the computationally intensive task of image segmentation.

Beyond image differentiation, optical filtering has a potential to apply other passive optical elements in order to achieve a variety of algorithms, e.g., integration [46, 47]. Algorithm specific metasurfaces can also be employed as convolutional elements for a variety of other image kernels including box blur, sharpen, or unsharpen masking. The size of the kernel will depend on the resolution of the metasurface relative to the image size. With the help of inverse design and machine learning to create metasurfaces, new and unique methods to achieve these computational systems are rapidly evolving [48].

The current trend for communication systems in vehicles is a demand to shuttle higher loads of data which is often due to the integration of higher resolution sensors (see Sect. 2). The need to transmit data at Gbs speeds, from sensors to their controllers (e.g., those for future automated driver assistance systems (ADAS)) is becoming a considerable communication hurdle. To comprehend the significance of this transition point from the traditional data communication approach to that of future automotive vehicles, we should first review the electronic ecosystem of the present-day vehicle. At present, most traditional OEMs continue to rely on the electronic control unit (ECU), which through a single module controls a set of processes such as those in the radar, cameras, powertrain, transmission, suspension and other systems. The ECU may simply process the raw data and transmit it to another hub or it may act as a subsystem for sensor fusion, processing and controlling data from a multitude of other ECUs. In 2021, there are vehicles with upwards of 150 ECUs; each ECU contains a microcontroller, memory, embedded software, and communication ports for the systems, power, and data communications. With so many control units, each with their own protocols, the OEMs have garnished the burden of driving the performance of these processors while extricating excess cabling and redundancy. This has been particularly difficult given the low bandwidth communication channels that most standard ECUs maintain; these are typically handled via automotive bus systems like LIN (Local Interconnect Network), CAN (Controller area Network), MOST (Media oriented system transport) and FlexRay. However, given the transition to more data heavy sensors like ADAS, higher bandwidths have become a requisite, which has led to the adoption of SerDes (Serializer-Deserializer), Automotive Ethernet, and HDBaseT Automotive. In particular, HDBaseT has allowed for communication over 15 m in length, with limited requirements for shielding for both point-to-point and daisy-chain connectivity [49]. It should be noted that while these technologies have enabled several Gbs transfer rates, there are signs that the future vehicle ecosystem will soon need to shuttle information in the form of bits structured as vectors.

Moreover, trends in machine learning are leaning towards understanding the data structure of an entire tensor, rather than operating a convolution on a single matrix of the tensor at a time. In order to achieve this in our current state, the data would need to be stored locally on RAM hardware after being shuttled bit by bit. In order to accelerate this process, Peripheral Component Interconnect Express (PCIe) cables are finding themselves as a greater necessity. A PCIe cable is a computer expansion bus that traditionally allowed for direct, short connections (on the order of cm) between a motherboard, graphics card, or solid-state drive. As of 2018, PCIe has found a means to integrate onto HDBaseT technology permitting signals such as audio and video, power, and controls to be transmitted over a single cable. Today PCIe is on its 6th generation with versions doubling roughly every few years. However, the future bus communication standard may be to move to fiber optic communications either via PCIe over fiber or standard fiber optical cabling [50, 51]. Given that many sensors utilize optical inputs and the growing desire to process optical inputs in the optical domain, the shuttling of information from sensors to hardware without the transduction to electronic sensors could prove fruitful. It is possible that OEMs will continue to try and process data at the point of occurrence thus declining the need for high bandwidth shuttling; however, the question that remains is for how long can this trend be managed successfully.

With the future of communication systems being heavily reliant on optical networking, there exists an opportunity for analogue processing pre-emptive to transduction to an electrical signal, where information would be traditionally processed on a standard CMOS chip. One possible application of this would be to utilize not only sensing processing but also other functions such as control in an optical architecture. The architecture would utilize a lidar or other light-based sensor with outputs to be fed into an optical neural network with model predictive control (MPC). The concept of using an MPC with a neuromorphic photonics processor for general non-linear programming was demonstrated in [52]. In this reference, nonlinear processing was demonstrated for the high-speed control application of tracking a moving target, e.g., in the case of missile targeting. A similar case could be envisioned for driving automation applications, such as path planning at an on-ramp, a traffic circle, or a parking lot. The light perception device—a sensing solution, directly informs a vehicle computer implemented as optical neural network (ONN) of the positions and the velocities of surrounding agents for the purpose of respecting their trajectories and avoiding collisions while the vehicle's computer plans to maneuver around on the road [53].

## 4 Conclusion

We overviewed the state of art and promising developments in the field of automotive sensing focusing on radar, lidar and optical processing for driving environment sensing exemplified by driving automation. Since the DARPA Challenges of the beginning of this century various enabling technologies have advanced by leaps and bounds. Phased-array radars and solid-state lidars are taking places of mechanically scanning devices, enabling more precise temporal snapshot of the data and delivering more accurate range and angular measurements for the ever growing number of mobility features. Advances in RF-CMOS and similarly silicon photonics IC for lidar, which is compatible with digital CMOS—the wide-spread technology for cameras, will pave the way for integration of sensing modalities at the level of hardware simplifying sensor fusion. In terms of optical processing, computational metasurfaces, i.e., devices specially designed to implement image differentiation, convolution and other functions of essence to AI algorithms, are on the rise. Similar to artificial neural network software years ago, we expect to live through another Renaissance in the field of optical processing focused on integrated photonics driven by the needs for smart sensing and ultra-low power consumption. The sensor advances described here will help developing the next generation of intelligent vehicles.

## References

1. DARPA Grand Challenges (2006 March). DARPA Report to Congress. https://en.wikipedia.org/wiki/DARPA_Grand_Challenge
2. DARPA (2007). Darpa Urban Challenge. https://www.darpa.mil/about-us/timeline/darpa-urban-challenge
3. Urmson C et al (2008) Autonomous driving in urban environments: boss and the urban challenge. J Field Robot 25(8):425–466
4. Markoff J (2010, October) Google cars drive themselves, in traffic.https://www.nytimes.com/2010/10/10/science/10google.html
5. Waymo (n.d.) Waymo driver. Retrieved 10 Jan 2022, from https://waymo.com/waymo-driver/
6. Tesla (n.d.) Artificial intelligence and autopilot. Retrieved 10 Jan 2022, from https://www.tesla.com/autopilotAI
7. Preserve Knowledge (2021, March 27) How AI powers self-driving tesla with Elon Musk and Andrej Karpathy. https://www.youtube.com/watch?v=FnFksQo-yEY
8. Karpathy A (2020, April 20) AI for Full-Self Driving at Tesla. https://www.youtube.com/watch?v=hx7BXih7zx8
9. Baum E (2004) What is Thought? The MIT Press
10. Prokhorov D (2019 July) Toward next generation of autonomous systems with AI. In: Proceedings of IJCNN 2019, Budapest, Hungary
11. Okumura B et al (2016) Challenges in perception and decision making for intelligent automotive vehicles: a case study. IEEE Trans Intell Veh 1(1):20–32
12. Iandolla F (2020, July 9) Is computer vision still improving… or has it Reached a Plateau? AutoSens 2020 Presentation
13. Li J et al (2016) Deep neural network for structural prediction and lane detection in traffic scene. IEEE Trans Neural Netw Learn Syst 28(3):690–703
14. Toyota (2022) Designed for safety. https://www.toyota.com/safety-sense/
15. Ross PE (2019, January 7) IEEE-spectrum news-transportation. CES 2019: Toyota Lifts the Veil on Its Guardian Driver-Assist System. https://spectrum.ieee.org/cars-that-think/transportation/self-driving/ces-toyota-lifts-veil-from-driver-assist-system
16. Li L, Ruan H, Liu C, Li Y, Shuang Y, Alù A, Cui, TJ et al (2019). Machine-learning reprogrammable metasurface imager. Nat Commun 10(1):1–8
17. Ku BH, Schmalenberg P, Inac O, Gurbuz OD, Lee JS, Shiozaki K, Rebeiz GM (2014) A 77–81-GHz 16-Element Phased-Array Receiver With ±50 deg. Beam Scanning for Advanced Automotive Radars. IEEE Trans Microw Theory Tech 62(11):2823–2832
18. Schmalenberg PD, Li M, Lee JS (2019) U.S. Patent No. 10,333,209. Washington, DC: U.S. Patent and Trademark Office
19. Harris P, Malkowsky S, Vieira J, Bengtsson E, Tufvesson F, Hasan W. B, Edfors O et al (2017) Performance characterization of a real-time massive MIMO system with LOS mobile channels. IEEE J Sel Areas Commun 35(6):1244–1253
20. Chen CY, Vaidyanathan PP (2008, May) Minimum redundancy MIMO radars. In: 2008 IEEE international symposium on circuits and systems. IEEE. pp 45–48
21. Li Y, Luo Y, Yang G (2017) 12-port 5G massive MIMO antenna array in sub-6GHz mobile handset for LTE bands 42/43/46 applications. IEEE Access 6:344–354
22. Heinemann B, Barth R, Bolze D, Drews J, Fischer GG, Fox A, Yamamoto Y (2010, December). SiGe HBT technology with f T/f max of 300GHz/500GHz and 2.0 ps CML gate delay. In: 2010 International Electron Devices Meeting. IEEE. pp 30–5
23. Ji Z, Prokhorov D (2008, June) Radar-vision fusion for object classification. In: 2008 11th International conference on information fusion. IEEE. pp 1–7
24. Prokhorov DV (2010, June) Road obstacle classification with attention windows. In: 2010 IEEE intelligent vehicles symposium. IEEE. pp 889–895
25. Prokhorov DV (2012) U.S. Patent No. 8,254,670. Washington, DC: U.S. Patent and Trademark Office

26. Nobis F, Geisslinger M, Weber M, Betz J, Lienkamp M (2019, October) A deep learning-based radar and camera sensor fusion architecture for object detection. In: 2019 sensor data fusion: trends, solutions, applications (SDF). IEEE. pp 1–7
27. Li X, Zhai W, Repeta M, Cai H, Ross T, Ansari K, Tong W (2021) A scalable 256-elements e-band phased-array transceiver for broadband communication. arXiv preprint arXiv:2106.10623
28. Halterman R, Bruch M (2010, May) Velodyne HDL-64E lidar for unmanned surface vehicle obstacle detection. In: Unmanned systems technology XII. International Society for Optics and Photonics, vol 7692. p 76920D
29. Fahrenkopf NM, McDonough C, Leake GL, Su Z, Timurdogan E, Coolbaugh DD (2019) The AIM photonics MPW: a highly accessible cutting edge technology for rapid prototyping of photonic integrated circuits. IEEE J Sel Top Quantum Electron 25(5):1–6
30. Poulton CV, Byrd MJ, Moss B, Timurdogan E, Millman R, Watts MR (2020, May) 8192-element optical phased array with 100 steering range and flip-chip CMOS. In CLEO: Applications and Technology. Optical Society of America. pp JTh4A-3
31. Lukashchuk A, Riemensberger J, Karpov M, Liu J, Kippenberg TJ (2021) Hardware-efficient megapixel per second coherent soliton microcomb ranging. arXiv preprint arXiv:2101.03952
32. Zhang C, Lindner S, Antolović IM, Pavia JM, Wolf M, Charbon E (2018) A 30-frames/s, 252 x 144 SPAD Flash LiDAR With 1728 Dual-Clock 48.8-ps TDCs, and pixel-wise integrated histogramming. IEEE J Solid-State Circuits 54(4):1137–1151
33. Continental Automotive (n.d.) High Resolution 3D Flash LIDAR. Retrieved January 14, 2022, from https://www.continental-automotive.com/en-gl/Passenger-Cars/Autonomous-Mobility/Enablers/Lidars/3D-Flash-Lidar
34. Rogers C, Piggott AY, Thomson DJ, Wiser RF, Opris IE., Fortune SA, Nicolaescu R (2021) A universal 3D imaging sensor on a silicon photonics platform. Nature 590(7845):256–261
35. Batet O, Dios F, Comeron A (2010, August) FMCW lidar for multiple-target sounding. In: Remote Sensing System Engineering III. International Society for Optics and Photonics, vol 7813. p 78130H
36. Yang X, Hao L, Wang H, Wang Y (2020) Spatial and temporal multiplexing array imaging lidar technique based on OOCDMA. Opt Lasers Eng 129:106066
37. Prokhorov D (2010) A convolutional learning system for object classification in 3D lidar data. IEEE Trans Neural Netw 21(5):858–863
38. Prokhorov D (2009) Object recognition in 3D lidar data with recurrent neural network. In: Proceedings of CVPR workshops, pp 9–15
39. Li Y et al (2020, May 20) Deep learning for LiDAR point clouds in autonomous driving: a review. https://arxiv.org/abs/2005.09830
40. Zhang J, Letaief KB (2020) Mobile edge intelligence and computing for the internet of vehicles. (Special Issue on Internet of Vehicles) Proc of IEEE 108(2):246–261
41. Rodrigues SP et al (2021) Weighing in on photonic-based machine learning for automotive mobility. Nat Photonics 15:66–67. https://doi.org/10.1038/s41566-020-00736-0
42. Cordaro A et al (2019) High-index dielectric metasurfaces performing mathematical operations. Nano Lett 19:8418–8423
43. Guo C, Xiao M, Minkov M, Shi Y, Fan S (2018) Photonic crystal slab Laplace operator for image differentiation. Optica 5:251–256
44. Zhou Y, Zheng H, Kravchenko I, Valentine J (2020) Flat optics for image differentiation. Nat Photonics 14:316–323. https://doi.org/10.1038/s41566-020-0591-3
45. Wang H, Guo C, Zhao Z, Fan S (2020) Compact incoherent image differentiation with nanophotonic structures. ACS Photonics 7:338–343
46. Estakhri NM, Edwards B, Engheta N (2021) Inverse-designed metastructures that solve equations. Science 363(6433):1333–1338
47. Silva A et al (2014) Performing mathematical operations with metamaterials. Science 343(6167):160–163. https://doi.org/10.1126/science.1242818
48. Ma W et al (2021) Deep learning for the design of photonic structures. Nat Photonics 15:77–90. https://doi.org/10.1038/s41566-020-0685-y

49. Embedded Computing (n.d.) PCIe in the connected car. Retrieved 17 January 2022, from https://www.embeddedcomputing.com/application/automotive/vehicle-networking/pcie-in-the-connected-car
50. Ciordia O (2020) Optical automotive ethernet for electrical and hybrid powertrains. Power Electronic News.https://www.powerelectronicsnews.com/optical-automotive-ethernet-for-electrical-and-hybrid-powertrains/
51. Kagami M (2005) Visible optical fiber communication. R&D Review of Toyota CRDL, 40(2). https://www.tytlabs.com/english/review/rev402epdf/e402_001kagami.pdf
52. Lima TF et al (2019) Machine learning with neuromorphic photonics. J Lightwave Technol 37(5):1515–1534
53. Dede E, et al (2021) Adaptable optical neural network system. US Patent Application No. 20210097378A1

# Automated Driving Decisions and Control

# Robust AI Driving Strategy for Autonomous Vehicles

**Subramanya Nageshrao, Yousaf Rahman, Vladimir Ivanovic, Mrdjan Jankovic, Eric Tseng, Michael Hafner, and Dimitar Filev**

**Abstract** There has been significant progress in sensing, perception, and localization for automated driving, However, due to the wide spectrum of traffic/road structure scenarios and the long tail distribution of human driver behavior, it has remained an open challenge for an intelligent vehicle to always know how to make and execute the best decision on road given available sensing/perception/localization information. In this chapter, we talk about how artificial intelligence and more specifically, reinforcement learning, can take advantage of operational knowledge and safety reflex to make strategical and tactical decisions. We discuss some challenging problems related to the robustness of reinforcement learning solutions and their implications to the practical design of driving strategies for autonomous vehicles. We focus on automated driving on highway and the integration of reinforcement learning, vehicle motion control, and control barrier function, leading to a robust AI driving strategy that can learn and adapt safely.

S. Nageshrao (✉) · Y. Rahman · V. Ivanovic · M. Jankovic · E. Tseng · M. Hafner · D. Filev
Ford Motor Company, Research and Advanced Engineering, Dearborn, MI 48124, USA
e-mail: snageshr@ford.com

Y. Rahman
e-mail: yrahman@ford.com

V. Ivanovic
e-mail: vivanovi@ford.com

M. Jankovic
e-mail: mjankov1@ford.com

E. Tseng
e-mail: htseng@ford.com

M. Hafner
e-mail: mhafner2@ford.com

D. Filev
e-mail: dfilev@ford.com

 
Y. L. Murphey et al. (eds.), *AI-enabled Technologies for Autonomous and Connected Vehicles*, Lecture Notes in Intelligent Transportation and Infrastructure,
https://doi.org/10.1007/978-3-031-06780-8_7

# 1 Introduction

Reinforcement learning (RL) is a key branch of Artificial Intelligence (AI) at the intersection of machine learning, decision making and control. It is a method of learning from interaction with the environment and it is inspired by the human learning process. RL gained success in the last few years as the right approach to learn how to make decisions in tasks that are too complex to explain or to learn from supervised examples.[1] In 2015 DeepMind pioneered a new approach to RL, Deep Q-Network (DQN) that approximates the Q-function [45] with a deep neural network [26, 32, 33, 43]. The key idea of the DQN algorithm is to store the agent's experiences in a replay buffer and then randomly sample and replay these experiences to provide diverse and decorrelated training data for learning the Q-function. The DQN concept was further extended to the other RL methods, e.g. Deep Deterministic Policy Gradient (DDPG) and Proximal Policy Optimization (PPO), resulting in a new subfield aptly termed as Deep Reinforcement Learning (DRL). Following the success of the DQN methodology, DRL has become one of the most used AI methods in the development of autonomous vehicles (AV).

DRL can be an essential part of the prototypical AV technology stack (Sensing, Perception, Localization, Driving Policy and Actuation) as a method for decision making in the hierarchical driving policy workstream Fig. 1.

The role of the decision making and motion planning module is to transform the information about the route and the state of the AV and surrounded traffic and environment into a high-level strategy—modulation of the speed set-point, merging, lane changing, car following, aborting the current maneuver, etc.—that is further implemented by the motion control system encompassing path planning, path following, and actuator control modules. Early AV research and prototypes utilized motion planning concepts derived from classical control techniques—finite state machines, rules, heuristic strategies and Model Predictive Control (MPC) [5, 6, 11, 21, 62]. Following the massive introduction of AI methods in AV development in the last years, we can observe a considerable growth in the share of the DRL based methods for motion planning [4]. The rationale behind this trend is the ability of the DRL to handle complex and ill-defined situations and environments that often occur in autonomous driving setting. Presently, alternative versions and implementations of the DRL are at the core of the AV motion planning methods (e.g., detailed surveys of the DRL applications to AV can be found in [4, 61]).

One of the challenging problems of the DRL based decision making and motion planning for AV is improving the robustness of the algorithms and preventing the possibility of unsafe actions and accidents. The essence of this problem is the probabilistic nature of the RL concept, which does not generally imply that the maximization of the reward function can guarantee the safety and repeatability of the solution.

One common approach to improving DRL robustness, we call it the probabilistic robustness, is focused on enhancing the exploration strategy during the training of the

[1] *Selected portions reprinted, with permission, from [34], ©2019 IEEE.

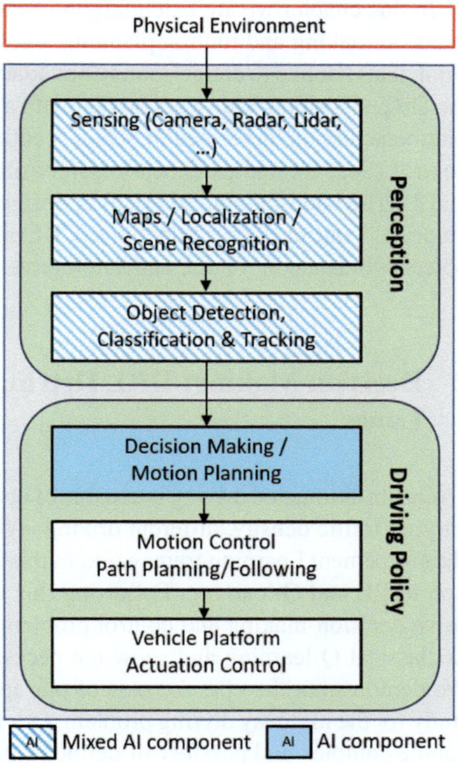


**Fig. 1** AI for AV: a prototypical technology stack

DRL algorithms. The final goal is enriching the training set by introducing, through adversarial disturbances and non-reward based exploration, unexpected scenarios and situations well beyond the space defined by the simulation model/distribution and the handcrafted reward function. The methods of this group include model-based exploration [38, 54, 63], adversarial perturbations of the state observations [27], infusing disturbances that are modulated by the capability of the control policy [28], generating socially acceptable perturbations [64], maximization of the rewards associated with the high risk trajectories [46], etc. In general, the outcome of these techniques is improved robustness of the algorithms and consequentially reducing the level of risk, but without providing deterministic guaranties for solution.

An alternative methodology proposed by Alshiekh et al. [2] introduces the idea of constraining ('shielding') the output of the RL algorithm within a safety envelope that is defined by a deterministic decision-making strategy. Human crafted rules improving the safety of the DRL algorithm are applied in [34, 35] and [31] in order to eliminate the unsafe actions produced by the DRL algorithm. On one side, the rules have the advantage of introducing human knowledge and experience in conjunction with the machine learning approach of developing the DRL. On the other side, the deterministic envelope defined by the rules might be too conservative and limited only to the specific situations considered by the designers.

In this chapter we are extending the 'shielding' concept by combining the DRL decision making and motion planning algorithm with a generic deterministic algorithm that is aimed to define a safety boundary, called the 'safety filter' at the output of the DRL algorithm. Our work started with a rule-based safety filter [35] with simple motion actuation, described in Sect. 2, continued with representative vehicle motion control (Sect. 3), and extended to safety filter for wide-ranging situations (Sects. 4 and 5). The wide-ranging safety filter design is inspired by the recent progress in the theory of Control Barrier Functions (CBF) as a generic and reliable methodology for object avoidance in robotic and automotive applications.

## 2 Decision Making: DRL Driving Strategy for Changing Lanes

To ensure automated driving is capable of operating in diverse environment including varying traffic density, different driving style and norms, we develop a novel Deep Reinforcement Learning framework. In this section we first provide a basic introduction to RL and Q-learning. Following this we elaborate on our hybrid approach to solve decision-making and control problems. We introduce essential modifications to classical Q-learning and show the necessity in incorporating basic safety rules. We demonstrate the effectiveness of our approach using a comparative simulation study for the highway driving problem. Our novel hybrid architecture can effectively handle complex and possibly ill-defined situations and environments.

### *2.1 Reinforcement Learning and Deep Reinforcement Learning an Introduction*

In this section we will provide a brief theoretical background on decision making using deep reinforcement learning. Additionally, we elaborate on the main motivation for using the hybrid control architecture presented in this chapter.

#### 2.1.1 Reinforcement Learning

In reinforcement learning (RL), an Agent learns an optimal policy for a given cost function by directly interacting with the environment. Broadly RL constitutes a set of algorithms that efficiently solve the sequential decision making problem for an underlying Markov decision process (MDP) [45]. An MDP is formally defined by a tuple $\langle \mathcal{S}, \mathcal{A}, \mathcal{T}, \mathcal{R} \rangle$. Where $\mathcal{S} \in \mathbb{R}^n$ is the state-space, $\mathcal{A}$ is the action space, $\mathcal{T} : \mathcal{S} \times \mathcal{A} \rightarrow \mathcal{S}$ is the state transformation function, and $\mathcal{R} : \mathcal{S} \times \mathcal{A} \times \mathcal{S} \rightarrow \mathbb{R}$ is the reward function. At each discrete time step $t$ the learning Agent selects an action $a_t$

according to some policy $\pi$ for a given system state $s_t$ , i.e., $a_t = \pi(s_t) \in \mathcal{A}$, where $\mathcal{A}$ is a set of feasible actions. On applying this action the system transitions to a new state $s_{t+1} \in \mathcal{S}$, and provides a scalar reward $r_{t+1}$. This process is repeated till the policy converges. It must be noted, both state transition $\mathcal{T}$ and policy $\pi$ can be stochastic.

The goal of an RL algorithm is to learn a policy $\pi$ so as to maximize the total cumulative reward termed as return $R_{\mathrm{t}} = \sum_{k=0}^{\infty} \gamma^k r_{t+k}$ where the scalar constant $\gamma \in (0, 1]$ is the discount factor. For discrete action space the optimal policy is obtained by solving for the optimal $Q$ function, termed as the action-value function. An optimal $Q$ function satisfies the Bellman equation:

$$Q^{\pi_{\mathrm{opt}}}(s_t, a_t) = Q^*(s_t, a_t) = \mathbb{E}_{s_{t+1}}\left[r_t + \gamma \max_{a_{t+1}} Q^*(s_{t+1}, a_{t+1})|(s_t, a_t)\right]. \quad (1)$$

For a given state $s_t$ any action $a_t$ that satisfies (1) is the optimal action.

### 2.1.2 Decision Making and Low-Level Actuation

Inspired by an example from [45] (see Sect. 1.2), we elaborate on the difference between decision making ($\mathcal{D}$'s) and low level actuation ($\mathcal{C}$'s) :

- Phil wants to have breakfast, he may chose between having cereal or a bagel ($\mathcal{D}_1$), after deciding he will either walk to cupboard or to the counter ($\mathcal{C}_1$).
- He will then either pick a choice of cereal ($\mathcal{D}_2$) and then walk to the fridge ($\mathcal{C}_2$) to get milk or chose a bagel ($\mathcal{D}_3$) and walk to the toaster ($\mathcal{C}_3$), etc.

This entire process involves a series of decision making followed by low-level actuation. High-level decisions (i.e., $\mathcal{D}$'s) are decided by reward function such as pleasure or getting enough nutrition etc.

Any RL methodology that tries to solve both decision making and actuation simultaneously may require a large amount of training data. In this work we simplify this process by having a clear hierarchy between high-level decision making and low level actuation. RL is used to solve the high-level decision making problem, while the classical feedback control methods are used for low-level actuation. Our approach has considerable overlap with hierarchical RL, where both decision making and control are learned simultaneously [12]. To learn optimal high-level decisions we use a modified Double DQN (DDQN) algorithm from [53].

### 2.1.3 Need for Robustness and Safety

The trade-off between exploration and exploitation is a key characteristic that distinguish RL from other forms of machine learning algorithms [45]. In an unknown situation an agent needs to be curious and explore new action. Hence during the initial learning phase the Agent will invariably explore all viable actions as the

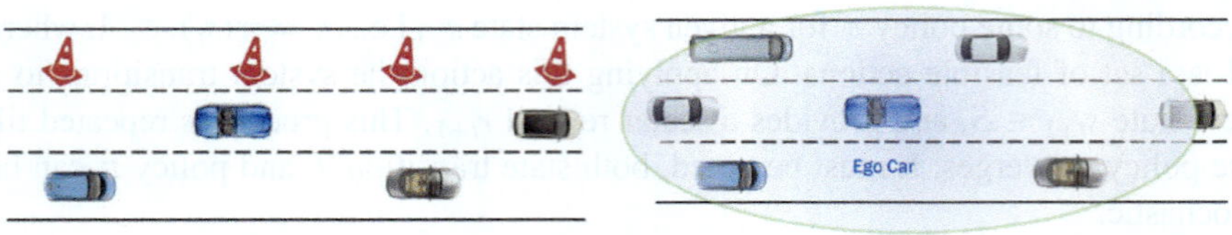


(a) Highway with barricaded right lane. (b) A three lane highway scenario.

**Fig. 2** An example for common sense road rule along with the ego vehicle perspective considered in this work (©2019 IEEE)

entire environment is unknown. Unfortunately this curiosity can be fatal and potentially expensive [13]. For example, during training, using unguided exploration could potentially lead to frequent collisions resulting in slow the training process. Additionally in the inference stage, due to perception noise, function approximation etc. a trained agent may still potentially recommend a non-safe maneuver. In order to address these issues, we augment the DDQN decision maker with an explicit *short-horizon safety check* that is used both during training as well as in the inference phase. Safe exploration or safe-RL is an active research topic, for a detailed survey see [14].

A standard DRL approach such as DDQN, may require a lot of samples before learning that certain action to be potentially dangerous in certain states. For example, consider a highway where one of the lane has been barricaded, see Fig. 2a. A standard epsilon-greedy algorithm may need to collide multiple times before learning that in certain situations going to the left-most lane, for e.g. when it is blocked, to be catastrophic. This may result in significant waste of learning effort as the Agent will spend considerable amount of time exploring irrelevant regions of the state and action space. Additionally due to function approximation there is a small probability where the trained DDQN Agent may still chose unsafe actions leading to a catastrophic outcome. This can be avoided by including an explicit short-horizon safety check that evaluates the action choice by the learning Agent and provides an alternative safe action whenever it is feasible [2].

Our first solution to this problem called the Rule Based Safety Filter [34] (see Sect. 2.2) uses a simple safety check based on the common sense road rules. This forces the Agent to avoid non-safe actions in dangerous situations resulting in a faster training. Our approach is similar to a teacher who provides corrective action when it is necessary [47]. Note that the safety filter may not be optimal, i.e., an expert in the task under consideration. Additionally, due to explicit safety check, new data can be obtained even in the inference phase which can then be used for continuous adaption of the learned network, this is further elaborated in Sect. 2.4.3.

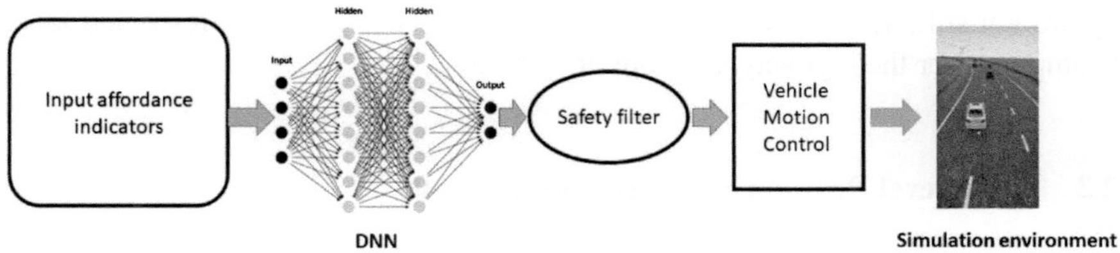


**Fig. 3** DRL agent control architecture

## 2.2 *DRL for Autonomous Driving*

The DRL architecture used in this work is given in Fig. 3.

Unlike the mediated perception method that relies on complete reconstruction of scene prior [16, 60], we use the concept of affordance indicators, i.e., state variables based on the direct perception approach from [7]. For a three lane highway scenario, the ego vehicle (EV) can be surrounded by up to six traffic vehicles (TV), see Fig. 2b.

### 2.2.1 Affordance Indicators

For Autonomous driving, the Agent uses information about its location within the road coordinate system and surrounding vehicles to select an action. This information is collectively termed as affordance indicators. The action taken by the Agent is then rewarded or penalized based on transitioned state. Cumulative reward is used by the Agent to learn safe and effective driving behavior. In this work, we use affordance indicators from 6 target vehicles in the vicinity of the ego vehicle. The 6 target vehicles are designated as Front Left, Front Center, Front Right, Rear Left, Rear Center and Rear Right.

To define the traffic vehicle's variable we use the following notation

$$State_{Lane_{Location\,\&\,Direction}} \tag{2}$$

where State can be distance $d$ or velocity $v$, Lane is either right $r$, center $c$, or left lane $l$. Location is either front $f$ or rear $r$ and Direction is either longitudinal $x$ or lateral $y$.

In total the following 24 indicators are used to represent the spatio-temporal information of the six nearest traffic vehicles (see Fig. 2b), they are formulated from the ego vehicle's perspective. Here the lane occupancy of a TV and the closest car to EV are obtained using the methodology presented in Zhang et al. [62]. In addition to the 24 traffic vehicle states we use longitudinal velocity $v_{e_x}$, lateral position $d_{e_y}$, and lateral velocity $v_{e_y}$ of the ego vehicle $e$. Since the affordance indicators are formulated w.r.t. ego vehicle we do not need a state corresponding to longitudinal position of the ego vehicle. These variables are minimal requirement for highway driving, however they are not sufficient for all highway driving tasks such as use of restricted lane, on-ramp to enter the highway, off-ramp to exit, etc.

### 2.2.2 High Level Decision and Feedback Control

A total of 27 affordance indicators, i.e., $s_t \in \mathbb{R}^{27}$ is used as an input to the deep $Q$-network. The $Q$-network is trained by using a modified double deep Q-learning algorithm from [53], see Algorithm 1. For highway driving decision making, we consider four longitudinal action choices:

1. *maintain*
2. *accelerate*
3. *brake*
4. *hard brake*

and three lateral action choices:

1. *keep lane*
2. *change lane to right*
3. *change lane to left.*

The combination results in set of 12 unique actions. For each of these action choice a numerical value can be assigned [25], for example for longitudinal acceleration four different discrete choices $\{a_1, 0, -a_1, -a_2\}$ may be considered. Alternatively, one can obtain the reference for the throttle or brake controller either using the intelligent driver model (IDM) [50] or adaptive cruise control [41, 52]. The reference for the steering controller is self-evident, it is either stay in lane or change lane to right or left. To obtain the front wheel angle we use a simple feedback controller (see Sect. 3.2 for details)

$$\kappa_{cmd} = f(k_{road}, e_{y_{off}}, e_{\psi}, T_{LC}, v_{e_x}) \tag{3}$$

where $k_{\text{road}}$ is the road curvature, $e_{\psi}$ is the heading angle offset, $T_{\text{LC}}$ is the desired time to complete a lane change, and the reference $e_{y_{off}}$ is the lateral offset to the desired position. When the DDQN Agent decides to perform a lane change, the absolute value of the lane offset $e_{y_{off}}$ will be set to the lane width. The inputs to the steering feedback controller (3) can be considered as additional affordance indicators which can be obtained from the perception module.

**Algorithm 1** A DRL based safe decision maker for autonomous highway driving

```
Initialize: Buf_S, Buf_C, Q(θ) and target network Q̂(θ̂) with θ̂ = θ
for episode =1,··· ,N_e, do
  Initialize: {1, ··· , N_T} cars randomly, obtain affordance indicator s_0
  for samples t =1,··· ,N_S, or Collision, do
    With ϵ select random action a_t, else a_t = arg max_a Q ((s_t, a, θ_t))
    For ego car: If a_t is not safe Then store (s_t, a_t, *, r_col) in Buf_C and replace a_t by safe action a_s
    Apply action, observe s_{t+1} and obtain r_{t+1} = ρ(s_t, s_{t+1}, a_t)
    if Collision then
      Store transition (s_t, a_t, *, r_col) in collision buffer Buf_C
    else
      Store transition (s_t, a_t, s_{t+1}, r_{t+1}) in safe buffer Buf_S
    end if
    Sample random minibatch (s_j, a_j, s_{j+1}, r_{j+1}) from Buf_S and Buf_C
    Set
```

$$y_j = \begin{cases} r_{j+1} & \text{if sample is from Buf}_\text{C} \\ r_{j+1} + \gamma \hat{Q}\left(s_{j+1}, \arg\max_a Q\left(s_{j+1}, a, \theta_t\right), \hat{\theta}_t\right) & \text{if sample is from Buf}_\text{S} \end{cases}$$

```
    Perform gradient descent on ||y_j − Q(s_j, a_j, θ_t)||_2 w.r.t. θ
    Every N_C episodes set Q̂ = Q
  end for
end for
```

### 2.2.3 Rule Based Safety Filter

As elaborated in Sect. 2.1.3, we use an explicit short-horizon safety check to validate the action choice by DDQN. For the current action choice, the safety filter verifies common sense but well known rules of the road such as ensuring a minimum relative gap to a TV based on relative velocity

$$d_{TV} - T_{min} \times v_{TV} > d_{TV_{min}} \tag{4}$$

where $d_{TV}$, $v_{TV}$ are the relative distance and velocity to a given traffic vehicle, $T_\text{min}$ is the minimum time to collision, $d_{TV_{min}}$ is the minimum gap which must be ensured before executing the action choice by the Agent. If this condition is not satisfied and when feasible, an alternate safe action will be provided by the short-horizon safety filter. In this example we use a simple variant of the intelligent driver model (IDM) [50] to provide safe alternative longitudinal action, it is formulated as

$$a_\text{s} = \begin{cases} \text{Hard brake} & \text{if } T_\text{C} \leq T_\text{HB} \\ \text{Brake} & \text{if } T_\text{HB} < T_\text{C} \leq T_\text{B} \\ \text{Maintain} & \text{if } T_\text{B} < T_\text{C} \end{cases} \tag{5}$$

where $a_\text{s}$ is the safe action, $T_\text{C}$ is the calculated time to collision. It is defined as

$$T_{\mathrm{C}} = \frac{d_{\mathrm{TV}}}{v_{\mathrm{TV}}} \tag{6}$$

[30], $T_{\mathrm{HB}}$ and $T_{\mathrm{B}}$ are the thresholds above which the decision made by the DDQN Agent is considered to be safe.

In particular we use the following safety check prior to performing an action given by DDQN:

1. Instead of the in-lane longitudinal action by DDQN, chose a safe action using (5) if (4) is not satisfied and the ego vehicle is faster than the preceding vehicle.
2. If the ego vehicle is in left most lane then *change lane to left* is not valid, similarly for the right lane.
3. For *change lane to left* continuously monitor (4) for center-front car and to the left-front and the left-rear car in the target lane. If condition (4) fails then lane change is either not initiated or aborted, similarly for the *change lane to right*.

After training, generally the trained Agent is frozen and used only in inference mode. However, in reality the Agent may encounter new information, additionally there can be a considerable variation between the training environment and the real-world experience. Also due of function approximation there can be a small probability of choosing an unsafe action even by the trained Agent, this can happen even after convergence and in the absence of any explicit exploration. In order to address these issues, in the implementation phase we augment the trained DDQN Agent with the short-horizon safety check that was used during learning. Any new safety violation data will be added to the collision buffer $\mathrm{Buf}_{\mathrm{C}}$, by using the training part of the Algorithm 1 (line 13 to 15) the learned Agent can be re-trained or adapted in a continuous manner. In the following section we apply the developed DRL based decision making Algorithm 1 for autonomous highway driving.

## *2.3 Vehicle Dynamics*

Throughout this chapter, we use different vehicle dynamics models based on the modeling requirements. In this section, we present the 3 models that are used in subsequent sections.

### 2.3.1 Point Mass Model

When using this model, each vehicle is modeled as a computationally efficient point-mass in discrete time. For longitudinal equations of motion we use a discrete-time double integrator, and for lateral motion we use a simple kinematic model.

$$
\begin{aligned}
x(t+1) &= x(t) + v_x(t)\Delta t, \\
y(t+1) &= y(t) + v_y(t)\Delta t, \\
v_x(t+1) &= v_x(t) + a_x(t)\Delta t,
\end{aligned} \tag{7}
$$

where $t$ is the time index, $\Delta t$ is the sampling time, $x \in \mathbb{R}$ is the longitudinal position, $y \in \mathbb{R}$ is the lateral position of the car, and $v_x \in \mathbb{R}$ is the longitudinal velocity of the vehicle. In (7), the external control inputs $a_x(t)$ and $v_y(t)$ represents the longitudinal acceleration and lateral velocity of the vehicle, respectively.

We assume $a_x(t)$ to vary from nominal acceleration, to hard brake and is discretized into four values, i.e., $a_x = \{a_1, 0, -a_1, -a_2\}$, with $a_1 = 2\,\text{m/s}^2$ and $a_2 = 4\,\text{m/s}^2$. Only in case of emergency hard braking of $a_x = -a_2$ is applied. The lateral velocity $v_y(t)$ provides a reference lane for the vehicle, we assume 5 s to complete a lane change action [48], with an option to abort at any sampling instance. In this work we use 1 $Hz$ sampling for the driving policy.

### 2.3.2 Dynamic Bicycle Model

In subsequent sections, we also use a continuous dynamic bicycle model for vehicle dynamics [41]. This model is required when simulating vehicle dynamics at higher frequencies.

$$
\begin{aligned}
\dot{x} &= v\cos(\phi + \beta), \\
\dot{y} &= v\sin(\phi + \beta), \\
\dot{v}_{\text{lon}} &= g\alpha - F_{\text{aero}} - mg\sin(\theta_r) \\
\dot{v}_{\text{lat}} &= \frac{F_{\text{f}} + F_{\text{r}}}{m} - v_{\text{lon}}\dot{\phi} \\
\ddot{\phi} &= \frac{F_{\text{f}}L_{\text{f}} - F_{\text{r}}L_{\text{r}}}{I_{\text{z}}},
\end{aligned} \tag{8}
$$

where $v$ is the velocity, $\phi$ is the heading angle, $\beta$ is the slip angle, $L_{\text{f}}$ is the distance of the front wheel to the center of mass, $L_{\text{r}}$ is the distance of the rear wheel to the center of mass, $\delta$ is the front wheel angle, $v_{\text{lon}}$ is the longitudinal velocity, $v_{\text{lat}}$ is the lateral velocity, $g$ is the acceleration due to gravity, $\alpha$ is the longitudinal acceleration request in $g$'s, $F_{\text{aero}}$ is the aerodynamic drag, $m$ is the mass of the vehicle, $\theta_r$ is the road grade, $F_{\text{f}}$ is the lateral tire force on the front tire, $F_{\text{f}}$ is the lateral tire force on the rear tire, and $I_{\text{z}}$ is the yaw moment of inertia of vehicle. The lateral tire forces are calculated as follows

$$
\begin{aligned}
F_{\mathrm{f}} &= 2C_{\mathrm{f}}(\delta - \theta_{v\mathrm{f}}),\\
F_{\mathrm{r}} &= 2C_{\mathrm{r}}(-\theta_{v\mathrm{r}}),\\
\theta_{v\mathrm{f}} &= \arctan\left(\frac{v_{\mathrm{lat}} + L_{\mathrm{f}}\dot{\phi}}{v_{\mathrm{lon}}}\right),\\
\theta_{v\mathrm{r}} &= \arctan\left(\frac{v_{\mathrm{lat}} - L_{\mathrm{r}}\dot{\phi}}{v_{\mathrm{lon}}}\right),
\end{aligned}
\tag{9}
$$

where $C_{\mathrm{f}}$ is the cornering stiffness of each front tire, $C_{\mathrm{r}}$ is the cornering stiffness of each rear tire, $\theta_{v\mathrm{f}}$ is the front tire velocity angle and $\theta_{v\mathrm{r}}$ is the rear tire velocity angle. In all simulations in Sect. 4, this model is used.

### 2.3.3 Simplified Bicycle Model

To aid in calculating the barrier dynamics in Sects. 4 and 5, we use the simplified decoupled system dynamics shown below.

$$
\begin{aligned}
\dot{x}_{\mathrm{T}} &= v_{\mathrm{T}}\cos(\phi_{\mathrm{T}}) - v_{\mathrm{H}}\cos(\phi_{\mathrm{H}}),\\
\dot{v}_{\mathrm{H}} &= g\alpha\\
\dot{y}_{\mathrm{T}} &= v_{\mathrm{T}}\sin(\phi_{\mathrm{T}}) - v_{\mathrm{H}}\sin(\phi_{\mathrm{H}}),\\
\dot{\phi}_{\mathrm{H}} &= \frac{v_{\mathrm{H}}}{L_{\mathrm{H}}}\delta,
\end{aligned}
\tag{10}
$$

where $x_{\mathrm{T}}$ and $y_{\mathrm{T}}$ are the relative $x$ and $y$ positions of the target center with respect to the ego center in road coordinates, $\phi_{\mathrm{T}}$ and $\phi_{\mathrm{H}}$ are the heading angles of the target and ego vehicles w.r.t. the road coordinate system, $v_{\mathrm{T}}$ and $v_{\mathrm{H}}$ are the absolute velocities of the target and ego respectively, $L_{\mathrm{H}}$ is the wheelbase of the ego vehicle, and $\delta$ is the front wheel angle of the ego vehicle.

## 2.4 Simulation Results

In this section we will show the applicability of our DRL based decision making Algorithm 1 for autonomous highway driving. First, we will elaborate on the training environment and evaluate the learned policy.

### 2.4.1 Training Environment

A schematic of the simulation environment used for training is given in Fig. 4. It is a three lane circular loop and is used to approximate an infinite stretch of straight highway. At the beginning of an episode, anywhere between $\{1, \ldots, N_{\mathrm{T}}\}$ number

**Fig. 4** A schematic of the simulation environment used for training

of cars are placed randomly within a distance of 250 m from the ego car. In this example we chose $N_T$ to be 30.

During learning stage the ego car (for e.g., white Fusion in Fig. 4) uses an $\epsilon$-greedy RL policy to make decisions, whereas for the traffic vehicles a combination of controllers from [25, 62] are used along with an IDM controller [50]. Additionally the traffic vehicles can randomly chose to perform lane change. For the traffic vehicles, the system parameters such as maximum velocity are randomly chosen. This is to ensure a diverse traffic scenario in training and evaluation. We assume that all the traffic vehicles take into account the relative distance and velocity to preceding vehicle before making a decision, i.e., they will not rear end the preceding car in the same lane. We use Algorithm 1 to train an Agent for decision making.

### 2.4.2 Reward Components for DRL Training

In order to train the policy $\pi$ we use a reward function $\rho$ that consists of a set driving goals for the ego car. It is formulated as a function of

- Desired traveling speed subject to traffic condition (11),
- Desired lane and lane offset subject to traffic condition (12),
- Relative distance to the preceding car based on relative velocity (13),

$$r_v = \mathrm{e}^{-\frac{(v_{\mathrm{ex}} - v_{\mathrm{des}})^2}{10}} - 1, \tag{11}$$

$$r_y = \mathrm{e}^{-\frac{\left(d_{\mathrm{ey}} - y_{\mathrm{des}}\right)^2}{10}} - 1, \tag{12}$$

$$r_x = \begin{cases} \mathrm{e}^{-\frac{(d_{\mathrm{lead}} - d_{\mathrm{safe}})^2}{10 d_{\mathrm{safe}}}} - 1 & \text{if } \mathrm{e}_x < d_{\mathrm{safe}} \\ 0 & \text{otherwise,} \end{cases} \tag{13}$$

where $v_{\mathrm{e_x}}$, $d_{\mathrm{e_y}}$, and $d_{\mathrm{lead}}$ are the ego velocity, lateral position, and the longitudinal distance to the lead vehicle respectively. Similarly, $v_{\mathrm{des}}$, $y_{\mathrm{des}}$, and $d_{\mathrm{safe}}$ are the desired speed, lane position, and safe longitudinal distance to the lead vehicle respectively.

Figure 5 gives an indicative plot of the reward functions (11)–(13), it is formulated assuming $v_{\mathrm{des}} = 30$ m/s which can be achieved in the center lane i.e., $y_{\mathrm{des}} = 3.8m$ with a minimum safe distance $d_{\mathrm{safe}} = 40$ m. The desired values are based on the traffic condition and can change depending on the scenario. For slow/fast moving traffic the peak in Fig. 5c will be adjusted based on the traffic condition. In this work we penalize the ego vehicle if it cannot maintain a minimum time headway of at least 1.3 s.

During learning, we evaluate the (partially) trained DRL controller every 100th episode. Figure 6 shows the average reward per decision during the training phase. It takes nearly 2000 episodes for the Agent to converge. We train the DRL Agent for a total of 10,000 episodes. Where each episode lasts until 200 samples or collision, whichever is earlier. Exploration is continuously annealed from 1 to 0.2 over first 7000 episodes and then kept constant for the remaining duration of learning. The $Q$-network is a deep neural network with 2 hidden layers each having 100 fully connected leaky ReLU's [29]. We train the network using Adam optimizer [22] with a fixed learning rate of $1e-4$.

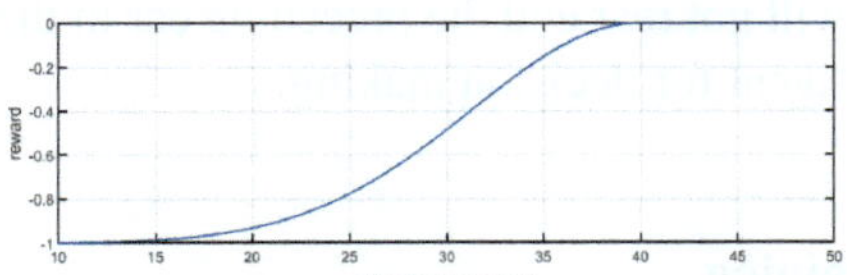


(a) Reward for desired relative distance.

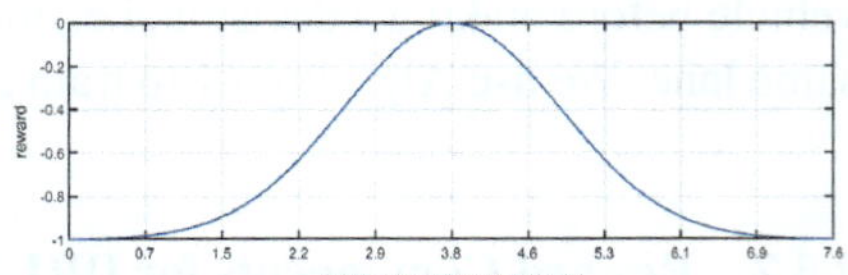


(b) Reward for desired lateral position.

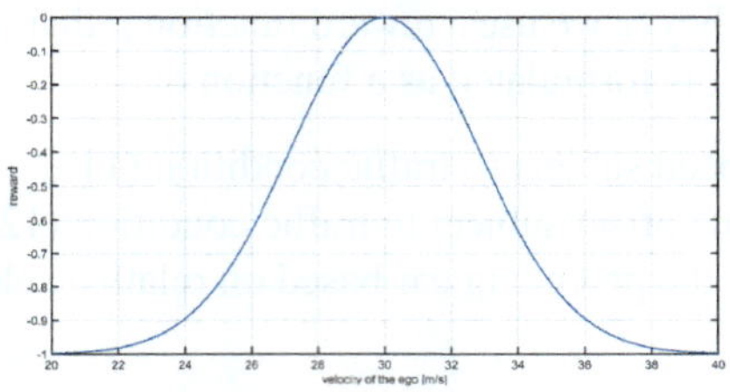


(c) Reward for desired ego speed.

**Fig. 5** Reward for the ego car based on traffic condition, sub goals are weighted equally when calculating the final reward

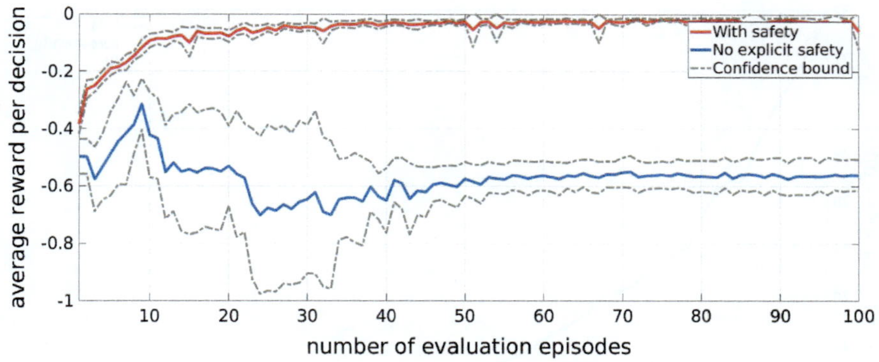


**Fig. 6** Average learning curve with confidence bound for with and without short horizon safety check in Algorithm 1

For the highway driving task, the safety filter was found to be a key component for learning a meaningful policy. Figure 6 shows the mean and confidence bound for training with and without safety filter over 200 training iterations of Algorithm 1. Training a standard DDQN Agent without explicit safety check could not learn a decent policy and always resulted in collision. Whereas DDQN with explicit safety check was able to converge to an optimal policy. Based on (11)–(13), the maximum reward an Agent can receive is zero per decision, the average reward per decision obtained by our trained DDQN Agent with safety check is around −0.025.

In Fig. 7 We evaluate our trained DDQN Agent to obtain average velocity with increase in traffic density. We compare this against modified safety filter from (5), the modification provides an acceleration command when the calculated time to collision $T_C$ is higher than $T_A$. This is referred as IDM in Fig. 7. It must be noted IDM controller from (5) cannot initiate lane change, in order to address this we integrate IDM with SUMO lane change decision making from [10, 20]. Figure 7, clearly demonstrates advantage of RL for high level decision making when compared to model-based approaches. With the increase in traffic density both the trained DDQN Agent and the model-based lane change controller converges to IDM controller. This is anticipated since lane change is neither safe nor advantageous in higher traffic density.

Use of two explicit buffers namely $Buf_S$ and $Buf_C$ in Algorithm 1 to store safe and non-safe transitions is simplified version of prioritized experience replay (PER) from [42]. Figure 8 shows the mean and confidence bound for training with two buffers and PER over 200 training iterations of Algorithm 1. For the highway driving example using two explicit buffers provides marginally better policy when compared to PER. This can be due to clear bifurcation of safe and non-safe transitions.

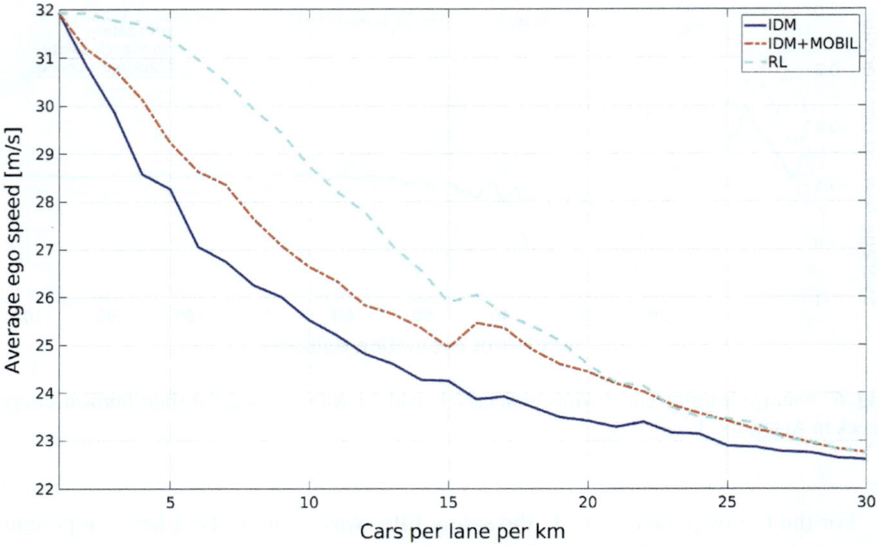


**Fig. 7** Average speed for simple IDM controller, with lane change, and trained RL Agent

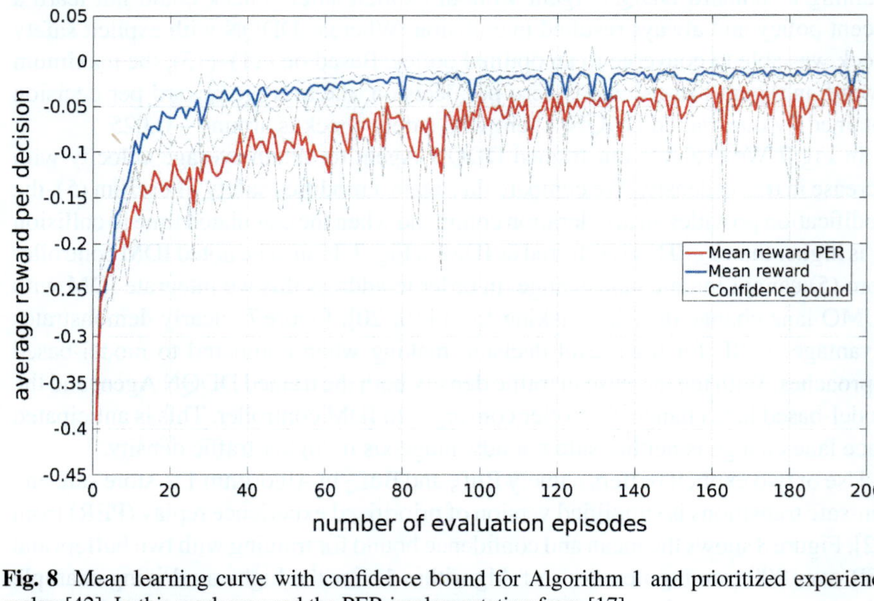


**Fig. 8** Mean learning curve with confidence bound for Algorithm 1 and prioritized experience replay [42]. In this work we used the PER implementation from [17]

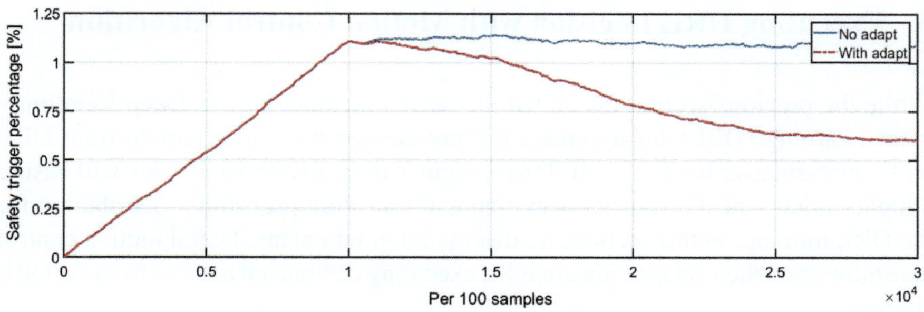


**Fig. 9** Comparison of number of safety trigger after learning with and without continuous adaptation

### 2.4.3 Continuous Adaptation

During the implementation phase, we replace the $\epsilon$-greedy policy, i.e., $\pi_\epsilon$ in Algorithm 1 line 5 by the learned policy $\pi$. Whenever the control decision by DDQN fails the short-horizon safety check, buffer $\mathrm{Buf}_\mathrm{C}$ is updated with additional data. Using a lower learning rate than the one used for training, $Q$-network can be retrained (line 13 until 16). Figure 9 shows the continuous adaptation result over 30K episodes and is obtained by averaging the data over 10k episodes using a moving average filter. Because of filtering, the mean number of safety trigger increases over first 10k episodes and stays constant for no adaptation scenario whereas it monotonically decreases to a smaller value thanks to continuous adaptation. Even with continuous adaptation the mean number safety trigger never converges to zero, this may be due to

1. Use of function approximation where a trained NN can potentially chose a non-safe action,
2. Use of rigid and static safety rules.

## *2.5 Summary*

In this section we provided an introduction to basics of RL, introduced our hybrid decision making—control architecture. We explained the need for including a safety filter during RL training and inference, we also showed its effectiveness in learning a driving policy. Although the rule-based safety filter is handcrafted and predetermined it significantly improved the learning speed. Additionally the number of safety interventions reduces as the agent learns to perform optimal actions.

## 3 Executing DRL Decision with Motion Control Algorithm

While the previous section illustrated the design methodology of Deep Reinforcement Learning (DRL) for a generic driving environment, it is best to train DRL with representative motion control that executes the DRL decision. This will ensure a higher fidelity of the vehicle—environment transition probability distribution for the DRL training. In this section, we discuss longitudinal and lateral motion control algorithms and their integration aimed at executing designated actions from AI DRL strategy.

### *3.1 Longitudinal Motion Control*

Longitudinal motion control systems need to provide tracking of the desired vehicle speed while maintaining a safety gap/range relative to the selected target vehicle. This functionality is provided by modern adaptive cruise control (ACC) systems which compute the longitudinal acceleration command based on the actual and set vehicle speed ($V_x$ and $V_{x.set}$), the actual and set range[2] ($R$ and $R_{set}$), and range rate[3] ($\dot{R}$)

$$a_{x.ACC.cmd} = f_{ACC}(V_{x.set}, R_{set}, V_x, R, \dot{R}) \tag{14}$$

The commanded acceleration is delivered by the actuation layer through modulating propulsion or brake torque. The ACC mapping function $f_{ACC}$ can be implemented in various ways. For example, [41] described an ACC mapping function that includes both car-following mode regulating steady state range/gap and speed-control mode responding to transient situations such as target vehicle cut-in or loss of its lead vehicle. The two modes were integrated through a calibrated range-range rate ($R - \dot{R}$) transition diagram. Treiber and Kesting [49] proposed the Intelligent Driver Model (IDM) which has been widely used in traffic simulation environments and represents human car-following behavior observed on freeway and urban traffic. Jin et al. [19] implemented state ($[R, \dot{R}]$) feedback controller imitating human driving behavior and experimented on connected automated vehicle. Treiber and Kesting [49] further developed Improved IDM (IIDM) and the ACC Model to avoid unrealistically large deceleration when desired speed suddenly changes, e.g., when entering a reduced speed zone or when a vehicle cuts in. For more variety of ACC a reader is referred to the survey [58] and references therein.

To execute DRL's lane change decision, we leverage our production ACC design. This will avoid the need of complicated spatial-temporal optimization [24]. By adding scenario dependent smart 'target selection' to existing ACC design, the vehicle speed will change during the lane change maneuvers for a smooth transition into a new

---

[2] Range $R$ is defined as the distance or gap between the ego and target vehicle, i.e., distance between the ego vehicle's front bumper and target vehicle's rear bumper.

[3] Range rate $\dot{R}$ is defined as relative speed between the ego and target vehicle.

gap and a new lane speed. The elements considered in the scenarios include the lane change decision, the range and range rate of surrounding vehicles, target lane speed, and the status/prediction of lateral motion. The 'target selection' refers to the selected lead vehicle whose range and range-rate of the ACC mapping function (Eq. 14) will be based on. For example, in case of staying in lane decision, which represents the base ACC functionality, the target vehicle selected is the lead vehicle of the current lane. In case of lane changing decision, the target selection is considered in two phases: before and after merging into target lane. In the first phase, both the lead vehicles in current lane and target lane will be fed into the ACC mapping function, with a calibratable desired range, and the commanded acceleration will take the minimum values of the two, thus ensuring safety. In the second phase (after merging into the target lane), the algorithm reverts to the base ACC functionality, i.e., selecting the lead vehicle of the current (new) lane as the target.

Therefore, with the scenario-based ACC always aware of the status of the lateral motion control (to be described next), smooth and human-driver-like motion control can be expected for lane change and lane keep decisions.

## 3.2 Lateral Motion Control

For lane centering and lane changing maneuvers, we first introduce the relevant vehicle-road kinematics. Figure 10 shows the definition of coordinate frames and variables for vehicle road/lane level localization. The vehicle motion dynamics is considered with respect to the curvilinear Frenet-Serret coordinate frame ($x_r$-$y_r$). The states of the vehicle-road kinematics include the travel distance along the curvilinear coordinate $s$, the path offset $e_y$, and the heading offset $e_\psi$, while the curvature of the road is considered as an exogenous input, denoted by $\kappa_{road}$. Vehicle motion—the longitudinal velocity $V_x$, lateral velocity $V_y$, and yaw rate $r$—are described in the Cartesian coordinate vehicle body frame $(x - y)$. Vehicle motion curvature $\kappa$ is defined as $\kappa = r/V_x$.[4]

Figure 11 shows the system block diagram. The lateral motion controller performs lane localization and computes vehicle motion curvature command. The plant includes actuator and vehicle dynamics, as well as vehicle-road kinematics/geometry.

Vehicle-Road kinematics in the Frenet-Serret coordinate frame is given by Werling et al. [56]

[4] Alternatively, the curvature can be computed based on the front wheel angle $\delta$ as $\kappa = \delta/(L + K_u V_x)$ where $L$ is the vehicle wheelbase, and $K_u$ is the vehicle understeer gradient in units of rad-s$^2$/m. The understeer gradient is defined as: $K_u = m_f/C_f - m_r/C_r$ where $m_f$ and $m_r$ are the front and rear axle mass, $C_f$ and $C_r$ are the front and rear axle cornering coefficients.

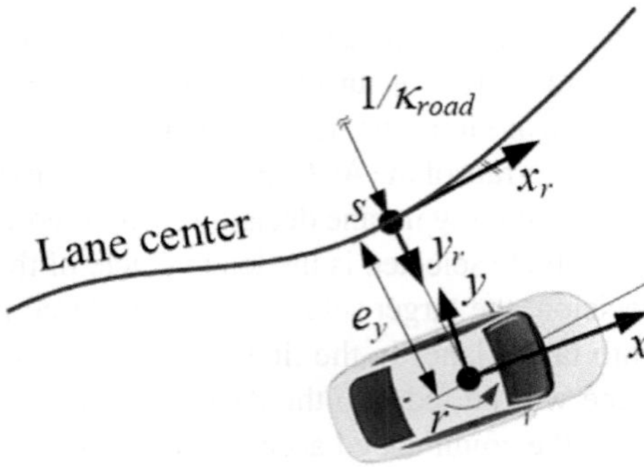


**Fig. 10** Coordinate frames and vehicle motion dynamics variables. Legend: $(x_r - y_r)$ is the curvilinear coordinate road frame, $(x - y)$ is the Cartesian coordinate vehicle body frame, $\kappa_{road}$ is road curvature, $e_y$ is path offset, $e_\psi$ is heading offset, $s$ is curvilinear coordinate, $r$ is vehicle yaw rate

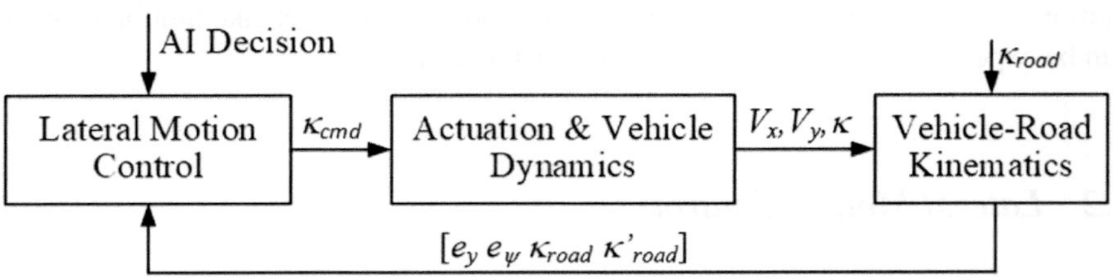


**Fig. 11** Block diagram of lateral motion control and plant. Legend: $\kappa_{cmd}$ is vehicle motion curvature command, $V_x$ and $V_y$ are vehicle longitudinal and lateral velocity, $\kappa$ is vehicle motion curvature, $\kappa_{road}$ is road curvature

$$\dot{e}_y = V_x \sin e_\psi - V_y \cos e_\psi \tag{15a}$$

$$\dot{e}_\psi = \kappa_{road}\dot{s} - r \tag{15b}$$

$$\dot{s} = \frac{1}{1 - \kappa_{road} e_y}(V_x \cos e_\psi - V_y \sin e_\psi). \tag{15c}$$

Since the value of heading offset, road curvature, and lateral velocity are relatively small during the lane centering and lane change maneuvers on highways, the vehicle-road kinematics can be further simplified to the following

$$\dot{e}_y = V_x e_\psi \tag{16a}$$

$$\dot{e}_\psi = V_x(\kappa_{road} - \kappa) \tag{16b}$$

$$\dot{s} = V_x. \tag{16c}$$

The path offset $(e_y)$ dynamics from $(\kappa_{road} - \kappa)$ are represented by a double integrator with crossover frequency of $|V_x|$. The vehicle motion curvature $\kappa$ is the system input, whereas the road curvature $\kappa_{road}$ is considered as a system disturbance.

In typical applications, the system states $e_y$ and $e_\psi$ and road curvature $\kappa_{road}$ are obtained through image processing of video frames taken from a front view camera and by fitting the lane markers to a 3rd order polynomial [55]. Coefficients of these lane marker polynomials are then fused to obtain the coefficients of the lane center polynomial with respect to vehicle coordinate ($x$-$y$), referred to as the path polynomial, where

$$y_{path} = a_0 + a_1 x + a_2 x^2 + a_3 x^3. \tag{17}$$

In this format, coefficients $a_0$ and $a_1$ are equivalent to the path offset ($e_y$) and heading offset respectively ($e_\psi$) (see Fig. 10). Coefficients $a_2$ and $a_3$ are related to the road curvature $\kappa_{road}$ and curvature rate $\kappa'_{road} = d\kappa_{road}/dx$. Based on vector differential calculus, curvature of a parametric curve $y = y(x)$ in the $x - y$ plane is given by Kreyszig [23]

$$\kappa(x) = |y''|/(1 + y'^2)^{3/2} \tag{18}$$

where $y' = dy/dx$. Computing $y'_{path}$ and $y''_{path}$ from Eq. (17) and replacing with $y'$ and $y''$ in (18) gives

$$\kappa(x) = \frac{2a_2 + 6a_3 x}{[1 + (a_1 + 2a_2 x + 3a_3 x^2)^2]^{3/2}}. \tag{19}$$

Assuming small heading offset $a_1$,[5] the curvature and curvature rate of the path (road) at the origin ($x = 0$) are equal to

$$\kappa_{road} = 2a_2 \tag{20a}$$

$$\kappa'_{road} = 6a_3 \tag{20b}$$

### 3.2.1 Lane Centering Control

Vehicle-Road kinematics for lane centering applications is given by Eqs. (16a) and (16b) where the road curvature is considered as an exogenous disturbance which can be measured. The proposed curvature command (Eq. 21) represent a generic lane centering algorithm which includes both a feedforward term of the road curvature and the feedback terms. As such, with different coefficient design, it can represent various well-known path following algorithms/approaches including Pure Pursuit [8], Bezier curves based planning [15], or lookahead point design [51]. Our contribution is to introduce a novel and heuristic calibration approach for any of the above path following approaches to achieve desired closed-loop system response. The generic control law is given by

[5] For highway driving nominal value of $a_1 \leq 5^\circ$ hence the error introduce by the small heading offset assumption is $\leq 1\%$.

$$\kappa_{cmd} = \underbrace{K_{ff}(\kappa_{road} + V_x T_{prev} \kappa'_{road})}_{\text{feedforward}} + \underbrace{K_y e_y + K_\psi e_\psi}_{\text{feedback}} \tag{21}$$

where $K_{ff}$ is the feedforward gain, $T_{prev}$ is the preview time that can be effectively utilized to compensate for actuation delays, $K_y$ is the path offset gain, and $K_\psi$ is the heading offset gain. Substituting $\kappa$ in (16b) with $\kappa_{cmd}$ from (21) and combining with (16a) gives the closed-loop dynamics

$$\begin{bmatrix} \dot{e}_y \\ \dot{e}_\psi \end{bmatrix} = V_x \begin{bmatrix} 0 & 1 \\ -K_y & -K_\psi \end{bmatrix} \begin{bmatrix} e_y \\ e_\psi \end{bmatrix} + V_x(1 - K_{ff}) \begin{bmatrix} 0 & 0 \\ 1 & V_x T_{prev} \end{bmatrix} \begin{bmatrix} \kappa_{road} \\ \kappa'_{road} \end{bmatrix}. \tag{22}$$

Eigenvalues of this 2nd order system are

$$\lambda_{1,2} = \frac{V_x K_\psi}{2} \pm V_x \sqrt{K_y} \sqrt{\frac{K_\psi^2}{4K_y} - 1}. \tag{23}$$

To design the state feedback gains in terms of desired closed-loop metrics such as the natural frequency and damping ratio, we compare eigenvalues (23) with eigenvalues of the 2nd order system expressed in terms of the natural frequency $\omega_n$ and damping ratio $\zeta$

$$\lambda_{1,2}^* = -\omega_n \zeta \pm \omega_n \sqrt{\zeta^2 - 1} \tag{24}$$

Derived feedback gains are given by

$$K_y = \frac{\omega_n^2}{V_x^2} \tag{25a}$$

$$K_\psi = \frac{2\zeta \omega_n}{V_x}. \tag{25b}$$

Instead of the natural frequency as the tuning parameter, for lane centering applications it is convenient to use alternative time response metrics such as the transient time to reach 95% of the target value or 5% of the initial value. This time is referred to as the response time and denoted by $T_r$. For practical damping ratios in range between 0.7 and 0.9, relation between the transient time, closed-loop frequency and damping ratio can be obtained by numerical optimization (with coefficient of determination $R^2 = 0.99$) and is given by

$$T_r \approx \frac{4.3\zeta}{\omega_n}. \tag{26}$$

Solving (26) for $\omega_n$ by using (25a) and (25b) gives the feedback gains in terms of the transient time as

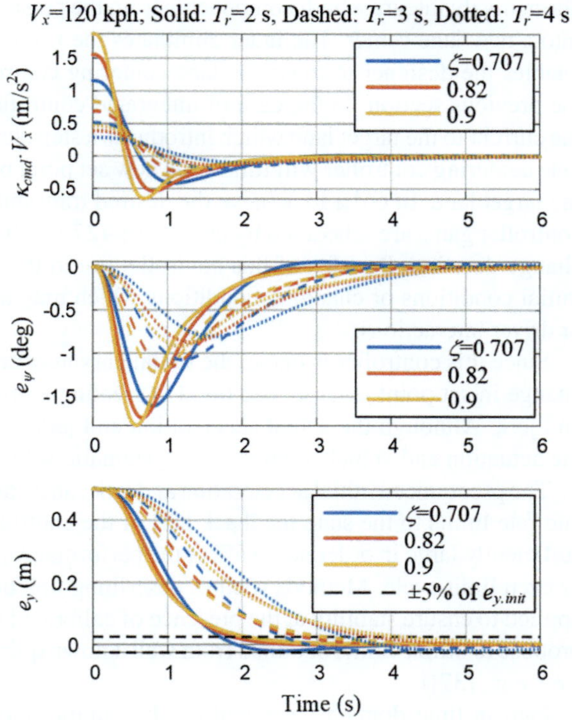


**Fig. 12** Initial condition response Lane change control for various damping ratios $\zeta$ and response times $T_r$. The initial conditions are: $e_{y.0} = 0.5$ m and $e_{\psi.0} = 0°$. Actuation dynamics is approximated by 2nd order term with $\omega_{n.act} = 5\omega_n$ and $\zeta_{act} = 0.6$

$$K_y = \frac{18.5\zeta^2}{T_r^2 V_x^2} \tag{27a}$$

$$K_\psi = \frac{8.6\zeta^2}{T_r V_x}. \tag{27b}$$

Figure 12 illustrates vehicle response with non-zero initial conditions using the proposed controller for different choice of the design parameters.

### 3.2.2 Lane Change Control

Lane change control systems need to provide a smooth lateral shift of a vehicle from current to target lane within specified time and with small or zero overshoot. Typical duration of lane changes[6] is in the range of 5 and 7 s with lateral accelerations amplitudes in the range of 0.4 and 0.8 m/s$^2$ (for making a lane change on straight roads). One can divide the problem into explicit path planning and path following, or

[6] Duration is defined as time required to reach path offset relative to the target lane center that is withing the range of nominal lane centering oscillations in range of 0.1–0.2 m. For nominal lane width of 3.4 m, this corresponds to approximately 5% of the lane width or the path offset at the lane change start.

alternatively integrate both components from the perspective of smooth transitioning into a new lane center. The latter eliminates the need of specifying way points and enables the designer to reuse the lane centering control methodology described in the previous section. In the case of integrated controller, the path is switched from the current to the target lane which introduces a step change in the path offset $e_y$. The lane centering controller will follow the new set point by driving the vehicle towards the target lane. In order to achieve the desired transient time and damping ratio, the controller gains are scheduled by using (26)–(27b), where $T_r$ corresponds to the lane change duration. Benefits of this controller are in the ease of tuning, adaptation to initial conditions or changing conditions which may arise from lane change aborts or driver interactions.

Since this controller is linear, the initial curvature command in response to a step change in set point is large (see Fig. 12) creating a pronounced lateral acceleration and jerk. Although the actual acceleration and jerk will be partially filtered out by the actuation and vehicle dynamics, a systematic solution is proposed below.

The pronounced initial acceleration and jerk can be overcome by adding amplitude and rate limits to the state feedback part of the control input. The limits need to be sufficiently large in order not to affect the performance, e.g., not to cause oscillations or even limit cycle. Methods such as describing function analysis [1, 9, 44] can be applied to ensure stability at the presence of calibrated rate and amplitude limits. We propose to utilize limits obtained from well-known quintic polynomial path planning (see, e.g., [37]).

Path in time domain described by the quintic polynomial is given by: $y(t) = a_0 + a_1 t + a_2 t^2 + a_3 t^3 + a_4 t^4 + a_5 t^5$. For given initial and final conditions ($t_f$ is final time): $y(0) = 0$, $\dot{y}(0) = 0$, $\ddot{y}(0) = 0$, $y(t_f) = y_f$, $\dot{y}(t_f) = 0$, $\ddot{y}(t_f) = 0$, the coefficients are equal to:$a_0 = 0$, $a_1 = 0$, $a_2 = 0$, $a_3 = 10y_f/t_f^3$, $a_4 = -15y_f/t_f^4$, and $a_5 = 6y_f/t_f^5$. From the second derivative of the polynomial, time instances at which lateral acceleration peaks occur are $t_{1,2} = (3 \pm \sqrt{3})/6t_f \Rightarrow t_1 \approx 0.21t_f$, $t_2 \approx 0.79t_f$. Accordingly, the acceleration maximum is

$$a_{y.max} = \frac{5.77y_f}{t_f^2} \tag{28}$$

Maximum jerk occurs at the beginning and end with magnitude are equal to

$$j_{max} = \frac{60y_f}{t_f^3} \tag{29}$$

Obtained lateral acceleration and jerk limits[7] provide comfortable maneuver yet are high enough to prevent oscillations or limit cycles which can be proved by nonlinear simulations or describing function analysis. Figure 13 shows simulation results with

[7] For lane width of 3.4 m and lane change duration of 6 s, the limits are $a_{y.max} = 0.54\,\text{m/s}^2$ and $j_{max} = 0.94\,\text{m/s}^3$.

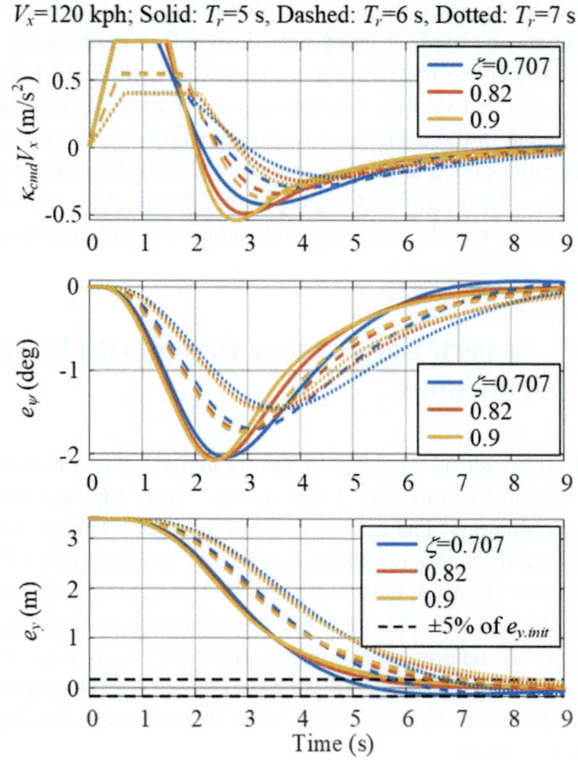


**Fig. 13** Lane change control with control input amplitude and rate limiting. Actuation dynamics is approximated by 2nd order term with $\omega_{n.act} = 5\omega_n$ and $\zeta_{act} = 0.6$

amplitude and rate limiting of the curvature command. We see that although there is a major effect on the initial transient response, the target metrics in terms of the duration and damping ratio are preserved.

### *3.3 Summary*

In this section, we presented longitudinal and lateral motion control, and how they work together to execute the decision made by higher level AI (DRL). For the longitudinal control, we introduced a smart 'target' selection logic to leverage production ACC design, and to facilitate the integration with lateral motion control for delivering smooth and human-driver-like lane change maneuvers. For the lateral control, we introduced a unified lane centering and lane changing algorithm with a general control algorithm formula that can be calibrated to represent various well-known path following design methodologies. We further introduced a new calibration approach that can automatically adjust the state feedback gains to achieve desired system damping, closed loop bandwidth, or 0–95% set point response time. In addition, for lane change maneuvers, we integrated both path planning and path following aspects.

Since the integrated control does not try to regulate the vehicle to specific waypoints, no explicit waypoints designation is required. It provides implicit, smooth, and seamless re-planning in case of unexpected vehicle motion which may arise from unintended steering movement, and in case of abrupt path offset change (set point) change in the event of lane change abort (to return to the original lane). We further recommended a way to analyze and calibrate the limit of lateral acceleration and jerk—the nonlinearities purposely added to the controller for comfort and actuation constraints—for assuring both subjective and objective performance.

## 4 Generic Safety Filter Design with Control Barrier Functions

In this section we develop a more generic safety filter than the heuristic handcrafted rules utilized in Sect. 2, because the latter may not be sufficient for avoid all types of collisions. The performance of an Agent depends on what it sees during training. Since the handcrafted heuristic rules are designed with certain imminent collision threats in mind, they could miss corner cases in the real-world, and even in a simulation world with long tail distribution where simulated target vehicles behaving beyond the engineers' imagination. For this reason, the rule-based safety filter can be 'brittle'. Hence a collision-avoidance intervention mechanism that is not specific to certain ego-target vehicle poses/situations is developed here to serve as a generic safety filter.

Control Barrier Functions (CBF) have been used as a method to compute minimally-invasive feedback control that satisfy various prescribed system constraints. For AVs, CBFs can provide a computationally efficient approach to avoid collision by correcting ego vehicle motion based on its relative position and velocity compared with surrounding road users. The CBF framework used here assumes that a nominal control signal $u_0$ is computed by the autonomous driver, in this case an Agent. That signal is then evaluated by the CBF for safety. If $u_0$ is safe, the CBF safety filter does not modify it. If it is deemed unsafe, the safety filter overrides the nominal control and calculates a minimally invasive safe control that can avoid collision. The advantage of using a CBF based collision avoidance system is that it provides a safe steering and longitudinal acceleration command that is as close as possible (in the Euclidean sense) to the nominal control. This results in a much less intrusive safety filter compared to the rule based safety filter from Sect. 2.

However, it is not straightforward to decide if the actuation of braking or steering should be used, (or how the actuation should be combined/split) to best correct vehicle motion when CBF intervenes to avoid a collision. To arbitrate between steering and/or braking, we use Contextual Selection of Decoupled CBF. This algorithm addresses the above issue by taking advantage of the structure and rules of the road, as well as the knowledge of steering/braking efficacy for different situations based on physics. Therefore, the proposed logic and resulted methodology, will determine steering and braking action required to effectively and decisively impose decoupled longitudinal and lateral CBFs in order to avoid collisions. The algorithm is designed to use the same input structure, i.e. affordance indicators as the DRL Agent. In addition to the affordance indicators used by the Agent, the CBF safety filter also uses the heading with respect to the road of the six surrounding traffic vehicles. The algorithm has been verified in extensive simulation with varying traffic and road users.

### 4.1 Control Barrier Functions

Barrier Functions have been used to enhance system robustness [39], while Control Barrier Functions (CBFs) have been used as a method to provide minimally-invasive feedback control that satisfies various constraints prescribed on that system [3, 18, 57, 59]. Here we mix the two concepts because we would like to apply them to systems with control inputs that are outside of the ego (the entity doing the calculations) control. Consider a control affine system

$$\dot{\mathbf{x}} = f(\mathbf{x}) + g(\mathbf{x})u, \tag{30}$$

where $\mathbf{x} \in \mathbb{R}^n$ and $u \in \mathbb{R}^m$ is the control input. We assume that a control input $u_0$ is computed by a driver (or an Agent/controller) to achieve some objective, and it is known. Let us also assume that an additional control objective is to keep the state of the system in a closed admissible set $\mathcal{C} \subset \mathbb{R}^n$ defined as

$$\begin{aligned}
\mathcal{C} &\triangleq \{\mathbf{x} \in \mathbb{R}^n : h(\mathbf{x}) \geq 0\}, \\
\partial\mathcal{C} &\triangleq \mathbf{x} \in \mathbb{R}^n : h(\mathbf{x}) = 0, \\
\text{Int}(\mathcal{C}) &\triangleq \mathbf{x} \in \mathbb{R}^n : h(\mathbf{x}) > 0,
\end{aligned} \tag{31}$$

where $h : \mathbb{R}^n \to \mathbb{R}$ is a twice continuously differentiable function. In addition, we assume that $\text{Int}(\mathcal{C}) \neq \emptyset$, where $\text{Int}(\mathcal{C})$. Figure 14 shows an example Barrier Function $h(\mathbf{x})$, the admissible set $\mathcal{C}$ and the boundary $\partial\mathcal{C}$. $h(\mathbf{x})$ is relative degree 1 with respect to the control inputs if $h_{\mathbf{x}} \cdot g(\mathbf{x}) \neq \emptyset$, where $h_{\mathbf{x}} = \frac{\partial h(\mathbf{x})}{\partial \mathbf{x}}$. For the system (30) with a given control input $\bar{u}$ a function $h(\mathbf{x})$ is a Barrier Function [3] with respect to the admissible set $\mathcal{C}$ if $h(\mathbf{x})$ is relative degree 1 with respect to the control inputs and

$$\dot{h}(\mathbf{x}, \bar{u}) + l_0 h(\mathbf{x}) \geq 0, \tag{32}$$

where $\dot{h}(\mathbf{x}, \bar{u}) = h_{\mathbf{x}} \cdot f(\mathbf{x}) + h_{\mathbf{x}} \cdot g(\mathbf{x})\bar{u}$. If a Barrier Function is relative degree 2 with respect to the control inputs (i.e. $h_{\mathbf{x}} \cdot g(\mathbf{x}) \equiv 0$, as is the case later in this chapter) we can—motivated by [36]—consider the second order barrier constraint: $\ddot{h}(\mathbf{x}) + l_1\dot{h}(\mathbf{x}) + l_0 h(\mathbf{x}) \geq 0$. The parameters $l_1, l_0$ are selected such that the roots of the polynomial $s^2 + l_1 s + l_0$ are negative real. The forward invariant set is $\{x : h >$

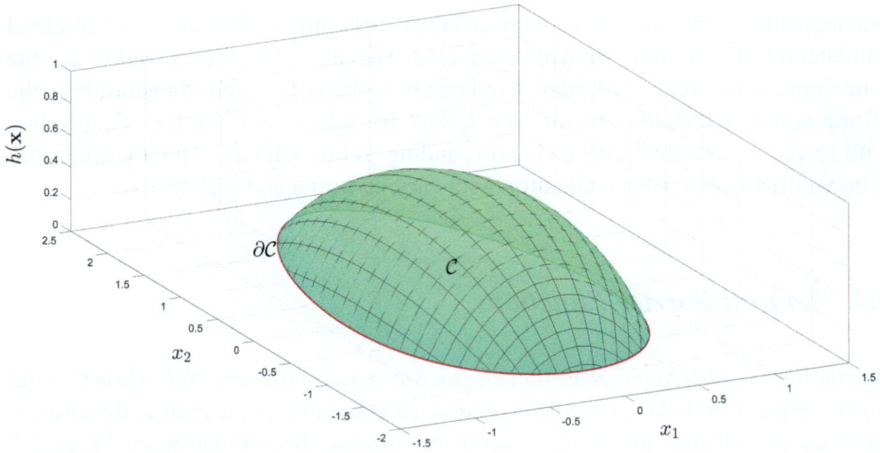


**Fig. 14** Example barrier function

0 and $h_{\mathbf{x}} \cdot f(\mathbf{x}) > -\lambda_i h\}$, where $\lambda_i$ is either one of the two roots of $s^2 + l_1 s + l_0 = 0$. A barrier function $h(\mathbf{x})$ is a CBF with respect to the admissible set $\mathcal{C}$ if $h$ is differentiable in Int($\mathcal{C}$), $h(\mathbf{x}) \to 0$ as $\mathbf{x} \to \partial\mathcal{C}$, and for $\mathbf{x} \in$ Int($\mathcal{C}$)

$$h_{\mathbf{x}} \cdot g(\mathbf{x})\bar{u} = 0 \implies h_{\mathbf{x}} \cdot f(\mathbf{x}) + l_0 h(\mathbf{x}) \geq 0 \tag{33}$$

if $h(\mathbf{x})$ is relative degree 1 with respect to the control inputs and

$$\frac{\partial(h_{\mathbf{x}} \cdot f(\mathbf{x}))}{\partial \mathbf{x}} \cdot g(\mathbf{x})\bar{u} = 0 \implies \frac{\partial(h_{\mathbf{x}} \cdot f(\mathbf{x}))}{\partial \mathbf{x}} \cdot f(\mathbf{x}) + l_1(h_{\mathbf{x}} \cdot f(\mathbf{x})) + l_0 h(\mathbf{x}) \geq 0 \tag{34}$$

if $h(\mathbf{x})$ is relative degree 2 with respect to the control inputs.

### 4.1.1 Decoupled Barrier Functions for Autonomous Driving

We define the longitudinal collision avoidance barrier functions as

$$\begin{aligned} h_{x,\mathrm{F}} &= x_{\mathrm{T}} - k_v v_{\mathrm{H}} - d_{x,\min} - \frac{L_{\mathrm{H}}}{2} - \frac{L_{\mathrm{T}}}{2}, \\ h_{x,\mathrm{R}} &= -x_{\mathrm{T}} - k_v v_{\mathrm{T}} - d_{x,\min} - \frac{L_{\mathrm{H}}}{2} - \frac{L_{\mathrm{T}}}{2}, \end{aligned} \tag{35}$$

where the subscript F is for the forward barrier and the subscript R is for the rearward barrier. Similarly, we define the lateral collision avoidance barrier functions as

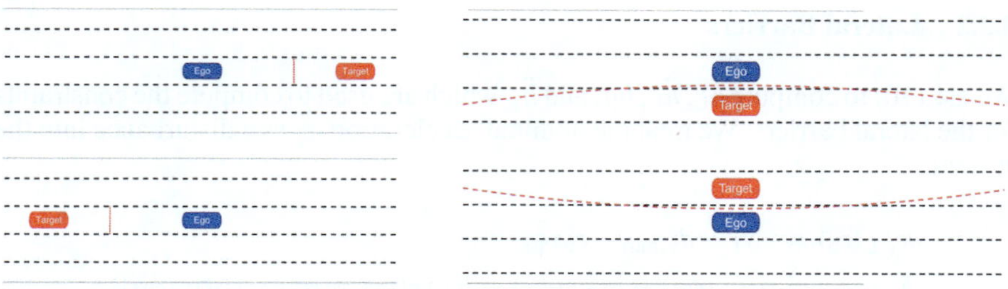


**Fig. 15** Longitudinal and lateral barriers

$$h_{y,\mathrm{L}} = -y_\mathrm{T} - d_{y,\min} + c_\mathrm{b} x_\mathrm{T}^2, \quad h_{y,\mathrm{R}} \;= y_\mathrm{T} - d_{y,\min} + c_\mathrm{b} x_\mathrm{T}^2, \tag{36}$$

where the subscript L is for the barrier to the left of the ego, the subscript R is for the barrier to the right of the ego vehicle, $L_\mathrm{T}$ is the length of the target vehicle, $d_{x,\min}$ is the minimum longitudinal distance allowed between vehicle bumpers, $d_{y,\min}$ is the minimum lateral distance allowed between vehicle centers, and $c_\mathrm{b}$ is a coefficient that determines the amount of bowing in the lateral barrier. The bowing is meant to reduce the steering effort required to satisfy the collision avoidance constraint when the target is far away from the ego. The minimum lateral distance of the barrier $d_{y,\min}$ is only enforced when the ego is driving alongside the ego (i.e. $x_\mathrm{T} = 0$). The possible longitudinal and lateral barriers are illustrated in Fig. 15. The dotted red line shows the position which the front/rear/left/right side of the ego vehicle must maintain in order to satisfy the barrier constraint.

## *4.2 Calculation of Barrier Constraints*

### 4.2.1 Longitudinal Barrier

We use the simplified bicycle model (10) to compute $\dot{h}_{x,F}, \dot{h}_{x,R}, \ddot{h}_{x,R}$, which are used to compute the constraints for the longitudinal barrier. Since $h_{x,\mathrm{F}}$ is relative degree 1 with respect to the control input $\alpha$, only $\dot{h}_{x,F}$ is required to calculate the constraints for vehicles ahead of the ego vehicle.

$$\begin{aligned}
h_{x,\mathrm{F}} &= x_\mathrm{T} - k_v v_\mathrm{H} - d_{x,\min} - \frac{L_\mathrm{H}}{2} - \frac{L_\mathrm{T}}{2}, \\
\dot{h}_{x,\mathrm{F}}(\alpha) &= -g k_v \alpha + \Big(v_\mathrm{T}\cos(\phi_\mathrm{T}) - v_\mathrm{H}\cos(\phi_\mathrm{H})\Big) \\
h_{x,\mathrm{R}} &= -x_\mathrm{T} - k_v v_\mathrm{T} - d_{x,\min} - \frac{L_\mathrm{H}}{2} - \frac{L_\mathrm{T}}{2}, \\
\dot{h}_{x,\mathrm{R}} &= \Big(v_\mathrm{T}\cos(\phi_\mathrm{T}) - v_\mathrm{H}\cos(\phi_\mathrm{H})\Big) \\
\ddot{h}_{x,\mathrm{R}}(\alpha) &= g\cos(\phi_\mathrm{H})\alpha
\end{aligned} \tag{37}$$

### 4.2.2 Lateral Barriers

We use (10) to compute $\dot{h}_{\mathrm{L}}, \ddot{h}_{\mathrm{L}}, \dot{h}_{\mathrm{R}}$, and $\ddot{h}_{\mathrm{R}}$ which are used to compute the constraints for the lateral barriers. We treat the nominal acceleration $\alpha_0$ as a disturbance into the system.

$$\begin{aligned} h_{y,\mathrm{L}}(c_{\mathrm{b}}) &= -y_{\mathrm{T}} - d_{y,\min} + c_{\mathrm{b}} x_{\mathrm{T}}^2, \\ \dot{h}_{y,\mathrm{L}}(c_{\mathrm{b}}) &= v_{\mathrm{H}}\sin(\phi_{\mathrm{H}}) - v_{\mathrm{T}}\sin(\phi_{\mathrm{T}}) + 2c_{\mathrm{b}}x_{\mathrm{T}}(v_{\mathrm{T}}\cos(\phi_{\mathrm{T}}) - v_{\mathrm{H}}\cos(\phi_{\mathrm{H}})) \\ \ddot{h}_{y,\mathrm{L}}(\alpha_0, \delta, c_{\mathrm{b}}) &= \Big(\cos(\phi_{\mathrm{H}}) + 2c_{\mathrm{b}}x_{\mathrm{T}}\sin(\phi_{\mathrm{H}})\Big)\frac{v_{\mathrm{H}}^2\delta}{L_{\mathrm{H}}} + \Big(\sin(\phi_{\mathrm{H}}) - 2c_{\mathrm{b}}x_{\mathrm{T}}\cos(\phi_{\mathrm{H}})\Big)g\alpha_0 \\ &\quad + 2c_{\mathrm{b}}\Big(v_{\mathrm{T}}\cos(\phi_{\mathrm{T}}) - v_{\mathrm{H}}\cos(\phi_{\mathrm{H}})\Big)^2, \end{aligned} \tag{38}$$

and

$$\begin{aligned} h_{y,\mathrm{R}}(c_{\mathrm{b}}) &= y_{\mathrm{T}} - d_{y,\min} + c_{\mathrm{b}} x_{\mathrm{T}}^2, \\ \dot{h}_{y,\mathrm{R}}(c_{\mathrm{b}}) &= v_{\mathrm{T}}\sin(\phi_{\mathrm{T}}) - v_{\mathrm{H}}\sin(\phi_{\mathrm{H}}) + 2c_{\mathrm{b}}x_{\mathrm{T}}(v_{\mathrm{T}}\cos(\phi_{\mathrm{T}}) - v_{\mathrm{H}}\cos(\phi_{\mathrm{H}})), \\ \ddot{h}_{y,\mathrm{R}}(\alpha_0, \delta, c_{\mathrm{b}}) &= \Big(2c_{\mathrm{b}}x_{\mathrm{T}}\sin(\phi_{\mathrm{H}}) - \cos(\phi_{\mathrm{H}})\Big)\frac{v_{\mathrm{H}}^2\delta}{L_{\mathrm{H}}} + \Big(-\sin(\phi_{\mathrm{H}}) - 2c_{\mathrm{b}}x_{\mathrm{T}}\cos(\phi_{\mathrm{H}})\Big)g\alpha_0 \\ &\quad + 2c_{\mathrm{b}}\Big(v_{\mathrm{T}}\cos(\phi_{\mathrm{T}}) - v_{\mathrm{H}}\cos(\phi_{\mathrm{H}})\Big)^2 \end{aligned} \tag{39}$$

### 4.2.3 Road Keeping Barriers

We also use road keeping barriers that prevent the vehicle from veering off the drivable area while avoiding a collision whenever possible

$$h_{\mathrm{RK}} = \begin{bmatrix} 3w_{\mathrm{l}} - \frac{W_{\mathrm{H}}}{2} - y_{\mathrm{H}} \\ y_{\mathrm{H}} - \frac{W_{\mathrm{H}}}{2} \end{bmatrix}, \quad \dot{h}_{\mathrm{RK}} = \begin{bmatrix} -v_{\mathrm{H}}\phi_{\mathrm{H}} \\ v_{\mathrm{H}}\phi_{\mathrm{H}} \end{bmatrix}, \quad \ddot{h}_{\mathrm{RK}}(\delta) = \begin{bmatrix} \frac{-v_{\mathrm{H}}^2\cos(\phi_{\mathrm{H}})\delta}{L_{\mathrm{H}}} \\ \frac{v_{\mathrm{H}}^2\cos(\phi_{\mathrm{H}})\delta}{L_{\mathrm{H}}} \end{bmatrix} \tag{40}$$

where $y_{\mathrm{H}}$ is the $y$-coordinate of the ego vehicle w.r.t. a lane attached frame where the y-coordinate of the right most lane line is 0, $L_{\mathrm{H}}$ is the length of the ego vehicle, and $w_{\mathrm{l}}$ is the lane width.

## *4.3 Contextual Selection of Decoupled CBF*

In this section, we describe a method that takes advantage of the structure and rules of the road and knowledge of steering/braking efficacy for different situations to select which barrier constraints to enforce. We assume $\alpha_{\mathrm{S}} = \alpha_0 + \alpha_{\mathrm{CBF}}$ where $\alpha_0$ is the nominal longitudinal acceleration that control provided by the virtual driver and $\alpha_{\mathrm{CBF}}$ is the correction computed by the collision avoidance algorithm. Similarly $\delta_{\mathrm{S}} = \delta_0 + \delta_{\mathrm{CBF}}$ where $\delta_0$ is the nominal front wheel angle and $\delta_{\mathrm{CBF}}$ is the correction computed by the collision avoidance algorithm. The goal is then to calculate $\alpha_{\mathrm{CBF}}$ and $\delta_{\mathrm{CBF}}$ given $\alpha_0$ and $\delta_0$ so that the constraints are not violated.

### 4.3.1 Provisional Selection of Decoupled CBF

The algorithm has two primary mechanisms for determining whether to use CBF constraints for a target vehicle. The first mechanism is to use the below constraints to determine whether a target vehicle is a threat to ego when only using the nominal acceleration and front wheel angle proposed by the DRL Agent

$$
\begin{aligned}
C_x &= \begin{cases} \dot{h}_{x,\mathrm{F}}(\alpha_0) + l_{0,x} h_{x,\mathrm{F}} \geq 0, & \text{if } x_{\mathrm{T}} \geq 0 \\ \ddot{h}_{x,\mathrm{R}}(\alpha_0) + l_{1,x} \dot{h}_{x,\mathrm{R}} + l_{0,x} h_{x,\mathrm{R}} \geq 0, & \text{otherwise} \end{cases} \\
C_y &= \ddot{h}_y(\alpha_0, \delta_0, 0) + l_{1,y} \dot{h}_y(0) + l_{0,y} h_y(0) \geq 0.
\end{aligned} \tag{41}
$$

These constraints arise out of the longitudinal and lateral CBF. We use the first order longitudinal barrier constraint if $x_{\mathrm{T}} \geq 0$ and the second order barrier constraint if $x_{\mathrm{T}} < 0$ because $h_{x,\mathrm{F}}$ is relative degree 1 with respect to $\alpha$ and $h_{x,\mathrm{R}}$ is relative degree 2. In (41), the threat assessment is performed with $\alpha_0$, $\delta_0$, and $c_b = 0$.

Once a vehicle has been determined as a threat, the next step is to determine whether to brake, steer, or to do both. This is determined by the relative distance and velocity of the obstacle as shown in Fig. 16. The $x$-axis shows the relative velocity and the $y$-axis shows the distance from the target. To arbitrate between steering and braking to avoid a collision, we use the the physical limits of the vehicle to determine whether to steer around and obstacle or to brake ahead of it. The minimum distances at which the ego vehicle can brake ahead of a target $d_{\mathrm{b}}$ and at which the ego vehicle can steer around a target $d_{\mathrm{s}}$ are given by

$$
d_{\mathrm{b}} = -\frac{v_{\mathrm{R}}^2}{2\mathrm{dec}_{\max}}, \quad d_{\mathrm{s}} = \sqrt{\frac{2d_{y,\min}}{a_{y,\max}}} v_{\mathrm{R}},
$$

where $v_{\mathrm{R}} = v_{\mathrm{H}} - v_{\mathrm{T}}$, $\mathrm{dec}_{\max}$ is the maximum deceleration and $a_{y,\max}$ is the maximum lateral acceleration of the ego vehicle. We define $v_{\mathrm{crit}}$ as the relative velocity where $d_{\mathrm{b}} = d_{\mathrm{s}}$. In the green shaded area we choose to brake, in the blue shaded area we choose to steer, and in the red shaded area we choose to do both. The boundary between the green and blue regions is determined by $v_{\mathrm{crit}}$. The second mechanism that determines whether to use CBF constraints is based on whether target vehicle is in a lane adjacent to the ego vehicle and within a predefined longitudinal threshold distance. For vehicles that are in either:

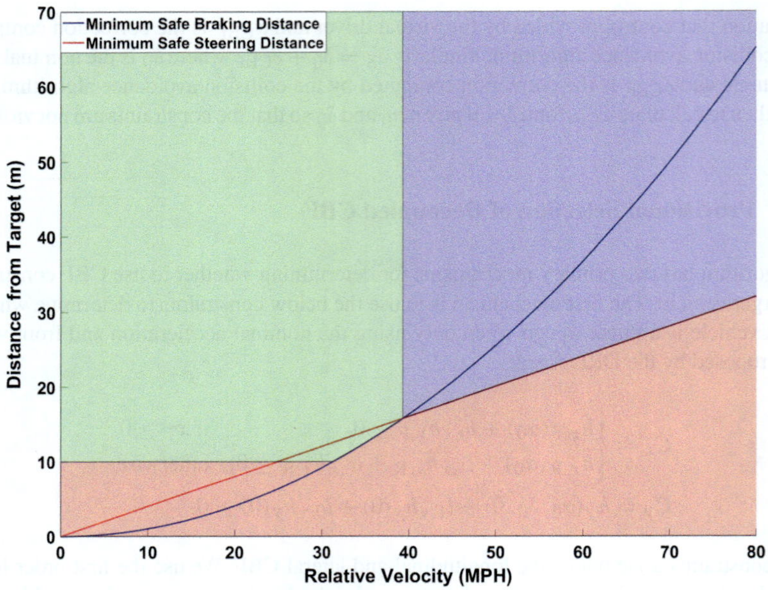


**Fig. 16** Initial selection of decoupled CBF

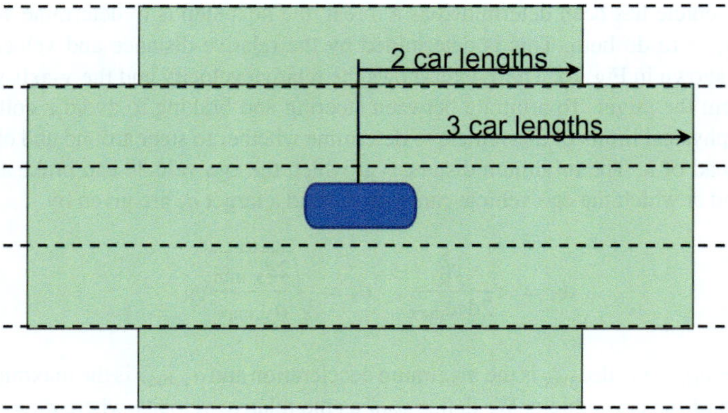


**Fig. 17** Geometric interpretation of preemptive lateral barriers

- the ego lane
- the lane immediately adjacent to the ego lane and within 3 car lengths from the center of the ego vehicle
- the lane two lanes adjacent to the ego lane and within 2 car lengths from the center of the ego vehicle.

We preemptively apply the lateral barriers. Figure 17 shows a geometric interpretation of the logic. If the center of a target vehicle falls within the shaded area, the appropriate lateral barrier is enforced.

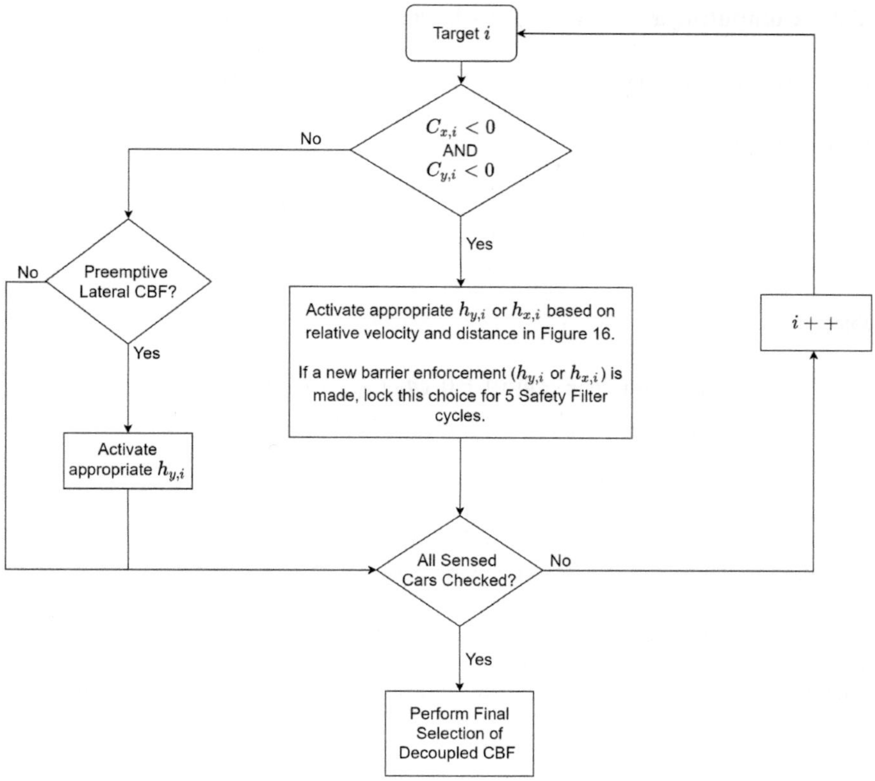


**Fig. 18** Initial selection of decoupled CBF

Using these two mechanisms, the barriers to be enforced are provisionally selected, as shown in Fig. 18. The process shown in Fig. 18 is performed for the closest front and rear targets in each lane. By considering all the vehicles nearby the ego, the collision avoidance algorithm can prevent a broad range of collisions and is not limited to forward collision avoidance. This barrier selection is provisional because in the case the selection cannot be enforced due to conflicting constraints, the algorithm systematically replaces lateral constraints with longitudinal constraints until a feasible solution is found.

Once the barriers have been provisionally selected, we designate the threat (i.e. barrier enforced through the threat assessment mechanism) with the minimum longitudinal distance as the *primary obstacle* $i_{\mathrm{PO}}$. For the primary obstacle, we provisionally activate both the left and right lateral barriers. This results in 2 distinct options, going to the right or the left of the primary obstacle. For all other lateral barriers, we choose the left or right barrier based on the location of the target w.r.t. the ego vehicle.

### 4.3.2 Computing $\alpha_{\text{CBF}}$ and $\delta_{\text{CBF}}$ with Quadratic Programs

We calculate the safe longitudinal acceleration $\alpha_{\text{S}} = \alpha_{\text{CBF}} + \alpha_0$ and the safe front wheel angle $\delta_{\text{S}} = \delta_{\text{CBF}} + \delta_0$ using 3 quadratic programs, $\text{QP}_{y\text{L}}$, $\text{QP}_{y\text{R}}$ and $\text{QP}_x$. $QP_{y\text{L}}$ and QP $_{y\text{R}}$: We define $\text{QP}_{y\text{L}}$ as

$$\underset{\delta_{\text{CBF,L}}, s_{\text{RK}}, s_{\text{saT}}}{\arg\min} \quad \begin{bmatrix} \delta_{\text{CBF,L}} & s_{\text{RK}} & s_{\text{saT}} \end{bmatrix} Q_y \begin{bmatrix} \delta_{\text{CBF,L}} \\ s_{\text{RK}} \\ s_{\text{saT}} \end{bmatrix}, \tag{42}$$

subject to:

$$\begin{aligned} \ddot{h}_{y,\text{L},i_{\text{PO}}}(\alpha_0, \delta_0 + \delta_{\text{CBF,L}}) + l_{1,y}\dot{h}_{y,\text{L},i_{\text{PO}}} + l_{0,y}h_{y,\text{L},i_{\text{PO}}} \geq 0, \\ \ddot{h}_{y,i}(\alpha_0, \delta_0 + \delta_{\text{CBF,L}}) + l_{1,y}\dot{h}_{y,i} + l_{0,y}h_{y,i} \geq 0, \quad \forall i \in \mathcal{Y}, \\ \ddot{h}_{\text{RK}}(\delta_0 + \delta_{\text{CBF,L}}) + l_{1,\text{RK}}\dot{h}_{\text{RK}} + l_{0,\text{RK}}h_{\text{RK}} + \begin{bmatrix} 1 \\ 1 \end{bmatrix} s_{\text{RK}} \geq 0, \\ \delta_{\min} - \delta_0 \leq \delta_{\text{CBF,L}} + s_{\text{saT}}, \\ \delta_{\text{CBF,L}} - s_{\text{saT}} \leq \delta_{\max} - \delta_0, \end{aligned} \tag{43}$$

where $\mathcal{Y}$ is the set of target ID's for which the lateral barrier is enforced excluding the primary obstacle, and $l_{1,y}$, $l_{0,y}$, $l_{1,\text{RK}}$, and $l_{0,\text{RK}}$ are tunable parameters that determine the sensitivity of the barriers. We define $J_{y\text{L}}$ as the optimal quadratic cost of $\text{QP}_{y\text{L}}$. Similarly, we define $\text{QP}_{y\text{R}}$ as

$$\underset{\delta_{\text{CBF,R}}, s_{\text{RK}}, s_{\text{saT}}}{\arg\min} \quad = \begin{bmatrix} \delta_{\text{CBF,R}} & s_{\text{RK}} & s_{\text{saT}} \end{bmatrix} Q_y \begin{bmatrix} \delta_{\text{CBF,R}} \\ s_{\text{RK}} \\ s_{\text{saT}} \end{bmatrix}, \tag{44}$$

subject to:

$$\begin{aligned} \ddot{h}_{y,\text{R},i_{\text{PO}}}(\alpha_0, \delta_0 + \delta_{\text{CBF,R}}) + l_{1,y}\dot{h}_{y,\text{R},i_{\text{PO}}} + l_{0,y}h_{y,\text{R},i_{\text{PO}}} \geq 0, \\ \ddot{h}_{y,i}(\alpha_0, \delta_0 + \delta_{\text{CBF,R}}) + l_{1,y}\dot{h}_{y,i} + l_{0,y}h_{y,i} \geq 0, \quad \forall i \in \mathcal{Y}, \\ \ddot{h}_{\text{RK}}(\delta_0 + \delta_{\text{CBF,R}}) + l_{1,\text{RK}}\dot{h}_{\text{RK}} + l_{0,\text{RK}}h_{\text{RK}} + \begin{bmatrix} 1 \\ 1 \end{bmatrix} s_{\text{RK}} \geq 0, \\ \delta_{\min} - \delta_0 \leq \delta_{\text{CBF,R}} + s_{\text{saT}}, \\ \delta_{\text{CBF,R}} - s_{\text{saT}} \leq \delta_{\max} - \delta_0, \end{aligned} \tag{45}$$

and we define $J_{y\text{R}}$ as the optimal quadratic cost of $\text{QP}_{y\text{R}}$. The distinction between $\text{QP}_{y\text{L}}$ and $\text{QP}_{y\text{R}}$ is that in $\text{QP}_{y\text{L}}$, the left barrier is enforced for the primary obstacle and in $\text{QP}_{y\text{R}}$ the right barrier is enforced for the primary obstacle. The slack variable $s_{\text{RK}}$ allows the algorithm to momentarily leave the road surface if such an action is necessary to avoid a collision and return when it is safe to do so. Similarly, $s_{\text{saT}}$ allows the algorithm to find a solution to the QPs even in the event of actuator saturation.

*Final Selection of Lateral CBF*

The procedure used for the final selection of the CBFs is shown in Fig. 19. After the provisional selection of the CBFs, we attempt to solve $\text{QP}_{y\text{L}}$ and $\text{QP}_{y\text{R}}$. We define one or both $\text{QP}_{y\text{L}}$ and $\text{QP}_{y\text{R}}$ as infeasible if any of the lateral constraints conflict (one lateral constraint requires $\delta > 0$ and another requires $\delta < 0$). If either $\text{QP}_{y\text{L}}$ or $\text{QP}_{y\text{R}}$ is infeasible, we replace the provisional $h_{y,i}$ of the target with the lowest value of $C_{x,i}$ with the longitudinal CBF $h_{x,i}$, and resolve. This procedure is repeated until either a solution is found or all the provisional lateral barriers have been replaced by a longitudinal barrier. Note that both $\text{QP}_{y\text{L}}$ and $\text{QP}_{y\text{R}}$ may still be infeasible after all provisional lateral barriers have been replaced. Once $\text{QP}_{y\text{L}}$ and $\text{QP}_{y\text{R}}$ have either been solved or deemed infeasible, we select $\delta_{\text{CBF}}$ based on the following criteria.

- If $\text{QP}_{y\text{L}}$ is infeasible, $\delta_{\text{CBF}} = \delta_{\text{CBF,R}}$
- If $\text{QP}_{y\text{R}}$ is infeasible, $\delta_{\text{CBF}} = \delta_{\text{CBF,L}}$
- If $\text{QP}_{y\text{L}}$ and $\text{QP}_{y\text{R}}$ are both feasible,
  - if $|J_{y\text{L}} - J_{y\text{R}}| < 1e-5$, $\delta_{\text{CBF}} = \delta_{\text{CBF,L}}$,
  - else if $J_{y\text{L}} < J_{y\text{R}}$ AND ($\delta_{\text{CBF,R}}$ was NOT selected in the last step OR $J_{y\text{L}} < \frac{J_{y\text{R}}}{2}$), $\delta_{\text{CBF}} = \delta_{\text{CBF,L}}$,
  - else if $J_{y\text{R}} < J_{y\text{L}}$ AND ($\delta_{\text{CBF,L}}$ was NOT selected in the last step OR $J_{y\text{R}} < \frac{J_{y\text{L}}}{2}$), $\delta_{\text{CBF}} = \delta_{\text{CBF,R}}$,
  - else if $\delta_{\text{CBF,L}}$ was selected in the last step, $\delta_{\text{CBF}} = \delta_{\text{CBF,L}}$,
  - else if $\delta_{\text{CBF,R}}$ was selected in the last step, $\delta_{\text{CBF}} = \delta_{\text{CBF,R}}$,
- If $\text{QP}_{y\text{L}}$ and $\text{QP}_{y\text{R}}$ are both infeasible, $\delta_{\text{CBF}}$ is infeasible. Skip 4.3.2 and perform max braking.

The logic above chooses the cheaper option and also prevents switching between choosing left or right in sequential steps. $\delta_{\text{CBF}}$ is saturated after its computation by the dual QPs.

*$QP_x$*: We solve $\text{QP}_x$ defined below, and saturate $\alpha_{\text{CBF}}$ based on the actuation limits of the ego vehicle.

$$\alpha_{\text{CBF}} = \arg\min_{\alpha_{\text{CBF}}} \alpha_{\text{CBF}}^2, \tag{46}$$

subject to:

$$\dot{h}_{x,i}(\alpha_0 + \alpha_{\text{CBF}}) + l_{0,x} h_{x,i}, \geq 0 \quad \forall i \in \mathcal{X}, \tag{47}$$

where $\mathcal{X}$ is the set of target ID's for which the longitudinal barrier is enforced and $l_{0,x}$ is a tunable parameter that determine the sensitivity of the barriers. For each target $i$, we use $h_{x,\text{F}}$ if $x_\text{T} >= 0$ and $h_{x,\text{R}}$ if $x_\text{T} < 0$. If $\text{QP}_x$ is infeasible due to conflicting constraints, we remove the conflicting rear constraints and resolve.

## 4.4 Examples

In all examples shown in this section, the parameters are set to the values shown in Table 1.

In the position plots shown in (Figs. 20, 23 and 26), the green vehicle is the ego with the collision avoidance activated and the dotted pink is the ego vehicle without the collision avoidance algorithm. All target vehicles are shown in blue. The dotted red lines show the active constraints at that time step.

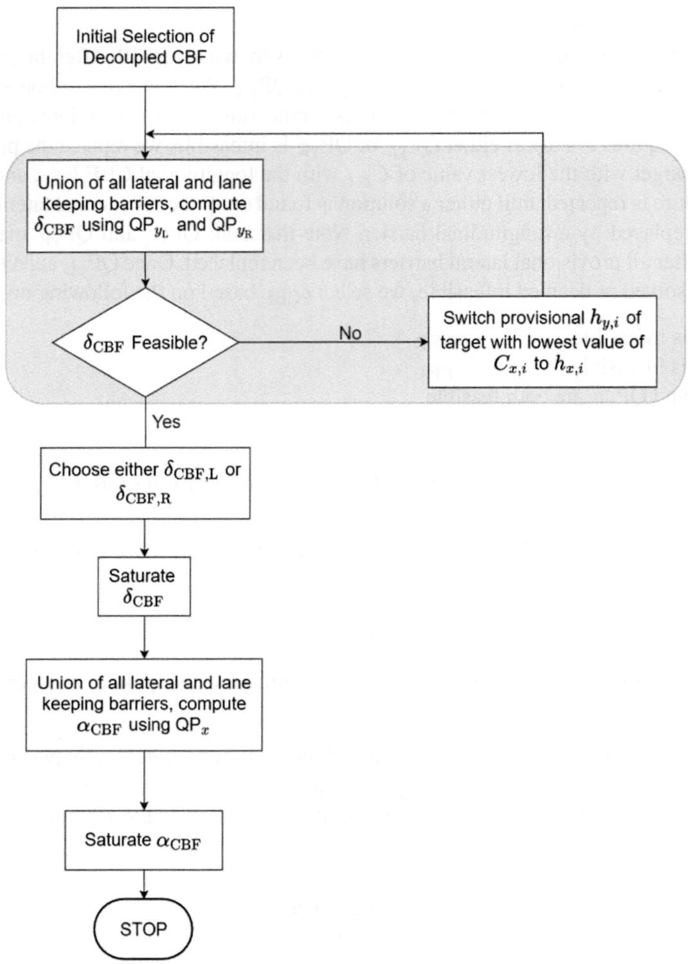


**Fig. 19** Final selection of decoupled CBF. The steps in the shaded area are performed for both $\text{QP}_{y\text{L}}$ and $\text{QP}_{y\text{R}}$

**Table 1** Tuning parameters

| Parameter | $k_v$(s) | $d_{x,\min}$(m) | $d_{y,\min}$(m) | $c_{\text{b}}$ | $l_{1,y}$ | $l_{1,y}$ | $l_{1,\text{RK}}$ | $l_{0,\text{RK}}$ | $l_{0,x}$ |
|---|---|---|---|---|---|---|---|---|---|
| Value | 1 | 6 | 3.15 | 0.0025 | 7 | 10 | 7 | 10 | $2\sqrt{\frac{0.4g}{\lvert x_{\text{T}}\rvert}}$ |

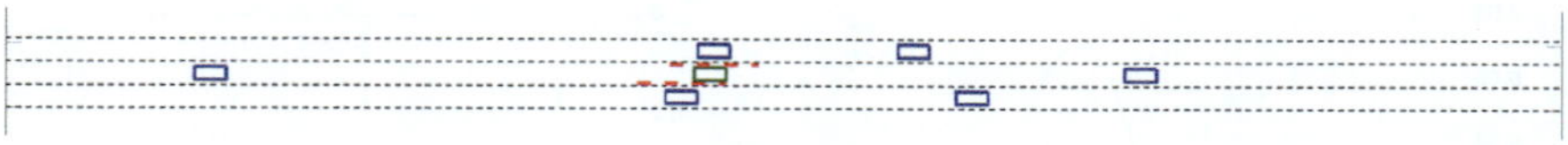

(a) $t = 0$ sec

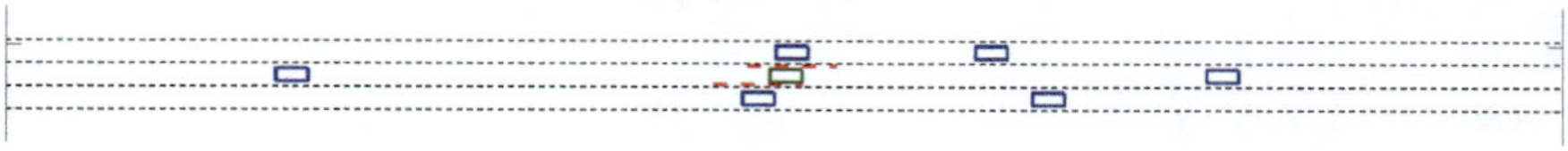

(b) $t = 0.5$ sec

(c) $t = 1$ sec

(d) $t = 1.5$ sec

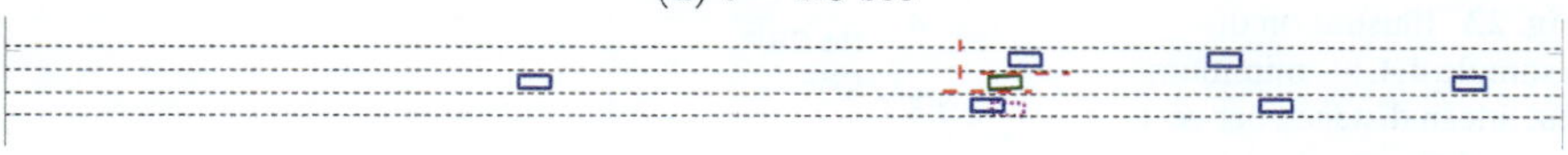

(e) $t = 2$ sec

(f) $t = 2.5$ sec

(g) $t = 3$ sec

**Fig. 20** Illustration of Example 4.4.1—position of the ego vehicle and the target vehicles at 0.5 s intervals

### 4.4.1 Aggressive/Erratic Ego Cut-in Prevention

In this example, the ego vehicle attempts to change lanes while a vehicle is in its blind spot. Figure 20 shows the position of the ego vehicle and the target vehicles at 0.5 s intervals. Figure 21a, b show the front wheel angle and acceleration of the ego vehicle and Fig. 22 shows the minimum Euclidean distance between the ego and the critical target vehicle (i.e. the one threatening to collide with the ego vehicle) edges. The CBF safety filter corrected the nominal steering actuation and prevented a potentially erratic right lane change—which is shown by the dashed-red rectangle in Fig. 20d–g.Similarly, it also prevented the ego vehicle from making a left lane change (shown by the dotted red line constraint).

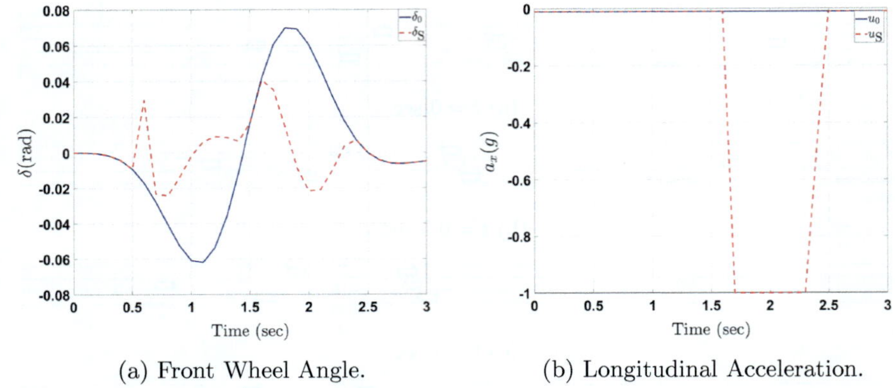


**Fig. 21** Illustration of Example 4.4.1—front wheel angle and longitudinal acceleration

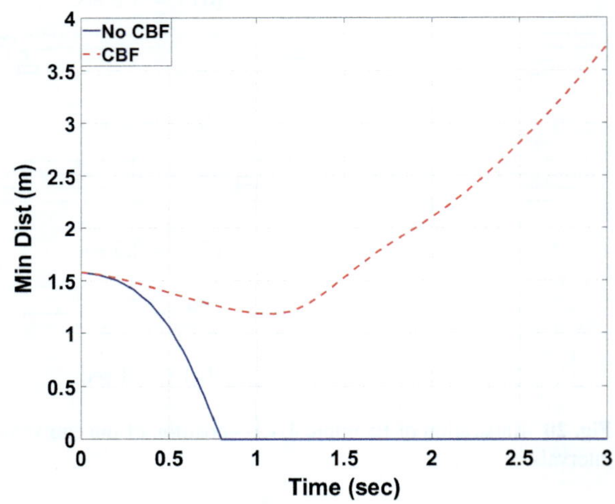


**Fig. 22** Illustration of Example 4.4.1—minimum Euclidean distance between ego and target vehicles

### 4.4.2 Target Cut-in Evasion

In this example, the ego vehicle evades a target vehicle that cuts-in ahead. Figure 23 shows the position of the ego vehicle and the target vehicles at 0.5 s intervals. Figure 24a, b show the front wheel angle and acceleration of the ego vehicle and Figure 25 shows the minimum Euclidean distance between the ego and the critical target vehicle edges. The safety filter avoids the vehicle cutting in by changing lanes, and also prevents the ego vehicle from driving off the highway due to the road keeping barriers.

### 4.4.3 Avoiding a Stationary Vehicle Ahead

In this example, the ego vehicle evades a potential collision with an obstacle that is stationary ahead. Figure 26 shows the position of the ego vehicle and the target vehicles (including the

(a) $t = 0$ sec

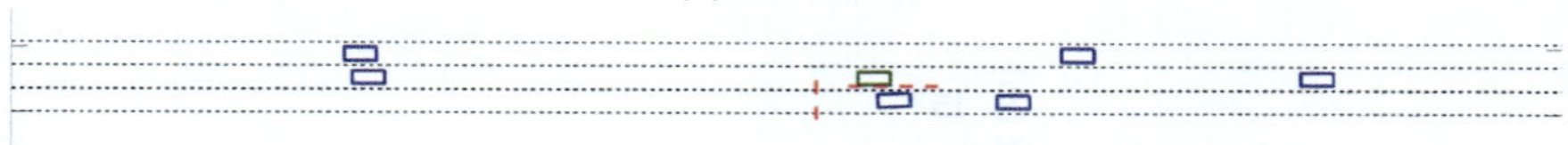

(b) $t = 0.5$ sec

(c) $t = 1$ sec

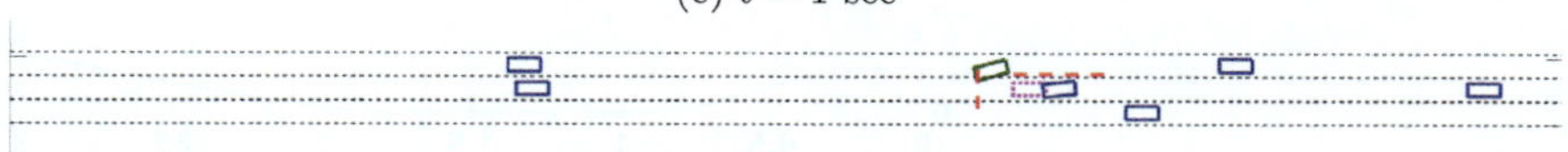

(d) $t = 1.5$ sec

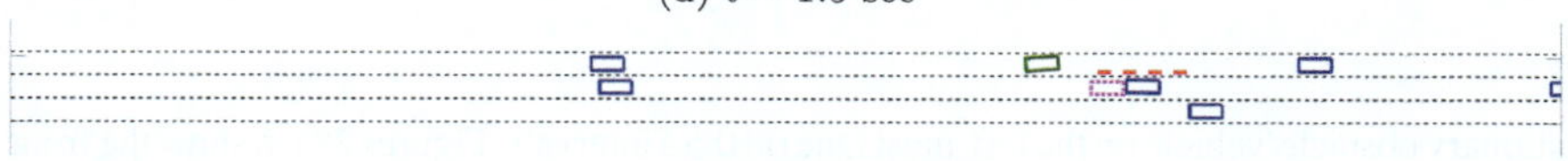

(e) $t = 2$ sec

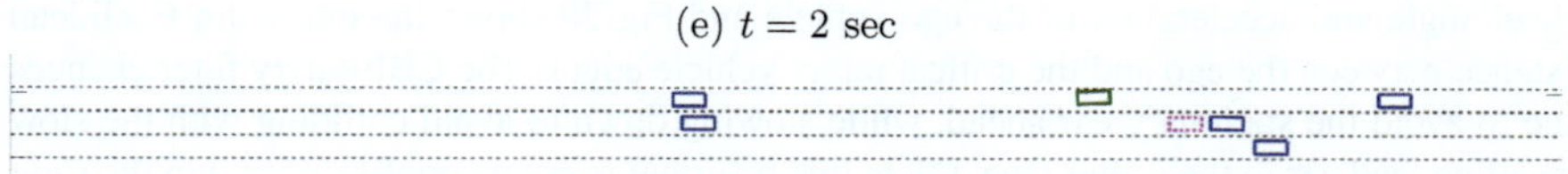

(f) $t = 2.5$ sec

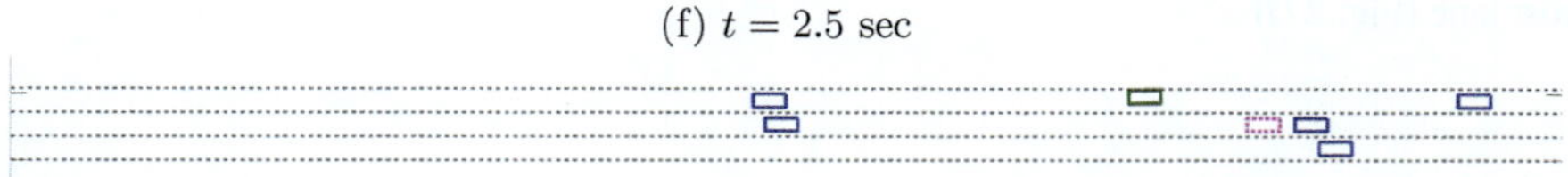

(g) $t = 3$ sec

**Fig. 23** Illustration of Example 4.4.2—position of the ego vehicle and the target vehicles at 0.5 s intervals

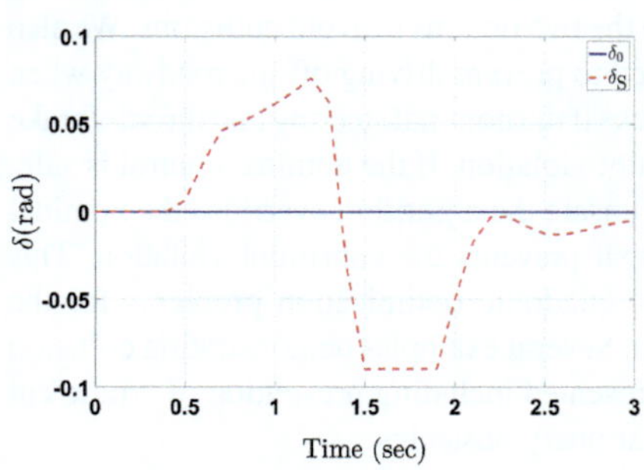


(a) Front Wheel Angle.

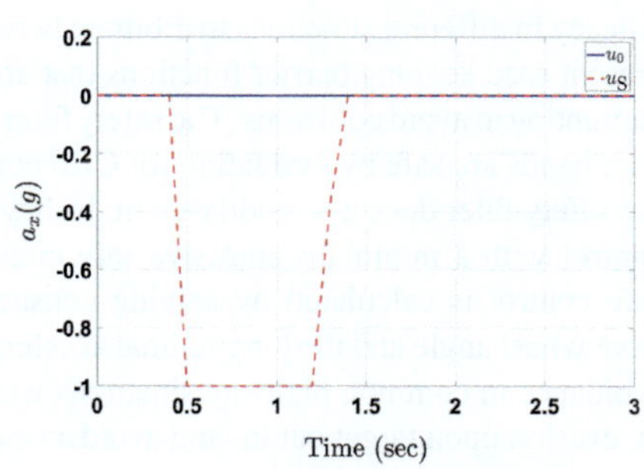


(b) Longitudinal Acceleration.

**Fig. 24** Illustration of Example 4.4.2—front wheel angleand longitudinal acceleration

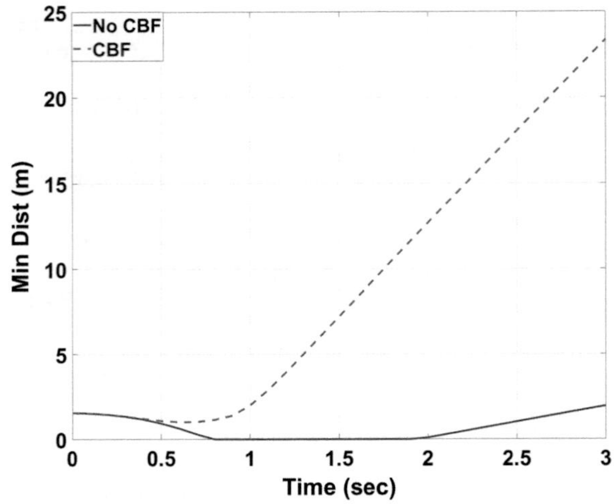


**Fig. 25** Illustration of Example 4.4.2—minimum Euclidean distance between ego and target vehicles

stationary obstacle/vehicle on the left-most lane) at 0.5 s intervals. Figures 28a, b show the front wheel angle and acceleration of the ego vehicle and Fig. 28 shows the minimum Euclidean distance between the ego and the critical target vehicle edges. The CBF safety filter changes lane to avoid the stationary car ahead, while slowing down to avoid colliding with the slow travelling lead car in the center lane. (Note that it further prevents erratic cut-in into the right most lane (Fig. 27))

## *4.5 Summary*

In this section, we introduced decoupled longitudinal and lateral Control Barrier Functions for collision avoidance, and a safety filter that utilizes contextual knowledge of steering/braking efficacy in different situations to arbitrate between the two options to avoid collisions. We also present road keeping barrier functions that are used to prevent driving off the roadway when attempting to avoid collisions. The safety filter checks if the nominal steering and throttle/brake commands are safe by evaluating for CBF constraint violation. If the nominal control is safe, the safety filter does not modify them. If they are unsafe the algorithm overrides the nominal control with a minimally intrusive safe control that prevents the constraint violation. This safe control is calculated by solving constrained quadratic optimization problems for the front wheel angle and the longitudinal acceleration. Several examples demonstrating collision avoidance in common highway situations were presented including prevention of erratic cut in, evasion upon target cut in, and avoidance of stationary obstacles.

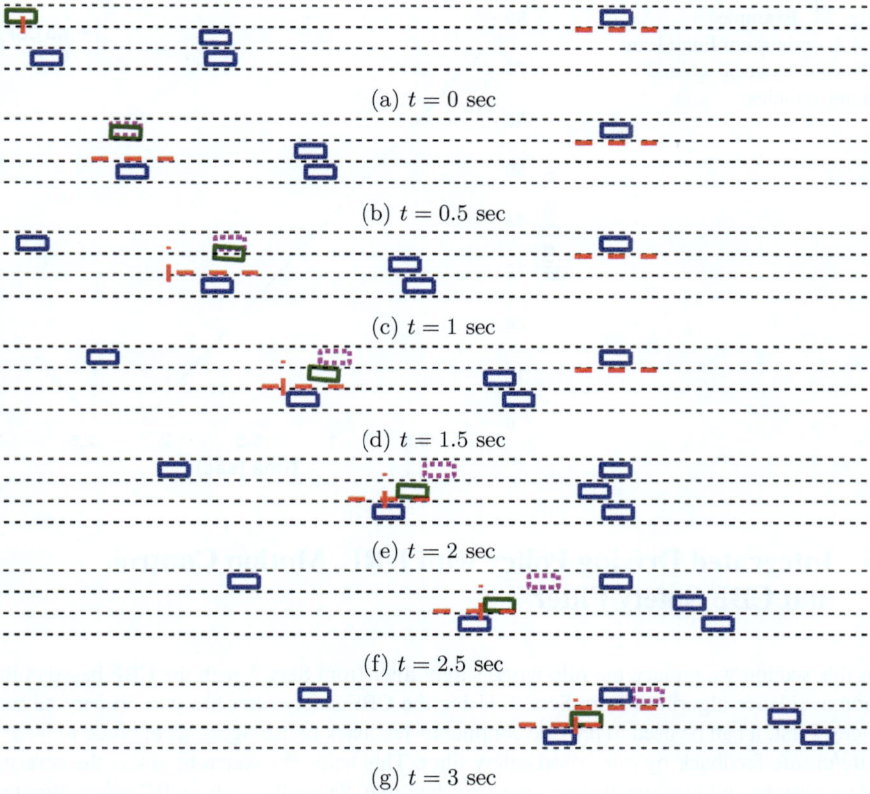


**Fig. 26** Example 4.4.3—position of the ego vehicle and the target vehicles at 0.5 s intervals

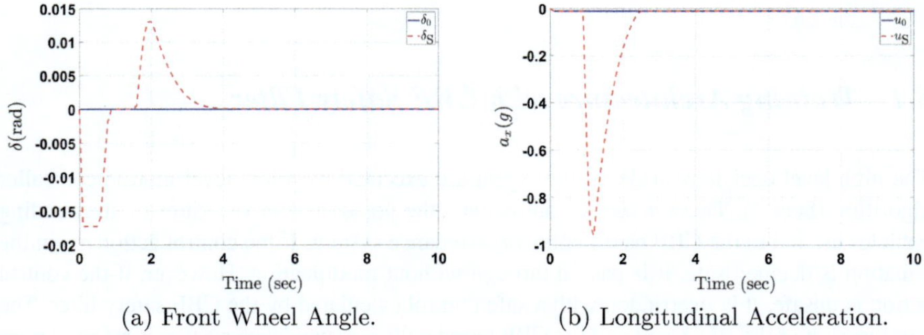


**Fig. 27** Example 4.4.3—front wheel angleand longitudinal acceleration

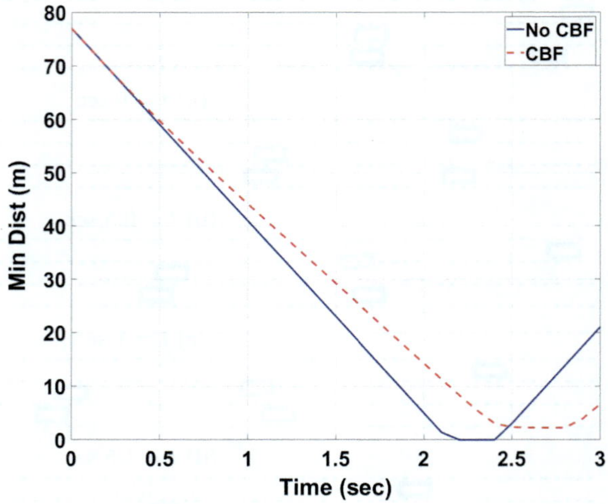


**Fig. 28** Example 4.4.3—minimum Euclidean distance between ego and target vehicles

# 5 Integrated Driving Policy with DRL, Motion Control, and CBF Safety Filter

In this section we replace the rule-based safety filter from Sect. 2 with the CBF based collision avoidance algorithm from Sect. 4. Using the CBF safety filter has two primary advantages. First, it can be used to provide continuous feedback to the Agent, as opposed to binary safe/unsafe feedback by rule-based safety filter. This helps the Agent to assess the severity of its mistake and prevents it from repeating the same. Secondly, with a CBF safety filter the Agent can adapt to different environments after its initial training without endangering the occupants of the AV or nearby motorists.

## *5.1 Training Architecture with CBF Safety Filter*

The high level decisions made by the Agent are executed by a low level motion controller algorithm (Sect. 3). These *nominal* controls and the positions and velocities of surrounding vehicles are fed to the CBF based collision avoidance system. If the control action given the situation is deemed safe, it is passed through without modification. However, if the control action is unsafe, it is overridden with a safe control calculated by the CBF safety filter. The combination of the RL Agent and the CBF based collision avoidance system is referred to as the Autonomous Driver. Figure 29 shows the system used for training the Agent. Similar to Sect. 2, sensors provide affordance indicators to the Agent, which then makes a decision in the form of a high level action. The high level action is then translated to a front wheel angle, throttle and brake command by a motion controller algorithm. These low level commands and affordance indicators are then provided to the CBF Safety Filter. The safety filter computes the safe front wheel angle and throttle/brake commands. If the original path is safe, these commands are unchanged. The safe front wheel angle and throttle/brake commands are then applied to the ego vehicle. The Action Translator block translates the safe low level commands

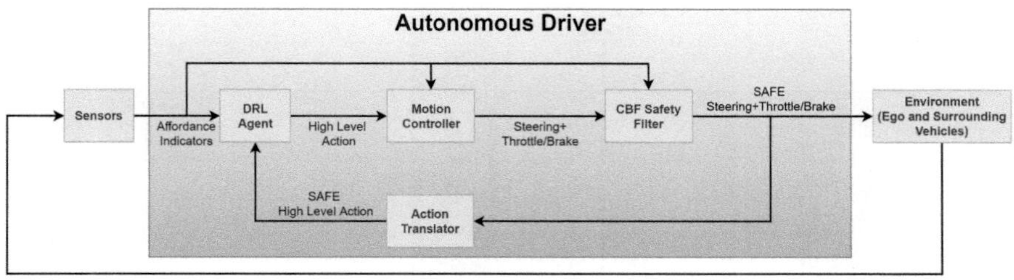


**Fig. 29** Training architecture

back into a safe high level action. This is used to provide feedback to the Agent about the action it has taken. If the safe action differs from the action selected by the Agent, it is penalized. In this work, the Agent makes a decision every second, and the motion controller and CBF updates the front wheel angle and throttle/brake every 10th of a second.

### 5.1.1 Action Translation and Penalizing Unsafe Actions

To provide effective feedback to the DRL Agent, the continuous safe front wheel angle and longitudinal acceleration signals must be translated back into the high level action space. For the longitudinal acceleration, this is straightforward, since both the high level action and the continuous signal have the same units (acceleration in $g$'s). For the front wheel angle $\delta$, we use a Barrier Function based intent prediction [40] to determine if the steering action is trying to prevent a lane change or triggering one.

We use the same reward function components (11)–(13) as in Sect. 2, with an additional component. The addition is a safety component that determines how safe the action is by comparing it to the safe action output by the CBF

$$r_s = f_x(a_x, \bar{a}_x) + f_y(a_y, \bar{a}_y), \tag{48}$$

where $a_x$ is the longitudinal action selected by the Agent, $\bar{a}_x$ is the safe longitudinal action by the CBF filter, $a_y$ is the lateral action selected by the Agent, $\bar{a}_y$ is the safe lateral action by the CBF filter, and $f_x$ and $f_y$ are functions that determine the size of the penalty for unsafe longitudinal and lateral actions respectively. The size penalty on unsafe actions depends on how different the unsafe action is compared to the safe action, and also how long the decision chosen by the Agent was deemed unsafe. Figures 30 and 31 show qualitatively how the size of the penalty is determined. In Fig. 30a, the RL action is to accelerate at $0.2g$, but the safe (determined by CBF Safety filter) action is to brake hard for nearly the entire duration of the RL Action. Such a case incurs a large penalty. In Fig. 30b, the RL action is again to accelerate at $0.2g$. In this case, the safe action is to brake near the end of the RL Action, and so this decision incurs a smaller penalty. Similarly, in Fig. 31a, the RL action is to maintain the lane. However, at the 6th time step, the safety filter must perform a left lane change in order to avoid a collision. The time duration at which RL action becomes unsafe determines how large the penalty is.

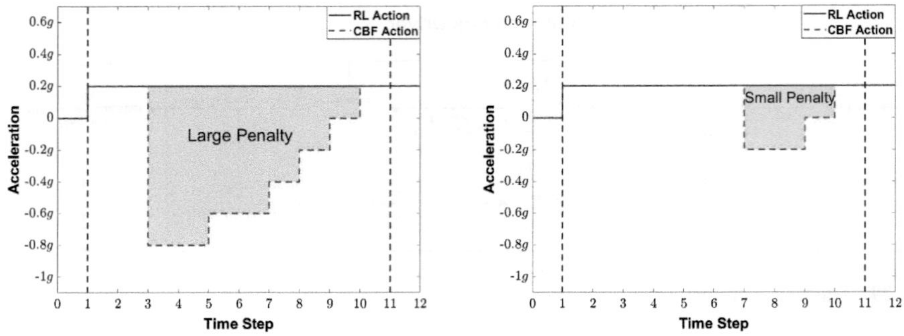


(a) Example large longitudinal penalty. (b) Example small longitudinal penalty.

**Fig. 30** Determination of penalty for unsafe longitudinal actions

### 5.1.2 Training Results

To train the Agent, we use episodes consisting of 200 s of highway driving. In each episode, the surrounding environment, i.e. the density, speed, location of the traffic is randomized. Figure 32 shows the training performance with and without CBF, and also using only the rule based safety filter from Sect. 2. Whilst learning how to drive without any safety filter, the Agent has many collisions initially, and does not learn how to drive safely. With the CBF, the time taken to learn to drive acceptable driving behavior is reduced significantly, the driving behavior is better as shown by the higher reward, and the driving behavior is safer, since CBF prevents collisions in case the Agent makes an unsafe decision. Without a safety filter, it is difficult to structure the reward function in a way that guides the Agent to drive safely and in a manner that is not exceedingly conservative.

Figure 33 shows the mean over 100 episodes of the number of safe Agent actions in each episode. In each episode, 20 actions are random exploration, so the maximum number of safe actions selected by the Agent is 180. The also shows that the Agent learns to drive safer as time progresses.

*Comparison of CBF Safety Filter Intervention with Rule-Based Safety Filter*

Figures 34 and 35 show the longitudinal and lateral interventions of the CBF and RB safety filters as the agent is trained. For both safety filters, the number of interventions and the severity of those interventions decrease over time, implying that the Agent is learning to be a safer driver, and since the reward is higher with the safety filter, this implies that the addition of the safety filter does not cause the Agent to become too conservative. The plots show that the longitudinal safety filter is more intrusive with its intervention, but this does not have an adverse effect on the reward. An advantage of using the CBF safety filter over the rule-based safety filter from Sect. 2 is that the CBF can be tuned to a desired degree of conservative behavior, whereas the rule-based safety filter can only choose between several predefined actions that may be either insufficient or unnecessarily excessive, which can reduce passenger confidence and also result in an uncomfortable ride.

*Collision avoidance with CBF Safety Filter Intervention with Rule-Based Safety Filter*

In this example, we show an use case where the intervention by the RB safety filter is insufficient. The AV and a target vehicle are driving at 70 mph. The RL policy initiates a lane change at $t = 0$ s. At $t = 2$ s, the black target vehicle decelerates at $0.35g$ for 3 s. Figure 36 shows the

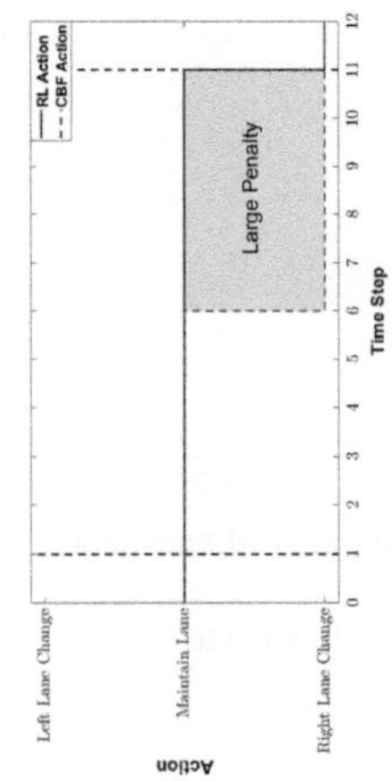


(a) Example large lateral penalty.

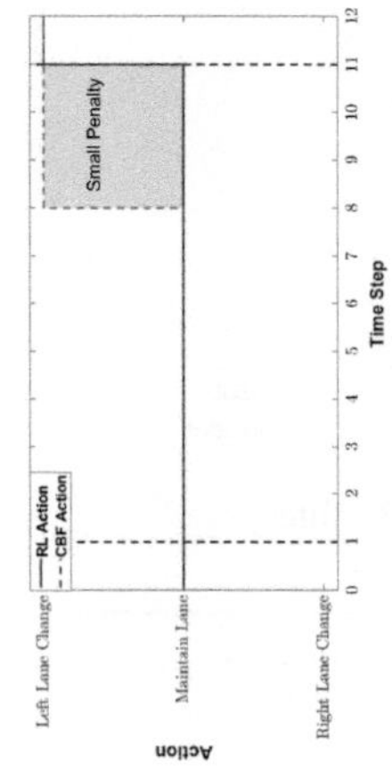


(b) Example small lateral penalty.

**Fig. 31** Determination of penalty for unsafe lateral actions

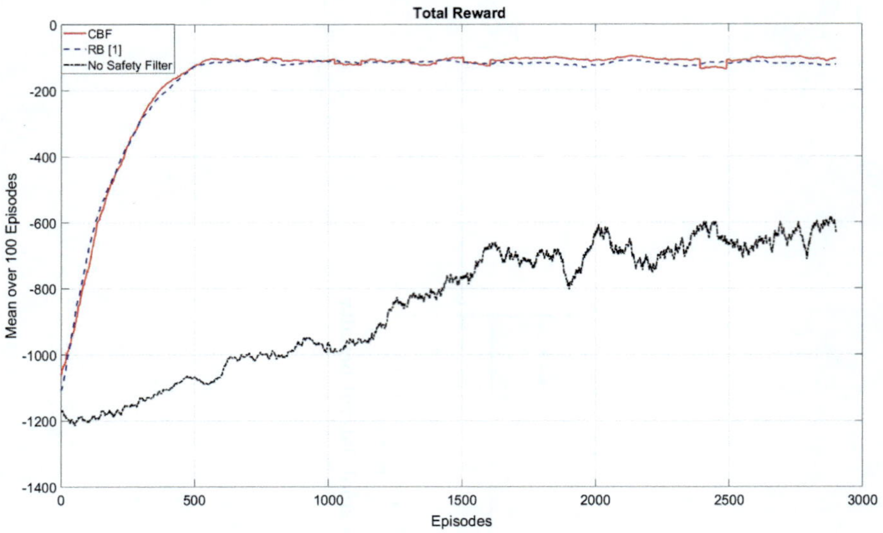


**Fig. 32** Training with and without the CBF filter

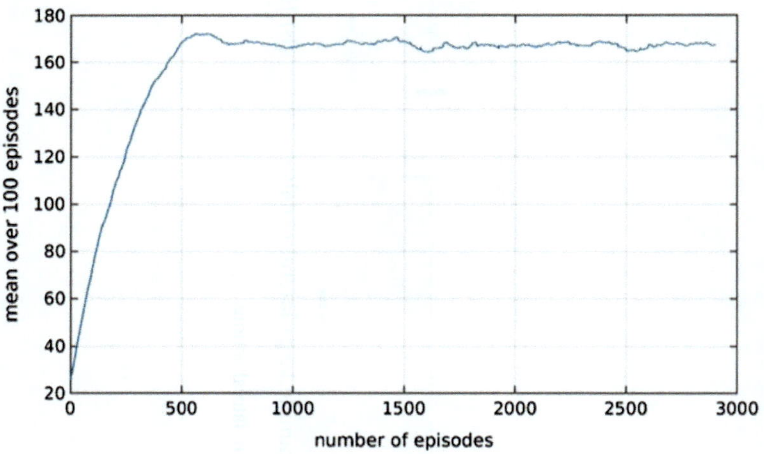


**Fig. 33** Number of safe agent actions in each episode

response of the RB safety filter and the CBF Safety Filter. The green vehicle is the AV with the CBF safety filter and the red vehicle is the AV with the RB safety filter. The RB safety filter reacts later to the decelerating target vehicle and is unable to avoid a collision.

*Retraining*

One of the key advantages of the proposed approach is that it can allow an already trained Agent to continue to adapt its driving policy once on the road. For example, acceptable driving behavior in different areas of a country are often slightly different. An Agent can potentially adapt in the field and optimize to its local conditions by retraining itself. The actual adaptation algorithm is a subject of ongoing research and is not discussed in this paper. However, without a

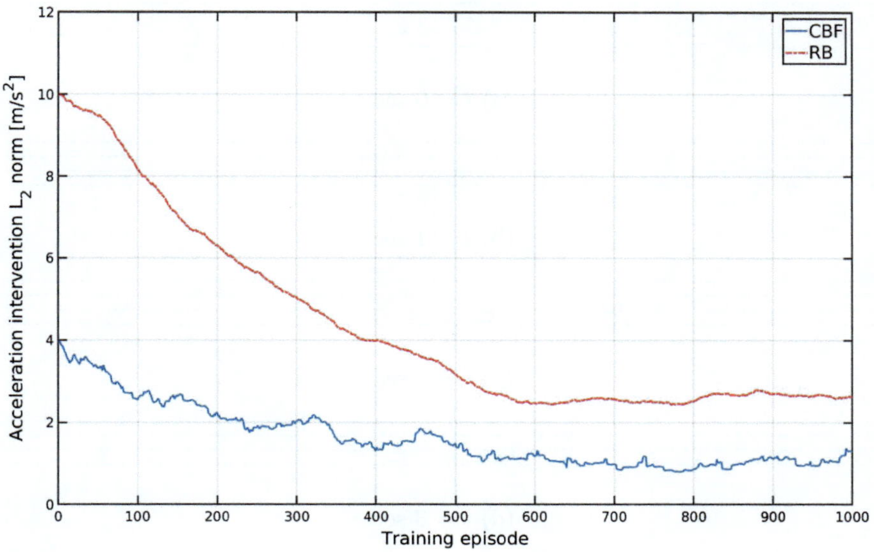


**Fig. 34** Comparison of CBF longitudinal intervention with RB safety filter

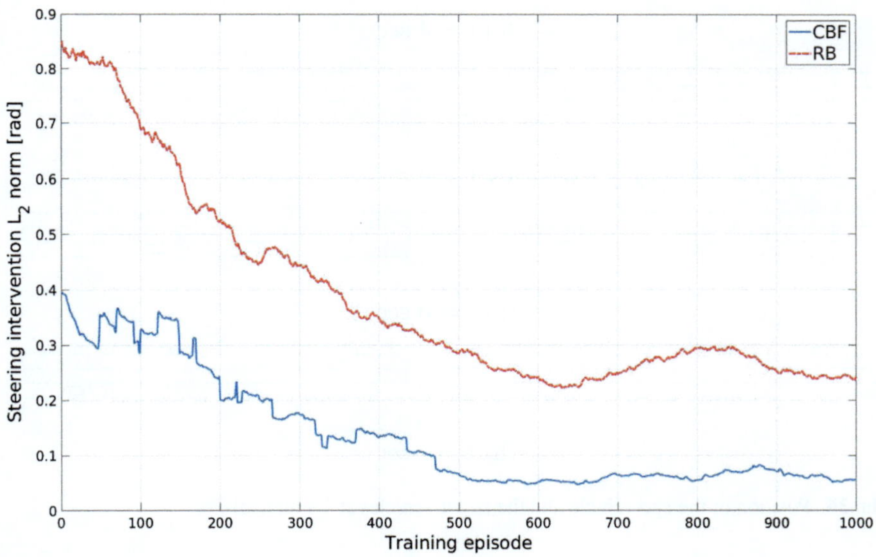


**Fig. 35** Comparison of CBF lateral intervention with RB safety filter

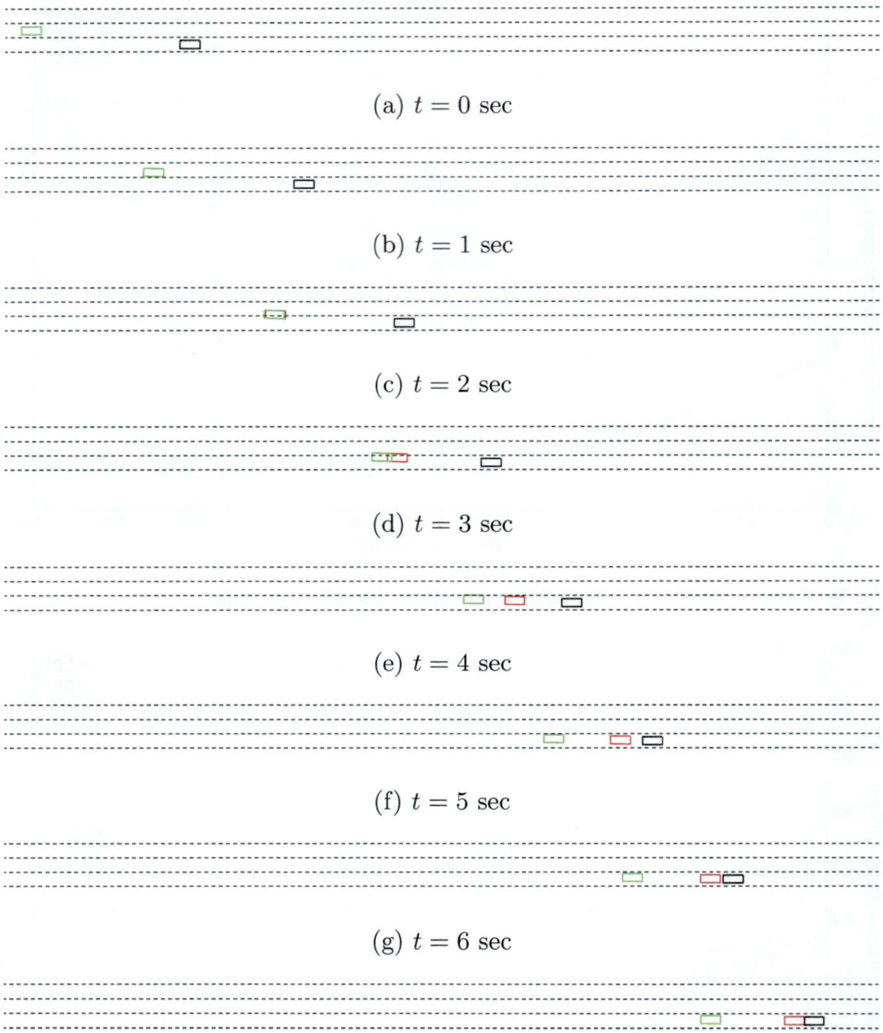

(a) $t = 0$ sec

(b) $t = 1$ sec

(c) $t = 2$ sec

(d) $t = 3$ sec

(e) $t = 4$ sec

(f) $t = 5$ sec

(g) $t = 6$ sec

(h) $t = 7$ sec

**Fig. 36** Position of the ego vehicle and the target vehicles at 1 s intervals

safety filter, an Agent making exploratory decisions may get itself into an unsafe situation. The benefit of the safety filter is twofold. If the Agent makes an unsafe decision during retraining, the decision is overridden and the safety of the AV is not compromised. Secondly, the Agent is provided immediate feedback that the decision was unsafe, therefore making it unlikely to make the same mistake again.

## 5.2 Summary

In this section we replaced the rule-based safety filter from Sect. 2 with the CBF based safety filter from Sect. 4. The amount of time taken for the Agent to learn a safe driving behavior is significantly reduced thanks to the efficient use of the Safety Filter. It overrides the agent's unsafe action and provides a safe alternative which acts as an additional feedback to the agent. To this end we modified the reward function from Sect. 2 to include an additional penalty based on the difference between the agent's nominal action and the safe action by the CBF. Training results show that the agent learns to drive safely and that the number of unsafe decisions reduces with training. The severity of the safety filter interventions also decrease as the training proceeds. The addition of the CBF safety filter helps to make the Autonomous Driver more robust to unsafe high level decisions and to aggressive traffic vehicles. In addition, the Autonomous Driver can adapt to new environments by learning online without forgoing on safety and comfort.

# 6 Summary and Conclusion

In this chapter we provide a practical approach to the design of RL-based driving policy for highway autonomous driving. Proposed driving policy system integrates a high level DRL-driven decision-making module with a path planning and path following strategy transforming the discrete DRL action space into a sequence of motion control commands. The benefits of the hierarchical decoupling of the RL decision logic from the algorithms for path formation and execution are the simplification of the RL algorithm design and training, and the opportunity to use versatile motion control algorithms. The latter facilitates the incorporation of engineering knowledge and experience in designing smooth human-like driving and safe lane change maneuvering. The engineers' field experience and knowledge includes the selection of lead target vehicle for longitudinal speed profile, the calibration of lateral acceleration profile and corresponding feedback control gains, the compliance of actuation constraints during lane change maneuvers, and the time required to perform a lane change. The chapter elaborates on the practical aspects of the design of the DRL decision-making strategy and motion control as critical components for real-world implementation of AI based decision policy for autonomous driving.

The chapter further focuses on addressing the overall robustness and safety of the output produced by the decision-making and motion control layers of the driving policy. We discuss the concept of a safety filter as an important means in autonomous driving applications of RL. The safety filter overrides the nominal front wheel angle and throttle/brake commands in case of imminent/unexpected threats while allowing the RL optimizer to focus on longer term goal. Two alternative safety filters defining the safety boundary of the produced control output are discussed. The first one uses handcrafted rules to constrain the DRL/motion control output within a predefined safety boundary. It accepts or rejects the RL decisions by enforcing traffic regulations and handcrafted rules for time headway and gap size requirement. The second solution introduces a CBF-based filter that provides a dynamic safety envelope safeguarding the control output along the vehicle trajectory. It adjusts the control actuation corresponding to RL decisions by leveraging the concept of invariant set to assess imminent threats and calculate minimal required adjustment (in steering/braking) for collision avoidance. The design and vehicle implementation of CBFs as a tool for collision avoidance and following road geometry constraints are discussed. The chapter concludes with a discussion on the pros and cons of both safety filters and their impact on the training and overall performance of the DRL algorithm, and the open opportunity of DRL—integrated with motion control and a safety filter—to adapt to new environments by learning online with corresponding safety and comfort assurance.

## References

1. Ackermann J, Bünte T (1999) Robust prevention of limit cycles for robustly decoupled car steering dynamics. Kybernetika 35(1):105–116
2. Alshiekh M, Bloem R, Ehlers R, Könighofer B, Niekum S, Topcu U (2017) Safe reinforcement learning via shielding. arXiv preprint arXiv:170808611
3. Ames AD, Xu X, Grizzle JW, Tabuada P (2016) Control barrier function based quadratic programs for safety critical systems. IEEE Trans Autom Control 62(8):3861–3876
4. Aradi S (2020) Survey of deep reinforcement learning for motion planning of autonomous vehicles. IEEE Trans Intell Transp Syst
5. Buehler M, Iagnemma K, Singh S (2007) The 2005 DARPA grand challenge: the great robot race, vol 36. Springer, Berlin
6. Buehler M, Iagnemma K, Singh S (2009) The DARPA urban challenge: autonomous vehicles in city traffic, vol 56. Springer, Berlin
7. Chen C, Seff A, Kornhauser A, Xiao J (2015) Deepdriving: Learning affordance for direct perception in autonomous driving. In: 2015 IEEE international conference on computer vision (ICCV). IEEE, pp 2722–2730
8. Coulter RC (1992) Implementation of the pure pursuit path tracking algorithm. Carnegie-Mellon UNIV Pittsburgh PA Robotics INST, Tech. rep
9. Duda H (1998) Flight control system design considering rate saturation. Aerosp Sci Technol 2(4):265–275
10. Erdmann J (2014) Lane-changing model in sumo. In: Proceedings of the SUMO2014 modeling mobility with open data, vol 24, pp 77–88
11. Falcone P, Tufo M, Borrelli F, Asgari J, Tseng HE (2007) A linear time varying model predictive control approach to the integrated vehicle dynamics control problem in autonomous systems. In: 2007 46th IEEE conference on decision and control. IEEE, pp 2980–2985
12. Frans K, Ho J, Chen X, Abbeel P, Schulman J (2017) Meta learning shared hierarchies. arXiv preprint arXiv:171009767
13. Garcia J, Fernández F (2012) Safe exploration of state and action spaces in reinforcement learning. J Artif Intell Res 45:515–564
14. Garcıa J, Fernández F (2015) A comprehensive survey on safe reinforcement learning. J Mach Learn Res 16(1):1437–1480
15. Giersiefer A, Dornhege J, Klein P, Klein C (2019) Driver assistance system. US Patent 20190071126A
16. Hecker S, Dai D, Van Gool L (2018) End-to-end learning of driving models with surround-view cameras and route planners. In: European conference on computer vision (ECCV)
17. Hill A, Raffin A, Ernestus M, Traore R, Dhariwal P, Hesse C, Klimov O, Nichol A, Plappert M, Radford A, Schulman J, Sidor S, Wu Y (2018) Stable baselines. https://github.com/hill-a/stable-baselines
18. Jankovic M (2018) Robust control barrier functions for constrained stabilization of nonlinear systems. Automatica 96:359–367
19. Jin IG, Avedisov SS, He CR, Qin WB, Sadeghpour M, Orosz G (2018) Experimental validation of connected automated vehicle design among human-driven vehicles. Transp Res Part C Emerg Technol 91:335–352
20. Kesting A, Treiber M (1999) Helbing D (2007) General lane-changing model mobil for car-following models. Transp Res Rec 1:86–94

21. Kim E, Kim J, Sunwoo M (2014) Model predictive control strategy for smooth path tracking of autonomous vehicles with steering actuator dynamics. Int J Automot Technol 15(7):1155–1164
22. Kingma DP, Ba J (2014) Adam: A method for stochastic optimization. arXiv preprint arXiv:14126980
23. Kreyszig E (1993) Advanced engineering mathematics. Wiley, New York, pp 482–488
24. Lee S, Tseng HE (2018) Trajectory planning with shadow trolleys for an autonomous vehicle on bending roads and switchbacks. In: 2018 IEEE intelligent vehicles symposium (IV). IEEE, pp 484–489
25. Li N, Oyler DW, Zhang M, Yildiz Y, Kolmanovsky I, Girard AR (2017) Game theoretic modeling of driver and vehicle interactions for verification and validation of autonomous vehicle control systems. IEEE Trans Control Syst Technol
26. Lillicrap TP, Hunt JJ, Pritzel A, Heess N, Erez T, Tassa Y, Silver D, Wierstra D (2015) Continuous control with deep reinforcement learning. arXiv preprint arXiv:150902971
27. Lin YC, Hong ZW, Liao YH, Shih ML, Liu MY, Sun M (2017) Tactics of adversarial attack on deep reinforcement learning agents. arXiv preprint arXiv:170306748
28. Ma X, Driggs-Campbell K, Kochenderfer MJ (2018) Improved robustness and safety for autonomous vehicle control with adversarial reinforcement learning. In: 2018 IEEE intelligent vehicles symposium (IV). IEEE, pp 1665–1671
29. Maas AL, Hannun AY, Ng AY (2013) Rectifier nonlinearities improve neural network acoustic models. In: Proceedings of ICML, vol 30, p 3
30. Minderhoud MM, Bovy PH (2001) Extended time-to-collision measures for road traffic safety assessment. Accid Anal Prev 33(1):89–97
31. Mirchevska B, Pek C, Werling M, Althoff M, Boedecker J (2018) High-level decision making for safe and reasonable autonomous lane changing using reinforcement learning. In: 2018 21st international conference on intelligent transportation systems (ITSC). IEEE, pp 2156–2162
32. Mnih V, Kavukcuoglu K, Silver D, Rusu AA, Veness J, Bellemare MG, Graves A, Riedmiller M, Fidjeland AK, Ostrovski G et al (2015) Human-level control through deep reinforcement learning. Nature 518(7540):529
33. Mnih V, Badia AP, Mirza M, Graves A, Lillicrap T, Harley T, Silver D, Kavukcuoglu K (2016) Asynchronous methods for deep reinforcement learning. In: International conference on machine learning, pp 1928–1937
34. Nageshrao S, Tseng HE, Filev D (2019) Autonomous highway driving using deep reinforcement learning. In: 2019 IEEE international conference on systems man and cybernetics (SMC). IEEE, pp 2326–2331
35. Nageshrao S, Tseng HE, Filev DP, Baker RL, Cruise C, Daehler L, Mohan S, Kusari A (2020) Vehicle adaptive learning. US Patent 10733510
36. Ngyuen Q, Sreenath K (2016) Exponential control barrier functions for enforcing high relative-degree safety-critical constraints. In: American control conference, pp 322–328
37. Papadimitriou I, Tomizuka M (2003) Fast lane changing computations using polynomials. In: Proceedings of the 2003 American control conference. IEEE, vol 1, pp 48–53
38. Pathak D, Agrawal P, Efros AA, Darrell T (2017) Curiosity-driven exploration by self-supervised prediction. In: International conference on machine learning, PMLR, pp 2778–2787
39. Prajna S, Jadbabaie A, Pappas GJ (2007) A framework for worst-case and stochastic safety verification using barrier certificates. Trans Autom Control 52(8):1415–1428
40. Rahman Y, Jankovic M, Santillo MA (2021) Driver intent prediction with barrier functions. In: American control conference
41. Rajamani R (2011) Vehicle dynamics and control. Springer Science & Business Media
42. Schaul T, Quan J, Antonoglou I, Silver D (2015) Prioritized experience replay. arXiv preprint arXiv:151105952
43. Silver D, Schrittwieser J, Simonyan K, Antonoglou I, Huang A, Guez A, Hubert T, Baker L, Lai M, Bolton A et al (2017) Mastering the game of go without human knowledge. Nature 550(7676):354
44. Slotine JJE, Li W et al (1991) Applied nonlinear control, vol 199. Prentice Hall, Englewood Cliffs

45. Sutton RS, Barto AG (1998) Reinforcement learning: an introduction, vol 1. MIT Press, Cambridge
46. Tamar A, Glassner Y, Mannor S (2015) Optimizing the cvar via sampling. In: Twenty-ninth AAAI conference on artificial intelligence
47. Thomaz AL, Breazeal C (2008) Teachable robots: understanding human teaching behavior to build more effective robot learners. Artif Intell 172(6–7):716–737
48. Toledo T, Zohar D (2007) Modeling duration of lane changes. Transp Res Rec J Transp Res Board 1999:71–78
49. Treiber M, Kesting A (2013) Traffic flow dynamics. Data, models and simulation. Springer, Berlin
50. Treiber M, Hennecke A, Helbing D (2000) Congested traffic states in empirical observations and microscopic simulations. Phys Rev E 62(2):1805
51. Tseng HE, Asgari J, Hrovat D, van Der Jagt P, Cherry A, Neads S (2002) Steering robot for evasive maneuvers-experiment and analysis. IFAC Proc 35(2):79–86
52. Vahidi A, Eskandarian A (2003) Research advances in intelligent collision avoidance and adaptive cruise control. IEEE Trans Intell Transp Syst 4(3):143–153
53. Van Hasselt H, Guez A, Silver D (2016) Deep reinforcement learning with double q-learning. AAAI 16:2094–2100
54. Vezzani G, Gupta A, Natale L, Abbeel P (2019) Learning latent state representation for speeding up exploration. arXiv preprint arXiv:190512621
55. Wang Y, Shen D, Teoh EK (2000) Lane detection using spline model. Pattern Recogn Lett 21(8):677–689
56. Werling M, Ziegler J, Kammel S, Thrun S (2010) Optimal trajectory generation for dynamic street scenarios in a frenet frame. In: 2010 IEEE international conference on robotics and automation. IEEE, pp 987–993
57. Wieland P, Allgöwer F (2007) Constructive safety using control barrier functions. IFAC Proc 40(12):462–467
58. Xiao L, Gao F (2010) A comprehensive review of the development of adaptive cruise control systems. Veh Syst Dyn 48(10):1167–1192
59. Xiao W, Belta C (2019) Control barrier functions for systems with high relative degree. In: IEEE 58th conference on decision and control (CDC), pp 474–479
60. Xu H, Gao Y, Yu F, Darrell T (2017) End-to-end learning of driving models from large-scale video datasets. In: Proceedings of the IEEE conference on computer vision and pattern recognition, pp 2174–2182
61. Ye F, Zhang S, Wang P, Chan CY (2021) A survey of deep reinforcement learning algorithms for motion planning and control of autonomous vehicles. arXiv preprint arXiv:210514218
62. Zhang M, Li N, Girard A, Kolmanovsky I (2017) A finite state machine based automated driving controller and its stochastic optimization. In: ASME 2017 dynamic systems and control conference, American Society of Mechanical Engineers, pp V002T07A002–V002T07A002
63. Zhang S, Peng H, Nageshrao S, Tseng E (2019) Discretionary lane change decision making using reinforcement learning with model-based exploration. In: 2019 18th IEEE international conference on machine learning and applications (ICMLA). IEEE, pp 844–850
64. Zhang S, Peng H, Nageshrao S, Tseng HE (2020) Generating socially acceptable perturbations for efficient evaluation of autonomous vehicles. In: Proceedings of the IEEE/CVF conference on computer vision and pattern recognition workshops, pp 330–331

# Artificially Intelligent Active Safety Systems

**Trevor Vidano, Francis Assadian, and Nihal Gulati**

**Abstract** To better connect the Artificial Intelligence and Vehicle Control Communities, the SAE set of terms and definitions for Active Safety Systems is used to discuss modern applications of artificial intelligence to Active Safety Systems. This chapter begins with an introduction to the technology enabling Active Safety Systems and its impact on improving on-road safety. A new dataset is introduced that captures the prevalence of Active Safety Systems in the United States (U.S.) automotive industry for the model year 2021. Three different analyses are performed on this dataset to demonstrate its potential value in studying the offered Active Safety Systems in the U.S. automotive industry. Finally, promising Artificial Intelligence applications to the automotive industry are presented from the fields of deep learning, reinforcement learning, and imitation learning. An example of reinforcement learning applied to Automatic Emergency Braking is provided and demonstrates that reinforcement learning agents can learn policies that are effective in avoiding a collision. This chapter concludes that Artificial Intelligence will play a critical role in the future of automotive safety systems.

## 1 Introduction

The United States (U.S.) automotive industry has seen enormous change in the last decade. Electrification is one major trend that is sweeping across nearly all vehicle manufacturers and brands. Some manufacturers have even promised to change their lineup of gasoline-powered passenger vehicles to use a fully electric powertrain in the next several decades. While this trend has gained momentum, a second and more immediately rewarding trend is taking place: the push for zero on-road vehicle collisions. This goal, generally shared by all automotive companies, is still decades,

T. Vidano · F. Assadian (✉) · N. Gulati
University of California-Davis, 1 Shields Avenue, Davis, CA 95616, USA
e-mail: fassadian@ucdavis.edu

T. Vidano
e-mail: tavidano@ucdavis.edu



Y. L. Murphey et al. (eds.), *AI-enabled Technologies for Autonomous and Connected Vehicles*, Lecture Notes in Intelligent Transportation and Infrastructure,
https://doi.org/10.1007/978-3-031-06780-8_8

if not centuries, in the future. Despite this distant goal, nearly all vehicle brands now offer some form of advanced safety system beyond the minimum safety systems imposed by the Federal Motor Vehicle Safety Standards (FMVSS). A secondary goal of these advanced safety systems is to reduce the damage of unavoidable collisions. New startup companies have entered the automotive market by striving to achieve this goal of zero collisions by developing Automated Driving Systems (ADS). This new section of the automotive market grew significantly following the Defense Advanced Research Projects Agency (DARPA) Grand Challenge in 2004.

Older original equipment manufacturers sometimes referred to as legacy manufacturers, have been acquiring or investing in some of these startup companies. Legacy automotive manufacturers and competing startups are approaching the same goal of zero collisions from different directions. The former seeks to achieve the goal with driver assistance, and the latter seeks the goal with ADS. However, the differences between the most sophisticated, advanced safety systems and ADS will eventually become negligible. These technologies will converge, and the industry may be very close to achieving its safety goal. However, the industry has not yet reached this point.

This chapter focuses on advanced safety systems, defined as Active Safety Systems, instead of on ADS. While ADS is discussed briefly at times, the scope of discussing Active Safety Systems in sufficient detail is substantial enough to fill this chapter. This chapter also emphasizes the forms of safety systems that directly influence the vehicle dynamics over those that augment the driver's perception of the surrounding environment. Finally, this chapter will not emphasize the current technologies for sensing, perception, or motion planning topics of ADS. Instead, it will focus on the underlying technologies of the control systems in Active Safety Systems.

This chapter intends to introduce Artificial Intelligence researchers to some current applications in Active Safety Systems. It also will introduce current Intelligent Vehicle researchers to new tools coming from the field of Artificial Intelligence. These tools present possible solutions to some of the most challenging problems in the field.

The remaining part of this section presents a unifying set of definitions that will facilitate clear and focused communication for the remainder of this chapter. Section 2 introduces these Active Safety Systems and their enabling technologies. Discussion of their safety benefits is reserved for Sect. 3, where the warning and intervention systems are the primary focus. Section 4 introduces a new data set that is collected directly from brand websites. This data set provides a snapshot of the Active Safety Systems on vehicles in the model year 2021.

The first three sections introduce the Active Safety Systems and supply enough background for Sect. 5. This background is a prerequisite to discussing applications of Artificial Intelligence (AI) to Active Safety Systems. Section 5 focuses on Deep Learning, Reinforcement Learning, and other AI applications. Finally, Sect. 6 closes this chapter with thoughts on the gaps and shortcomings of current safety systems and how these problems might be addressed in the future.

## 1.1 SAE Definitions

Before discussing advanced technologies in the automotive field, this section introduces the Society of Automotive Engineers (SAE)'s cohesive set of terms and definitions [1, 2]. This chapter uses this internationally recognized terminology to constrain this chapter's scope and to facilitate a discussion that is understandable to both automotive and AI researchers.

### 1.1.1 SAE's Driving Automation Systems

The SAE provides a Taxonomy and Definitions for driving automation technology [2] that this chapter follows as best as possible. This section only introduces a subset of definitions, but if the reader wishes to know more, refer to [2].

Committee [2] defines 6 levels (beginning with 0) of automation for ground vehicles. No detailed definitions are provided for these as the names of each level are sufficiently self-describing. Important differences will be explained as specific implementations of these levels are introduced.

SAE Level 0 No Driving Automation
SAE Level 1 Driver Assistance
SAE Level 2 Partial Driving Automation
SAE Level 3 Conditional Driving Assistance
SAE Level 4 High Driving Automation
SAE Level 5 Full Driving Automation

The following definitions are abbreviated: (for full definitions refer to [2]):

**Active Safety System** Systems that sense the inside and outside conditions of a vehicle, and can either issue a warning to the driver or actuate vehicle subsystems that modify vehicle dynamical responses for a *limited* time. This encompasses SAE Levels 0–2.

**Dynamic Driving Task (DDT)** The real-time functions required to operate a vehicle such as lateral and longitudinal motion control, perception, and planning. Think of all the functions a driver in the 1950s would perform when operating a vehicle.

**Automated Driving System (ADS)** The vehicle itself and the software required to perform the full DDT on a *sustained* basis. This encompasses SAE Levels 3–5. This term excludes Active Safety Systems because Active Safety Systems perform on a *momentary* basis, where ADS perform on a *sustained* basis.

**Operational Design Domain (ODD)** The set of environmental and situational conditions in which a feature belonging to SAE Levels 1–5 is designed to operate.

### 1.1.2 SAE's Active Safety Systems

As the field of Driver Assistance is rapidly evolving and growing both in academic and industry communities, it is increasingly important to use a cohesive set of terms to discuss these technologies. The development of new technologies is inevitable. The customer, researcher, regulator, and Original Equipment Manufacturer (OEM) must be specific when discussing these new systems. These various stakeholders have different values when defining terms. These values can disagree in a way that can unintentionally confuse the other stakeholders.

***Example: Conflicting Definitions***
Ford offers a system called *BlueCruise*, which performs similar functions to Mercedes Benz's *Distronic Plus with Steering Assist* and Tesla's *Autopilot*. A problem occurs when these OEMs do not consider their competing systems equivalent in performance and market them differently. Without a cohesive set of definitions, the consumer will struggle to compare each technology. Furthermore, these technologies must be categorized based on some functional classifications so that regulators can ensure these systems meet a base level of safety. Adopting the SAE categories solves these problems. The SAE categorizes *BlueCruise*, *Distronic Plus with Steering Assist*, and *Autopilot* into Active Driving Assistance [1]. Using a united set of definitions, consumers can compare performances between products, regulators can enforce safety standards, and researchers can better improve the performance and robustness of these systems.

This chapter considers the industry and literature term Advanced Driver Assistance Systems (ADAS) to refer to the same systems as defined by Active Safety Systems. Note that this consideration implies that ADAS is not included in ADS. This chapter prefers the term *Active Safety Systems* over ADAS because it is part of the specific, internally consistent SAE terms and definitions.

Section 2 discusses in detail the technology that enables each system type in these subcategories. However, before proceeding, it should be pointed out that the problem of terminology is not entirely solved by using the SAE definitions. This is because of the accelerating development of these technologies. The industry is seeing an accelerating pace towards the adoption and development of new systems that improve the safety and control of ground vehicles.

***Example: Quickly Evolving Systems***
Tesla's *Autopilot* can be considered an Active Driving Assistance system according to the SAE's definition of the term [1]. However, its recent upgrade to include Tesla's *Active Lane Change Assist* arguably exceeds the SAE's definition. This upgrade extends beyond Active Driving Assistance systems because not only is the system maintaining a safe distance from surrounding vehicles (laterally and longitudinally), but it is also planning and deciding on maneuvers that seek to achieve the driver's set speed. While this additional functionality may not be substantial enough to consider this new feature an example of SAE Level 3, it is unfortunate that there is not a

separate SAE definition to include this unique functionality. There are several other examples of similar conflicts from competing brands. In the interest of choosing a single set of Terms and Definitions, Tesla's newest upgrade of *Autopilot* is still considered an Active Driving Assistance system in this chapter.

There is no doubt that in the future, the SAE's definitions must be modified and new definitions are added. This is important to both reduce any friction to innovation and to support focused discussion on safety systems.

## 2 Active Safety Technology

This section does not paraphrase the SAE Terms and Definitions of Active Safety Systems. For the full definitions, the reader is referred to [1]. Instead, this section introduces the technology that enables these Active Safety Systems. The six categories of Active Safety Systems are Collision Warning, Collision Intervention, Driving Control Assistance, Parking Assistance, and Other Driving Assistance Systems.

### *2.1 Collision Warning*

Active Safety Systems categorized as *Collision Warning* are systems that do not directly control the dynamic response of the vehicle. Instead, they act through warnings that the driver can quickly interpret and respond accordingly. The SAE defines five subcategories of Collision Warning systems: Forward Collision Warning, Blind Spot Warning, Lane Departure Warning, Parking Collision Warning, and Rear Cross Traffic Warning [1]. These subcategories are discussed in this order.

Most warning systems must:

1. Measure the environment outside the host vehicle
2. Identify objects
3. Track objects
4. Assess the threat of collision with objects
5. Efficiently inform the driver

Forward Collision Warning

Forward Collision Warning (FCW) systems measure objects directly in front of the vehicle and issue warnings when the vehicle has a high chance of colliding with one or more objects.

In some implementations, the system will also infer the driver's intention based on the driving commands. This method projects the host vehicle's predicted trajectory and then assesses the location of objects relative to that trajectory.

FCW systems commonly use radar to identify objects. Radar can distinguish most objects in adverse weather conditions but often struggles to differentiate between two close objects on a curved road if the radar does not have sufficient angular resolution [3]. This shortcoming can result in false positives. After several false positives, the driver might grow to distrust the system and disable it, ending any potential safety benefits.

Laser-based sensors show great promise as an alternative to radar. Raster-scanned laser rangefinders provide high performance for low-cost [3].

An increasingly common sensor used in detecting forward objects is the forward-facing optical camera. These cameras are easy to spot in parking lots because they typically are found between the rear-view mirror and the windshield. This location makes it easy to keep the field of view clear of debris without obstructing the driver's point of view.

The most promising methods are redundant sensors that can provide two or more different estimations of the same states. This redundancy helps reduce false alarm rates. Therefore, it is likely that many FCW systems use two radars and a forward-facing camera to detect forward objects [4]. An additional sensor that helps reduce collisions is a rear-facing radar. This radar can provide feedback on the location of vehicles behind the host vehicle. The braking torque can then be adapted to reduce the chance of being rear-ended [4].

The threat of collision is commonly assessed by two types of algorithms [3]:

1. time-based, which computes the time to collision
2. distance-based, which computes the distance to stop

Both methods are susceptible to false alarm rates, and the timing of the alarm is critical. If the alert is issued too early, the driver may ignore it, or worse, become irritated and turn off the system entirely. If the alarm is issued too late, then the driver is unable to avoid or mitigate the collision.

Finding a warning system that gets the driver's attention without annoying the driver is a heavily researched area. Some methods tested in research are:

- visual icons
- visual text
- digitized speech
- earcons and tones
- tactile (seat vibration, steering wheel vibration, or belt tension).

Zador et al. [3] investigates many combinations of these. The method that is most effective at getting the driver's attention is auditory warnings. This finding is the outcome of human trials. However, this is also the most annoying warning method. Instead, the most preferred alert among test participants is yellow icons displayed on a Heads Up Display.

Further complicating the issue of designing an appropriate warning system is the finding that drivers are inaccurate at estimating their temporal headway [5]. They are also highly sensitive to false positives. This combination of driver behavior can be very challenging to overcome. One promising finding suggests that a training period in which the FCW system trains the driver to maintain a safe headway improves the driver's performance in the short term and long term (6 months after training) [5]. Training may also help encourage the driver to be more tolerant of the FCW system's false positives.

FCW stacks, meaning the hardware and software components that compose the system, vary widely between brands. Not all FCW systems are equal in performance, but they all must meet a base level of performance. Improvements beyond this base level performance include the ability to [3]:

1. detect objects at long range on rolling terrain
2. reduce the rate of false alarms on roadside objects
3. reduce the rate of false alarms at long range when near cargo trucks.

To achieve these future improvements, the tools in machine vision, sensing, and planning must improve. All three of these areas can benefit from the quickly advancing area of Artificial Intelligence.

Blind Spot Warning

Blind Spot Warning (BSW) is very similar, in terms of system requirements, to FCW systems. The fundamental roles: sensing, object detection, and warning, must be performed. The difference is the location of the sensors and the preferred location of the alerts.

Radars used to detect vehicles in the host vehicle's blind spot are typically located towards the vehicle's rear and face radially outward of the vehicle center. Cameras, which replace radars, are placed either on the side view mirrors or on the beam that supports the car's roof and separates the front and rear side doors.

Almost all warning systems on vehicles are icons colored red or yellow and appear on the side-view mirrors. Some systems also deliver an audible warning, and some only issue this audible warning when the turn signal is on and when there is a vehicle in the driver's blind spot.

The decision of when to issue a warning is often made with less complexity than the forward collision warning. This simplicity is because the system does not need to compute the distance or time to collision with that object. Instead, it only needs to detect the presence of an object in a bounded area. This detection is done differently depending on the type of sensor.

Lane Departure Warning

Lane departure warning (LDW) systems have almost an identical stack compared to FCW systems. LDW systems typically prefer a forward-facing camera instead of radar or laser-based sensors. Some additional components in the stack are lane modeling, lane detection, and lane tracking [6].

Machine vision is one of the crucial technologies to developing a successful LDW system. A drawback to using cameras is their inability to detect the lane in adverse conditions. To mitigate this shortcoming, LDW systems leverage lane models to improve detection. The model and sensor measurements are fused in a variant of a Kalman filter or particle filter. These techniques are the most common ways to combine model information and sensor measurements [6, 7].

The development of the lane model has numerous trade-offs that are typical in model-based solutions. The most basic models, linear models, assume the system is linear, thereby constraining the system's Operational Design Domain (ODD). As lane models increase in complexity, they improve in prediction accuracy but become more susceptible to sensor inaccuracies and misclassifications [6]. Lane model robustness is an ongoing research pursuit, one that deep learning has shown to be increasingly dominant. It is also common for these models to misclassify lane-like objects, such as traffic signs and guard rails, as lane markers. A step called Lane Verification is therefore commonly used to remove these falsely classified objects [6].

Many camera-based sensors use grey visuals to reduce both computational load and sensor costs. The black and white contrast between the road and lane marker is typically sufficient to detect the lane. However, color cameras are becoming increasingly common [7]. Laser-based systems offer a more accurate sensor alternative. The laser-based systems are composed of several sensors that are underneath the front bumper and look directly down. However, due to the low latency imposed by this sensor array's field of view, these types of systems cannot be used for assistance, only for warning the driver [7]. As a result, these systems are fairly uncommon. Some other alternatives use special infrastructures, such as magnetic nails and underground cables that emit radio-frequency to mark lanes. While these are not expected to scale widely, they are promising in areas with heavy snow and ice. Finally, Global Positioning Systems (GPS) can be used, but are currently limited by position and map accuracy [7].

The greatest challenges to the LDW system's ability to identify lanes are [6]:

1. lane marker variations
2. road surface material
3. shadows (especially long linear shadows from trees and building)
4. lighting conditions
5. night conditions
6. rain
7. fog.

A final challenge to LDW systems is when there are no lanes or the lanes are obstructed from view [6]. Naturally, these issues are nearly impossible to overcome, and as a result, most LDW systems deactivate themselves in these scenarios.

The challenges for LDW systems to get the driver's attention are similar to the other Collision Warning Systems. [8] found that LDW systems that used steering wheel vibrations were more likely to be turned on than those that used only auditory warning. Other systems use visual alerts on the host vehicle's digital dashboard.

Parking Collision Warning

Parking Collision Warning (PCW) systems are less complex than the previously discussed Collision Warning Systems. These systems use sonar sensors typically placed on the rear bumper to measure the distance between the host vehicle and an object. Most of these systems only activate when the vehicle is in reverse gear. They also issue an audible warning. Some systems issue different warnings that correspond to the distance between the host vehicle and an object. For example, slow repetitive beeps correspond to a large distance. Fast repetitive beeps correspond to short distances.

Rear Cross Traffic Warning
Rear Cross Traffic Warning (RCTW) systems are very similar to parking collision warning systems. They typically use sonar sensors placed on the rear corners of the vehicle to detect vehicles that are in the blind spot of the host vehicle. If the host vehicle has blind-spot warning systems, the existing sensors that enable blind-spot warning systems are used instead of the corner sonar sensors. The warning methods are similar to Parking Collision Warning. However, they typically issue a single, urgent beep instead of repetitive beeps.

## *2.2 Collision Intervention*

The natural evolution of Collision Warning systems is to go beyond warning the driver and issue actuator commands on behalf of the driver. After this extension, they are called Collision Intervention systems. The SAE defines 4 subcategories for Collision Intervention systems: Automatic Emergency Braking, Automatic Emergency Steering, Reverse Automatic Emergency Braking, and Blind Spot Collision Intervention [1]. Nearly every Collision Warning system is extended to become a Collision Intervention system. Forward Collision Warning extends to Automatic Emergency Braking and Automatic Emergency Steering. Parking Collision Warning systems and Rear Cross Traffic Warning combine and extend to Reverse Automatic Emergency Braking systems, and Blind Spot Warning systems extend to Blind Spot Collision Intervention. The requirements imposed on Collision Warning systems are insufficient for Collision Intervention systems because the acceptance of false positives and false negatives is far more stringent. All intervention systems issue commands to actuators to reduce the potential damage of collisions. The actions that the actuators take can eliminate the potential damage by avoiding the collision. However, avoiding a crash is not always possible. The SAE definitions make this distinction very clear [1].

The tires are the only contact between the vehicle and the road. Therefore all actuators that change the vehicle motion must act through the tires. Steering actuators modify the direction of tires to primarily generate more or less lateral forces, thereby changing the vehicle's heading (yaw angle). Brakes and motors change the torque on the wheels to control the longitudinal tire forces. Taking this perspective exposes the limitations of every ground-vehicle control: tire friction limits. All Collision Intervention systems should make careful considerations of these limits. Saturating tire forces (demanding a tire force beyond the current tire friction limits) can lead to dangerous instabilities. For instance, saturating the front tires on a typical passenger vehicle results in the inability to steer. Furthermore, these friction limits change drastically with the environmental conditions. Road-tire limits are one of many challenges for Collision Intervention systems.

There are two actions that forward Collision Intervention systems can take on behalf of the driver: steer away from the collision or stop the vehicle. These two actions are realized in Automatic Emergency Steering (AES) and Automatic Emergency Braking (AEB), respectively. It is important to note that these systems do not control both steering and braking simultaneously. Therefore, AES and AEB are Level 1: Driver Assistance systems.

Both AEB and AES have great safety potential. In 2014, 32% of all police-reported crashes were rear-ending collisions. From analyzing data, AEB appears to be most effective at preventing the host vehicle from rear-ending another vehicle at speeds of 40–45 mph [9]. These collisions can have substantial energy and therefore increase the likelihood of severe or fatal injuries.

The only actuator used to extend Parking Collision Warning (PCW) and Reverse Cross Traffic Warning (RCTW) to perform Collision Intervention is the brake system. A Reverse Automatic Emergency Braking (RAEB) system uses brakes as the actuator to avoid or lessen the severity of a collision when the host vehicle is backing up. RAEB controls longitudinal vehicle dynamics intermittently and therefore is classified as SAE Level 1.

Blind Spot Warning (BSW) is extended to Blind Spot Collision Intervention (BSCI) by actuating brakes and steering to avoid or mitigate a collision. However, not all implementations of BSCI systems control both longitudinal and lateral actuators. Despite some forms of this system using brakes and steering, all Collision Intervention systems are SAE Level 1. By definition, Collision Intervention systems do not perform their task on a sustained basis. BSCI activates to aid in avoiding a collision with a vehicle in the driver's blind spot and only remains active while there exists a threat of collision.

These Collision Intervention systems must have a lower rate of false positives and false negatives than their Collision Warning counterparts. The risks created by actuating the vehicle on behalf of the driver are far more severe than simply issuing a warning. If a warning is issued when there is no probable collision, the driver will likely not change their behavior. With more false alarms, the driver will lose trust in the system and will turn it off. Besides the driver acting dangerously in response to a false alarm, this is the worst-case scenario for warning systems. While this is not preferred, it is far better than the risk of the vehicle, for instance, braking for a non-existent imminent collision. Such a maneuver may increase the likelihood of a crash. Therefore, most systems are tuned to be conservative in identifying potential collisions.

***Example: AEB False Negative***
If an AEB system acts on a false positive, at best, the driver will be shocked. At worst, the driver will react irrationally and create a collision. A third alternative is that the driver does not act to counter the emergency braking, and the vehicle following the host vehicle must react in time to avoid a collision. If the following driver does not, they will collide with the host vehicle. Though not legally at fault, the AEB system created the potential for a crash that otherwise would not have occurred.

The consequences of a false negative are obvious. These consequences demand more stringent requirements imposed on Collision Intervention systems. This lower tolerance typically drives up the cost, development time, and imposed regulation of these systems. Furthermore, improvements in reducing false positive and false negative rates could have significant impacts on increasing safety. This is further discussed in Sect. 3.

## 2.3 Driving Control Assistance

Collision Intervention systems are properly named because they only intervene until the threat of collision is gone. A system that provides longer momentary support and replaces a portion of the Dynamic Driving Task (DDT) is classified as Driving Control Assistance. Driving Control Assistance still requires the human driver to continually supervise the active system. This is why Driving Control Assistance is not yet considered ADS (SAE levels 3–5). In ADS the human driver does not need to be continually supervising, but is required to be ready to take over the DDT should some system failure occur. A typical example of a system failure that would trigger such a fallback is the mechanical failure of some vehicle subsystem, such as the suspension system [2]. At this level of support, the lines between Level 1 and Level 2 blur because a vehicle may be equipped with two or more Level 1 systems that, when working together, appear to be providing Level 2 functionality. The difference is in which part of the DDT each system performs. Level 1 systems perform only longitudinal or lateral control on a sustained basis, while Level 2 systems provide both longitudinal and lateral control on a sustained basis. The SAE defines three subcategories within Driving Control Assistance: Adaptive Cruise Control, Lane Keeping Assistance, and Active Driving Assistance [1].

Adaptive Cruise Control
Adaptive Cruise Control (ACC) is a Level 1: Driving Assistance system. ACC systems typically use forward-facing radar placed on the vehicle's nose (commonly placed behind the brand's logo). A common alternative to using radar is to use stereo cameras. A single-camera system is uncommon because of the difficulty in estimating distances. Stereo cameras use two cameras, typically grey-scale to reduce computational load, spaced a known distance apart. Knowing this distance allows

the software to compare both images and compute the distances to objects. However, this is only possible in the overlapping field of views of each camera.

These sensor configurations have to compensate for the road's curvature in both the horizontal plane and the vertical plane. Additional challenges to achieving accurate estimates are the variable vehicle speed and external object density. Cameras must have a high frame rate, which requires a large amount of processing to detect objects at highway speeds. Control systems are also challenging to design in the full range of the vehicle's velocity. Some ACC systems are automatically disengaged below a set speed to ensure the validity of their design assumptions. Almost all systems are designed to operate in highway ODDs, which provide drivers with significant comfort and safety benefits. Object density, such as the number of vehicles surrounding the host vehicle, mostly limits the sensing capabilities. However, a high density of objects can also force tighter requirements on controller performance.

A high-level challenge posed to the designers of ACC systems is striking the right balance between providing some comfort to the driver and providing too much comfort. In other words, ACC must not replace too much of the driver's mental workload of performing the DDT. Doing so might create a more distracted or a more tired driver [4].

Lane Keeping Assistance

While ACC controls longitudinal dynamics, Lane Keeping Assistance (LKA) systems control lateral dynamics. These systems typically use the same sensors as LDW systems since the sensing requirements are similar. There are several different techniques to assisting the driver [7]:

1. **Loose Guidance**: Torque is applied only when the vehicle is within a close tolerance of the lane border. This performs like an intervention system, acting only at elevated risk.
2. **Tight Guidance**: Torque is applied based on a linear scale, starting with no torque in the lane center and maximum torque at the lane border.
3. **Comfort Oriented Guidance**: A piece-wise or non-linear function is used to provide no torque at the lane center, low amounts of torque until a set tolerance around the lane border, and maximum torque at the lane border.

It is important to note that all three guidance techniques stop applying torque to the steering wheel when the host vehicle has crossed the lane border as defined by some tolerance around the lane border. The tolerance is necessary because it helps make the system robust against sensor noise. If the function were to stop applying torque immediately after crossing the lane border, then the entire system would be highly dependent on the estimation of the lane border. Stopping the application of torque after entering a new lane is also necessary because if the system does not recognize the driver's intention to change lanes, it may try to steer the vehicle back to the previous lane.

Most systems use the lane change indicator, colloquially called a "blinker", to determine the driver's intent to change lanes. However, the driver is still capable of changing lanes without the blinker by overcoming the assisted torque. As a safety measure, all systems do not apply more force than the typical human can counter. The human must always be capable of performing the DDT should the LKA system malfunction or disengage.

LKA systems are often complex in deciding when to deactivate. For instance, many systems have capacitor-based sensors that detect when the driver is holding the steering wheel. After issuing a warning, these systems will deactivate. This detection is critical to ensuring that the driver is attentive and closely monitoring the outside environment. Many LKA systems also detect quick avoidance maneuvers and deactivate to minimize the steering resistance in performing such a maneuver [7].

While most systems use electric power steering, hydraulic power steering is common in larger vehicles. In addition to power steering, the brakes can also be used, similar to how they are used in electronic stability control systems, to create a yaw moment [7].

The overall system must have tighter requirements than those for LDW systems. It is common to use Kalman Filter variants and particle filters to smooth out lane detection measurements. Smoothing out these measurements prevents oscillating between engagements and disengagements [7]. Such behavior may be considered a nuisance by the driver.

Active Driving Assistance

Active Driving Assistance (ADA) systems combine both ACC and LKA. These systems rely on the same sensor and actuator types that ACC and LKA use, except that ADA can control longitudinal and lateral vehicle dynamics. Some brands have a single system classified as ADA, while others offer separate ACC and LKA systems that, when engaged together, perform the same functions as an ADA system.

## 2.4 *Parking Assistance*

Parking assistance is the subcategory of active safety systems that aid the driver in the task of parking. This aid can be provided as different visualizations of the environment directly behind and surrounding the vehicle. These visualizations are provided to the driver by Backup Camera systems and Surround View Cameras. They mostly require moderately-high definition cameras and displays. They also employ some image processing technologies. In these systems, the driver is in full control of the vehicle and is only aided visually or through audible alerts.

More complex systems can take control of the vehicle to automatically parallel park or perpendicular park. These systems are called Active Parking Assistance. They mostly rely on short-range sensors such as sonar proximity sensors and cameras. Some systems control steering, throttle, and braking, while others control only one or more of these actuators. Additionally, the driver may be relied upon to change gears. These systems have looser requirements than ADA systems, but some are considered Level 2: Partial Driving Automation since they control longitudinal and lateral dynamics in a limited ODD.

More sophisticated systems completely remove the driver's physical interactions, allowing the vehicle to operate without a passenger. These systems are considered Remote Parking Assistance systems. This operation is controlled and monitored remotely by the driver. For instance, in some versions of this system, the driver uses a controller attached to the car keys to command the vehicle forward or backward. Some systems parallel park or perpendicularly park without the need for remote driver input. Other systems require all actuator commands to be communicated remotely from the driver. Some systems that operate at Level 2: Partial Driving Automation can pull out of a parking space and navigate a parking lot to arrive at the driver's location. This system usually operates in a parking lot or driveway. The driver hails the vehicle from outside, and the car navigates from its parking space to the driver.

The last type of Parking Assistance system, Trailer Assistance, aids the driver in attaching a trailer to the hitch of a vehicle. These systems are similar to Backup Cameras, except the camera is oriented to provide visualization of the host vehicle's hitch and the trailer being approached. Some systems aid in steering control, and some systems supply visualizations to the driver while driving with a trailer. The enabling technologies are very similar to Surround View Cameras.

## *2.5 Other Driving Assistance*

The SAE captures miscellaneous Active Safety Systems in the subcategory Other Driving Assistance. This chapter ignores the technology behind these systems to better focus on vehicle control. Nevertheless, they are listed here for completeness, and they have self-explanatory titles:

- Automatic High Beams
- Driver Monitoring
- Head-Up Display
- Night Vision
- Speed Warning.

## *2.6 Beyond Assistance*

This chapter does not try to provide comprehensive coverage of the technology that enables each Active Safety System. Entire books have been written to cover this topic. If the reader wants deeper coverage of these systems they are referred to [10, 11].

Active Safety Systems attempt to aid the driver in avoiding collisions, and when avoidance is not possible, reduce the potential damage. The primary problem of avoiding collisions in all possible driving scenarios is not yet solved. The secondary problem of mitigating the damage of such a collision in all possible driving scenarios is a more general problem, and therefore also unsolved. Due to the difficulty of solving the second problem, it is likely the case that the behavior of Active Safety Systems does not change with respect to whether or not the collision is avoidable. For instance, it is likely that an Automatic Emergency Brake (AEB) system will apply maximal brake force to both avoid a collision and to reduce the potential damage of the collision. However, suppose that the host vehicle can avoid rear-ending the car in front, but doing so will require the host vehicle to brake so quickly that the vehicle behind the host vehicle will rear-end the host vehicle. Some AEB systems may not consider the probability of being rear-ended. To consider this probability requires the ability to sense approaching vehicles behind the host vehicle as well as the ability to determine which type of collision is preferable: the host vehicle rear-ending or the host vehicle being rear-ended.

The literature that studies Automated Driving Systems (ADS)s has begun to propose frameworks and other strategies to help formulate and evaluate solutions to this type of dilemma. One common approach is to avoid having to solve the problem altogether by only generating trajectories for the ADS that prevent possible collisions. The concept of Inevitable Collision States, proposed in [12], is one example of such an approach. Inevitable Collision States are a formal definition of conditions in which a robot cannot avoid a collision with a static or dynamic obstacle. This concept helps to generate plans that will avoid collisions with objects that are known a priori as well as objects that suddenly appear. Despite this example of a theoretical solution, it is likely that ADS will encounter a situation in which a collision is inevitable and the system will have to choose between different types of collisions. This is where the fields of philosophy and ethics enter into the study of ADS. Some ethical concerns that are raised by challenging situations that an ADS might end up in are analyzed by [13]. One example of an ethical framework is the one proposed by [14], which attempts to minimize a cost function that quantifies risk and attempts to achieve equal treatment of people. The intersection of ethics and ADS is critical to study because even if an ethical framework is not used in the design of planning algorithms, ethical choices are being programmed into the ADS. For instance, consider an ADS that has to decide whether to brake and be rear-ended or to collide with some obstacle in front. If the system is designed to always brake without considering the ethics of that choice, then despite the lack of consideration, an ethical decision is being made: the choice to favor being rear-ended over rear-ending.

## 3 Active Safety Potential

The first brand to introduce an FCW system in the United States was Mercedes-Benz in the model year 2000. Soon after, other brands offered FCW and AEB. By 2016, 54% of U.S. vehicles offered either FCW or AEB. Twenty automakers, 99% of the U.S. automotive market, have committed to making FCW and AEB standard on almost all passenger vehicles by 2022 [9]. With these promising trends, it is natural to expect Active Safety Systems to have significant potential in preventing collisions and reducing the damage resulting from collisions. This inclination is accurate for FCW and AEB.

An analysis of insurance claims shows that vehicles with FCW alone are associated with 7–22% reductions in property damage and 4–25% reductions in bodily injury. When FCW couples with AEB, the number improves to a 10–16% reduction in property damage and 14–32% reductions in bodily injury. Based on collision data, vehicles with FCW and AEB were 50% less likely to strike a vehicle in front of the host vehicle [9].

However, this successful result of FCW and AEB is offset slightly by the increased probability of being rear-ended. [9] found that vehicles with both FCW and AEB were 20% more likely to be struck in a rear-end collision than vehicles without this technology. The consumer may interpret this as a net improvement because they receive a decrease in collisions in which they are likely to be considered at-fault while suffering an increase in collisions in which they are not likely to be considered at-fault. The regulator and brands, however, might view this as more of a problem. The solution may come with time because if the automakers fulfill their promise, after 2022 almost all new vehicles will have FCW and AEB standards. This will gradually reduce the rear-end collisions and consequently the chances of being rear-ended. Unfortunately, it is expected that it will take 35 years or more for 95% of all vehicles on roads in the U.S. (not just new vehicles) to have AEB [15]. AEB systems should therefore improve their ability to reduce the chances of not only rear-ending a vehicle but also of being rear-ended. This improvement should be made instead of waiting for all vehicles on the road to have AEB.

Benefits from Active Safety Systems are not only confined to FCW and AEB. If all vehicles had been equipped with BSW in 2015, about 50,000 collisions could have been prevented in that year [16].

Furthermore, [17] showed that all warning systems provide an overall decrease in insurance "collision" (property damage of at-fault vehicles). Systems that use parking sensors showed a reduction in insurance "collisions", property damage liability, medical payment, and personal injury protection [17]. According to [17], RAEB is the most effective in reducing property damage liability when compared to FCW, AEB, and many other Active Safety Systems.

These safety improvements also extend to pedestrians and cyclists. In 2019, two out of three AEB systems in the U.S. included pedestrian detection, and the EuroNCAP has been rating pedestrian detection since 2016 [15].

## 4 Systems on the Road Today

The Insurance Institute for Highway Safety (IIHS) collected data from numerous brands by asking them to report how their customers used and liked the Active Safety Systems installed on their cars. [8] found that 49% of vehicles had systems relating to lane maintenance (LKA and LDW) turned off. The researchers also found that if the lane maintenance system can be disabled easily, it is more likely to be disabled than those versions that are more difficult to be disabled. [8] also found that 93% of systems that helped avoid front collisions, AEB and FCW, were turned on. Consumers also preferred these systems over the lane maintenance systems.

These findings indicate the amount of usage of the Active Safety Systems. Unfortunately, they do not show how many Active Safety Systems are available to consumers. There is a lack of publicly available data that records the level of the automotive industry's offering of Active Safety Systems. This section presents a new data set[1] that captures the level of adoption of automotive brands. The data set records the Active Safety Systems offered on each brand's non-commercial vehicles in 2020 (not the model year). The data set covers the top six brands based on total vehicle sales, by volume, in the United States. Subaru, which is not in the top six, is also included in this list. The data set also contains the top three luxury vehicle brands, based on total vehicle sales, to capture the luxury vehicle market's adoption of Active Safety Systems. In total, this covers 67% of all new passenger vehicle transactions in the year 2020.

The intended purpose of this data set is to capture the state of the industry adoption of Active Safety Systems. This data is rich with information, and this section does not fully explore all the possible analyses. Future efforts can perform new analyses and extend this data set to include more features and all vehicle brands. However, these additional efforts diverge too far from the intended scope of this chapter. Therefore the reader is encouraged to explore the raw data and to use it to test their hypotheses.

After introducing this data set, this section investigates trends relating to the Manufacturers Suggested Retail Price (MSRP) and the number of Active Safety Systems on both specific vehicle trims and the industry as covered by this data set. This data set may not encompass enough of the U.S. passenger vehicle market to comprehensively study market-wide trends. Conclusions drawn from this data set about the market as a whole are preliminary investigations. The reader is also cautioned against interpreting this data as evidence that one vehicle is better than another based on MSRP, safety, and or any other features included in this data. Quantifying the number of safety features may mislead the reader to assume that more Active Safety Systems cause the vehicle to be safer or more valuable than its competition. These sorts of arguments are outside the scope of this chapter and this data set.

## *4.1 Methods*

The collection of this data began as a very preliminary investigation of the most efficient method to collect this data. Initially, this study selected a comprehensive

[1] The Google Sheet as a *.ods file and the web scraper are available on The Future Mobility Lab's GitHub at https://github.com/FutureMobilityLab/ADASVehicleScraper3000.

**Table 1** Top 6 vehicle brands[a]

| Brand name | % of U.S. vehicle sales by volume |
|---|---|
| Ford | 13.13 |
| Toyota | 12.50 |
| Chevrolet | 11.77 |
| Honda | 8.16 |
| Nissan | 5.70 |
| Jeep | 5.41 |

[a]Brands whose cheapest model and trim level is less than $30,000

**Table 2** Top 3 luxury vehicle brands[a]

| Brand name | % of U.S. Vehicle sales by volume |
|---|---|
| Mercedes-Benz | 2.21 |
| Tesla | 1.99 |
| BMW | 1.91 |

[a]Brands whose cheapest model and trim level is more than $30,000

list of vehicle brands in the U.S. The study then began collecting data randomly. Subaru was one of the first vehicle brands to be studied before it became clear that the scope of this data collection was too vast for the allocated resources. The list of brands was subsequently constrained to include a subset of all vehicle brands that sold passenger vehicles in the U.S. in 2020. This subset is the current list of vehicle brands (Tables 1, 2).

Subaru, which covers about 4.16% of U.S. passenger vehicle sales by volume, is included to avoid throwing out already collected data. In total, these ten vehicle brands account for 66.95% of all vehicle sales by volume in the U.S. Furthermore, most new cars sold in 2020 were either the 2020 model or the 2021 model. This study focuses on the 2021 model year to capture the most recent models.

As previously discussed in Sect. 1.1, there is an inconsistency between most vehicle brands and the terminology used for Active Safety Systems. A dictionary that maps the brand packages and features names to the SAE Active Safety Systems helps overcome this problem. This dictionary is defined manually and reviewed several times to ensure accuracy and consistency. Despite this, there is still an opportunity for arguments over which brand terms should correspond to specific SAE terms.

A web scraper extracts the data from the brand websites, and then this dictionary translates the brand packages and terms assigned to each trim level of each model. The data collected is in a binary format except for the features that describe the Vehicle Information:

1. Brand
2. Model Name
3. Trim Level
4. Model Year
5. Base MSRP for the 95616 zip code.

The fuel type available for each trim level is also collected, but not analyzed in this chapter. Finally, the raw data contains some additional safety and comfort features such as Active Driving Assistance with Lane Change and Cruise Control, respectively. The result is a data array of 713 rows (one for each trim level of each model) long and 35 columns (including the extra features previously discussed) wide. However, in this chapter, only the Active Safety Systems are analyzed.

Aggregating the data into vectors that capture industry-wide data is done in two calculations. The first calculation is summing the number of SAE Active Safety Features installed on each trim of each model. The second calculation counts the number of trims of each vehicle with each SAE Active Safety Feature.

This data set covers all SAE Active Safety Features except for Lane Departure Warning. This feature is excluded because of the difficulty to differentiate it from Lane Keeping Assist for each brand. Future work may revise the dictionary used to collect this data set to include Lane Departure Warning separate from Lane Keeping Assist.

## *4.2 Results*

This section uses three different methods to extract results that convey the current state of the industry's adoption of Active Safety Systems. These methods intend to present a preliminary view of the industry from three different perspectives. Together they are more holistic, but should not be considered comprehensive. The ultimate goal is to show how this data can be used in further research.

### 4.2.1 Aggregating Data per Brand

This first analysis shows if a relationship exists between Models and the number of Active Safety Systems available in those Models. It also provides aggregated data enabling the formation of a hypothesis about the form of this relationship. The process begins by computing the number of Active Safety Systems per Trim level for every model. The Trim Levels that offer the lowest number of Active Safety Systems for each Model can then be identified. These Trim Levels are given the name Minimum Safety Trim Levels (MSTL)s. These MSTLs are used to represent each Model. The data is reduced further by identifying the MSTL with the lowest number of Active Safety Systems for that brand. This information is used to compare the Vehicle Type or the Base MSRP to the remaining MSTLs. To clarify this process, it is performed for Ford below:

***Example: Ford Data Reduction*** Beginning with the Ford Escape, the following is found: the S, SE, and SEL Trim Levels have 9 Active Safety Systems, the Titanium trim level has 11 Active Safety Systems. Since the S, SE, and SEL Trim Levels have the lowest number of Active Safety Systems for the Ford Escape, they are the MSTLs for the Ford Escape. In this situation either all three can remain, or the cheapest Trim can be considered the MSTL. Repeating this for the Ford Ecosport, the S Trim Level is found to be the MSTL because it has only one Active Safety System. The MSTLs can be found for the remaining Ford models by repeating this process for each model. A comparison of all the MSTLs for Ford reveals that the Ecosport S has the lowest number of Active Safety Systems. Having found this, the other information for this Trim Level can be used to make other observations. For example, this is also the cheapest Trim Level and Model offered by Ford.

Carrying out this process for all brands identifies the following Models and Trim Levels with the lowest amount of Active Safety Systems of all the MSTLs for each brand:

- Ford Ecosport S
- BMW i3
- Chevrolet Spark LS Manual
- Toyota GR Supra 2
- Honda Civic Type R
- Nissan GT-R Premium and NISMO
- Jeep Wrangler 4xe Sahara and Rubicon
- Subaru BRZ Limited and tS
- Mercedes-Benz C-Class Cabriolet and Coupe, and A-Class, and CLA
- Tesla Model S Long Range and Plaid.

By looking at the MSRP, and the style of advertising of each of these models, they all fall into three categories: the most affordable model offered by the brand, an all-terrain model, or a high-performance model.

### 4.2.2 Active Safety System Prevalence

A second analysis is performed to identify the least and the most prevalent Active Safety Systems across all of the vehicle Trim Levels collected. This is done by comparing the total number of the Trim Levels that contain the specific Active Safety Systems. The results are shown in Tables 3 and 4.

**Table 3** Top 5 most prevalent active safety features on all brands

| Active safety system | Number of trims with system |
|---|---|
| Backup camera | 711 |
| Automatic emergency braking | 552 |
| Parking collision warning | 474 |
| Blind spot collision warning | 459 |
| Forward collision warning | 452 |

**Table 4** Top 5 least prevalent active safety features on all brands

| Active safety system | Number of trims with system |
|---|---|
| Night vision | 4 |
| Remote parking assist | 17 |
| Automatic emergency steering | 18 |
| Blind spot collision intervention | 27 |
| Trailer assist | 48 |

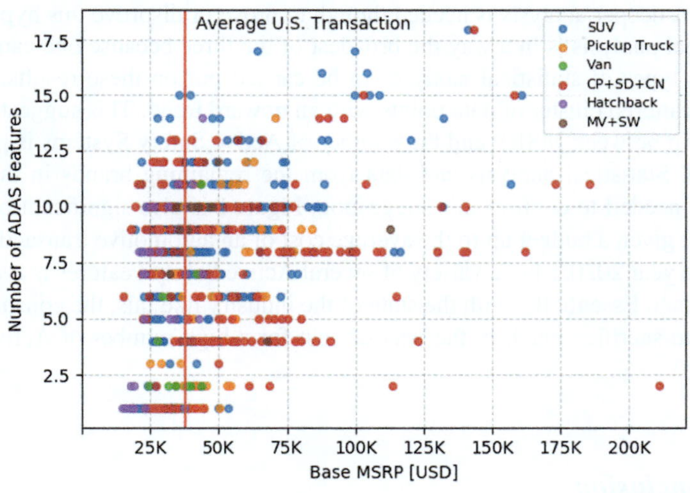


**Fig. 1** Active safety systems for each trim versus the MSRP of that trim with vertical line for average US transaction cost

### 4.2.3 Scatter Plot Analysis

A third analysis is performed to investigate the relationship between MSRP and the number of safety features for each Trim Level. The average U.S. transaction cost [18] is also included as the red vertical line in Fig. 1.

The cluster of data points in Fig. 1, although quite scattered, appear to show an up and right-ward trend. However, this is a qualitative observation and statistical analysis should be performed to determine its significance. Furthermore, by grouping Coupe, Sedan, and Convertible Vehicle Types into "CP+SD+CN" and grouping Minivan and Station Wagon Vehicle Types into "MV+SW" the reader is also able to view Vehicle Types as a third variable.

## 4.3 Discussion

These three analyses hardly span the possible methods of extracting meaningful results from this data. More related to the scope of this chapter, the first analysis helps show that not all models offered by vehicle brands use the same amount of Active Safety Systems.

The second analysis is important when put in the context of the previous safety and technology discussions of Sects. 2 and 3. It can be hypothesized that the most prevalent systems are either the systems most proven to be effective in improving safety or the cheapest systems to offer convenience to the driver. Backup Cameras and Parking Collision Warning fall in the second category, while the rest fall in the first category. A deeper analysis is needed though to prove or disprove this hypothesis.

The final analysis is in a way the broadest of the three because one can imagine numerous types of statistical analyses to be carried out on these results. Figure 1 shows a scattered cluster of data points with an upward trend. This suggests that the relationship between MSRP and the number of Active Safety Systems is positively correlated. Statistical analysis and data from the remaining brands in the United States are needed to confirm this suggestion. Figure 1 is also significant because it shows that given a budget up to the average cost of an automotive transaction in the U.S. in the year 2020, a huge variety of several Active Safety Features is available to the consumer. Essentially, with the state of the industry like this, the consumer does not need to sacrifice much in the way of cost for a high number of Active Safety Systems.

## 4.4 Conclusion

This newly collected data and three preliminary analyses capture most of the current state of the automotive industry's adoption of Active Safety Systems. It does not, however, capture the success of individual brands' technology, their implementations of the technology, the past trends, or the future projections of the industry. Despite this limited scope, the reader is encouraged to investigate this data and perform their analysis or to extend the scope of the data set. Through this collaboration a better understanding of where research efforts should be targeted is achievable. In this way, the potential of Active Safety Systems can be fully realized in a shorter time and more efficiently.

# 5 Promising AI Applications

Artificial Intelligence (AI) has proven to be an effective tool in many applications ranging from biology to machine vision. The ability to make computer systems identify patterns and interpret meaning at a proficiency approaching that of humans will only improve over time. For this reason, it is crucial to learn to apply AI techniques and technology to the automotive industry.

## *5.1 Deep Learning*

Deep learning is one of the most promising fields in AI. Deep learning refers to learning a mapping between inputs and outputs with a deep Artificial Neural Network (ANN). An ANN is considered "deep" when it has a large number of layers in between the first (known as the input layer) and last (known as the output) layers. The layers between the input and output layers are referred to as hidden layers. The number of hidden layers is called the depth of the network, giving rise to the popular term *deep* to refer to networks that have many hidden layers. A layer is composed of nodes, or perceptrons, that perform simple computations. The number of perceptrons in a layer is called the width. A perceptron performs two linear operations before passing the output through a nonlinear function. The two linear operations are (1) multiplication of the input with a weighted value and (2) addition of that product with a bias term. The sum is then passed to a nonlinear function, called an activation function. This architecture is shown in Fig. 2 which shows a shallow ANN with 2 input nodes, 1 output node, two hidden layers, and each hidden layer's width is 3 nodes. Figure 2 also zooms in on a node to show the perceptron in a block diagram representation. On its own, a perceptron is very simple. However, organizing perceptrons into layers and then fully connecting those layers (where the outputs of the first layer are the inputs to the second layer and so on) gives the network the ability to approximate continuous functions. Informally, the universal approximation theorem states that an ANN with one or more hidden layers and an unlimited number of perceptrons in those hidden layers can approximate any continuous function with any desired amount of accuracy [19]. This is the power behind ANNs. Most research in deep learning has been spent studying methods that most efficiently, in terms of computing time and memory size, approximate a given function to some desired accuracy. Two excellent resources to learn this topic, both theoretically and practically, are [20] and [21].

The automotive research literature is starting to show the promise of deep learning in vehicle applications. Nearly all problems in vehicle applications that are solved are being revisited in a new perspective. Furthermore, some of the most difficult unsolved problems in the automotive field are being approached with deep learning.

A guideline in modeling is to develop the simplest model that can explain the dynamic phenomenon under study. Vehicle dynamic researchers rely on the funda-

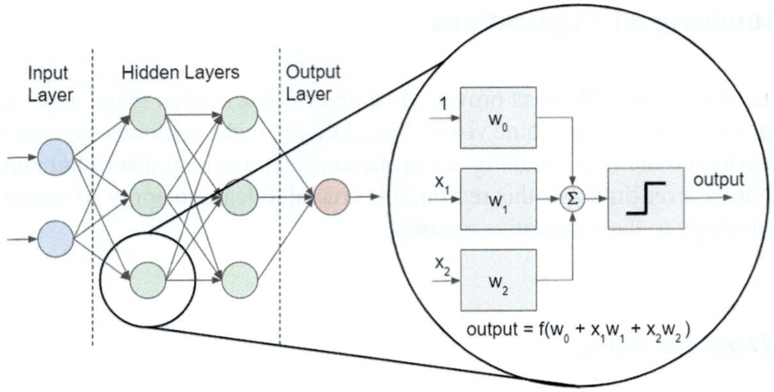


**Fig. 2** A fully connected artificial neural network with an expansion showing the perceptron in the network

mental technique of constructing ordinary differential equations (ODE)s to capture dynamic phenomena exhibited by various vehicle types. The bicycle model, also known as a single-track model, is one such model and is covered in nearly every vehicle dynamics textbook.

Recent applications of ANNs to model vehicle dynamics suggest that control systems that leverage the approximating power of ANNs can outperform controllers that rely on physics-based models. [22] compares the performance of a control system designed using traditional methods and using an ANN. The control system is a feedforward-feedback steering controller. The traditional design method uses a bicycle-car model, and the machine learning method uses an ANN. The ANN is trained on simulated data and real data collected from vehicle measurements. Their results show that the control system designed with the ANN outperforms the physics-based control system. Furthermore, they show that the ANN vehicle model predicts the vehicle's response more accurately than the bicycle car model when the simulation runs on different road conditions. However, the ANN model must be properly trained with diverse data [22].

A common complaint about ANNs is their black-box nature, which is an important concern that must be addressed by further research. Control design typically employs models developed from physics-based first principles because they are reasonably interpretable and can be studied analytically. For most vehicle models analytical methods are used to understand the different properties of the model. For instance, analytical equations can be derived from the bicycle car model to define the speed at which a vehicle transitions from understeer to oversteer. Attempting such an analytical study on an ANN vehicle model is impractical. However, this does not mean that designing models and controllers with ANNs requires blind trust. Instead, experimentation can be used to study ANN models of dynamical systems. This is a topic that requires significant progress before ANNs become a commonly used tool of control engineers.

James et al. [23] suggests that using ANNs as data-driven models can be better than a physics-derived model in studying longitudinal vehicle dynamics. They demonstrate that greater model accuracy is achieved by using a data-driven ANN model over both a nonlinear, physics-derived model and a linear state-space model. Although not emphasized as much as the model accuracy, they also propose that their ANN model leads to better model interpretation than physics-derived models. This is because the typical process of fitting a physics-derived model (such as a linearized bicycle model) may result in one or more specific model parameters that are unreasonably large or small. These unreasonable parameters lead to doubt in the model's accuracy and reliability. These unreasonable values can occur when the data exhibits non-linear behavior, and the fitting procedure converges on a less-than-reasonable value for some model parameters. For instance, a vehicle that weighs 2000 kg may be fit to data such that a mass value of 1500 kg results in the closest fit to the collected data. This may be the result of the unmodeled or nonlinear kinematics or dynamics of the vehicle, the data collection methods, noise in sensor measurements, or numerous other sources of error. Linear models are more susceptible to this issue than nonlinear models. The ANN model constructed in this paper provides more reasonable model parameter values than the linear physics-based model [23].

More closely related to Active Safety Systems is the application of Machine Learning algorithms such as Support Vector Machines and Hidden Markov Models to capture driving styles and driver behavior. These methods, and many more, have shown significant promise in their ability to identify drivers, detect drowsiness, aid warning systems, and monitor road condition and vehicle aging [24].

More cutting-edge control algorithms such as model-reference adaptive control, model predictive control, and nonlinear internal model control are finding benefits from applications of ANNs [25]. Learning-based model predictive control appears to be one of these control algorithms with great promise for robustness and performance [26].

### 5.1.1 Datasets

The challenge any deep learning researcher faces is always data. If using supervised learning techniques, the amount of labeled data is crucial. The quality of those labels is also crucial. This subsection briefly mentions some datasets to let readers know that extensive datasets exist.

- KITTI Vision Benchmark Suite [27, 28]
- Caltech Lanes Dataset [29]
- EuroCity Persons Dataset [30]
- The UAH-DriverSet [31]
- Berkeley BDD100k [32].

There are many more datasets available for aiding the task of training and testing a machine vision algorithm. Fairly up-to-date and comprehensive coverage of datasets for Machine Learning in ADS can be found in the fourth chapter of [33]. Most datasets used for deep learning are images and videos. There are no known audio datasets, such as recording audio outside the vehicle, or numerical datasets designed to improve Active Safety Systems or Automated Driving Systems (ADS).

***Example: Audio Dataset Uses*** One of the more obvious uses of an audio dataset for ADS would be to help identify approaching emergency vehicles. However, a less obvious use would be to record the host vehicle's road noise, which enables the computer systems to estimate road-tire conditions. Conditions such as road-tire coefficient of friction and low tire pressure are useful in estimating the physical limits of the tire forces. An example of a machine learning method applied to some of this is [34]. A third audio dataset example might be to collect data in the cabin of the car for monitoring driver distractions. Loud music, shouting, or soft sleepy music might indicate driver distraction levels.

***Example: Numerical Dataset Uses*** There is one numerical dataset that does not exist yet: vehicle dynamics input-output data. There is no known open-source dataset that provides a vehicle's actuator inputs (gas pedal displacement, steering wheel angle, and brake pedal pressure) and dynamic response (longitudinal and lateral). In other words, there is no open-source dataset that provides plant input-output data for a vehicle on the road, much less on various road conditions performing various maneuvers. Such data is typically closely held by brands for vehicle model validation. This secrecy is most likely due to the high cost of data collection. Obtaining this data may provide a brand an advantage over a competitor because it could improve virtual sensing technology, vehicle modeling methods, and control system design. It could also help improve predictions of how other vehicles will act. For instance, several generic vehicles representing most of the vehicles encountered on-road can be created in simulation. Then, deep learning models can be trained on each vehicle. While detecting these other vehicles, they can be classified as being one of these several modeled, generic vehicles. This would allow the prediction algorithm to assign a dynamics model to each vehicle in its vicinity and more accurately predict the future trajectories of those vehicles. More generally, this theoretical dataset will aid the development of deep learning methods applied to the Active Safety Systems discussed in this chapter.

There are many other gaps in publicly available data sets that can aid Active Safety Systems. Future studies are needed to identify automotive areas that can collect large amounts of data at a low cost. Labeling this data will most likely be costly. For instance, a new data set that allows ANNs to model tire dynamics will need to be experimentally collected using accurate measurement devices. Because this proposed data set needs to be experimentally collected, it will be expensive to obtain. The cost of high-quality data is a great challenge to apply Deep Learning to the automotive industry. However, this is a common challenge that has been approached in many ways, and in many applications. For instance, the fields of biology and medicine have benefited greatly from collecting large datasets and applying Deep Learning models to model that data. There may be techniques that mitigate the cost of data collection that can be applied to the automotive field. The goal must be to create large, accurate data sets at a low cost.

### 5.1.2 Computer Vision

The most prevalent and successful application of AI to ADS is in the field of Computer Vision. The intersection of Computer Vision and ADS includes many problems that the community is actively working to solve. In this brief overview of Computer Vision, only a subset of these problems is discussed: Object Detection, Object Tracking, and Ego-motion estimation.

Object Detection

The need to detect objects around the ego vehicle (the vehicle with the Active Safety System) is an obvious requirement. In addition to just locating the presence of obstacles, this problem typically includes classifying those objects. This is important because the vehicle's reaction to a collision with a paper box should be very different from its reaction to a collision with a pedestrian.

This problem is very challenging because of the variety of objects that the ego vehicle may encounter. The variety of ways in which the ego vehicle may encounter those objects further complicates this problem. For instance, an object may be partially hidden from the ego vehicle's sensors, making detection and classification challenging.

Cameras are common in object detection and classification. The visible light spectrum is the most popular in daylight and night-time scenarios, but the infrared spectrum is also rising in popularity for night-time applications. Both sensors are typically passive and therefore require low power and are not cost-prohibitive to many passenger vehicles. Laser scanning systems are also used, but they are active systems and require more power and typically cost significantly more than cameras [33].

At some point in the detection pipeline (the series of machine learning techniques), modern solutions to object detection and classification employ a Convolutional Neural Network (CNN). A CNN is a variant of an ANN that has shown significantly better performance on image processing than ANNs. Surprisingly, these deep and often very complex CNN-based pipelines can classify objects in real-time on modern computing systems. These computers are also small enough to be installed in passenger vehicles. The latency of classification is short enough to make object detection and classification feasible in Active Safety Systems and Automated Driving Systems (ADS)s [33].

Object Tracking

Once an object has been detected and most likely classified, the state of that object is of significant importance. Nearly all Active Safety Systems can benefit from object tracking, even though it is not necessary. This is because knowing the state of an object and the predictions of where that object will be in the future can inform the decision-making of Active Safety Systems and ADSs.

Tracking an object typically utilizes an object's temporal data to predict where that object will move to in the future. The previous states help inform the possible future trajectories of that object. Extended Kalman and particle filters are the traditional techniques used to solve this problem. However, now deep learning methods have begun to propose new and more accurate solutions to this problem [33].

Ego-vehicle Estimation

Ego-vehicle position and orientation estimation is a critical aspect of both driver assistance systems and ADSs. This technology locates the ego vehicle in space and time. Traditionally, estimators use GPS, wheel-encoders, and steering wheel angle sensors to estimate the position and orientation of the ego vehicle. However, these often require corrections for nonlinearities (wheel slip, uneven roads, suspension travel) and sensor errors. A more robust method now is to use visual odometry and laser scanning odometry. Extracting information from the visual and laser scanning sensors typically uses deep learning methods. This technique requires referencing this extracted information to a map of the surrounding environment. Obtaining such a map is a significant undertaking. It is likely that many driver assistance systems do not make use of these new estimation methods because of this challenge. However, many ADSs, which are produced by companies willing to overcome this challenge, make use of this new method because of its increased robustness and accuracy [33].

Suggested Reading

The intersection of Computer Vision and Active Safety Systems is vast and advancing every day. It is challenging to enter this space as a beginner because of the recent and substantial changes from the traditional methods. These new methods came with the rise of deep learning and breakthroughs in the cost of computation. An excellent source to get started in this intersection is [33]. It covers nearly every major research problem being studied currently in this intersection of two fields. The book also provides an excellent starting point for finding more specific research literature. Most importantly, it is recent enough to contain state-of-the-art methods and algorithms.

## *5.2 Reinforcement Learning*

Reinforcement Learning (RL) is another AI field that has been increasingly used to solve classical control problems. Reinforcement learning is fundamentally a method that maximizes a reward signal through trial-and-error search. RL frameworks consist of an agent that is able to observe, either partially or entirely, some environment and then interact with that environment through actions. For each action the agent takes, the environment presents new situations. The environment also provides the agent with some reward, which is typically a function defined by the AI researcher that captures the desired behavior of the agent. The agent learns a policy, a mapping between observed states and actions, to maximize the reward the agent receives from the reward signal throughout the whole episode. An episode is the set of actions, observations, and rewards between the initialization time of the environment and the time at which some end state is reached. Sometimes the agent also learns a model of the environment and leverages it to better explore the environment and tune its policy. A long-term concept of rewards is the value function, which defines the value of each state. The value of a state is the total expected reward that the agent can receive when starting at that state. This value function is the most important aspect of reinforcement learning because this function is not directly accessible to the agent. Instead, it must be explored and approximated. This estimated value function is what should be used by the agent to maximize rewards throughout an entire episode [35]. For a deeper and more holistic introduction to reinforcement learning the reader is referred to the popular book: [35].

***Example: Training a Dog*** Consider a dog and its owner. The owner wishes to train the dog, the agent, to sit. The dog can see that the owner is holding their hand just above the dog's nose. It can also hear the owner say, "sit." These are the agent's observations. The dog begins exploring its environment randomly. It sits, it lays down, and it walks away. These are the agent's actions. When the dog sits, the owner gives the dog a treat, which is a reward. After several repetitions, the dog comes to realize that when the owner holds their hand up and says "sit," and the dog sits, it receives a reward. If the dog lays down or walks away when the owner issues the sit command, it does not receive a reward. The agent is estimating the value function by mapping the observation of the sit command and the action of sitting to receive maximal rewards. This is one of many biological examples of training formulated as an RL problem.

The communities supporting RL have made the technology open-source friendly. Numerous RL environments simulate autonomous vehicles, robotics, and classical control problems. One Application Programming Interface (API) that is particularly useful for learning RL, as well as sharing new environments with the research community, is Open AI Gym [36]. An open-source simulator that works well with Open AI Gym is CARLA, which enables RL agents to control a passenger vehicle in a photo-realistic simulation [37]. An alternative open-source simulator is Microsoft's Airsim, which supports ground-vehicle simulation despite it being mostly oriented toward drone RL research [38].

One particular simulation project that is well-suited for testing RL algorithms on highway driving scenarios is the Highway-Env [39]. This project uses Open-AI Gym, making benchmarking and comparing RL algorithms on this environment easy and repeatable. The Highway-Env is a set of scenarios oriented towards allowing RL algorithms to perform high-level control of Automated Driving Systems (ADS). In each scenario, RL algorithms control one or more automobiles, allowing for the study of a single ADS or a fleet of ADS. Existing scenes in this package include straight and continuous stretches of highways, on-ramp merging scenarios, roundabouts, and parking lots. However, the environment, and therefore all the scenarios previously mentioned, use low-fidelity vehicle models. The lateral motion is modeled with a kinematic bicycle model, and the longitudinal motion is modeled as a point particle.

### 5.2.1 Reinforcement Learning for Automatic Emergency Braking

The goal of this subsection is to present an example of how Artificial Intelligence provides an alternative method of developing an Active Safety System. This section introduces a new scenario in the Highway-Env to show that RL can learn a policy that performs similar functions to an Automatic Emergency Braking (AEB) system. This new environment models a continuous straight stretch of a three-lane highway. The simulation approximates realistic highway driving within fixed proximity to the host vehicle. This can be conceptualized as a control volume drawn around the host vehicle and placed in the host vehicle's inertial reference frame. This control volume is shown in Fig. 3, where the orange vehicle is the host vehicle or ego vehicle. This method reduces computational load by lowering the number of vehicles other than the ego vehicle (uncontrolled vehicles).

This environment simulates 20 uncontrolled vehicles to make the road similar to moderate traffic levels on a highway. While the ego vehicle is initialized at the same highway velocity for each episode, the uncontrolled vehicles are initialized at randomly selected highway velocities. This ensures that the relative velocities at initialization time between the ego vehicle and the uncontrolled vehicles are different for each episode. The uncontrolled vehicles are randomly spaced in front of and behind the ego vehicle. The ego vehicle also starts on a randomly selected lane. This randomization significantly increases the likelihood that each episode is unique, and therefore each imminent collision is also unique. The uniqueness allows the RL agent to train on diverse episodes.

These uncontrolled vehicles can also enter and leave the control volume. The control volume encompasses less than 100 m and all three lanes of the highway. It is also important to note that the uncontrolled vehicles exhibit behaviors approximately similar to human-driving behavior on the highway. The uncontrolled vehicles' longitudinal behavior is determined by the Intelligent Driver Model (IDM), which is developed for time-continuous microscopic studies of traffic on the highway [40]. The Minimizing Overall Braking Induced by Lane change (MOBIL) model, which governs when the uncontrolled vehicle changes lanes, determines the lateral behavior [41]. Both the IDM and MOBIL models control the uncontrolled vehicles by commanding discrete meta-actions to the low-level controllers. These meta-actions are defined as (1) change to the left lane, (2) do not change lanes or change velocity, (3) change to the right lane, (4) accelerate, and (5) decelerate.

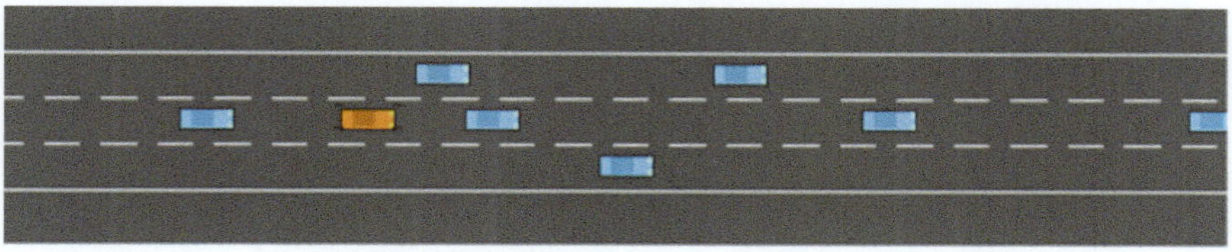

**Fig. 3** Rendering of the automatic emergency braking scenario control volume

A significant gap exists between the intended usage of the Highway-Env and the goal of this new scenario. The Highway-Env is designed for continuous control of the ego vehicle. However, AEB systems are not always controlling the vehicle for the driver. Instead, the AEB system always passively monitors both the environment and the driver's actions to determine if a collision is imminent. When a collision is determined as imminent, the AEB system activates and supplies sufficient brake torque to either stop the ego vehicle or to reduce the velocity of the ego vehicle. To bridge this gap, the environment is modified such that the agent is more likely to collide with other uncontrolled vehicles. This is done by randomly selecting uncontrolled vehicles and commanding them to brake as hard as possible to come to a stop. This behavior increases the chances that an uncontrolled vehicle in front of the host vehicle will suddenly stop, forcing the intervention system, in this case, the RL agent, to execute a collision-avoidance maneuver. A state machine models the intervention system's passive and active states. This state machine constrains the RL agent to either be inactive, active, or transitioning from active to inactive. When the agent is inactive, it cannot act, observe, or receive rewards. The agent is inactive when a collision is not imminent, as determined by a computed time-to-collision. The time-to-collision is computed in Eq. 1. When the agent is active, it can provide actions and receive rewards in response to observations. The agent is active when a collision is imminent. The agent is transitioning when the previous state was active, but there is no current imminent collision. The transitioning state keeps the agent active until the end of the episode, which is defined by either a collision or a set time duration.

$$t_{collision} = \begin{cases} \infty, & \text{if } v_{rel} \geq 0 \\ \frac{(d_{rel}+l_{veh})}{v_{rel}}, & \text{otherwise} \end{cases} \tag{1}$$

where $v_{rel}$ is the relative velocity between the controlled vehicle and the vehicle immediately in front, $d_{rel}$, is the relative distance between the controlled vehicle and the vehicle immediately in front, and $l_{veh}$ is the length of the host vehicle.

This environment created to induce collisions is also highly tunable with the following tuning parameters:

| | |
|---|---|
| Duration | The length of time of each episode |
| Road friction | The road-tire coefficient of friction |
| Time to intervene | The set time-to-collision threshold in which the RL agent becomes active |
| Vehicle density | The average space between each vehicle when first initialized on the highway |
| Vehicles count | The number of uncontrolled vehicles on the highway |
| Stopping vehicles count | The number of uncontrolled vehicles that will suddenly brake |

The host vehicle offers the agent discrete actions that directly influence the vehicle actuators: steering wheel and brakes. The possible actions are (0) do nothing, (1) 50% maximum braking, (2) 100% maximum braking, (3) 15 ° (ground-wheel) steering to the left, and (4) 15 ° (ground-wheel) steering to the right. It is important to provide multiple actions to ensure that the action space allows the agent to sufficiently explore the environment. Steering is allowed to be commanded by the agent because the agent may learn to steer away from a potential collision instead of just applying the brakes. The host vehicle can observe the world around it with measurements of its current (1) longitudinal velocity, (2) distance to the vehicle in front, (3) distance to the vehicle behind, and (4) the current lane number. These observations are reasonably realistic for a vehicle with Active Driving Assistance. The first observation can be captured with wheel speed sensors, sensors that measure the output of the motor, or GPS. The second and third observations can be measured with forward-facing radar and rear-facing radar, respectively. The fourth observation can be thought of as the vehicle pose in the world, which can come from GPS or even forward-facing cameras that can detect which lane the current vehicle occupies. With three lanes, the options for the current lane are lane zero, lane one, or lane two.

The environment is made more realistic by modeling the controlled vehicle's lateral dynamics with a nonlinear, dynamic bicycle model. The longitudinal dynamics are modeled as a single wheel with a single mass on top. The tires are modeled using the magic formula, presented in Eq. 2 with constant tire parameters. While the coupled vehicle dynamics are decoupled, the coupled tire forces are limited by Eq. 3. The specific characteristic curves for this tire model used on the host vehicle are shown in Fig. 4.

$$F_{x,y}(i) = F_z \times \mu \times sin(C \times atan(B \times i - E \times (B \times i - atan(B \times i)))) \quad (2)$$

where $F_{x,y}$ is either the tire longitudinal force or lateral force, $F_z$ is the tire normal force, $\mu$ is the tire-road coefficient of friction, $i$ is either the tire longitudinal slip or lateral slip, and $B$,$C$, and $E$ are parameters that control the shape of the tire force curve. Because the lateral and longitudinal tire forces are treated separately and as decoupled, it is helpful to define a simple method of limiting the tire force so that tire saturation is more realistic. Geometrically, this can be thought of as all the possible tire forces of $F_x$ and $F_y$ spanning a rectangle in 2D space. In reality, they should span an ellipse that fits within that rectangle because there is some amount of coupling between longitudinal and lateral forces. Idealizing a little to simplify computation, they can be limited to a circle with Eq. 3, where $F_{\max}$ is the radius of this circle.

$$\begin{aligned} &\text{if } F_y^2 + F_x^2 > F_{\max}^2, \text{ then:} \\ &F_x = \frac{F_x}{||F||} \times F_{\max} \\ &F_y = \frac{F_y}{||F||} \times F_{\max} \end{aligned} \tag{3}$$

where $F_y$ is the lateral tire force, $F_{\max}$ is the maximum tire force (defined arbitrarily to create realistic coupling effects), $F_x$ is the longitudinal tire force, and $F$ is the force vector consisting of both the longitudinal and lateral tire forces. This equation is only active when the magnitude of the coupled tire forces, normalized by $F_{\max}$, exceeds 1. This creates realistic tire saturation effects when the vehicle is both steering and braking.

A single-wheel model is used to capture the phenomenon of longitudinal vehicle dynamics. The tire model in Eq. 2 allows a maximum longitudinal force to be achieved at a specific slip value. The single-wheel model approximates the vehicle as one wheel with rotational inertia and a motor or brake torque applied at the wheel center. This wheel is connected to the chassis, which is modeled with a single mass. The equations of motion are given in Eq. 4.

$$\begin{aligned} \dot{v}_x &= \frac{F_x - F_{drag}}{m} \\ \dot{\omega} &= \frac{T - r_t \times F_x}{J} \end{aligned} \tag{4}$$

where $v_x$ is longitudinal velocity, $F_x$ is the longitudinal tire force, $F_{drag}$ is the vehicle drag force, $m$ is the chassis mass, $T$ is brake and motor torque applied to the wheel, $r_t$ is the tire radius, and $J$ is the wheel rotational inertia.

The result of these modeling efforts is a host vehicle that has longitudinal and lateral dynamics that capture the important dynamical phenomenon of ground vehicles. To ensure that the uncontrolled vehicles behave reasonably when $\mu$ varies, the uncontrolled vehicles are limited in deceleration and acceleration. This limit is in the form of a simple linear equation fit to the deceleration of the host vehicle. To keep the model simple, there is no Anti-Lock Braking System modeled and so the host vehicle locks its wheel when braking with full brake torque. This results in the vehicle decelerating at approximately 0.16 g's when $\mu = 0.2$ and at 0.6 g's when $\mu = 1.0$, representing icy and dry asphalt road conditions respectively.

The Deep Q Learning algorithm provided by the package StableBaselines 3 [42] is used to reduce the development time of an RL agent that can perform similar functions to an Automatic Emergency Braking (AEB) system. This scenario is designed to be extensible to Automatic Emergency Steering functionalities as well, but these functions were not the target of this example. Instead, AEB is the goal functionality.

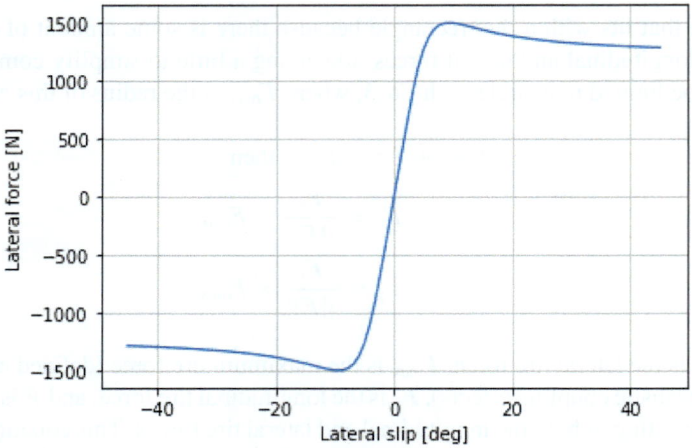


(a) Lateral Tire Curve

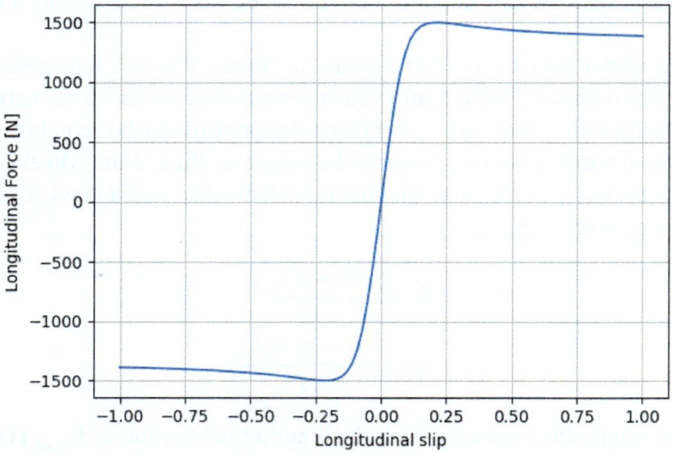


(b) Longitudinal Tire Curve

**Fig. 4** Tire characteristic curves when normal force is 1500 N N and $\mu$ is 1.0

The challenge in training an RL agent in a new, unproven environment is defining a reward function that encourages the desired behavior without allowing the agent to exploit unintended options. For example, when a reward is given at the end of an episode to the agent for not ending in a collision, then the agent learns to drive the vehicle off the road. By driving off the road, the vehicle cannot collide with another vehicle since the uncontrolled vehicles are constrained to the road. This is unintended behavior that comes out of poorly defined reward functions and environment constraints. For this reason, when the vehicle leaves the road, it is considered a collision and ends the episode.

Another challenge is deciding whether to provide positive rewards or negative rewards (penalties). Penalties for colliding with another vehicle might encourage the vehicle to end the episode as soon as possible. By ending the episode as soon as possible, the agent reduces the chance of receiving a penalty. Positive rewards might encourage the vehicle to extend the duration of an episode as long as possible. This is a desirable outcome because the best way to extend the duration of an episode is to avoid a collision.

Defining positive rewards is challenging though. A first attempt provides a reward every time an action does not result in a collision. This proved too challenging for the Deep Q Learning algorithm to learn in a reasonable amount of timesteps, approximately 20,000. The agent was unable to perform significantly better than random actions. This reward function comes from interpreting the reward engineering problem as giving a treat to a dog. The poor performance of the agent suggests this is the wrong interpretation of the reward engineering problem.

A second interpretation considers the problem of designing a reward function as being similar to the natural tendency of all things to be in a lower energy state. In terms of positive rewards, it is the inverse of this: the reward function should resemble a maximal energy state. Interpreting the problem like this inspires the need for the reward function to be continuous and differentiable. Applying this interpretation to the AEB scenario, the reward function should provide rewards based on the current host vehicle velocity. Since the goal of AEB is typically to bring the vehicle to a full stop, the target velocity of this reward function is zero. The function is therefore defined as $reward = 1 - V_{current}/V_{initial}$. When the current velocity is zero, the maximum reward per action is 1, and when the current velocity has not reduced from the initial velocity, no reward is provided.

The results of training for 20,000 timesteps with this reward are shown in Fig. 5. Training is performed for 180 episodes, and each timestep is $\frac{1}{15}$ of a second in simulation time. The average reward per episode is computed by averaging the reward received per step over each episode. The average length of the reward is computed by averaging the number of steps taken in each episode. Because both plots look similar in shape, the positive reward appears to encourage the agent to extend the episode as long as possible.

When evaluating this agent over 50 episodes for more granular assessments of the agent's performance, and using the same random number generator seed as the one used when training the agent, there were a total of 39 collisions. This startlingly high number of collisions initially indicates that there is a significant amount of development work left. However, looking deeper it is found that only 13 of those collisions occurred because the ego vehicle rear-ended another vehicle in front. In other words, the RL agent caused 13 out of the 39 collisions. Looking more closely at not-at-fault collisions reveals that nearly all of the collisions occur because another vehicle rear-ends the ego-vehicle. The smart driver models are not tuned to avoid collisions. Instead, they are tuned to provide reasonable approximations to human highway driving behavior. Collision avoidance requires more aggressive behavior, which is not captured with the non-aggressive smart vehicle models. A result is a high number of uncontrolled vehicles rear-ending the host vehicle because the uncontrolled vehicle cannot appropriately behave in emergencies.

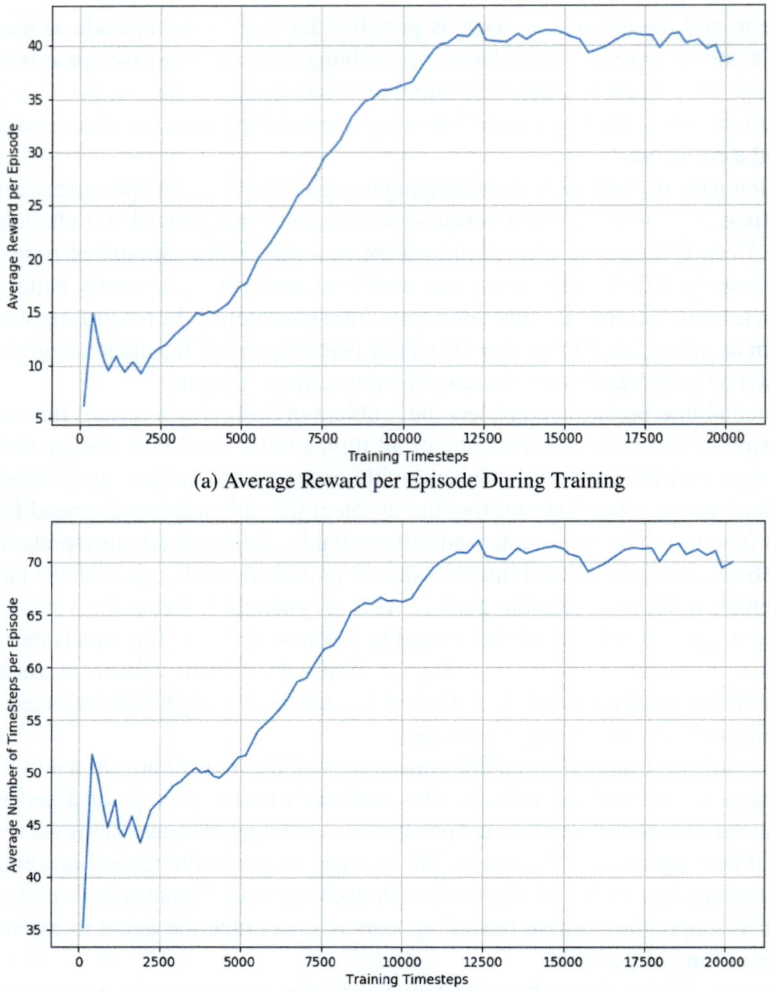


(a) Average Reward per Episode During Training

(b) Average Timesteps per Episode During Training

**Fig. 5** Deep Q learning training over training timesteps

Over those 50 evaluation episodes, all actions taken were action 2: 100% maximum braking. It was believed that by not modeling Anti-Lock Braking systems, the RL agent would learn to maximize the brake force by using a combination of 50% maximum braking and 100% maximum braking. By using a combination of these two actions it might be possible to control the slip such that it stays close to a value that maximizes longitudinal tire forces. Referring to Fig. 4, the magnitude of the optimal slip value is approximately 0.2. Unfortunately, the agent did not learn this policy. One possible reason why this RL agent does not achieve this hoped-for behavior is that it is provided with only two brake torque actions: 100 and 50%.

This limited number of options makes it difficult to finely tune a control signal that can maintain the optimal brake torque. Furthermore, Fig. 4 shows a fairly shallow drop off after the maximal longitudinal braking force. This means that the maximal brake torque generates a longitudinal brake force that is not significantly less than the optimal longitudinal brake force. It is possible that this insignificant change in longitudinal brake force when using 100% brake torque may explain why the agent does not attempt to oscillate between both possible brake actions. It should also be noted that the RL agent is able to steer. This ability is given to test the agent's ability to converge on a braking maneuver. If the agent were only allowed to brake, then random actions would achieve very good performance. However, since the agent can steer, random actions can create more collisions than a braking policy.

Further Work

Of course, numerous other traditional controllers can be used to deliver a brake torque command to the vehicle. A comprehensive comparison of the most promising traditional controllers for this and this RL agent is beyond the scope of this chapter. In general, these types of comparisons have not been investigated fully. The reason is the immense variety of control applications, RL algorithms, and traditional control architectures and algorithms.

This RL scenario is highly extensible. Further research with this scenario might apply this trained agent to all uncontrolled vehicles to investigate the question of what happens when all vehicles on the road have AEB. A possible reward function to do this might be to connect the reward to the tire longitudinal force. The ability to change the road coefficient of friction could also help train agents that are robust to environmental conditions. Another possible use of this environment to design a new type of control system is to create an RL agent that performs continuous Anti-lock Braking. Modern Anti-lock Braking Systems often scare drivers because the brake pressure is being released in a discrete fashion. If there could be brake systems that could deliver continuous control to each wheel, an RL agent might be able to learn to control these actuators continuously to achieve maximum brake force. Finally, this scenario should allow the training of an agent that can both steer and brake to avoid a collision. This agent would combine Automatic Emergency Steering and Automatic Emergency Braking. Defining a proper reward function for this will be challenging, but worth the effort.

Furthermore, AEB systems are not only designed for highway driving. Significant work is required to design a new scenario in the Highway-Env to allow for the design of an RL agent that operates in low-speed environments. Furthermore, an additional environment might be desirable to combine both low-speed and high-speed environments. Listing the requirements of these two environments exposes the problem of selecting the intervention system's operational design domains. This is an important problem that must be considered carefully before designing any Active Safety System because it imposes critical assumptions and constraints on the system design.

Using a state machine is also not the only way to design an RL agent to perform an AEB function. Instead, the ego vehicle can be given the same human driver models as the uncontrolled vehicles. The agent's observations can then be extended so that the agent can observe the driver. The agent is then tasked with deciding when to act and how to act with the driver to avoid a collision. In this new scenario, the RL agent should only be given the ability to brake, since an AEB system is the intended functionality. The challenge with this new setup is the design of an appropriate reward function. This is because the RL agent must decide when and how to act while the driver is also acting. An ideal RL agent should not override the driver's demands unless a collision is imminent. Also if the driver's demands are not going to avoid or optimally reduce the damage of that collision the agent should also step in and act on the driver's behalf. Such an ideal behavior might be challenging to achieve with an RL algorithm.

These considerably large extensions of this work are just a sample of the possible extensions of RL to AEB systems. One can further imagine a set of applications of RL to other intervention systems such as Automatic Emergency Steering. Furthermore, it might be possible to use RL algorithms to design highly robust and adaptive control systems for more generic vehicle applications. This brief discussion of future work indicates the vast amount of research opportunities in the application of RL to Active Safety Systems.

Other AI Applications

Autonomous Driving Systems appear to be an increasing interest in research applications of Artificial Intelligence. [43] showed that not only can Imitation Learning control a vehicle end-to-end, but it also can do so with cheaper onboard sensors than a vehicle that uses Model Predictive Control. The authors trained an Imitation Learning algorithm using an off-road vehicle controlled by high-quality sensors and Model Predictive Control. When given control of the vehicle, the Imitation Learning algorithm was able to perform similarly with lower-cost sensors than the original vehicle that used model predictive control.

## 6 Final Thoughts

One great challenge that is not discussed in this chapter is the verification and validation of these Active Safety Systems. While it is true that several standards exist to guide engineers in this process such as ISO 26262, ISO 21448, IEC 61508, and many others, this challenge is unique to every OEM and perhaps unique to each model and trim. To integrate AI applications into vehicles, verification and validation must be done. Currently, there are many efforts to do this efficiently and cost-effectively.

The common way to verify the safety of Automated Driving Systems (ADS) is to collect huge amounts of data. The general idea is that if the ADS can achieve a lower statistical likelihood of causing a collision than a human in the desired Operational Design Domain (ODD), then the ADS is valid to operate in that ODD. As Active Safety Systems become more complex and assume more of the Dynamic Driving

Task (DDT), verification and validation challenges grow larger and more difficult to overcome. The solution to this will probably become similar to the solution used for ADS: lots of data. In addition to collecting all this data for collisions, most companies also collect sensor data from their extensive sensor arrays. This data is used to improve different parts of the autonomous software stack, such as machine vision components and decision-making components. One area that is not using this vast amount of data is vehicle dynamic modeling and control.

Never before in the history of the automotive industry has there been the opportunity to collect data from every vehicle with Active Safety Systems, which is a significant portion of all vehicles on the road as Sect. 4 indicates. As a result of this newly available data, there must be ways to leverage its information to revisit current solutions and improve on them and solve new, unsolved problems.

***Example: Self-Learning Vehicles*** Suppose a new vehicle manufacturer enters the market and immediately can put thousands of vehicles on the road, driven by consumers. These vehicles have all the Driving Control Assistance systems covered in Sect. 2. However, these vehicles have an additional technology on-board: machine learning models, controllers, and estimators. Huge amounts of data are collected nearly every mile from these vehicles. Now consider that each of these thousands of vehicles is manufactured with tolerances. No two vehicles are the same. Future work might be training deep learning models that can capture this vehicle-specific uncertainty. If the data collected from measurements and estimators can be used to train a deep learning model of the vehicle, this model can be improved as more data is collected. Imagine controllers that leverage this continually collected data. These controllers could improve the performance and safety of the vehicle the longer the vehicle is operating. As the vehicle changes over time, (such as wear and tear, tire replacements, or new batteries) these data-driven models, controllers, and estimators also adapt. It is therefore conceivable that there are many applications of these models, such as alerting drivers to when systems might fail in the future due to mechanical wear or system faults. These data-driven models and controllers can also overcome the challenge of variation between the same model of car, as each car builds and trains separate models, controllers, and estimators. In a sense, the vehicle learns how to better observe itself, better predict how it will behave, and better control itself over time.

While verification and validation are some of the many topics not covered in this chapter, AI and its sub-fields may help solve new challenges or improve on existing solutions. Not only this, but the safety benefits of leveraging the available data from these Active Safety Systems are too great to ignore. The industry and the research communities must learn to collect and use this data in anonymous, secure, and beneficial ways.

AI is not the only technology that offers huge improvements to Active Safety Systems and Autonomous Driving Systems. Connected vehicles, that is vehicles that can wirelessly communicate to one another, are the next natural step for the automotive industry. However, AI is not separate from this new field of study. Instead, it too is offering huge gains to connected vehicles.

The future of the automotive industry can be unlocked by applying AI to improve existing solutions and find solutions to new problems. The goal of having zero collisions on the road is a worthy goal to orient the research and development of Active Safety Systems and ADS. This chapter shows that integrating AI technology and data-driven components, where possible, is helping the industry approach this goal.

## References

1. Committee ASSS (2021) Active safety systems terms and definitions. https://doi.org/10.4271/J3063_202103
2. Committee, O.R.A.D.O (2021) Taxonomy and definitions for terms related to driving automation systems for on-road motor vehicles. https://doi.org/10.4271/J3016_202104
3. Zador P, Krawchuk S, Voas R (2000-8) Final report—automotive collision avoidance system (acas) program (2000-8)
4. Maurer M (2012) Forward collision warning and avoidance. Springer London, London, pp 657–687. https://doi.org/10.1007/978-0-85729-085-4_25
5. Ben-Yaacov A, Maltz M, Shinar D (2002) Effects of an in-vehicle collision avoidance warning system on short-term and long-term driving performance. Hum Fact 44(2):335–342. https://doi.org/10.1518/0018720024497925PMID: 12452277
6. Narote SP, Bhujbal PN, Narote AS, Dhane DM (2018) A review of recent advances in lane detection and departure warning system. Pattern Recogn 73:216–234. https://doi.org/10.1016/j.patcog.2017.08.014
7. Gayko J (2012) Lane departure and lane keeping. Springer London, London, pp 657–687. https://doi.org/10.1007/978-0-85729-085-4_26
8. Reagan IJ, Cicchino JB, Kerfoot LB, Weast RA (2018) Crash avoidance and driver assistance technologies—are they used? Transp Res Part F Traffic Psychol Behaviour 52:176–190. https://doi.org/10.1016/j.trf.2017.11.015
9. Cicchino JB (2017) Effectiveness of forward collision warning and autonomous emergency braking systems in reducing front-to-rear crash rates. Accid Anal Prev 99:142–152. https://doi.org/10.1016/j.aap.2016.11.009
10. Eskandarian A (ed) (2012) Handbook of intelligent vehicles, 1 ed. Springer, London. https://doi.org/10.1007/978-0-85729-085-4
11. Jiménez F (2018) Chapter 6—driver assistance systems and safety systems. In: Jiménez F (ed) Intelligent vehicles. Butterworth-Heinemann, pp. 209–226. https://doi.org/10.1016/B978-0-12-812800-8.00006-0
12. Fraichard T, Asama H (2004) Inevitable collision states-a step towards safer robots? Adv Rob 18(10):1001–1024
13. Millán-Blanquel L, Veres SM, Purshouse RC (2020) Ethical considerations for a decision making system for autonomous vehicles during an inevitable collision. In: 2020 28th mediterranean conference on control and automation (MED). IEEE, pp 514–519
14. Geisslinger M, Poszler F, Betz J, Lütge C, Lienkamp M (2021) Autonomous driving ethics: from trolley problem to ethics of risk. Philos Technol 1–23
15. Wakeman K, Moore MJ, Zuby DS, Hellinga LA (2019) Effect of Subaru eyesight on pedestrian-related bodily injury liability claim frequencies. In: Proceedings of the 26th international technical conference on the enhanced safety of vehicles. National Highway Transportation Safety Administration
16. Cicchino JB (2018) Effects of blind spot monitoring systems on police-reported lane-change crashes. Traffic Inj Prevention 19(6):615–622. https://doi.org/10.1080/15389588.2018.1476973PMID: 29927678

17. Bischak C (2021) Compendium of hldi collision avoidance research. Bulletin 37(12). https://www.itskrs.its.dot.gov/node/209273. Accessed 16 July 2021
18. Average new-vehicle prices up nearly 3% year-over-year in february 2020, according to kelley blue book (2021). URL https://mediaroom.kbb.com/2020-03-03-Average-New-Vehicle-Prices-Up-Nearly-3-Year-Over-Year-in-February-2020-According-to-Kelley-Blue-Book. Accessed 16 July 2021
19. Cybenko G (1989) Approximation by superpositions of a sigmoidal function. Math Cont Signals Syst 2(4):303–314
20. Goodfellow I, Bengio Y, Courville A (2016) Deep learning. MIT Press. http://www.deeplearningbook.org
21. Géron A (2019) Hands-on machine learning with Scikit-Learn, Keras, and TensorFlow: concepts, tools, and techniques to build intelligent systems. O'Reilly Media
22. Spielberg NA, Brown M, Kapania NR, Kegelman JC, Gerdes JC (2019) Neural network vehicle models for high-performance automated driving. Sci Robot 4(28). https://doi.org/10.1126/scirobotics.aaw1975
23. James SS, Anderson SR, Lio MD (2020) Longitudinal vehicle dynamics: a comparison of physical and data-driven models under large-scale real-world driving conditions. IEEE Access 8:73714–73729. https://doi.org/10.1109/ACCESS.2020.2988592
24. Meiring GAM, Myburgh HC (2015) A review of intelligent driving style analysis systems and related artificial intelligence algorithms. Sensors 15(12):30653–30682. https://doi.org/10.3390/s151229822
25. Aly A, Salem F (2013) Vehicle suspension systems control: a review. Int J Control Autom Syst 2:2
26. Hewing L, Wabersich KP, Menner M, Zeilinger MN (2020) Learning-based model predictive control: toward safe learning in control. Ann Rev Control Rob Auton Syst 3(1):269–296. https://doi.org/10.1146/annurev-control-090419-075625
27. Geiger A, Lenz P, Stiller C, Urtasun R (2013) The kitti vision benchmark suite. http://www.cvlibs.net/datasets/kitti/. Accessed 21 July 2021
28. Geiger A, Lenz P, Stiller C, Urtasun R (2013) Vision meets robotics: the kitti dataset. Int J Rob Res (IJRR)
29. Aly M (2008) Caltech lanes dataset. http://www.mohamedaly.info/datasets/caltech-lanes. Accessed: 21 July 2021
30. Braun M, Krebs S, Flohr F, Gavrila D (2018) The eurocity persons dataset: a novel benchmark for object detection. ArXiv abs/1805.07193
31. Romera E, Bergasa LM, Arroyo R (2016) Need data for driver behaviour analysis? Presenting the public UAH-DriveSet. In: 2016 IEEE 19th international conference on intelligent transportation systems (ITSC), pp 387–392. https://doi.org/10.1109/ITSC.2016.7795584
32. Yu F, Chen H, Wang X, Xian W, Chen Y, Liu F, Madhavan V, Darrell T (2020) Bdd100k: a diverse driving dataset for heterogeneous multitask learning
33. Janai J, Güney F, Behl A, Geiger A (2020) Computer vision for autonomous vehicles: problems, datasets and state of the art. IEEE
34. Doğan D (2017) Road-types classification using audio signal processing and svm method. In: 2017 25th signal processing and communications applications conference (SIU), pp 1–4. https://doi.org/10.1109/SIU.2017.7960154
35. Sutton RS, Barto AG (2018) Reinforcement learning: an introduction. MIT Press
36. Brockman G, Cheung V, Pettersson L, Schneider J, Schulman J, Tang J, Zaremba W (2016) Openai gym
37. Dosovitskiy A, Ros G, Codevilla F, Lopez A, Koltun V (2017) CARLA: an open urban driving simulator. In: Proceedings of the 1st annual conference on robot learning, pp 1–16
38. Shah S, Dey D, Lovett C, Kapoor A (2017): Airsim: high-fidelity visual and physical simulation for autonomous vehicles. In: Field and service robotics. https://arxiv.org/abs/1705.05065
39. Leurent E (2018) An environment for autonomous driving decision-making. https://github.com/eleurent/highway-env

40. Treiber M, Hennecke A, Helbing D (2000) Congested traffic states in empirical observations and microscopic simulations. Phys Rev E 62:1805–1824
41. Kesting A, Trieber M, Helbing D (2007) General lane-changing model MOBIL for car-following models. Transp Res Rec J Transp Res Board 1999:86–94
42. Raffin A, Hill A, Ernestus M, Gleave A, Kanervisto A, Dormann N (2019) Stable baselines3. https://github.com/DLR-RM/stable-baselines3
43. Pan Y, Cheng CA, Saigol K, Lee K, Yan X, Theodorou EA, Boots B (2019) Imitation learning for agile autonomous driving. Int J Rob Res 39:286–302. https://doi.org/10.1177/0278364919880273

# Model Predictive Control for Safe Autonomous Driving Applications

**Ivo Batkovic, Mario Zanon, and Paolo Falcone**

**Abstract** Although Model Predictive Control is widely used in motion planning and control for autonomous driving applications, accommodating closed-loop stability with respect to an *arbitrary* reference trajectory and avoidance of pop-up or moving obstacles is still an open problem. While it is well-known how to design a closed-loop stable MPC with respect to a reference trajectory that satisfies the system dynamics, this chapter discusses how to guarantee stability of a vehicle motion planner and controller when a user-provided *arbitrary* reference is used. Furthermore, the proposed MPC scheme enables recursive collision-avoidance constraint satisfaction in the presence of pop-up or moving obstacles (e.g., pedestrians, cyclists, human-driven vehicles), provided that their predicted future motion trajectory is available together with some uncertainty bound and satisfies some mild requirement. The proposed motion planner and controller is demonstrated through simulations.

## 1 Introduction

In order to fully deploy highly automated driving technologies, vehicles need to be able not only to reliably sense their surrounding environment, but also to *safely* interact with it. To that end, challenging problems need to be solved that span from

I. Batkovic (✉)
Zenseact AB, Lindholmspiren 2, Gothenburg 417 56, Sweden

Chalmers University of Technology, Chalmersplatsen 4, Gothenburg 412 96, Sweden
e-mail: ivo.batkovic@zenseact.com

M. Zanon
IMT School for Advanced Studies Lucca, Piazza S.Francesco, 19, 55100 Lucca, LU, Italy
e-mail: mario.zanon@imtlucca.it

P. Falcone
Engineering Department, Università di Modena e Reggio Emilia, Via Universitá, 4, 41121 Modena, MO, Italy
e-mail: paolo.falcone@unimore.it; falcone@chalmers.se

Chalmers University of Technology, Chalmersplatsen 4, Gothenburg 412 96, Sweden

 
Y. L. Murphey et al. (eds.), *AI-enabled Technologies for Autonomous and Connected Vehicles*, Lecture Notes in Intelligent Transportation and Infrastructure,
https://doi.org/10.1007/978-3-031-06780-8_9

robust and reliable sensors design (e.g., cameras, lidars, radars, GPS, HD-maps) to the development of robust perception and motion planning and control algorithms. While *safe* autonomous driving in complex environments still remains an open problem, research is progressing in the fields of localization [1, 2], object detection and tracking [3–5], and planning [6–8], to move beyond the current state of the art.

In this chapter, we focus on the vehicle motion planning and control problems for autonomous driving applications. We build our results on the Model Predictive Control (MPC) technique, as it has been proven to be a convenient design tool for self-driving applications including, e.g., optimal coordination [9–11], energy consumption minimization [12–14], and planning [15–22]. The theory of MPC is equipped with well-known, well-established tools to enforce closed-loop stability w.r.t. a reference trajectory, while ensuring constraint satisfaction [23, 24]. However, these results build upon assumptions that can be challenging, or impossible, to satisfy in practical autonomous driving applications.

Such challenges include: (a) enforcing closed-loop stability w.r.t. reference trajectories that do not satisfy the system dynamics; and (b) providing recursive feasibility guarantees in uncertain environments (e.g., in presence of pop-up or moving obstacles). Indeed, if the MPC controller is provided with a reference trajectory that *does not* satisfy the system dynamics, then the well-known results for asymptotic stability of the closed-loop system do no longer hold. Furthermore, in an autonomous driving setting, the vehicle needs to interact with other road users that may appear almost anywhere and at any point in time within the sensor range. Ensuring that the vehicle motion planner can *persistently* be able to avoid collisions with other road users *at all future times* in such uncertain settings still remains an open problem. To address a) we resort to Input-to-State Stability (ISS) analysis to prove closed-loop stability also when *infeasible* (in the sense that they do not satisfy the system dynamics) references are used; and to address (b) we provide a scheme which enables recursive collision-avoidance constraint satisfaction in the presence of pop-up or moving obstacles, provided that the uncertainty on their predicted motion is *not growing* as new information is available about the environment and the road users therein.

This chapter is structured as follows. In Sect. 2 we introduce the problem of safely planning and controlling the vehicle within an uncertain environment, while Sect. 3 outlines the Model Predictive Flexible trajectory Tracking Control (MPFTC) framework. In Sect. 4 we show how stability can be ensured when an infeasible reference trajectory is used, and in Sect. 5 we provide recursive feasibility guarantees for uncertain settings with suddenly appearing (pop-up) obstacles. Finally, in Sect. 6 we provide clarifying examples that illustrate the results derived in Sects. 4 and 5.

## *1.1 Notation*

We denote a discrete-time nonlinear system by

$$\mathbf{x}_{k+1} = f(\mathbf{x}_k, \mathbf{u}_k), \tag{1}$$

where $\mathbf{x}_k \in \mathbb{R}^{n_\mathbf{x}}$ and $\mathbf{u}_k \in \mathbb{R}^{n_\mathbf{u}}$ are the state and input vectors at time $k$, respectively. The state and inputs are subject to two categories of constraints: *a-priori known* constraints $h(\mathbf{x}, \mathbf{u}) : \mathbb{R}^{n_\mathbf{x}} \times \mathbb{R}^{n_\mathbf{u}} \to \mathbb{R}^{n_h}$; and *a-priori unknown* constraints $g(\mathbf{x}, \mathbf{u}) : \mathbb{R}^{n_\mathbf{x}} \times \mathbb{R}^{n_\mathbf{u}} \to \mathbb{R}^{n_g}$, i.e., the state and inputs must satisfy $h(\mathbf{x}, \mathbf{u}) \leq 0$ and $g(\mathbf{x}, \mathbf{u}) \leq 0$, where the inequalities are defined element-wise.

We use the notation $g_{n|k}(\mathbf{x}, \mathbf{u})$ to denote function $g$ at time $n$, given the information available at time $k$. Moreover, we will denote by $g_n(\mathbf{x}, \mathbf{u}) := g_{n|\infty}(\mathbf{x}, \mathbf{u}) = g_{n|k}(\mathbf{x}, \mathbf{u})$, $\forall\, k \geq n$ the actual constraint, while in general $g_{n|k}(\mathbf{x}, \mathbf{u}) \neq g_n(\mathbf{x}, \mathbf{u})$, $\forall\, k < n$. Note that for a-priori known constraints $h_{n|k}(\mathbf{x}, \mathbf{u}) := h_n(\mathbf{x}, \mathbf{u})$ holds $\forall\, k$ by definition. We apply the same notation to state and inputs, e.g., $\mathbf{x}_{n|k}$ and $\mathbf{u}_{n|k}$ denote the predicted state and input at time $n$ given the information available at the current time $k$. In addition, to denote a set of integers, we use $\mathbb{I}_a^b := \{a, a+1, ..., b\}$.

## 2 Problem Description

In this section we formulate the problem of *safely* planning and controlling the motion of a vehicle within an environment with static and moving obstacles, as a *Model Predictive Control Problem.*

The dynamical model of the vehicle, and the state and input constraints it is subject to, are denoted as in Sect. 1.1. Function $h$ includes actuator limitations, design and safety (e.g., distance from the lane boundaries) constraints and is known beforehand. Function $g$ models a-priori unknown constraints such as, e.g., pop-up (suddenly appearing) or moving obstacles, whose exact future motion trajectories are unknown. Our aim is then to control the vehicle motion described by (1) such that both known constraints $h_k(\mathbf{x}, \mathbf{u}) \leq 0$ and a-priori unknown constraints $g_k(\mathbf{x}, \mathbf{u}) \leq 0$ are satisfied at all times $k$.

Our first and essential objective is to guarantee safety of (1), which we define formally as follows.

**Definition 1** (*Safety*) A controller is said to be safe in a given set $\mathcal{S} \subseteq \mathbb{R}^{n_\mathbf{x}}$ if $\forall\, \mathbf{x} \in \mathcal{S}$ it generates control inputs $\mathbf{U} = \{\mathbf{u}_0, ..., \mathbf{u}_\infty\}$ and corresponding state trajectories $\mathbf{X} = \{\mathbf{x}_0, \mathbf{x}_1, ..., \mathbf{x}_\infty\}$ such that $h_k(\mathbf{x}_k, \mathbf{u}_k) \leq 0$ and $g_k(\mathbf{x}_k, \mathbf{u}_k) \leq 0$, $\forall\, k \geq 0$.

Our second objective is to control the system such that the state and input $\mathbf{x}_k$, $\mathbf{u}_k$ track a parameterized reference trajectory $\mathbf{r}(\tau) := (\mathbf{r}^\mathbf{x}(\tau), \mathbf{r}^\mathbf{u}(\tau))$ *as closely as safety allows*. If the reference parameter $\tau$ is selected to be time, its natural dynamics are given by

$$\tau_{k+1} = \tau_k + t_\mathrm{s}, \tag{2}$$

where $t_\mathrm{s}$ is the sampling time for sampled-data systems and $t_\mathrm{s} = 1$ in the discrete-time framework. Given the presence of nonlinear dynamics and constraints, we frame the problem in the context of MPC. Note that if $\tau$ is forced to follow its natural dynamics (2), then the reference tracking problem in the absence of a-priori unknown constraints $g$ is a standard MPC problem and, therefore, inherits all stability guarantees, but also a possibly aggressive behavior when the initial state is far from the reference. In order to tackle that issue and be able to deal with a-priori unknown constraints, we adopt next the concept of *Model Predictive Flexible Trajectory Tracking Control* introduced in [25].

# 3 Model Predictive Flexible Trajectory Tracking Control

The main idea in Model Predictive Flexible trajectory Tracking Control (MPFTC) is to avoid aggressive behaviors by adapting the dynamics of the reference trajectory by means of the parameter $\tau$, which acts as a fictitious time for the reference, through relaxed dynamics given by

$$\tau_{k+1} = \tau_k + t_\mathrm{s} + \upsilon_k, \tag{3}$$

where $\upsilon$ is an additional auxiliary control input and $\tau$ becomes an auxiliary state. Note that the system dynamics are unchanged and the fictitious time $\tau$ makes only the reference dynamics deviate from the natural ones.

We formulate the MPFTC problem as the following MPC problem

$$\begin{aligned}
V(\mathbf{x}_k, \tau_k) := \min_{\substack{\mathbf{x},\mathbf{u}\\ \tau,\upsilon}} & \sum_{n=k}^{k+N-1} q_\mathbf{r}(\mathbf{x}_{n|k}, \mathbf{u}_{n|k}, \tau_{n|k}) + w{\upsilon_{n|k}}^2 \\
& + p_\mathbf{r}(\mathbf{x}_{k+N|k}, \tau_{k+N|k}) && \text{(4a)}\\
\text{s.t. } \mathbf{x}_{k|k} &= \mathbf{x}_k, \quad \tau_{n|k} = \tau_k \quad , && \text{(4b)}\\
\mathbf{x}_{n+1|k} &= f(\mathbf{x}_{n|k}, \mathbf{u}_{n|k}), \quad n \in \mathbb{I}_k^{k+N-1}, && \text{(4c)}\\
\tau_{n+1|k} &= \tau_{n|k} + t_\mathrm{s} + \upsilon_{n|k}, \quad n \in \mathbb{I}_k^{k+N-1}, && \text{(4d)}\\
h_n(\mathbf{x}_{n|k}, \mathbf{u}_{n|k}) &\le 0, \quad n \in \mathbb{I}_k^{k+N-1}, && \text{(4e)}\\
g_{n|k}(\mathbf{x}_{n|k}, \mathbf{u}_{n|k}) &\le 0, \quad n \in \mathbb{I}_k^{k+N-1}, && \text{(4f)}\\
\mathbf{x}_{k+N|k} &\in \mathcal{X}_\mathbf{r}^\mathrm{f}(\tau_{k+N|k}), && \text{(4g)}
\end{aligned}$$

where $k$ is the current time, $N$ is the prediction horizon, and $w > 0$ is the weight defining the cost associated with the auxiliary input $\upsilon_{n|k}$. In tracking MPC, typical choices for the stage and terminal costs are

$$q_\mathbf{r}(\mathbf{x}_{n|k}, \mathbf{u}_{n|k}, \tau_{n|k}) = \begin{bmatrix} \Delta\mathbf{x}_{n|k} \\ \Delta\mathbf{u}_{n|k} \end{bmatrix}^\top W \begin{bmatrix} \Delta\mathbf{x}_{n|k} \\ \Delta\mathbf{u}_{n|k} \end{bmatrix}, \tag{5}$$

$$p_\mathbf{r}(\mathbf{x}_{k+N|k,\tau_{k+N}}) = \Delta\mathbf{x}_{k+N|k}^\top P \Delta\mathbf{x}_{k+N|k}, \tag{6}$$

$$\Delta\mathbf{x}_{n|k} := \mathbf{x}_{n|k} - \mathbf{r}^\mathbf{x}(\tau_{n|k}), \quad \Delta\mathbf{u}_{n|k} := \mathbf{u}_{n|k} - \mathbf{r}^\mathbf{u}(\tau_{n|k}),$$

where the matrices $W \in \mathbb{R}^{(n_\mathbf{x}+n_\mathbf{u})\times(n_\mathbf{x}+n_\mathbf{u})}$ and $P \in \mathbb{R}^{n_\mathbf{x}\times n_\mathbf{x}}$ are symmetric positive-definite and $\mathbf{r}(\tau_{n|k}) = (\mathbf{r}^\mathbf{x}(\tau_{n|k}), \mathbf{r}^\mathbf{u}(\tau_{n|k}))$ is a user-provided reference trajectory. Note that the cost functions $q_\mathbf{r}$ and $p_\mathbf{r}$ depend on $\tau_{n|k}$ only through the reference trajectory and that the cost is built with convex quadratic forms just for simplicity, while the proposed framework can accommodate more general cost definitions. The predicted state and controls are defined as $\mathbf{x}_{n|k}$, $\tau_{n|k}$, and $\mathbf{u}_{n|k}$, $v_{n|k}$ respectively, and are subject to constraints (4b)–(4f). Constraint (4b) initializes the state prediction to the current system state $\mathbf{x}_k$, (4c)–(4d) impose that the predicted states generated from the system dynamics and the controls $\mathbf{u}_{n|k}$, (4e) enforces constraints stemming from, e.g., actuator physical limitations and reference trajectory bounds, while constraint (4f) forces the predicted state and controls to satisfy constraints imposed to avoid the collision with obstacles detected by a perception layer, hence, not known a-priori. Finally, (4g) is a terminal set constraint, which, differently from standard formulations, depends on the auxiliary state $\tau_{k+N}$ relative to the reference parameter. Note that, while the introduction of one additional state and control results in an increased computational complexity, such increase is typically small, since these variables have decoupled linear dynamics.

The stability proof for MPFTC has been provided in [25]. However, in Sect. 4 we will recall the necessary assumptions and state the stability theorem for completeness. We ought to stress that, in the absence of a-priori unknown constraints, the necessary assumptions for stability reduce to those commonly used to prove stability in standard MPC schemes. In the presence of a-priori unknown constraints, these assumptions may become too restrictive leading to the lack of recursive feasibility. Hence, in [25] we proposed to relax the standard assumptions while introducing new ones which can be summarized as follows. We first require one to "be able to predict a (possibly conservative) worst-case scenario" for the a-priori unknown constraints. Since the uncertainty typically becomes too large in rather short times, we need a second assumption which postulates the existence of a safe set, consisting of states for which the a-priori unknown constraints can be neglected. We will provide a more detailed discussion on this aspect in Sect. 5, where we will also discuss how to enforce these assumptions for autonomous driving.

## 4 Tracking an Infeasible Reference

We recall that our objective is to control the system (1), such that the state $\mathbf{x}_k$ tracks a *user-provided parameterized reference trajectory* $\mathbf{r}(\tau) = (\mathbf{r}^\mathbf{x}(\tau), \mathbf{r}^\mathbf{u}(\tau))$ as closely as possible. For the remainder of the chapter, we assume that the reference trajectory is parameterized with the time parameter $t$, and that it follows the natural dynamics (2). Furthermore, we will refer to any time dependence of the reference using the notation $(\mathbf{r}^\mathbf{x}_k, \mathbf{r}^\mathbf{u}_k) := (\mathbf{r}^\mathbf{x}(\tau_k), \mathbf{r}^\mathbf{u}(\tau_k))$, where $\tau$ is the fictitious time introduced in (3).

In order to prove stability, we recall the following standard assumptions, see, e.g., [24, 26].

**Assumption 1** (*System and cost regularity*) The system model $f$ is continuous, and the stage cost $q_{\mathbf{r}} : \mathbb{R}^{n_\mathbf{x}} \times \mathbb{R}^{n_\mathbf{u}} \times \mathbb{R} \rightarrow \mathbb{R}_{\geq 0}$, and terminal cost $p_{\mathbf{r}} : \mathbb{R}^{n_\mathbf{x}} \times \mathbb{R} \rightarrow \mathbb{R}_{\geq 0}$, are continuous at the origin and satisfy $q_{\mathbf{r}}(\mathbf{r}_k^\mathbf{x}, \mathbf{r}_k^\mathbf{u}, \tau_k) = 0$, and $p_{\mathbf{r}}(\mathbf{r}_k^\mathbf{x}, \tau_k) = 0$. Additionally, $q_{\mathbf{r}}(\mathbf{x}_k, \mathbf{u}_k, \tau_k) \geq \alpha_1(\|\mathbf{x}_k - \mathbf{r}_k^\mathbf{x}\|)$ for all feasible $\mathbf{x}_k$, $\mathbf{u}_k$, and $p_{\mathbf{r}}(\mathbf{x}_k, \tau_k) \leq \alpha_2(\|\mathbf{x}_k - \mathbf{r}_k^\mathbf{x}\|)$, where $\alpha_1$ and $\alpha_2$ are $\mathcal{K}_\infty$-functions.

This assumption is common in MPC [24, 26] and can be relaxed in case one wants to account for so-called "economic" costs, see, e.g., [27–34] for a generic theory and [12, 31] for applications to autonomous driving. The MPFTC framework can be extended to yield an approximate economic MPC, similar to the one proposed in [35–38].

Assumptions 2, 3 are introduced as for standard MPC formulations, where no difference is made between a-priori known $h$ and unknown constraints $g$. Nevertheless, we distinguish the case where the assumption is required to hold for $h$ only from the case where the assumptions hold for both constraints, which is too restrictive in practice.

**Assumption 2** (*Reference feasibility*) The reference is feasible for the system dynamics, i.e., $\mathbf{r}^\mathbf{x}(t + t_\mathrm{s}) = f(\mathbf{r}^\mathbf{x}(t), \mathbf{r}^\mathbf{u}(t))$, and:

(a) the reference satisfies the known constraints (4e), i.e., $h_n(\mathbf{r}^\mathbf{x}(t_n), \mathbf{r}^\mathbf{u}(t_n)) \leq 0$, for all $n \in \mathbb{I}_0^\infty$;
(b) the reference satisfies the unknown constraints (4f), i.e., $g_{n|k}(\mathbf{r}^\mathbf{x}(t_n), \mathbf{r}^\mathbf{u}(t_n)) \leq 0$, for all $n, k \in \mathbb{I}_0^\infty$.

Assumption 2b is a strong assumption since it assumes that the reference is feasible for the unknown constraints for all future times, i.e., at time $k$ the constraint $g_{n|k+1}$ is also assumed to be satisfied. This is clearly unrealistic for autonomous driving, since pedestrians and other vehicles may at some time cross the road and make the reference infeasible. Therefore, Assumption 2a serves as a relaxed version which is more realistic and will be used later, while dropping Assumption 2b.

**Assumption 3** (*Stabilizing Terminal Conditions*) There exists a parametric stabilizing terminal set $\mathcal{X}_\mathbf{r}^\mathrm{f}(t)$ and a terminal control law $\kappa_\mathbf{r}^\mathrm{f}(\mathbf{x}, t)$ yielding:

$$\mathbf{x}_+^\kappa = f(\mathbf{x}, \kappa_\mathbf{r}^\mathrm{f}(\mathbf{x}, t)), \qquad\qquad t_+ = t + t_\mathrm{s},$$

such that $p_\mathbf{r}(\mathbf{x}_+^\kappa, t_+) - p_\mathbf{r}(\mathbf{x}, t) \leq -q_\mathbf{r}(\mathbf{x}, \kappa_\mathbf{r}^\mathrm{f}(\mathbf{x}, t), t)$, and

(a) $\mathbf{x} \in \mathcal{X}_\mathbf{r}^\mathrm{f}(t) \Rightarrow \mathbf{x}_+^\kappa \in \mathcal{X}_\mathbf{r}^\mathrm{f}(t_+)$, and $h_n(\mathbf{x}, \kappa_\mathbf{r}^\mathrm{f}(\mathbf{x}, t)) \leq 0$, for all $n, k \in \mathbb{I}_0^\infty$;
(b) $\mathbf{x} \in \mathcal{X}_\mathbf{r}^\mathrm{f}(t) \Rightarrow g_{n|k}(\mathbf{x}, \kappa_\mathbf{r}^\mathrm{f}(\mathbf{x}, t)) \leq 0$, for all $n, k \in \mathbb{I}_0^\infty$.

Similarly to Assumption 2b, Assumption 3b is also difficult to verify due to the unknown constraints. Hence, the milder Assumption 3a, which is standard in MPC settings, will be used later on whereas Assumption 3b will be dropped.

In order to track the reference $(\mathbf{r}_k^{\mathbf{x}}, \mathbf{r}_k^{\mathbf{u}})$, we temporarily consider the following tracking MPC Problem

$$V(\mathbf{x}_k, \tau_k) := \min_{\mathbf{x},\mathbf{u}} \sum_{n=k}^{k+N-1} q_{\mathbf{r}}(\mathbf{x}_{n|k}, \mathbf{u}_{n|k}, \tau_n) + p_{\mathbf{r}}(\mathbf{x}_{k+N|k}, \tau_{k+N}) \tag{7a}$$

$$\text{s.t. } \mathbf{x}_{k|k} = \mathbf{x}_k, \tag{7b}$$

$$\mathbf{x}_{n+1|k} = f(\mathbf{x}_{n|k}, \mathbf{u}_{n|k}), \qquad n \in \mathbb{I}_k^{k+N-1}, \tag{7c}$$

$$h(\mathbf{x}_{n|k}, \mathbf{u}_{n|k}) \leq 0, \qquad n \in \mathbb{I}_k^{k+N-1}, \tag{7d}$$

$$\mathbf{x}_{k+N|k} \in \mathcal{X}_{\mathbf{r}}^{\mathrm{f}}(\tau_{k+N}). \tag{7e}$$

where, as opposed to Problem (4), $\tau$ follows the natural dynamics (2) instead of (3) and $g_{n|k}$ is removed, i.e., we consider a setting where there are no pop-up or other moving obstacles, or we make the restrictive assumption that these constraints cannot become active.

Assumptions 1–3 make it possible to derive the following standard (i.e., for feasible references) stability result.

**Proposition 1** (Nominal Asymptotic Stability *[39]*) *Suppose that the constraints $g_{n|k}$ are inactive, the Assumptions 1, 2a, and 3a hold, and that the initial state* $(\mathbf{x}_k, \tau_k)$ *at time k belongs to the feasible set of Problem* (7)*. Then the system* (1) *in closed loop with the solution of* (7) *applied in receding horizon is an asymptotically stable system.*

***Proof*** Since constraints $g_{n|k}$ are inactive by assumption, disregarding them in Problem (7) does not jeopardize feasibility. The rest of the proof follows from standard arguments, see, e.g., [23, 24]. □

Proposition 1 recalls the known stability results from the existing literature, which apply to tracking MPC schemes. We emphasize that, the design procedure resulting from Proposition 1 requires precomputing a feasible reference trajectory $(\mathbf{r}_k^{\mathbf{x}}, \mathbf{r}_k^{\mathbf{u}})$ that satisfies Assumption 2. However, in practice, it may be convenient to use a reference trajectory that is infeasible w.r.t. the system dynamics, yet simpler to define. For example, in the design of a motion planner and controller for an autonomous vehicle driving on public roads, it would be convenient to just use as a reference trajectory an easily available lane centerline, which in general would not be feasible for kinematic or dynamic vehicle models.

While in standard MPC settings the stability with respect to an unreachable set point has been studied in [40], the approach therein applies to time-invariant infeasible references. In order to overcome such a limitation, we consider a setting where the reference can be time-varying and does not need to satisfy Assumption 2, and the terminal conditions (7e) do not need to hold at the reference trajectory, but in a neighborhood. To lay down the main result of this section (Theorem 2), we need to first introduce a few preliminary results.

Consider the optimal state and input trajectories obtained as the solution of the optimal control problem (OCP)

$$(\mathbf{x}^{\mathrm{r}}, \mathbf{u}^{\mathrm{r}}) := \lim_{M\to\infty} \arg\min_{\boldsymbol{\xi},\boldsymbol{\nu}} \sum_{n=0}^{M-1} q_{\mathbf{r}}(\boldsymbol{\xi}_n, \boldsymbol{\nu}_n, \tau_n) + p_{\mathbf{r}}(\boldsymbol{\xi}_M, \tau_n) \tag{8a}$$

$$\text{s.t. } \boldsymbol{\xi}_0 = \mathbf{x}_0, \tag{8b}$$

$$\boldsymbol{\xi}_{n+1} = f(\boldsymbol{\xi}_n, \boldsymbol{\nu}_n), \qquad n \in \mathbb{I}_0^{M-1}, \tag{8c}$$

$$h(\boldsymbol{\xi}_n, \boldsymbol{\nu}_n) \leq 0, \qquad n \in \mathbb{I}_0^{M-1}. \tag{8d}$$

Note that, since constraints $g_{n|k}$ are considered to be inactive in this section, we do not include them in the OCP formulation in order to simplify the following analysis. Let $\mathbf{y}^{\mathrm{r}} := (\mathbf{x}^{\mathrm{r}}, \mathbf{u}^{\mathrm{r}})$ denote the solution of (8) and its optimal multipliers as $\boldsymbol{\lambda}^{\mathrm{r}}, \boldsymbol{\mu}^{\mathrm{r}}$. Hereafter, we will refer to the reference $\mathbf{y}^{\mathrm{r}}$ as the *feasible reference*, as it satisfies Assumption 2.

The result in Theorem 2 builds upon the stability theory for *economic MPC schemes*, where the cost is not of the *tracking type*. While the interested reader is referred to [27, 32, 41, 42] for an in-depth understanding of the stability analysis tools for economic MPC schemes, in this chapter we just recall that the main difference between economic and tracking MPC schemes is in the cost function, which satisfies

$$q_{\mathbf{r}}(\mathbf{x}_k^{\mathrm{r}}, \mathbf{u}_k^{\mathrm{r}}, \tau_k) = 0, \; q_{\mathbf{r}}(\mathbf{x}_k, \mathbf{u}_k, \tau_k) > 0, \; \forall \, \mathbf{x}_k \neq \mathbf{x}_k^{\mathrm{r}}, \; \mathbf{u}_k \neq \mathbf{u}_k^{\mathrm{r}}, \tag{9}$$

in tracking schemes but not in economic ones. While the MPC scheme built on a reference trajectory satisfying Assumption 2 is of a tracking type, an infeasible reference trajectory yields an economic scheme. Hence, in order to retrieve a tracking cost from the economic one, we introduce the following *rotated costs*

$$\begin{aligned}\bar{q}_{\mathbf{r}}(\mathbf{x}_{n|k}, \mathbf{u}_{n|k}, \tau_n) &:= q_{\mathbf{r}}(\mathbf{x}_{n|k}, \mathbf{u}_{n|k}, \tau_n) - q_{\mathbf{r}}(\mathbf{x}_n^{\mathrm{r}}, \mathbf{u}_n^{\mathrm{r}}, \tau_n) \\ &\quad + \boldsymbol{\lambda}_n^{\mathrm{r}\top}(\mathbf{x}_{n|k} - \mathbf{x}_n^{\mathrm{r}}) - \boldsymbol{\lambda}_{n+1}^{\mathrm{r}\top}(f_n(\mathbf{x}_{n|k}, \mathbf{u}_{n|k}) - f_n(\mathbf{x}_n^{\mathrm{r}}, \mathbf{u}_n^{\mathrm{r}})),\end{aligned} \tag{10}$$

$$\bar{p}_{\mathbf{r}}(\mathbf{x}_{n|k}, \tau_n) := p_{\mathbf{r}}(\mathbf{x}_{n|k}, \tau_n) - p_{\mathbf{r}}(\mathbf{x}_n^{\mathrm{r}}, \tau_n) + \boldsymbol{\lambda}_n^{\mathrm{r}\top}(\mathbf{x}_{n|k} - \mathbf{x}_n^{\mathrm{r}}), \tag{11}$$

which are commonly used in economic MPC. The rotated cost essentially shifts the stage and terminal costs $q_{\mathbf{r}}$ and $p_{\mathbf{r}}$, so that the minimum is attained at the *feasible reference* $(\mathbf{x}^{\mathrm{r}}, \mathbf{u}^{\mathrm{r}})$, i.e., $\bar{q}_{\mathbf{r}}(\mathbf{x}_k^{\mathrm{r}}, \mathbf{u}_k^{\mathrm{r}}, \tau_k) = 0$, $\bar{p}_{\mathbf{r}}(\mathbf{x}_k^{\mathrm{r}}, \tau_k) = 0$. However, in order to ensure that the rotated costs remain positive definite, we must assume that the system dynamics are linear time-varying, i.e., that Assumption 4 (which we introduce next) must hold.

The issue of tracking an infeasible time-varying reference has been studied in [39], under the additional assumption of linear time-varying (LTV) system dynamics. This assumption is technical and in practice we expect that the results can be extended to the fully nonlinear case, which will be the subject of future research.

**Assumption 4** The system dynamics $f$ are linear time-varying, i.e.,

$$\mathbf{x}_{k+1} = f_k(\mathbf{x}_k, \mathbf{u}_k) = A_k \mathbf{x}_k + B_k \mathbf{u}_k. \tag{12}$$

In order to construct a problem that tracks the *feasible reference* obtained from (8), we formulate the following *ideal* formulation

$$V^{\mathrm{i}}(\mathbf{x}_k, \tau_k) = \min_{\mathbf{x},\mathbf{u}} \sum_{n=k}^{k+N-1} q_{\mathbf{r}}(\mathbf{x}_{n|k}, \mathbf{u}_{n|k}, \tau_n) + p_{\tilde{\mathbf{y}}^{\mathrm{r}}}(\mathbf{x}_{k+N|k}, \tau_{k+N}) \tag{13a}$$

$$\text{s.t. } (4b) - (4e), \ \mathbf{x}_{k+N|k} \in \mathcal{X}^{\mathrm{f}}_{\mathbf{y}^{\mathrm{r}}}(\tau_{k+N}), \tag{13b}$$

where

$$\tilde{\mathbf{y}}^{\mathrm{r}}_k := \arg\min_{\mathbf{x}} p_{\mathbf{y}^{\mathrm{r}}}(\mathbf{x}, \tau_k) - \boldsymbol{\lambda}_k^{\mathrm{r}\top}(\mathbf{x} - \mathbf{x}^{\mathrm{r}}_k). \tag{14}$$

We refer to this formulation as being *ideal* since the terminal conditions are in general not known, unless one solves OCP (8).

**Theorem 1** *Suppose that*

1. *Assumption 1 holds,*
2. *Problem* (8) *is feasible,*
3. *Assumption 3 holds for $\bar{q}_{\mathbf{r}}$ and $\bar{p}_{\tilde{\mathbf{y}}^{\mathrm{r}}}$, with terminal set $\mathcal{X}_{\mathbf{y}^{\mathrm{r}}}$,*
4. *Assumption 4 holds for the system dynamics.*

*Then, the system* (1) *in closed-loop with the ideal MPC* (13) *is asymptotically stabilized to the optimal trajectory* $\mathbf{x}^{\mathrm{r}}$. □

Theorem 1 establishes that an MPC problem can be formulated using an *infeasible reference*, which stabilizes system (1) to the *feasible reference* obtained from Problem (8), provided that appropriate terminal conditions are used. The remaining issue, however, is to express the terminal constraint set as a positive invariant set containing $\mathbf{x}^{\mathrm{r}}$, and a terminal control law that stabilizes the system to $\mathbf{x}^{\mathrm{r}}$. To that end, one needs to have prior knowledge of the *feasible reference*, i.e., Problem (8) needs to be solved. Instead, we consider expressing terminal conditions that are based on an *approximately feasible* reference. In that case, asymptotic stability in the sense of Proposition 1 cannot be proven. We will therefore resort to input-to-state stability for the closed-loop system, where the input will be a terminal reference $\mathbf{y}^{\mathrm{f}}$ satisfying the following assumption.

**Assumption 5** (*Approximate feasibility of the reference*) The reference $\mathbf{y}^{\mathrm{f}}$ satisfies the constraints (4e), i.e., $h(\mathbf{x}^{\mathrm{f}}_n, \mathbf{u}^{\mathrm{f}}_n) \leq 0$, $n \in \mathbb{I}_k^{k+N-1}$, for all $k \in \mathcal{N}^+$. Additionally, recursive feasibility holds for both Problem (7) and (13) when the system is controlled in closed-loop using the feedback from Problem (7).

Assumption 5 sets a rather mild requirement from a practical standpoint. Using an infeasible reference for simplicity, or approximating system dynamics to capture the most relevant dynamics of the system ($\|\mathbf{x}^{\mathrm{f}}_{n+1} - f_n(\mathbf{x}^{\mathrm{f}}_n, \mathbf{u}^{\mathrm{f}}_n)\| \leq \epsilon$, for some small $\epsilon$) is not uncommon in practice. In particular, in a practical setting we can select $\mathbf{y}^{\mathrm{f}} = \mathbf{r}(t_{k+N})$, or in an ideal setting $\mathbf{y}^{\mathrm{f}} = \mathbf{y}^{\mathbf{r}}(t_{k+N})$. To that end, we define the following closed-loop dynamics

$$\mathbf{x}_{k+1}(\mathbf{y}^{\mathrm{f}}) = f_k(\mathbf{x}_k, \mathbf{u}_{\mathrm{MPC}}(\mathbf{x}_k, \mathbf{y}^{\mathrm{f}})) = \bar{f}_k(\mathbf{x}_k, \mathbf{y}^{\mathrm{f}}), \tag{15}$$

where $\mathbf{u}_{\mathrm{MPC}}$ is obtained as $\mathbf{u}^{\star}_{k|k}$ solving Problem (7) in case $\mathbf{y}^{\mathrm{f}} = \mathbf{r}$; and as $\mathbf{u}^{\mathrm{i}}_{k|k}$ solving the *ideal* Problem (13) in case $\mathbf{y}^{\mathrm{f}} = \mathbf{y}^{\mathrm{r}}$.

We are now ready to state the main results of this section.

**Theorem 2** *Suppose that*

1. *Problem* (8) *is feasible,*
2. *Assumptions 1 and 3 hold for the reference* $\mathbf{y}^{\mathrm{r}}$ *with costs* $\bar{q}_{\mathbf{r}}$ *and* $\bar{p}_{\mathbf{y}^{\mathrm{r}}}$ *and terminal set* $\mathcal{X}_{\mathbf{y}^{\mathrm{r}}}$,
3. *Problem* (7) *and Problem* (13) *are feasible at time k with initial state* $(\mathbf{x}_k, t_k)$,
4. *the reference* $\mathbf{y}^{\mathrm{f}}$, *with terminal set* $\mathcal{X}^{\mathrm{f}}_{\mathbf{y}^{\mathrm{f}}}$, *satisfies Assumption 5.*

*Then, system* (15) *obtained from* (1) *in closed-loop with MPC formulation* (7) *is ISS.*

□

This theorem proves that if an infeasible reference is used, system (1) does not converge exactly to the (unknown) optimal trajectory from OCP (8), but to a neighborhood around it which depends on how inaccurate the terminal reference is. We note, however, that the effect of the terminal condition on the closed-loop trajectory decreases as the prediction horizon $N$ increases [32, 42].

This section has so far considered the a-priori unknown constraint $g_{n|k}$ to be inactive, i.e., no road users or other obstacles are present, in order to simplify the analysis. In the next section, we consider settings where $g_{n|k}$ may not be ignored, e.g., one has to ensure collision-avoidance w.r.t other road users.

## 5 Safety-Enforcing MPC

The aim of this section is to tackle the issues posed by the presence of the a-priori unknown constraints (4f), which in Sect. 4 have been assumed to always remain inactive. While we cast the problem in the framework of MPFTC, we stress that the developments proposed to enforce safety are independent of the specific tracking scenario, i.e., flexible trajectory, path, setpoint, etc., and can also be deployed in the context of Model Predictive Path Following Control (MPFC) proposed in [43, 44].

We introduce the following assumption, imposing some structure on $g$ that is needed in order to ensure that the feasibility of a solution is preserved between consecutive time instances.

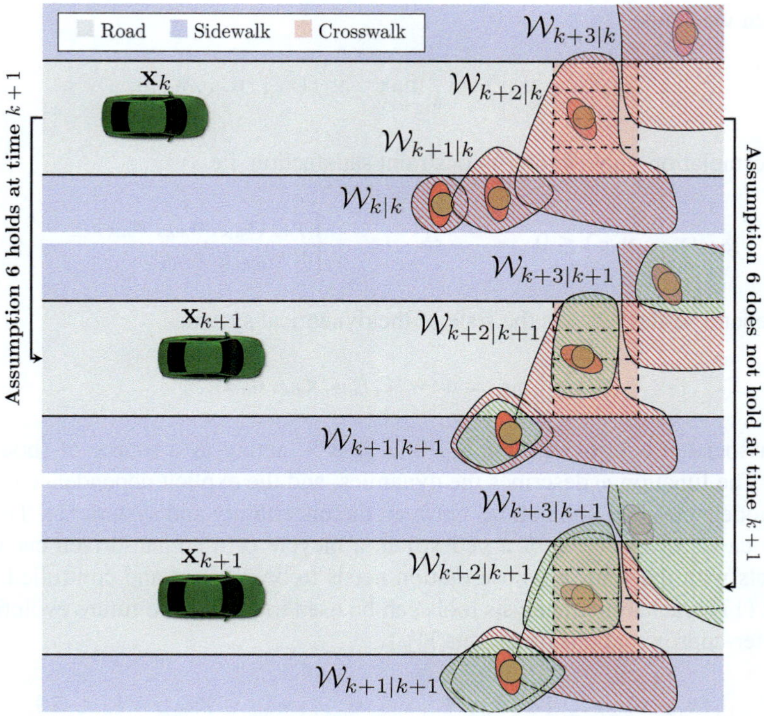


**Fig. 1** The top panel shows an initial prediction for a pedestrian at time $k$, while the middle and bottom panels show different predictions at time $k+1$. The middle panel illustrates a model that satisfies (18), i.e., satisfying Assumption 6, while the bottom panel does not satisfy (18), hence, Assumption 6 is also not satisfied

**Assumption 6** (*Unknown constraint dynamics*) The a-priori unknown constraints satisfy $g_{n|k+1}(\mathbf{x}_{n|k}, \mathbf{u}_{n|k}) \leq g_{n|k}(\mathbf{x}_{n|k}, \mathbf{u}_{n|k})$, for all $n \geq k$.

While Fig. 1 gives a visual interpretation of the requirement in Assumption 6, at this stage it is natural to wonder how function $g$ should be constructed such that Assumption 6 is satisfied. In order to answer this question, we will first provide a formal description of how $g$ can be constructed, and then provide some examples which are relevant to autonomous driving to provide more intuition on the meaning of Assumption 6.

We introduce the function $\gamma(\mathbf{x}, \mathbf{u}, \mathbf{w}) : \mathbb{R}^{n_\mathbf{x}} \times \mathbb{R}^{n_\mathbf{u}} \times \mathbb{R}^{n_\mathbf{w}} \rightarrow \mathbb{R}^{n_g}$ and the uncertain variable $\mathbf{w}_{n|k} \in \mathcal{W}_{n|k} \subseteq \mathbb{R}^{n_\mathbf{w}}$ whose bounded support is a subset of set $\mathcal{W}_{n|k}$, lumping all the uncertainty related to the a-priori unknown constraints. Note that we only require knowledge on a superset of the support of $\mathbf{w}_{n|k}$, which can in principle even be deterministic, in which case its probability distribution is a Dirac and the superset $\mathcal{W}_{n|k}$ can be any set of measure 1.

Then we define

$$g_{n|k}(\mathbf{x}_{n|k}, \mathbf{u}_{n|k}) := \max_{\mathbf{w}_{n|k} \in \mathcal{W}_{n|k}} \gamma_{n|k}(\mathbf{x}_{n|k}, \mathbf{u}_{n|k}, \mathbf{w}_{n|k}). \tag{16}$$

This formulation implies robust constraint satisfaction, i.e.,

$$g_{n|k}(\mathbf{x}_{n|k}, \mathbf{u}_{n|k}) \leq 0 \qquad \Leftrightarrow \qquad \begin{cases} \gamma_{n|k}(\mathbf{x}_{n|k}, \mathbf{u}_{n|k}, \mathbf{w}_{n|k}) \leq 0, \\ \forall\, \mathbf{w}_{n|k} \in \mathcal{W}_{n|k}. \end{cases}$$

In a general setting, $\mathbf{w}_{n|k}$ is the state of the dynamical system

$$\mathbf{w}_{n+1|k} = \omega(\mathbf{w}_{n|k}, \xi_{n|k}, \mathbf{x}_{n|k}, \mathbf{u}_{n|k}), \tag{17}$$

with associated control variable $\xi_{n|k} \in \Xi \subseteq \mathbb{R}^{m_\xi}$, acting as a source of (bounded) noise. The function $\omega$ describes the dynamics, and the explicit dependence on $\mathbf{x}_{n|k}$, $\mathbf{u}_{n|k}$ models possible interactions between the uncertainty and system (1). This can be the case, for example, of a pedestrian, a bicycle or a human-driven car which interacts with the vehicle whose motion needs to be planned and controlled. With model (17) reachability analysis tools can be used to predict the future evolution of the outer-approximations of the sets $\mathcal{W}_{n|k}$.

$$\mathcal{W}_{n+1|k}(\mathbf{x}_{n|k}, \mathbf{u}_{n|k}) :\supseteq \{\, \omega(\mathbf{w}_{n|k}, \xi_n, \mathbf{x}_{n|k}, \mathbf{u}_{n|k}) \mid \mathbf{w}_{n|k} \in \mathcal{W}_{n|k},\ \forall\, \xi_n \in \Xi \,\}, \tag{18}$$

for some initial $\mathcal{W}_{k|k} = \mathbf{w}_{k|k}$.

The introduction of the uncertainty sets (18) allows one to model road users which (a) are detectable by the sensors, and (b) are either beyond the onboard sensors range or hidden by other obstacles, as depicted in Fig. 2. For type (a), the model (16), (17) does not underestimate the set of future states that can be reached by the road users. For type (b), the uncertainty model must predict the possibility that a road user appears at any time either at the boundary of the sensor range or from behind an obstacle.

We can now state the following result.

**Lemma 1** *Suppose that $g_{n|k}$ is defined according to* (16) *with $\mathcal{W}_{n|k}$ satisfying* (18). *Then, Assumption 6 holds.*

Note that this lemma amounts to assuming that the uncertainty in $g_{n|k}$ cannot increase as additional information becomes available (from either the onboard sensors or through communication links). Furthermore, a direct consequence of Assumption 6 is $g_{n|k}(\mathbf{r}^{\mathbf{x}}(t_n), \mathbf{r}^{\mathbf{u}}(t_n)) \leq 0 \implies g_{n|k+1}(\mathbf{r}^{\mathbf{x}}(t_n), \mathbf{r}^{\mathbf{u}}(t_n)) \leq 0$. We provide the following clarifying illustrations in order to better understand Lemma 1 and Assumption 6.

Figure 1 shows the difference between having a model that satisfies Lemma 1, and one that does not. The middle panel shows that the predictions made at time $k+1$ belong to a subset of the previous predictions made at time $k$, i.e., $\mathcal{W}_{n|k+1} \subseteq \mathcal{W}_{n|k}$,

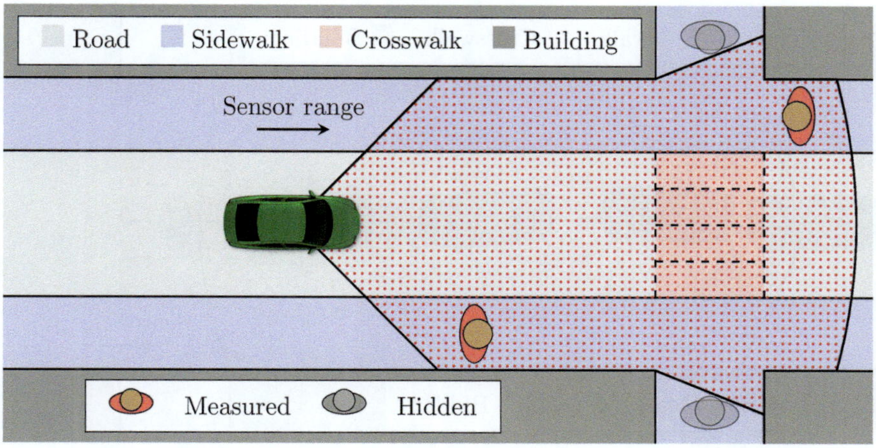


**Fig. 2** Due to limited sensing capabilities, it is not possible to directly measure all road users in the environment. Therefore, one must assume that hidden road users may appear outside the sensor range at all times

which satisfies Lemma 1. The bottom panel on the other hand illustrates predictions made at time $k + 1$, where Lemma 1 does not hold. Figure 2 illustrates the fact that road users in the environment may be occluded due to limited sensing capabilities. In that case, in order to satisfy Assumption 6 at all times, one needs to model the possibility that a road user might appear at the boundary of the sensor range.

We recall that, in the context of autonomous driving, the constraints $g_{n|k}$ could enforce avoiding collisions with obstacles (e.g., other road users) detected by the sensors, whose behavior can just be predicted to some limited extent. Hence, Assumption 6 amounts to assuming that the uncertainty on, e.g., position and velocity estimates of the detected objects at a specific time instance cannot increase as additional information becomes available. We note however that limited sensor range makes it impossible to detect obstacles which are too far away. To ensure satisfaction of Assumption 6 one can adopt a worst-case approach which ensures that the predicted trajectory $\mathbf{x}_{n|k}$ may never leave the sensor range, and by also assuming that new obstacles appear at the boundary of the sensor range at all times. A visual example of this shown in Fig. 3, where the planned trajectory is forced to remain within the sensor range at all times.

Enforcing *safety* of the controller (4) according to Definition 1 requires the controller to be *recursively feasible*. Hence, the terminal constraint (4g) needs to be designed in order to guarantee its satisfaction despite of the presence of the constraints (4f), which can grow unbounded in time, such that no state is safe, see Fig. 4. We therefore design the terminal conditions by assuming the existence of a *safe set*, where the constraints (4f) are guaranteed to be satisfied *regardless*. This will allow us to rely on standard approaches in MPC [23, 45, 46] which are based on the existence of a robust invariant set.

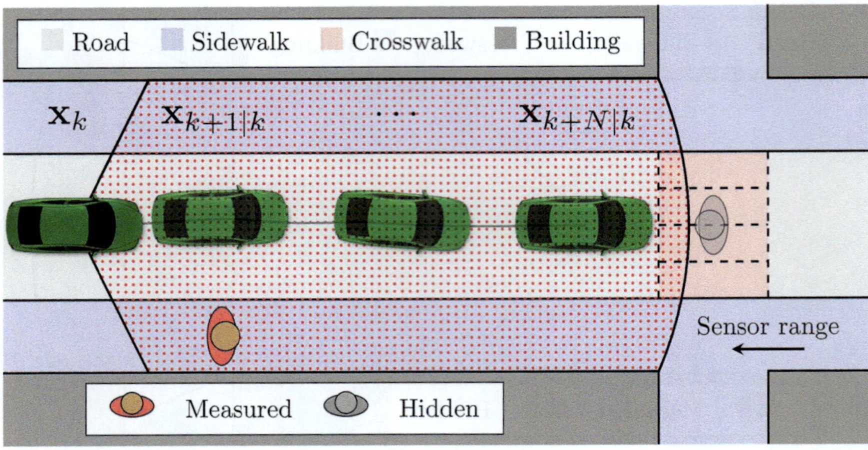


**Fig. 3** Since it is impossible to know what lies ahead of the sensing range, the planned trajectory $\{\mathbf{x}_{k|k}, \mathbf{x}_{k+1|k}, \ldots, \mathbf{x}_{k+N|k}\}$ must be forced to remain within the sensor range

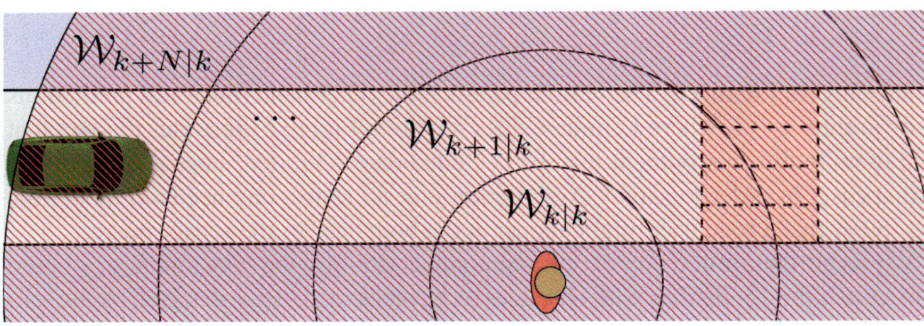


**Fig. 4** As the uncertainty of the pedestrian predictions grow in time, the collision-free region of the vehicle drastically shrinks. Hence, after some time, any model will predict that a pedestrian can be anywhere, such that no state is safe

**Assumption 7** There exists a robust invariant set denoted $\mathcal{X}_{\text{safe}}(\tau_{n|k}) \subseteq \mathbb{R}^{n_x}$ such that for all $\mathbf{x}_{n|k} \in \mathcal{X}_{\text{safe}}(\tau_{n|k})$ there exists a safe control set $\mathcal{U}_{\text{safe}}(\mathbf{x}_{n|k}, \tau_{n|k}) \subseteq \mathbb{R}^{n_u+1}$ entailing that $f(\mathbf{x}_{n|k}, \mathbf{u}_{\text{safe}}) \in \mathcal{X}_{\text{safe}}(\tau_{n|k} + t_{\text{s}} + v_{\text{safe}})$, and $h_n(\mathbf{x}_{n|k}, \mathbf{u}_{\text{safe}}) \leq 0$, for all $(\mathbf{u}_{\text{safe}},\ v_{\text{safe}}) \in \mathcal{U}_{\text{safe}}(\mathbf{x}_{n|k}, \tau_{n|k})$ and for all $n \geq k$. Moreover, for all $\mathbf{x}_{n|k} \in \mathcal{X}_{\text{safe}}(\tau_{n|k})$ the a-priori unknown constraints can never be violated, i.e., by construction $g_{n|k}(\mathbf{x}_{n|k}, \mathbf{u}_{\text{safe}}) \leq 0$ for all $\mathbf{x}_{n|k} \in \mathcal{X}_{\text{safe}}(\tau_{n|k})$ and $(\mathbf{u}_{\text{safe}},\ v_{\text{safe}}) \in \mathcal{U}_{\text{safe}}(\mathbf{x}_{n|k}, \tau_{n|k})$.

While this assumption might seem strong, it only postulates the existence of *known* safe configurations for system (1). However, if no such configurations exist, then the controller based on Problem (4) is intrinsically unsafe. On the other hand, if such configurations do exist for (1), then the safe set $\mathcal{X}_{\text{safe}}$ is non-empty and invariant. Note that the safe configuration depends on system (1), the problem setting, and must be known a-priori.

***Example 1*** Many practical settings where safety is emphasized consider a system to be safe at steady state, in which case the safety set $\mathcal{X}_{\text{safe}}$ can be formulated as

$$\mathcal{X}_{\text{safe}}(\tau_k) := \{\mathbf{x} \,|\, \mathbf{x} = f(\mathbf{x}, \mathbf{u}),\ h_k(\mathbf{x}, \mathbf{u}) \le 0,\ m_k(\mathbf{x}, \mathbf{u}) \le 0\}, \tag{19}$$

where function $m_k$ defines additional constraints which might be needed in the set definition. A notable example for automotive settings is that a vehicle parked in a safe configuration, e.g., a parking lot, emergency lane or any other safe environment that can be modeled by $m_k$, is not responsible for collisions with other road users. This reasoning can be applied to the setting illustrated in Fig. 4, where the uncertainty grows such that the only safe thing to do is to force the vehicle to a complete stop.

The introduction of Assumption 7, allows us to drop Assumption 3b, so that we can build our approach based on standard strategies in MPC [23, 45, 46], i.e., we rely on stabilizing terminal control laws $\kappa_{\mathbf{r}}^{\text{s}}(\mathbf{x}, t)$ and sets $\mathcal{X}_{\mathbf{r}}^{\text{s}}(t)$ satisfying Assumption 3a. In order to obtain recursive feasibility also with respect to a-priori unknown constraints, we rely on the safe set $\mathcal{X}_{\text{safe}}$ to introduce the following terminal set

$$\begin{aligned}
\mathcal{X}_{\mathbf{r}}^{\text{f}}(\tau_{k+N|k}) := \{\mathbf{x}_{k+N|k} \mid \exists\, \mathbf{u}_{n|k}, v_{n|k}, \quad & (20a)\\
\tau_{n+1|k} = \tau_{n|k} + t_{\text{s}} + v_{n|k}, \quad & (20b)\\
\mathbf{x}_{n+1|k} = f(\mathbf{x}_{n|k}, \mathbf{u}_{n|k}), \quad & (20c)\\
h_n(\mathbf{x}_{n|k}, \mathbf{u}_{n|k}) \le 0, \quad & (20d)\\
g_{n|k}(\mathbf{x}_{n|k}, \mathbf{u}_{n|k}) \le 0, \quad & (20e)\\
\mathbf{x}_{n|k} \in \mathcal{X}_{\mathbf{r}}^{\text{s}}(\tau_{n|k}), \quad & (20f)\\
\mathbf{x}_{k+M|k} \in \mathcal{X}_{\text{safe}}(\tau_{k+M|k}) \subseteq \mathcal{X}_{\mathbf{r}}^{\text{s}}(\tau_{k+M|k}), \quad & (20g)\\
(20a) - (20f),\ \forall n \in \mathbb{I}_{k+N}^{k+M-1}\}, \quad & (20h)
\end{aligned}$$

where $M \ge N$ is a degree of freedom. Note that the construction of (20) implies that $\mathcal{X}_{\text{safe}}(\tau) \subseteq \mathcal{X}_r^{\text{s}}(\tau),\ \forall \tau \ge 0$. If (20) is a non-empty set we are guaranteed that for all $\mathbf{x} \in \mathcal{X}_{\mathbf{r}}^{\text{f}}(\tau)$ a terminal control law exists, which steers the states to the safe set.

In order to provide a practical approach to design the terminal control law, we propose to first design a control law $\kappa_{\text{r}}^{\text{s}}$ as one would do in standard MPC formulations, i.e., by ignoring a-priori unknown constraints $g$ and by forcing the time in the reference to evolve according to its true dynamics. We can then define the terminal control law $(\kappa_{\mathbf{r}}^{\text{f}}(\mathbf{x}_{k+N|k}, \tau_{k+N|k}), \nu_{\mathbf{r}}^{\text{f}}(\mathbf{x}_{k+N|k}, \tau_{k+N|k}))$ by using $\kappa_{\text{r}}^{\text{s}}$, as the solution of

$$\begin{aligned}
\min_{\mathbf{u},\nu}\ & \|\mathbf{u} - \kappa_{\text{r}}^{\text{s}}(\mathbf{x}_{k+N|k}, \tau_{k+N|k})\|_2 + \nu^2 && (21a)\\
\text{s.t.}\ & f(\mathbf{x}_{k+N|k}, \mathbf{u}) \in \mathcal{X}_{\mathbf{r}}^{\text{f}}(\tau_{k+N|k} + t_{\text{s}} + \nu), && (21b)\\
& h_{k+N}(\mathbf{x}_{k+N|k}, \mathbf{u}) \le 0, && (21c)\\
& g_{k+N|k}(\mathbf{x}_{k+N|k}, \mathbf{u}) \le 0. && (21d)
\end{aligned}$$

The idea behind the terminal set (20) is to ensure safety by forcing the system to be able to reach a safe set $\mathcal{X}_{\text{safe}}$ in a finite amount of time $M - N \geq 0$, while always remaining inside a stabilizing set $\mathcal{X}_{\text{r}}^{\text{s}}(t)$ around the reference. Note that $M$ is a parameter which can be used to tune the stabilizing terminal safe set and, consequently, the NMPC scheme (4). If $M = N$, then the terminal set coincides with the safe set, possibly limiting the capabilities of the terminal control law, i.e., $\kappa_{\mathbf{r}}^{\text{f}}(\mathbf{x}, \tau) \neq \kappa_{\mathbf{r}}^{\text{s}}(\mathbf{x}, \tau)$ and $\nu_{\mathbf{r}}^{\text{f}}(\mathbf{x}, \tau) \neq 0$. On the other hand, if $M \gg N$, the computational complexity of $\mathcal{X}_{\mathbf{r}}^{\text{f}}$ can become excessive.

**Theorem 3** (Recursive Feasibility) *Suppose that Assumptions 1, 2a, 3a, 6, and 7 hold, and that Problem* (4) *is feasible for the initial state* $(\mathbf{x}_k, \tau_k)$*, with terminal set and terminal controllers given by* (20) *and* (21)*, respectively. Then, system* (1)-(3) *in closed loop with the solution of* (4) *applied in receding horizon is safe (recursively feasible) at all times.*

While Theorem 3 only proves recursive feasibility, the presence of obstacles makes it more difficult to discuss closed-loop stability. We note, however, that if the a-priori unknown constraints become inactive, then the proposed formulation yields nominal asymptotic stability.

Next we consider two simulation examples to illustrate the theory from Sects. 4 and 5.

## 6 Simulations

In this section we present two simulations related to autonomous driving settings in order to illustrate the theory from Sects. 4 and 5. In Sect. 6.1 we first show how the closed-loop behavior is affected when an infeasible reference is used. Then, in Sect. 6.2 we introduce moving and pop-up obstacles to show how recursive feasibility is ensured through Theorem 3.

For both simulations we consider the single-track vehicle model with kinematics

$$\begin{bmatrix} \dot{x} \\ \dot{y} \\ \dot{\psi} \\ \dot{\delta} \\ \dot{v} \end{bmatrix} = \begin{bmatrix} v\cos(\psi) \\ v\sin(\psi) \\ \frac{v}{l}\tan(\delta) \\ \omega \\ a \end{bmatrix}, \tag{22}$$

where $x$, $y$ are the position coordinates in a global frame, $v$ is the velocity, $\psi$ is the orientation angle, $l$ is the wheelbase length, $\delta$ is the steering wheel angle, and $a$ and $\omega$ denote the acceleration and steering wheel angle rate, respectively. Since one of the objectives is to track a user-defined reference $\mathbf{r}$, it is possible to geometrically derive the following vehicle kinematics in the frame of the reference path [47]

$$\begin{bmatrix} \dot{s} \\ \dot{e}_y \\ \dot{e}_\psi \\ \dot{\delta} \\ \dot{v} \end{bmatrix} = \begin{bmatrix} v\cos(e_\psi)(1-\kappa^{\mathrm{r}}(s)e_y)^{-1} \\ v\sin(e_\psi) \\ vl^{-1}(\tan(\delta)-\tan(\delta^{\mathrm{r}}(s))) \\ \omega \\ a \end{bmatrix}, \ \mathbf{x} = \begin{bmatrix} e_y \\ e_\psi \\ \delta \\ v \end{bmatrix}, \ \mathbf{u} = \begin{bmatrix} a \\ \omega \end{bmatrix} \tag{23}$$

where $s$ is the longitudinal position along the path, $\kappa^{\mathrm{r}}$ is the path curvature, $e_y$ is the lateral displacement error, $e_\psi$ is the yaw error with respect to the reference $\mathbf{r}$ and $\delta^{\mathrm{r}}$ is the reference steering angle. Note that we consider $s$ to be an auxiliary state since we are only interested in tracking the velocity and not the longitudinal position. For both simulations, we assume that system (23) is subject to the following known constraints

$$\|e_y\| \leq 0.4, \ \|e_\psi\| \leq 0.16, \ \|\delta\| \leq 0.53,$$
$$0 \leq v \leq 50/3.6, \ -5 \leq a \leq 3, \ \|\omega\| \leq 0.35,$$

while the stage cost matrices in (5) are

$$W = \text{blockdiag}(Q, R), \ Q = \text{diag}(1, 10, 1, 1), \ R = \text{diag}(1, 1). \tag{24}$$

In order to compute the stabilizing terminal set $\mathcal{X}_{\mathrm{s}}^{\mathrm{f}}(t)$, we decouple the longitudinal and lateral kinematics. Using an an LQR controller with costs $Q^{\mathrm{lon}} = 1$, $R^{\mathrm{lon}} = 50$ one can then obtain the feedback gain $K = 0.14$ a terminal cost $P^{\mathrm{long}} = 72.63$ with corresponding terminal set

$$\mathcal{X}_s^{\mathrm{lon}}(\tau) := \{v \mid -5 \leq -K(v - \mathbf{r}^{\mathbf{x}_v}(\tau)) \leq 3\}. \tag{25}$$

For the terminal set of the lateral kinematics, we consider the velocity to be an uncertain parameter and linearize the system (23) on the reference to obtain the Linear Parameter Varying (LPV) system $(A(v), B(v)) \in \mathbb{R}^{3\times 3} \times \mathbb{R}^3$. Then, by considering a nominal velocity $v^{\mathrm{nom}} = 28/3.6$ m/s, we use the feedback gain $K$, obtained from an LQR controller with tuning $Q^{\mathrm{lat}} = \text{blockdiag}(1,\ 100,\ 2)$ and $R^{\mathrm{blockdiag}} = 1$, to stabilize the LPV system by considering the following polytopic system

$$\Gamma := \{(A, B) \in \mathbb{R}^{3\times 3} \times \mathbb{R}^3 : A = A(v), B = B(v), v \in [1, 50/3.6]\}, \tag{26}$$

for velocities $v \in [1, 50/3.6]$ m/s. We then use the MPT toolbox [48] to compute the terminal set [49] for the polytopic system (26) and obtain that

$$X_{\mathbf{r}}^{\mathrm{lat}}(\tau) := \{H([e_y, e^\psi, \delta]^\top - [0, 0, \delta^{\mathrm{ref}}(\tau)]^\top) \leq b\}. \tag{27}$$

We note that, while we compute the terminal set for the polytopic system (26), one can also use the methods presented in [50, 51] to compute low-complexity invariant sets for linear parameter-varying models. By solving the linear matrix inequalities that satisfy the Lyapunov equations for the polytopic system, we also obtain the terminal cost

$$P^{\text{lat}} = \begin{bmatrix} 402.37 & 839.90 & 323.00 \\ 839.90 & 7272.88 & 2781.57 \\ 323.00 & 2781.57 & 2556.74 \end{bmatrix} \tag{28}$$

Finally, the terminal set $\mathcal{X}_{\text{s}}^{\text{f}}(\tau)$ can then be constructed as

$$\mathcal{X}_{\text{s}}^{\text{f}}(\tau) := \{\mathbf{x} \mid [e_y, e_\psi, \delta] \in \mathcal{X}_{\mathbf{r}}^{\text{lat}}(\tau),\ v \in \mathcal{X}_{\text{r}}^{\text{long}}(\tau)\}, \tag{29}$$

with terminal cost $P = \text{blockdiag}(P^{\text{lat}}, P^{\text{lon}})$.

The following simulations were implemented in Matlab and interfaced with Acados [52] and CasADi [53], together with solvers IPOPT [54] and HPIPM [55].

## 6.1 ISS: Stability with Infeasible Reference

In this section we show the closed-loop behavior of Problem (7) when using an infeasible reference. To that end, we consider a reference with a discontinuous curvature, i.e., the steering angle reference is discontinuous,

$$\mathbf{r}^{\mathbf{x}}(\tau) = [0,\ 0,\ \delta^{\text{r}}(\tau),\ 50/3.6]^\top, \mathbf{r}^{\mathbf{u}}(\tau) = [0,\ 0]^\top, \tag{30}$$

where

$$\kappa(\tau) = \begin{cases} 0.0452 & \text{if } 7.5 \le \tau < 12.5, \\ 0 & \text{otherwise.} \end{cases} \tag{31}$$

In order to obtain the feasible reference $\mathbf{y}^{\text{r}} = (\mathbf{x}^{\text{r}}, \mathbf{u}^{\text{r}})$, we linearize model (23) to obtain an LTV system and approximate the infinite horizon Problem (8) with a prediction horizon of $M = 600$ and sampling time $t_{\text{s}} = 0.05$ s. For the closed-loop simulations, we use the control input obtained from formulations (7) and (13) with horizon $N = 10$ and sampling time $t_{\text{s}} = 0.05$ s.

Figure 5 shows the closed-loop trajectories for the initial condition $t_0 = 0$ and $\mathbf{x}_0 = [0.1,\ 0.02,\ 0,\ 50/3.6]^\top$, where the gray lines denote the infeasible reference $\mathbf{r}$, and the black lines denote the optimal reference $\mathbf{y}^{\text{r}}$ from (8). The blue lines show the closed-loop evolution of each state for *ideal* MPC formulation (13), i.e., when the terminal conditions are based on the *feasible* reference $\mathbf{y}^{\text{f}} = \mathbf{y}^{\text{r}}$. The orange lines, on the other hand, show the closed-loop trajectories for the practical MPC formulation (7), which has the terminal conditions based on the infeasible reference, i.e., $\mathbf{y}^{\text{f}} = \mathbf{r}$. The bottom right plot of Fig. 5 shows the closed-loop error with respect to $\mathbf{y}^{\text{r}}$ for the two MPC formulations. It is visible that for times $t \le 5$ s, the reference trajectory is feasible and both formulations manage to stabilize towards the optimal trajectory. For 5 s $\le t \le 15$ s, the discontinuity of the reference $\mathbf{r}$ affects how the two

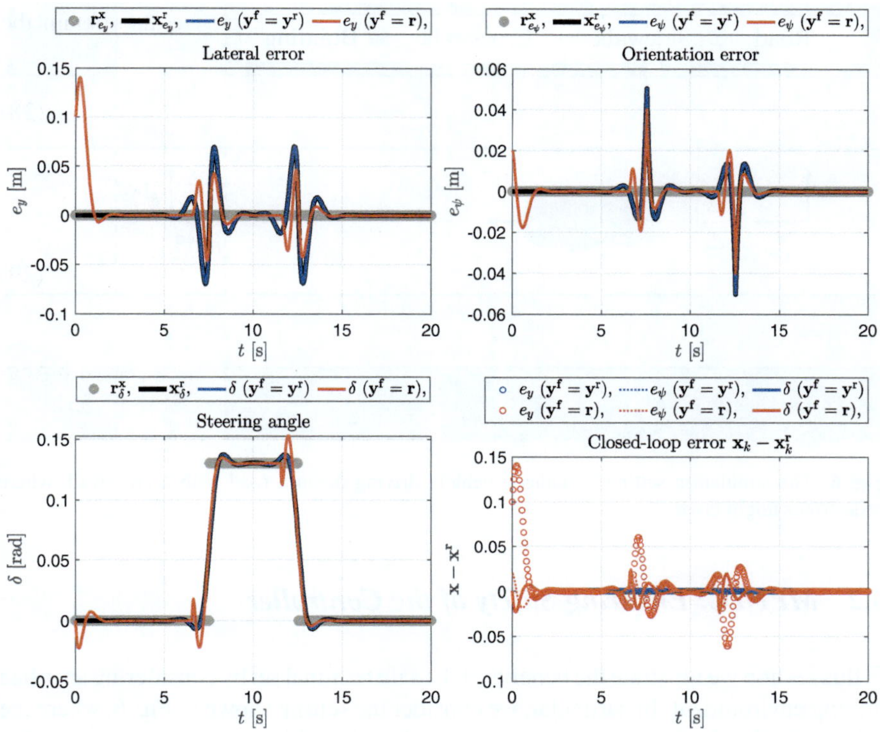


**Fig. 5** Closed-loop simulation with initial conditions $\mathbf{x}_0 = [0.1,\ 0.02,\ 0,\ 50/3.6]^\top$ and initial time $t_0 = 0$. The gray trajectories show the infeasible reference $\mathbf{r} = (\mathbf{r}^{\mathbf{x}}, \mathbf{r}^{\mathbf{u}})$, while the black trajectories show the optimal trajectory $\mathbf{y}^{\mathrm{r}} = (\mathbf{x}^{\mathrm{r}}, \mathbf{u}^{\mathrm{r}})$ obtained from Problem (8). The orange trajectories show the closed-loop behavior for the practical MPC Problem (7), while the blue trajectories show the closed-loop behavior for the *ideal* MPC Problem (13)

formulations behave. The *ideal* formulation manages to track the optimal reference $\mathbf{y}^{\mathrm{r}}$ (black trajectory) since it is formulated with proper terminal conditions. The practical formulation (orange trajectory) on the other hand tries to track the infeasible reference $\mathbf{r}$, and therefore deviates from the optimal trajectory and ideal formulation. After the discontinuity, the rest of the reference trajectory is feasible, and both formulations are asymptotically stabilizing.

In order to verify in simulation the theoretical result of [32, 42] stating that the effect of the terminal condition on the closed-loop trajectory decreases as the prediction horizon $N$ increases, we ran the same simulation with $N = 60$ and observed that the closed-loop trajectory obtained with the infeasible reference becomes indistinguishable from the optimal one by eye inspection.

While we have shown how an infeasible reference affects the closed-loop system in section, we consider next a setting where the reference trajectory becomes infeasible due to pop-up obstacles in the environment. To that end, in the next section we show recursive feasibility guarantees from Theorem 3 are enforced.

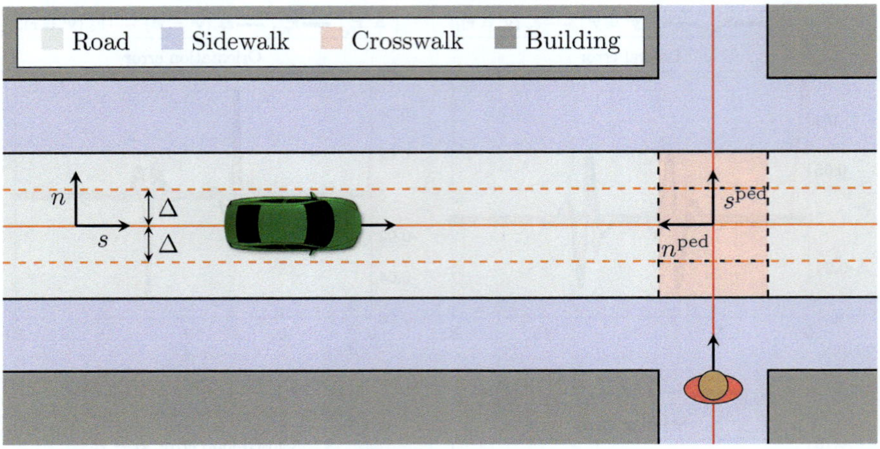


**Fig. 6** The simulation setting includes a vehicle driving down a road with a crosswalk where pedestrians might cross

## 6.2 MPFTC: Ensuring Safety of the Controller

In this section we illustrate the benefits of the safe terminal set by considering an urban driving environment. In particular, we consider the setting shown in Fig. 6, where the vehicle needs to safely navigate the road and avoid collisions with pedestrians that may be occluded by the environment. To that end, we will show next that when the conditions required our safe framework, i.e., Theorem 3, are satisfied, we are able to avoid collisions with suddenly appearing road users.

For simplicity and ease of illustration, we consider the following reference trajectory

$$\mathbf{r}^{\mathbf{x}}(\tau) = [v^{\text{ref}}\tau,\ 0,\ 0,\ 0,\ v^{\text{ref}}]^\top,\ \mathbf{r}^{\mathbf{u}}(\tau) = [0,\ 0]^\top, \tag{32}$$

which models a constant velocity trajectory with no turning, i.e., the road center line from Fig. 6. In order show constraint satisfaction of our framework while tracking this reference, we need to introduce a model that predicts the future motion of other road users. Hence, to simplify the analysis and presentation of the results, we consider only pedestrians that may appear from the bottom side of the crosswalk in Fig. 6. Therefore, we model the pedestrian dynamics as following the red straight line in Fig. 6. By defining the uncertain variable related to the pedestrian position as $\mathbf{w}_k := [s_k^{\text{ped}}, n_k^{\text{ped}}]$, we formulate the pedestrian kinematics as two single integrators

$$\mathbf{w}_{k+1} = \omega(\mathbf{w}_k, \boldsymbol{\zeta}) = \begin{bmatrix} 1 & 0 \\ 0 & K \end{bmatrix} \mathbf{w}_k + \begin{bmatrix} 1.3 \\ 0 \end{bmatrix} t_\text{s} + t_\text{s}\boldsymbol{\zeta}, \tag{33}$$

where the lateral states $n_k^{\text{ped}}$ are stabilized if $K < 1$, and $\boldsymbol{\zeta} \in \mathcal{Z}$, with $\mathcal{Z} = \{\boldsymbol{\zeta} \mid \|\boldsymbol{\zeta}\|_\infty \leq 0.75\}$, relates to some bounded uncertainty in the pedestrian movement. We stress that we use this model for simplicity, and for real autonomous driving scenarios, model (33) is far from realistic and one should instead consider more advanced pedestrian models like the ones proposed in, e.g., [56–59].

With model (33), it is straightforward to follow the steps from (18) in Sect. 5 and propagate the future uncertainty from an initial measurement $\mathbf{w}_k$ as

$$\mathcal{W}_{n+1|k} = \{\omega(\mathbf{w}_{n|k}, \boldsymbol{\zeta}) \mid \mathbf{w}_{n|k} \in \mathcal{W}_{n|k}, \forall \boldsymbol{\zeta} \in \mathcal{Z}\}, \tag{34}$$

with $\mathcal{W}_{k|k} = \mathbf{w}_k$. However, we note that even though prediction model (34) ensures that the predicted sets satisfy $\mathcal{W}_{n|k+1} \subseteq \mathcal{W}_{n|k}$, we are still faced with the problem of accounting for pedestrians that are outside of our sensing range, e.g., occluded by buildings or other vehicles. To that end, the sensor-suite must provide information, or have an understanding, where pedestrians might appear. This becomes a crucial part in satisfying Assumption 6, which is needed in Theorem 3 for recursive feasibility. We therefore assume for this simple example, that the sensor-suite has access to information on locations where pedestrians might appear, e.g., from a map. With the information of each such "hidden" location, we expect that a pedestrian might be hidden, and as such, propagate the uncertainty model of (34) for the hidden pedestrian as well. To that end, we must propagate the uncertainty for all measured pedestrians, and potentially "hidden" pedestrians. Hence, if we can measure $i$ pedestrians, and have $j$ hidden locations, we predict the sets $\mathbf{w}_{n|k}^i \in \mathcal{W}_{n|k}^z$, $\forall z \in \mathbb{I}_1^{i+j}$, and construct $g_{n|k}$ as

$$g_{n|k}^{(\mathbf{x}_{n|k}, \mathbf{u}_{n|k})} = \max_{\mathbf{w}_{n|k}{}^i \in \mathcal{W}_{n|k}^i,\ \forall i \in \mathbb{I}_1^{i+j}} \begin{cases} s_{n|k} + n_{n|k}^{\text{ped},i} + s^{\text{inter}} + r & \text{if } \|s_{n|k}^{\text{ped},i}\| \leq \Delta \\ 0 & \text{otherwise} \end{cases} \leq 0,$$

where $s^{\text{inter}}$ is the intersecting point of the reference trajectory and the walkable path, i.e., $(s, n) = (s^{\text{inter}}, 0)$ and $(s^{\text{ped}}, n^{\text{ped}}) = (0, 0)$ map to the same point in the global frame, $r$ is an additional safety distance, and $\Delta$ denotes the distance threshold when the pedestrian should be considered for collision avoidance.

Having formulated the collision-avoidance constraints, we now turn to formulating the safe terminal set. In a similar light to Example 1 in Sect. 5, we consider the safe set to be given when the vehicle is fully stopped, i.e.,

$$\mathcal{X}_{\text{safe}}(\tau) = \{\mathbf{x} \mid v = 0\}. \tag{35}$$

We note that this may not be a suitable safe set for general autonomous driving settings, and that it in general can vary for different cases. However, in order to avoid further technicalities, we consider this safe set to be sufficient for the simplified setting that we use for illustration.

For the simulations we use the terminal set defined in (20), using the sets (35) and (29), with $N = 40$ and $M = 80$. We use the MPFTC formulation (4) with sam-

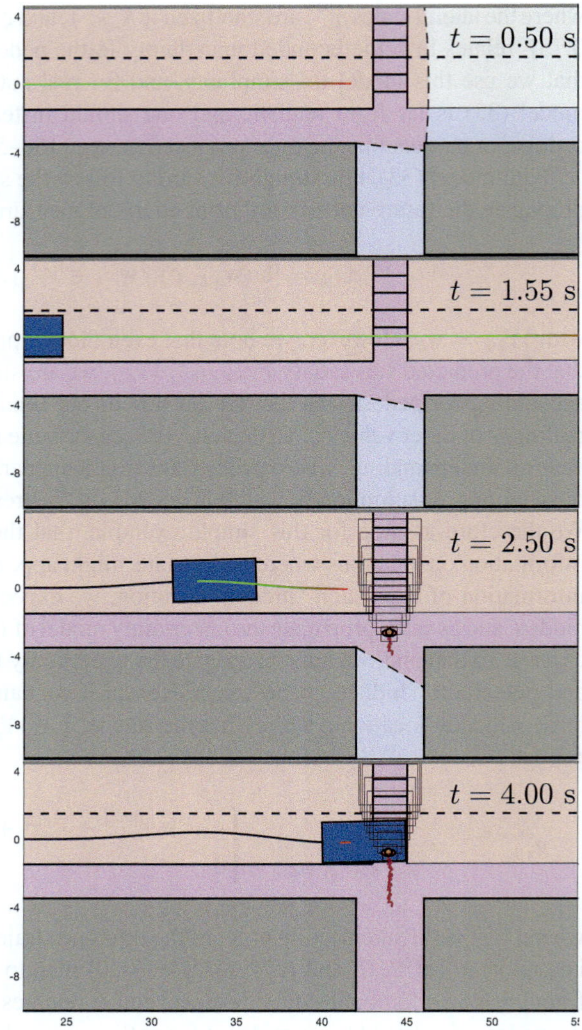


**Fig. 7** Four different time instances of the simulation environment. The two top panels show that the sensors (shaded region) cannot see behind a wall, and that the vehicle as such plans a trajectory within the sensing range. The two last panels show that a pedestrian, who was not visible for the sensors, shows up and forces the vehicle to perform an emergency braking

pling time $t_s = 0.05$ s and nonlinear system model (23). In order to illustrate the benefits of our proposed safe framework, we implement two controllers: one that satisfies Theorem 3; and one where Assumption 6 is not satisfied.

Figure 7 shows four time instances of the MPC controller which only reacts to what it can sense directly, i.e., the controller does not satisfy Assumption 6. It is visible that the vehicle believes that it can safely cross the intersection in the first two frames. In the last two frames the vehicle has moved close enough to the intersection, such that the sensors can now detect the pedestrian. However, in this case the vehicle velocity is too high, so that it causes a collision with the pedestrian. The closed-loop trajectories are shown in Fig, 8. Here it is visible that the vehicle sees the

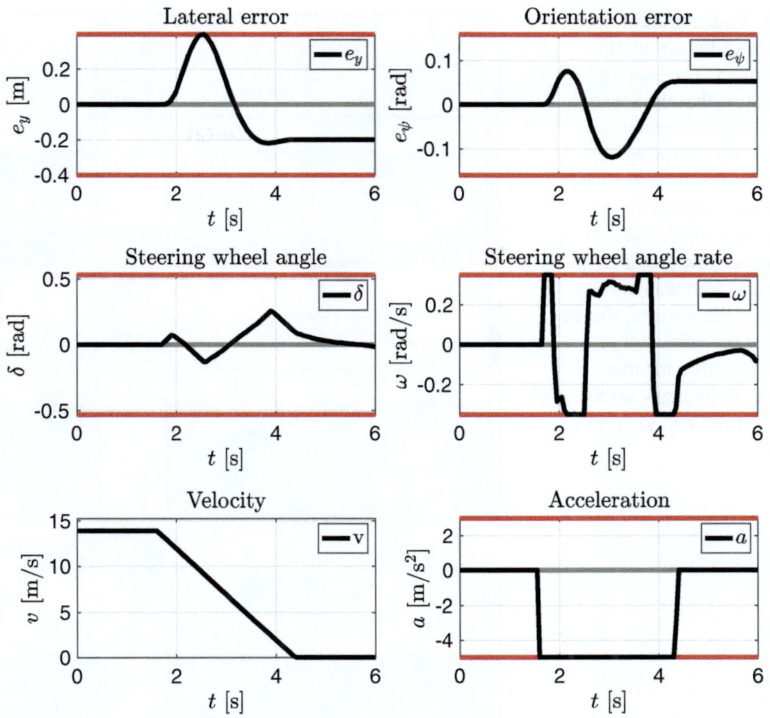


**Fig. 8** Closed-loop evolution of the unsafe MPC controller shown in Fig. 7. Just before $t \leq 2$ s, a pedestrian appears and forces the vehicle to perform an emergency braking

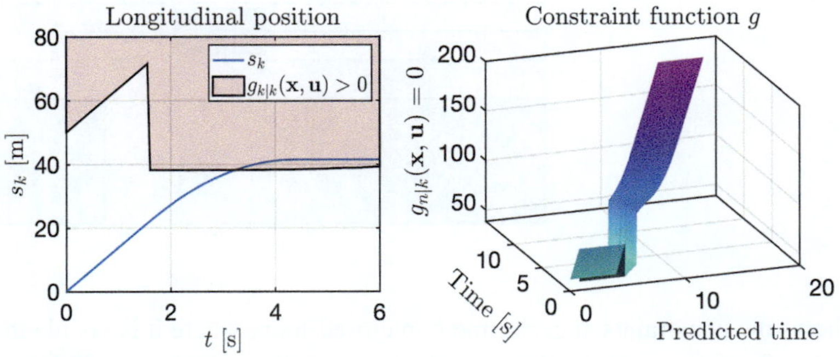


**Fig. 9** Constraint evolution for the unsafe MPC controller. The left-hand side shows the constraint $g_{k|k}$, while the right-hand side shows the time evolution of the predicted constraint $g_{n|k}$

pedestrian just before $t < 2$ s, and applies full braking. From the lateral error, and orientation error, we realize that the MPC controller in fact tries to avoid a collision, by marginally maximizing the traveled distance by actively steering. Figure 9 shows that Assumption 6 is not satisfied since constraint $g_{n|k+1} \nleq g_{n|k}$ for all $n > k$, which indeed is needed for safety, i.e., guaranteeing that Theorem 3 holds. The left-hand

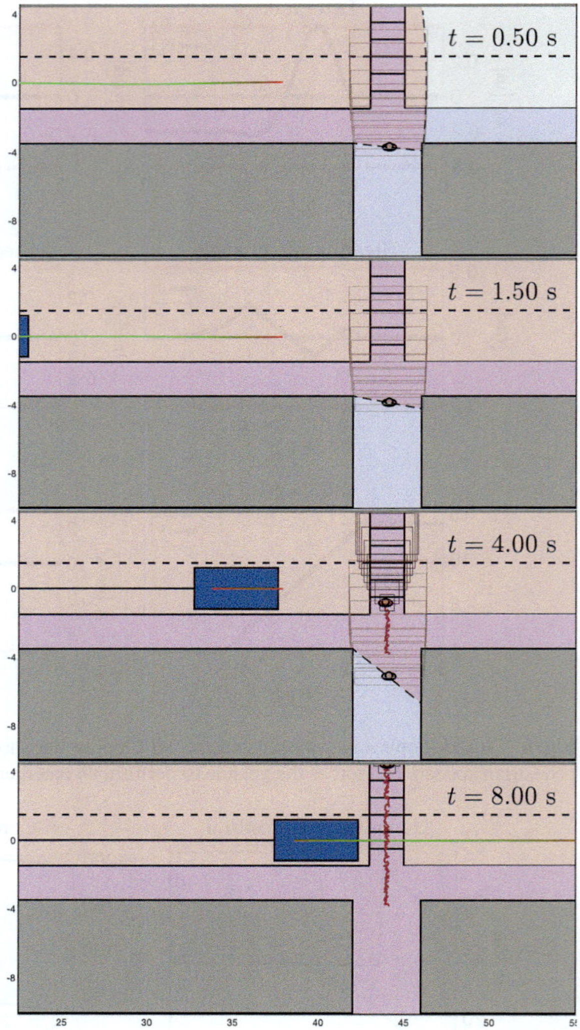


**Fig. 10** Four different time instances of the simulation environment. The two top panels show that the sensors (shaded region) cannot see behind a wall, however, the vehicle plans a trajectory as if there were a pedestrian behind the corner. The two last panels show that a pedestrian, who was not visible for the sensors, shows up. Since the vehicle was already prepared for this situation, it manages safely adjust it speed and yield to the pedestrian

side shows the constraints at each time $k$ in closed-loop, where it is visible that just before $t < 2$ s the constraint shrinks, which causes the vehicle to collide with the pedestrian.

Figure 10 shows how the safe MPC controller behaves when Assumption 6 is satisfied. As opposed to the unsafe controller shown in Fig. 7, the safe controller approaches the intersection more cautiously, as any rational human driver would do. By adjusting the speed, it anticipates that a moving obstacle may appear behind the wall. This can be seen in the two last frames. As time moves on, the pedestrian can safely pass, and the vehicle moves close enough to see that there are no more remaining pedestrians, so that it can safely accelerate to pass the intersection.

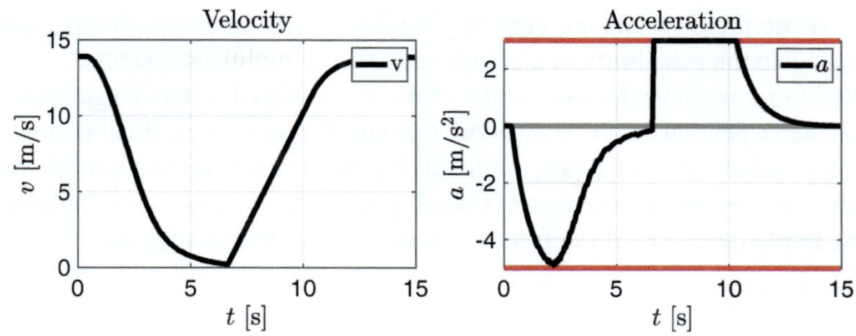


**Fig. 11** Closed-loop evolution of the safe MPC controller shown in Fig. 10. The vehicle approaches the intersection by reducing its velocity. After $t = 6$ s the pedestrian passes, and the vehicle is free to accelerate again

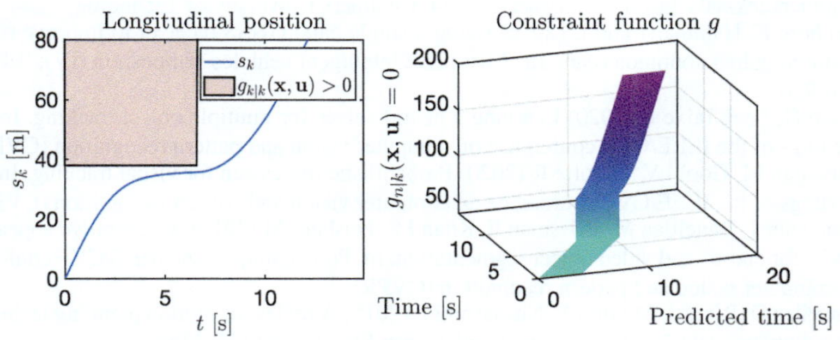


**Fig. 12** Constraint evolution for the safe MPC controller in Fig. 10. The left-hand side shows the constraint $g_{k|k}$, while the right-hand side shows the time evolution of the predicted constraint $g_{n|k}$

Figure 11 shows the closed-loop velocity and acceleration trajectories. It is worth noting how the vehicle slows down earlier than the unsafe controller in Fig. 8, since it anticipates that a pedestrian might appear behind the corner. Note that, the states $(e_y, e_\psi, \delta)$ have been omitted since their dynamics essentially remain unchanged. Finally, Fig. 12 shows that the constraint $g$ is monotonic, and hence, Assumption 6 is satisfied. The "jump" in the right plot illustrates the point in time when the pedestrian is no longer predicted to block the intersection, and the vehicle is free to accelerate again.

## 7 Conclusions

The possibility of using *infeasible reference trajectories* is of great interest in MPC-based motion planning and control algorithms, due to the convenience and simplicity they offer. In this chapter, we have discussed how such reference trajectories affect the closed-loop behavior of the system, and proposed conditions sufficient for stability to hold. While these results are currently limited to LTV systems, future research will investigate the possibility to also include general nonlinear systems.

Furthermore, we have discussed safety for autonomous driving in a general sense and presented a new safe MPC framework that enables recursive collision-avoidance constraint satisfaction at all times, while relying on assumptions that can be verified in practice on the perception system. Ongoing research is focusing on the practical real-time implementation of the framework in a full-scale test vehicle.

## References

1. Stenborg E (2020) Long term localization for self driving cars. Doktorsavhandlingar vid Chalmers tekniska högskola. Ny serie: 4844. Chalmers University of Technology
2. Stenborg E, Hammarstrand L (2016) Using a single band GNSS receiver to improve relative positioning in autonomous cars. In: 2016 IEEE intelligent vehicles symposium (IV). IEEE, pp 921–926
3. Braso G, Leal-Taixe L (2020) Learning a neural solver for multiple object tracking. In: Proceedings of the IEEE/CVF conference on computer vision and pattern recognition (CVPR)
4. Danelljan M, Gool LV, Timofte R (2020) Probabilistic regression for visual tracking. In: Proceedings of the IEEE/CVF conference on computer vision and pattern recognition (CVPR)
5. Johnander J, Danelljan M, Brissman E, Khan FS, Felsberg M (2019) A generative appearance model for end-to-end video object segmentation. In: Proceedings of the IEEE/CVF conference on computer vision and pattern recognition (CVPR)
6. González D, Pérez J, Milanés V, Nashashibi F (2015) A review of motion planning techniques for automated vehicles. IEEE Trans Intell Transp Syst 17(4):1135–1145
7. Paden B, Čáp M, Yong SZ, Yershov D, Frazzoli E (2016) A survey of motion planning and control techniques for self-driving urban vehicles. IEEE Trans Intell Veh 1(1):33–55
8. Shalev-Shwartz S, Shammah S, Shashua A (2018) On a formal model of safe and scalable self-driving cars
9. de Campos GR, Falcone P, Hult R, Wymeersch H, Sjöberg J (2017) Traffic coordination at road intersections: autonomous decision-making algorithms using model-based heuristics. IEEE Intell Transp Syst Mag 9(1):8–21
10. Campos GR, Falcone P, Wymeersch H, Hult R, Sjöberg J (2014) Cooperative receding horizon conflict resolution at traffic intersections. In: 53rd IEEE conference on decision and control. IEEE, pp 2932–2937
11. Hult R, Zanon M, Gros S, Falcone P (2018) Optimal coordination of automated vehicles at intersections: theory and experiments. IEEE Trans Control Syst Technol 27(6):2510–2525
12. Hult R, Zanon M, Gros S, Falcone P (2018) Energy-optimal coordination of autonomous vehicles at intersections. In: 2018 European control conference (ECC), pp 602–607
13. Uebel S, Murgovski N, Bäker B, Sjöberg J (2019) A two-level MPC for energy management including velocity control of hybrid electric vehicles. IEEE Trans Veh Technol 68(6):5494–5505
14. Zanon M (2020) A Gauss-Newton-Like Hessian approximation for economic NMPC. IEEE Trans Autom Control
15. Batkovic I, Rosolia U, Zanon M, Falcone P (2020) A robust scenario MPC approach for uncertain multi-modal obstacles. IEEE Control Syst Lett 5(3):947–952
16. Batkovic I, Zanon M, Ali M, Falcone P (2019) Real-time constrained trajectory planning and vehicle control for proactive autonomous driving with road users. In: Proceedings of European Control Conference (ECC)
17. Cesari G, Schildbach G, Carvalho A, Borrelli F (2017) Scenario model predictive control for lane change assistance and autonomous driving on highways. IEEE Intell Transp Syst Mag 9(3):23–35

18. Chen Y, Rosolia U, Ubellacker W, Csomay-Shanklin N, Ames AD (2021) Interactive multi-modal motion planning with branch model predictive control. arXiv preprint arXiv:2109.05128
19. Gros S, Zanon M, Quirynen R, Bemporad A, Diehl M (2020) From linear to nonlinear MPC: bridging the gap via the real-time iteration. Int J Control 93(1):62–80
20. Gutjahr B, Gröll L, Werling M (2017) Lateral vehicle trajectory optimization using constrained linear time-varying MPC. IEEE Trans Intell Transp Syst 18(6):1586–1595. https://doi.org/10.1109/TITS.2016.2614705
21. Lima PF, Pereira GC, Martensson J, Wahlberg B (2018) Experimental validation of model predictive control stability for autonomous driving. Control Eng Pract 81:244–255. https://doi.org/10.1016/j.conengprac.2018.09.021
22. Nair SH, Govindarajan V, Lin T, Meissen C, Tseng HE, Borrelli F (2021) Stochastic MPC with multi-modal predictions for traffic intersections. arXiv preprint arXiv:2109.09792
23. Borrelli F, Bemporad A, Morari M (2017) Predictive control for linear and hybrid systems. Cambridge University Press
24. Rawlings JB, Mayne DQ, Diehl M (2017) Model predictive control: theory, computation, and design, vol 2. Nob Hill Publishing, Madison
25. Batkovic I, Ali M, Falcone P, Zanon M (2020) Safe trajectory tracking in uncertain environments. IEEE Trans Autom Control (submitted). Available on arXiv:2001.11602
26. Grüne L, Pannek J (2011) Nonlinear model predictive control. Springer, London
27. Amrit R, Rawlings J, Angeli D (2011) Economic optimization using model predictive control with a terminal cost. Ann Rev Control 35:178–186
28. Faulwasser T, Grüne L, Müller M (2018) Economic nonlinear model predictive control: stability, optimality and performance. Found Trends Syst Control 5(1):1–98. https://doi.org/10.1561/2600000014
29. Grüne L (2013) Economic receding horizon control without terminal constraints. Automatica 49:725–734
30. Müller MA, Angeli D, Allgöwer F (2015) On necessity and robustness of dissipativity in economic model predictive control. IEEE Trans Autom Control 60(6):1671–1676
31. Zanon M (2021) A Gauss-Newton-Like Hessian approximation for economic NMPC. IEEE Trans Autom Control 66(9):4206–4213
32. Zanon M, Faulwasser T (2018) Economic MPC without terminal constraints: gradient-correcting end penalties enforce asymptotic stability. J Process Control 63:1–14
33. Zanon M, Gros S, Diehl M (2013) A Lyapunov function for periodic economic optimizing model predictive control. In: Proceedings of the 52nd conference on decision and control (CDC), pp 5107–5112
34. Zanon M, Grüne L, Diehl M (2017) Periodic optimal control, dissipativity and MPC. IEEE Trans Autom Control 62(6):2943–2949
35. De Schutter J, Zanon M, Diehl M (2020) TuneMPC—a tool for economic tuning of tracking (N)MPC problems. IEEE Control Syst Lett 4(4):910–915. https://github.com/jdeschut/tunempc
36. Zanon M, Gros S, Diehl M (2014) Indefinite linear MPC and approximated economic MPC for nonlinear systems. J Process Control 24:1273–1281
37. Zanon M, Gros S, Diehl M (2016) A tracking MPC formulation that is locally equivalent to economic MPC. J Process Control 45:30–42
38. Zanon M, Gros S, Diehl M (2017) A periodic tracking MPC that is locally equivalent to periodic economic MPC. In: Proceedings of the 2017 IFAC world congress, vol 50, issue no 1, pp 10711–10716
39. Batkovic I, Ali M, Falcone P, Zanon M (2020) Model predictive control with infeasible reference trajectories. IEEE Trans Autom Control (submitted). Available on arXiv:2109.04846
40. Rawlings J, Bonne D, Jorgensen J, Venkat A, Jorgensen S (2008) Unreachable setpoints in model predictive control. IEEE Trans Autom Control 53:2209–2215
41. Diehl M, Amrit R, Rawlings JB (2010) A Lyapunov function for economic optimizing model predictive control. IEEE Trans Autom Control 56(3):703–707

42. Faulwasser T, Zanon M (2018) Asymptotic stability of economic NMPC: the importance of adjoints. In: Proceedings of the IFAC nonlinear model predictive control conference
43. Faulwasser T, Findeisen R (2016) Nonlinear model predictive control for constrained output path following. IEEE Trans Autom Control 61(4):1026–1039
44. Faulwasser T, Kern B, Findeisen R (2009) Model predictive path-following for constrained nonlinear systems. In: Proceedings of the 48th IEEE conference on decision and control. CDC. IEEE, pp 8642–8647
45. Kerrigan EC (2001) Robust constraint satisfaction: Invariant sets and predictive control (Order No. U142244). Available from ProQuest Dissertations & Theses Global: The Sciences and Engineering Collection. (301571506)
46. Yu S, Maier C, Chen H, Allgöwer F (2013) Tube MPC scheme based on robust control invariant set with application to Lipschitz nonlinear systems. Syst Control Lett 62(2):194–200
47. Lima PF, Martensson J, Wahlberg B (2017) Stability conditions for linear time-varying model predictive control in autonomous driving. In: 2017 IEEE 56th annual conference on decision and control (CDC). IEEE, pp 2775–2782
48. Herceg M, Kvasnica M, Jones C, Morari M (2013) Multi-parametric toolbox 3.0. In: Proceedings off the European control conference. Zürich, Switzerland, pp 502–510. http://control.ee.ethz.ch/~mpt
49. Blanchini F, Miani S (2008) Set-theoretic methods in control. Springer, Berlin
50. Gupta A, Köroglu H, Falcone P (2019) Computation of low-complexity control-invariant sets for systems with uncertain parameter dependence. Automatica 101:330–337
51. Gupta A, Mejari M, Falcone P, Piga D (2020) Computation of parameter dependent robust invariant sets for LPV models with guaranteed performance. arXiv preprint arXiv:2009.09778
52. Verschueren R, Frison G, Kouzoupis D, Frey J, van Duijkeren N, Zanelli A, Novoselnik B, Albin T, Quirynen R, Diehl M (2020) acados: a modular open-source framework for fast embedded optimal control. Math program comput 14:147–183
53. Andersson JA, Gillis J, Horn G, Rawlings JB, Diehl M (2019) Casadi: a software framework for nonlinear optimization and optimal control. Math Program Comput 11(1):1–36
54. Wächter A, Biegler LT (2006) On the implementation of an interior-point filter line-search algorithm for large-scale nonlinear programming. Math Program 106(1):25–57
55. Frison G, Diehl M (2020) HPIPM: a high-performance quadratic programming framework for model predictive control. IFAC-PapersOnLine 53(2):6563–6569
56. Batkovic I, Zanon M, Lubbe N, Falcone P (2018) A computationally efficient model for pedestrian motion prediction. In: 2018 European control conference (ECC), pp 374–379. https://doi.org/10.23919/ECC.2018.8550300
57. Koschi M, Althoff M (2020) Set-based prediction of traffic participants considering occlusions and traffic rules. IEEE Trans Intell Veh 6(2):249–265
58. Koschi M, Pek C, Beikirch M, Althoff M (2018) Set-based prediction of pedestrians in urban environments considering formalized traffic rules. In: 2018 21st international conference on intelligent transportation systems (ITSC). IEEE, pp 2704–2711
59. Pek C, Manzinger S, Koschi M, Althoff M (2020) Using online verification to prevent autonomous vehicles from causing accidents. Nat Mach Intell 2(9):518–528

# Energy-Efficient Autonomous Driving Using Cognitive Driver Behavioral Models and Reinforcement Learning

**Huayi Li, Nan Li, Ilya Kolmanovsky, and Anouck Girard**

**Abstract** Autonomous driving technologies are expected to not only improve mobility and road safety but also bring energy efficiency benefits. In the foreseeable future, autonomous vehicles (AVs) will operate on roads shared with human-driven vehicles. To maintain safety and liveness while simultaneously minimizing energy consumption, the AV planning and decision-making process should account for interactions between the autonomous ego vehicle and surrounding human-driven vehicles. In this chapter, we describe a framework for developing energy-efficient autonomous driving policies on shared roads by exploiting human-driver behavior modeling based on cognitive hierarchy theory and reinforcement learning.

## 1 Introduction

With recent advances in sensing technologies and artificial intelligence, there has been a rapidly growing interest in connected and autonomous vehicles (CAVs) [12, 43]. Such vehicles are expected to improve the safety and mobility of transportation and to alleviate traffic congestion.

Another expected benefit of CAVs is a reduction in fuel/energy consumption [38, 41]. Since 2016, the U.S. Department of Energy has awarded more than $50 million in funding for studies by the Advanced Research Projects Agency-Energy's (ARPA-E) Next-Generation Energy Technologies for Connected and Automated On-Road Vehicles (NEXTCAR) program for which the goal is to reduce the energy

H. Li (✉) · N. Li · I. Kolmanovsky · A. Girard
Department of Aerospace Engineering, University of Michigan, Ann Arbor, MI, USA
e-mail: huayil@umich.edu

N. Li
e-mail: nanli@umich.edu

I. Kolmanovsky
e-mail: ilya@umich.edu

A. Girard
e-mail: anouck@umich.edu

 
Y. L. Murphey et al. (eds.), *AI-enabled Technologies for Autonomous and Connected Vehicles*, Lecture Notes in Intelligent Transportation and Infrastructure,
https://doi.org/10.1007/978-3-031-06780-8_10

**Table 1** Summary of selected literature on using CAV technologies to improve energy efficiency

| References | Energy consumption reduction by (%) | Methods | Traffic conditions |
|---|---|---|---|
| [1] | 5.4 | Thermal management (air conditioning) | Driving cycles |
| [42] | 3.1 | Thermal management (battery) | Driving cycles |
| [8] | 41 | Thermal management (air conditioning and engine) | Driving cycles |
| [32] | 50 | Eco-driving | Intersections |
| [14] | 12 | Eco-routing | City |
| [20] | 57.8 | Platooning, cooperative merging | Ramps on highway |
| [2] | 8.5 | Anticipative lane change | City and highway |
| [4] | (Up to) 59 | Cooperative lane change | Highway |
| [30] | 47 | Cooperative driving | Intersections and roundabouts |

consumption of vehicles in all classes by more than 20% via CAV technologies [40]. Table 1 shows selected results by this program. This is not intended to be a comprehensive review, but the collection illustrates the variety of methods and traffic conditions being explored in some of the most recent studies to achieve energy efficiency improvements using CAV technologies.

To realize this encouraging potential in real-world driving circumstances, we observe that at least the following two problems remain to be studied. Firstly, interactions of the ego vehicle with surrounding vehicles in traffic need to be considered. In the foreseeable future, autonomous vehicles will operate together with human-driven vehicles in traffic. Thus, it is necessary to consider the different vehicle actions and reactions caused by different types of human driving styles. In [18, 26, 28, 39], level-$k$ game theory is used to model the interactions with the focus on different driving scenarios. Researchers such as of [10, 35, 36] have utilized traffic-in-the-loop models and closed-loop control to achieve simultaneous optimization for safety and fuel economy. However, only longitudinal control is considered in these studies. Indeed, the vast majority of recent studies on improving energy efficiency using CAV technologies assume that the ego vehicle is driven in single-lane traffic. Thus, the second problem worth investigating is the simultaneous longitudinal control and lateral control (such as lane changes) of AVs, which increases the dimension of the problem but provides additional possibilities to save energy. A more detailed discussion on lane changes for better energy efficiency is given in Sect. 2.

There has been a rich set of research on machine learning (ML) methods for automotive applications to improve energy efficiency and emissions by modeling and control of the powertrain system (see, e.g., [15, 22, 25, 29, 34, 46]). In particular, to meet increasingly stringent fuel economy and emissions regulations, the powertrain systems of hybrid electric vehicles (HEVs) have become more and more complex.

Consequently, commonly studied model-based control methods for the energy management system (EMS), such as dynamic programming (DP) and model predictive control (MPC) [13], are facing growing difficulties, as they rely on models with good accuracy and many control-oriented models are physics-based. In comparison, machine learning methods can handle this challenge well. For example, for HEVs with small electrical energy storage, there is a significant potential to utilize recurrent neural networks to learn driving patterns and improve energy efficiency [9]. In the CAV domain, example applications of ML include perception and localization, route/path planning/optimization, and motion control, despite challenges such as computation, safety, and adaptability/generalizability that are actively being studied [11, 19, 21]. Studies such as [6, 45] use ML to inform energy-efficient acceleration/braking of electric vehicles. The authors previously developed a level-$k$ game theory-based traffic simulator in [26] (following the methodology originally proposed in [47]). The simulator is based on cognitive driver behavioral models trained by reinforcement learning (RL).

In this chapter, we describe a novel framework for developing energy-efficient AV control policies, including both longitudinal (speed) and lateral (lane change) controls, through RL. We focus our attention on highway driving and autonomous battery electric vehicles (BEVs), as BEVs are getting increasingly popular due to their environmental benefits [50]. A BEV powertrain model is developed to calculate the energy consumption over trips. To enable the AV control policy to properly respond to the interactions with human-driven vehicles on shared roads, the game-theoretic traffic model developed in [26, 28] is used as the RL training environment. Reference [23] is a preliminary conference version of this chapter. Extensions to other powertrain types and traffic environments are possible [27, 39] but are left to future work.

The remainder of this chapter is organized as follows. Firstly, further background on lane changes for energy-efficient AV driving is discussed in Sect. 2. Then, we begin the development by building a BEV model and validating it in Sect. 3. In Sect. 4, the control development is detailed, and another control policy trained by RL and the finite-state-machine (FSM) controller from [28] are introduced to be subsequently used for comparison. Section 5 presents results on RL convergence and on performance of the developed control policy in simulations. Finally, concluding remarks are made in Sect. 6.

## 2 Lane Changes for Energy-Efficient AV Driving

Including lateral actions such as lane changes may further improve the energy efficiency [31], though the survey results of [41] show that this is still an emerging area that remains to be studied. Anticipating lane selection has been proposed, such as in [17, 37]. Furthermore, instead of focusing on an individual vehicle, cooperative lane change [3] is expected to benefit the neighboring vehicles and harmonize the surrounding traffic. However, these studies assume having connected vehicle technologies, such as vehicle-to-vehicle and vehicle-to-infrastructure communications.

A general observation is that the energy efficiency can be improved by reducing the change of speed and acceleration. In contrast, our work uses the position and velocity information of the immediate neighboring vehicles detected by the sensors of the ego vehicle, and a powertrain model is used to accurately predict the energy consumption.

There are two major challenges in combining longitudinal and lateral actions for safe energy-efficient driving. Firstly, considering both longitudinal and lateral actions increases the problem complexity. In particular, there are subtle trade-offs between safety and energy efficiency. For instance, in scenarios such as a sudden cut-in by a slow vehicle, changing lanes rather than hard braking preserves vehicle momentum and avoids energy loss, but if the traffic density is high, performing a lane change may not be safe or feasible.

Another challenge is that, unlike safety constraint violation scenarios (e.g., collisions) where events typically occur within seconds, energy efficiency evaluation requires longer time horizons of several hundreds of seconds. Hence, optimization-based control algorithms that simultaneously address driving safety and energy efficiency requirements need to account for both short-term and long-term objectives. One approach is to define a terminal cost function for the short horizon optimization reflective of long-term rewards. However, how to practically determine such a terminal cost function is often a priori unclear. In special cases, the problem can be reformulated to focus on maintaining component operation in more efficient regions [7] for which short-horizon optimization is sufficient. However, such reformulations are not always feasible and the performance with such an approach could be suboptimal. In particular, Stackelberg policies and the decision tree policies considered in [5, 48, 49] rely on rewards being evaluated short-term. It may not be straightforward to extend these to account for energy efficiency requirements. For our AV control policy, since the RL algorithm updates the value functions with not only the one-step/instantaneous reward but also the average reward over time, it is able to handle multiple optimization objectives that need to be evaluated over different time horizons.

## 3 Powertrain Modeling for Battery Electric Vehicles

### *3.1 Model Description*

Accounting for energy efficiency in the AV controller design requires a longitudinal powertrain model. As an RL process is generally computationally intensive, a powertrain model used in RL should have low computational footprint but sufficiently high accuracy.

Figure 1 shows the layout of the BEV model considered in this work. The powertrain system consists of a battery pack, a motor/generator (MG), a drivetrain with a single-speed final drive, the wheels and tires, and the powertrain control unit (PCU).

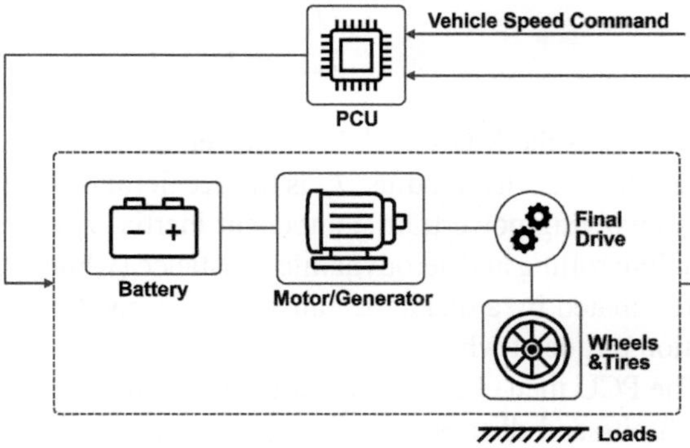


**Fig. 1** Battery electric vehicle powertrain system layout

The powertrain model is of the backward type [13, 33, 44]. (See Remark 1 below for details.) It has two states, state of charge ($SOC$) and total energy consumption. High-fidelity dynamics, such as transient responses of the MG and drivetrain, are not considered. Maps (i.e., look-up tables) are used to represent component operating characteristics as described below. A benefit of using a map-based approach is that it can facilitate the change of component specifications so that potential extensions, such as component sizing or fleet energy efficiency studies, are possible.

*Remark* 1 A backward powertrain model assumes that the actual vehicle speed is always equal to the reference speed command. Rotational speeds of components are coupled/scaled by the gear ratios and wheel radius based on this command. The torque required at the wheels to meet the acceleration demand is first calculated and then translated component by component to the actuators such as the motor and the brakes. Backward models typically entail low computational costs and are suitable for (approximate) energy consumption evaluations. In comparison, forward models involve a driver model, typically modeled as a PID controller, that commands the motor/brake torque in order to track the vehicle speed reference. The torque is then translated to the wheels through drivetrain components. The speed of each component, as well as the actual vehicle speed, is calculated by integrating the acceleration produced by the torque. Consequently, forward models can better capture the component dynamics, at the cost of higher computational complexity than backward models. They are typically used for more detailed (e.g., componentwise) energy efficiency analysis as well as drivability-related simulations [13, 33, 44]. Also, the forward model is not as suitable as the backward model in our case considering safe driving since there is a tracking error between the commanded and actual vehicle speed the value of which depends on the tuning of the control parameters.

In the BEV model, the MG speed and traction torque at the wheels are first calculated based on the vehicle speed command according to

$$\omega_{mg} = \frac{V \cdot g}{r}, \tag{1}$$

$$T_a = \dot{V} \cdot M \cdot r, \tag{2}$$

$$T_l = a + b \cdot V + c \cdot V^2, \tag{3}$$

$$T_{whl} = T_a + T_l, \tag{4}$$

where $\omega_{mg}$ is the MG speed, $V$ is the vehicle speed, $g$ is the final drive gear ratio, $r$ is the effective wheel radius, $T_a$ is the acceleration torque, $M$ is the effective vehicle mass involving powertrain component inertia, $T_l$ represents lumped external loads including rolling and aerodynamic resistance (while the grade is assumed to be zero) approximated by a quadratic function with coefficients $a$, $b$, and $c$, and $T_{whl}$ is the traction torque at wheels.

The PCU then checks whether the traction torque demand exceeds the battery or MG limits and distributes the torque command to the MG and the friction brake, using the following logic,

$$T_{mgPos} = \begin{cases} 0, & \text{if } T_{whl} < 0, \\ T_{whl} \cdot \frac{1}{g}, & \text{if } \ 0 \leq T_{whl} \leq T_{mgMax} \cdot g, \\ T_{mgMax}, & \text{if } \ T_{whl} > T_{mgMax} \cdot g, \end{cases} \tag{5}$$

$$T_{whlBrk} = \begin{cases} 0, & \text{if } \ -T_{whl} < 0, \\ -T_{whl}, & \text{if } \ 0 \leq -T_{whl} \leq T_{brkMax}, \\ T_{brkMax}, & \text{if } \ -T_{whl} > T_{brkMax}, \end{cases} \tag{6}$$

$$T_{mgReg} = \begin{cases} F_{reg} \cdot T_{whlBrk} \cdot \frac{1}{g}, & \text{if } \ T_{whlBrk} \leq T_{mgRegLim} \cdot g, \\ F_{reg} \cdot T_{mgRegLim}, & \text{if } \ T_{whlBrk} > T_{mgRegLim} \cdot g, \end{cases} \tag{7}$$

$$T_{mg} = \begin{cases} T_{mgPos}, & \text{if } \ T_{whl} > 0, \\ -T_{mgReg}, & \text{if } \ T_{whl} \leq 0, \end{cases} \tag{8}$$

$$T_{mechBrk} = -(T_{whlBrk} - T_{mgReg}), \tag{9}$$

where $T_{mgMax}$ is the maximum MG torque limit, $T_{mgPos}$ is the positive portion of the MG torque limited by $T_{mgMax}$, $T_{whlBrk}$ is the negative portion of the traction torque at wheels limited by a constant brake torque limit denoted by $T_{brkMax}$, $T_{mgRegLim}$ is the MG regeneration torque limit, $F_{reg}$ is the regeneration factor, $T_{mgReg}$ is the negative portion of the MG torque limited by $T_{mgRegLim}$, $T_{mg}$ is the MG torque, and $T_{mechBrk}$ is the torque demand assigned to the friction brakes. We obtain the values of $T_{mgMax}$, $T_{mgRegLim}$, and $F_{reg}$ through maps, where both $T_{mgMax}$ and $T_{mgRegLim}$ depend on $\omega_{mg}$, and $F_{reg}$ is a function of $V$ (to reduce the regenerative braking at low vehicle speeds) and the battery $SOC$ (to reduce the regenerative braking at high $SOC$).

The power drawn by the MG is then obtained from

$$P_{mg} = \begin{cases} T_{mg}\omega_{mg}/\eta, & \text{if } T_{mg}\omega_{mg} \geq 0, \\ T_{mg}\omega_{mg} \cdot \eta, & \text{if } T_{mg}\omega_{mg} < 0, \end{cases} \tag{10}$$

where $P_{mg}$ is the MG electric power, and $\eta \in (0, 1)$ is the MG efficiency as a function of the MG torque and speed, as given by a map.

To calculate the $SOC$ and cumulative energy consumption, the battery is modeled as follows:

$$\begin{aligned} P_{mg} &= (V_{oc} - \frac{1}{N_p} \cdot I \cdot R) \cdot N_s \cdot I \\ &= N_s \cdot V_{oc} \cdot I - \frac{N_s}{N_p} \cdot R \cdot I^2, \end{aligned} \tag{11}$$

which can be re-arranged as

$$\frac{N_s}{N_p} \cdot R \cdot I^2 - N_s \cdot V_{oc} \cdot I + P_{mg} = 0, \tag{12}$$

where $V_{oc}$ and $R$ are, respectively, the open circuit voltage and the resistance of a single battery cell, $I$ is the battery pack current, $N_s$ is the number of battery cells in series, and $N_p$ is the number of battery cells in parallel. Here, $V_{oc}$ and $R$ are acquired through maps, and both variables depend on $SOC$, with the assumption that the battery temperature is constant. Then, we can solve for the battery current as

$$I = \frac{N_s \cdot V_{oc} - \sqrt{(N_s \cdot V_{oc})^2 - 4\frac{N_s}{N_p} \cdot R \cdot P_{mg}}}{2\frac{N_s}{N_p} \cdot R}, \tag{13}$$

and the battery dynamics are given by

$$\dot{SOC} = \frac{-I}{C_{max} \cdot N_p \cdot 3600} \cdot 100, \tag{14}$$

where $C_{max}$ is the maximum battery capacity.

The total discharged electric energy $E_{batt}$ is computed by integrating the battery power $P_{batt}$ based on

$$\dot{E}_{batt} = P_{batt} = N_s \cdot I \cdot V_{oc}, \tag{15}$$

and the energy consumption can be determined from

$$MPGe = \frac{x}{E_{batt}} \cdot \gamma, \tag{16}$$

where $MPGe$ stands for miles per gallon equivalent, $x$ is the total distance traveled, and $\gamma$ represents the unit conversion coefficient.

## 3.2 Model Calibration and Validation

The BEV powertrain model described above is calibrated using the BEV reference model in the Powertrain Blockset Toolbox (PTBS) version 1.5 developed by MathWorks. The PTBS reference model is a forward model and includes more detailed component controls and dynamics than our model. Our model uses some maps and parameter values from the PTBS reference model, and the other model parameters are hand-tuned to reduce errors between the two models.

After calibration, we validate our model by testing and comparing the $MPGe$ of our model and that of the PTBS reference model for different driving cycles. The $MPGe$ mismatches between the two models for the Urban Dynamometer Driving Schedule (UDDS), the Highway Fuel Economy Test (HWFET), and the US06 driving cycles are 5.94%, 5.90%, and 7.95%, respectively. Figure 2 shows the time histories of powertrain signals for the BEV driving through the UDDS cycle, where the blue curves correspond to our model after calibration and the red curves correspond to the PTBS reference model. It can be observed that the signals of our model closely match those of the PTBS reference model. These results validate that our model (1)–(16) after calibration can be used to produce sufficiently accurate energy consumption estimates (accurate in terms of matching the estimates produced by the high-fidelity PTBS reference model). Note that our model entails much lower computational footprint than the PTBS reference model, and is thus suitable for RL purposes.

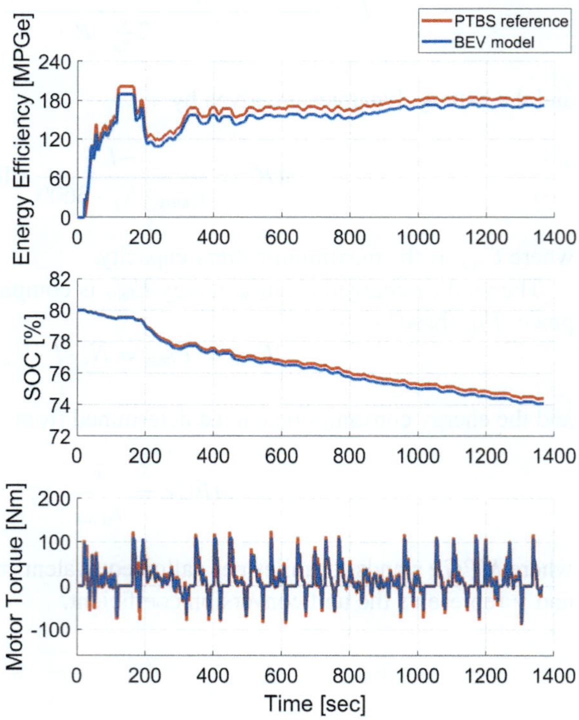


**Fig. 2** Time histories of powertrain signals for the UDDS cycle

The BEV powertrain model (1)–(16) is then converted to discrete-time, assuming a 1-s sampling period, and integrated with the traffic simulator of [26, 28], used for the RL-based development of energy-efficient autonomous vehicle control policy.

# 4 Controller Design

## 4.1 Game-Theoretic Traffic Environment

In order to train an autonomous vehicle control policy offline, we use an interactive traffic simulator, following the approach described in Li et al. [26], as the training environment. This level-$k$ game-theory based simulator assumes that the traffic consists of human-driven vehicles that can be modeled by cognitive behavioral models with $k$ levels. Studies such as [39] show that human reasoning process rarely exceeds three steps, so $k = 0, 1, 2$ is used in this work. The level-0 vehicles use a hand-crafted policy that commands one of the three actions, "maintain speed", "decelerate", or "hard decelerate", based on the range and range rate with the vehicle in its front to represent vehicle behavior under minimal rationality; level-1 and level-2 vehicles use policies trained by RL assuming that the surrounding vehicles are all level-0 and level-1, respectively. Moreover, the level-1 and 2 vehicles have a larger action space consisting of seven actions, including acceleration, deceleration, and lane change. Overall, the results of [26] indicate that the level-0 drivers/vehicles have the most conservative behaviors, while level-1 vehicles behave the most aggressively such as driving faster and frequently making lane changes, and the aggressiveness of level-2 vehicles falls between level-0 and level-1.

*Remark* 2 Similarities and differences between the setup considered in this work and that in [26] are highlighted in Table 2, in terms of models, reward functions, surrounding traffic, observation space and action space, RL algorithm, and training process. Details are given in the subsequent subsections.

**Table 2** Comparison summary on control development for the level-$k$ ($k = 1, 2$) policies and the autonomous vehicle policy considered in this work (AV)

| | Level-$k$ for $k = 1, 2$ | AV |
|---|---|---|
| BEV model | Not included | Included |
| Reward function | $R_1$ | $R_1 + R_2$ |
| Surrounding traffic during RL | Level-($k - 1$) | A mixture of level-0, level-1, and level-2 with a certain ratio |
| Observation space | 11 observations | 11 observations plus $V$ and $SOC$ |
| Action space | 7 actions | Same as level-$k$ |
| RL algorithm | Jaakkola RL algorithm | Same as level-$k$ |
| After training, assign level-0 policy to states visited fewer than $n$ times | $n = 20$ | $n = 40$ |

## 4.2 Observation and Action Spaces

The observation space, i.e., input space, for our autonomous vehicle control policy is extended from the observation space for the level-$k$ driver policies. The observation space for the level-$k$ policies has 11 observations, including the range and range rate of the ego vehicle to the vehicles in its front, front right, front left, rear right, and rear left, as well as the lane index of the ego vehicle. The range is categorized by "far", "nominal", or "close", and the range rate is categorized by "moving away", "stable", and "approaching". The simulator is configured for a three-lane highway. As a result, the total number of possible states in the level-$k$ policy is $3^{11}$. Here, a state means a unique combination of observations.

To incorporate considerations of energy efficiency, it is necessary to enlarge the observation space by including additional observations. Since vehicle speed and battery $SOC$ are critical factors that affect the PCU decision for the regenerative power distribution, as well as the component efficiencies of the battery and the MG, they are added to the observation space, each being categorized by "high", "medium", and "low". Consequently, the total number of possible states for our autonomous vehicle control policy increases to $3^{13}$.

The action space, i.e., output space, for the proposed control design is the same as that for the level-$k$ driver policies. It includes the following seven actions: (1) maintain speed, (2) accelerate, (3) decelerate, (4) hard accelerate, (5) hard decelerate, (6) move left (if the vehicle is not in the left-most lane) and (7) move right (if the vehicle is not in the right-most lane).

The exact definitions of the observation and action spaces for the level-$k$ driver policies are given based on the parameters including the relative longitudinal position and speed thresholds, acceleration rates, deceleration rates, and lane change rates, whose values are the same as in [26], so they are not repeated here. For the two additional observations of the autonomous vehicle control policy, i.e., the vehicle speed and battery $SOC$, the thresholds that divide the three categories are, respectively, 17.22 m/s and 22.22 m/s, and 70% and 80%, chosen by trial and error.

## 4.3 Reward Function

The reward function used for RL training is as follows,

$$\mathcal{R} = R_1 + R_2, \tag{17}$$

where

$$R_1 = w_1 \cdot c + w_2 \cdot v + w_3 \cdot h + w_4 \cdot u, \tag{18}$$

$$R_2 = w_5 \cdot e. \tag{19}$$

Here, $w_i > 0$, $i = 1, \ldots, 5$ are weights, and $c$, $v$, $h$, $u$, and $e$ are reward features. In particular, the reward function consists of two parts: The first part $R_1$ with the terms $c$, $v$, $h$, and $u$ accounts for the safety, performance, and comfort requirements and shares the same setup as for the level-$k$ policies. The second part $R_2$ is an additional term that accounts for energy efficiency.

The reward features and their corresponding weights are chosen based on engineering insight and tuning by simulation as follows:

- $c$ accounts for constraint violations,

$$c = \begin{cases} -1, & \text{if a collision occurs to the ego vehicle,} \\ 0, & \text{otherwise,} \end{cases} \tag{20}$$

  with $w_1 = 10{,}000$.

- $v$ accounts for travel speed,

$$v = \frac{V - v_n}{a}, \tag{21}$$

  where $V$ is the speed of the ego vehicle in the longitudinal direction, and the constants $v_n$, a nominal speed, and $a$, a nominal acceleration rate, are used to scale this term to the same order of magnitude of the other terms, with $w_2 = 5$.

- $h$ accounts for headway, encouraging the ego vehicle to keep a reasonable distance from preceding vehicles,

$$h = \begin{cases} 1, & \text{if headway} \in \text{“far”,} \\ 0, & \text{if headway} \in \text{“nominal”,} \\ -1, & \text{if headway} \in \text{“close”,} \end{cases} \tag{22}$$

  with $w_3 = 1$. Here, “headway” means the range of the ego vehicle to the vehicle in its immediate front.

- $u$ accounts for control effort,

$$u = \begin{cases} 0, & \text{if action} = \text{“maintain speed”,} \\ -1, & \text{if action} = \text{“accelerate” or “decelerate”,} \\ -3, & \text{if action} = \text{“move left” or “move right”,} \\ -5, & \text{if action} = \text{“hard accelerate” or “hard decelerate”,} \end{cases} \tag{23}$$

  with $w_4 = 1$.

- $e$ is for energy efficiency, defined by the time derivative of $MPGe$ as

$$e = \frac{dMPGe}{dt} = \frac{V \cdot E_{batt} - x \cdot P_{batt}}{E_{batt}^2} \cdot \gamma, \tag{24}$$

with $w_5 = 5$.

## 4.4 Training Algorithm

The goal of training is to find a control policy that maximizes the reward averaged over an infinite horizon. Using the settings described above, we formulate this problem as a partially observable Markov decision process (POMDP) problem since only certain observations are available to the ego vehicle. For example, if there are multiple vehicles in front of the ego vehicle in the same lane, the ego vehicle can only observe the relative range and range rate of the vehicle immediately in front of it. Thus, the algorithm used for training the control policy should guarantee convergence of the average reward with respect to POMDP problems. We choose to use the Jaakkola RL algorithm [16] since, under suitable assumptions, this algorithm guarantees convergence of the average reward to a local maximum for POMDP problems. The proof of such a convergence guarantee can be found in Appendix A of [24].

A summary of the Jaakkola RL algorithm is given in [26] and Sect. 1.2.7 of [24]. The key variables and equations are reviewed here. The algorithm iterates with two steps at every simulation time step $t$. First, the one-step reward $R_t$ is evaluated based on the results of the simulation following the current policy $\pi_t$. Then, for each observation state $o \in \mathcal{O}$, and state and action pair $(o, a) \in \mathcal{O} \times \mathcal{A}$, the state-value functions $V(o|\pi_t)$ and the action-value functions $Q(o, a|\pi_t)$, also called Q-values, are updated based on the difference of $R_t - \bar{R}(\pi_t)$ where $\bar{R}(\pi_t)$ is the average reward for an infinite duration with the policy $\pi_t$. The state-value $V(o|\pi_t)$ represents the expected cumulative reward starting at state $o$ following policy $\pi_t$, while the Q-value $Q(o, a|\pi_t)$ represents the expected cumulative reward if the state starts at $o$, we take action $a$ first, and then follow policy $\pi_t$ afterward. Specifically, the state-value functions and Q-values are updated with equations given as

$$\beta_t^o(o) = \left(1 - \frac{\chi_t^o(o)}{K_t^o(o)}\right)\gamma_t \beta_{t-1}^o(o) + \frac{\chi_t^o(o)}{K_t^o(o)}, \tag{25}$$

$$V(o|\pi_t) = \left(1 - \frac{\chi_t^o(o)}{K_t^o(o)}\right)V(o|\pi_{t-1}) + \beta_t^o(o)\Big(R_t - \bar{R}(\pi_t)\Big), \tag{26}$$

$$\beta_t^a(o, a) = \left(1 - \frac{\chi_t^a(o, a)}{K_t^a(o, a)}\right)\gamma_t \beta_{t-1}^a(o, a) + \frac{\chi_t^a(o, a)}{K_t^a(o, a)}, \tag{27}$$

$$Q(o, a|\pi_t) = \left(1 - \frac{\chi_t^a(o, a)}{K_t^a(o, a)}\right)Q(o, a|\pi_{t-1}) + \beta_t^a(o, a)\Big(R_t - \bar{R}(\pi_t)\Big), \tag{28}$$

where $\chi$ represents a binary (0 or 1) indicator function that equals to one if $o$ or $(o, a)$ is visited at the current time step, $K$ is a function that counts how many times $o$ or $(o, a)$ has been visited, and $\gamma_t$ is a time-dependent discount factor that takes a value between zero and one and converges to one as $t$ goes to infinity. In the second step, the policy is updated by the following equation

$$\pi_{t+1}(o, a) = (1 - \epsilon)\pi_t(o, a) + \epsilon\hat{\pi}_t(o, a), \quad \forall(o, a) \in \mathcal{O} \times \mathcal{A}, \tag{29}$$

where $\epsilon \in (0, 1)$ is the learning rate and $\hat{\pi}_t$ is the greedy policy that maximizes

$$J_t(\pi, o) = \sum_{a \in \mathcal{A}} \pi(o, a)\Big(Q(o, a|\pi_t) - V(o|\pi_t)\Big), \quad \forall o \in \mathcal{O}. \tag{30}$$

The process then moves on to the next time step and the iteration of the above two steps continues.

Note that the Jaakkola RL algorithm updates the value functions and Q-values at each time step using both the immediate one-step reward $R_t$ and the infinite-horizon average reward $\bar{R}$. The one-step reward can make the policy update respond instantly to the large penalty of a safety constraint violation. The average reward, which in actual implementation is estimated by averaging all past instant rewards, eventually contains the weighted energy efficiency $MPGe$ computed over a long horizon. In this way, objectives with different time horizons are handled simultaneously.

## 4.5 Training Process

The level-$k$ policies with $k = 1$ and 2 are obtained following Algorithms 1 and 2 of [26] (similar to Algorithm 1 below) with the reward function $R_1$ described above, and thus, they do not account for energy efficiency. We then use vehicles operating with level-$k$ policies to provide the traffic environment for the RL training of our autonomous vehicle controller that considers energy efficiency.

The training process for the proposed autonomous vehicle control policy is summarized by the pseudo-code in Algorithm 1. Each training episode corresponds to a simulation trajectory with a duration of 200 s. This RL training process is similar to the process described by Algorithms 1 and 2 of [26] with the following major differences: (1) When initializing a training episode, we initialize the ego vehicle battery $SOC$ randomly according to a uniform distribution in the interval [15%, 90%]. For Algorithm 2 of [26], however, since $SOC$ is not a state of the level-$k$ models, this initialization step does not exist; (2) A traffic environment consisting of a mixture of level-0, 1, and 2 vehicles with a ratio of 15%, 55%, and 30% is used to train our autonomous vehicle control policy, while when training a level-$k$ policy, by definition, vehicles in the environment are all level-$(k - 1)$; (3) Due to the increase of the size of the observation space, the total number of possible observation combinations is 9 times greater than for the level-$k$ policies of [26]. With the same number of training episodes (i.e., 50,000, determined/limited by the affordable computational resources such as training time duration), it is more likely that some states are not sufficiently visited during the training. Thus, an increased value of the parameter $n$ is used in the last step for the autonomous vehicle policy training, where $n$ represents the number of times a state has to be visited during training, lest it be assigned the level-0 policy.

---

**Algorithm 1:** Training process

---

Initialize the ego car's policy with equal action probabilities for every state.<br>
*episode* $\leftarrow 0$.<br>
**while** *episode* $\leq 50,000$ **do**<br>
    Randomly select the number of surrounding cars, $n_c \in [21, 30]$.<br>
    Initialize surrounding cars with level-$k$ policies with probabilities corresponding to 15%, 55% and 30% for $k = 0$, 1 and 2.<br>
    Initialize the ego car with $SOC \in [15\%, 90\%]$.<br>
    **while** $t \leq 200$ **do**<br>
        Run simulation and evaluate the ego car's policy with the reward function $\mathcal{R}$.<br>
        Update the ego car's policy.<br>
        **if** *a collision occurs to the ego vehicle* **then**<br>
            Terminate the current episode.<br>
        **end if**<br>
    **end while**<br>
    *episode* $\leftarrow$ *episode* $+ 1$.<br>
    Assign the level-0 policy to states visited less than $n = 40$ times.<br>
**end while**

---

## 4.6 Autonomous Vehicle Control Policy for Benchmarking

For comparison purposes, a second RL-based policy is trained in the mixed traffic environment described above. This benchmark policy uses only $R_1$, as defined in (18), as its reward function, and allows one to study the differences between considering the fuel economy or not, in similar traffic conditions.

In addition to policies trained by RL, the FSM-based policy described in [28] is adopted for comparison. The FSM-based policy is a rule-based controller with three modes including cruise control, adaptive cruise control, and lane change control. Switches between modes are triggered when certain traffic conditions are satisfied. The FSM-based policy is calibrated to optimize safety and performance, while the energy efficiency is not being considered.

# 5 Results

## 5.1 *Training for RL-Based Policies*

As discussed above, each RL-based policy is trained for 50,000 episodes. Figure 3 shows the values of the average reward as the training progresses for the level-1 policy, level-2 policy, the proposed autonomous vehicle control policy with the energy efficiency consideration (AV w/ $e$), and the benchmark policy that does not account for the energy efficiency (AV w/o $e$). It can be observed that the average rewards all converge smoothly, suggesting the success of the RL procedures.

The value of the converged average reward of each policy is a combined result of the different reward features in $\mathcal{R}$. For example, the converged average reward of the level-2 policy is higher than those of the other policies. This is because, among the level-0, 1, and 2 policies, the level-1 policy is the most aggressive as concluded in [26], tending to make many lane changes to pursue higher travel speeds. Since the traffic environment for training the level-2 policy is composed of purely level-1 vehicles, the level-2 policy is relatively conservative and collisions are less likely. Moreover, the overall faster traffic flow allows a higher travel speed of the ego vehicle. Thus, the coupled effect of these factors leads to a higher converged average reward for the level-2 policy.

## 5.2 *Control Performance*

### 5.2.1 Evaluation Process

The control policies are evaluated based on simulations using the process described by the pseudo-code in Algorithm 2. In particular, for each policy and each traffic density (represented by the number of surrounding vehicles in traffic, ranging from 0 to 30), 10,000 simulation episodes are run, each with a duration of 200 s. Then, the policy is evaluated with respect to four evaluation metrics, including:

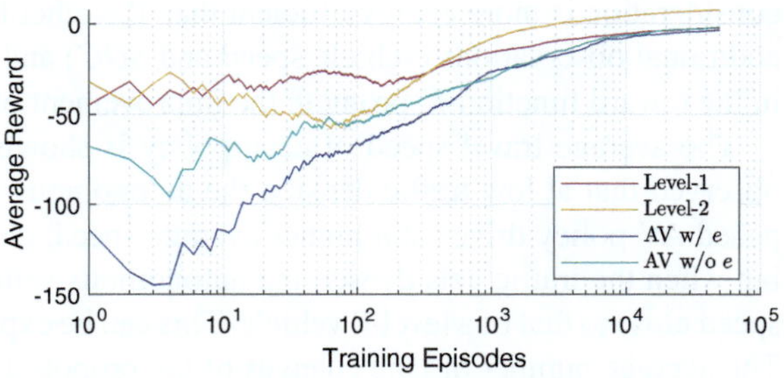


**Fig. 3** Average reward evolution during RL

- Constraint violation rate, defined as the percentage of simulation episodes where a collision occurs to the ego vehicle;
- Average number of lane changes per simulation episode;
- Average $MPGe$;
- Average travel speed.

**Algorithm 2:** Evaluation process

**for** $n_c = 0 : 30$ **do**
  *episode* $\leftarrow 0$.
  **while** *episode* $\leq 10,000$ **do**
    Initialize the ego car with $SOC \in [15\%, 90\%]$ and the control policy to be evaluated.
    Initialize surrounding cars with level-$k$ policies randomly with probabilities of 15%, 55% and 30% for $k = 0$, 1 and 2.
    Simulate and record variables relevant to the evaluation metrics.
    *episode* $\leftarrow$ *episode* $+ 1$.
  **end while**
  Compute and output the evaluation metric values.
**end for**

### 5.2.2 Simulation Result Analysis

Proposed AV control policy

Figures 4, 5 and 6 show the results of different policies in regards to the four aforementioned evaluation metrics as functions of the traffic density. Figures 4a and 5 show the constraint violation rate, and Figs. 4b and 6a show the average number of lane changes. It can be observed that, when driving in the mixed traffic environment (vs. Mix), the proposed policy (AV w/ *e*) has the lowest constraint violation rate among all policies that can perform lane changes.

Figure 4c compares the average $MPGe$ among the three AV control policies, i.e., the proposed policy (AV w/ *e*), the RL-based benchmark policy (AV w/o *e*), and the FSM-based policy. It shows that the proposed policy with the energy efficiency consideration is more energy-efficient than the other two policies, verifying that the additional observations (vehicle speed and $SOC$) and the energy efficiency term $R_2$ in the reward function $\mathcal{R}$ promote the improvement of energy efficiency.

The average travel speed of each policy is shown in Figs. 4d and 6b. It can be observed that at low traffic density, the autonomous vehicle controlled by the proposed AV policy drives at a higher average speed, close to that of level-2 vehicles; but when the traffic gets denser, the autonomous vehicle slows down to an average speed close to that of a level-0 vehicle. This can be explained with the help of Fig. 4b: The average number of lane changes of the proposed policy varies only slightly for different traffic densities, since when the traffic density is low, there is not much need to change lanes, while when the traffic density gets high, it may be neither safe nor

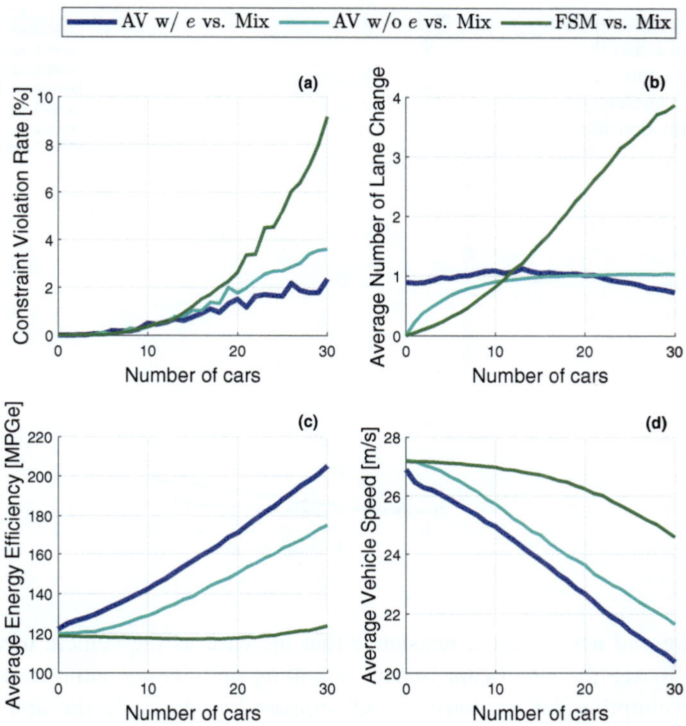


**Fig. 4** Evaluation results for AV control policies in traffic environments of different traffic densities: **a** constraint violation rate, **b** average number of lane changes per simulation episode, **c** average $MPGe$, **d** average travel speed

energy-efficient to perform lane changes. This feature distinguishes the behavior of the proposed policy from the level-1 policy and the FSM-based policy that prefer higher travel speeds and thus make many lane changes to achieve them.

Miscellaneous Observations

The results indicate that the effects of the different evaluation aspects, that are closely related to the five features in the reward function, are not decoupled. For example, for the level-$k$ policies, although the level-0 policy has the lowest constraint violation rate, it is a very conservative policy that does not allow lane changes (as illustrated in Fig. 6a) and has the lowest average vehicle speed (as shown in Fig. 6b). When driving in the level-0 environment (vs. level-0), consequently, the level-1 policy also has a very low constraint violation rate. However, when driving in the mixed traffic environment, where there are other level-1 cars and level-2 cars, the level-1 policy has the highest constraint violation rate.

It is worth highlighting some additional observations in the simulation results. Firstly, for policies trained by RL, the average $MPGe$ increases as the average travel speed decreases in denser traffic. This is attributed to the fact that for highway driving, the energy efficiency at the vehicle level is affected largely by energy losses

**Fig. 5** Constraint violation rates for level-$k$ and the proposed AV control policies in traffic environments of different traffic densities

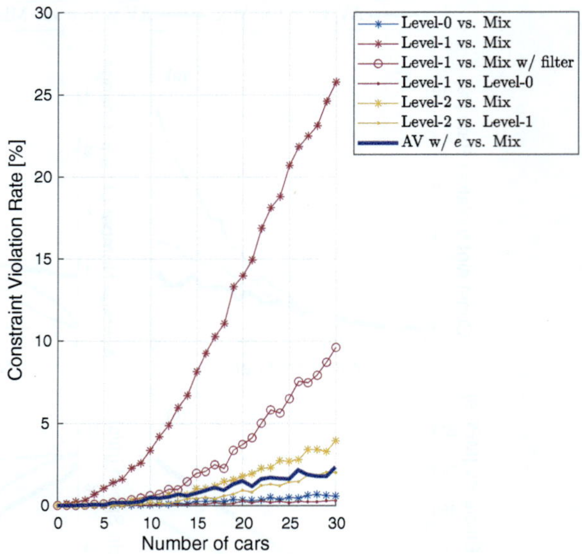


from rolling and aerodynamic resistance that increase as the vehicle travel speed increases. To see the significant impact of rolling and aerodynamic resistance on energy consumption, let us consider and compare two cases: In the first case, the vehicle is driving at a constant speed of 20.5 m/s. In the second case, the vehicle is driving at 24.5 m/s. These two cases roughly represent, respectively, the average longitudinal behavior of the proposed AV policy and that of the FSM-based policy with 30 surrounding vehicles, shown in Fig. 4d. Here, we ignore the differences in the MG efficiency by assuming a constant value of $\eta_0$. Then, we have that the MG power consumed by the rolling and aerodynamic resistance can be calculated as

$$\begin{aligned} P_l &= (T_l \cdot \frac{1}{g})\, \omega_{mg}/\eta_0 \\ &= (a + b \cdot V + c \cdot V^2)V \,/\, (r\eta_0), \end{aligned} \tag{31}$$

depending cubically on the vehicle speed $V$. Without considering the discrepancy in the battery and MG efficiency, we then use (31) to estimate the difference in the MG power used to counteract the rolling and aerodynamic resistance. We obtain that the discrepancy between the two cases is about 33%. This contributes to the difference in the *MPGe* shown in Fig. 4b, where the change is about 64%. It is within a reasonable range according to Table 1 (e.g., [4, 20]).

Note, however, that the proposed policy does not always operate the autonomous vehicle at a low speed, as the reward function represents several different objectives besides energy efficiency. In general, energy efficiency depends on the traffic scenario, the powertrain type, as well as the component specifications. This fact high-

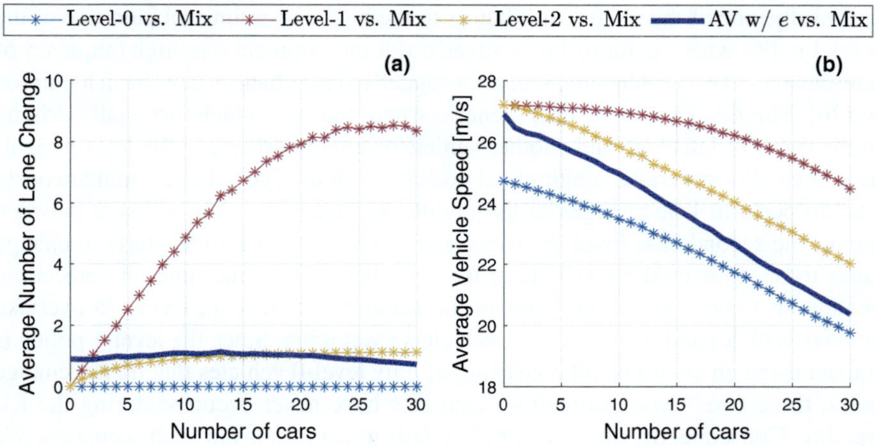


**Fig. 6** Average number of lane changes per simulation episode and average travel speed for level-$k$ and the proposed AV control policies in traffic environments of different traffic densities

lights the benefit of modeling the powertrain system using maps, so that components can be easily sized or swapped.

Secondly, it is observed in Fig. 4b that when there is no other vehicle in traffic ("zero-traffic"), the autonomous vehicle controlled by the proposed policy makes on average one lane change. This is because the policy by RL converges to a solution where when there is no other vehicle in the immediate vicinity of the ego vehicle, the ego vehicle tends to change to and stay in the right-most lane to reduce the possibility of having interactions/conflicts with other vehicles that may degrade its safety and energy efficiency in the future. Note also that such a solution may only be locally optimal (i.e., not globally optimal), as the Jaakkola RL algorithm used to train the policy guarantees only convergence to a local optimum (and not necessarily the global optimum) [16].

Intuitively speaking, vehicles need not change lanes when there are no slower vehicles in their front blocking their ways. This is the case for most policies shown in Figs. 4b and 6a except for the proposed policy for which the average number of lane changes in the zero-traffic environment is close to but slightly less than one. For most cases, when initialized in the middle lane, the autonomous vehicle controlled by the proposed policy immediately makes a lane change to the right, as explained above. However, for some states with high SOC that were not visited enough during RL training, the policy was overridden by the level-0 policy that does not perform lane changes (see the last line of Algorithm 1). This caused the average number of lane changes to be slightly less than one. Note also that the training is conducted only for traffic environments with 21–30 surrounding vehicles, i.e., not covering the zero-traffic environment. Such sub-optimal behavior in the zero-traffic environment of the trained policy indicates that it might be beneficial to conduct training for a wider range of traffic environments if computational resources allow.

Thirdly, one of the major contributors to the high constraint violation rate of the level-1 policy when operating in the mixed traffic environment is its high frequency of lane changes. Two problematic scenarios related to lane changes have been identified in [26]. The first case involves a scenario where the ego vehicle originally driving in the right (or left) lane and another vehicle originally driving in the left (or right) lane in an almost parallel longitudinal position with the ego vehicle simultaneously start to perform lane changes to the middle and lead to a side collision between them. The second case involves a scenario where the ego vehicle starts to change lanes trying to overtake a vehicle in its front, but at the same time, the preceding vehicle also starts to change lanes in the same direction (e.g., trying to overtake another vehicle) and blocks the ego vehicle's overtaking. Since the level-1 policy is trained using an environment consisting of only level-0 vehicles that do not change lanes, these two "unrare-in-reality" scenarios have never occurred during the RL training. Consequently, the level-1 policy fails to learn to avoid such scenarios. We have identified all constraint violation cases in the Level-1 vs. Mix data that belong to these two scenarios and computed the constraint violation rate after filtering out these cases. The result is plotted in Fig. 5, called Level-1 versus Mix w/ filter. It can be seen that the constraint violation rate of the level-1 policy after this filtering is significantly reduced.

If such an issue happens when developing autonomous vehicle control algorithms, where problematic scenarios can be clearly identified, they can be handled by specific add-on mechanisms. For example, the autonomous vehicle may be commanded to go back to its original lane when either of the above two cases is detected.

## 6 Conclusions

In this chapter, an autonomous vehicle control policy is developed focusing on energy efficiency optimization while safety, performance, and comfort are balanced. We first discuss the potential of autonomous vehicle (AV) controls for energy-efficient driving and the major challenges to develop such control policies. Then, we show the powertrain model built to capture the energy consumption of a battery electric vehicle (BEV), integrated with the highway traffic simulator consisting of cognitive driver behavioral models based on level-$k$ game theory. An AV control policy is trained by reinforcement learning (RL) for this BEV and compared with two benchmark policies as well as the level-$k$ policies from different evaluation perspectives.

Analysis of the results indicates that the addition of the energy efficiency term in the RL reward function, in addition to the expanded observation space to include the vehicle speed and $SOC$, is effective in improving the energy efficiency while maintaining low collision rates. Through analysis of the BEV powertrain model, the increase of the energy efficiency represented by $MPGe$ is likely dominated by the reduction of the average vehicle speed that lowers the rolling and aerodynamic resistance. However, this does not make the vehicle always travel at the lower speed limit, which highlights the capability of the RL-based approach that does a good job in balancing travel speeds, safety, and efficiency. The results also imply the potential to further extend and explore the control design in terms of higher computational efficiency and advanced RL algorithms for control performance improvement. In the future, our AV policy may serve as a baseline control strategy for more advanced autonomous driving control development.

**Acknowledgements** This research was supported by Mcity, University of Michigan. This research was also supported in part through computational resources and services provided by Advanced Research Computing at the University of Michigan, Ann Arbor.

## References

1. Amini MR, Kolmanovsky I, Sun J (2020) Hierarchical MPC for robust eco-cooling of connected and automated vehicles and its application to electric vehicle battery thermal management. IEEE Trans Control Syst Technol 29(1):316–328
2. Aoki S, Jan LE, Zhao J, Bhat A, Chang CF et al (2021) Multicruise: eco-lane selection strategy with eco-cruise control for connected and automated vehicles. arXiv preprint arXiv:2104.11959
3. Awal T, Murshed M, Ali M (2015) An efficient cooperative lane-changing algorithm for sensor- and communication-enabled automated vehicles. In: 2015 IEEE intelligent vehicles symposium (IV). IEEE, pp 1328–1333
4. Chen R, Cassandras CG, Tahmasbi-Sarvestani A, Saigusa S, Mahjoub HN, Al-Nadawi YK (2020) Cooperative time and energy-optimal lane change maneuvers for connected automated vehicles. IEEE Trans Intell Transp Syst
5. Claussmann L, Carvalho A, Schildbach G (2015) A path planner for autonomous driving on highways using a human mimicry approach with binary decision diagrams. In: 2015 European control conference (ECC). IEEE, pp 2976–2982
6. Delnevo G, Di Lena P, Mirri S, Prandi C, Salomoni P (2019) On combining big data and machine learning to support eco-driving behaviours. J Big Data 6(1):64
7. Di Cairano S, Liang W, Kolmanovsky IV, Kuang ML, Phillips AM (2012) Power smoothing energy management and its application to a series hybrid powertrain. IEEE Trans Control Syst Technol 21(6):2091–2103
8. Doshi N, Hanover D, Hemmati S, Morgan C, Shahbakhti M (2019) Modeling of thermal dynamics of a connected hybrid electric vehicle for integrated HVAC and powertrain optimal operation. In: Dynamic systems and control conference, vol 59155. American Society of Mechanical Engineers, p V002T23A005
9. Feldkamp L, Abou-Nasr M, Kolmanovsky IV (2009) Recurrent neural network training for energy management of a mild hybrid electric vehicle with an ultra-capacitor. In: 2009 IEEE workshop on computational intelligence in vehicles and vehicular systems. IEEE, pp 29–36
10. Gamage HD, Lee JB (2017) Reinforcement learning based driving speed control for two vehicle scenario. In: Australasian transport research forum (ATRF), 39th, 2017, Auckland, New Zealand
11. Grigorescu S, Trasnea B, Cocias T, Macesanu G (2020) A survey of deep learning techniques for autonomous driving. J Field Robot 37(3):362–386
12. Guanetti J, Kim Y, Borrelli F (2018) Control of connected and automated vehicles: State of the art and future challenges. Annu Rev Control 45:18–40
13. Guzzella L, Sciarretta A et al (2007) Vehicle propulsion systems, vol 1. Springer, Berlin
14. Houshmand A, Cassandras CG, Zhou N, Hashemi N, Li B, Peng H (2020) Combined eco-routing and power-train control of plug-in hybrid electric vehicles in transportation networks. arXiv preprint arXiv:2004.05161

15. Hu X, Liu T, Qi X, Barth M (2019) Reinforcement learning for hybrid and plug-in hybrid electric vehicle energy management: recent advances and prospects. IEEE Ind Electron Mag 13(3):16–25
16. Jaakkola T, Singh SP, Jordan MI (1995) Reinforcement learning algorithm for partially observable markov decision problems. In: Advances in neural information processing systems, pp 345–352
17. Kamal MAS, Taguchi S, Yoshimura T (2015) Efficient vehicle driving on multi-lane roads using model predictive control under a connected vehicle environment. In: 2015 IEEE intelligent vehicles symposium (IV). IEEE, pp 736–741
18. Karimi S, Vahidi A (2020) Receding horizon motion planning for automated lane change and merge using Monte Carlo tree search and level-k game theory. In: 2020 American control conference (ACC). IEEE, pp 1223–1228
19. Kiran BR, Sobh I, Talpaert V, Mannion P, Al Sallab AA, Yogamani S, Pérez P (2021) Deep reinforcement learning for autonomous driving: a survey. IEEE Trans Intell Transp Syst
20. Kumaravel SD, Malikopoulos AA, Ayyagari R (2020) Decentralized cooperative merging of platoons of connected and automated vehicles at highway on-ramps. arXiv preprint arXiv:2002.04826
21. Kuutti S, Bowden R, Jin Y, Barber P, Fallah S (2020) A survey of deep learning applications to autonomous vehicle control. IEEE Trans Intell Transp Syst 22(2):712–733
22. Li H, Butts K, Zaseck K, Liao-McPherson D, Kolmanovsky I (2017) Emissions modeling of a light-duty diesel engine for model-based control design using multi-layer perceptron neural networks. Technical report, SAE technical paper
23. Li H, Li N, Kolmanovsky I, Girard A (2020) Energy-efficient autonomous vehicle control using reinforcement learning and interactive traffic simulations. In: 2020 American control conference (ACC). IEEE, pp 3029–3034
24. Li N (2021) Game-theoretic and set-based methods for safe autonomous vehicles on shared roads. Ph.D. thesis, University of Michigan
25. Li N, Han K, Kolmanovsky I, Girard A (2021) Coordinated receding-horizon control of battery electric vehicle speed and gearshift using relaxed mixed-integer nonlinear programming. In: IEEE transactions on control systems technology. https://doi.org/10.1109/TCST.2021.3111538
26. Li N, Oyler DW, Zhang M, Yildiz Y, Kolmanovsky I, Girard AR (2017) Game theoretic modeling of driver and vehicle interactions for verification and validation of autonomous vehicle control systems. IEEE Trans Control Syst Technol 26(5):1782–1797
27. Li N, Yao Y, Kolmanovsky I, Atkins E, Girard AR (2020) Game-theoretic modeling of multivehicle interactions at uncontrolled intersections. IEEE Trans Intell Transp Syst
28. Li N, Zhang M, Yildiz Y, Kolmanovsky I, Girard A (2019) Game theory-based traffic modeling for calibration of automated driving algorithms. In: Control strategies for advanced driver assistance systems and autonomous driving functions. Springer, pp 89–106
29. Liu C, Murphey YL (2014) Power management for plug-in hybrid electric vehicles using reinforcement learning with trip information. In: 2014 IEEE transportation electrification conference and expo (ITEC). IEEE, pp 1–6
30. Mahbub AI, Malikopoulos AA, Zhao L (2020) Decentralized optimal coordination of connected and automated vehicles for multiple traffic scenarios. Automatica 117:108958
31. McDonough K, Kolmanovsky I, Filev D, Szwabowski S, Yanakiev D, Michelini J (2014) Stochastic fuel efficient optimal control of vehicle speed. In: Optimization and optimal control in automotive systems. Springer, pp 147–162
32. Meng X, Cassandras CG (2020) Eco-driving of autonomous vehicles for nonstop crossing of signalized intersections. IEEE Trans Autom Sci Eng
33. Mohan G, Assadian, F., Longo, S.: Comparative analysis of forward-facing models vs backwardfacing models in powertrain component sizing. In: IET hybrid and electric vehicles conference 2013 (HEVC 2013). IET, pp 1–6
34. Murphey YL, Park J, Chen Z, Kuang ML, Masrur MA, Phillips AM (2012) Intelligent hybrid vehicle power control-Part I: machine learning of optimal vehicle power. IEEE Trans Veh Technol 61(8):3519–3530

35. Qiu S, Qiu L, Qian L, Abdollahi Z, Pisu P (2018) Closed-loop hierarchical control strategies for connected and autonomous hybrid electric vehicles with random errors. IET Intel Transport Syst 12(10):1378–1385
36. Rajendran AV, Hegde B, Ahmed Q, Rizzoni G (2017) Design and development of traffic-in-loop powertrain simulation. In: 2017 IEEE conference on control technology and applications (CCTA). IEEE, pp 261–266
37. Schildbach G, Borrelli F (2015) Scenario model predictive control for lane change assistance on highways. In: 2015 IEEE intelligent vehicles symposium (IV). IEEE, pp 611–616
38. Sciarretta A, Vahidi A et al (2020) Energy-efficient driving of road vehicles. Springer, Berlin
39. Tian R, Li N, Kolmanovsky I, Yildiz Y, Girard AR (2020) Game-theoretic modeling of traffic in unsignalized intersection network for autonomous vehicle control verification and validation. IEEE Trans Intell Transp Syst
40. U.S. Department of Energy: next-generation energy technologies for connected and automated on-road vehicles. https://arpa-e.energy.gov/technologies/programs/nextcar. Accessed: 2021-07-15
41. Vahidi A, Sciarretta A (2018) Energy saving potentials of connected and automated vehicles. Transp Res Part C Emerg Technol 95:822–843
42. Wang H, Amini MR, Song Z, Sun J, Kolmanovsky I (2019) Combined energy and comfort optimization of air conditioning system in connected and automated vehicles. In: Dynamic systems and control conference, vol 59148. American Society of Mechanical Engineers, p V001T08A001
43. Waschl H, Kolmanovsky I, Willems F (2019) Control strategies for advanced driver assistance systems and autonomous driving functions. Springer, Berlin
44. Wipke KB, Cuddy MR, Burch SD (1999) ADVISOR 2.1: a user-friendly advanced powertrain simulation using a combined backward/forward approach. IEEE Trans Veh Technol 48(6):1751–1761
45. Wu G, Ye F, Hao P, Esaid D, Boriboonsomsin K, Barth MJ et al (2019) Deep learning-based eco-driving system for battery electric vehicles. Technical report, Institute of Transportation Studies, UC Davis
46. Xu B, Hu X, Tang X, Lin X, Li H, Rathod D, Filipi Z (2020) Ensemble reinforcement learning-based supervisory control of hybrid electric vehicle for fuel economy improvement. IEEE Trans Transp Electrification 6(2):717–727
47. Yildiz Y, Agogino A, Brat G (2014) Predicting pilot behavior in medium-scale scenarios using game theory and reinforcement learning. J Guid Control Dyn 37(4):1335–1343
48. Yoo JH, Langari R (2013) Stackelberg game based model of highway driving. In: ASME 2012 5th annual dynamic systems and control conference joint with the JSME 2012 11th motion and vibration conference. American Society of Mechanical Engineers Digital Collection, pp 499–508
49. Yoo JH, Langari R (2014) A stackelberg game theoretic driver model for merging. In: ASME 2013 dynamic systems and control conference. American society of mechanical engineers digital collection
50. Zheng G, Peng Z (2021) Life cycle assessment (LCA) of BEV's environmental benefits for meeting the challenge of ICExit (internal combustion engine exit). Energy Rep 7:1203–1216

# Self-learning Decision and Control for Highly Automated Vehicles

**Jianyu Chen, Jingliang Duan, Yang Guan, Qi Sun, Yuming Yin, and Shengbo Eben Li**

**Abstract** The decision and control module plays a key role for autonomous driving, which is responsible for generating appropriate control commands that navigate the autonomous vehicles safely and efficiently. Existing decision and control modules for automated vehicles are mainly using a rule-based hand-engineered approach. Although working well in a number of specialized scenarios, such method shows its limitation when dealing with highly automated driving tasks such as dense urban scenarios. Recent advances in artificial intelligence have inspired a line of works about self-learning based decision and control, which enable self-reinforcement of the control policy to potentially super-human performance. In this chapter, we will introduce how to appropriately apply such techniques to automated vehicles. The chapter will begin with the motivations and basics, followed by the key challenges and recent achievements of self-learning decision and control for automated vehicles, focusing on the following key aspects: scalability, performance, interpretability, mixed-model, and emergency handling.

## 1 Introduction

Most of today's autonomous driving decision and control systems are using a rule-based modularized hand-engineered approach [43]. Even though working well in a few driving tasks, this rule-based framework starts to touch its performance limitation in urban driving scenarios because (1) too much human heuristics can lead to conservative driving policies; (2) it is hard to generalize as we might need to redesign

J. Chen
Institute for Interdisciplinary Information Sciences, Tsinghua University, Beijing 100084, China

J. Duan · Y. Guan · Q. Sun · S. E. Li (✉)
School of Vehicle and Mobility, Tsinghua University, Beijing 100084, China
e-mail: lishbo@tsinghua.edu.cn

Y. Yin
Department of Mechanical Engineering, Zhejiang University of Technology, Hangzhou 310023, China

 
Y. L. Murphey et al. (eds.), *AI-enabled Technologies for Autonomous and Connected Vehicles*, Lecture Notes in Intelligent Transportation and Infrastructure,
https://doi.org/10.1007/978-3-031-06780-8_11

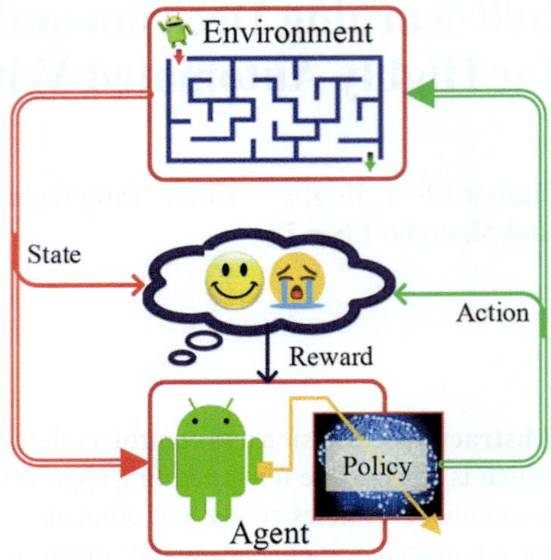


**Fig. 1** RL with agent-interaction environment

the heuristics for each new scenario and task; (3) the computation time is relatively high, and (4) these modules are strongly entangled with each other, and the whole system becomes expensive to scale and maintain.

Those limitations might be avoided with self-learning approaches, where a driving policy can be learned and generalized to new tasks without much hand-engineered involvement. Reinforcement learning (RL) [12] provides a principled framework to obtain self-learning decision and control policies by trial-and-error interaction with the environments. In general, the essence of RL relies on the following three elements: (1) state-action samples from the environment and the agent outputs, (2) a policy mapping states to actions, (3) reward signals defining how well the policy behaves in the environment. The concept of RL is illustrated in Fig. 1, in which an agent interacts with the environment to generate state, action, and reward signals, and its policy is then adjusted for another round of interaction to achieve better reward. Approaches for solving RL problems can be mainly divided into two categories: (1) Indirect RL based on dynamic programming, and (2) direct RL based on policy gradient [12]. More details can be found in RL textbooks such as [31].

Combined with deep learning techniques, RL already shows its power on tackling complex decision making and control problems, bringing a series of breakthroughs in recent years. Agents trained with deep RL techniques achieve super-human-level performance in game playing [34] and robotics [19]. Related deep RL algorithms range from value based methods such as DQN [34], actor-critic based methods such as DDPG [32] and TD3 [10], policy optimization based methods such as TRPO [40], and maximum entropy RL methods such as SAC and DSAC [5, 19]. With RL, a policy can be learned automatically without any hand-engineering or expert data. It can explore various kinds of possible cases including some dangerous ones, and then learn useful skills. It also has the potential to achieve superhuman performance.

Researchers have been trying to apply deep RL to the domain of autonomous driving. Lillicrap et al. [32] proposed a continuous control deep RL algorithm which learned a deep neural network policy that was able to drive the autonomous car on a simulated racing track. Chen et al. [3] proposed a hierarchical deep RL framework to solve driving scenarios with complex decision making such as traffic light passing. There are also some other related works not mentioned here. However, existing works are still unsatisfying in terms of the following important properties when applying RL to autonomous driving:

1. **Scalability**: There are various scenarios and tasks for autonomous driving, such as intersections, roundabouts, ramps, lane changing, overtaking, etc. Standard RL requires designing different reward functions and retraining for each of the scenarios, which largely limits the scalability.
2. **Performance**: Although existing RL algorithms have achieved successes in a variety of challenging domains from games to robotic control, autonomous driving puts forward much higher requirements for policy performance due to its high degree of dynamics, randomness and sophistication.
3. **Interpretability**: For safety-critical autonomous driving applications, good interpretability is necessary as we need to understand how the driving system is reasoning and making decisions. Existing RL algorithms often learn an end-to-end deep neural network policy, making it hard to figure out how it understands the scenarios and makes decisions.
4. **Mixed Model**: Standard model-free RL methods suffer from low sample efficiency with slow convergence. For autonomous driving, some parts of the model such as the ego vehicle's dynamics can be easily obtained. By appropriately mixing such prior models with the data, RL can yield high sample efficiency while maintaining high training performance.
5. **Emergency Handling**: The improvement of traffic efficiency and economy generally implies maintaining a relatively high vehicle speed, which would increase the potential of causing severe traffic collision accidents, especially in emergent and extreme scenarios when the tire-road contact condition is near the road adhesion limit. Therefore, when applied to autonomous driving, it is crucial that the RL algorithm should be able to handle such emergent scenarios.

In this chapter, we will introduce how to address the above properties with extended RL techniques for applying to autonomous driving.

## 2 Scalability

Decision and control are core functionalities of high-level automated vehicles. Current rule-based methods split the decision and control functionality into several submodules [41], such as scene understanding [16], prediction [22, 29], behavior selection [43], trajectory planning [45] and control [30]. The modularized design is interpretable, and is relatively easy to generalize to different driving scenarios. However, it yields several disadvantages as described in the above introduction part.

In contrast, reinforcement learning (RL) learns a control policy, usually carried out by neural networks (NNs), from trial-and-error in real-world or a high-fidelity

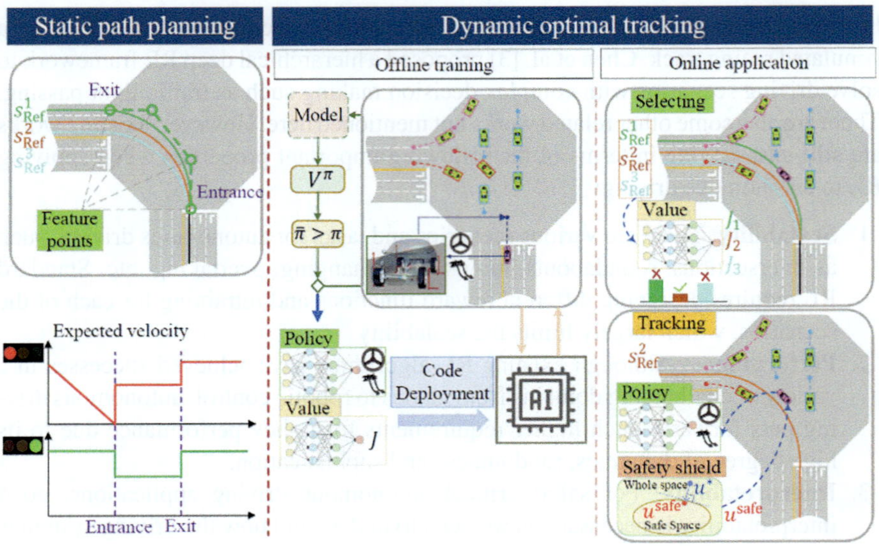


**Fig. 2** Illustration of the integrated decision and control framework

simulator [12]. This policy can be used to realize the decision and control functionality in an end-to-end manner, i.e., computing the expected control commands directly from inputs given by perception module. Such method has been applied in certain driving scenarios, such as highway [6, 13], intersection [14], roundabout [4] and ramp [25], etc. But nevertheless, they are mostly task-specific and in each task, a set of complicated reward functions is required to provide guidance for policy optimization (e.g., distance traveled towards a destination, collisions with other road users or scene objects, maintaining comfort and stability while avoiding extreme acceleration, braking or steering). This is non-trivial and needs a lot of human efforts, causing poor scalability among various driving scenarios and tasks. Although meta-RL is considered to be an effective means to help agents adapt to a new task from just a few examples, it requires similarities between old and new tasks, which is not suitable for diverse driving tasks and environments [9].

To tackle the above challenges, Guan et al. [15] proposed an integrated decision and control framework (IDC) for automated vehicles, which combines common road and traffic rules with RL-based methods to significantly improve scalability among different driving scenarios and tasks. The framework is illustrated in Fig. 2, which consists of two layers: static path planning and dynamic optimal tracking. Different from existing schemes, the upper layer aims to generate multiple candidate paths only considering static information such as road structure, speed limit, traffic signs and lights. Note that these paths do not include time-related information, though each is attached with an expected velocity determined by traffic rules.

The lower layer considers the dynamic information such as surrounding vehicles, pedestrians and bicycles. For each candidate path specified by the upper layer, a constrained optimal control problem (OCP) is formulated and solved, where the cost function is to minimize the accumulative future tracking errors within a finite horizon while satisfying the constraints characterizing safety requirements. In each time step, the optimal trajectory is chosen as the one with the lowest cost and thereafter tracked by a low-level controller. Such scheme allows replacing all the expensive online computation modules with two light-weight NNs trained offline by RL. To do that, instead of solving these OCPs with online optimization, a multi-task RL problem considering all the candidate paths is formulated. Then, a model-based RL algorithm is developed to solve this problem to obtain a policy network that is capable of tracking different target paths efficiently and safely. Meanwhile, a value network is obtained which estimates the optimal cost of choosing each path, which is used for online path selection.

This framework has several advantages over conventional methods. First, it has high online computing efficiency. The upper layer only needs to deal with static path information and compare the outputs given by the value network. On the other hand, the lower layer utilizes trained networks to compute the optimal path and control command, which is also time saving since online inference of NNs is very efficient. Second, it can be easily transferred among different driving scenarios without a lot of human design. The only thing we need to do is to generate different static paths in the upper layer, defined by the road topology of various scenes such as intersections, roundabouts and ramps. Besides, the lower layer always formulates the same tracking problem with safety constraints no matter which path it chooses, which can be solved by the same model-based RL solver, avoiding designing separate reward functions for different tasks.

As shown in Fig. 3, the IDC framework is verified in an intersection of two-way streets, where the east-west street has an eight-lane dual carriageway from both directions, while the north-south street has an only four-lane dual carriageway. The experiment vehicle is Chang'an CS55 equipped with RTK GPS, which enables precise localization of the ego vehicle. In each time step, the ego states gathered from the CAN bus and the RTK are mapped into the SUMO traffic simulation to obtain the current virtual traffic information including the states of surrounding vehicles and traffic signals. Then they both are sent to the on-board computer, where the trained policy and value networks are stored. The computer is a KMDA-3211 with a 2.6 GHz Intel Core I5-6200U CPU. The computed final control commands including the steering angle and the expected acceleration are then delivered from the CAN bus to the real vehicle.

A demo is visualized by snapshotting its featured time steps shown in Fig. 4. The key parameters are shown in Fig. 5. At the beginning, the ego pulls up before the stop line, waiting for the traffic light (t = 0). When the light turns green, the ego accelerates into the intersection area (t = 28s). In the center of the intersection, it encounters a straight-going vehicle with high speed from the opposite direction. In order to avoid collision, the ego slows down and switches to the adjacent path, with which it is able to bypass the vehicle from back (t = 34s). However, another straight-going vehicle comes over after the previous one passes through with a relatively low speed. This time, the ego chooses to accelerate to pass the vehicle. As the vehicle approaches, the optimal path is also automatically adjusted to minimize the tracking errors (t = 34s and t = 36s) and the ego finally passes the intersection successfully. The computation time for each step is within 15 ms.

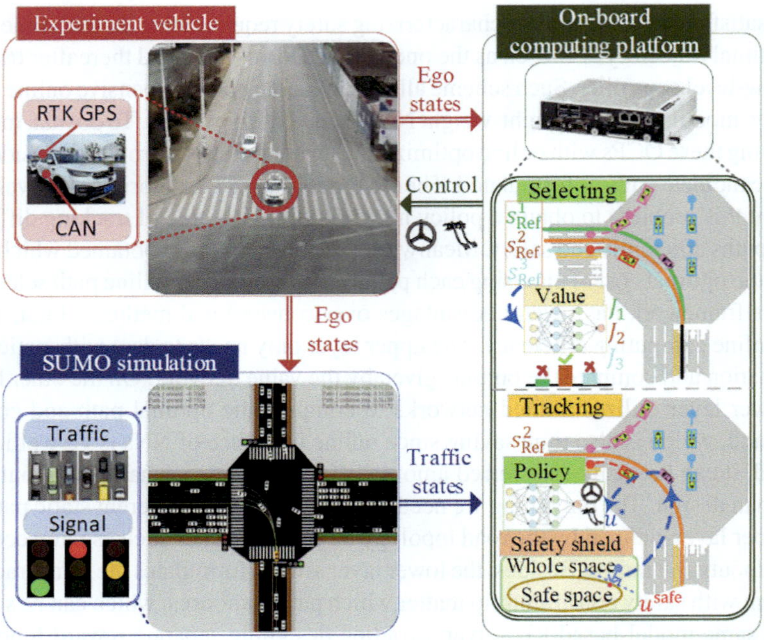


**Fig. 3** Illustration of the real-world road test

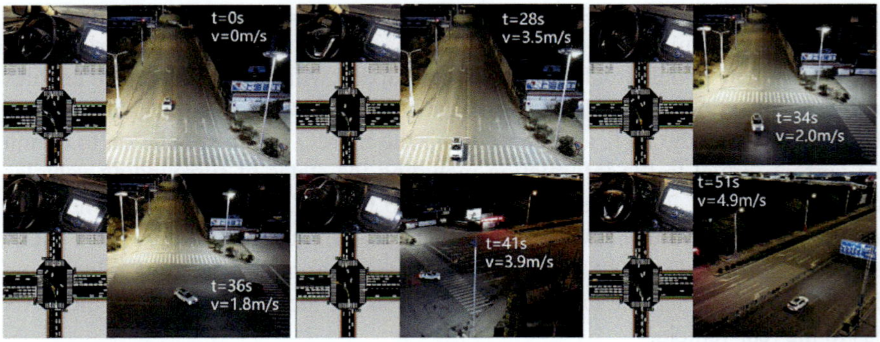


**Fig. 4** Featured time steps in real world test

# 3 Performance

The above section provides an IDC framework that is scalable to diverse driving scenarios; however, an RL algorithm is still required to learn a driving policy. The optimality of RL algorithms directly determines the performance of the learned policy. Although existing RL algorithms have achieved successes in a variety of challenging domains ranging from games to robotic control, autonomous driving puts forward much higher requirements for policy performance due to its high degree of dynamic, randomness and sophistication [6]. The performance of RL algorithms, especially

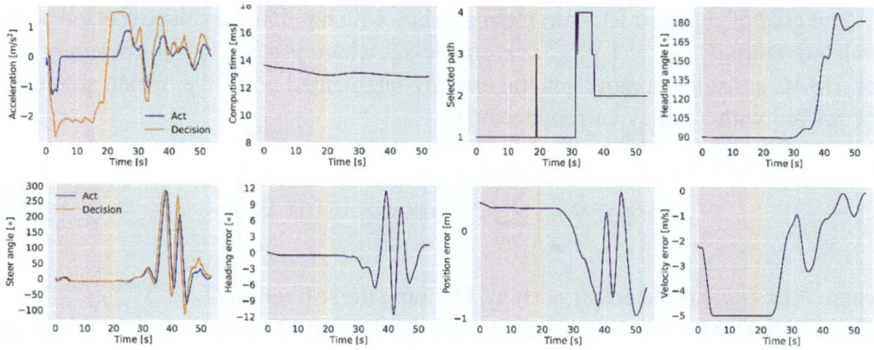


**Fig. 5** Key parameters in real world test

for Deep RL, may be damaged by many difficulties, such as non-iid sequential data, easy divergence, overestimation, sample and exploration inefficiency.

The first successful attempt to improve RL performance attributes to DQN [34], which introduces two critical techniques, i.e., experience replay and target network, to provide nearly iid training samples and stabilize the learning process, achieving human-level performance in almost half of Atari games. However, it can only handle discrete action spaces. Inspired by DQN, Lillicrap et al. proposed an off-policy RL algorithm for continuous control settings, called DDPG, by concurrently learning a Q-function and a deterministic policy using samples from the replay buffer [32]. A common failure mode for DDPG is that the learned Q-function still overestimates, which leads to a brittle policy. To address this issue, Fujimoto et al. (2018) proposed the TD3 algorithm by updating the Q-network in a clipped double Q-learning way [10]. SAC embeds such technique in the maximum entropy RL framework to simultaneously mitigate overestimations and improve exploration efficiency [19]. Duan et al. (2020) proposed Distributional SAC (DSAC) to further improve the value estimation accuracy by learning a return distribution instead of the standard Q-value [5, 39]. Besides, DSAC incorporates the off-policy and experience replay to improve sample efficiency and nearly iid training samples, and the maximum entropy framework to encourage exploration, resulting in state-of-the-art performance. Therefore, in the following, we will give a brief introduction to the DSAC algorithm and its applications to autonomous driving.

The autonomous driving problem can be modeled as a Markov Decision Process (MDP), defined by the tuple $(\mathcal{S}, \mathscr{A}, \mathscr{R}, p)$. At each time step $t$, the vehicle receives a state $s_t \in \mathcal{S}$ and selects an action $a_t \in \mathcal{A}$ according to the policy $\pi(a_t|s_t)$. In return, the vehicle receives the next state $s_{t+1} \in \mathcal{S}$ and a scalar reward $r_t \sim R(s_t, a_t)$. The unknown state transition probability $p(s_{t+1}|s_t, a_t) : \mathcal{S} \times \mathcal{A} \rightarrow \mathcal{P}(s_{t+1})$ maps a given $(s_t, a_t)$ to the probability distribution over $s_{t+1}$. For the sake of simplicity, the current and next state-action pairs are also denoted as $(s, a)$ and $(s', a')$, respectively. This section will use $\rho_\pi(s)$ and $\rho_\pi(s, a)$ to denote the state and state-action distribution induced by policy $\pi$.

The goal of standard RL is to learn a policy which maximizes the expected accumulated return, $\mathbb{E}_{(s,a)\sim\rho_\pi}\left\{\sum_{i=0}^{+\infty}\gamma^i r_{t+i}|s_t=s\right\}$, where $\gamma\in[0,1)$ is the discount factor. DSAC considers a more general entropy-augmented objective, which augments the reward with a policy entropy term $\mathcal{H}$,

$$J_\pi=\mathbb{E}_\pi\left\{\sum_{i=0}^{+\infty}\gamma^i\left[r_{t+i}+\mathcal{H}(\pi(\cdot|s_i))\right]\right\} \tag{1}$$

where $\mathcal{H}(\pi(\cdot|s))$ is the policy entropy. Defining the soft return $G_t=\sum_{i=t}^{\infty}\gamma^{i-t}[r_i-\alpha log\pi(a_i|s_i)]$ with the balance parameter $\alpha$, the soft Q-value of policy $\pi$ can be expressed as

$$Q^\pi_{soft}(s_t,a_t)=r_t+\mathbb{E}_\pi\left\{G_{t+1}\right\} \tag{2}$$

which describes the expected soft return for selecting $a_t$ in state $s_t$ and thereafter following policy $\pi$. Given the Q-estimate $Q^\pi_{soft}$, DSAC improves $J_\pi$ by directly maximizing

$$\pi_{new}=arg\max_\pi\mathbb{E}_{s\sim\rho_\pi,a\sim\pi}\left\{Q^\pi_{soft}(s,a)-\alpha log\pi(a|s)\right\} \tag{3}$$

Compared with most RL algorithms that directly learn the expected return $Q^\pi_{soft}(s_t,a_t)$, DSAC chooses to learn the return distribution to reduce the overestimation. To learn the distribution, Duan et al. first defines the soft state-action return of policy $\pi$ from $(s_t,a_t)$ as

$$Z^\pi(s_t,a_t)=r_t+\gamma G_{t+1}, \tag{4}$$

which is usually a random variable due to the randomness of the system and policy. Then, they define $\mathcal{Z}^\pi(Z^\pi(s,a)|s,a):\mathcal{S}\times\mathcal{A}\to\mathcal{P}(Z^\pi(s,a))$ as a mapping from $(s,a)$ to a distribution over soft state-action returns, and call it the soft state-action return distribution whose expectation is Q-value. The distributional variant of the Bellman operator in the maximum entropy framework can be derived as

$$\mathcal{T}^\pi_\mathcal{D}Z^\pi(s,a)\overset{D}{=}r+\gamma(Z^\pi(s',a')-\alpha\log\pi(a'|s')) \tag{5}$$

where $s'\sim p, a'\sim\pi$, and $A\overset{D}{=}B$ denotes that random variables $A$ and $B$ have equal probability laws. Let $\mathcal{T}^\pi_\mathcal{D}\mathcal{Z}(\cdot|s,a)$ denote the distribution of $\mathcal{T}^\pi_\mathcal{D}Z^\pi(s,a)$, i.e., $\mathcal{T}^\pi_\mathcal{D}Z^\pi(s,a)\sim\mathcal{T}^\pi_\mathcal{D}\mathcal{Z}(\cdot|s,a)$. To implement (5), DSAC directly updates the return distribution by

$$\mathcal{Z}_{new}=arg\min_\mathcal{Z}\mathbb{E}_{s\sim\rho_\pi,a\sim\pi}\left\{d\left(\mathcal{T}^\pi_\mathcal{D}\mathcal{Z}(\cdot|s,a),\mathcal{Z}(\cdot|s,a)\right)\right\} \tag{6}$$

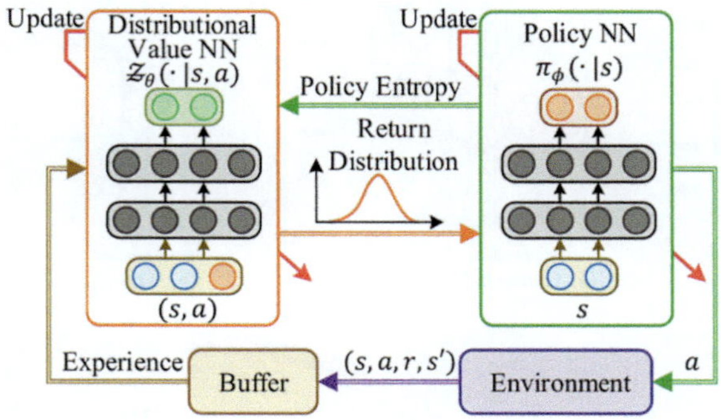


**Fig. 6** DSAC diagram

where $d$ is some metric to measure the distance between two distributions. It has been proved that DSAC which alternates between (6) and (3) also leads to policy improvement with respect to the maximum entropy objective (1).

For practical applications, both the return distribution and policy can be represented by any parameterized functions, such as neural networks. Figure 6 shows the diagram of the DSAC algorithm. DSAC first updates the distributional value network based on the samples collected from the buffer, then the output of the value network is used to guide the update of the policy network.

Before applying to autonomous driving, the performance of DSAC is first evaluated on five MuJoCo continuous control tasks. The learning curves of DSAC and 9 baselines, including DDPG, TRPO, PPO, D4PG, TD3, SAC, TD4, Double-Q SAC and Single-Q SAC, are shown in Fig. 7. Results show that the DSAC algorithm outperforms or matches all other baselines across all benchmark tasks in terms of the final performance. For example, compared with famous RL algorithms such as SAC, TD3, PPO, and DDPG, DSAC gains 20.0%, 63.8%, 39.8%, 97.6% improvements on the most complex Humanoid-v2 task, respectively. This indicates that the final performance of DSAC on these benchmarks exceeds the state-of-the-art. Therefore, the results demonstrate that the return distribution learning can greatly improve policy performance by mitigating overestimations.

A DSAC-based self-driving policy learning system is then constructed to learn a policy for autonomous driving on multi-lane roads [5, 7, 25]. To verify the learned policy, the autonomous driving decision-making experiment is carried out based on Chang'an CS55 and two-lane park road. Figure 8 shows the experimental framework and results. Results show that the learned policy can smoothly complete maneuvers such as lane-keeping, lane-changing and overtaking, so as to realize autonomous driving in response to different surrounding vehicles.

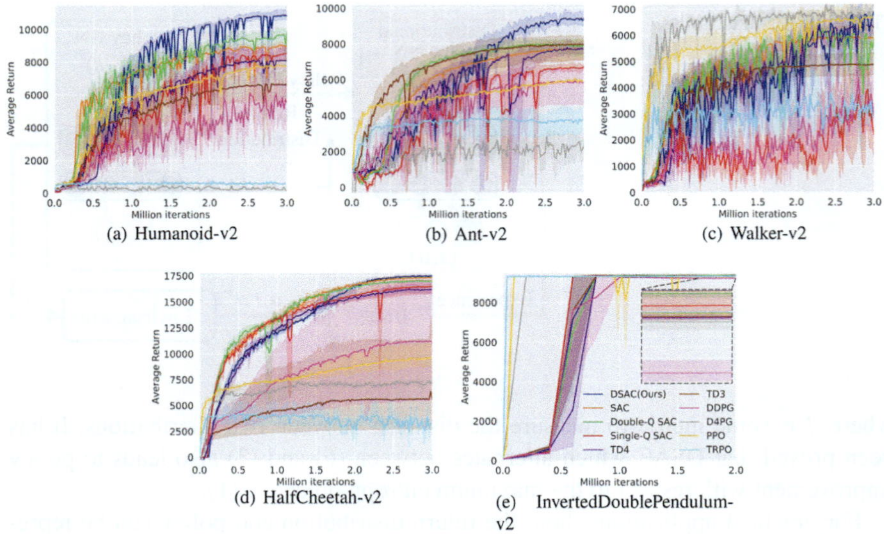


(a) Humanoid-v2 (b) Ant-v2 (c) Walker-v2

(d) HalfCheetah-v2 (e) InvertedDoublePendulum-v2

**Fig. 7** Training curves on continuous control benchmarks. The solid lines correspond to the mean and the shaded regions correspond to 95% confidence interval over 5 runs

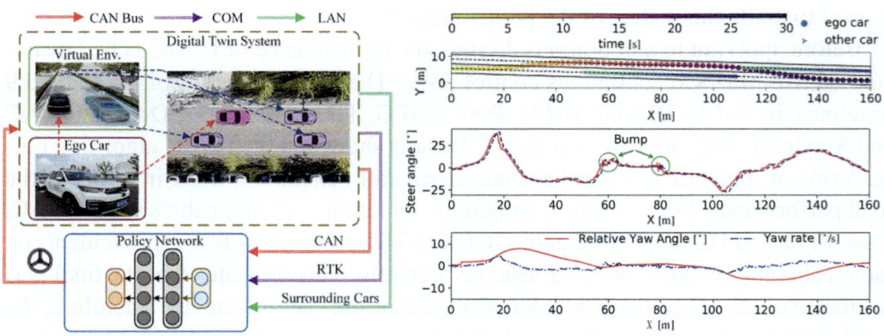


(a) Experiment framework (b) Driving results

**Fig. 8** Experimental framework and results

# 4 Interpretability

Interpretability is one of the most important properties for RL when applied to autonomous driving. For such safety-critical applications, good interpretability is necessary as we need to understand how the driving system is reasoning and makes decisions. Thus, we can debug when things go wrong and prevent future failures. However, existing learning-based approaches for autonomous driving are lacking of interpretability. Although RL algorithms, such as DSAC, enable us to learn a policy with good performance, the learned end-to-end deep neural network policy is like a black box. This indicates that we can never know whether or how it understands the environments and makes decisions.

Some works have made efforts in augmenting the interpretability of learning-based autonomous driving. Bojarski et al. [1] visualized NVIDIA's deep neural network based driving system by extracting the convolutional layer feature maps and highlighting the salient objects. Kim et al. [24] used a visual attention model with a causal filter to visualize the attention heatmap. However, the interpretable information they provide—mostly just which part of the observed image is within attention—is rather weak.

The traditional modularized approach for autonomous driving, on the other hand, is interpretable in its nature. This is because they have well-defined module structures as well as their interfaces. However, if we use end-to-end learning with a black-box neural network policy, we would lose those structures and well-defined interfaces, resulting in a lack of interpretability. Nevertheless, as illustrated in the introduction section, the modular approach lacks adaptivity, since these modules are strongly entangled with each other. Therefore, how to find a balance between the interpretability of modular approach and the adaptivity of end-to-end learning approach is essential but quite challenging.

Probabilistic graphical model (PGM) is a generic and powerful tool to formulate many machine learning problems. One of its most attractive properties is that it is flexible to impose desired structures, while preserving the end-to-end learning mechanism. Recently, the sequential latent model [27] is one of the extensions of PGM that is very relevant to sequential decision making problems such as autonomous driving. Some recent works also propose to integrate sequential latent model learning and reinforcement learning [21]. Such methods show great potential in building structural learning frameworks which preserves both structural modular and end-to-end learning style.

Inspired by the PGM based methods, Chen et al. [2] introduce the latent reinforcement learning method to build an interpretable end-to-end autonomous driving system. The latent space is employed to encode the complex urban driving environment, which is learned jointly with the maximum entropy reinforcement learning process. The introduced latent space resembles the interface between modules, which enables an interpretable explanation of how the policy reasons about the environment by decoding the latent state to a semantic birdeye mask. Meanwhile, the latent space provides a much more compact state representation, which significantly reduces the sample complexity of learning the driving policy.

The latent reinforcement learning approach for interpretable end-to-end driving framework is shown in Fig. 9. The agent takes multi-modal sensor inputs from the driving environment, and then outputs control commands to drive the car in urban scenarios. In the meantime, the agent generates a semantic mask to interpret how it understands the current driving situation. To learn this interpretable end-to-end driving agent, a probabilistic graphical model (PGM) is designed as shown in Fig. 10, where $z_t$ represents for the latent state, $a_t$ for action, $O_t$ for the optimality variable, $x_t$ for the sensor inputs, and $m_t$ for the birdeye semantic mask. Edges in this PGM are represented using deep neural networks, including the policy $p(a_t|z_t)$, the latent

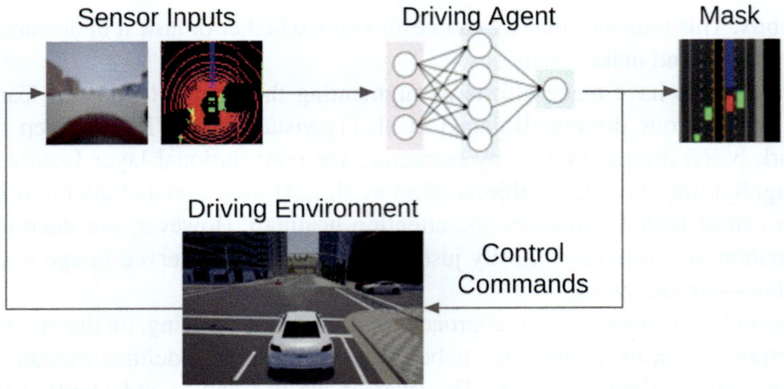


**Fig. 9** The interpretable end-to-end autonomous driving agent

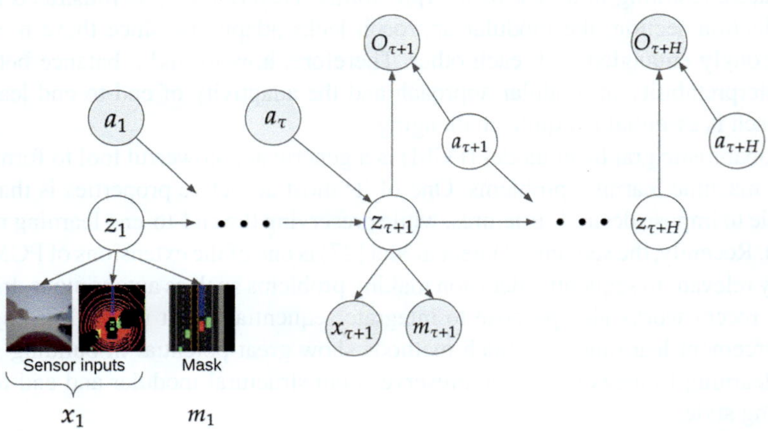


**Fig. 10** The probabilistic graphical model for interpretable end-to-end autonomous driving

dynamics $p(z_{t+1}|z_t, a_t)$, the filter model $p(z_{t+1}|x_{t+1}, a_t, z_t)$, and the generative models $p(x_t|z_t)$, $p(m_t|z_t)$. The whole PGM can be learned by maximizing the following log likelihood:

$$\log \prod_{(\overrightarrow{x}, \overrightarrow{m}, \overrightarrow{a}, \overrightarrow{r}) \in \mathcal{D}} p(\overrightarrow{x}, \overrightarrow{m}, \overrightarrow{O} \,|\, \overrightarrow{a}) \tag{7}$$

where $\mathcal{D}$ is the dataset and $\overrightarrow{x} = x_{1:\tau+1}$, $\overrightarrow{m} = m_{1:\tau+1}$, $\overrightarrow{a} = a_{1:\tau+1}$, $\overrightarrow{r} = r_{1:\tau}$, $\overrightarrow{O} = O_{\tau+1:\tau+H}$. With variational inference, an evidence lower bound (ELBO) can be derived to be optimized. Furthermore, the ELBO can be decoupled into two parts, a sequential latent environment model [20, 26, 27] part and a maximum entropy reinforcement learning [17, 18, 28] part. Details of the derivations can be found in [2].

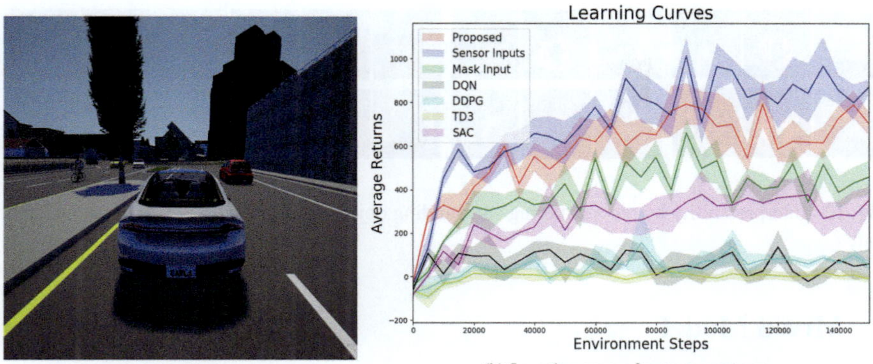


(a) Virtual town

(b) Learning curves for average returns

**Fig. 11** **a** Simulation view of autonomous driving experiments. **b** Training learning curves of average returns for different methods

The model is trained and evaluated on the CARLA simulator, as shown in Fig. 11a, which includes various urban driving scenarios such as intersections and roundabouts. 100 vehicles are running autonomously in the virtual town to simulate a multi-agent environment. The vehicles will randomly choose a direction at intersections, then follow the route, while slowing down for front vehicles and stopping when the front traffic light becomes red.

The ego vehicle is asked to learn from scratch to drive through this virtual city following a predefined route. It is spawned at a random feasible point in the city at the start of each episode. Several variations of the interpretable end-to-end driving method, as well as the state-of-the-art baseline RL algorithms including DQN [34], DDPG [32], TD3 [10] and SAC [18] are implemented and compared. The performance (Average return vs environment steps) comparison is shown in Fig. 11b. We can see that all variants of the interpretable end-to-end driving ("Proposed", "Sensor Inputs", and "Mask Input") are significantly better than the baselines ("DQN", "DDPG", "TD3", and "SAC").

Besides the performance, the interpretable end-to-end driving method also has a significant advantage in terms of interpretability by decoding a semantic mask from the latent state. By doing so, the interpretable explanations about how the agent reasons about the environments can be provided. Figure 12 shows random sampled frames of the sensor inputs, ground truth masks, and reconstructions during running with the learned model and policy. For each sample, the first row contains the raw sensor inputs and ground truth mask (left to right: camera, lidar, bird-view mask). The second row contains the corresponding reconstructed images from the latent state. Note here only the raw sensor inputs are observed, the ground truth bird-view image is displayed only for comparison. From the reconstructed bird-view mask, we can see that it can accurately locate the ego car and decode the map information (e.g., drivable areas and road markings), even though there is no direct information from the raw sensor inputs indicating the ego car is in an intersection. We can also see that the model can accurately detect the surrounding vehicles (green boxes) given raw camera and lidar observations.

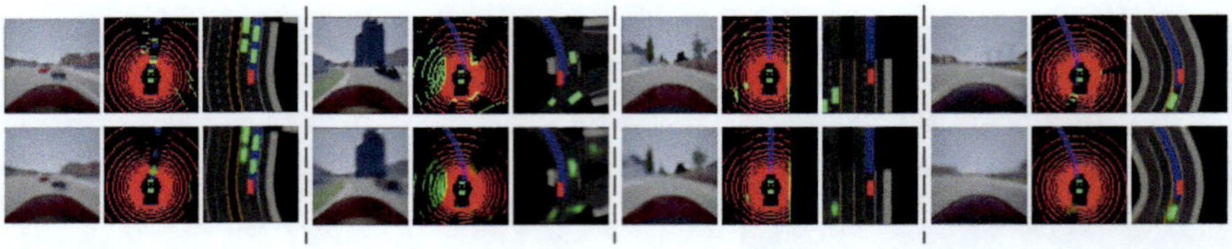

**Fig. 12** Interpretability by reconstructing surrounding objects and road conditions from the latent state

## 5 Mixed Model

Besides the performance and interpretability, how to improve sample efficiency is also very important for RL-based autonomous driving. Model-free RL methods are known to be sample inefficient since it has no prior knowledge and thus requires extensive trial-and-error experiences. Actually, human beings can learn the optimal policy and achieve goals in a complex environment without much interaction since they can abstract prior knowledge from the physical world to construct a model. This mechanism is similar to the principle of model-based RL, which searches for the optimal policy with a known environmental model. Prior works can be divided into Dyna-like algorithms [42], back-propagation through model [37, 38], and sampling-based planning [21]. Although the model-based approaches have achieved significant progress over the past few years, their performance inevitably suffers from the inherent modeling errors and the time-varying characteristics. Such inaccuracy issue usually results in a locally optimal and causes an unstable training process, severely limiting the applicability of model-based RL approaches.

To overcome these challenges, Mixed Actor-Critic (MAC) [35, 36] utilizes the dual representations of the environmental dynamics to improve both the learning accuracy and the training speed. Briefly speaking, the empirical model is used as the prior information to reduce the difficulty of model learning and avoid overfitting, while the model inaccuracy is iteratively compensated by the interaction data using Bayesian estimation.

Consider a discrete-time environment with additive stochastic uncertainty:

$$x_{t+1} = f(x_t, u_t) + \xi_t, \quad \xi_t \sim N(\mu, \mathcal{K}) \tag{8}$$

where $x_t$ is the state, $u_t$ is the action, $f(x_t, u_t)$ is the deterministic part of environmental dynamics, $\xi_t$ is the additive stochastic uncertainty with unknown mean $\mu$ and covariance $\mathcal{K}$. Assume that the additive stochastic uncertainty follows the Gaussian distribution.

The objective of MAC is to minimize the expectation of cumulative cost under the distribution of additive stochastic uncertainty $\xi$ :

$$\min_{\pi} V(x_t) = \mathbb{E}_{\xi}\left[\sum_{k=0}^{\infty} \gamma^k l(x_{t+k+1}, u_{t+k})\right], \quad \forall x_t \in \mathcal{X} \tag{9}$$

where $\pi$ is policy, $V(\cdot)$ is the state value function, $l(\cdot, \cdot)$ is the utility function, and $E_{\xi}(\cdot)$ is the expectation with respect to the additive stochastic uncertainty $\xi$. The policy is a deterministic mapping: $u_t = \pi(x_t)$.

Like other indirect RL problems, MAC aims to find the optimal policy by minimizing cost (9) subject to the environmental dynamics model constraint (8). Such a problem can be solved by iteratively applying the Bellman equation. The performance of the trained policy depends on the accuracy of the environmental model. Either an analytical model or state-action samples can be a useful representation, which corresponds to the so-called model-based RL and model-free RL, respectively. In fact, the analytical model in model-based RL is usually inaccurate due to environmental uncertainties, which impairs the optimality of the generated policy. The state-action samples in model-free RL, on the other hand, have low sampling efficiency which slows down the training process by a wide margin.

In MAC, the environmental dynamics model is dually represented by combining an analytical model $\mathcal{M}$ with the state-action data $\mathcal{D}$. The former represents the empirical knowledge about the environmental dynamics. The latter directly collect the data of state-action pairs during training. This dual representation is generally more accurate than $\mathcal{M}$ by estimating the uncertain part in the analytical model. As a consequence, it benefits from the accelerated training compared to purely model-free RL while achieving better policy satisfaction than purely model-based approaches.

The Bayesian estimator is adopted to fuse the distribution information of the additive stochastic uncertainty $\xi_t^{\mathcal{M}}$ from both model $\mathcal{M}$ and data $\mathcal{D}$, by maximizing the posterior probability $p(\mu, \mathcal{K}|\mathcal{D})$. In general, $p(\mu)$ and $p(\mathcal{K})$ are the prior distribution of $\mu$ and $\mathcal{K}$. The maximum likelihood problem becomes

$$\max_{\mu,\mathcal{K}}\{p(\mu, \mathcal{K}|\mathcal{D})\} \Leftrightarrow \max_{\mu,\mathcal{K}}\{p(\mathcal{D}|\mu, \mathcal{K})\, p(\mu)\, p(\mathcal{K})\} \tag{10}$$

Under the assumption that data D is i.i.d., (10) can be rewritten into iterative form:

$$\max_{\mu,\mathcal{K}}\{p(\xi_k^{\mathcal{D}}|\mu, \mathcal{K})\, p(\mathcal{D}_{k-1}|\mu, \mathcal{K})\, p(\mu)\, p(\mathcal{K})\}, \quad \mathcal{D}_{k-1} = \xi_1^{\mathcal{D}}, \ldots, \xi_{k-1}^{\mathcal{D}} \tag{11}$$

Therefore, an iterative Bayesian estimator $IBE(\cdot)$ can be derived,

$$\begin{bmatrix} \mu_k \\ \mathcal{K}_k \end{bmatrix} = IBE(\mu_{k-1}, \mathcal{K}_k, \xi_k^{\mathcal{D}}) \tag{12}$$

Existing indirect RL algorithms find the optimal policy via the Bellman equation, including Policy Evaluation (PEV) and Policy Improvement (PIM) steps. While MAC consists of three alternating steps, IBE, PEV and PIM, as shown in Fig. 13. IBE estimates the mean and covariance of the additive stochastic uncertainty iteratively.

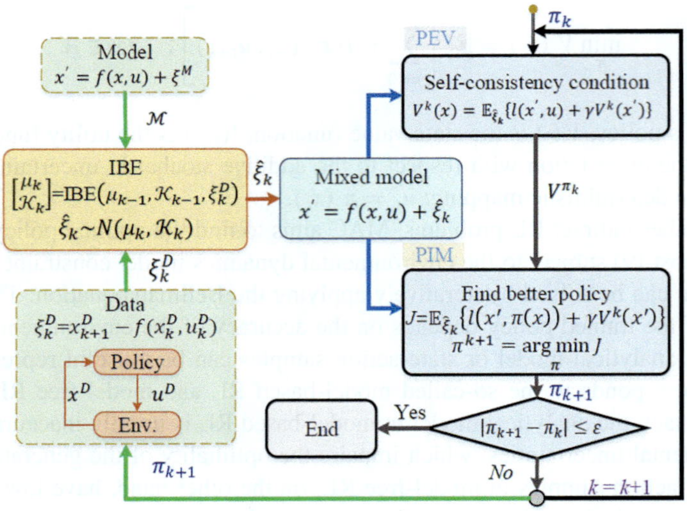


**Fig. 13** The framework of MAC

PEV seeks to numerically solve a group of algebraic equations governed by the self-consistency condition under the current policy, and PIM is to search for a better policy by minimizing a "weak" Bellman equation.

MAC is applied to a lateral and longitudinal control task of an automated vehicle with stochastic disturbance. The vehicle is subjected to random longitudinal interference force $F_{dis}$ in the tracking process and the vehicle dynamics is:

$$\dot{x} = \begin{bmatrix} \frac{F_{yf}cos\delta + F_{yr}}{m} - v_x r \\ \frac{aF_{yf}cos\delta - bF_{yr}}{I_z} \\ a_x + v_y r - \frac{F_{yf}sin\delta}{m} + \frac{F_{dis}}{m} \\ r \\ v_x sin\phi + v_y cos\phi \end{bmatrix} \tag{13}$$

where the state $x = [v_y \; r \; v_x \; \phi \; y]^T$ includes the lateral speed, the yaw rate, the longitudinal speed difference, the yaw angle and the distance between vehicle's centroid and the target trajectory. The control input $u = [\delta \; a_x]^T$ is the front wheel angle and the longitudinal acceleration. A double-lane change scenario is established to compare different RL algorithms. The task is to track the desired trajectory in the lateral direction while maintaining the desired longitudinal velocity under the longitudinal interference $F_{dis}$. The optimal control problem with discretized stochastic system equation is given by:

**Fig. 14** Convergence rate comparison between MAC, model-driven RL, and data-driven RL

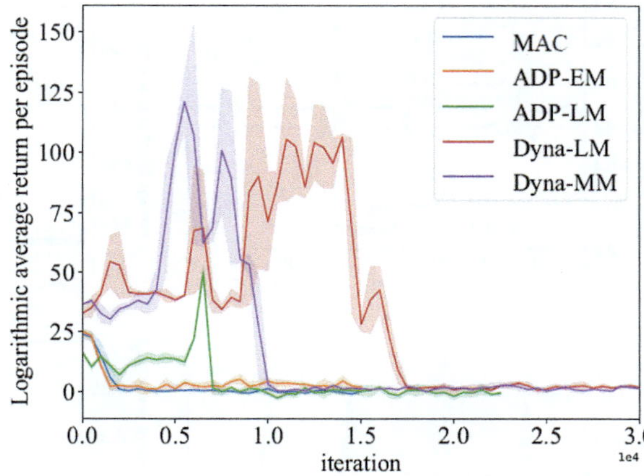


$$\min_{u} \sum_{t=0}^{\infty} \gamma^t \left(45v_x^2 + 60y^2 + u^T \begin{bmatrix} 800 & 0 \\ 0 & 1 \end{bmatrix} u\right) \tag{14}$$

$$\text{s.t. } x_{t+1} = f\left(x_t, u_t\right) + \xi_t \tag{15}$$

$$\xi_t = F_{dis}T/m \tag{16}$$

To demonstrate the effectiveness of MAC, its performance is compared with the widely used model-based RL methods, including dyna algorithm with the learned model (Dyna-LM), dyna algorithm with the mixed model (Dyna-MM), adaptive dynamic programming with the learned model (ADP-LM) and adaptive dynamic programming with the empirical model (ADP-EM). As shown in Fig. 14, The MAC and adaptive dynamic programming methods (i.e., MAC, ADP-EM and ADP-LM) converge faster than the Dyna-like algorithms (i.e., Dyna-LM, Dyna-EM), which demonstrates the advantage of analytical gradient given by the dynamic model. Moreover, MAC converges almost twice faster than ADP-LM without oscillation. In ADP-LM the mismatch of the data distributions between two adjacent iterations and the switching characteristics of the system leads to difficulties to learn an accurate model purely from data. ADP-EM achieves a similar convergence rate as the MAC. The above results confirm the effectiveness of the designer's knowledge embedded in MAC.

However, the control performance of ADP-EM is impaired by the model inaccuracy. As shown in Fig. 15, MAC has the minimum longitudinal speed error and lateral position error. In contrast, because of the model inaccuracy, the policies generated by ADP-EM have the highest speed error. MAC outperforms the other five benchmark methods.

In summary, MAC exhibits the fastest convergence speed during the training process and superior control performance in the given double-lane change task. The ADP-EM has a similar convergence speed as the MAC, but has a higher tracking error due to the model mismatch. Although the ADP-LM compensates the model

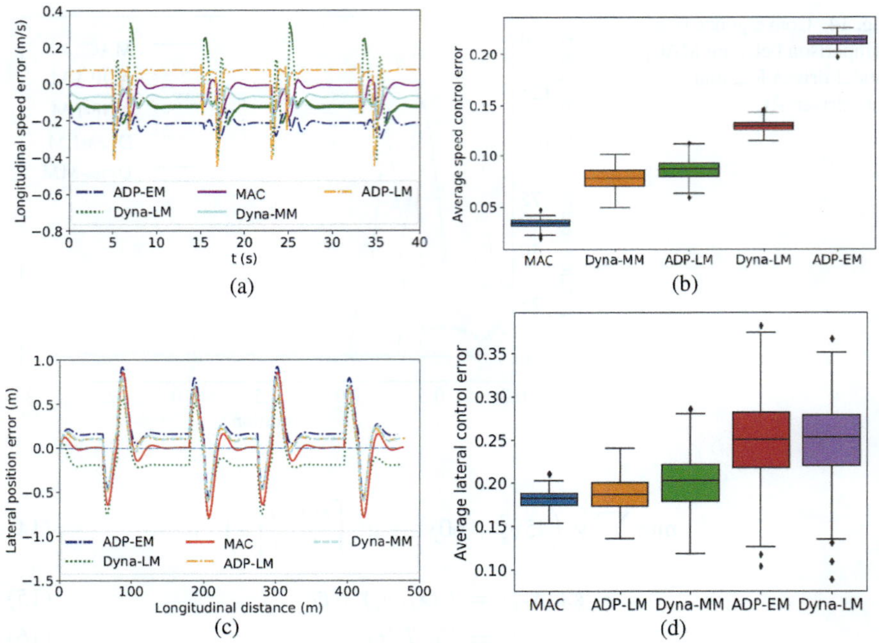


**Fig. 15** Lateral and longitudinal error during 20 tests

inaccuracy by iteratively updating the dynamic model, it converges slower than the MAC and ADP-EM due to the difficulties to learn an accurate enough model purely from data. The Dyna-like algorithms have a slower convergence rate than the MAC, due to the difficulties in finding the optimal policy only by state-action data.

# 6 Emergency Handling

The improvement of traffic efficiency and economy generally implies maintaining a relatively high vehicle speed, which would increase the potential of causing severe traffic collision accidents, especially in emergent and extreme scenarios when the tire-road contact condition is near the road adhesion limit. Therefore, when applied to autonomous driving, it is crucial that the RL algorithm should be able to handle such emergent scenarios.

Most of the existing literature on autonomous driving emergency handling are using model-based control. Lu et al. [33] proposed a centralized control strategy for multiple vehicles to minimize the impact of multiple-vehicle collision based on vehicle-to-vehicle communication techniques. Model predictive control (MPC) framework is used to formulate and solve the problem. Zhang et al. [47] proposed a path planning and motion control framework to plan and track a reference drift

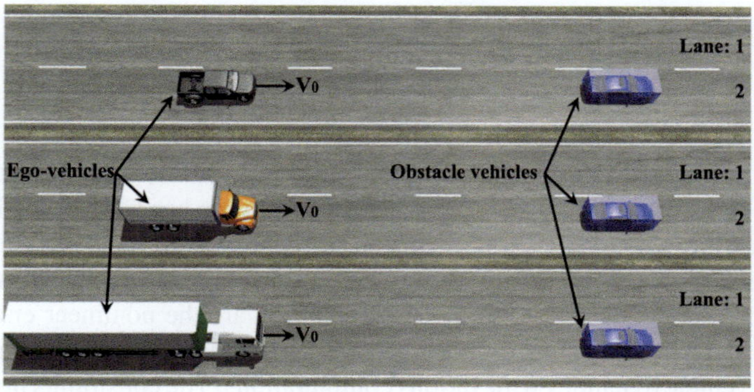


**Fig. 16** Emergent collision avoidance scenarios

trajectory along a sharp bend in a track. The path planner divides the path horizon into three regions to generate the reference trajectory. However, model-based control is usually limited by its heavy calculation burden, especially for nonlinear systems with a long planning horizon, and it requires accurate modeling.

Recently, RL-based methods have been applied to obstacle avoidance control of vehicles. Kahn et al. [23] proposed a model-based learning algorithm using camera image as input, which could enable a prototype vehicle to avoid obstacles with uncertainty. Emuna et al. [8] proposed a data-driven learning algorithm in order to imitate human driver's collision avoidance behaviors. These learning-based methods could adapt to various environment uncertainties and extreme scenarios [46], even for different vehicle configurations of steering, driving and braking systems.

In this section, the considered emergent collision avoidance scenario for three vehicle configurations is demonstrated in Fig. 16. The selected vehicle configurations are typical light pickup with 2 axles, single unit truck with 3 axles, and heavy semi-trailer with 2 vehicle units and 6 axles. These three can cover the commonly used vehicle types for passenger and cargo transportation. An obstacle vehicle in the front of the ego vehicle is making emergent braking, which has an initial speed of 100km/h and a constant deceleration of 4.5 m/s$^2$. The selected three ego vehicles are set to travel with an initial high speed of 120 km/h. On all the three roads, lane 1 is free without obstacle vehicles. In this emergent scenario, the ego vehicle will collide with the obstacle vehicle in lane 2, if only emergent braking is applied since their distance is too close. Alternatively, the ego vehicle can change to the nearby empty lane 1, so as to avoid the accident. The involved lane-change decision and collision avoidance path-following control is still challenging and requires optimization of nonlinear/continuous vehicle dynamics and discrete lane selection.

The decision and control process is integrated into a mix-integer constrained optimization problem, which can be expressed in the form of:

$$\max_{j,a}\left\{J_j=\int_t^{t+T} r\left(s(\tau), s_{ref}^j(\tau), a(\tau)\right) d\tau\right\} \tag{17}$$

$$\text{s.t.}\dot{s}=f(s,a),\ h(s(\tau))\leq 0 \tag{18}$$

where $J_j$ is the objective function of lane $j$, $j=1,2$ in this study; $t$, $\tau$ and $T$ are the initial time, current time and collision avoidance time horizon, respectively; $s$, $a$ and $r(\cdot)$ are the ego-vehicle's state, decision and control action, and reward function, respectively; $s_{ref}^j(\tau)$ is the reference state when selecting lane $j$, which is the lateral position of lane center in this study; $f(\cdot)$ and $h(\cdot)$ are the nonlinear environment dynamics and constraints functions, respectively. The environment dynamics include nonlinear vehicle dynamics with various configurations. The constraints in this study include the safety distance constraint with the obstacle vehicle and the road curbs of lane 1 and lane 2.

Although the vehicle dynamic responses of different configurations are significantly different, they are combined as a single dynamics function. The state, action and reward function of the ego vehicles are considered as:

$$s=\left[x\ \ y\ \ \varphi\ \ \dot{x}\ \ \dot{y}\ \ \dot{\varphi}\ \ x_{obs}\ \ y_{obs}\right] \tag{19}$$

$$a=\left[j_{ref}\ \ \delta_s\ \ Thr\ \ P_b\right] \tag{20}$$

$$r=-w_s\left|s-s_{ref}^j\right|-w_a\left|a\right|-P_{obs} \tag{21}$$

where $x$, $y$, $\varphi$, $\dot{x}$, $\dot{y}$, $\dot{\varphi}$ are the longitudinal position, lateral position, yaw angle, longitudinal speed, lateral speed, yaw rate of ego-vehicle, respectively; $x_{obs}$ and $y_{obs}$ are the longitudinal and lateral position of obstacle vehicle; $j_{ref}$ is the reference lane threshold; $\delta_s$ is the steering wheel angle; $Thr$ is the throttle position; $P_b$ is the braking pressure; $w_s$ and $w_a$ are weighting matrix of state and action; $|\cdot|$ indicates the absolute value; and $P_{obs}$ is the designed penalty function for the collision avoidance constraints.

The mix-integer constrained optimal decision and control problem described above is solved using deep reinforcement learning. The objective function is reformed and discretized as the accumulative return $G_t=\sum_{i=t}^{T-1}\gamma^{i-t}r_i$, which needs to be maximized at each time $t$, where $\gamma$ is the discounting factor, $i$ is discretized time step, $r_i$ is instant reward at ith time step, and $T$ is the total time steps of an RL training episode. The training curve of RL for this emergency collision avoidance task is shown in Fig. 17. The total return tends to converge after about 2500 episode iterations.

In order to verify the effectiveness of the method, the learned neural network policy at convergence is then deployed on the ego vehicles. A test scenario similar to the demonstration in Fig. 16 is used. The trajectories of obstacle vehicles and the three ego-vehicles during the emergent collision avoidance test are shown in Fig. 18(a) and the variations in vehicle speed are shown in Fig. 18(b). Note that the centerline's lateral positions of lane 1 and 2 are 2m and -2m, respectively. We can see that the obstacle vehicle is braking until fully stopped on lane 2 as designed. All

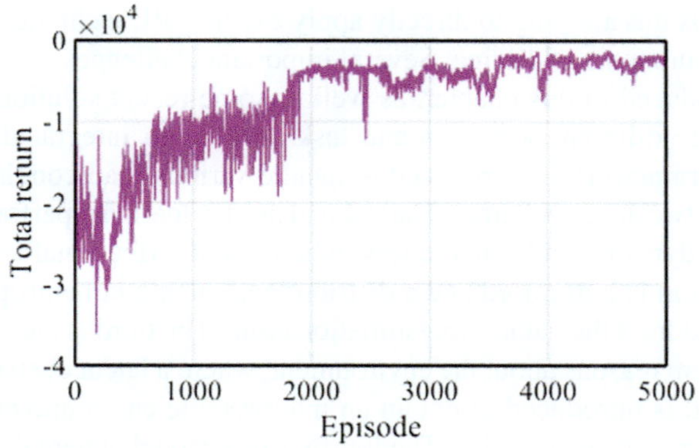


**Fig. 17** Episode total return during RL training

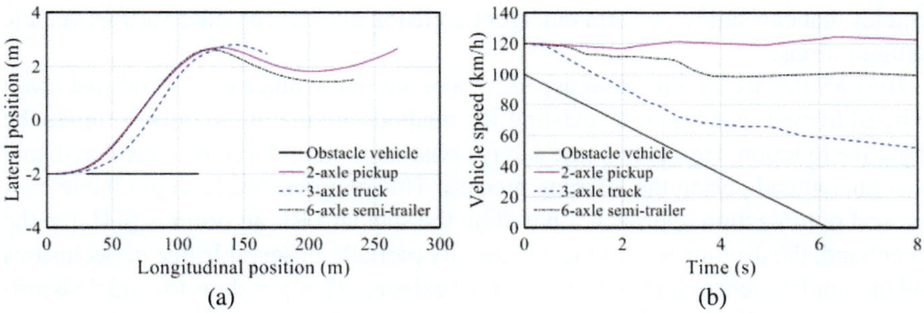


**Fig. 18** Emergent collision avoidance test demonstration: **a** vehicle trajectory and **b** speed

the trajectories of the three ego-vehicles change to lane 1 to avoid colliding with the obstacle. While the ego-vehicle trajectories are still close between different types of vehicles, the speed profiles are significantly different, as seen in Fig. 18(b). This is because the target vehicle speed is not considered in the reward function and thus different vehicle configurations yield different speeds.

## 7 Conclusion

Although widely used in existing automated vehicle systems, rule-based decision and control methods have limitations because of the lack of adaptability and extensive need for human-engineered heuristics. Self-learning based methods, mainly inspired by the recent advances in reinforcement learning, have become a new trend for designing the decision and control systems for highly automated vehicles. Such method enables the driving policy to self-adapt by interacting with the environment, and potentially results in super-human level performance. Minimum human efforts are required through the development process, and the online computation time can easily meet real time requirements.

However, it is unsatisfying to directly apply existing RL techniques from the AI community to autonomous driving. Several important challenges need to be solved, which are introduced in this chapter, as well as some recent solutions. The first is the scalability to different scenarios and tasks, where an integrated decision and control (IDC) framework is introduced to handle various road conditions using an interconnected two-layer scheme. The second is the learning performance under high degree of dynamic and randomness, where the distributional soft actor-critic (DSAC) algorithm is introduced and a distributional maximum entropy RL method is proposed to address the value overestimation issue. The third is the interpretability of how the system reasons about the environment, where a latent deep reinforcement learning method is introduced to obtain an interpretable environment model while learning the optimal driving policy. The fourth is mix model, where the mixed actor-critic (MAC) algorithm is introduced to combine the vehicle model information with interaction data. The last is emergency handling, where RL is applied to learn driving policies that can safely perform emergent collision avoidance under various vehicle configurations.

Besides the above important aspects, there are many other issues that still await to be addressed. For example, existing RL methods are mainly to only optimize the objective function. However, some safety constraints should also be considered, and even guaranteed during the learning process. This requires developing reliable constrained optimization approaches based on the standard RL algorithms [44]. On the other hand, the driving environment is usually partially observable due to occlusions and incomplete sensor inputs. Extensions of existing RL algorithms to model historical information should be designed to handle the partial observability [11]. Moreover, interactions with human drivers under unknown intentions, and connected control of networked automated vehicles are also important topics.

## References

1. Bojarski M, Yeres P, Choromanska A, Choromanski K, Firner B, Jackel L, Muller U (2017) Explaining how a deep neural network trained with end-to-end learning steers a car. arXiv preprint arXiv:1704.07911
2. Chen J, Li SE, Tomizuka M (2021) Interpretable end-to-end urban autonomous driving with latent deep reinforcement learning. IEEE Trans Intell Transp Syst
3. Chen J, Wang Z, Tomizuka M (2018) Deep hierarchical reinforcement learning for autonomous driving with distinct behaviors. In: 2018 IEEE intelligent vehicles symposium (IV). IEEE, pp 1239–1244
4. Chen J, Yuan B, Tomizuka M (2019) Model-free deep reinforcement learning for urban autonomous driving. In: 2019 IEEE intelligent transportation systems conference (ITSC). IEEE, pp 2765–2771
5. Duan J, Guan Y, Li SE, Ren Y, Sun Q, Cheng B (2021) Distributional soft actor-critic: off-policy reinforcement learning for addressing value estimation errors. IEEE Trans Neural Netw Learn Syst
6. Duan J, Li SE, Guan Y, Sun Q, Cheng B (2020) Hierarchical reinforcement learning for self-driving decision-making without reliance on labelled driving data. IET Intell Transp Syst 14(5):297–305

7. Duan J, Yu D, Li SE, Wang W, Ren Y, Lin Z, Cheng B (2021) Fixed-dimensional and permutation invariant state representation of autonomous driving. arXiv preprint arXiv:2105.11299
8. Emuna R, Borowsky A, Biess A (2020) Deep reinforcement learning for human-like driving policies in collision avoidance tasks of self-driving cars. arXiv preprint arXiv:2006.04218
9. Finn C, Abbeel P, Levine S (2017) Model-agnostic meta-learning for fast adaptation of deep networks. In: International conference on machine learning. PMLR, pp 1126–1135
10. Fujimoto S, van Hoof H, Meger D (2018) Addressing function approximation error in actor-critic methods. arXiv preprint arXiv:1802.09477
11. Gu Z, Yang Y, Duan J, Li SE, Chen J, Cao W, Zheng S (2021) Belief state separated reinforcement learning for autonomous vehicle decision making under uncertainty. In: 2021 IEEE 24th international conference on intelligent transportation systems (ITSC), pp 1–7
12. Guan Y, Li SE, Duan J, Li J, Ren Y, Sun Q, Cheng B (2019) Direct and indirect reinforcement learning. Int J Intell Syst
13. Guan Y, Li SE, Duan J, Wang W, Cheng B (2018) Markov probabilistic decision making of self-driving cars in highway with random traffic flow: a simulation study. J Intell Connected Veh
14. Guan Y, Ren Y, Li SE, Sun Q, Luo L, Li K (2020) Centralized cooperation for connected and automated vehicles at intersections by proximal policy optimization. IEEE Trans Veh Technol 69(11):12597–12608
15. Guan Y, Ren Y, Sun Q, Li SE, Ma H, Duan J, Dai Y, Cheng B (2021) Integrated decision and control: towards interpretable and computationally efficient driving intelligence. arXiv preprint arXiv:2103.10290
16. Guo J, Kurup U, Shah Mohak (2019) Is it safe to drive? An overview of factors, metrics, and datasets for driveability assessment in autonomous driving. IEEE Trans Intell Transp Syst 21(8):3135–3151
17. Haarnoja T, Tang H, Abbeel P, Levine S (2017) Reinforcement learning with deep energy-based policies. In: Proceedings of the 34th international conference on machine learning, vol 70, pp 1352–1361. (JMLR-organization)
18. Haarnoja T, Zhou A, Abbeel P, Levine S (2018) Soft actor-critic: off-policy maximum entropy deep reinforcement learning with a stochastic actor. arXiv preprint arXiv:1801.01290
19. Haarnoja T, Zhou A, Hartikainen K, Tucker G, Ha S, Tan J, Kumar V, Zhu H, Gupta A, Abbeel P et al (2018) Soft actor-critic algorithms and applications. arXiv preprint arXiv:1812.05905
20. Hafner D, Lillicrap T, Fischer I, Villegas R, Ha D, Lee H, Davidson J (2018) Learning latent dynamics for planning from pixels. arXiv preprint arXiv:1811.04551
21. Hafner D, Lillicrap T, Fischer I, Villegas R, Ha D, Lee H, Davidson J (2019) Learning latent dynamics for planning from pixels. In: International conference on machine learning, pp 2555–2565. PMLR
22. Hou L, Xin L, Li SE, Cheng B, Wang W (2019) Interactive trajectory prediction of surrounding road users for autonomous driving using structural-LSTM network. IEEE Trans Intell Transp Syst 21(11):4615–4625
23. Kahn G, Villaflor A, Pong V, Abbeel P, Levine S (2017) Uncertainty-aware reinforcement learning for collision avoidance. arXiv preprint arXiv:1702.01182
24. Kim J, Canny J (2017) Interpretable learning for self-driving cars by visualizing causal attention. In Proceedings of the IEEE international conference on computer vision, pp. 2942–2950
25. Kong Y, Guan Y, Duan J, Li SE, Sun Q, Nie B (2021) Decision-making under on-ramp merge scenarios by distributional soft actor-critic algorithm. arXiv preprint arXiv:2103.04535
26. Krishnan RG, Shalit U, Sontag D (2015) Deep kalman filters. arXiv preprint arXiv:1511.05121
27. Lee AX, Nagabandi A, Abbeel P, Levine S (2019) Stochastic latent actor-critic: deep reinforcement learning with a latent variable model. arXiv preprint arXiv:1907.00953
28. Levine S (2018) Reinforcement learning and control as probabilistic inference: tutorial and review. arXiv preprint arXiv:1805.00909
29. Li G, Li SE, Cheng B, Green P (2017) Estimation of driving style in naturalistic highway traffic using maneuver transition probabilities. Transp Res Part C Emerg Technol 74:113–125

30. Li Shengbo, Li Keqiang, Rajamani Rajesh, Wang Jianqiang (2010) Model predictive multi-objective vehicular adaptive cruise control. IEEE Trans Control Syst Technol 19(3):556–566
31. Li SE (2020) Reinforcement learning and control. Tsinghua University: Lecture Notes. http://www.idlab-tsinghua.com/thulab/labweb/publications.html
32. Lillicrap TP, Hunt JJ, Pritzel A, Heess N, Erez T, Tassa Y, Silver D, Wierstra D (2015) Continuous control with deep reinforcement learning. arXiv preprint arXiv:1509.02971
33. Lu X-Y, Wang J, Li SE, Zheng Y (2014) Multiple-vehicle longitudinal collision mitigation by coordinated brake control. Math Probl Eng 2014
34. Mnih V, Kavukcuoglu K, Silver D, Rusu AA, Veness J, Bellemare MG, Graves A, Riedmiller M, Fidjeland AK, Ostrovski G et al (2015) Human-level control through deep reinforcement learning. Nature 518(7540):529
35. Mu Y, Li SE, Liu C, Sun Q, Nie B, Cheng B, Peng B (2020) Mixed reinforcement learning with additive stochastic uncertainty. arXiv preprint arXiv:2003.00848
36. Yao Mu, Baiyu Peng, Ziqing Gu, Shengbo Eben Li, Chang Liu, Bingbing Nie, Jianfeng Zheng, and Bo Zhang. Mixed reinforcement learning for efficient policy optimization in stochastic environments. In: 2020 20th international conference on control, automation and systems (ICCAS). IEEE, pp 1212–1219
37. Peng B, Mu Y, Duan J, Guan Y, Li SE, Chen J (2021) Separated proportional-integral lagrangian for chance constrained reinforcement learning. arXiv preprint arXiv:2102.08539
38. Peng B, Mu Y, Guan Y, Li SE, Yin Y, Chen J (2020) Model-based actor-critic with chance constraint for stochastic system. arXiv preprint arXiv:2012.10716 2020
39. Ren Y, Duan J, Li SE, Guan Y, Sun Q (2020) Improving generalization of reinforcement learning with minimax distributional soft actor-critic. In: 2020 IEEE 23rd international conference on intelligent transportation systems (ITSC). IEEE, pp 1–6
40. Schulman J, Levine S, Abbeel P, Jordan M, Moritz P (2015) Trust region policy optimization. In: International conference on machine learning, pages 1889–1897
41. Shengbo LI, Yang G, Lian HOU, Hongbo GAO , Jingliang DUAN , Shuang LIANG , WANG Yu, CHENG Bo, LI Keqiang, REN Wei et al (2019) Key technique of deep neural network and its applications in autonomous driving. J Autom Saf Energy 10(2):119
42. Sutton RS, Szepesvári C, Geramifard A, Bowling MP (2012) Dyna-style planning with linear function approximation and prioritized sweeping. arXiv preprint arXiv:1206.3285
43. Urmson C, Anhalt J, Bagnell D, Baker C, Bittner R, Clark MN, Dolan J, Duggins D, Galatali T, Geyer C et al (2008) Autonomous driving in urban environments: boss and the urban challenge. J Field Robot 25(8):425–466
44. Wen L, Duan J, Li SE, Xu S, Peng H (2020) Safe reinforcement learning for autonomous vehicles through parallel constrained policy optimization. In: 2020 IEEE 23rd international conference on intelligent transportation systems (ITSC). IEEE, pp 1–7
45. Xin L, Kong Y, Li SE, Chen J, Guan Y, Tomizuka M, Cheng B (2021) Enable faster and smoother spatio-temporal trajectory planning for autonomous vehicles in constrained dynamic environment. Proc Inst Mech Eng Part D J Autom Eng 235(4):1101–1112
46. Yin Y, Li SE, Li K, Yang J, Ma F (2020) Self-learning drift control of automated vehicles beyond handling limit after rear-end collision. Transp Saf Environ 2(2):97–105
47. Zhang F, Gonzales J, Li SE, Borrelli F, Li K (2018) Drift control for cornering maneuver of autonomous vehicles. Mechatronics 54:167–174

# Advanced Driver Assistant Systems

# MAGMA: Mobility Analytics Generated from Metrics on ADAS

**Jeremy Lerner and Dina Tayim**

**Abstract** Modern Advanced Driver Assistance Systems (ADAS) have complex logic for determining when and where feature use is appropriate, generally based on geolocation and the vehicle's sensor suite. This variability can lead to a problem of how to meaningfully measure customer experience from ADAS feature usage data. To provide a broad understanding of customer experience and where the feature should have been active but was not, the data must be viewed relative to the feature availability map. The feature activation and availability experienced by a driver is dependent on numerous design decisions (such as the map and map previewing logic), which may affect the process of understanding the raw data. Therefore, it is critical to compare customer ADAS feature usage data to what the vehicle could have previewed in the best-case customer experience scenario by simulating the feature availability based just on the feature design logic, the vehicle's location, and the map. This enhanced understanding of customer experience allows for the discovery of corner cases and enables improved feature design. In short, customer ADAS feature usage data can be better understood in the appropriate context, where offline simulations of the designed feature logic provide an appropriate normalization factor.

**Keywords** Connectivity · ADAS · Driver assistance systems · Driver behavior · Analysis methodologies · Big data · Autonomous vehicles · Automated vehicles · Connected vehicle data

J. Lerner (✉) · D. Tayim
Ford Motor Company, Dearborn, USA
e-mail: Lernerjn@gmail.com

D. Tayim
e-mail: dtayim@ford.com



Y. L. Murphey et al. (eds.), *AI-enabled Technologies for Autonomous and Connected Vehicles*, Lecture Notes in Intelligent Transportation and Infrastructure,
https://doi.org/10.1007/978-3-031-06780-8_12

## Definitions/Abbreviations

| | |
|---|---|
| ADAS | Advanced driver assistance systems assume some level of control over the driving task, assisting the driver in the driving task |
| ODD | Operational Design Domain (ODD) defines the domain over which the automated vehicle can operate |
| Expected feature experience | Quantitative metrics that predict the ultimate customer's interaction with a feature |
| Actual feature experience | Quantitative metrics that measure the customer's interaction with a feature |
| Map matching | Offline determination of a vehicle's position on a map based on a full route's reported GNSS coordinates |
| Localization | Real-time determination of a vehicle's position on a map based on current and past GNSS coordinates |
| Map | The virtual representation of the geospatial features (e.g. lane markings and traffic sign positions) that can be used for autonomous driving applications to localize a vehicle or control ADAS feature availability |
| GNSS | Global Navigation Satellite System |

## 1 Introduction

Advanced Driver Assistance Systems (ADAS) features are designed to automate portions of the driving task. The SAE definitions of the six levels of vehicle autonomy are generally well-known in the automotive industry. Levels 0, 1, and 2, commonly known as ADAS, assist the driver but the driver must remain aware of the vehicle and its surroundings. Examples of ADAS technologies include automatic emergency braking, automatic lane centering, and adaptive cruise control. Whereas for Levels 3, 4, and 5, the driver may be required to resume control (for Levels 3 and 4) but the vehicle autonomously handles the driving task and understands its limitations well enough to request that the driver resume control when necessary [1]. All levels of autonomy include some form of object detection based on the vehicle's sensing suite and may involve path planning and localizing to a map as well [2].

One of the fundamental problems of vehicle automation is that the system is exposed to numerous edge-cases, which a rule-based approach cannot account for. Machine learning models can learn from these edge cases and generalize them for overall operational improvement. There are two major steps in machine learning: learning and inference. Many machine learning applications that run on vehicles are only capable of inference on the embedded hardware. For example, embedded computer vision and/or signal processing models utilizing cameras and other sensors (e.g. LiDAR) allow vehicles to perceive and make inferences about their environments based on pre-trained models [3]. To improve the performance of machine learning models, one must collect sizable datasets and retrain the model frequently. There are federated learning approaches that involve training a machine learning model on each vehicle and combining the results (rather than the raw data) into a single model [4]. The learning or retraining step typically involves collecting these sizeable datasets from a fleet of connected vehicles. In this chapter, one of the main goals is to introduce measurable properties from complex ADAS time series data that can be utilized by supervised or unsupervised machine learning algorithms [5].

Questionnaires are currently the most common method of assessing customer experience with ADAS features on mass scale [6]. Here, we will address quantifying customer experience with ADAS features in an automated manner. Data collected from connected vehicles running ADAS features can be utilized to estimate the customer experience for more vehicles and on a more consistent basis than questionnaires. Below, we will introduce metrics that are intended to track driver satisfaction with ADAS features as well as the operation of those features relative to a known feature availability.

An ADAS system may or may not have a defined Operational Design Domain (ODD) which can be dependent on geolocation. Typically, the real-time determination of feature availability is based primarily on a computer vision system. Many vision systems rely on convolutional neural networks to create a model of the road to enable ADAS features [7]. The result and confidence of these machine learning models generally dictates the real-time availability of the ADAS feature.

In this chapter, a map shall be defined as the virtual representation of physical geospatial features (e.g. lane markings and traffic sign positions) used for autonomous driving applications to localize a vehicle and/or control ADAS feature availability [8, 9]. When ADAS features are available in pre-mapped locations on a feature availability map [10], the vehicle's local sensing suite may still not allow for feature activation (e.g. due to poor visibility of lane markings). In addition, vehicles localize to roadways based on real-time geolocation, which may have significant noise factors and limited accuracy. Map-indicated feature availability is necessary but not sufficient for ADAS operation. The real-time vehicle sensing ultimately determines if ADAS features are available or not and may request that the driver resume the driving task, which impacts driver experience.

If the data is evaluated in the cloud (to enable centralized analysis of fleet data), the data that reaches the cloud may not be in a usable format (perhaps due to poor cellular connectivity). In a hypothetical scenario, a particular signal used locally on the vehicle may have a native frequency of 50 Hz, and the data transmitted from connected vehicles occurs at 1 Hz. In effect, all the signal state changes may not be captured in the data received by a remote server and therefore that data may be incomplete. Additionally, signals on the vehicle may be overridden with higher priority signals for specific vehicle events (e.g. on a Controller Area Network bus) so some events or full datasets may be unavailable for data collection [11].

Utilizing ADAS data in an offline simulation of feature availability allows for deeper understanding of feature operation. Simulating individual vehicle operation can assist in validation of the vehicle design [5, 12]. Here, we focus on macro-scale simulation to predict and understand vehicle operation for an entire fleet. This simulation can create the best-case customer experience scenario that the ADAS feature availability map could have provided based on an offline recreation of the feature's expected behavior. This simulation methodology can further take advantage of advanced offline map matching algorithms that utilize the entire set of GNSS coordinates to effectively determine the route taken by a vehicle [13].

For example, feature usage data may show that a driver had 20 miles of hands-free activation on their route, but this must be normalized relative to the total availability provided by the map. If the total availability was close to 20 miles, the feature performed almost exactly as expected by the map, but if the total availability was significantly greater than 20 miles, then the ability to use the feature was not well aligned with the pre-mapped availability. Similarly, continuous feature usage is a strong indicator of customer experience. For example, for a route with 90 miles of feature availability, the customer could have experienced continuous availability for 90 miles or unavailability for one out of every ten miles.

Further, in correlating the feature activation to the map, anomalous roads and vehicles can be detected by low feature utilization. For example, if one vehicle out of many does not activate hands-free driving where the map indicates hands-free availability, there may be a sensor fault, but if many vehicles do not activate hands-free driving on a roadway, that road may represent types of corner cases to be assessed. There may be road layouts that cause vehicles to occasionally localize to the incorrect roadway and therefore preview the incorrect ADAS feature availability. These locations could be discovered in the mismatch between the vehicle-created ADAS availability data and that of the simulation.

## 2 Constraints

Many fields have experienced an explosion of available labeled and unlabeled data and have been utilizing machine learning at ever-increasing scale to enable advanced applications. However, while the automotive industry is generally doing this at scale, there are many obstacles to overcome due to legacy components that were not designed for data collection. The challenge in retrofitting these legacy systems for data collection is the need for contextual information from the vehicle. Obtaining this contextual information is complex and the methods required can differ with each platform and model year. Most modern technology companies have been able to design their products with data collection at the forefront of the design process, while automotive OEMs have long legacies and long lead time for incorporating new features in vehicles. While vehicles can be specially configured to record a great deal of the data, it is not yet practical nor generally necessary to capture the massive volume of data generated in each vehicle on each drive [2].

## 2.1 Data Collection Costs

Data collection from large fleets of connected vehicles has a non-trivial cost component [2]. Most connected vehicles can connect to Wi-Fi networks, but so few drivers complete this setup, that most connected vehicle data must be offloaded over a cellular network. Therefore, it is pertinent to estimate and design for the cellular costs when planning connected vehicle data collection. In an upcoming work, the authors introduce the Leveraging Aggregated Vehicles Analytics (LAVA) methodology, which uses historical connected vehicle data to rigorously estimate performance metrics with various simulated parameters [14]. The LAVA methodology is particularly relevant in this context in estimating the volume of cellular data that would be transmitted from connected vehicles when data collection is dependent on specific parameters, such as collecting data only while the vehicle is on the highway.

### 2.1.1 Determining Trip Routes from Fleet Data

As existing connected vehicle data collection may be triggered only by specific events (e.g. key-on and key-off events), the complete routes of connected vehicles may not be obvious to an OEM analyzing the data. Therefore, when only origin and destination pairs are collected, a routing engine can be used to create full routes. Routing engines typically return multiple options for routes, so the ground truth data for the trip can be used to probabilistically determine the true route. That is, the route that minimizes the following objective function is the most likely to be the true route taken by the vehicle:

$$f(d_{route}, t_{route}) = w_1 \frac{|d_{true} - d_{route}|}{d_{true}} + w_2 \frac{|t_{true} - t_{route}|}{t_{true}}$$

where $w_1 \geq 0$ and $w_2 \geq 0$ are weights for this multi-objective optimization problem, $d_{true}$ is the odometer change of the route, $t_{true}$ is the time change from the start to the end of the trip, $d_{route}$ is the length of a proposed route, and $t_{route}$ is the estimated duration of the proposed route.

Additionally, depending on the trigger for data collection, some routes may be fundamentally undiscoverable. For example, if a driver makes a stop without turning their vehicle off prior to arriving at the final destination, the detour will result in a greater than expected odometer change (based on just the origin and destination). To avoid improbable route matches, routes whose length is a certain percentage over the odometer change are excluded as possibilities. In other words, the odometer change can be significantly larger than the route length, but not vice versa.

Using this methodology on 341,792 connected vehicle origin–destination pairs, we were able to match 324,916 (95.01%) trips to routes, where only 9198 (2.69%) trips had a trip length greater than the odometer change (average length over odometer change for those trips was 2.29 km, and average percentage over odometer change was 2.43%). Therefore, we have sufficiently high confidence in the accuracy of these routes and are certainly comfortable using them in aggregate to discover overall trends (where the accuracy of individual routes is not critical).

Note that additional connected vehicle data can be utilized in the routing process. For example, fuel usage provides valuable information about the route. The average fuel usage per road type (per vehicle type) could be estimated from higher frequency connected vehicle data to validate routes chosen based on an expected fuel consumption for an entire route. The expected fuel usage could be summed over the roads included in a route and compared against the actual fuel usage which can be determined with only collecting data at the origin and destination. Expected fuel usage can be determined based on crowdsourcing connected vehicles fuel usage at relatively high frequency (e.g. when entering and exiting each road edge) and using the fuel usage, weather information, and road information (e.g. slope) as labelled training data for a machine learning model to predict fuel usage for any road type in various weather conditions [15].

Other considerations could include vehicle weight distribution, vehicle closure status at origin and destination, seat occupancy detection, passenger identities [16], and trailer status. Routed trips can be sanity checked against this data. For example, if the trailer status (i.e. towing a trailer or not), cabin occupancy, or weight distribution have changed from origin to destination, there must have been a stop in between them which will make extrapolating the full route nearly impossible.

However, a thorough examination of a vehicle's location history may provide waypoints that explain such trips. That is, waypoints whose addition into the route makes the most sense of the ground truth vehicle data can be chosen from the vehicle's location history based on distance from the origin and destination. For example, as in Fig. 1, if the origin is point A, the destination is point C, and the odometer change is 10 km, then the distance from A to C does not explain the odometer change. However, if the vehicle has frequently visited point B before or after point A or point C and the length from point A to point C via point B is 10 km, then the trip likely included point B as a waypoint.

A machine learning model can utilize known routes to more accurately determine a true trip route from origin and destination information. These known routes can be created from higher frequency data collection from connected vehicles, as data

**Fig. 1** Example routing options with associated distances between three points of interest

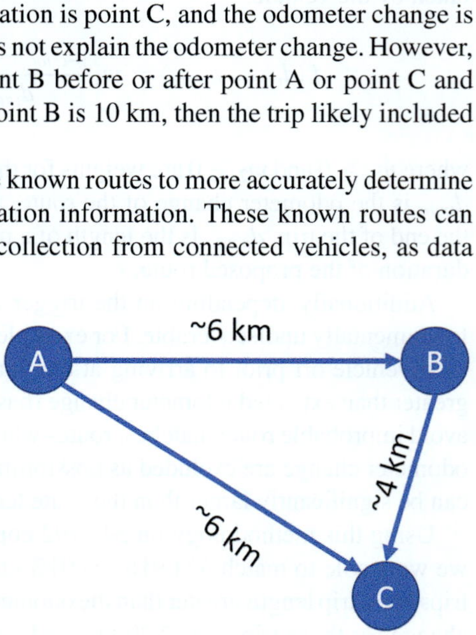

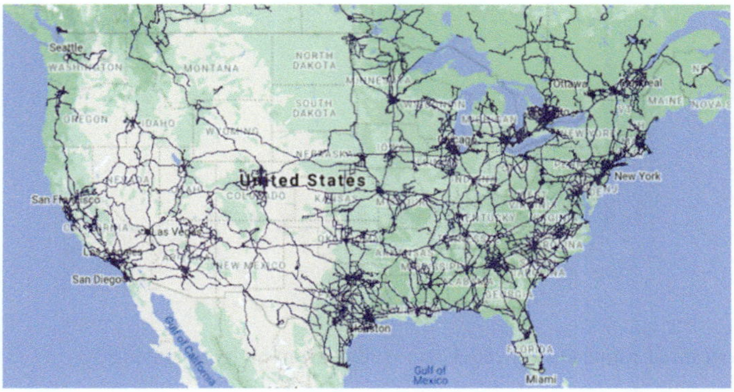


**Fig. 2** Map Data ©2021 Google. Set of 128,682 routes from 303 vehicles over 6 months

collection efforts scale up. For example, a long short-term memory recurrent neural network can be used to determine the order that road edges were used. Essentially, the label would be the order of road edges actually taken and the input data would be ground truth data such as origin and destination GNSS location, trip length (i.e. odometer change), trip duration, total fuel used for the entire trip, etc. Then for trips where only origin and destination information is known, data from other connected vehicles can help fill in the entire route [17].

### 2.1.2 Data Collection Cost Estimation

Figure 2 shows a dataset of all 128,682 trips taken by 303 unique vehicles over six months. Note that we have narrowed the scope from all 324,916 available routes (over 15 months) to avoid excessive data processing times. This subset of data can be used to estimate the total number of miles driven by a large set of vehicles on various road types and create data-driven estimates for the volume and cost of cellular data transmission.

When these routes are created and map matched to a road network, the total number of highway miles can be summed. Most data collection (and native signal behavior) occurs at a time-driven frequency, so the total number of highway miles can be converted to the time dimension based on the roadways' speed limits or even a global estimate for highway speeds (e.g. 65 miles per hour). Figure 3 shows a histogram of an estimated data volume collected by connected vehicles over six months, where the total data volume would be 4203.54 Megabytes. This assumes that 0.1 kB are generated every one second and the vehicles are travelling at 65 miles per hour on average.

Given a cost per megabyte and by assuming this subset of vehicles is representative of the wider fleet's behavior, this analysis can be used to estimate the aggregate data cost. Let us assume cellular data transmission costs $\$x$ per megabyte, which results in a total cost of $\$x * 4203.54$ for this fleet of 303 vehicles over the course of 6 months. If

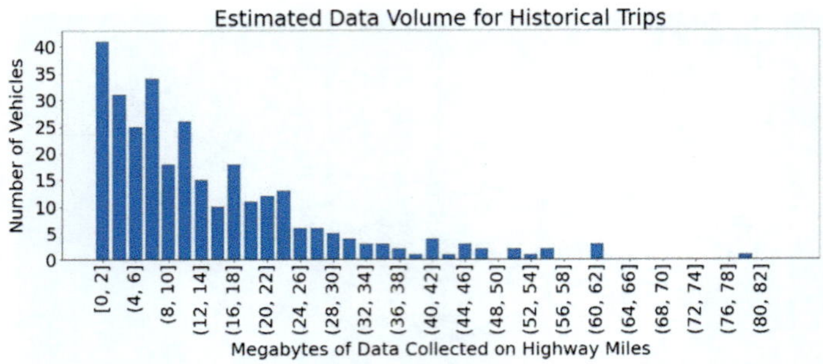


**Fig. 3** Histogram of estimated collected data volume from highway driving

we assume that the fleet from which data will be collected contains 100,000 vehicles, then the cost for 12 months of data transmission is given by: $\frac{\$x*4203.54}{303\text{vehicles}\times 6\text{ months}} * 12 \text{ months} * 100{,}000 = \$x * 2{,}774{,}613.90$, or roughly $\$x *$ 13.87 per vehicle per year for an extremely small sample of data (0.1 kB per second). The cost of data collection will also likely continue for the life of the vehicle.

Therefore, the cost of data collection must be considered when designing centralized data analyses, and the solution is generally not to collect all signals at the native frequency. It is important to collect the appropriate signals in an intelligent manner (e.g. a signal indicating the ADAS feature's status), but it may be prohibitively difficult to design large-scale in-vehicle data aggregation and is demonstrably expensive to transmit over a cellular network.

## 2.2 *Data Generation and Collection Issues*

Significant work is required to ensure collected data is representative of vehicle operation, but current vehicle architecture rightfully prioritizes in-vehicle performance over data collection efforts. Reusing the hypothetical situation mentioned above, most signals on vehicles refresh at a relatively high frequency (e.g. 50 Hz) and are not coordinated to occur at the same time. The frequency at which signals are captured, and, if not captured at native frequency, how signals are aggregated has a drastic impact on the ability to process the data in the cloud. Figure 4 shows sample vehicle data where the signals at native frequency are shown as blue dots, downsampled signals (to 1 Hz) are shown as green stars, and times where native frequency signal state changes are missed in the downsampled data are shown as red triangles.

In this example, signal_1 carries information regarding causes of state changes in signal_2. For a roughly 2.5 h drive, there are 52 state changes in signal_1 that would not be captured by a low frequency down-sampling (in this case 1 Hz subsample with no aggregation). Since the data was not originally designed to be understood outside of the in-vehicle context, a downsampled data collection scheme would miss

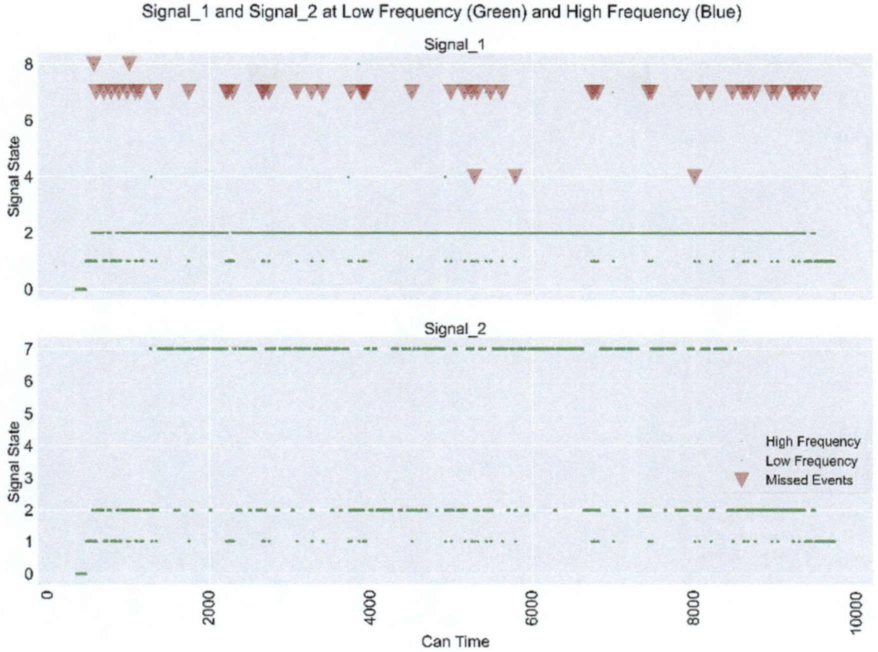


**Fig. 4** Sample missed states in high frequency data

important signal states and would be insufficient to understand the state changes in signal_2.

Figure 5 shows a zoomed-in view of the data in Fig. 4 where it is visually obvious that the state change in signal_2 (from state 2 to 1) occurs in time with the state change in signal_1 from 2 to 4 to 1. Without appropriate aggregation or higher frequency data sampling, the state change in signal_1 would not be captured by a downsampled data collection process. Therefore, if an algorithm were designed to automatically process data coming from connected vehicles, then the state changes in signal_1 that is expected and explains the state change in signal_2 would not be captured. Other collected signals whose state values persist longer may be useful in piecing together the information in signal_1, but the in-vehicle evaluation would be lost without the appropriate data aggregation design.

## 3 Determination of Actual Feature Experience

In order to quantify the actual feature experience, we will use a single signal that indicates the status of an in-development ADAS feature, which automates some portions of the driving task. Figure 6 shows a development drive (with non-production map and software) color-coded by the ADAS feature status only. This data represents

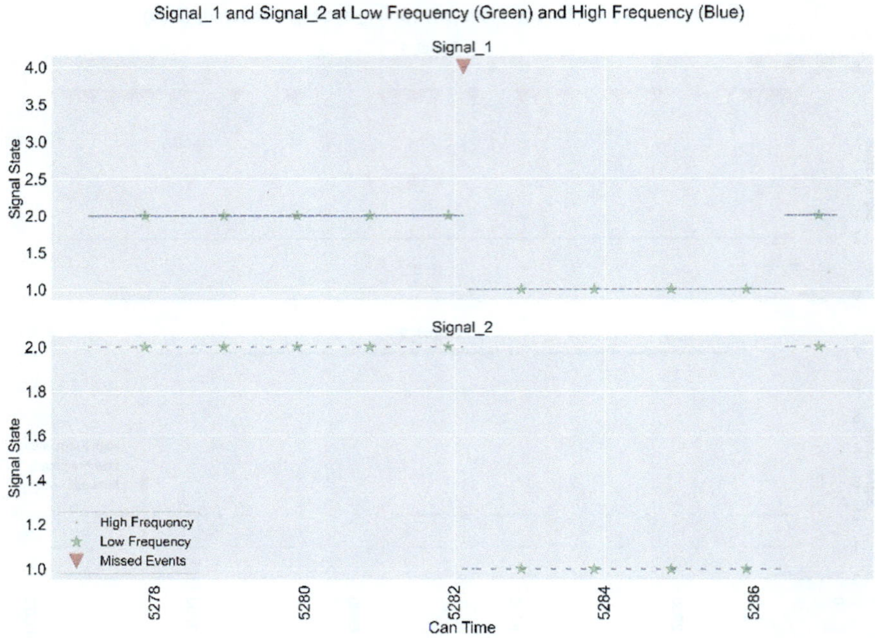


**Fig. 5** Zoomed in sample missed states in high frequency data

**Fig. 6** Map Data ©2021 Google. Actual feature experience for a development drive

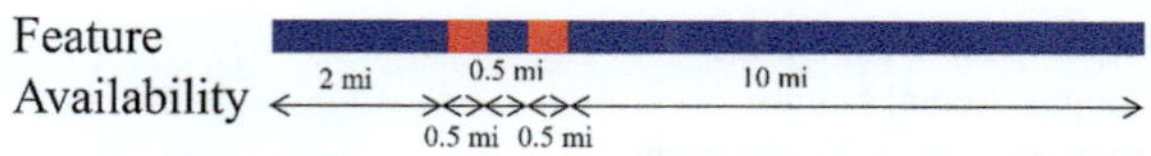


**Fig. 7** Minimum length of feature activation

3294.055 miles of total driving, where the feature was active for 1845.921 miles. This gives a high-level view into the usage of the ADAS feature but is not normalized properly. That is, the total length of the drive is not indicative of the total length of feature availability. The route could include unsupported road types and could include situations where the feature is not available by design.

To answer the question of where the feature was available but was not active and to give an appropriate normalization factor to the total number of miles the feature was used, the vehicle-generated signals can be utilized directly (assuming there is a properly collected signal that indicates the availability of the feature).

Many ADAS features have complex logic regarding how ADAS feature availability maps are previewed and used in-vehicle [8, 10]. Therefore, it is key that any analysis of the vehicle data is done in that context. For example, some ADAS features may disallow activation in otherwise acceptable conditions due to an upcoming roadway that does not allow for feature usage (perhaps that roadway will be encountered in the next *y* miles). Figure 7 shows an example of this situation where the map has a 0.5-mile section of feature availability (blue) surrounded by feature unavailability (red). Due to the example 0.5-mile area of feature availability being sufficiently short, the vehicle may not practically allow for the feature to be used there as feature usage in this location would result in an interrupted customer experience. While the map and the vehicle both reflect that the current roadway is technically available for ADAS feature usage, the vehicle nonetheless inhibits activation.

Figure 8 shows the same data as in Fig. 6 but now evaluated relative to the vehicle's preview of the feature availability map. That is, the feature status and the classification of the road on the feature availability map are considered in the color-coding and evaluating the vehicle data. Note that there are several possible reasons for the feature to be inactive on roads where the map indicates it is available, such as how the vehicle localizes to the map in real-time, vehicle-sensed concerns with the roadway (e.g. lane lines are not visible), or driver preference. These categories can and should be separated, but that can be difficult based on the downsampled vehicle data mentioned above. Therefore, the production data collection strategy must ensure that the data collected is able to differentiate between these types of events.

This view into the ADAS feature usage data now contains a normalization factor. We can calculate the percentage of available roads where the feature determined it could have been available but was not active. This leads to the *Utilization* of ADAS features, which is given by

$$\text{Utilization} = \frac{\text{Feature Active}}{\text{Feature Available}}$$

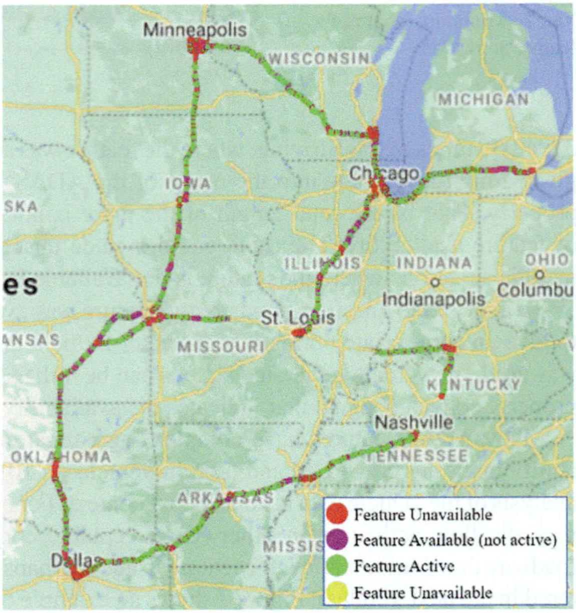


**Fig. 8** Map Data ©2021 Google. Actual feature experience for a development drive relative to the ADAS feature availability map

where the numerator and denominator can be in units of length or time. Utilization ranges between zero and one, where larger values indicate better customer experience. Note that the locations where the feature was available but was not active can be separated into two distinct groups: one where the driver decides to not activate the feature even though it is available and two, where the vehicle's local sensing suite determines that the feature should not be active even though the map indicates that the feature should be available. The latter group is of more interest in understanding the feature operation and capabilities, while the former reflects individual choice.

In the case of the development drive in Figs. 6 and 8, the utilization is $\frac{1845.921\text{ mi}}{2694.981\text{mi}} = 0.685$, because the ADAS feature was active for 1845.921 miles and the vehicle previewed feature availability for 2694.981 miles (where the total trip length was 3294.055 miles). That is, the feature was utilized (i.e. active) about 68.5% of the total length that it was available.

## 4 Determination of Expected Feature Experience

Using the vehicle-generated signals relies on the data generation and collection process fully capturing the nuances in the determination of feature availability. While this may capture feature availability and usage sufficiently well, it may be difficult to analyze the granular reasons for the feature not being active. For example, based on downsampled data, it may be impossible to determine whether the feature was not active due to real-time road conditions as determined by the vision system [7], due to the driver's condition (e.g. eyes off the road for a Level 2 SAE ADAS feature), whether the driver cancelled voluntarily, or due to the vehicle's location on the map.

Given the above concerns with vehicle-created data, it can be beneficial to validate the vehicle-determined feature availability in the *best-case customer experience scenario* with an offline simulation [12]. That is, the GNSS coordinates (paired with the feature status) of a route are map matched to the feature availability map and the feature availability logic is simulated offline. This firstly allows for more advanced methods of map matching [13] that utilize the entire set of GNSS points in a route, ensuring more accurate map matching to the road network. Secondly, this process provides a sanity check against the vehicle-generated signals.

The purpose of a feature availability map could be to maximize the warning time given to a driver prior to an area incompatible with ADAS feature usage. However, the in-vehicle sensor suite would generally be the most reliable source of real-time road information, so the map may indicate feature availability while the vehicle disallows feature activation. Significant discrepancies between the map-based expectation of feature availability and actual feature usage are cause for examination of the logic behind the feature availability map. Therefore, it is prudent to examine how the purported feature availability (provided by the map) directly compares to the actual feature usage by a customer. It is also important to examine the discrepancy between the vehicle-determined feature availability and an offline simulation to determine how accurately the vehicle can localize to the road in real-time, whether the vehicle is previewing the upcoming roadways in the correct manner, and whether the vehicle data is capturing all the relevant information for analysis. This further opens the door to simulating what the customer experience would have been with changes in the design of the map and map previewing logic.

Further, it can be important to understand the feature's ideal capabilities without consideration to transitory issues. For example, a driver may have attempted to use an ADAS feature in a downpour and be told it was unavailable. It would be relevant to know that the map indicated that the feature should have been available as well as the reason why the feature was not allowed to be activated. A driver could also deactivate the feature despite it being available. An offline simulation can indicate where the feature could have been active but was deactivated. This would provide concrete metrics regarding the total usage relative to the total availability, aggregated by different reasons for non-usage when the feature should have been available.

Figure 9 shows the result of the offline simulation of the best-case customer experience scenario on the development drive data given in Figs. 6 and 8. That is, Fig. 9 utilizes offline map-matching and provides a simulation of the feature availability indicated by the map as it should be previewed by the vehicle. These results show what the logic behind the ADAS feature availability map and the in-vehicle process previewing the map determined to be the maximum feature offering for customers.

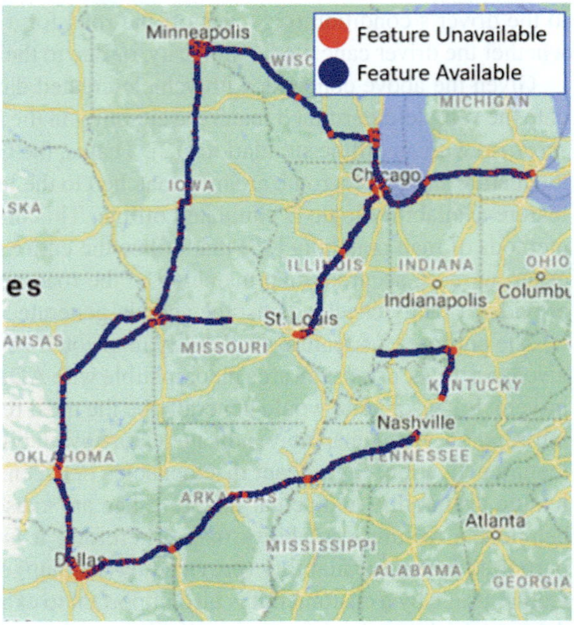


**Fig. 9** Map Data ©2021 Google. Expected feature experience for a development drive relative to the ADAS feature availability map

## 5 ADAS Feature Customer Experience Metrics

*Utilization* can still be defined as above, however, we can now slightly alter the definition of "Feature Available" to include the availability as given by the ADAS feature availability map, rather than only the in-vehicle determination of availability. Using the offline simulation as the normalization factor can expose areas where the map indicates that the feature should be available but the vehicles travelling there frequently encounter a situation that is not conducive for feature activation, for example the lane lines may not be detectable by the vision system. It is also important to exclude areas where the driver voluntarily cancels the ADAS feature (or consider these events separately). If the driver simply does not want the feature to be active (perhaps they are about to exit an available roadway and are preparing to resume control) then the feature is behaving as expected by the map and by the feature design logic.

Further, utilization can be calculated for a subset of vehicles or roads. That is, certain sets of vehicles may have unexpectedly low feature utilization wherever they drive, perhaps due to improperly calibrated sensors. Figure 10 shows an example roadway where localization errors could cause most vehicles to determine that the feature is unavailable, when an offline map match determines that it should have in fact been available. The discrepancy between the vehicle-created data showing that the feature is unavailable and the offline simulation showing availability would be an important flag for discovering roadways where localization errors are common.

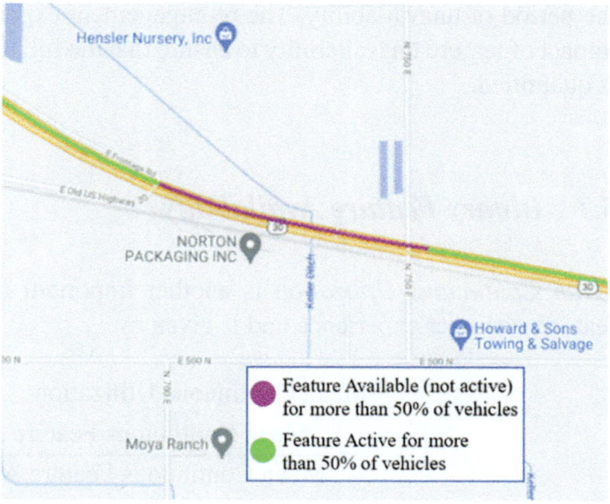


**Fig. 10** Map Data ©2021 Google. Example feature utilization at the road level

Further, the effectiveness of the re-engagement strategy used before and after unavailability can be evaluated through *Re-Engagement Experience*, as in

$$\text{Re - Engagement Experience} = \frac{\text{Feature Inactive} - \text{Feature Unavailable}}{\text{Feature Inactive}},$$

where the denominator only includes roadways where the feature is unavailable and immediately precedes and/or follows a roadway where the feature is available. Re-Engagement Experience ranges between zero and one where smaller values indicate better customer experience. In Fig. 11, where blue indicates feature usage and availability, red indicates feature unavailability, and orange indicates that the feature is available but not in use, then the Re-Engagement Experience would be $\frac{450-350}{400} = \frac{100}{400} = 0.25$. That is, the interruption to feature usage was 25% longer than the interruption to feature availability. This view of customer experience should be considered both in designing the feature availability map and how it is used by the vehicle to capture the ultimate impact of feature unavailability on feature usage. That is, when the map indicates that features are unavailable, there will be an extended effect on the ability of a driver to utilize the feature, in that it will take time for the feature to re-engage manually or automatically before and/or after

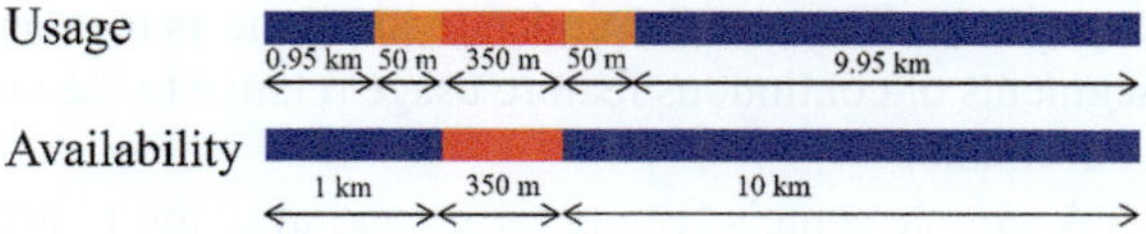


**Fig. 11** Visualization of feature re-engagement

the period of unavailability. The re-engagement experience captures this expanded impact of feature unavailability to ensure that the full impact on customer experience is quantified.

## 5.1 Binary Feature Availability

*Mean Continuous Utilization* is another important metric in determining ADAS feature customer experience and is given by

$$\text{Mean Continuous Utilization} \\ = \frac{\text{Mean Continuous Feature Active}}{\text{Mean Continuous Feature Available}}$$

where again the numerator and denominator can be in units of length or time. Mean Continuous Utilization ranges between zero and one, where larger values indicate better customer experience. This metric gives an important view into how interrupted the feature usage was relative to the optimal continuous feature offering. It is important to note that the mean length that the feature was continuously active and the mean length that the feature was available are themselves important metrics, where larger values indicate better customer experience. The mean continuous utilization compares the feature usage to the best-case customer experience scenario given by an offline simulation using the feature availability map and feature design logic. This best-case customer experience scenario is an important method to evaluate feature availability maps, but it is critical to quantify how well it truly represents the feature's capabilities.

Another relevant metric is the comparison between the number of continuous feature activation usage events and the number of continuous feature availability roadways, which we call the *Interruptedness*, as in

$$\text{Interruptedness} \\ = \frac{\text{Count of Feature Active} - \text{Count of Feature Available}}{\text{Count of Feature Active}}.$$

This metric tracks how frequently the feature activation was interrupted relative to the best-case customer experience scenario (given by feature availability). Interruptedness ranges between zero and one, where smaller values indicate better customer experience. This can be understood as the percentage increase in the number of segments of continuous feature usage relative to the number of segments the feature is available.

A sample route's feature usage and availability are given by Fig. 12, where red shows where the feature is unavailable, orange shows where the feature should be

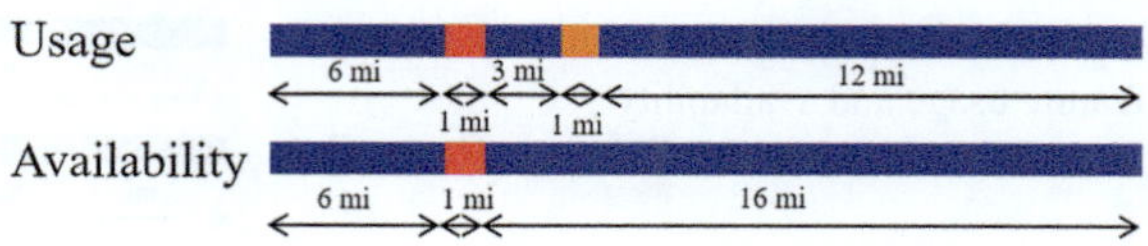


**Fig. 12** Example feature usage versus availability

available but was not active, and blue shows where the feature was active on the top and available on the bottom. For this route, the mean continuous feature usage is $\frac{6+3+12}{3} = 7$ miles and the mean continuous feature availability is $\frac{6+16}{2} = 11$ miles. Therefore, for the feature usage and feature availability given by Fig. 12, the mean continuous utilization is $\frac{7}{11} \approx 0.636$. Further, the utilization is $\frac{6+3+12}{6+16} = \frac{21}{22} \approx 0.955$. Lastly, the interruptedness would be given by $\frac{3-2}{3} \approx 0.333$, as the 16-mile-long portion of the route that was available was broken into two segments of feature usage.

## 5.2 *Nuanced Feature Availability*

When ADAS feature availability is extended outside of current implementations, measuring the customer experience becomes more subtle. This is due to an expectation of differentiated feature operation in new road situations. In cases where the feature availability map has multiple options for feature availability (e.g. Feature Mode 1, Feature Mode 2, etc.), then the continuous usage of a single feature mode may not be a good indicator of customer experience. If the feature usage is aligned with the map's availability, then the feature is performing as expected, so the *Mean Continuous Operation* could be given by

$$\text{Mean Continuous Operation} \\ = \frac{\text{Mean Continuous Feature Usage as Expected by Map}}{\text{Mean Continuous Feature Availability}},$$

where the feature availability would essentially distill the entire nuanced availability map into available or not, and the continuous spans of feature usage would not be broken up by changes in the feature mode of operation changes. Mean Continuous Operation ranges between 0 and 1, where larger values indicate better customer experience. For example, in Fig. 13, for Route A, the Mean Continuous Operation would be $\frac{((5+3)+(14))/2}{(5+3+1+14)/1} = \frac{(22)/2}{(5+3+15)/1} = \frac{11}{23} \approx 0.478$. The denominator is divided by one because the entire roadway in Fig. 13 is a single span of feature availability, where part is available for Feature Mode 1 and part is available for Feature Mode 2. The numerator is divided by two to reflect the two distinct spans of feature usage interrupted by a one-mile span of feature inactivity. For Route B, the Mean Continuous Operation would be 1.00, as the feature was behaving according to the map provided availability for the entire route.

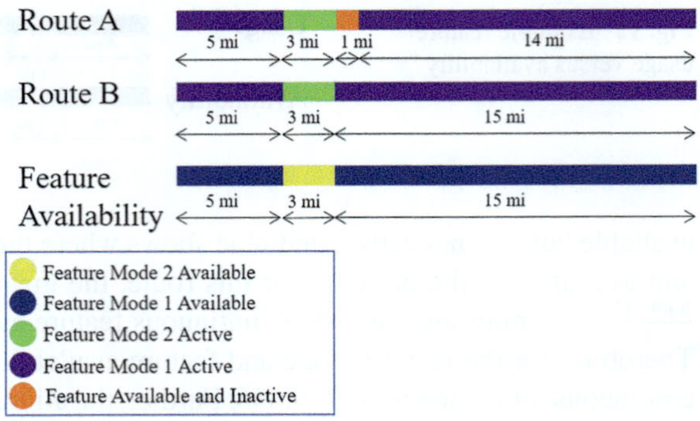


**Fig. 13** Example nuanced feature usage and availability

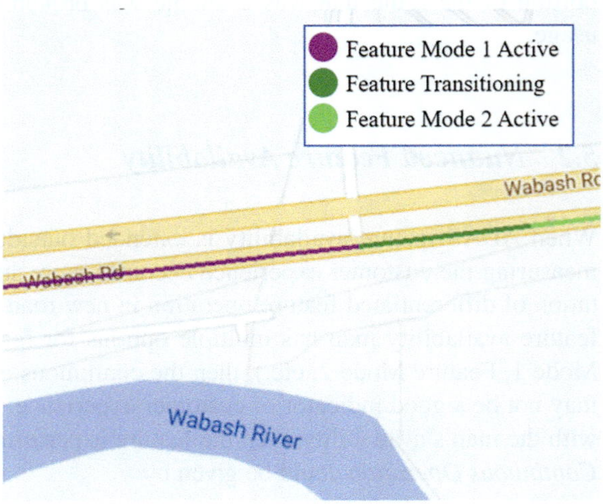


**Fig. 14** Map Data ©2021 Google. Example nuanced feature usage

This metric could optionally be modified for adjustments in feature operation. For example, as shown in Fig. 14, there may a noticeable distance where the feature mode is transitioning and neither mode is active. Such transitions may require input from the driver and exist by design. Therefore, if these transitions are sufficiently short, it may be appropriate to count these sections as feature usage as expected by the map (without breaking up the Feature Mode 1 and Feature Mode 2 active sections). These may, for example, occur due to a required driver re-engagement (e.g. transitioning from Level II to Level I SAE ADAS feature).

## 5.3 *Clustering for Outlier Discovery*

Unsupervised machine learning can also be employed with a subset of the above metrics to identify anomalous roads, drivers, or vehicles [18, 19]. Road edges (from node to node, where intersections, exits, and entrances would be considered nodes) or longer portions of roadways (e.g. 10 mile stretches of highways) could be considered.

For example, the input features to a clustering algorithm for discovering outlier portions of highways could include (a) the mean continuous feature operation vehicles experience while on a road, (b) the percentage of vehicles utilizing the feature when it is available on a road, (c) the percentage of drivers who manually deactivate the feature on a road, (d) the percentage of trips where the feature deactivates the feature on a road due to the vehicle's sensing, (e) interruptedness for trips that include a road, and (f) re-engagement experience when a road has some feature unavailability. The portion of highway could be determined by junctions with other highways. Urban and rural highways would likely end up in two separate clusters, and anomalous roadways in terms of feature capabilities should be outliers. The time of day may also play a role in the formation of clusters, as the behavior and availability of ADAS features may vary significantly between heavy and light traffic situations.

Anomalous vehicles (or drivers) can also be identified based on a similar set of features to find clusters of vehicles that have been unable to sustain feature usage. These clusters would likely be broken up along the lines of driver versus feature deactivation and non-use. Sets of vehicles that do not utilize the feature (especially if the cause for deactivation was due to the feature and not driver preference) can be found and, for certain problems such as camera calibration issues, potential software fixes can be delivered via over-the-air updates.

## 6 Conclusion

Connected vehicle data can be utilized to determine customer experience with ADAS features. Typically, these features utilize a combination of the host vehicle's sensor suite and pre-mapped information about roadways in the form an ADAS feature availability map. This map is used to ensure that the feature is used only in appropriate areas. In this work, we have proposed several metrics for grounding the vehicle data to improve understanding of customer experience. Specifically, there must be a normalizing factor to understand total customer usage of the feature, which can be based on the vehicle data as well as on offline simulation of the expected feature experience. By simulating the design of feature availability and allowance of activation, the vehicle's performance can be checked against a ground truth. That is, the total ADAS feature usage can only be properly understood within the context of where it *could have* been active. The simulation results can also take advantage of advanced map matching algorithms (that utilize the entire route) to discover and diagnose real-time localization issues on vehicles. In this manner, customer experience with ADAS features can be concretely measured and potential over-the-air updates to ADAS feature offerings meaningfully compared.

Effective use of large scale connected vehicle data enables a powerful new feedback loop in the iterative vehicle design process. As discussed in this chapter, customer experience with ADAS features can be measured to drive post-production improvements to ADAS technology via over-the-air updates to connected vehicles. We have introduced several measurable properties (also called features in the machine learning community) of ADAS feature usage data. These properties utilize data from connected vehicle fleets to optimize the data-driven engineering process. Now that customer experience can be measured directly and objectively, machine learning algorithms can be utilized to discover anomalies and variations in customer experience.

## References

1. Takács Á, Drexler DA, Galambos P, Rudas IJ, Haidegger T (2018) Assessment and standardization of autonomous vehicles. In: 2018 IEEE 22nd international conference on intelligent engineering systems (INES). IEEE, pp 000185–000192
2. McQueen B (2017) Big data analytics for connected vehicles and smart cities. Artech House
3. Gao H, Cheng B, Wang J, Li K, Zhao J, Li D (2018) Object classification using CNN-based fusion of vision and LIDAR in autonomous vehicle environment. IEEE Trans Industr Inf 14(9):4224–4231
4. Ye D, Yu R, Pan M, Han Z (2020) Federated learning in vehicular edge computing: a selective model aggregation approach. IEEE Access 8:23920–23935
5. Vishnukumar HJ, Butting B, Müller C, Sax E (2017) Machine learning and deep neural network—artificial intelligence core for lab and real-world test and validation for ADAS and autonomous vehicles: AI for efficient and quality test and validation. In: 2017 Intelligent systems conference (IntelliSys). IEEE, pp 714–721
6. Harms IM, Bingen L, Steffens J (2020) Addressing the awareness gap: a combined survey and vehicle registration analysis to assess car owners' usage of ADAS in fleets. Transp Res Part A: Policy Pract 134:65–77
7. Stein Gdeon, Blumenthal I, Shaag N, Moskowitz J (2020) Vehicle environment modeling with a camera. U.S. Patent Number 10,872,433. U.S. Patent and Trademark Office, Washington, DC
8. Matthaei R, Bagschik G, Maurer M (2014) Map-relative localization in lane-level maps for ADAS and autonomous driving. In: 2014 IEEE Intelligent vehicles symposium proceedings. IEEE, pp 49–55
9. Okuda R, Kajiwara Y, Terashima K (2014) A survey of technical trend of ADAS and autonomous driving. In: Technical Papers of 2014 International Symposium on VLSI Design, Automation and Test. IEEE, pp 1–4
10. Groh A, Thomas C, Lerner J, Gifford J et al (2019) Location-based vehicle operation. U.S. Patent Application No. 20210063167(A1). U.S. Patent and Trademark Office, Washington, DC
11. Johansson KH, Törngren M, Nielsen L (2005) Vehicle applications of controller area network. In: Handbook of networked and embedded control systems. Birkhäuser Boston, pp 741–765
12. Fling T, Filev D, Kristinsson J, Wisniewski J et al (2017) Vehicle mode determination. U.S. Patent No. 9,796,388. U.S. Patent and Trademark Office, Washington, DC
13. Newson P, Krumm J (2009) Hidden Markov map matching through noise and sparseness. In: Proceedings of the 17th ACM SIGSPATIAL international conference on advances in geographic information systems, pp 336–343
14. Lerner J, Tayim D, Pervez N, Zwicky T, LAVA: a methodology for leveraging aggregated vehicle analytics. SAE Int J CAV 6(1):2023. https://doi.org/10.4271/12-06-01-0003
15. Zeng X, Schmotzer J, Poll M, Raval S, Bandi P (2020) Systems and methods for providing predictive distance-to-empty for vehicles. U.S. Patent Application No. 16/353,922. U.S. Patent and Trademark Office, Washington, DC

16. Lerner J, Issac J, Lim JH (2020) Automated vehicle profile differentiation and learning. U.S. Patent Application No. 16/441,613. U.S. Patent and Trademark Office, Washington, DC
17. Liu S, Qu Q (2016) Dynamic collective routing using crowdsourcing data. Transp Res Part B: Methodol 93:450–469
18. Chen Y, Zhou M, Zheng Z, Huo M (2019) Toward practical crowdsourcing-based road anomaly detection with scale-invariant feature. IEEE Access 7:67666–67678
19. Hall MA (1999) Correlation-based feature selection for machine learning (Doctoral Dissertation). Retrieved from: https://www.cs.waikato.ac.nz/~ml/publications/1999/99MH-Thesis.pdf

# Driver Assistance Systems and Safety—Assessment and Challenges

**Jinwei Zhou, Pavlo Tkachenko, Daniel Adelberger, and Luigi del Re**

**Abstract** Safety assessment of Highly Automated Vehicles, including Advanced Driver Assistance Systems and Advanced Driving Functions, is of paramount importance for the acceptance and diffusion of these technologies. On-road testing alone is no option due to the enormous time requirements, so virtual testing is generally considered to be a necessary complement. While more time efficient than on road testing, also virtual testing cannot be performed for all possible situations. Moreover, virtual testing can be even misleading if the considered scenarios are not realistic or do not include the critical situations which can occur in the intended real traffic use. Against this background, we discuss different options and challenges as well as outlooks.

## 1 Introduction

Highly Automated Vehicle (HAV) technologies have advanced dramatically in recent years and offer an impressive potential to transform ground mobility in the future. Over the past few decades, considerable efforts have been made to develop automated transportation technology and have manifested tremendous technological advances in numerous areas. The advancement of HAVs can bring massive social and economic benefits in terms of time saving [1], fuel economy and reduction of harmful emissions and greenhouse gas [2], traffic capacity [3] etc.

J. Zhou (✉)
Kontrol GmbH, 4020 Linz, Austria
e-mail: j.zhou@kontrol.tech

P. Tkachenko · D. Adelberger · L. Re
Johannes Kepler University Linz, 4040 Linz, Austria
e-mail: pavlo.tkachenko@jku.at

D. Adelberger
e-mail: daniel.adelberger@jku.at

L. Re
e-mail: luigi.delre@jku.at

 
Y. L. Murphey et al. (eds.), *AI-enabled Technologies for Autonomous and Connected Vehicles*, Lecture Notes in Intelligent Transportation and Infrastructure,
https://doi.org/10.1007/978-3-031-06780-8_13

A major interest is related to the expected increase of safety. Safety hazards can have many causes, like sensor or actuator faults, but also by faults in the software which plays an increasingly important role in transportation. Accidents can arise from a wrong decision based on correct detection—as in the Arizona Uber case—or even be induced by hackers [4]. Still, according to *The National Motor Vehicle Crash Causation Survey* (NMVCCS) [5], up to 94 % of crashes are due to drivers' errors. In view of this, a vision of delegating driving to computers is appealing. Indeed, this has been happening to some extent for long in the background, for instance by active safety systems, like Anti-lock Braking System (ABS), Electronic Stability Program (ESP), collision warning/avoidance or intelligent speed adaptation. However, all these systems are not meant to replace but to "enhance" the driver.

Moving up in terms of automation[1] implies that the HAV should take over decisions traditionally reserved to the driver. Even so, some errors are bound to happen, and this is strongly affecting the attitude of potential buyers, in spite of the fact that the accident rates are very low, lower than conventional vehicles, and improving. For instance, the National Highway Traffic Safety Administration (NHTSA)'s report on Tesla, has pointed out that crash rates involving Tesla cars have dropped by almost 40 percent since the wide introduction of *Autopilot* system [7], which can be classified as somewhere between levels 2 and 3 under Society of Automotive Engineers (SAE)'s automation level definitions [6].

Accordingly, it is important both for commercial but also for regulatory reason to be able to assess the safety of the automated driving functions (ADFs), separately or jointly, which make up the HAV and finally the fully automated vehicle (AV), level 6 in the SAE scaling. Safety will never be absolute, so it boils down to estimate the real probability of accidents due to a wrong behavior of ADF.

The standard approach of the automotive industry—road testing—is not viable, due to the much higher complexity and the resulting requirements. For instance, [8] has predicted that more than 100 Mio km of road driving would be required for the thorough validation of an automated vehicle. Only if these extensive tests have been done, it can be shown—within an acceptable confidence interval—that the automated vehicle is at least as safe as a manually driven car.

Against this background, virtual testing, by simulation, has become a necessity. Virtual testing encompasses two aspects: the modeling of the vehicle so that the results are representative, and the choice of the test conditions.

In the following we shall concentrate on the evaluation procedure. Of course simulation programs and vehicle models are needed as well, but there are several commercially available options (like PTV VISSIM,[2] IPG Carmaker[3] and others), so that we shall not discuss them.

---

[1] There is no unique definition of level of automation in vehicles, the most common used one being the SAE [6], where level 0 stands for not automated and level 5 stand for fully automated vehicle.

[2] https://www.ptvgroup.com/en/solutions/products/ptv-vissim/.

[3] https://ipg-automotive.com/products-services/simulation-software/carmaker/.

# 2 The Scenario Approach

## *2.1 The Idea*

In general terms, a safety assessment could follow a grid approach, *i.e.* defining all possible situations, then picking out a number of combinations chosen in order to cover approximately the whole test space and then performing a very large number of simulations for "all" these possible conditions. In practice, this would be as impossible as finding the right book in the Library of Babel [9]. With a very coarse discretization, an exhaustive search might become feasible, but still very inefficient, as its combinatorial process would generate many candidates that are not relevant or of little interest.

The sensible alternative consists in fixing a limited number of traffic situations, the so called scenarios, which are considered especially representative for the average traffic conditions. Indeed, in UN Regulation of Automated Lane Keeping Systems (ALKS) [10], known also as UNECE R157, 3 scenarios are defined for the performance and safety assessment of an Automated Lane Keeping System. For example, looking in the accident database of [7], it turns out that 96% of the highway accidents occurred in 23 specific situations, so that evaluating safety for these cases is a good indicator of the general safety.

## *2.2 Elements of a Scenario Based Evaluation*

In order for the results of the scenario based method to be representative, some elements need to be chosen carefully:

1. The definition of the scenarios
2. Their parametrization
3. The way the surrounding traffic is represented
4. The measure of safety.

Definition of scenarios

Basically, two main lines can be followed: scenarios can be constructed or extracted from data. The first line offers the possibility of being more general, but the scenarios could be unrealistic or hardly ever occur in real life. In the second case, the scenarios are definitely realistic, but they will represent only the conditions for which they were measured—e.g. if only data from freeways were used, the safety evaluation will probably be wrong for country roads. Of course, combinations are possible.

Parametrization of scenarios

The purpose of the evaluation defines also the parametrization. There are (at least) two different situations. If the aim of the assessment is to evaluate an average accident probability, it will be sensible to use a parametrization corresponding to the expected situation, albeit with some strategy to reduce computational effort. If the aim is to assess a comparative advantage, it will be sensible to look for a limit boundary, which leads to accidents for one option and not for the other.

Surrounding traffic

It is well known that many accidents are prevented by a correcting reaction of other traffic participants. Some situations, like merging, frequently are associated to a good informal cooperation—when it fails, traffic becomes more difficult—and dangerous.

So to have a realistic assessment of safety, the reactions of other participants should be considered as well, at least as long as we consider mixed traffic with automated and not automated vehicles. Unfortunately, the reactions of a human drivers are not deterministic, so we should consider a full set of possibilities, which would make the evaluation very difficult–with an unknown advantage in terms of quality of the assessment. For example, modeling the human driver's reaction to cut-in maneuvers to validate safety of the automated lane change function in terms of crash probability need large amount of measurements of not only normal (safe and proper) reaction but (unsafe) reactions that result in near-crash or crashes as well. Moreover, it is still unknown if the cut-in maneuver of automated driving function identical with human driver. And thus, it still unknown if the human reaction model, derived from the measured reaction to other human drivers, applicable for ADAS/ADF. In practice, most evaluations are done without considering changes in the trajectories of other vehicles.

Safety metrics

The most simple way to measure safety would be through the boolean indicator crash/non-crash. However, such indicator is similar to using black and white colors to describe the world. So, it is sensible to look for metrics that are able to describe the "shades" of safety. Unfortunately, there is no single metric reflecting the safety of a HAV under all conditions. Combined metrics have been proposed.

In the following, we discuss these topics more in detail.

## 3 Scenario Generation and Selection

The first step consists in fixing the so called Operational Design Domain (ODD). Thus reducing the set of the test cases to a manageable size and thus enable the so-called scenario based engineering.

The ODD defines the environmental condition of ADAS/ADF. SAEJ3016:2021 provides the taxonomy to define the ODD as "operating conditions under which a given driving automation system or feature thereof is specifically designed to function including, but not limited to, environmental, geographical, and time-of-day restrictions, and/or the requisite presence or absence of certain traffic or roadway characteristics." AV Safety Consortium has published best practice for defining the ODD [11]. Thus, fixing ODD can put limits on road environment, behavioral specification of ADAS/ADF, and the state of the vehicle. Thus reducing the set of the test cases to a manageable size and thus enable the so-called scenario based engineering. For example, the research project PEGASUS has worked on the ODD of a highway chauffeur function to define the validation process and scenarios of the L3 autonomous driving function [12]. The latest UN Regulation of ALKS [10] defines that the maximum operational velocity of the ALKS is $v_{\max} = 60\,\text{km/h}$ and road environment is highway so that the total test cases are limited to 4 logic scenarios [13]. An alternative way to specify the ODD is introduced [14], where the risks are assessed so that the ODD of ADAS/ADF is limited in a low risk area. In [15], describes the relationship among ODD, scenarios, and crashes/near-crashes. Once the environmental condition is fixed, ODD relevant scenarios can be generated for testing and validation. Up to very special cases, testing is done on the basis of a scenario catalogue, not of single scenarios.

### *3.1 Scenario Generation*

As already stated, there are different phylosophies: scenarios can be constructed (or generated) or extracted from measured data. Following the first line, stochastic modeling and similar methods are usually applied for massive generation of test scenarios. In [16, 17], the authors build up stochastic models using real measurements and produce various test scenarios through randomized sampling, searching for the situations with fault behavior, such as a crash. Nevertheless, they are usually applied to some simple traffic situations, e.g. emergency braking in a car-following situation. In the case of complex situations, e.g. modeling the large scale traffic [18], it is normally time-consuming, due to the low occurrence of critical events, e.g. cases in which the specified test scenarios result in a crash.

To overcome these limitations, [19–21] have proposed a methodology for scenario generation in a deterministic way, through the combination of various components, e.g. the weather, road type, driving maneuver, etc. In [22–24] the surrounding area of the ego-vehicle is divided into various observation cell according to the possible relative position of other vehicles to the ego. Depending on the cell occupancy and vehicle location, the static combination of possible behavior of the ego and/or surrounding vehicles form the concrete scenarios. However, this method greatly relies on engineers' expertise knowledge to single out the realistic scenario and to test the critical situations.

### *3.2 Scenarios from Crash Databases*

It is worthy mention that within this work the terminology of scenarios introduced in [13] is applied, namely three levels of abstraction for scenarios: functional sce-

**Table 1** Assignment of critical causes of NMVCCS crashes

| Critical cause attributed to | Estimated percentage± standard deviation[a] |
|---|---|
| Drivers | 94% ± 2.2% |
| Vehicles | 2% ± 0.7% |
| Environment | 2% ± 1.3% |
| Unknown Critical Causes | 2% ± 1.4% |
| **Total** | 100% |

[a] ±95% confident limits
*Source* Singh [26]

narios, logical scenarios, and concrete scenarios. Functional scenarios depict the most abstract level of scenario representations, which are represented by language to ensure that human experts can easily understand existing scenarios, e.g. Tables 2, 4. Logical scenarios depict a detailed representation of functional scenarios with the help of state space variables and the value ranges of the state space variables via a formal notation, see Fig. 8 panel (1). Every logical scenario can be converted to a concrete scenario by selection of a concrete value from a parameter range.

Manual driving is the main cause of road traffic accidents. As mentioned in a study conducted by **NMVCCS**[4] from 2005 to 2007 [25], it is estimated that 94% of the direct causes of NMVCCS collision events can be attributed to driver action, and the vehicle and driving environment are estimated to contribute only by 4% of direct causes. Relevant statistics on key causes are detailed in Table 1. Thus, by analyzing the accident data, we can find out the challenging driving situations that the system under test (SUT) can encounter in real traffic, as it is expected to have a mixed traffic in the next decade, consisting of autonomous driving and human drivers.

It is worth noting that crashes due to human errors as those summarized in [25] result from the pre-crash actions. They usually belong to the same set of actions (e.g. accelerating) which are perfectly safe in the utmost majority of cases. So the key question is: what made a specific action dangerous? If we can systematically describe these situations through a limited number of scenarios, we can limit the overall cases and make the test feasible.

To make the distinction more clear between accident cause and pre-crash action, let us consider as an example the non-performance error "sleep" in [25]. Figure 1 panel (1) shows the actions resulting from a sleeping driver that leads to collision with another vehicle, whereas in panel (2) it leads to crash with road barrier. In the case of panel (1), the HAV is required to react appropriately to avoid the crash. If we switch both vehicles, as in panel (2), the HAV does not need to react. Technically, as we consider only accidents of the HAV, only the pre-crash action of the red vehicle in panel (1) represents a relevant scenario for safety testing of HAV. Analyzing the accident data in a similar way, we can find out the relevant test scenarios. Table 2 shows some exemplary scenarios derived from analysis of SHRP2ND crash data,[5] in which the collision threat is identified when surrounding vehicle execute various

[4] National Motor Vehicle Crash Causation Survey (NMVCCS).

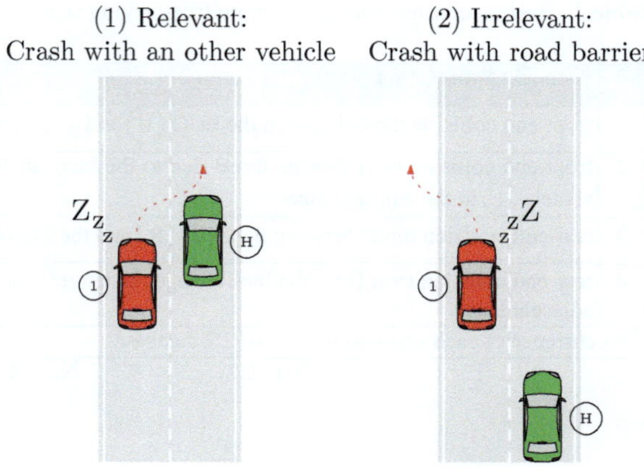


**Fig. 1** Relevance of the crashes in terms of testing of HAVs

maneuvers. For example, in Table 2 rear-end collision between EGO and vehicle 1 may happen if EGO does not react to braking maneuver (No.1) or cut-in maneuver (No.2) of the vehicle 1 in time and behavior as expected. It is worthy to mention that the abstraction level of such kind of scenarios, according to the terminology of scenarios in [13], is functional scenario, which are represented by language to ensure that human experts can easily understand existing scenarios. The functional scenario needs further parametrization to form the logic scenario with the help of state space variables and their value ranges via a formal notation, e.g. Figure 8 panel (1), and converted to thousands of concrete scenarios by selection of various concrete values from a parameter range of logic scenario [13, 27]. Further (functional) scenarios are provided in Appendix.7.2.

### *3.3 Automated Scenario Catalogue Learning*

As it was mentioned before, the scenarios for virtual testing can be generated or learned from measurements of the real-world environment. The generation of scenarios has the advantages of high flexibility as well as a low initial investment. Complex traffic situations can thus be created precisely. Unfortunately, with increasing functionalities and the continuous operation of modern ADAS, an a priori defined set of relevant scenarios within this approach can be hardly found. In this context, an alternative validation methodology consists in combining real drivings for data acquisition and further extraction of the scenarios from collected data.

The overall architecture of such methods is represented by Fig. 2. In the Data Collection step the data is recorded and the generation of additional information relevant to the use-case is performed. Here, the data can be enriched with information

[5] The Second Strategic Highway Research Program (SHRP 2) Naturalistic Driving Study(NDS), known to be the largest study of naturalistic driving behaviors available to date [26].

**Table 2** Basic scenarios: the ego is in the free-driving state

| No. | Scenario Description | Occ.† |
|---|---|---|
| 1 | front-end collision threat between the EGO (H) and a preceding vehicle (1) | 309 |
| 2 | front-end collision or sideswipe threat due to the lane change maneuver of a front vehicle (1) in the adjacent lane | 108 |
| 3 | rear-end collision threat between the EGO (H) and the following vehicle (1) | 10 |
| 4 | rear-end collision treat from the lane change maneuver of a rear vehicle (1) on the adjacent lane | 0 |

† Occurrence of each scenario in the SHRP2 database.

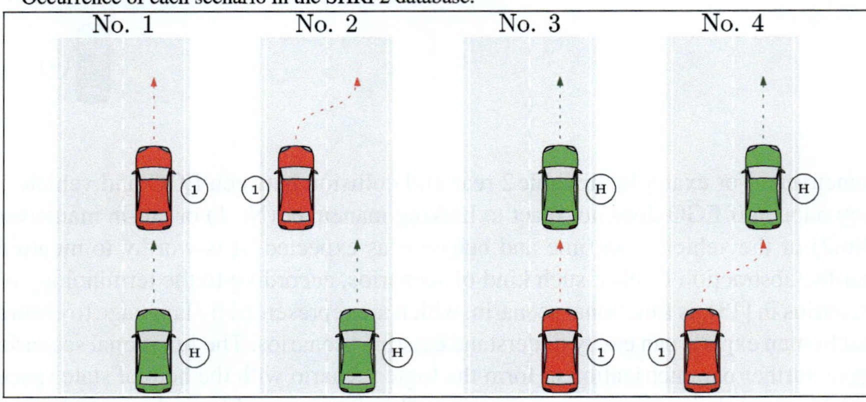


that is commonly not directly found in the recorded data like the TTC [28]. Then, the processed data is transferred to a Data Clustering block responsible for finding similarities between collected signals and grouping then into potential scenarios. The resulting clusters are then seen as scenarios and are saved into a scenario database. Finally, when new data come it is compared to the already learned catalogue of scenarios to decide whether these data belongs to already known case or should extend the database by a new scenario. If the new data can be assigned to an already learned scenario, then the meta-information of this scenario (appearance frequency, parameter values, ranges or distributions etc.) is updated without creating a new entry in the catalogue. The latter one allows keeping the catalogue compact.

The choice of the clustering and classification algorithms then defines the concrete approach. For example, in [29] the authors use the extended (Modified) Unsupervised Random Forest (MURF) for clustering and Random Forest (RF) algorithm for classification. In [30], the clustering is done using the variant of on-line k-means approach and the classification part is delegated to the nearest centroid method.

To show how the approach can be used in practice, we illustrate it through the experiments from [30]. In the experiments the highway driving and the vehicles in front of the ego car (150 m ahead at most) were considered. Further restriction to maneuvers in lateral direction is assumed. The goal will be to learn the scenarios of 5.5 s duration.

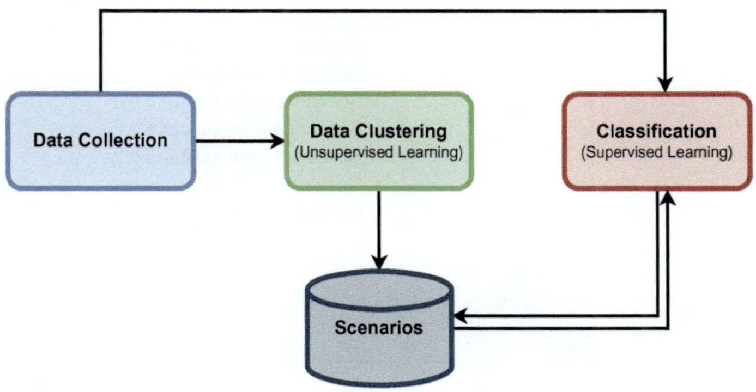


**Fig. 2** Architecture of automated scenario learning approaches

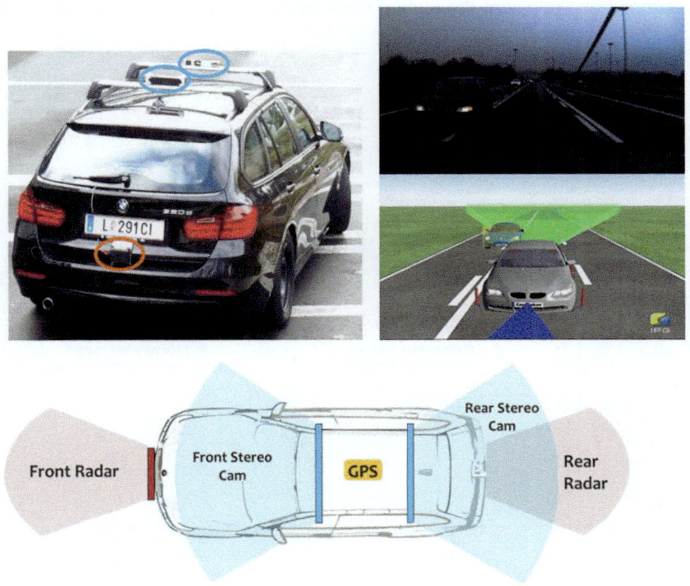


**Fig. 3** The experiment vehicle equipped with 2 radars and a screenshot of the replayed data from IPG CarMaker

To collect the sensor data, a production standard BMW 320d equipped with 2 Radar sensors (front, back) and 2 Stereo cameras (front, back) was used. The sampling time of the sensors is $t_s = 0.5\ s$. A trip from Spittal an der Drau via Villach (AT) to Malborghetto (IT) has been recorded, out of which 50 min highway driving were extracted and post processed in order to obtain lateral displacements between the ego car and other vehicles, see Fig. 3. Only the radar signals were used for the purpose of testing, whereas the camera recordings were used to label the detected objects.

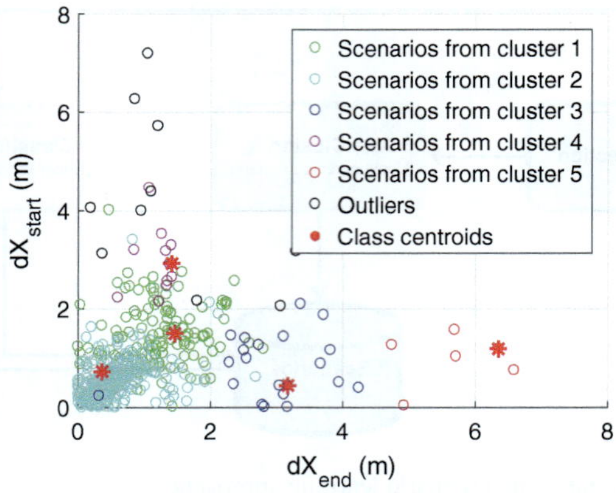


**Fig. 4** The start and end points of the motion trajectories and their centroids (patterns) belonging to different clusters

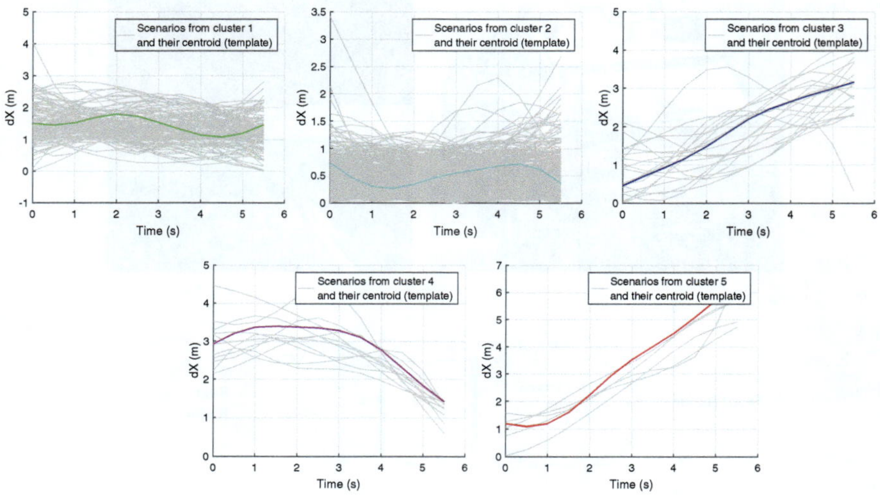


**Fig. 5** The 5 learned clusters, corresponding to 5 subfigures, the time series assigned to them (grey lines) and the centroids for each class (colored lines)

In our experiment we test the approach over the limit case—an empty catalogue at the beginning of testing. We run the learning algorithm on the sensor data replayed on-line.

For illustration purposes, we plot the start $dX_{start}$ and end $dX_{end}$ positions (relative lateral distances of the cars with respect to ego) of the learned maneuvers in Fig. 4 as well as each of the learned clusters together with its cluster centroid (in terms of Dynamic Time Warping (DTW) average) in more details in Fig. 5. Moreover, the algorithm detected some outliers as well, in our set-up the clusters with less than 5 cases.

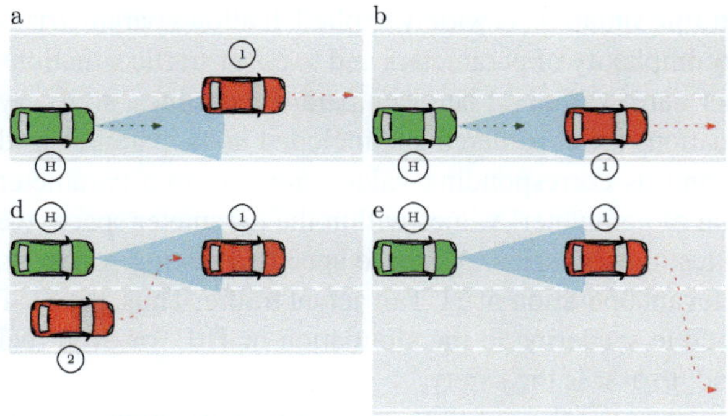


**Fig. 6** Schematic representation of the learned scenarios

The results show that the algorithm was able to learn 5 scenarios. Figure 5 illustrates that they clearly distinguish from each other, even though some misclassification is present, in particular in the first and second cluster. For the presented example, the scenarios are: transition phase between two scenarios corresponding to in-between lane driving, the following case, cut-out maneuver, cut-in maneuver, and a scenario when during the cut-out the preceding vehicle changes 2 lanes fast on a 3-lane highway. The schematic representation of these scenarios corresponding to learned clusters is shown in Fig. 6.

It should be commented that the first cluster contains somewhat unclear cases, including the above mentioned transitions, but also very noisy data as well as cases which are not detectable by the radar. To the latter ones we can include the changes in the highway topology resulting in changing the relative distances between vehicles, e.g. highway narrowing before tunnels.

Depending on the use case, different outcomes of the learning process can be used. For example, for safety testing not the clusters themselves are of interests, since they provide common driving models from data, but the outliers (see Fig. 4) that represent in some sense rare scenarios that are usally safety critical more likely. On the other hand, for evaluating whether the simulation environment is realistic or not, one may analyze the clusters of typical driving behaviors by comparing the learning outcomes from real data against the simulated ones.

## 4 Scenario Parametrization

### *4.1 Numerical Assessment Methods*

The parametrization of the scenario is one of the key steps towards the scenario-based validation. Both experts knowledge based and real measurement based approaches are applicable. Specifically, using Field-Operational-Test (FOT) measurements to determine the suitable mathematical description (model) with corresponding parameter set of a scenarios so that the resulting mathematical model (parametrization of a scenario) represents the real traffic situation, is widely applied. It allows parametrization to effectively reduce the complexity of parameters and to cover traffic situations in the real world with a rather narrow interval between parameter values, so to say only relevant and realistc situations are considered and included in the parametrized scenario. The parameter set and its corresponding value ranges form a parameter space. Thus, through variation of parameter' values within the parameter space, we can obtain various specified testing scenarios. All these specified testing scenarios together describe the most relevant operation of SUT in actual traffic. Thus, the SUT can be tested with these concrete scenarios in the simulation or HiL, or even real driving test on proving ground, to assess the safety.

Figure 7 illustrates a exemplary workflow of FOT-based parametrization and virtuall testing of safety.

Because the parameter space of a parameterized scenario is continuous, is can result again infinite many concrete scenarios by varying the parameter values. It is necessary to apply various sampling methods to generate concrete scenario for testing and evaluation, which result in various safety outcome

Depending on sampling method, various descriptions in terms of safety can be obtained. It can be obtained as the final outcome in terms of the safety, a boundary in the parameter space separating the safe or unsafe state or a statistic indicator like crash rate.

### 4.1.1 Grid Search

As already mentioned, grid search is a traditional way of performing the space exploration, which is simply an exhaustive searching. The parameter space is normally discreted evenly. Each grid point (a set of parameter values) will sampled to generate the concrete scenario to test the safety of SUT. Finally, a map can be obtained that indicates the safety and unsafety region of the SUT in the parameter space for the testing scenario. Grid search is simply to implement but it suffers from the curse of dimensionality. The resulting total samples depends on the numbers of the parameters and discretization accuracy.

The resulting safe and unsafe regions are normally refers to boundary that exactly separate the safe state from unsafe. This boundary represents the safety limit of the SUT. We can evaluate the SUT in certain scenario by comparing its safety limit with human's.

Figure 8 shows the virtual tests results for an ACC function, obtained through a full scale grid searching for a cut-in scenario. The scenario is parametrized through initial longitudinal distance $\Delta y_0 \in [2, 45]$ m between EGO and vehicle 1 (cut-in vehicle), initial velocity difference $\Delta v_0 \in [-6.5, -1.5]$ m/s between them, and the lane change duration $T \in [1, 15]s$ of the vehicle 1. The lane change duration characterized the aggressiveness of the cut-in maneuver, which is modeled through the sigmoid function, see [31, 32].

Three different simulation results, that is, collision and two safe situations, are identified and shown in Fig. 8 panel (2). The red crosses represent the parameter

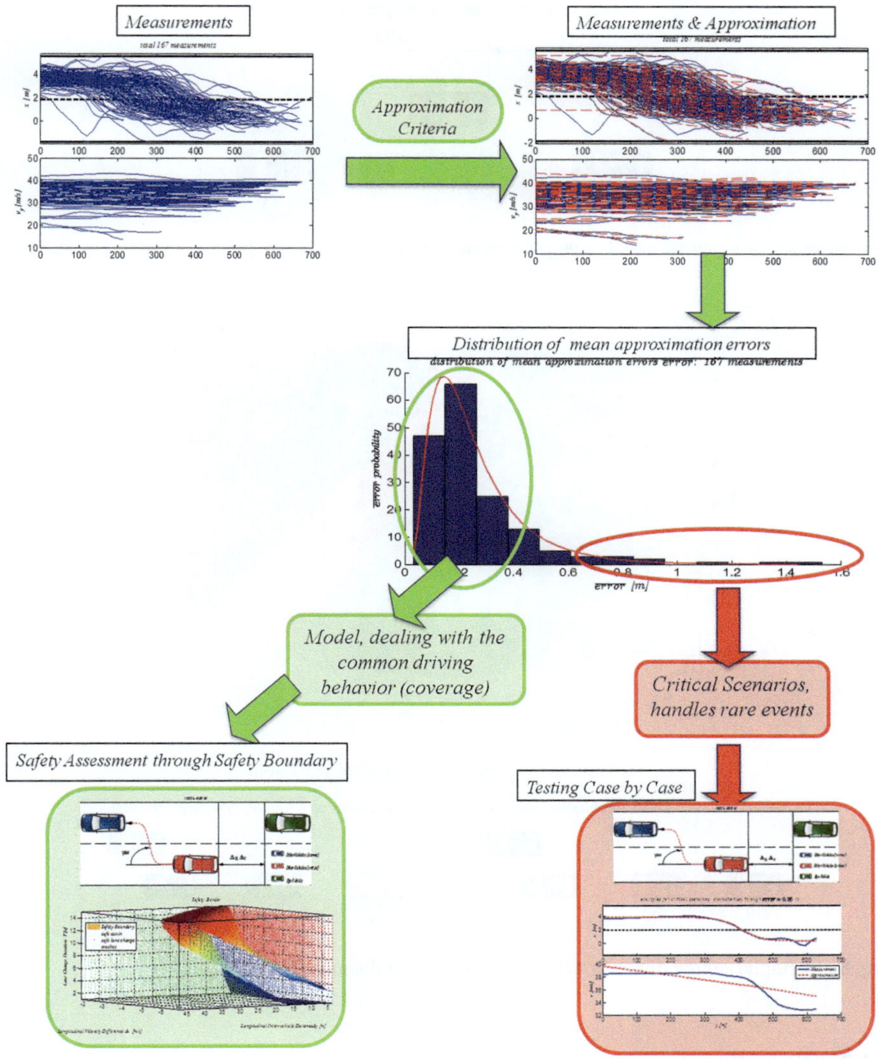


**Fig. 7** Exemplary workflow of FOT measurements-based parametrization of scenario and numerical methods based assessment

combinations that leads to front-end collision or sideswipe between the ego-vehicle and the cut-in vehicle. The green circles represent the parameter combinations that the ACC controller reacts to the cut-in vehicle properly and avoids the crash successfully [(see panel (3) variant (1)]. The blue dots represent the situation that the cut-in vehicle successfully changes the lane without cutting in between the ego-vehicle and the preceding vehicle but behind the ego-vehicle, see panel (3) variant (2). The light blue hull that envelops the safe situations in panel (3) is the collision-free boundary. That indicates the safety performance limit of the HAV in terms of the cut-in situation.

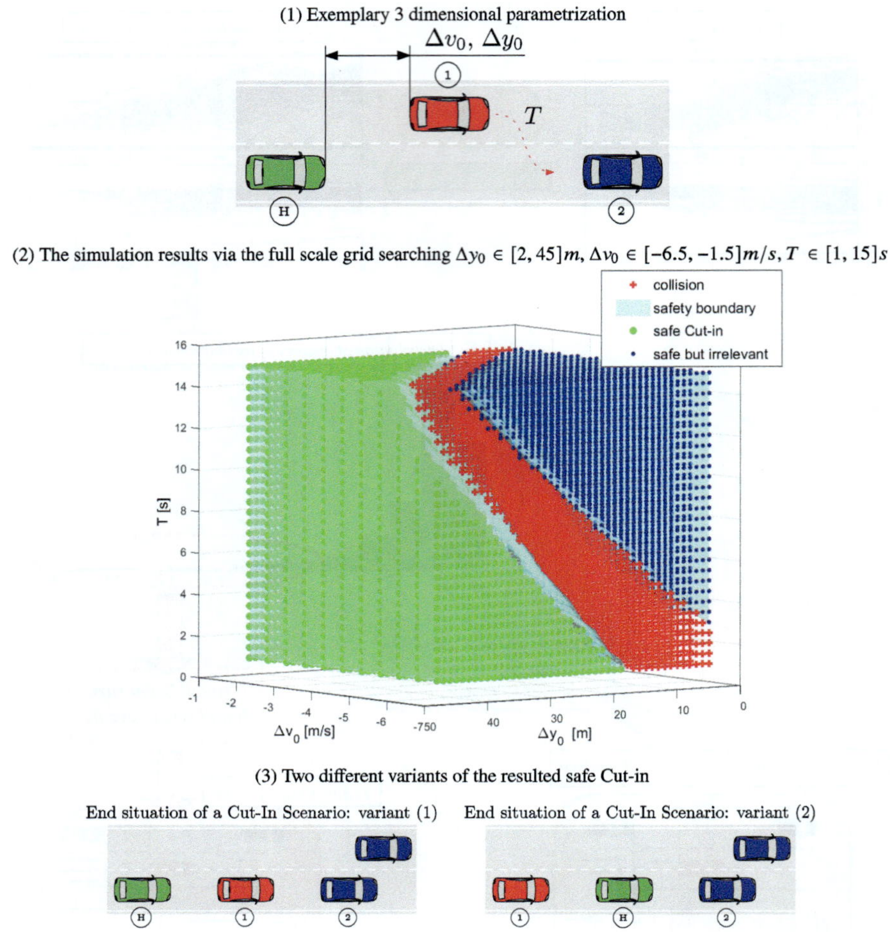


**Fig. 8** Exemplary three dimensional parametrization and simulation results

### 4.1.2 Sampling, Importance Sampling

Random sampling is another widely used method to generate parameter values from a parameter space. It is assumed that the distribution of each parameter is known. By applying Monta-Carlo test, various massive concrete scenarios will be generated and evaluated for the SUT. This results at the end a failure probability, or say crash rate (collision as unsafe indicator). Based on the crash rate, people can evaluate the safety of SUT by comparing the crash rate with the statistic of human drivers.

However, Random sampling may face a problem that in the FOT data (the parameter distribution), the critical situations or unsafe situations of the SUT belong to rare situations, and a reliable evaluation of the probability of crash requires a large number of tests. To reduce the efforts put on testing of safe situations, importance sampling techniques are usually applied to reduce the number of tests and obtain a reliable test result.

The basic idea of applying importance sampling in test amount reduction is to replace the original distribution density function $f(x)$ by a new one $f^*(x)$ to generate critical situation $x$, which leads to a higher probability of occurrence of high-risk events. And then the risk calculation function is modified to obtain the safety benefits of ADAS/ADF/HAVs [33].

### 4.1.3 DOE Based Approaches

Another method widely investigated method for space exploration is Design-of-Experiment DOE based approach or learning based method [34]. DOE methods are developed to speed up the process of boundary searching. The most common approaches are the methods based on convex hull or minimizing the parameters' variance [35, 36]. However, for the parametrization of a scenario, the collision free boundary can be non-convex. A Gaussian Process Classifier (GPC) and a Support Vector Machine (SVM) based DOE strategy, proposed in [37] and [38], respectively, can find and describe the non-convex boundaries iteratively. In [32], the Gaussian Process is extended to accelerate the boundary searching in ADAS/ADF/HAV testing. A criteria is proposed to evaluate the approximation quality of the resulted boundary during the iteration and thus, to stop and exit the iteration.

Figure 9 panel (1) describes the work-flow of the accelerated boundary searching using GPC based DOE strategy, taking an exemplary parametrization of the cut-in scenario with 3 dimensional inputs $\boldsymbol{\theta} = [\Delta v_0, \Delta y_0, T]^T$ and binary outputs $y \in \{+1, -1\}$ as the example, as shown in panel (2). The blue circles are tested safe states whereas the red crosses are tested crash states. The green surface is the resulting safe boundary separating the safe from unsafe states. As we can find that most tested states are cumulated close to the boundary and only limited samples are in the areas that are apart from boundary.

It is worthy mention that the proposed DOE methods normally works for scenario-based testing and validation of ADAS/ADF without stochastic characters. To including stochastic characters further adaption is necessary.

## 5 Representation of the Surrounding Traffic

Safety in its own is a static concept—did a crash occur? In practice, a crash analysis will be done only after a crash. However, such a black/white classification delivers a limited information, and is also difficult to analyze, because most data records do not include crashes, even though they include situations which could have easily led to a crash. Against this background it may be more appropriate to evaluate risk, additionally to crashes, and this requires estimating the future trajectories of the involved vehicles. While for virtual testing the ground truth of all surroundings

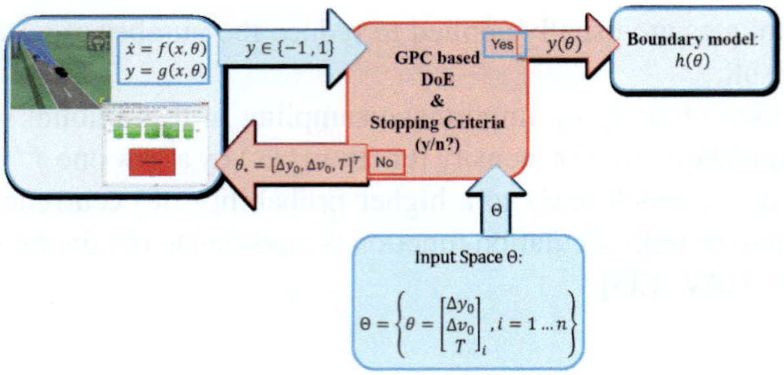


(1) Boundary searching using GPC based Design-of-Experiment

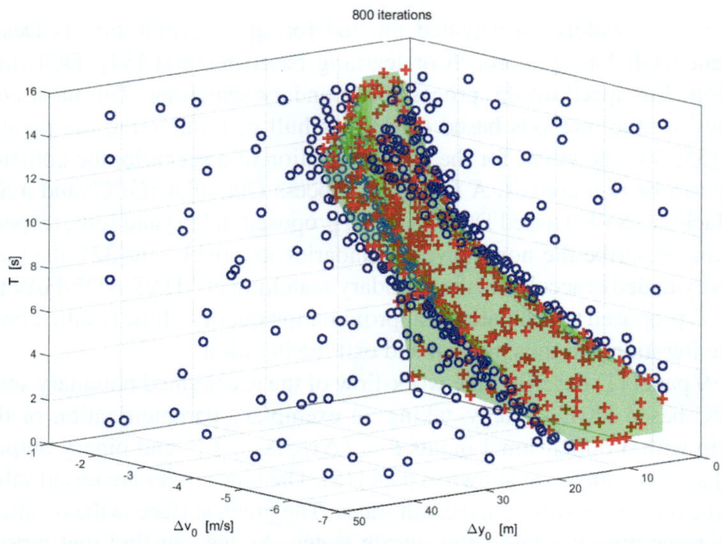


(2) Resulting safety boundary by applying GPC-based DOE for the cut-in scenario described in Fig.8

**Fig. 9** The workflow of a GPC based DOE and testing example

needs to be known, the respective tested system typically has no access to said ground truth. Based on this fact, the topic of prediction and, subsequently, threat assessment (from an ADAS perspective) will be briefly presented in the following subchapters.

## *5.1 Trajectory Prediction*

The prediction of other traffic participants' behavior represents a discipline in its own, as a huge variety of different facets and approaches exist [39–41]. The key challenge is, that there is not only a high inter-group variability between different types of traffic participants (i.e. a car will have a completely different behavior than

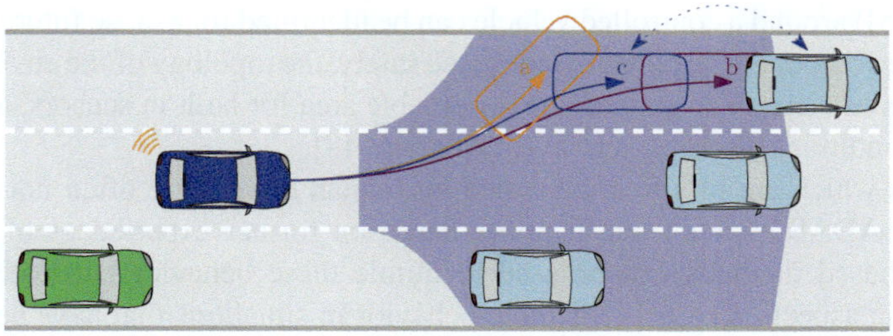


**Fig. 10** Exemplary visualization of different prediction methods. **a** first principles based **b** maneuver based **c** interaction based.eps

a bicycle), but also a notable intra-group variability which is intuitive, as one cannot for example expect drivers to react similar and also in all other groups there is a huge variation. The behavior is again depending highly on the environment, as the same driver will act different when comparing the behavior on highways to the behavior in urban areas, especially due to the fact that the respective environment allows only a limited number of maneuvers [42]. The fewer maneuvers are possible and the more clearly they can be assigned, the easier it is to make a prediction which is also a reason why the first autonomous driving functions are being used primarily on highways.

Prediction methods for cars can be mainly split into approaches that are based on first principles (usually used for short prediction horizons), maneuver based prediction methods (usually used for longer prediction horizons), as well as hybrid approaches combining these methods—as exemplarily shown in Fig. 10 [40].

However, in mixed traffic, a focus on the vehicle to be predicted is not sufficient as the interaction with other traffic participants plays a significant role. For this reason, methods from the field of game theory, such as the Level-k theory [18], or probability-based approaches, such as Markov decision processes [43], are often used. The more realistic the behavior should be modeled, the more challenging and complex the problem gets. The result of a prediction can also be given in different ways, either as a single trajectory, a possible area of occupancy, or a probabilistic estimation where the object might be most likely. Such predictions may also incorporate the response to legal restrictions, such as speed limits or sources of danger in the surrounding area, such as tight curves or roads that merge and also common behavior like not changing lanes without a reason. A physically reachable set based on the vehicle model often defines the outer limits here [44], but is usually of no direct practical significance since the result is typically too conservative for direct use in regular road traffic.

## 5.2 *Threat Assessment*

Based on the prediction of other traffic participants' behavior possible threats (or possibly occupied areas—see [44], which can also be used for tactical decision making [45]) around a controlled vehicle can be identified for a close future. But this is only one component required to guarantee safety, the topology of the environment, restrictions in view, or in general the observable area for built in sensors, as well as weather conditions also play a relevant role [46–48].

Threats which are intuitively handled by human drivers are often not trivial to model for ADAS as this often happens based on former experiences of a driver. Learning based methods can be used to mimic these behavior but lead to other problematic aspects, like unpredictable behavior in situations that have never been experienced before which can be very critical. As previously mentioned, a technically safe handling of a situation can provoke other traffic participants to dangerous maneuvers like cutting in in front of the ego-vehicle if the gap to the preceding vehicle (which is just safe from a physical perspective) appears too big for a human driver who may have incorrectly assessed the situation. This way, a cautious handling of an unsafe situation might lead to an even more critical situation, which is quite common in mixed traffic.

Before autonomous vehicles are on the road in large numbers, it would therefore seem to make sense to tackle those kind of misperceptions—which are often due to people's wrong assessment of situations—in road traffic to increase safety in mixed traffic. The idea is, that the behavior of a human driver is only observed and the system intervenes if it becomes apparent that a situation would lead to an accident without intervention. Thereby, it is necessary to model the environment and potential changes as precise as possible, but also the modeling of the ego-vehicle itself, especially when it comes to physical restrictions, like maximum possible lateral forces or velocity depending acceleration [43, 49].

## 6 Metrics of Risk

For the design of a safety system, the determination of the risk function plays a central role since it influences a variety of performance indicators. As elaborated in [50], on the one hand, large inter-vehicle distances or headways improve safety in terms of avoiding rear end collisions. On the other hand, traffic capacity decreases as shown in [51]. In addition, large inter-vehicle distances may pursue vehicles on the adjacent lane to merge which might decrease the acceptance. Hence the choice of the headway policy is crucial for ADAS development and is closely linked with safety and comfort. To guarantee acceptance of ADAS, the risk function should reflect to some extent the driver's risk perception.

Risk perception by human drivers and the consequences on driving behavior is an extensively studied field [52–54]. It was shown, that the main factors are the gender (there is a difference between men and women in evaluating the risk), age, driving experience (an experienced driver has in general a better risk appreciation compared to a young driver), and cultural aspects (e.g. risk evaluation of Indian drivers can be very different from German ones). For autonomous driving, however, the driver is not responsible anymore, meaning that the risk assessment is shifted to a vehicle. For the

**Table 3** Risk functions for vehicle following scenarios

| Name | Meaning | Formula and nomenclature |
|---|---|---|
| Distance | A minimum inter-vehicle distance | – |
| Time headway (TH) | The time that passes when the controlled vehicle reaches the position of the obstacle on the path | $t_h^{(n)}(t) = \frac{\Delta x^{(n)}(t)}{v(t)}$<br>$\Delta x^{(n)}(t)$ – relative distance<br>$v(t)$—absolute speed of the ego |
| Time to collision (TTC) | The time required for two vehicles to collide if they continue at their present speeds and the same path | $t_c^{(n)}(t) = \frac{\Delta x^{(n)}(t)}{\Delta v^{(n)}(t)}, \quad v(t) > v_x^{(n)}(t)$<br>$\Delta v^{(n)}(t)$—relative velocity |
| Modified TTC (MTTC) | TTC with current accelerations to predict the collision timeif both vehicles continue with the status quo | $t_{mc}^{(n)}(t) = \frac{-\Delta v^{(n)}(t) \pm \sqrt{v^{(n)2} + 2\Delta a \Delta x^{(n)}(t)}}{\Delta a}$<br>$\Delta a$—relative acceleration |
| Deceleration rate to avoid a crash (DRAC) | The minimum required deceleration rate which a vehicle has to apply to avoid a crash with a leading vehicle | $t_d^{(n)}(t) = \frac{\Delta v^{(n)2}(t)}{\Delta x^{(n)}(t) - L}$<br>$L$—length of the front vehicle |
| Modified DRAC (MDRAC) | DRAC and taking into account perception reaction time (PRT) | $t_{md}^{(n)}(t) = \frac{\Delta v^{(n)}(t)}{2(t_c^{(n)} - R)}$<br>$R$ is the PRT |

latter, the examined risk functions are typically defined for longitudinal control in a vehicle following scenario. However, the choice of a suitable risk function that defines the minimum headway is not trivial. For a human driver, the choice of the headway might be related to its risk perception and some authors describe the driving process as a result of risk homeostasis [55], although the definition of a the risk function is not clear. In this context, often the inverse of the headway indicator is considered to represent a drivers risk perception, see e.g. [56].

Some of the most popular (longitudinal) headway policies and risk functions are listed in Table 3 and briefly explained in the latter.

Basically, by imposing a minimum inter-vehicle distance as constraint of an automated driving algorithm, collisions could be avoided in car following situations. However, human drivers, as well as automated driving applications have a reaction time (where no action can be expected) which is greater than zero. This can become dangerous e.g. when it comes to a heavy braking maneuver of a preceding vehicle. For higher velocities, the driven distance during the reaction time could be large enough to lead to a violation of the minimum headway or to a collision. Imposing constraints on distances are most extensively applied for maneuvering with low velocity as for instance trajectory planning for automated parking control [57]. For higher velocities, it seems reasonable to use a risk function that depends on the velocity of the vehicle.

Instead of keeping a minimum distance to a preceding vehicle, many authors propose to track a time headway (TH) [58]. An example for an automated driving algorithm that utilizes a constant TH policy is conventional ACC. Beside the main goal of ACC, namely tracking the driver's desired velocity, many algorithms track a constant TH if a preceding vehicle on the same lane exists and drives slower. Obviously, by constraining the minimum TH to a vehicle, the resulting inter-vehicle distance must increase with the velocity of the ego vehicle. However, the minimum headway could lead to very small distances for low velocities of the ego vehicle. To keep a minimum distance in such situations, a constant term can be added to the TH, which ensures a minimum distance when it comes to stop and go traffic, etc.

Observations of real traffic situation show that human drivers often approach (e.g. dense traffic) very close to the preceding vehicle with TH values significantly lower than those typically found in ACC functions. This indicates that the relative velocity between controlled vehicle and obstacle plays an important role in human risk perception.

A suitable metric that serves as indicator in many collision avoiding systems is the time to collision (TTC). It assumes a constant speed difference over the time and requires that the path of the ego-vehicle is the same as the preceding vehicle. Several studies point out that it is beneficial to directly use TTC for decision making in automated driving applications. TTC is considered to meet a drivers expectations with respect to a collision avoidance system and it is shown that drivers decisions as when to start braking in a critical situation may be well based on TTC. It can be seen that TTC > TH, and TTC and TH are equal in case if the velocity of the obstacle is zero. The difference between the risk as measured by TH and TTC can be interpreted as follows. A constraint on TTC allows in general lower inter-vehicle distances for low relative speed compared to a constraint on the minimum TH. Whereas TH captures rather the potential danger of a situation, TTC reflects the imminent danger. In fact, low TH may not require any action of the system but low TTC signify that a reaction is required to avoid an accident.

As one could see, TH and TTC have their drawbacks. On the one hand, the TH is suitable to decrease the potential risk of a collision but it can be observed that many drivers allow much lower TH in some situations. In many cases, tracking of TH is not meaningful, e.g. when a vehicle crosses the lane and is driving faster or traffic is dense. On the other hand, TTC depends on the relative speed of a vehicle and can lead to very small headways to the preceding vehicle in steady state conditions.

In order to improve the risk assessment, some modifications and/or different indexes can be found in the literature. In the case of modified TTC (MTTC), the classical TTC is modified to include the current acceleration and thus predict the time of collision if both vehicles would continue with the status quo [59].

By using Deceleration Rate to Avoida Crash (DRAC), one defines the minimum required deceleration rate which a vehicle has to apply to avoid a crash with the leading vehicle [60], whereas modified DRAC (DRAC) considers an important factor in the analysis of safety—the minimum time required for the drivers to react, also called Perception Reaction Time (PRT).

# 7 Outlook

## *7.1 Induced Effects on Other Vehicles*

By considering the safety of the ego vehicle itself, it is usually assumed that the other vehicles do not change their behavior as a consequence of the actions of the ADAS. There are very good reasons for that, but in practice the reactions of other drivers usually contribute to reduce the probability of an accident. A good example of such contribution is a highway merging scenario, especially in a dense traffic. Very often a vehicle driving on the main lane assists the merging vehicle in performing it's maneuver, either by reducing the own speed or by changing the lane to left, if enough space is available. Therefore, the real safety tends to be even higher compared to that estimated under the assumption that the surrounding vehicles do not change their behavior.

On the other side, let us imagine the situation when an autonomous vehicle makes a lane change in front of another human-driven vehicle. If the velocity of the ego car is larger than the second vehicle's, no rear-end collision is possible between these two cars however close the merging has been. However, the human driver may be scared by a possibly very small distance between the vehicles and can have a completely different reaction when facing the safety critical situation. For instance, is driver may perform a hard braking leading to the rear-end collision not with the ego car, but with the car behind. As a result, the formal safety of the ADAS was fulfilled, but the traffic accident still happened between two human driven vehicles as a result of the ego's action.

The importance of the latter is to some extent included in many national regulation documents. For example, in German traffic rules [(Straßenverkehrsordnung (StVO)] it is stated: *When changing lane to the left lane during overtaking, no following road user shall be endangered.* That means: whenever a vehicle changes to the left lane to overtake another road user, the driver has to ensure that those on the left lane will not be endangered. In particular, if a vehicle that approaching on the left lane could be endangered in any way, overtaking is prohibited.

This can be interpreted in a more generic way by the so-called **defensive driving.** The "Safe Practices for Motor Vehicle Operations" defines the defensive driving as *driving to save lives, time, and money, in spite of the conditions around you and the actions of others*. Its aim is to reduce the risk of collision by anticipating dangerous situations, despite adverse conditions or the mistakes of others.

Accordingly, the defensive paradigm means the development of an ADAS that take in account the possible behavior so to minimize the risk that *any other vehicle is forced to slow down, speed up, or change lanes to avoid collision.*

## 7.2 Shields and Emergency Systems

The risk methods mentioned above can be used also to improve safety. In this context, there are two developments which share many common aspects and are already partly enforced, and partly under study.

The first approach are emergency which systems already exist (in different levels of complexity), and are in some cases also required by law. For example, new trucks and buses in the European Union have had to be equipped with Autonomous Emergency Braking (AEB) assistants since 2015. The relevant regulations are summarised in UN/ECE R131. Taking a look into the future, it can be assumed that this type of emergency assistance system will be standard for newly developed cars in a few years' time. To support this kind of development, AEB/AES-systems are getting more and more into focus in the European New Car Assessment Programme (NCAP) rating of the upcoming years, as listed in [61].

The second aspect, safety-shields, are concerned with an additional, independent layer which checks the plausibility of any decision by the ADAS and adapts the system inputs if a critical state would be reached, as shown e.g. in [62]. This concept is utilized both in the field of controller synthesis [63, 64] with automotive applications [62, 65, 66] as well as in the field of motion planning [67–69], including reachability analysis of surrounding traffic participants. The key idea is shown in Figs. 11 and 12. Of course, the approach can be extended to include more elements, e.g. road borders. To this end, the environment must be detected, the options available must be evaluated, and if the case arises that without intervention each option would lead

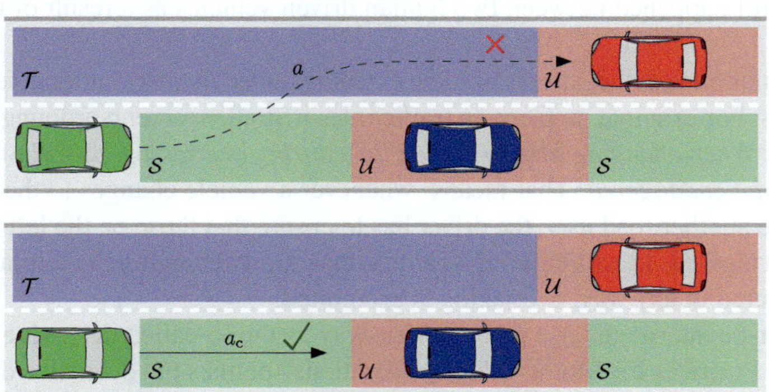


**Fig. 11** Exemplary visualization of safe ($\mathcal{S}$), temporary safe ($\mathcal{T}$) and unsafe ($\mathcal{U}$) sets of states including an unsafe action $a$ and a safe action $a_c$. For simplicity only temporal sets of states are shown

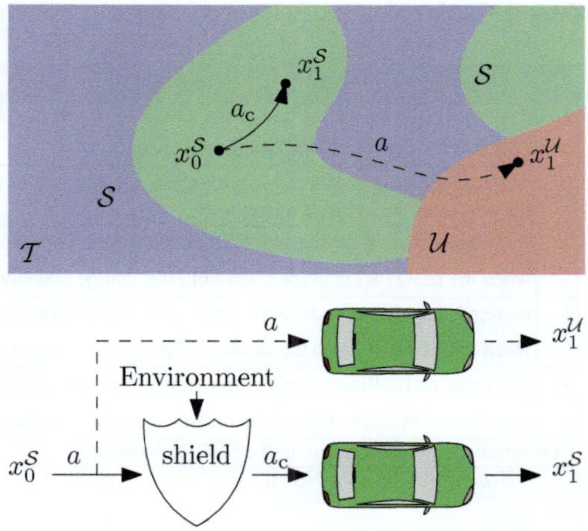


**Fig. 12** Exemplary concept of a safety-shield. Starting from a safe state $x_0^{\mathcal{S}}$ the shield utilizes environmental knowledge to prevent and adapt unsafe actions $a$ which would result in an unsafe state $x_1^{\mathcal{U}}$ into safe actions $a_c$ resulting in a safe state $x_1^{\mathcal{S}}$

to an accident, the system must intervene to bring the vehicle into a safe state. The challenges thereby are manifold, especially in mixed traffic, and simulation-based validation methods are urgently needed before such systems should be used in road traffic.

# Appendix

See Tables 4, 5, 6, 7, 8.

**Table 4** Basic scenarios: EGO in the course of lane change maneuver

| No. | Scenario Description | Occ.[†] |
|---|---|---|
| 5 | front-end collision threat due to the action of the front vehicle on the adjacent lane when the EGO (H) is in the course of lane change maneuver | 32 |
| 6 | rear-end collision or sideswipe with a rear vehicle (1) on the adjacent lane when the EGO (H) is in the course of lane change maneuver | 20 |
| 7 | front-end collision or sideswipe threat due to the lane change maneuver of a preceding vehicle (1) when the EGO (H) is in the course of lane change maneuver | 8 |
| 8 | rear-end collision or sideswipe threat due to the lane change maneuver of a following vehicle (1) when the EGO (H) is in the course of lane change maneuver | 1 |
| 9 | front/rear-end collision or sideswipe threat from lane change maneuver of the vehicle (1) on the most outer lane when the EGO (H) is in the course of lane change maneuver | 18 |

[†] Occurrence of each scenario in the SHRP2 database.

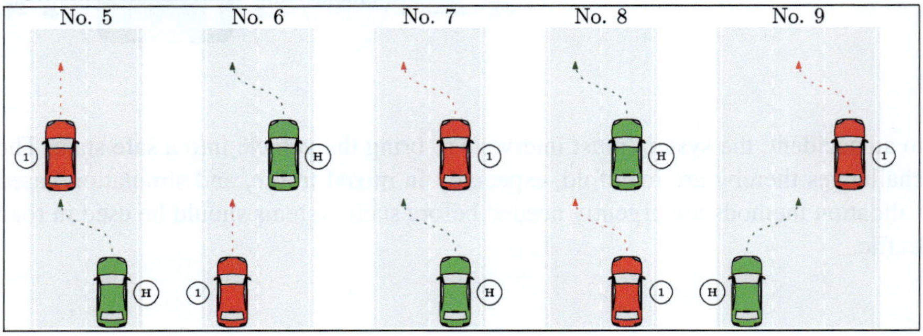


**Table 5** Basic scenarios: EGO in the car-following situation

| No. | Scenario Description | Occ.[†] |
|---|---|---|
| 10 | Rear-end collision or sideswipe threat due to the lane change maneuver (cut-in) of a front vehicle (2) on the adjacent lane when the EGO (H) is following a preceding vehicle (1) | 63 |
| 11 | front-end collision threat due to the action of the preceding vehicle (2), resulted from the lane change maneuver (pull-out) of the preceding vehicle (1) that reveals the preceding vehicle (2) when the EGO (H) is following the preceding vehicle (1) | 5 |

[†] Occurrence of each scenario in the SHRP2 database.

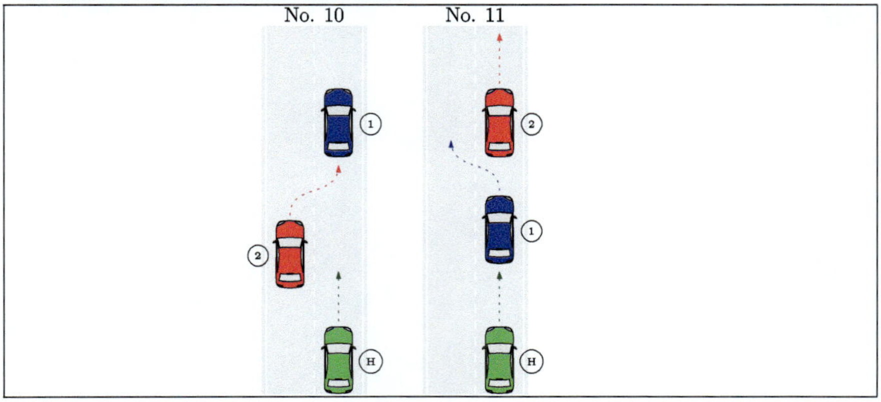

**Table 6** Complex test scenarios: via braking evasive action

| No. | Scenario Description | Occ.[†] |
|---|---|---|
| 12 | (Secondary) rear-end collision threat from a following vehicle (2), resulted from the braking evasive action of the EGO (H) to avoid the (Primary) front-end collision threat from a preceding vehicle (1) | 20 |
| 13 | (Secondary) rear-end collision threat from the lane change maneuver of the rear vehicle (2) on the adjacent lane, resulted from the braking evasive action of the EGO (H) to avoid the (Primary) front-end collision threat from a preceding vehicle (1) | 1 |
| 14 | (Secondary) rear-end collision threat from a following vehicle (2), resulting from the braking evasive action of the EGO (H) to avoid the (Primary) front-end collision or sideswipe threat due to the lane change maneuver of a front vehicle (1) on the adjacent lane. | 1 |

[†] Occurrence in SHRP2 database

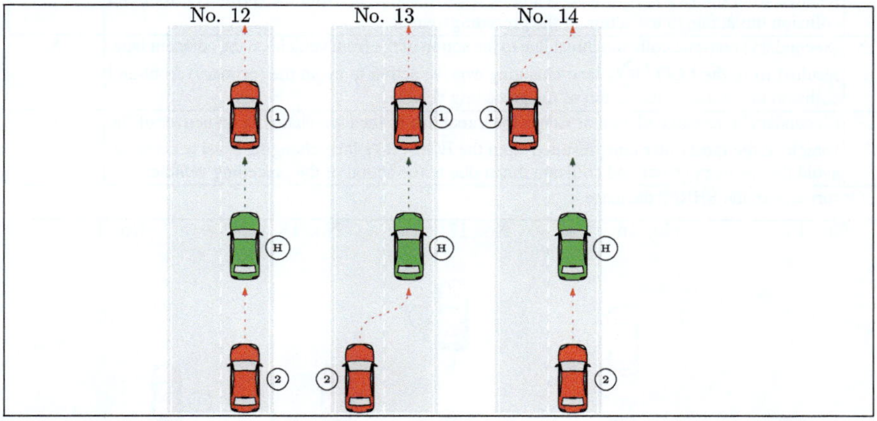

**Table 7** Complex test scenarios: via lane-changing evasive action

| No. | Description | Occ.[†] |
|---|---|---|
| 15 | (Secondary) rear-end collision or sideswipe threat due to the lane-changing maneuver of the preceding vehicle when EGO (H) is in the course of lane-changing maneuver, resulted from the EGO (H)'s lane-changing evasive action to avoid the (primary) front-end collision threat due to the action of the preceding vehicle | 4 |
| 16 | (Secondary) rear-end collision or sideswipe threat due to the lane-changing maneuver of the preceding vehicle (2) when EGO (H) is in the course of lane-changing maneuver, resulted from the EGO (H)'s lane-changing evasive action to avoid the (primary) front-end collision threat due to the action of the preceding vehicle (1) | 2 |
| 17 | (Secondary) rear-end collision or sideswipe threat due to a rear vehicle on the adjacent lane, resulted from the EGO (H)'s lane-changing evasive action to avoid the (primary) front-end collision threat due to the action of the preceding vehicle | 11 |
| 18 | (Secondary) rear-end collision threat due to the action of the front vehicle on the adjacent lane, resulted from the EGO (H)'s lane-changing evasive action to avoid the (primary) front-end collision threat due to the action of the preceding vehicle | 5 |
| 19 | (Secondary) rear-end collision or sideswipe threat due to the lane-changing maneuver of the vehicle in the most outer lane, resulted from the EGO (H)'s lane-changing evasive action to avoid the (primary) front-end collision threat due to the action of the preceding vehicle. | 7 |

[†] Occurrence in the SHRP2 database

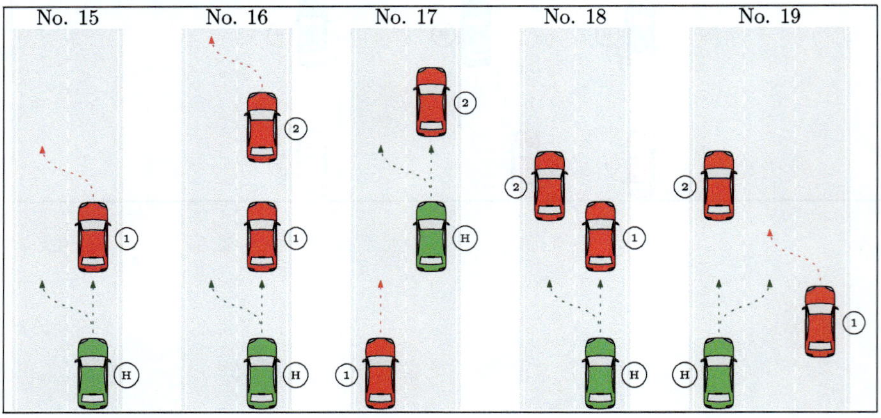

**Table 8** Complex test scenarios: via lane-changing evasive action

| No. | Scenario Description | Occ.[†] |
|---|---|---|
| 20 | (Secondary) rear-end collision threat due to the action of the front vehicle ② on the adjacent lane when the EGO Ⓗ is in the course of lane-changing maneuver, resulted from the EGO Ⓗ's lane-changing evasive action to avoid the (primary) front-end collision or sideswipe threat due to the lane-changing maneuver of another front vehicle ① on another adjacent lane | 1 |
| 21 | (Secondary) rear-end collision or sideswipe threat due to a rear vehicle ② on the adjacent lane, resulted from the EGO Ⓗ's lane-changing evasive action to avoid the (Primary) front-end collision or sideswipe threat due to the lane-changing maneuver of a front vehicle ① on another adjacent lane | 3. |
| 22 | (Tertiary) rear-end collision threat due to the action of the vehicle ② on the most outer lane when the EGO Ⓗ is in the course of lane-changing maneuver, resulted from the EGO Ⓗ's second lane-changing evasive action to avoid (Secondary) front-end collision or sideswipe threat due to the lane-changing maneuver of the preceding vehicle ① when the EGO Ⓗ is in the course of lane-changing maneuver, resulted from the EGO Ⓗ's first lane-changing evasive action to avoid the (Primary) front-end collision threat due to the action of the preceding vehicle ① | 2 |

[†] Occurrence in the SHRP2 database

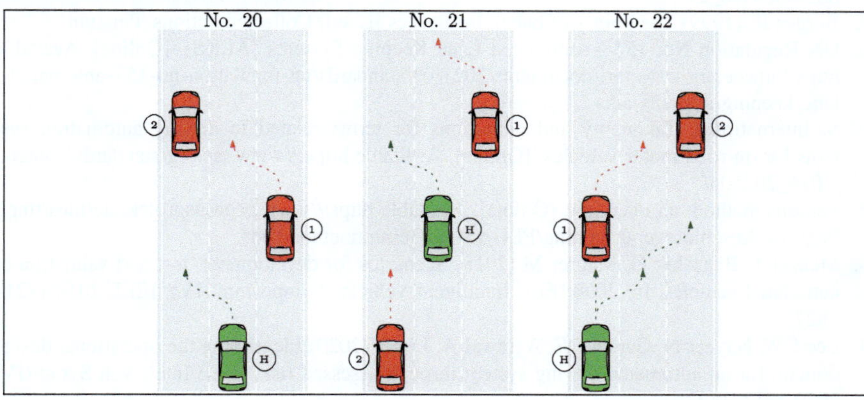

## References

1. Travel monitoring and traffic volume (2014). [Online]. Available https://www.fhwa.dot.gov/policyinformation/travel_monitoring/14augtvt/page2.cfm
2. Igliński H, Babiak M (2017) Analysis of the potential of autonomous vehicles in reducing the emissions of greenhouse gases in road transport. In: 12th international scientific conference of young scientists on sustainable, modern and safe transport, Procedia Engineering, vol 192, pp 353 – 358. [Online]. Available http://www.sciencedirect.com/science/article/pii/S1877705817326073
3. Friedrich B (2016) The effect of autonomous vehicles on traffic. Springer Berlin Heidelberg, Berlin, Heidelberg, pp 317–334. [Online]. Available https://doi.org/10.1007/978-3-662-48847-8_16
4. Greenberg A (2015) Hackers remotely kill a jeep on the highway-with me in it (2015). https://www.wired.com/2015/07/hackers-remotely-kill-jeep-highway
5. Singh S (2018) Critical reasons for crashes investigated in the national motor vehicle crash causation survey
6. Taxonomy and Definitions for Terms Related to Driving Automation Systems for On-Road Motor Vehicles (2018, June). [Online]. Available https://doi.org/10.4271/J3016_201806
7. National Highway Traffic Safety Administration, U.S. Department of Transportation (2006). [Online]. Available https://static.nhtsa.gov/odi/inv/2016/INCLA-PE16007-7876.PDF
8. Wachenfeld W, Winner H (2016) The release of autonomous vehicles. Berlin, Heidelberg: Springer Berlin Heidelberg, pp 425–449. [Online]. Available: https://doi.org/10.1007/978-3-662-48847-8_21
9. Borges JL (1999) The library of babel. In: Borges JL (ed) Collected fictions. Penguin
10. UN Regulation No. 157—Automated Lane Keeping Systems (ALKS). [Online]. Available https://unece.org/transport/documents/2021/03/standards/un-regulation-no-157-automated-lane-keeping-systems-alks
11. S. International, Taxonomy and definitions for terms related to driving automation systems for on-road motor vehicles. [Online]. Available https://www.sae.org/standards/content/j3016_202104/
12. Pegasus method: an overview. [Online]. Available https://www.pegasusprojekt.de/files/tmpl/Pegasus-Abschlussveranstaltung/PEGASUS-Gesamtmethode.pdf
13. Menzel T, Bagschik G, Maurer M (2018) Scenarios for development, test and validation of automated vehicles. In: 2008 IEEE Intelligent Vehicles Symposium (IV). IEEE 2018:1821–1827
14. Lee CW, Nayeer N, Garcia DE, Agrawal A, Liu B (2020) Identifying the operational design domain for an automated driving system through assessed risk. IEEE Intell Veh Symp (IV) 2020:1317–1322
15. Czarnecki K (2018) Operational design domain for automated driving systems
16. Zhao D, Lam H, Peng H, Bao S, LeBlanc DJ, Nobukawa K, Pan CS (2016)Accelerated evaluation of automated vehicles based on importance sampling techniques. CoRR. abs/1605.04965
17. Gietelink O, De Schutter B, Verhaegen M (2005) Probabilistic approach for validation of advanced driver assistance systems. Transp Res Rec J Transp Res Board 1910:20–28
18. Li N, Oyler DW, Zhang M, Yildiz Y, Kolmanovsky I, Girard AR (2017) Game theoretic modeling of driver and vehicle interactions for verification and validation of autonomous vehicle control systems. IEEE Trans Control Syst Technol
19. Schaedler O, Mueller S, Gruendl M (2013) Entwicklung eines verfahrens und der dazugehoerigen hard-und software zur bewertung und verbesserung der controllability von fahrerassistenzsystemen (confas). VDI-Berichte, no. 2205
20. Schuldt F, Saust F, Lichte B, Maurer M, Scholz S (2013) Effiziente systematische testgenerierung für fahrerassistenzsysteme in virtuellen umgebungen, AAET2013-Automatisierungssysteme. Assistenzsysteme und eingebettete Systeme für Transportmittel, Braunschweig

21. Domsch C, Negele H (2008) Einsatz von referenzfahrsituation bei der entwicklung von fahrerassistenzsystemen. In: 3. Tagung Aktive Sicherheit durch Fahrerassistenz
22. Huang L, Xia Q, Xie F, Xiu HL, Shu H (2018) Study on the test scenarios of level 2 automated vehicles. In: 2018 IEEE intelligent vehicles symposium (IV). IEEE 49–54
23. Huy Do Q, Tehrani H, Egawa M, Muto K, Yoneda K, Mita S (2015) Distance constraint model for automated lane change to merge or exit. In: FAST-zero'15: 3rd international symposium on future active safety technology toward zero traffic accidents
24. Bagschik G, Menzel T, Maurer M (2018) Ontology based scene creation for the development of automated vehicles. In 2018 IEEE intelligent vehicles symposium (IV). IEEE, pp 813–1820
25. Singh S (2015) Critical reasons for crashes investigated in the national motor vehicle crash causation survey. Technical Report
26. Hankey JM, Perez MA, McClafferty JA (2016) Description of the SHRP 2 naturalistic database and the crash, near-crash, and baseline data sets. Technical Report, Virginia Tech Transportation Institute
27. Menzel T, Bagschik G, Isensee L, Schomburg A, Maurer M (2019) From functional to logical scenarios: Detailing a keyword-based scenario description for execution in a simulation environment. In: 2019 IEEE intelligent vehicles symposium (IV). IEEE, pp 2383–2390
28. Pütz A, Zlocki A, Bock J, Eckstein L (2017) System validation of highly automated vehicles with a database of relevant traffic scenarios. Situations 1:E5
29. Kruber F, Wurst J, Morales ES, Chakraborty S, Botsch M (2019) Unsupervised and supervised learning with the random forest algorithm for traffic scenario clustering and classification. In: IEEE intelligent vehicles symposium (IV) 2019, pp 2463–2470
30. Tkachenko P, Zhou J, del Re L (2019) Unsupervised clustering of highway motion patterns. In IEEE intelligent transportation systems conference (ITSC) 2019, pp 2337–2342
31. Zhou J, del Re L (2017) Identification of critical cases of ADAS safety by FOT based parameterization of a catalogue. In: Control conference (ASCC) (2017) 11th Asian. IEEE, pp 453–458
32. Zhou J, del Re L (2018) Safety verification of adas by collision-free boundary searching of a parameterized catalog. In: American control conference (ACC) 2018. IEEE
33. Xu Y, Zou Y, Sun J (2018) Accelerated testing for automated vehicles safety evaluation in cut-in scenarios based on importance sampling, genetic algorithm and simulation applications. J Intell Connected Veh
34. Beglerovic H, Ravi A, Wikström N, Koegeler H-M, Leitner A, Holzinger J (2017) Model-based safety validation of the automated driving functio highway pilot. In: 8th international Munich Chassis symposium 2017. Springer, Berlin, pp 309–329
35. Renninger P, Aleksandrov M (2005) "Rapid hull determination: a new method to determine the design space for model based approaches. Design of Experiments (DoE) in Engine Development II. Expert-Verlag, Renningen
36. Pukelsheim F (2006) Optimal design of experiments, classics in appl
37. Oyama H, Yamakita M, Sata K, Ohata A (2016) Identification of static boundary model based on gaussian process classification. IFAC-PapersOnLine 49(11):787–792
38. Kampmann G, Kieft N, Nelles O (2012) Support vector machines for design space exploration. In: Proceedings of the world congress on engineering and computer science, vol 2, pp 1116–1121
39. Houenou A, Bonnifait P, Cherfaoui V, Yao W (2013) Vehicle trajectory prediction based on motion model and maneuver recognition. In: IEEE/RSJ international conference on intelligent robots and systems. IEEE, vol 2013, pp 4363–4369
40. Lefèvre S, Vasquez D, Laugier C (2014) A survey on motion prediction and risk assessment for intelligent vehicles. ROBOMECH J 1(1):1–14
41. Deo N, Trivedi MM (2018) Convolutional social pooling for vehicle trajectory prediction. In: Proceedings of the IEEE conference on computer vision and pattern recognition workshops, pp 1468–1476
42. Reisinger S, Adelberger D, del Re L (2021) A two-layer switching based trajectory prediction method. Eur J Control

43. Althoff M, Mergel A (2011) Comparison of Markov chain abstraction and monte Carlo simulation for the safety assessment of autonomous cars. IEEE Trans Intell Transp Syst 12(4):1237–1247
44. Söntges S, Althoff M (2018) Computing the drivable area of autonomous road vehicles in dynamic road scenes. IEEE Trans Intell Transp Syst 19(6):1855–1866
45. Manzinger S, Althoff M (2018) Tactical decision making for cooperative vehicles using reachable sets. In: Proceedings of the 21st IEEE international conference on intelligent transportation systems
46. Karaduman O, Eren H, Kurum H, Celenk M (2016) Road-geometry-based risk estimation model for horizontal curves. IEEE Trans Intell Transp Syst 17(6):1617–1627
47. Damerow F, Puphal T, Li Y, Eggert J (2017, June) Risk-based driver assistance for approaching intersections of limited visibility. In: 2017 IEEE international conference on vehicular electronics and safety (ICVES), pp 178–184
48. Adelberger D, del Re L (2021) Robust handling of diffuse hazard in country road traffic. In: American control conference (ACC). IEEE, pp 418–423
49. Bichiou Y, Rakha HA (2019) Developing an optimal intersection control system for automated connected vehicles. IEEE Trans Intell Transp Syst 20(5):1908–1916
50. Darbha S, Rajagopal K (1999) Intelligent cruise control systems and traffic flow stability. Transp Res Part C Emerg Technol 7(6):329–352
51. Barcelo J et al (2010) Fundamentals of traffic simulation, vol 145. Springer, Berlin
52. Deery HA (2000) Hazard and risk perception among young novice drivers. J Saf Res 30(4):225–236
53. Siren A, Kjær MR (2011) How is the older road users' perception of risk constructed? Transp Res Part F Traffic Psychol Behav 14(3):222–228
54. Crundall D, Chapman P, Trawley S, Collins L, van Loon E, Andrews B, Underwood G (2012) Some hazards are more attractive than others: drivers of varying experience respond differently to different types of hazard. Accid Anal Prev 45:600–609
55. Wilde GJ (1982) The theory of risk homeostasis: implications for safety and health. Risk Anal 2(4):209–225
56. Svensson Å (1998) A method for analysing the traffic process in a safety perspective. Ph.D. dissertation, Lund University
57. Schildbach G, Borrelli F (2016) A dynamic programming approach for nonholonomic vehicle maneuvering in tight environments. In: IEEE intelligent vehicles symposium (IV). IEEE, vol 2016, pp 151–156
58. Swaroop D, Rajagopal KR (2001) A review of constant time headway policy for automatic vehicle following. In: Proceedings of IEEE intelligent transportation systems, pp 65–69
59. Charly A, Mathew TV (2017) Estimation of modified time to collision as surrogate for mid-block crashes under mixed traffic conditions
60. Gettman D, Head L (2003) Surrogate safety measures from traffic simulation models. Transp Res Rec 1840(1):104–115
61. NCAP E (2018) Euro NCAP Rating Review 2018. Technical Report, Euro NCAP
62. Wu M, Wang J, Deshmukh J, Wang C (2019) Shield synthesis for real: enforcing safety in cyber-physical systems. In: Formal Methods in Computer Aided Design (FMCAD). IEEE 2019:129–137
63. Bloem R, Chatterjee K, Greimel K, Henzinger TA, Hofferek G, Jobstmann B, Könighofer B, Könighofer R (2014) Synthesizing robust systems. Acta Informatica 51(3–4):193–220
64. Bloem R, Könighofer B, Könighofer R, Wang C (2015) Shield synthesis. In: International conference on tools and algorithms for the construction and analysis of systems. Springer, Berlin, pp 533–548
65. Raman V, Donzé A, Sadigh D, Murray RM, Seshia SA (2015) Reactive synthesis from signal temporal logic specifications. In: Proceedings of the 18th international conference on hybrid systems: computation and control, pp 239–248
66. Nilsson P, Hussien O, Balkan A, Chen Y, Ames AD, Grizzle JW, Ozay N, Peng H, Tabuada P (2016) Correct-by-construction adaptive cruise control: two approaches. IEEE Trans Control Syst Technol 24(4):1294–1307

67. Pek C, Koschi M, Althoff M (2019) An online verification framework for motion planning of self-driving vehicles with safety guarantees. In: AAET—Automatisiertes und vernetztes Fahren, pp 260–274
68. Pek C, Althoff M (2019) Ensuring motion safety of autonomous vehicles through online fail-safe verification. In: Robotics: science and systems—Pioneers workshop
69. Gruber F, Althoff M (2018) Anytime safety verification of autonomous vehicles. In: IEEE conference on intelligent transportation systems, pp 1708–1714

# Factors Influencing Driver Behavior and Advances in Monitoring Methods

**Shahzeb Ansari, Haiping Du, Fazel Naghdy, and David Stirling**

**Abstract** Monitoring driver behavior in real-time is a challenging task as there are several factors that can influence the driver to commit unpredictable mistakes while driving. These factors mainly involve inattentive driver state, absent mind, unreliable cornering, and speeding, resulting in fatal accidents. This chapter identifies the factors that affect driver behavior and performance, and provides an in-depth analysis of various deployed scientific monitoring methods and proposes solutions for early and efficient real-time monitoring of driver behavior. The chapter also reviews real-time smart detection algorithms deployed for the classification of driver state. In addition, the chapter proposes an unsupervised deep learning neural network model that can be deployed in classifying driver states and actions.

**Keywords** Human factors · Environmental factors · Driver behavior · Intelligent monitoring methods · Driver states classification · Unsupervised deep learning

## 1 Introduction

Driving mistakes can result in injuries and fatalities. According to Transport for New South Wales (NSW), Australia [1], over the period 2015–2020, there were 1688 traffic accidents, resulting in more than 1100 fatalities, including both drivers and passengers. Figure 1 shows accident statistics and the casualty rate over this five year period. While there are various factors involved in such accidents, speeding, fatigue, and alcohol are considered as the most serious causes of catastrophic situations [1], as illustrated in Fig. 2.

Various categories of driving styles have been proposed in the literature [2–6]. Generally, driving behavior can be categorized as: (1) Aggressive, (2) Moderate, and (3) Defensive. An aggressive driving style is characterized by thrill seeking, risky and adventurous driving. Moderate driving behavior is the optimum driving style,

---

S. Ansari (✉) · H. Du · F. Naghdy · D. Stirling
School of Electrical, Computer and Telecommunications Engineering, University of Wollongong, Wollongong, NSW, Australia
e-mail: sa345@uowmail.edu.au

 
Y. L. Murphey et al. (eds.), *AI-enabled Technologies for Autonomous and Connected Vehicles*, Lecture Notes in Intelligent Transportation and Infrastructure,
https://doi.org/10.1007/978-3-031-06780-8_14

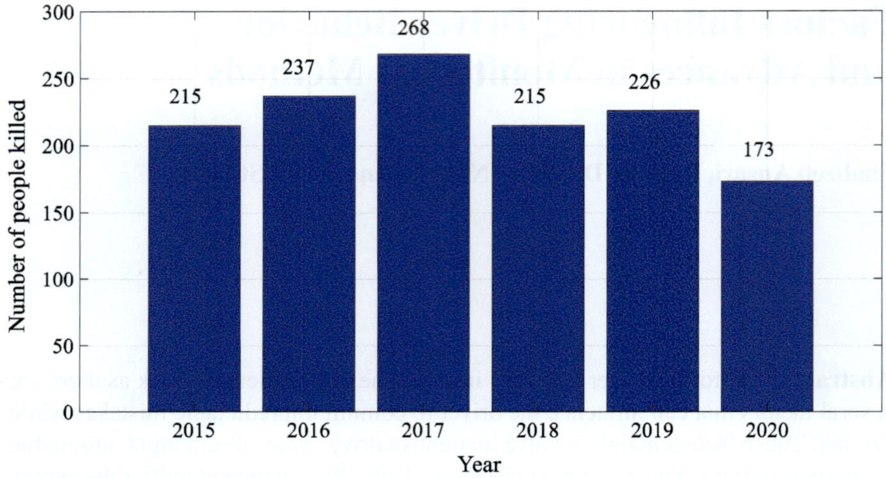


**Fig. 1** Year-wise fatality rate in NSW, Australia [1]

**Fig. 2** Fatality rate of three dangerous human factors [1]

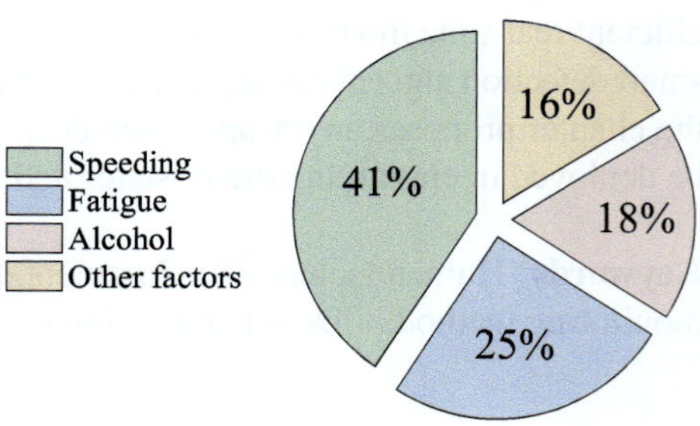


where drivers safely maneuver the vehicle and cope with difficult road situations. In this state, the driver is patient and avoids causing any trouble for other road users, as well as prioritizing safe and vigilant driving. In the defensive driving style, the drivers remains overly-patient and operates the vehicle when they feel confident and assume safe road environment. It can be considered as slow driving style, where drivers apply frequent braking and perform under-steering. Hence, this can cause irritation in surrounding vehicle drivers, and can slow the flow of traffic. Furthermore, this can also induce over-load mental fatigue in drivers [7].

Driving style can vary over time. For example, a young driver may exhibit an aggressive driving style at the start of the trip on an empty street by over accelerating the vehicle and engaging in unstable vehicle cornering. After some time though, the driver then adopts a defensive style on a crowded road by managing the vehicle speed within required speed limits, as well as leveraging the vehicle maneuverability within a safe handling envelope (to maintain a stable vehicle sideslip angle and yaw rate) [8].

Driver's performance is affected by several factors, among which fatigue, drowsiness, aggression, and misjudgment are the most common factors [2]. For instance, a driver experiencing fatigue and drowsiness can lose concentration and steer the vehicle in a wrong direction [9–11]. Similarly, younger drivers are more prone to risky and aggressive driving and can end up in fatal crashes [12]. According to statistics provided by NSW Transport [1], 168 young drivers between the ages 17 and 25 were involved in various crashes because of speeding. Figure 3 provides statistics about the type of crashes reported over the period of 2015–2020 and the factors involved.

Numerous methods have been proposed in the literature to understand driver behavior. In most approaches, camera-based methods are deployed to track the driver's facial and body features in real-time [10]. Such methods are called *direct driver measurement*. Alternatively, in *indirect driver measurement* [13], the driver's state is measured indirectly through vehicle dynamics by detecting motifs or anomalies in the time-series data.

In this chapter, factors that influence driver behavior are discussed using both direct and indirect methods. An in-depth analysis of different methods proposed to effectively monitor the driver in real-time is conducted. Furthermore, the chapter discusses the implementation of various machine learning algorithms in developing the intelligent driver models. Finally, the chapter proposes an unsupervised deep learning classifier known as *deep learning auto-encoder* that can be deployed to predict the different driver states.

The remainder of the chapter is structured as follows. Section 2 discusses the main factors affecting driving. Section 3 provides an in-depth analysis of driver monitoring methods that can be deployed to detect the factors in real-time. Section 4 provides an overview of several algorithms proposed to monitor the states of both the driver and vehicle. Section 5 presents an analysis of deep learning classifiers. Finally, some conclusions are drawn in Sect. 6.

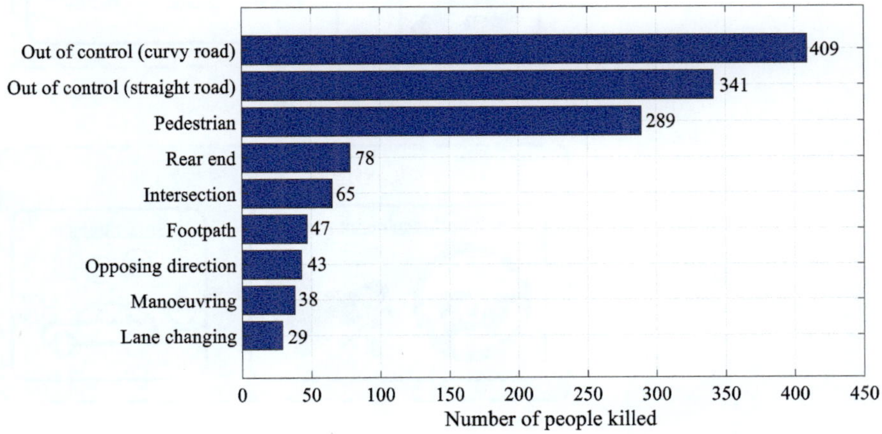


**Fig. 3** Types of crashes [1]

## 2 Factors Influencing Driver Behavior

Driver's state generally depends on the mental state and actions of the driver, as well as the application of the steering wheel, acceleration, and braking pedals through the vehicle's human-machine interface (HMI). Figure 4 shows the impact of different factors on driver behavior that result in leveraging the vehicle operation.

The most common factors that influence the driver are shown in Fig. 5. These factors can be categorized as *human factors* and *environmental factors*. Human factors include the driver's mental condition, personality, and ergonomic posture. Environmental factors are associated with the surrounding environment, such as emerging obstacles.

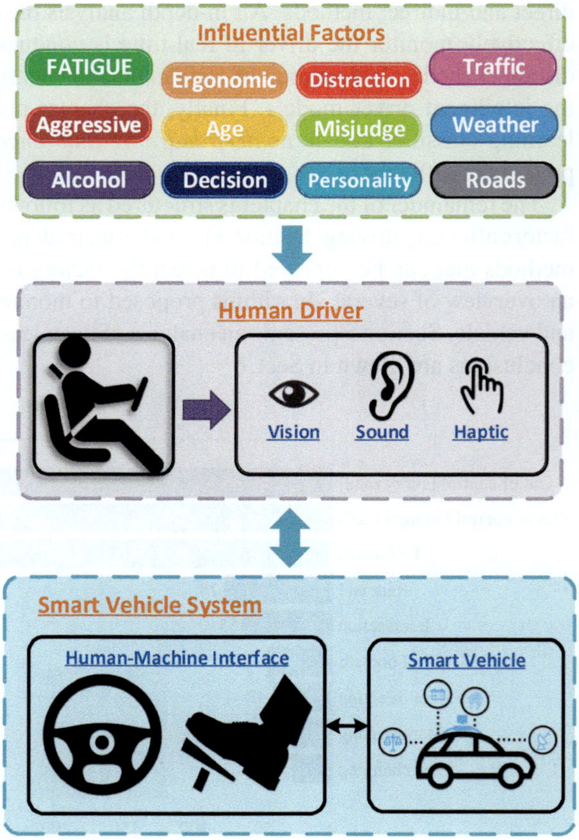


**Fig. 4** Impact of different factors on vehicle

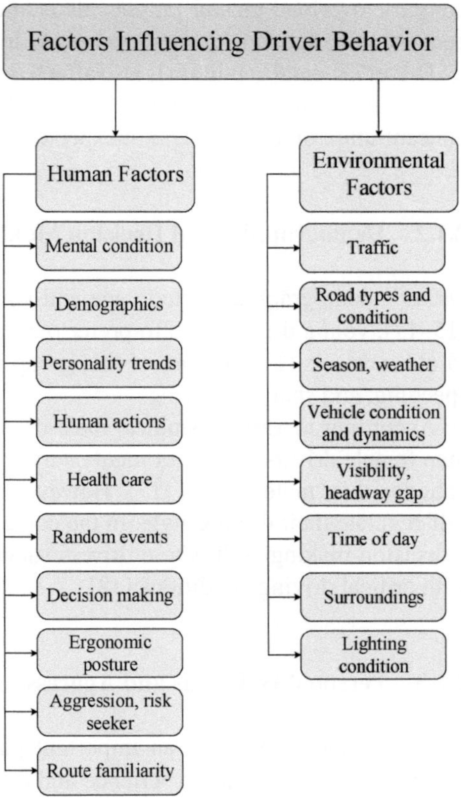


**Fig. 5** Factors influencing the driver behavior

## 2.1 *Types of Human Factors*

There are six basic human factors that influence the driver performance and decision-making process, as described below.

### 2.1.1 Mental Condition

Driver's mental condition is affected by fatigue, drowsiness, and sleepiness. According to [14], driver fatigue, drowsiness and sleepiness are similar in meaning. Generally, driver fatigue is of two types: (1) sleep-related (SR) fatigue and (2) task-related (TR) fatigue. SR fatigue is caused by sleep deprivation, inadequate sleep cycle, and circadian times.

TR fatigue is further divided into two groups: underload TR fatigue and overload TR fatigue. Underload mental or cognitive fatigue is induced because of extended hours of driving on long highways, and intervention of driver assistance systems in less dense traffic environments. It is also known as passive TR mental fatigue. On the contrary, overload mental fatigue, also called active TR mental fatigue, is induced because of demanding driving operations in dense traffic environments.

Driver's mental condition is also affected by workload, tiredness, and stress, resulting in absent-mindedness while driving. Poor mental state contributes to loss of concentration on the road and unexpected vehicle maneuvers.

### 2.1.2 Demographics and Decision Making

Driver's demographics include age and gender. Based on the research reported in [15–17], young drivers tend to prefer pleasure seeking and risky driving compared to older drivers. The driving of this group is characterized by higher acceleration, speeding, and sharp cornering.

According to various studies, male drivers are more prone to high risk driving than female drivers. A questionnaire survey showed that female drivers have higher patience than male drivers [17]. However, male drivers possess a better ability to self-regulate their driving style. In the context of driving behavior, self-regulation is a decision-making skill where drivers suddenly change their driving style to cope with critical driving conditions [18].

### 2.1.3 Personality Trends and Aggression Potential

Driver's personality plays an important part in driving performance. Personality depends on both driving experience and driving style [15]. Driving style relates to the driver's years of experience. Driving style generally falls into three categories: (1) Aggressive, (2) Moderate, and (3) Defensive.

According to the survey conducted in [17, 18], driver's aggressive nature reflects riskier and less experienced driving. Drivers with 2–3 years of driving experience can adopt aggressive driving for excitement, which results in higher likelihood of accidents. Drivers with more than three years of experience prefer a more moderate driving style, which is considered as an optimal driving style, where drivers are aware of the surrounding environment and can predict the consequences of the vehicle operation. A defensive driving style mainly falls in the category of the novice drivers and older drivers. Such drivers prefer lower speed and frequently apply braking in coping with road situations for a safe driving experience.

### 2.1.4 Driver Actions

Driver distraction often results in deterioration in driving performance. Distractions can include driver actions such as, talking with passengers, interacting with infotainment, inattention to the road, using a cell phone, and eating or drinking. Moreover, actions such as yawning, nodding or shaking the head while driving are also considered as distractive activities as they indicate fatigued mental state [19, 20].

#### 2.1.5 Health Care and Ergonomic Posture

The health of the driver can play a major role in the quality of driving. Medical problems, such as back or neck pain contribute to poor ergonomic posture. Driving under the effects of alcohol is more prone to accidents [21–23]. The breath alcohol concentration (BAC) test is conducted to scale the levels of alcohol and drug consumption. Drivers with high BAC levels tend to steer erratically, as well as apply erratic amount of pressure on the acceleration and brake pedals, resulting in lane exceedance on a curvy road.

#### 2.1.6 Route Familiarity

Route familiarity reduces the occurrence of distractive tasks, such as looking at a navigation screen, resulting in reduced risk for accidents. However, unfamiliarity with the route requires more concentration on the road, thereby increasing and potentially overloading mental tasks, such as, interacting with traffic signs, surrounding vehicles, pedestrians, and road obstacles.

### 2.2 *Types of Environmental Factors*

As described below, there are five basic environmental factors that can influence the performance and judgment of the driver.

#### 2.2.1 Traffic and Surrounding Environment

Traffic signals and signs are used to ensure a safe and smooth flow of traffic. As reported in [17], frequent changes of traffic signals and complex road signs result in frustration and uncertainty for the driver. Traffic violations can occur because of driver anger and anxiety [24]. The most important and potentially dangerous features in the surrounding environment are the presence of neighboring vehicles and pedestrians. The driver's most pressing task is to avoid collisions with these obstacles by selecting appropriate vehicle functions for a safe journey.

#### 2.2.2 Road Variations

Variations in the road can cause unexpected interference of the non-linear longitudinal, lateral and tire forces, resulting in vehicle rollover during high speed cornering [25]. Furthermore, road complexity level caused by lane variations and roundabouts can also affect driver behavior. It is observed that drivers prefer high speeds on long monotonous highways, thus violating speed limits [25].

#### 2.2.3 Seasons, Weather, and Reduced Visibility

Bad weather including rainy, wet, snow, windy, and fog conditions can impede driving performance and are considered as extreme driving conditions. These environmental factors result in challenging driving conditions such as slippery roadways and reduced visibility [26–28].

Technically, these factors reduce the tire-road friction co-efficient. Lower friction can cause the vehicle skid and roll. Similarly, reduced visibility can make the driving difficult for the driver to maintain a safe distance from other vehicles.

#### 2.2.4 Vehicle Condition and Dynamics

Vehicle parameters, such as mass, tire cornering stiffness ratios, and moment of inertia, are responsible for the feel that the driver experiences while driving the vehicle. Variations in these parameters can significantly affect vehicle performance. According to the study conducted in [29], it is observed that with an increase in the number of passengers in a vehicle, its overall mass increases, resulting in heavy load on the steering wheel and requiring more effort for the driver to turn the steering wheel on curvy high-inclined roads. Similarly, varying the tire stiffness ratio can induce arbitrary self-aligning torque that ultimately results in changing the feel of steering wheel.

#### 2.2.5 Time of Day and Lighting Condition

Night-time driving is more challenging because of the reduced lighting condition, making pedestrians, lane and road lines harder to see. On the other hand, excessive amounts of sunshine can reduce the driver's visibility and can increase the risk of accidents.

### *2.3 Driver Profile*

#### 2.3.1 Driver Speed Profile

A *driver speed profile* provides a way to characterize the driver's use of the acceleration and brake pedals. Various methods have been proposed to construct such a profile. In the study reported in [30], the driver's pedal behavior is modelled based on the information of the pedal's position, vehicle speed and acceleration, and road speed limits and signs. Another method discussed in [4], deployed to classify driving styles based on the headway gap, pedal position and vehicle speed. The classification is categorized as aggressive, moderate, and defensive styles. Aggressiveness is represented by fast speed, lower headway gap and higher pedal movements, whereas defensiveness is the opposite.

### 2.3.2 Driver Steering Profile

The driver steering behavior can be measured in terms of vehicle yaw, as well as lateral and longitudinal dynamics. Generally, a 2-DOF bicycle dynamic model is employed to estimate the vehicle lateral and yaw dynamics, as shown in Fig. 6. The state equation can be written as:

$$\begin{bmatrix} \dot{\beta} \\ \dot{\gamma} \end{bmatrix} = \begin{bmatrix} \frac{-C_f - C_r}{mV} & -1 + \frac{l_r C_r - l_f C_f}{mV^2} \\ \frac{l_r C_r - l_f C_f}{I_z} & \frac{-l_r^2 C_r - l_f^2 C_f}{I_z V} \end{bmatrix} \begin{bmatrix} \beta \\ \gamma \end{bmatrix} + \begin{bmatrix} \frac{C_f}{mv} \\ \frac{l_f C_f}{I_z} \end{bmatrix} \delta_{rw} \quad (1)$$

where, $\beta$ is the vehicle sideslip angle. $\gamma$ is the yaw rate, $C_f$ and $C_r$ are cornering stiffness coefficients of front and rear tires, respectively. $V$ is the vehicle speed, $l_f$ and $l_r$ are the distances from front and rear wheels to the center of gravity, respectively. $m$ is the vehicle mass, $I_z$ is the vehicle inertia and $\delta_{rw}$ is the road wheel angle. During cornering, lateral position and lateral acceleration determine the lateral force acting on the road wheels resulting in self-aligning torque ($\tau_a$) [31] as shown in Equation (2).

$$\tau_a = -C_f(t_m + t_p)(\beta + \frac{\gamma l_f}{V} - \delta_{rw}) \quad (2)$$

where, $t_m$ and $t_p$ are the tire mechanical and pneumatic trails, respectively.

**Fig. 6** 2-DOF bicycle model

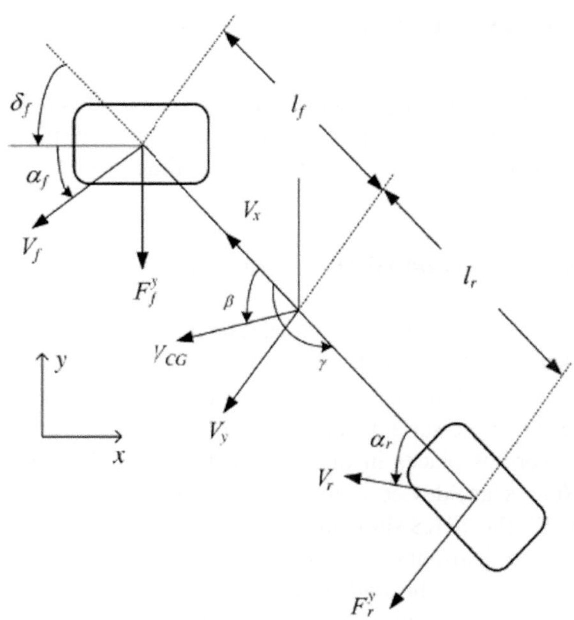

**Table 1** Intrusive measuring methods

| Physiological signals | Monitoring methods |
|---|---|
| Brain or cognitive | Electroencephalogram (EEG) |
| Heart beat or pulse | Electrocardiogram (ECG) |
| Eye's blinking frequency | Electrooculogram (EOG) |
| Respiration rate | Wearable sensors, accelerometers |
| Skin conductance and temperature | Wearable sensors |

# 3 Driver Behavior Monitoring Methods

Generally, the method used to assess driver behavior is either subjective or objective. In subjective approaches, also known as qualitative methods, the driver's behavior is assessed by performing pre-drive and post-drive interviews or answering questionnaires. Typical assessment questions focus on driver's vigilance, such as whether the driver failed to notice pedestrians in a desired path, or missed traffic signs, as well as underestimated the speed of oncoming vehicles during overtaking. In objective methods, driver behavior is assessed by measuring the physiological parameters as well as vehicle parameters, such as longitudinal and lateral vehicle dynamics. In this section, driver's objective real-time monitoring methods are described.

## *3.1 Intrusive Measuring Methods*

Driver monitoring based on intrusive methods are also known as direct measurement approaches. Table 1 presents the important intrusive physiological monitoring methods commonly deployed in real-time.

### 3.1.1 EEG Based Monitoring System

Monitoring the driver's mental state based on brain signals through EEG is considered as the gold standard for physiological assessment. A brain computer interface (BCI) device, such as the one shown in Fig. 7 is used to record a set of electrical signals (each a time-series) emanating from the brain, as recorded from sensors positioned in various places in the BCI. In Fig. 7, a 32-channel Quick-cap (Compumedics-Neuroscan) device is mounted on the driver's head, where a subject is driving a MATHWORKS simulated vehicle in a virtual environment [32].

The distribution of EEG electrodes over the brain regions is shown in Fig. 8. According to the studies reported in [33], the spectral power in frequency bands of the EEG channels/electrodes increases as the driver feels tired, stressed or fatigued.

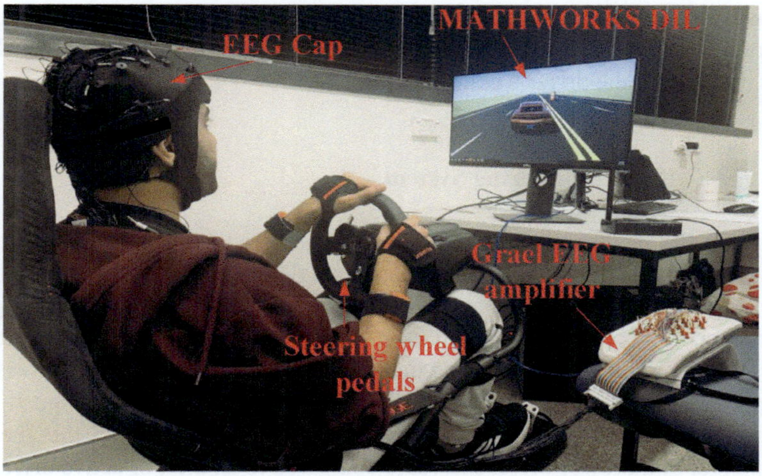


**Fig. 7** Subject driving MATHWORKS simulated vehicle wearing a 32-Channel EEG device

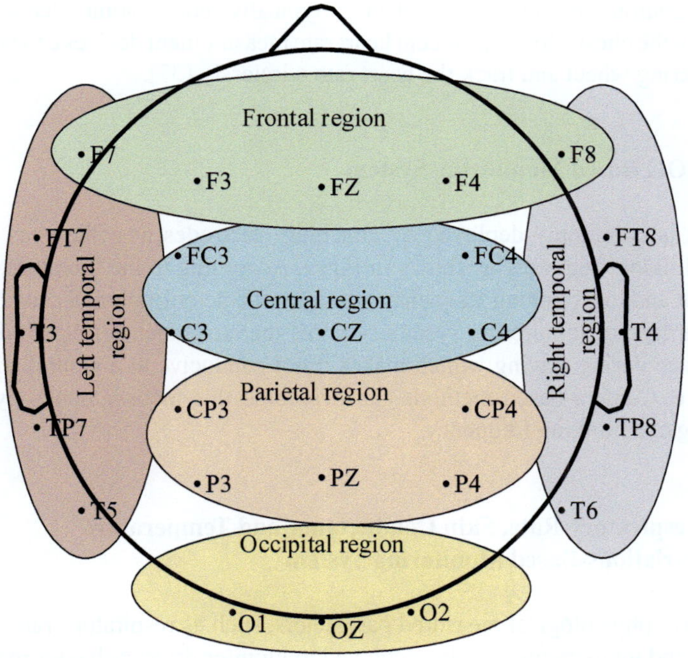


**Fig. 8** Distribution of EEG channels with respect to brain region

The frontal and parietal regions are the most active parts of the brain, from where driver's vigilance and absenteeism can best be tracked [34]. The trend in time-series spectrum of respective parts of the brain is highly visible so that motifs or anomalies can easily be detected. EEG-based monitoring systems provide an accurate assessment of driver state. However, wearing the intrusive cap consisting of several electrodes can disturb the driver's comfort and focus.

### 3.1.2 ECG Based Monitoring System

Monitoring the heart pulse and rate is also considered as an effective method to detect driver states. In the heart rate variability (HRV) analysis, the time variation in the heartbeats is measured. It is an evaluation technique that assesses the peripheral autonomic nervous system (ANS). It is an indicator of human health capability that needs to be maintained stable during daily life routine activities [35]. The research conducted in [35] demonstrates that the heart rate for a drowsy or fatigued driver tends to reduce when the level of sleepiness increases. This implies that the HRV time interval of a drowsy driver increases compared to an alert or active driver.

There are several devices available that can monitor the real-time heart activity features, such as Alive Heart Activity Monitor [36] that provides ECG, heart rate, and acceleration variations. These devices typically require some electrodes to be attached to the chest. However, recent heart rate measurement devices can be attached to the steering wheel and track the heart rate wirelessly [37].

### 3.1.3 EOG Based Monitoring System

EOG test is commonly deployed by attaching electrodes near the eyes to record the eye blinking frequency as shown in Fig. 9. According to the literature, slow eye movement and low blinking frequency are signs of a deterioration in driver alertness [33]. This type of monitoring system requires the attachment of electrodes on the driver's face during driving, which makes it not conducive to a naturalistic driving experience. Camera-based methods can overcome this inconvenience by visually measuring eye blinking frequency.

### 3.1.4 Respiratory Rate, Skin Conductance and Temperature Variations-Based Monitoring System

Other direct physiological measured parameters, such as respiratory rate, skin conductance and temperature are also deployed to monitor driver behavior in real-time. Wearable sensors have been extensively deployed, such as smartwatches, smart glasses, smart clothes, and fitness wristbands are some examples of such applications. A good example is Base Peak Watch that measures the heart rate, GSR (Galvanic Skin Response) and skin temperature [38]. Moreover, radar-based contact-less devices have also proven effective in tracking the respiratory rate during driving [37].

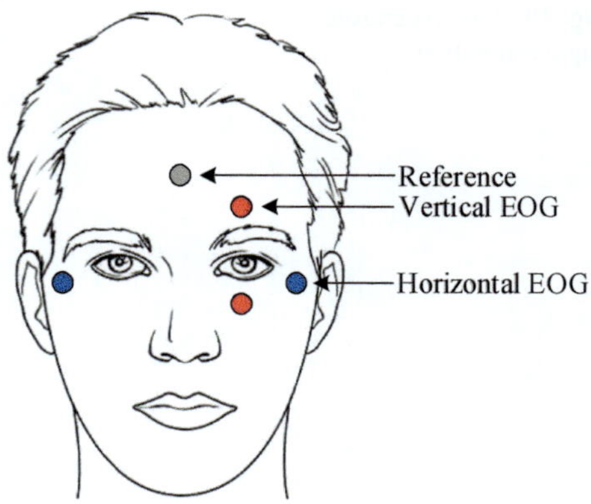


**Fig. 9** Attachment of EOG electrodes. The grey electrode is the reference, whereas, red and blue electrodes representing vertical and horizontal voltage potentials, respectively

## *3.2 Camera Based Measuring Methods*

Camera based monitoring methods offer a cost effective way to monitor driver behavior. Generally, face features are captured through cameras mounted inside the vehicle and analyzed to detect abnormalities and vigilance in the driver behavior. The features used in the analysis include degree of eyelid opening, percentage of eye rate closure (PERCLOS), eye pupil tracking, and mouth opening.

First, the images produced by the camera are processed to crop the face area of the driver. Next, the eye pupils are tracked in successive image frames. According to work reported in [39], a Kalman filter works effectively in estimating the position of the moving objects. It is important to accurately determine the location of eye pupils, as well as region of interest to be investigated in the next frame.

Another feature used is the percentage of number of frames with closed eyes $(N_{close})$ to the number of frames with open eyes $(N_{open})$ in a fixed time frame [10]. The so-called PERCLOS measure is given in Equation (3), and is mainly deployed to detect the level of drowsiness in the subject.

$$f_{PERCLOS} = \frac{N_{close}}{N_{open}} \times 100\% \tag{3}$$

For detecting the behaviors such as yawning, singing, and talking to passengers, the amount of mouth opening can be measured. To detect whether the mouth is open or closed, the subject's lips are segmented based on an eccentricity analysis, where the lips region is assumed to be like an ellipsoid with eccentricity equal to 1 when the mouth is closed. Furthermore, if the whole lip region is not detected then the mouth centroid point is estimated through the eye pupil center points using Eq. (4), as shown in Fig. 10. The mouth region is cropped from the image frames and then processed to assess the frequency of mouth opening $(FMO)$, which represents the percentage of number of frames with closed mouth $(M_{close})$ to the number of frames with open mouth $(M_{open})$, as given in Equation (5).

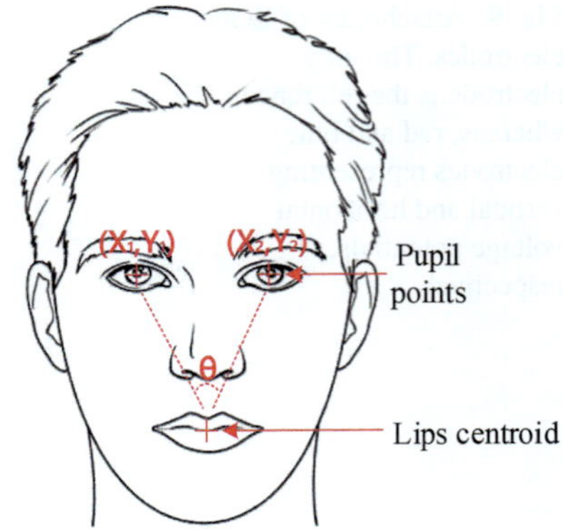


**Fig. 10** Mouth centroid angle estimation

$$\theta = \tan^{-1}\frac{|y_2 - y_1|}{|x_2 - x_1|} \tag{4}$$

$$f_{FMO} = \frac{M_{close}}{M_{open}} \times 100\% \tag{5}$$

### 3.3 Ergonomic and Body Posture Based Measuring Methods

Driver's ergonomic and body posture behavior have direct impact on the driving performance [7]. A normal ergonomic posture often indicates sound driving. When seated, the position of the head should be upright with gaze towards the road. The arms should be relaxed by adjusting the position of the steering wheel. Car mirrors should be adjusted so that they can be viewed without restraining the neck or twisting the body. The thighs should be straight, and the knees should be levelled with the hips in a way that the feet are approached easily when operating the pedals. A comparison of alert and inattentive ergonomic posture captured using XSENS motion capture system [19] is illustrated in Fig. 11.

The work reported in [19, 20], studied the effects of mental fatigue through head posture using motion capture suit. The study revealed that the driver's nodding and head shaking behavior are considered as distracted behavior due to influence of fatigue. Figure 12 shows the head posture inclination and nodding posture. The study conducted in [9], developed a radio frequency based smart hat that can be deployed in real-time driver behavior monitoring system to track driver's head movements. Furthermore, an inertial motion tracker based system was also deployed in [32] to track the driver's foot trajectory.

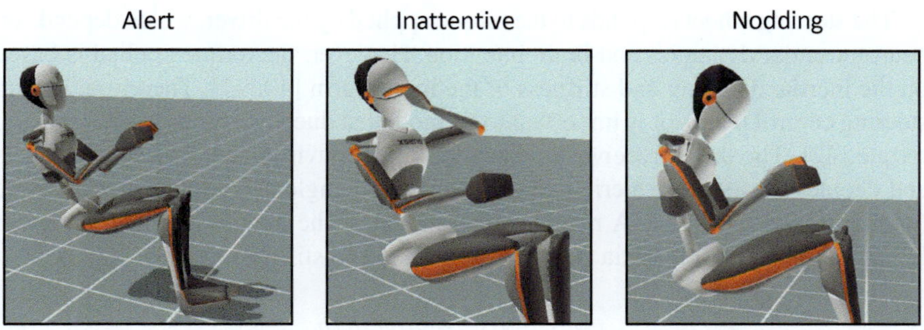


**Fig. 11** Driver's ergonomic posture [19, 20]

**Fig. 12** Head posture

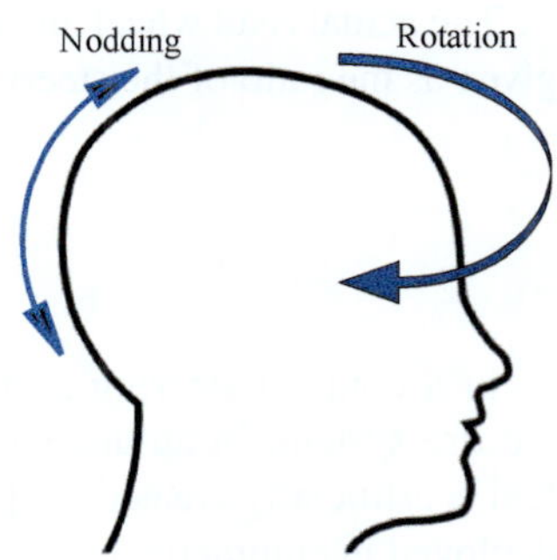


## *3.4 Vehicle Dynamics Based Measuring Methods*

Driver behavior monitoring based on vehicle parameters is an example of indirect driver measurement approach. Vehicle parameters such as speed, steering deviation, lane deviation, roll rate, mass variations etc. are monitored to detect any unusual activity or motif. In addition to driver behavior monitoring, this type of measurement approach is also used to detect the environmental factors discussed in Sect. 2.2.

Steering angle ($\delta_{sw}$)monitoring is a common method deployed to detect driver condition. The steering wheel subsystem model is deployed shown in Eq. (6) to detect the time-series anomalies during vehicle operation. A steering wheel subsystem includes the steering wheel, steering column, steering angle sensor, and a feedback motor. The inputs to a steering subsystem are driver's torque ($\tau_d$) and self-aligning torque ($\tau_a$). $J_s$, $C_s$ and $B_s$ are the moment of inertia, torsional stiffness, and viscous friction co-efficient.

$$\begin{bmatrix} \dot{\delta}_{sw}(t) \\ \ddot{\delta}_{sw}(t) \end{bmatrix} = \begin{bmatrix} 0 & 1 \\ -\frac{C_s}{J_s} & -\frac{B_s}{J_s} \end{bmatrix} \begin{bmatrix} \delta_{sw}(t) \\ \dot{\delta}_{sw}(t) \end{bmatrix} + \begin{bmatrix} 0 \\ 1 \end{bmatrix} (\tau_d(t) - \tau_a(t)) \tag{6}$$

The steering wheel responds to the torque applied by the driver, which depends on neuromuscular dynamics and brain intention. However, the torque applied is based on the inertia, damping and stiffness of the human arm [40, 41]. Therefore, human steering control behavior is important to be modelled due to its effect on the reaction torque [42]. The driver receives information of the environment and vehicle states and decides to steer the steering wheel at a desired angle and applies it as a torque ($\tau_{in}$) as shown in Fig. 13. A mathematical model of the human arm is provided in Equation (7) with arm inertia ($I_a$), damping ($B_a$) and stiffness ($K_a$).

$$\tau_{in} = I_a\ddot{\delta}_{sw} + B_a\dot{\delta}_{sw} + K_a\delta_{sw} \tag{7}$$

The actual road wheel angle depends on the steering ratio ($N$) of the system. It is given as the ratio of the steering wheel ($\delta_{sw}$) to the road wheel ($\delta_{rw}$).

$$\begin{aligned} \delta_{rw} &= \tfrac{1}{N}\delta_{sw} \\ \dot{\delta}_{rw} &= \tfrac{1}{N}\dot{\delta}_{sw} \end{aligned} \tag{8}$$

In a coupled steering system, the steering wheel is linked mechanically with the tire system. In an uncoupled system, this linkage is removed and the steering feel is artificially created [43]. Steer-by-Wire systems are the applications currently deployed in commercial cars such as Infiniti Q50.

Lane deviation monitoring while driving is also considered as an effective method to detect driver errors and mistakes. In this regard, vehicle lateral model shown in Equation (2) is usually deployed according to the handling envelope [8]. The handling envelope shown in Fig. 14 represents the relationship between vehicle sideslip angle and yaw rate. It shows the stable vehicle maneuverability when vehicle retains its lateral dynamics within the envelope.

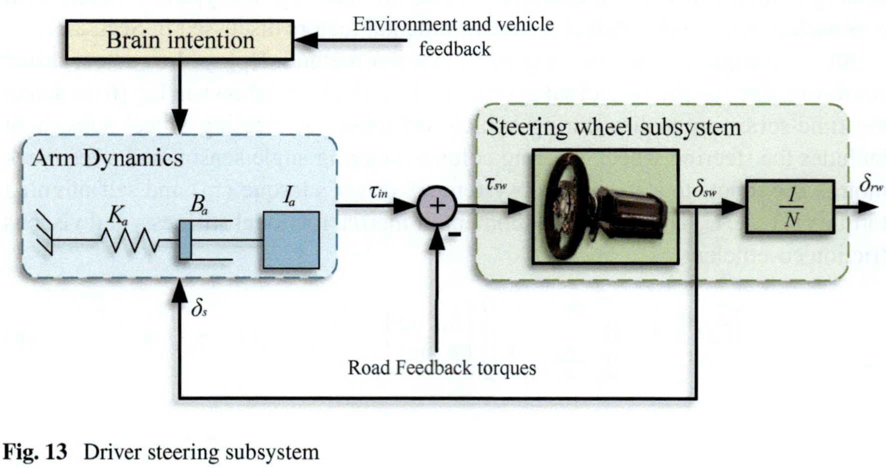


**Fig. 13** Driver steering subsystem

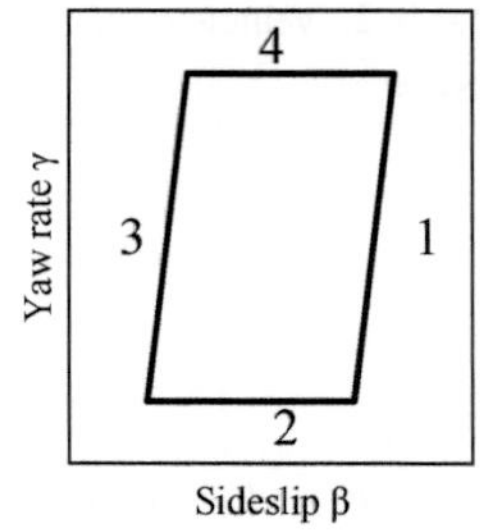


**Fig. 14** Handling envelope (from Balachandran et al. [8])

To monitor the driver speed profile, a dual-track 3-DOF nonlinear vehicle model is deployed with effects of the non-linear dynamics of the lateral tire forces, induced due to unreliable steering of the driver. The 3-DOF vehicle model consists of longitudinal speed, lateral velocity and yaw rate of the vehicle based on the non-linear Brush tire model described by Pacejka [44]. This non-linear tire model can effectively provide the in-vehicle parameters to identify the longitudinal and lateral driver behavior through tire longitudinal and lateral forces [45, 46]. The dynamics of the vehicle model is presented here, and the nomenclature of the variables is shown in Table 2.

$$\dot{v}_x = v_y\gamma + \frac{F_{xfl} + F_{xfr} + F_{xrl} + F_{xrr} + F_{xext}}{m} \tag{9}$$

$$\dot{v}_y = -v_x\gamma + \frac{F_{yfl} + F_{yfr} + F_{yrl} + F_{yrr} + F_{yext}}{m} \tag{10}$$

$$\dot{\gamma} = \frac{a(F_{yfl} + F_{yfr}) - b(F_{yrl} + F_{yrr}) + M_{zext} + \frac{b_f(F_{xfl}-F_{xfr})+b_r(F_{xrl}-F_{xrr})}{2}}{I_z} \tag{11}$$

Tire longitudinal and lateral forces are given as:

$$F_{xi} = F_{xit}\cos(\delta_i) - F_{yit}\sin(\delta_i) \tag{12}$$

$$F_{yi} = F_{xit}\sin(\delta_i) + F_{yit}\cos(\delta_i) \tag{13}$$

where, $i = fl, fr, rl, rr$, represents the front left, front right, rear left, and rear right wheel, respectively.

If external longitudinal velocity is applied, then $\dot{v}_x = 0$ under quasi steady-state condition. In this case, the longitudinal tire forces $F_{xit}$ also become zero. Hence,

**Table 2** Vehicle dynamics nomenclature

| Symbols | Description |
|---|---|
| $v_x$ | Vehicle longitudinal speed |
| $v_y$ | Vehicle lateral velocity |
| $\gamma$ | Yaw rate |
| $a, b$ | Distance of front and rear wheels from center |
| $b_f, b_r$ | Front and rear track widths |
| $F_{xfl}, F_{xfr}, F_{xrl}, F_{xrr}$ | Longitudinal forces applied on each wheel |
| $F_{yfl}, F_{yfr}, F_{yrl}, F_{yrr}$ | Lateral forces applied on each wheel |
| $F_{xext}, F_{yext}$ | Longitudinal and lateral external forces |
| $M_{zext}$ | External moment about vehicle center of gravity on yaw axis |
| $\delta_{fl}, \delta_{fr}, \delta_{rl}, \delta_{rr}$ | Steering angle of each road wheel |
| $F_t$ | Traction force on tire |
| $F_s$ | Side force on tire |
| $C_i$ | Longitudinal cornering stiffness |
| $C\alpha$ | Lateral cornering stiffness |
| $\alpha_i$ | Tire slip angle |
| $s$ | Slip ratio |
| $\mu$ | Tire-road friction co-efficient |
| $F_z$ | Normal tire force |
| $g$ | Gravity |
| $h$ | Vehicle height |

based on the Fiala's Brush tire model [44], non-linear lateral forces due to slip angle applied on each wheel can be calculated as follows:

$$F_{yit} = -C_{\alpha i}\tan(\alpha_i) + \frac{C_{\alpha i}^2}{3\mu_i F_{zi}}\tan(\alpha_i)|\tan(\alpha_i)| - \frac{C_{\alpha i}^3}{27(\mu_i F_{zi})^2}\tan^3(\alpha_i) \tag{14}$$

Tire slip angle of each wheel is defined as:

$$\alpha_{fl} = \tan^{-1}\left(\frac{v_y + a\gamma}{v_x + \gamma\frac{b_f}{2}}\right) - \delta_{fl} \tag{15}$$

$$\alpha_{fr} = \tan^{-1}\left(\frac{v_y + a\gamma}{v_x - \gamma\frac{b_f}{2}}\right) - \delta_{fr} \tag{16}$$

$$\alpha_{rl} = \tan^{-1}\left(\frac{v_y - b\gamma}{v_x + \gamma\frac{b_r}{2}}\right) - \delta_{rl} \tag{17}$$

$$\alpha_{rr} = \tan^{-1}\left(\frac{v_y - b\gamma}{v_x - \gamma\frac{b_r}{2}}\right) - \delta_{rr} \tag{18}$$

Normal tire forces are computed as follows:

$$F_{zfl} = \frac{bmg - (\dot{v}_x - v_y\gamma)mh}{a+b} + (mh\left(\dot{v}_y + v_x\gamma\right))\frac{2}{b_f} \tag{19}$$

$$F_{zfr} = \frac{bmg - (\dot{v}_x - v_y\gamma)mh}{a+b} + (-mh\left(\dot{v}_y + v_x\gamma\right))\frac{2}{b_f} \tag{20}$$

$$F_{zrl} = \frac{amg + (\dot{v}_x - v_y\gamma)mh}{a+b} + (mh\left(\dot{v}_y + v_x\gamma\right))\frac{2}{b_r} \tag{21}$$

$$F_{zrr} = \frac{amg + (\dot{v}_x - v_y\gamma)mh}{a+b} + (-mh\left(\dot{v}_y + v_x\gamma\right))\frac{2}{b_r} \tag{22}$$

The cornering stiffness ratio ($C_{\alpha i}$) and surface friction coefficient ($\mu_i$) are the design parameters which determine the road friction conditions of each tire.

## 3.5 *Hybrid Measuring Methods*

Hybrid driver monitoring methods deploy the multiple detection systems in real-time by integrating direct as well as indirect measurement methods. Hybrid approaches deploy physiological or camera-based measuring methods to monitor the driver's condition and then combine the information with vehicle parameters to estimate the state of both the driver and vehicle. One of the common approaches deployed is to combine the driver's eye gaze information with the road scene information [47]. Figure 15 shows the methodology of deployment of intelligent hybrid system for detecting driver states. Following are the typical hybrid driver monitoring systems:

1. Physiological system + vehicle dynamics.
2. Camera based systems + vehicle systems.
3. Wearable/Motion tracker sensors + vehicle systems.

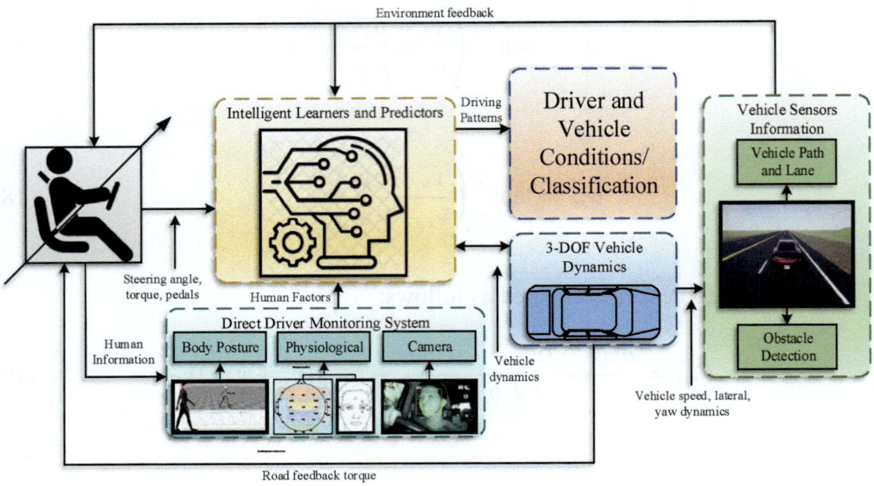


**Fig. 15** Hybrid driver monitoring system

In the first type of system, the driver's physiological features, such as EEG, ECG, and EOG are analyzed simultaneously with the information of vehicle dynamics. However, deployment of an intrusive approach is not practical in real-time because of several electrodes that are required to be attached to the subject. In this regard, alternative devices based on wearable sensors can monitor the physiological parameters, such as heart rate, skin conductance and temperature variations in real-time, and can be deployed simultaneously with vehicle dynamics information. The motion trackers/sensors record different parameters such as acceleration, velocity, orientation angles (Euler angles), and angular motion/displacement of body posture [19, 20]. The sensors need to be attached on the respective body segment whose data needs to be monitored. On the contrary, non-intrusive systems such as camera-based devices can be deployed in real-time in conjunction with vehicle sensors and dynamics. MIT human-centered smart car is an example of a hybrid approach that tracks the driver's activities and vehicle dynamics simultaneously [48].

## 4 Smart Detection Algorithms

The techniques used in real-time driver behavior monitoring fall into two categories. In the first category, a personalized driver behavior model is designed based on the mathematical parametric relationship such as driver's arm model as shown in Equation (7) [49–51]. The mathematical models are hard-coded based on specific threshold-rule strategies. Hence, they cannot be considered as general driver behavior model. In the second category, driver behavior is analyzed using pattern classification learning [52, 53]. The classification models are machine learning models that

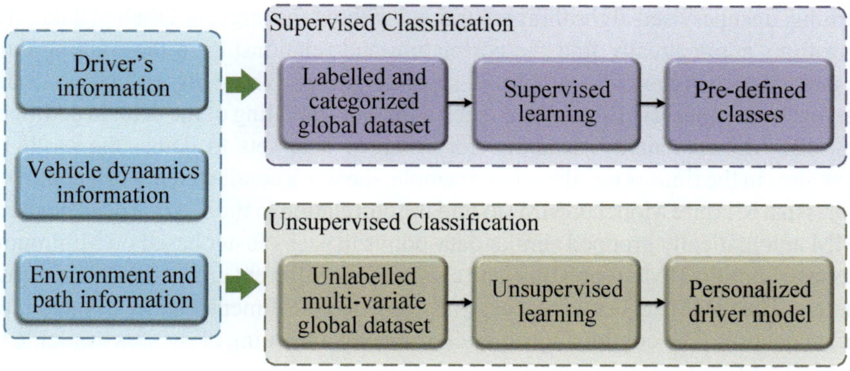


**Fig. 16** Illustration of supervised and unsupervised methods

are trained over large data sets of human and vehicle information under different scenarios and conditions. The classification methods can be divided into two groups: (1) *Supervised classification* and (2) *Unsupervised classification* as shown in Fig. 16.

## 4.1 Supervised Classification

In supervised classification, the prediction of an unknown input is carried out based on the label set. The supervised machine learning algorithms are trained over labelled dataset of pre-defined classes. The dataset is labelled by human experts or validated according to other benchmarks. These methods accurately detect the new data, and are easily implementable for real-time driver and vehicle states classification. However, the uncertainty in assessing driver's performance due to different factors can lead to incorrect labelling. Moreover, large complex multivariate datasets require extensive labelling effort and time, resulting in subjective or human errors. Following are the efficient supervised algorithms available in MATLAB, and commonly used for categorical time-series data classification [54]:

1. Decision trees
2. Support vector machines
3. K-nearest neighbors (KNN)
4. AdaBoost Algorithm
5. RUSBoosted trees.

## 4.2 Unsupervised Classification

In unsupervised classification, the prediction of unknown data is carried out based on the patterns or groups formed from the training of unlabeled data. Unlike supervised learning, unsupervised algorithms are trained without the target or labelled data. The algorithms automatically find the correlation and relationship in the training data set and cluster similar data-samples in separate groups. These algorithms avoid the drawbacks of supervised algorithms, such as manual labelling of the data. In addition, they offer dimensionality reduction for complex data sets to detect the motif or anomalies in the time-series data. For example, the work conducted in [19], deployed a Gaussian Mixture Model (GMM) to find the correlation in the driver's head posture. GMM automatically grouped similar data-points in each cluster based on Minimum-Message Length (MML) encoding, and clustered the different driver states in separate clusters. The GMM sequences successfully reduced the dimensionality of the multi-variate data set. The common unsupervised machine learning classifiers available in the MATLAB are listed below [55]:

1. Principal component analysis
2. Hierarchical clustering
3. K-Means clustering
4. Gaussian mixture models
5. Hidden Markov models
6. Self-organizing maps.

## 5 Deep Learning Neural Network Classifiers

Deep learning neural networks (DLNN) consist of neural network architectures with several hidden layers, and have been proved to offer high performance in many different types of classification problems. The architecture represents a more human-like artificial intelligence, and is based on different layers that produce an artificial neural network (ANN), and make decisions based on learning through hidden neurons. The decision is processed through several non-linear processing units. The major advantage of DLNN over traditional machine learning algorithms is that it takes the raw input data and automatically extracts the features from the data.

The most common types of DLNN are convolutional neural networks (CNN) and recurrent neural networks (RNN). CNNs are mostly deployed in image processing, particularly for monitoring the driver behaviour using Camera-based approaches, while RNNs are used for temporal sequence classification applications, where driver's direct measuring signals, such as EEG, ECG, and accelerometer, are used for training the model and detecting the driver status in real-time. Unlike to Feed-Forward Neural Network, RNNs are feedback-learners and can predict the long-term dependencies [56, 57]. In this chapter, a special type of RNN known as long-short term memory (LSTM) classifier is discussed.

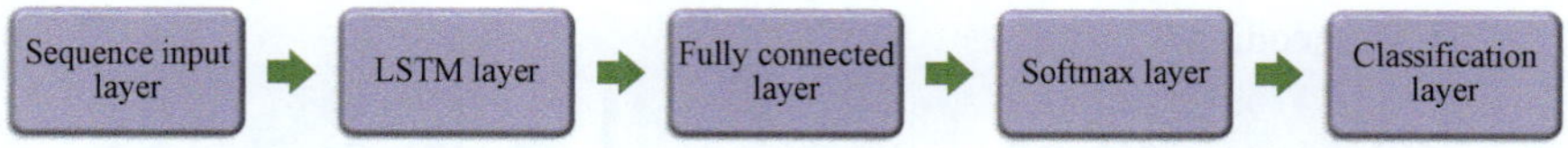


**Fig. 17** Supervised LSTM classifier architecture

## *5.1 Supervised LSTM Classifier*

LSTM is a type of RNN and is deployed for temporal sequence classification applications. It can learn the long-term dependencies between the temporal sequence-to-sequence data. The architecture of a supervised sequence-to-sequence LSTM classifier is shown in Fig. 17.

The raw data is passed through the *Sequence Input Layer* to the network that declares the dimension (number of input features) of the dataset. The data is then processed by the *LSTM Layer*, which learns the long-term dependencies between the time-series data and performs the additive interactions that can improve the gradient flow during training based on the hidden neurons. The data is then multiplied by the neurons weight matrix and added by a bias vector in *Fully Connected Layer*. It creates the $k$ fully connected layers based on the number of output classes. The non-linear *SoftMax Layer* then applies the SoftMax function to the data and sends to the *Classification Layer*. This final layer then computes the cross-entropy loss function and infers the number of output classes from the output size of previous layers. The mathematical relationships among the layers can be seen in [20].

## *5.2 Unsupervised LSTM Classifier*

The unsupervised LSTM classifier works as an auto-encoder network classifier that encodes the input signal and then decodes it based on the learning achieved through the LSTM layer. An auto-encoder LSTM network learns the dataset without the labelled or target set. It reconstructs the input signal based on the feature extraction by learning from the training dataset. The architecture of the unsupervised LSTM auto-encoder is shown in Fig. 18. It has a simple model architecture with low complexity that results in short training time [9].

The LSTM auto-encoders work as a binary unsupervised classifier that are trained on a specific type of data. When, the model is tested on the new data of $n$ samples that is similar to its training data, then model reconstructs the similar data of $n$ samples with higher similarity score (high accuracy). However, when the test input is completely different from its training data, then the similarity score drops considerably, representing low detection accuracy.

In the example of detecting different driver states based on head posture as discussed in [19], a novel architecture of the unsupervised classifier can be deployed as shown in Fig. 19, where multiple auto-encoders can be used for detecting each

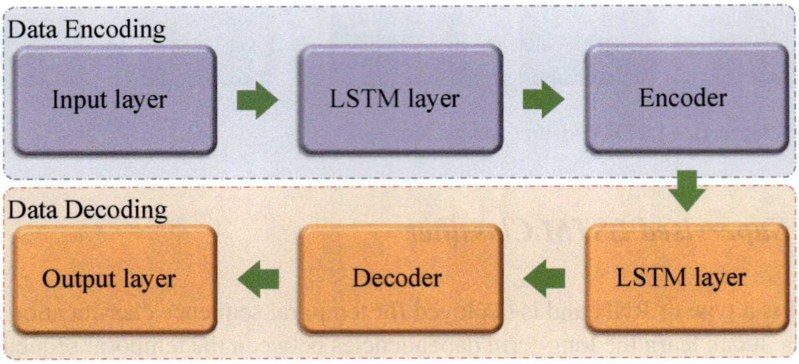


**Fig. 18** LSTM auto-encoder architecture

**Fig. 19** Driver states classification using LSTM auto-encoders

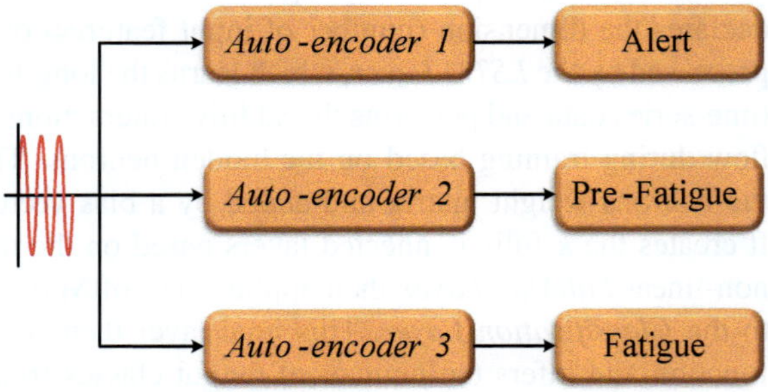


driver state, such as alert, pre-fatigue and fatigue. Each auto-encoder can be trained over data of a particular driver state. For example, in Fig. 19, the *auto-encoder1* can be trained over driver's alert information provided in [19]; the *auto-encoder2* and *auto-encoder3* can be trained over driver pre-fatigue and fatigued driving datasets, respectively. In such a case, if the input data corresponds to the alert data-points, then the matching score of *auto-encoder1* will be high compared to the other auto-encoders. There are various similarity or matching score techniques available [58]. However, *Euclidean distance* or *dynamic time warping (DTW)* as given in Equation (23) is generally deployed.

$$dist(x, y) = \sqrt{\sum_{i=1}^{k} (x_i - y_i)^2} \tag{23}$$

Here, $x$ and $y$ are the test and predicted samples. If the $dist(x, y)$ is low, then the accuracy will be considered high and vice versa.

## 6 Summary

With rapid advancements in the development of intelligent vehicles, there is a need for smart, fast, and intelligent systems to predict the state of the driver. The next step in the advancement of smart vehicles is to predict the driver behavior efficiently and to provide assistance in coping with complex driving scenarios. Various intelligent algorithms and methods will be deployed in commercial vehicles to identify the intention of the driver in real-time and provide active shared control while driving.

In this chapter, several factors that influence driver behavior were presented, with discussion on their effects and detection methods. The factors were divided into two categories *human factors* and *environmental factors*. The core content of the chapter was to discuss the possible methods in detecting the driver behavior in real-time. The methods were divided into either *direct measurement* or *indirect measurement* methods. The advantages and disadvantages of all the presented methods were discussed considering practical deployment in commercial cars.

More specifically, the importance of hybrid driver monitoring methods in real-time applications was discussed and various smart detection algorithms were introduced. In addition, the chapter proposed an unsupervised deep learning model based on encoder-decoder LSTM network known as *LSTM auto-encoder*, which could perform the multi-level classification for detecting different states of the driver. The algorithm deploys separate auto-encoders for each driver state classification and can efficiently detect the various driver conditions by evaluating the performance of each *auto-encoder* using a *Euclidean distance-based* similarity score.

## References

1. Art Galley of NSW (2021) Nsw interactive crash statistics. Accessed 18 May 2021. [Online]. Available https://roadsafety.transport.nsw.gov.au/statistics/interactivecrashstats/index.html
2. Martinez CM, Heucke M, Wang F-Y, Gao B, Cao D (2017) Driving style recognition for intelligent vehicle control and advanced driver assistance: a survey. IEEE Trans Intell Transp Syst 19(3):666–676
3. Xie Z, Yu D, Yang J (2018) Analysis on lane changing trajectory for different style drivers. In: 2018 10th international conference on measuring technology and mechatronics automation (ICMTMA). IEEE, pp 198–203
4. Feng Y, Pickering S, Chappell E, Iravani P, Brace C (2018) Driving style modelling with adaptive neuro-fuzzy inference system and real driving data. In: International conference on applied human factors and ergonomics. Springer, Berlin, pp 481–490
5. Su C, Deng W, Sun H, Wu J, Sun B, Yang S (2017) Forward collision avoidance systems considering driver's driving behavior recognized by gaussian mixture model. In: IEEE intelligent vehicles symposium (IV). IEEE 2017:535–540
6. Xu L, Hu J, Jiang H, Meng W (2015) Establishing style-oriented driver models by imitating human driving behaviors. IEEE Trans Intell Transp Syst 16(5):2522–2530
7. Ansari S, Du H, Naghdy F, Stirling D (2021) Application of fully adaptive symbolic representation to driver mental fatigue detection based on body posture. In: 2021 IEEE international conference on systems, man, and cybernetics (SMC). IEEE, pp 1313–1318

8. Balachandran A, Brown M, Erlien SM, Gerdes JC (2015) Predictive haptic feedback for obstacle avoidance based on model predictive control. IEEE Trans Autom Sci Eng 13(1):26–31
9. Yang C, Wang X, Mao S (2020) Unsupervised drowsy driving detection with RFID. IEEE Trans Veh Technol 69(8):8151–8163
10. Savaş BK, Becerikli Y (2020) Real time driver fatigue detection system based on multi-task connn. IEEE Access 8:12,491–12,498
11. Kaplan S, Guvensan MA, Yavuz AG, Karalurt Y (2015) Driver behavior analysis for safe driving: A survey. IEEE Trans Intell Transp Syst 16(6):3017–3032
12. Braitman KA, Braitman AL (2017) Patterns of distracted driving behaviors among young adult drivers: exploring relationships with personality variables. Transp Res Part F Traffic Psychol Behav 46:169–176
13. Sikander G, Anwar S (2018) Driver fatigue detection systems: a review. IEEE Trans Intell Transp Syst 20(6):2339–2352
14. May JF, Baldwin CL (2009) Driver fatigue: the importance of identifying causal factors of fatigue when considering detection and countermeasure technologies. Transp Res Part F Traffic Psychol Behav 12(3):218–224
15. Witt M, Wang L, Fahrenkrog F, Kompaß K, Prokop G (2018) Cognitive driver behavior modeling: influence of personality and driver characteristics on driver behavior. In: International conference on applied human factors and ergonomics. Springer, Berlin, pp 751–763
16. Wishart D, Somoray K, Evenhuis A (2017) Thrill and adventure seeking in risky driving at work: the moderating role of safety climate. J Saf Res 63:83–89
17. Machin MA, Sankey KS (2008) Relationships between young drivers' personality characteristics, risk perceptions, and driving behaviour. Accid Anal Prev 40(2):541–547
18. Kostyniuk LP, Molnar LJ (2008) Self-regulatory driving practices among older adults: health, age and sex effects. Accid Anal Prev 40(4):1576–1580
19. Ansari S, Du H, Naghdy F, Stirling D (2020) Unsupervised patterns of driver mental fatigue state based on head posture using gaussian mixture model. In: 2020 IEEE symposium series on computational intelligence (SSCI). IEEE 2020:2699–2704
20. Ansari S, Naghdy F, Du H, Pahnwar YN (2021) Driver mental fatigue detection based on head posture using new modified relu-bilstm deep neural network. IEEE Trans Intell Transp Syst. https://doi.org/10.1109/TITS.2021.3098309
21. Armstrong KA, Watling CN, Davey JD (2018) Deterrence of drug driving: the impact of the act drug driving legislation and detection techniques. Transp Res Part F Traffic Psychol Behav 54:138–147
22. Zakletskaia LI, Mundt MP, Balousek SL, Wilson EL, Fleming MF (2009) Alcohol-impaired driving behavior and sensation-seeking disposition in a college population receiving routine care at campus health services centers. Accid Anal Prev 41(3):380–386
23. Li Z, Li X, Zhao X, Zhang Q (2019) Effects of different alcohol dosages on steering behavior in curve driving. Hum Factors 61(1):139–151
24. Precht L, Keinath A, Krems JF (2017) Identifying the main factors contributing to driving errors and traffic violations-results from naturalistic driving data. Transp Res Part F Traffic Psychol Behav 49:49–92
25. Rudin-Brown CM, Edquist J, Lenné MG (2014) Effects of driving experience and sensation-seeking on drivers' adaptation to road environment complexity. Saf Sci 62:121–129
26. Zhao Y, Andrey J, Deadman P (2018) Whether conversion and weather matter to roundabout safety. J Saf Res 66:151–159
27. Fylan F, Hughes A, Wood J, Elliott DB (2018) Why do people drive when they can't see clearly? Transp Res Part F Traffic Psychol Behav 56:123–133
28. Ahmed MM, Ghasemzadeh A (2018) The impacts of heavy rain on speed and headway behaviors: an investigation using the shrp2 naturalistic driving study data. Transp Res Part C Emerg Technol 91:371–384
29. Zhang C, Hu J, Qiu J, Yang W, Sun H, Chen Q (2018) A novel fuzzy observer-based steering control approach for path tracking in autonomous vehicles. IEEE Trans Fuzzy Syst 27(2):278–290

30. Zeng X, Wang J (2017) A stochastic driver pedal behavior model incorporating road information. IEEE Trans Hum Mach Syst 47(5):614–624
31. Pfeffer P, Braess H-H (2017) Basics of lateral vehicle dynamics. In: Steering handbook. Springer, Berlin, pp 91–120
32. Ansari S, Du H, Naghdy F (2020) Driver's foot trajectory tracking for safe maneuverability using new modified reLU-BiLSTM deep neural network. In: 2020 IEEE international conference on systems, man, and cybernetics (SMC). IEEE, pp 4392–4397
33. Hu X, Lodewijks G (2020) Detecting fatigue in car drivers and aircraft pilots by using non-invasive measures: the value of differentiation of sleepiness and mental fatigue. J Saf Res 72:173–187
34. Min J, Xiong C, Zhang Y, Cai M (2021) Driver fatigue detection based on prefrontal EEG using multi-entropy measures and hybrid model. Biomed Signal Process Control 69:102857
35. Jung S-J, Shin H-S, Chung W-Y (2014) Driver fatigue and drowsiness monitoring system with embedded electrocardiogram sensor on steering wheel. IET Intell Transp Syst 8(1):43–50
36. Alive Technologies (2021) Alive bluetooth heart and activity monitor. Accessed 5 Nov 2021. [Online]. Available https://www.alivetec.com/pages/alive-bluetooth-heart-activity-monitor
37. Rong Y, Dutta A, Chiriyath A, Bliss DW (2021) Motion-tolerant non-contact heart-rate measurements from radar sensor fusion. Sensors 21(5):1774
38. Reeder B, David A (2016) Health at hand: a systematic review of smart watch uses for health and wellness. J Biomed Inform 63:269–276
39. Azim T, Jaffar MA, Mirza AM (2014) Fully automated real time fatigue detection of drivers through fuzzy expert systems. Appl Soft Comput 18:25–38
40. Jiang Y, Deng W, Wu J, Zhang S, He R (2018) Study on the stability of the steering torque feedback system considering the time delay and the system characteristics. Proc Inst Mech Eng Part D J Autom Eng 232(5):707–721
41. Cole DJ (2012) A path-following driver-vehicle model with neuromuscular dynamics, including measured and simulated responses to a step in steering angle overlay. Veh Syst Dyn 50(4):573–596
42. Kim N, Cole DJ (2011) A model of driver steering control incorporating the driver's sensing of steering torque. Veh Syst Dyn 49(10):1575–1596
43. Marcano M, Díaz S, Pérez J, Irigoyen E (2020) A review of shared control for automated vehicles: theory and applications. IEEE Trans Hum Mach Syst 50(6):475–491. https://doi.org/10.1109/THMS.2020.3017748
44. Pacejka HB (2012) Semi-empirical tire models (Chapter 4). In: Pacejka HB (ed) Tire and vehicle dynamics, 3rd edn. Butterworth-Heinemann, Oxford, pp 149–209. ISBN: 9780080970165. https://doi.org/10.1016/B978-0-08-097016-5.00004-8. https://www.sciencedirect.com/science/article/pii/B9780080970165000048
45. Li B, Du H, Li W (2014, January 3) A novel method for side slip angle estimation of omni-directional vehicles. SAE Int J Passeng Cars Electron Electr Syst 7(3):471–480
46. Smith DE, Starkey JM (1995) Effects of model complexity on the performance of automated vehicle steering controllers: model development, validation and comparison. Vehicle System Dynamics 24(2):163–181
47. Dong Y, Hu Z, Uchimura K, Murayama N (2010) Driver inattention monitoring system for intelligent vehicles: a review. IEEE Trans Intell Transp Syst 12(2):596–614
48. Fridman L (2018) Human-centered autonomous vehicle systems: principles of effective shared autonomy. *arXiv preprint* arXiv:1810.01835
49. Ro JW, Roop PS, Malik A, Ranjitkar P (2017) A formal approach for modeling and simulation of human car-following behavior. IEEE Trans Intell Trans Syst 19(2):639–648
50. Qu T, Chen H, Cao D, Guo H, Gao B (2014) Switching-based stochastic model predictive control approach for modeling driver steering skill. IEEE Trans Intell Transp Syst 16(1):365–375
51. Plöchl M, Edelmann J (2007) Driver models in automobile dynamics application. Veh Syst Dyn 45(7–8):699–741

52. Wang Z, Liao X, Wang C, Oswald D, Wu G, Boriboonsomsin K, Barth MJ, Han K, Kim B, Tiwari P (2020) Driver behavior modeling using game engine and real vehicle: A learning-based approach. IEEE Trans Intell Veh 5(4):738–749
53. Khairdoost N, Shirpour M, Bauer MA, Beauchemin SS (2020) Real-time driver maneuver prediction using LSTM. IEEE Trans Intell Veh 5(4):714–724
54. MATLAB (2021) Matlab classification learner application. Accessed 30 May 2021. [Online]. Available https://au.mathworks.com/help/stats/classificationlearner-app.html
55. MATLAB (2021) Matlab unsupervised classification. Accessed 30 May 2021 [Online]. Available https://au.mathworks.com/discovery/unsupervised-learning.html
56. Hochreiter S, Schmidhuber J (1997) Long short-term memory. Neural Comput 9(8):1735–1780
57. Monner D, Reggia JA (2012) A generalized LSTM-like training algorithm for second-order recurrent neural networks. Neural Netw 25:70–83
58. Serra J, Arcos JL (2014) An empirical evaluation of similarity measures for time series classification. Knowl Based Syst 67:305–314

# Connected Autonomous Vehicles, Mobility, and Security

# Towards Learning-Based Control of Connected and Automated Vehicles: Challenges and Perspectives

**Alexander Katriniok**

**Abstract** The exploitation of communication technologies enables connected and automated vehicles (CAVs) to operate more collaboratively, that is, by exchanging or even negotiating future trajectories and control actions. That way, CAVs (or agents) can establish a networked control system such as to safely automate road traffic in a collaborative fashion. A rich body of literature is available, e.g., on intersection automation, automated lane change or lane merging scenarios. These control concepts, though, are most tailored to the particular application and are in general not applicable to multiple scenarios. This chapter conveys the challenges and perspectives of modeling and optimization-based control techniques for the safe coordination of multiple connected agents in road traffic scenarios. Along these lines, the perspective of generalizing controller design to serve multiple use cases simultaneously instead of designing separate controllers for every use case is discussed. Moreover, the opportunities of learning-based control in case of model uncertainties and mixed-traffic scenarios, involving connected and non-connected agents, are outlined.

## 1 Introduction

Automated vehicles (AVs) perceive their environment through instantaneous sensor measurements from radar, camera or LiDAR sensors and predict motion trajectories of other road users, e.g., by utilizing machine learning methods [8]. These mostly uncertain predictions are then leveraged in the AV control strategy to act more proactively [7]. When AVs are additionally equipped with communication technologies, these are referred to as connected and automated vehicles (CAVs). Compared to AVs, CAVs are even able to explicitly exchange or even negotiate future trajectories and control actions [27]. That way, the uncertainty related to future motion trajectories of other CAVs in the vicinity can significantly be reduced. Moreover, CAVs have the ability to establish a collaborative networked control system to safely automate traffic

A. Katriniok (✉)
Ford Research and Innovation Center (RIC), Süsterfeldstr. 200, 52072 Aachen, Germany
e-mail: de.alexander.katriniok@ieee.org

 
Y. L. Murphey et al. (eds.), *AI-enabled Technologies for Autonomous and Connected Vehicles*, Lecture Notes in Intelligent Transportation and Infrastructure,
https://doi.org/10.1007/978-3-031-06780-8_15

scenarios—such as intersections without traffic signs or lights [43], lane changing or lane merging maneuvers [2].

There is a rich body of literature which addresses collaborative maneuvers of CAVs, see [27, 43] for a comprehensive survey on intersection automation or [2] for an overview of automated lane changing and lane merging maneuvers. Control system architectures for such collaborative networked control systems can generally be categorized into centralized, distributed and hybrid approaches. In centralized control schemes, the CAVs (in the remainder referred to as *agents*) communicate with a central node, that is, the agents transmit their current state to the central node and in turn receive an optimized control action or (exclusive) access to a certain road section [16, 21, 47, 59]. Conversely, in distributed schemes the agents only communicate with each other, independent of a central node, and solve their part of the control problem locally [3, 18, 23, 25, 36]. Hybrid concepts are a combination of the previous two. Also decentralized control schemes are a prominent alternative in the literature, however, these do not involve communication and are therefore excluded from subsequent considerations. In terms of methodology, such kind of multi-agent coordination problems have, amongst others, been addressed in various ways, see [2, 27, 43] for a comprehensive review.

This chapter is predominantly devoted to optimization-based control schemes, particularly to model predictive control (MPC) [35], as these allow to explicitly accommodate constraints and exploit future trajectories of connected agents. That said, the main contribution of this chapter is to convey the challenges and perspectives related to modeling and optimization-based control for the safe coordination of connected agents in road traffic scenarios—such as intersections, lane changing or lane merging maneuvers. Along these lines, the author's own work forms the narrative but with sufficient and broad reference to the related literature. Given that control concepts in the literature are mostly designed to address a particular use case, this chapter discusses perspectives of generalizing controller design such as to serve multiple use cases simultaneously. Starting with a centralized control scheme, the chapter outlines the challenges that come along with the transition to a distributed optimization regime. This step is even more cumbersome if the centralized problem is already nonconvex, mostly due to collision avoidance constraints. Finally, the potentials of exploiting machine learning methods for such kind of problems are highlighted. Especially, model uncertainties or the presence of non-connected vehicles (also known as mixed-traffic scenarios [11]) give rise to control schemes which learn from data and account for the associated uncertainty in the resulting (stochastic) optimal control problem (OCP).

In the remainder, the definition of the considered multi-agent coordination problem and its fundamental assumptions are introduced in Sect. 2. Then concepts for modeling agent dynamics are conveyed in Sect. 3 before Sect. 4 proceeds with an outline of centralized and distributed control concepts. Extensions towards learning-based control are discussed in Sect. 5 before Sect. 6 concludes the chapter.

## 2 Multi-Agent Coordination Problem

**Notation** In the remainder, $x_{k+j|k}$ refers to the prediction of variable $x$ at the future time step $k + j$ given information up to time $k$. Moreover, the interval $[a, b] \subset \mathbb{N}$ with $a < b$ is denoted as $\mathbb{N}_{[a,b]}$, while $\mathbb{N}^+$ and $\mathbb{R}^+$ indicate the set of positive integers and real numbers greater than zero. Finally, $A^\top$ denotes the transpose of a matrix $A \in \mathbb{R}^{m\times n}$.

We refer to the problem of safely coordinating multiple connected agents in road traffic scenarios as the *multi-agent coordination problem*, which is generally defined as follows.

**Problem Statement**

The problem of interest is to automate multiple connected agents in a certain section of the road network and utilize vehicle-to-everything (V2X) communication for that purpose. The respective set of automated agents is denoted as $\mathcal{A} \triangleq \{1, \ldots, N_\mathcal{A}\}$ where $N_\mathcal{A}$ refers to the total number of agents. An agent is said to be automated if the associated control system is at least manipulating one control input. Starting at the current time $t_k$, every agent travels from its initial position $p_k \triangleq p(t_k) \triangleq (X(t_k), Y(t_k))$ along an a priori known route $\mathcal{R}$. In general, the route $\mathcal{R}$ rather prescribes a desired driving direction than the particular lane in the road network (see Fig. 1b). Along these lines, the agents commonly pursue individual

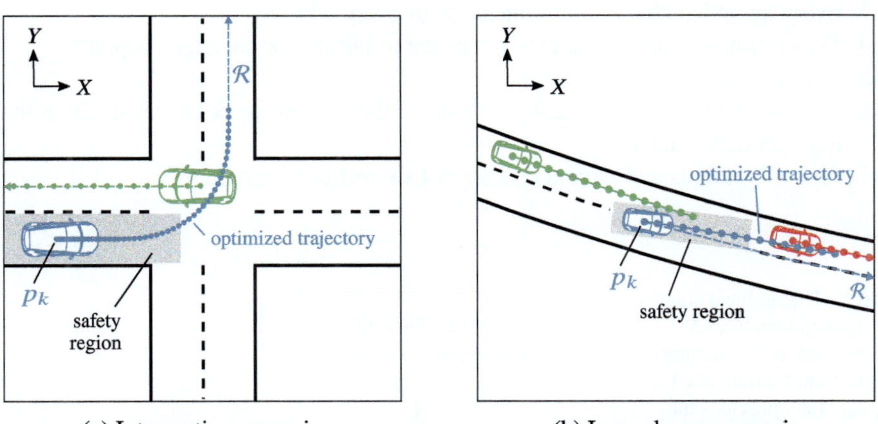


**Fig. 1** Two examples of multi-agent coordination problems. (Left) An unsignalized intersection with two agents. (Right) A lane change scenario on a road section with two lanes. (Both) Every agent moves in the direction of the route $\mathcal{R}$ (dashed line) while its actual optimized trajectory (solid line with dots, which indicate future time steps) does not have to coincide with $\mathcal{R}$. The safety region (hatched gray area) shall not be violated by the shape of the other agents. Fig. 1b Derived from [22] ©2020 IFAC with permission

objectives such as tracking a desired speed or ensuring comfortable and efficient driving. Most importantly, collisions between agents have to be avoided and agents must not leave their designated lane(s) along the route $\mathcal{R}$.

Two prominent and frequently researched examples of such multi-agent coordination problems are unsignalized intersections [18, 25, 29] and lane change scenarios [22] as shown in Fig. 1a, b, respectively. In both use cases, the hatched gray area indicates the safety area around the blue agent which must not be violated by the shape of other agents—an example realization of collision avoidance. In the literature, many different approaches have been pursued to mathematically formalize collision avoidance, see Sect. 4.2. Along these lines, Fig. 1a shows the blue agent crossing the intersection after the green agent, although it could have decided to cross first. These type of choices challenge coordination schemes by means of nonconvexity as outlined in the subsequent sections. Moreover, Fig. 1b gives an example that the route $\mathcal{R}$ does not necessarily have to coincide with the agent's optimized trajectory.

To further limit the scope of such multi-agent coordination problems, the following fundamental assumptions are made.

**Fundamental Assumptions**

A1. The desired route $\mathcal{R}$ of every agent is determined by a route planning algorithm, which resides on a higher control level (see Fig. 2).
A2. Every agent is equipped with V2X communication capabilities.
A3. Every agent is able to communicate with every other agent.
A4. The communication channel is not prone to failures or package dropouts.
A5. Agent clocks are synchronized.
A6. Every agent has access to a digital map of the road network to reason about the road geometry ahead.
A7. Agent dynamics and parameters are not subject to uncertainty.

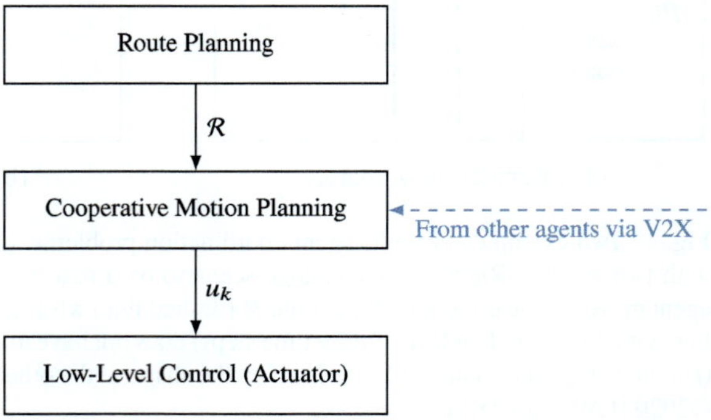


**Fig. 2** Hierarchical control system architecture. A high-level route planning algorithm, running at a low update rate, provides the desired route $\mathcal{R}$ to the cooperative motion planning algorithm. The vehicle actuators are finally manipulated by low-level controls

Assumptions A2–A7 are common in the literature to reduce complexity of the control problem at hand [17, 29]. Moreover, the use of a high-level route planning algorithm, as postulated in A1, is prevailing in AV control system architectures [30]. Along the lines of A1, a control system architecture as given in Fig. 2 is considered. More precisely, a high-level route planner determines every agent's route $\mathcal{R}$ which then serves as an input to the cooperative motion planning algorithm—the controller for multi-agent coordination, which is connected to other agents via V2X communication. With a central node in the infrastructure, the agents predominantly exploit vehicle-to-infrastructure (V2I) communication while vehicle-to-vehicle (V2V) communication is prevalent in distributed control schemes. Finally, low-level control either (i) further refines control input and state trajectories $(u_{\cdot|k},\ x_{\cdot|k})$ of the motion planning layer before forwarding the resulting control actions to the actuators [40] or (ii) represents the interface to the actuators which consume the (optimized) control action $u_k$ [18, 24]. As illustrated in Fig. 2, this chapter contemplates the latter.

## 3 Modeling Agent Dynamics

To solve the multi-agent coordination problem at hand within an optimal control framework, a dynamic agent model is required which is conducive to control. That said, it should be simple but describe agent dynamics in terms of input to state or input to output behavior sufficiently well. In the literature, the applied model inherently depends on the considered application. Referring to the use cases in Fig. 1, intersection coordination schemes mostly utilize a simple double integrator model (sometimes accompanied by first order drivetrain dynamics) while lane change or lane merging control regimes apply more complex (kinematic) bicycle models—as discussed subsequently. The reader should be aware, though, that there might be further alternative models which are not addressed in this chapter for the sake of brevity.

### 3.1 Double Integrator Model

The simplest model that is frequently utilized in the literature is the double integrator model [17, 23, 29, 42]. Every agent is represented by a reference point (mostly the geometric center) which is assumed to travel along a route coordinate $s$, see Fig. 3a. That said, the time evolution of the agent's velocity $v$ and route coordinate $s$ is prescribed through the integration of its longitudinal acceleration $a_x$ and its velocity $v$ respectively, i.e.,

$$\dot{v} = a_x, \quad \dot{s} = v. \tag{1}$$

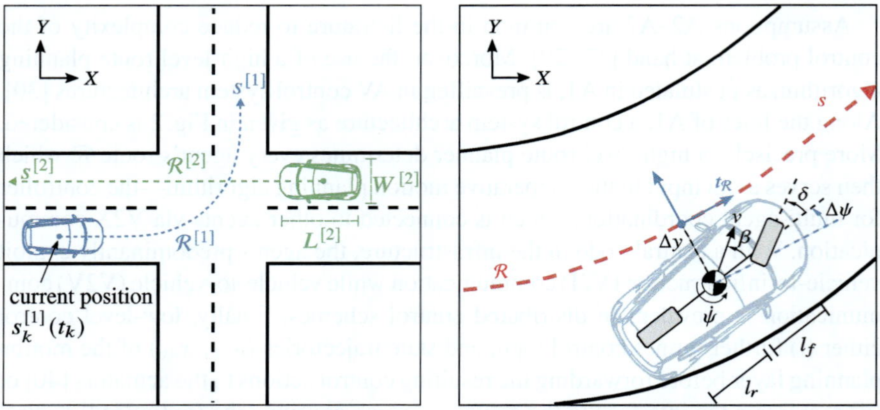


(a) Double integrator model (b) Kinematic bicycle model

**Fig. 3** Control-oriented models for motion planning. (Left) Double integrator model which is utilized in an intersection use case. The route $\mathcal{R}$ and the actual travel path are the same. Every agent's dimension is given by its length $L$ and its width $W$. The superscript indicates the respective agent. (Right) Kinematic bicycle model in a lane change use case. With an additional degree of freedom (steering), the travel path of the agent may deviate from $\mathcal{R}$. Fig. 3b Derived from [22] ©2020 IFAC with permission

The advantage of such linear model is its simplicity. By applying model (1), though, the underlying motion planning problem is one-dimensional. Thus, every agent moves along its a priori fixed route $\mathcal{R} : s \mapsto (X, Y)$ (provided by the route planner) where $X \in \mathbb{R}$ and $Y \in \mathbb{R}$ denote the respective coordinates in the global Cartesian frame in dependence of the route coordinate $s \in \mathbb{R}$, see Fig. 3a. In other words, the model does not allow the agents to deviate from the given route $\mathcal{R}$ and as such to involve lateral motion control. With (1), the resulting state space model $\dot{x} = f(x, u)$ with $x \in \mathcal{X} \subseteq \mathbb{R}^{n_x}$, $u \in \mathcal{U} \subseteq \mathbb{R}^{n_u}$ and vector field $f : \mathcal{X} \times \mathcal{U} \rightarrow \mathbb{R}^{n_x}$ exhibits the state vector $x = [s, v]^\top$ and the control input $u = a_x$. In this context, $\mathcal{X}$ and $\mathcal{U}$ are the admissible state and input sets, respectively.

## 3.2 Bicycle Model

To accommodate lateral vehicle dynamics in motion planning, a frequently pursued approach is to utilize a bicycle model [33, 46, 49], which may either be purely kinematic [46, 49] or is governed by Newton-Euler equations [33]. For a majority of motion planning problems, though, kinematic models are prevalent in the literature [2, 27, 43] as they are much less dependent on vehicle and tire parameters, while still being sufficiently accurate for the considered application. For this reason, these type of models are considered to be sufficient for the coordination problem at hand.

In the literature, the time evolution of the agent's position is commonly described either with respect to global coordinates $(X, Y)$ [49] or along a given path (or route) [22, 46]. We will follow the latter approach, appearing to be more suitable for our coordination problem, and devise the agent's motion along the given route $\mathcal{R}$ in a curvilinear reference frame [22, 46]. Compared to Sect. 3.1, the route $\mathcal{R} : s \mapsto (X, Y)$ with path coordinate $s$ and global Cartesian coordinates $X$ and $Y$ does not necessarily match the agent's actual trajectory but rather reflects the road geometry—usually defined by the road centerline. In the curvilinear Frenet frame [22, 46], as illustrated in Fig. 3b, the agent's position at time $t$ relative to the route $\mathcal{R}$ is given in terms of its coordinate $s$ along $\mathcal{R}$ and its perpendicular distance to $\mathcal{R}$. By considering the route coordinate $s$, the lateral distance $\Delta y$, the heading angle $\Delta\psi$ relative to the route tangent $t_{\mathcal{R}}$ and the velocity $v$ as agent states, a state space model of the form

$$\dot{s} = v\cos(\Delta\psi + \beta)\left(\frac{1}{1 - \Delta y\,\kappa(s)}\right) \tag{2a}$$

$$\Delta\dot{y} = v\sin(\Delta\psi + \beta) \tag{2b}$$

$$\Delta\dot{\psi} = \frac{v}{l_r}\sin(\beta) - v\cos(\Delta\psi)\left(\frac{\kappa(s)}{1 - \Delta y\,\kappa(s)}\right) \tag{2c}$$

$$\dot{v} = a_x \tag{2d}$$

with the vehicle sideslip angle

$$\beta = \arctan\big(\tan(\delta)\,l_r/(l_f + l_r)\big) \tag{2e}$$

is gained. In (2), $l_f$ and $l_r$, respectively, refer to the distance of the front and rear axle to the center of gravity. The route curvature $\kappa(s)$ is assumed to be provided as a parameterized curve by the high-level route planner (see Assumption A1). The devised state space model $\dot{x} = f(x, u)$ exhibits the state vector $x = [s,\ \Delta y,\ \Delta\psi,\ v]^\top$ along with an increased input vector $u = [a_x,\ \delta]^\top$. Compared to the double integrator model (1), the wheel steering angle $\delta$ adds an additional degree of freedom to accommodate lateral motion control.

## 3.3 Further Extensions

The control-oriented state space models in Sects. 3.1 and 3.2 can be augmented by first order drivetrain dynamics [25] such as to account for the time lag until the longitudinal acceleration $a_x$ reaches its desired set point $a_{x,\mathrm{ref}}$, i.e.,

$$\dot{a}_x = -\frac{1}{T_{a_x}}a_x + \frac{1}{T_{a_x}}a_{x,\mathrm{ref}} \tag{3}$$

where $T_{a_x} \in \mathbb{R}^+$ is the respective time constant. The inclusion of such dynamics is not necessarily required in simulations. However, real-world experiments on the test track, carried out by the author, have shown that these dynamics can generally not be neglected [24]. More precisely, time constants greater than 0.5 s have been recognized, which are even dependent on the current state $x$ and input $u$.

Likewise, in applications which involve lateral motion control, it may be required to accommodate steering system dynamics. Especially, if the time constant of the steering system is sufficiently large compared to the sample time of the controller, its dynamics may have to be contemplated during controller design. Assuming first order dynamics, such dynamic model can be represented as

$$\dot{\delta} = -\frac{1}{T_\delta}\delta + \frac{1}{T_\delta}\delta_{\text{ref}} \tag{4}$$

where $T_\delta \in \mathbb{R}^+$ is the respective time constant.

Independent of the particular model choice, a continuous-time state space model of the form $\dot{x} = f(x, u)$. is obtained. For the purpose of (direct) numerical optimization within an optimal control framework, we discretize these dynamics by using zero-order hold discretization. If $f(x, u)$ is nonlinear, a suitable numerical integration scheme, such as a fourth-order Runge-Kutta method [44], has to be applied. In the linear case, that is, if $f(x, u) = Ax + Bu$ with $A \in \mathbb{R}^{n_x \times n_x}$ and $B \in \mathbb{R}^{n_x \times n_u}$, we determine the exact time-discretization $x_{k+1} = A_d x_k + B_d u_k$ with $A_d = e^{AT_s}$ and $B_d = \int_{\tau=0}^{T_s} \mathrm{e}^{A\tau} d\tau B$ where $T_s \in \mathbb{R}^+$ denotes the controller sample time. In both case, the following discrete-time agent dynamics are obtained

$$x_{k+1} = f_d(x_k, u_k) \tag{5}$$

which can be exploited for numerical optimization purposes.

## 3.4 Challenges and Perspectives

The inclusion of nonlinear system dynamics (2) renders numerical optimization more complex, thus having to solve a nonlinear programming (NLP) problem. Utilizing the linear model (1) is much simpler, however, it restricts the degrees of freedom by only allowing for the manipulation of longitudinal dynamics. That said, the challenge is to devise an appropriate model which is conducive to control but preserves the real-time capability of the underlying optimization-based control scheme, see Sect. 4. Recent advances in solving NLPs in real-time, though, have made the accommodation of nonlinear dynamics tractable [15, 54].

To generalize the applicability of a single controller to multiple road traffic scenarios (for multi-agent coordination), it is essential to determine a suitable control-oriented model. Although the reader might conclude that the double integrator model is not appropriate to solve lane chance scenarios, the authors in [40] have demonstrated that an additional decision variable for the lane choice can indeed solve the problem. In that case, a low-level controller takes care of lateral vehicle dynamics control. As a first step towards a generalized control scheme, Sect. 4 will further discuss the integration of intersection and lane change scenarios, as depicted in Fig. 1.

Exogenous disturbances, unmodeled dynamics or parametric uncertainties pose additional challenges to control. Contemplating (1) and (2), it can be recognized that both models are simplified kinematic representations of the underlying vehicle dynamics, that is, resistance forces (such as aerodynamic drag and rolling resistances) or tire forces are not accommodated by the model. Moreover, model parameters may be time-varying and as such subject to uncertainty. A particular example is the drivetrain time constant as indicated in Sect. 3.3. We will further elaborate on this topic when discussing the perspective of learning-based control in Sect. 5.

## 4 Optimal Agent Coordination

Having devised a suitable state space model in Sect. 3, this section continues with a discussion on how to design centralized and distributed optimization-based control schemes for multi-agent coordination. Along these lines, the focus is placed on MPC-based control schemes which solve a finite horizon OCP at every time $t_k$ over a prediction horizon of $N \in \mathbb{N}^+$ time steps. After optimization, only the first control input is applied to the plant and optimization is repeatedly executed over a shifted horizon at the subsequent time $t_{k+1}$ [35].

### *4.1 Local Objectives and Constraints*

**Objectives** For multi-agent coordination problems, as defined in Sect. 2, the underlying optimization problem is usually separable with respect to every agent's individual (local) objectives and constraints, whereas collision avoidance imposes a coupling in terms of joint constraints. For the intersection automation problem, e.g., local agent objectives involve the tracking of a reference velocity $v_{\text{ref}}$ as well as ensuring comfortable and efficient driving [18, 25, 29, 42], where the latter generally translates into a regularization term on the inputs. The lane change use case adds additional objectives such as tracking the center of the target lane or minimizing the relative heading between the route $\mathcal{R}$ and the agent [22]. All these objectives can be summarized as a quadratic stage cost $\ell_j^{[i]} : \mathbb{R}^{n_x} \times \mathbb{R}^{n_u} \rightarrow \mathbb{R}$ for every agent $i \in \mathcal{A}$ (indicated by the superscript $[i]$) at time step $k + j$, $j \in \mathbb{N}_{[0,N-1]}$ over the prediction horizon, i.e.,

$$\begin{aligned}\ell_j^{[i]}(x_{k+j|k}^{[i]}, u_{k+j|k}^{[i]}) \triangleq (x_{k+j|k}^{[i]} - x_{k+j|k}^{[i],\text{ref}})^\top Q^{[i]}(x_{k+j|k}^{[i]} - x_{k+j|k}^{[i],\text{ref}}) \\ + (u_{k+j|k}^{[i]})^\top R^{[i]} u_{k+j|k}^{[i]}\end{aligned} \tag{6}$$

with reference state $x_{k+j|k}^{[i],\text{ref}}$ and positive semidefinite weighting matrices $Q^{[i]} \succeq 0$, $R^{[i]} \succeq 0$. At time step $k + N$, a terminal cost $\ell_N^{[i]} : \mathbb{R}^{n_x} \rightarrow \mathbb{R}$ such as

$$\ell_N^{[i]}(x_{k+N|k}^{[i]}) \triangleq (x_{k+N|k}^{[i]} - x_{k+N|k}^{[i],\text{ref}})^\top Q_N^{[i]}(x_{k+N|k}^{[i]} - x_{k+N|k}^{[i],\text{ref}}) \tag{7}$$

with $Q_N^{[i]} \succeq 0$ is applied, which does not depend on the control inputs. Mostly, the terminal cost is designed such as to guarantee closed-loop stability [35].

**Constraints** Alongside with agent $i$'s objectives, its states and inputs are constrained by polyhedral sets over the prediction horizon, that is,

$$x_{k+j+1|k}^{[i]} \in \mathcal{X}_{k+j+1|k}^{[i]}, \quad u_{k+j|k}^{[i]} \in \mathcal{U}_{k+j|k}^{[i]}, \quad j \in \mathbb{N}_{[0,N-1]}. \tag{8}$$

For instance, the agent's speed is frequently bounded from below by zero to avoid driving backwards and bounded from above by the speed limit of the road section [18, 25, 42]. Another example is to constrain the lateral displacement from the lane center to avoid that agents leave their designated lane [22]. Constraints on the inputs originate from actuator limitations [18, 29]. Additionally, the terminal state $x_{k+N|k}^{[i]}$ is oftentimes forced to be located in the terminal set $\mathcal{X}_{k+N|k}^{[i]}$, that is,

$$x_{k+N|k}^{[i]} \in \mathcal{X}_{k+N|k}^{[i]} \tag{9}$$

to ensure stability (together with terminal cost (7)) and/or recursive feasibility [33, 35]. In the intersection use case [25], for instance, a terminal constraint is applied to ensure that every agent has either passed the intersection at time step $k + N$ or waits at the stop line before entering the intersection. That way, unexpected collisions are avoided at the intersection, even if the prediction horizon is short. As the admissible state and input sets are polyhedrons, (8) can be rewritten as

$$P_x^{[i]} x_{\cdot|k}^{[i]} + P_u^{[i]} u_{\cdot|k}^{[i]} + q_{xu}^{[i]} \leq 0 \tag{10}$$

with matrices and vectors $P_x^{[i]}$, $P_u^{[i]}$, $q_{xu}^{[i]}$ of appropriate dimension. Moreover, $x_{\cdot|k}^{[i]} \triangleq \{x_{k+j|k}^{[i]}\}_{j=1}^{N}$ and $u_{\cdot|k}^{[i]} \triangleq \{u_{k+j|k}^{[i]}\}_{j=0}^{N-1}$ refer to the state and input trajectories over the entire prediction horizon. Terminal constraint (9) may be nonlinear (and even nonconvex), we therefore state it in the form

$$h_N(x_{k+N|k}^{[i]}) \leq 0. \tag{11}$$

To account for all crucial system constraints, in many cases the linear constraint (10) is not sufficient. For instance, the intersection use case [41] and the lane change scenario [22] require to constrain the lateral and total acceleration. Both quantities cannot be represented as a linear combination of states and inputs. Therefore, nonlinear constraints are added over the prediction horizon, which depend on the state $x$ and input $u$, i.e.,

$$h_{xu}(x_{\cdot|k}^{[i]}, u_{\cdot|k}^{[i]}) \leq 0. \tag{12}$$

## 4.2 Collision Avoidance

While objectives and constraints in Sect. 4.1 refer to the individual agent, collision avoidance between agents $i$ and $l$ introduces coupling constraints of the form

$$h_{CA}(x_{k+j|k}^{[i]}, x_{k+j|k}^{[l]}) \leq 0 \quad (13)$$

with $j \in \mathbb{N}_{[1,N]}$ and $h_{CA} : \mathbb{R}^{n_x} \times \mathbb{R}^{n_x} \to \mathbb{R}$. The constraint is imposed for every agent $l$ which belongs to the set $\mathcal{C}(i) \subseteq \mathcal{A} \setminus \{i\}$ of agents that may potentially collide with agent $i$. In the literature, there is a variety of options how to select $h_{CA}$ as outlined subsequently. An important factor is whether a certain *order* is imposed on the agents, e.g., agents are ordered in their lane (agent 1 is ahead of agent 2, agent 2 is ahead of agent 3, etc.) or an intersection is crossed in a prescribed sequence.

In the context of intersection automation, the authors in [17, 18] apply a central coordinator to determine intersection entry and exit times for a given intersection crossing order. Then, for the purpose of collision avoidance they impose linear constraints on the agents' position over the prediction horizon to meet the respective schedule and as such ensure safety. A similar principle of separating time slot allocation and motion planning is pursued in [29, 36, 37]. For a fixed crossing order, an alternative approach is to lower bound the distance between two agents along a given route through a linear constraint function $h_{CA}$ [42] (see Fig. 4a). The same principle can be applied to rear-end collision avoidance as the frontal vehicle is known (according to the given order). So, mostly $h_{CA}$ can be kept convex (or even linear) if a certain order of the agents is prescribed, which has also been shown for the use case of automated lane change maneuvers [22].

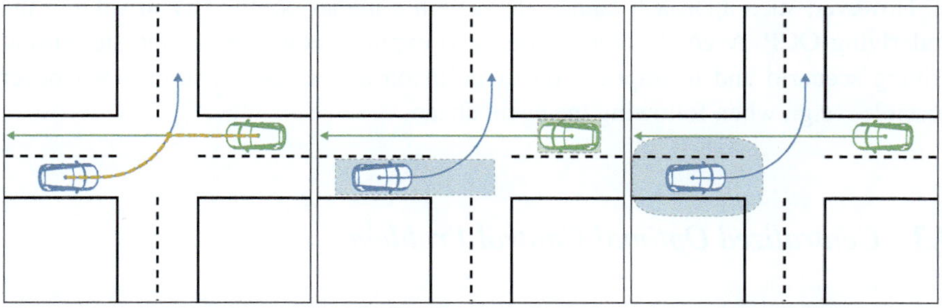

(a) Distance along path (b) Safety area ∩ bounding box (c) Safety superellipse

**Fig. 4** Examples of collision avoidance formulations at intersections. (Left) Distance along travel path (dashed orange) can be convex (crossing order fixed) or nonconvex (crossing order not fixed). (Middle) Agents are safe if safety area (blue) and bounding box (green) do not overlap. (Right) Agents are safe if the center point of the green agent is located outside the blue superellipse

Whenever the order of the agents is not prescribed, collision avoidance constraints generally become nonconvex. For instance, to relax the fixed crossing order in [42], the linear minimum distance constraints can, e.g., be formulated as nonconvex quadratic constraints [23], where the nonconvexity relates to agent $i$'s choice to cross the intersection before or after agent $l \in \mathcal{C}(i)$. With the same reasoning, collision avoidance constraints become nonconvex when claiming that agent $l \in \mathcal{C}(i)$ must not violate a certain safety area around agent $i$. This safety area can, e.g., be a rectangle [25] (see Fig. 4b) or a (super)ellipse [10] (see Fig. 4c), where the size of the rectangle and the (super)ellipse encodes the geometry of agents $i$ and $l$ as well as their desired safety distance to each other.

The challenge that arises with the choice of $h_{CA}$ is to ensure that the resulting optimization problem can be solved in real-time. Especially when $h_{CA}$ is nonconvex, the convergence towards a (local) optimal solution may take significantly longer than in the convex case. The properties of the cost and constraint functions (such as smoothness, Lipschitz continuity, etc.) as well as the choice of the numerical solver have significant influence on convergence guarantees and the time to solve the problem [44]. Our research work has shown that whenever $h_{CA}$ is nonconvex, there is unfortunately no best practice or recommendation on how to choose $h_{CA}$. Nevertheless, in many traffic scenarios the cardinality of $\mathcal{C}(i)$ and as such the number of collision avoidance constraints can significantly be reduced when contemplating that only a subset of $\mathcal{C}(i)$ in the vicinity of agent $i$ may pose a collision risk. For instance, agents $l$ that already have crossed the intersection and are then driving in a different direction than agent $i$ can be removed from $\mathcal{C}(i)$.

When aiming to generalize control schemes for multi-agent coordination such as to serve multiple scenarios simultaneously (without explicit notion of the driving context), the most intuitive way appears to be the definition of safety areas around the vehicle that must not be violated by other agents (see the two examples in Fig. 4b, c). However, such approach comes along with a higher complexity in solving the underlying OCP. A good compromise is to exploit knowledge about the current driving scenario and to impose convex collision avoidance constraints whenever possible—e.g., when following the agent ahead.

## *4.3 Centralized Optimal Control Problem*

When combining the individual cost terms (6), (7) for agent $i \in \mathcal{A}$ subject to agent constraints (10)–(12), agent dynamics (5) and collision avoidance constraints (13) over a prediction horizon of $N$ time steps, we can summarize the multi-agent coordination problem as the following centralized OCP with initial condition (14f)

$$\underset{\{u^{[i]}_{\cdot|k}, x^{[i]}_{\cdot|k}\}_{i=1}^{N_\mathcal{A}}}{\text{minimize}} \sum_{i=1}^{N_\mathcal{A}} \Big[ \ell_N(x^{[i]}_{k+N|k}) + \sum_{j=0}^{N-1} \ell_j(x^{[i]}_{k+j|k}, u^{[i]}_{k+j|k}) \Big] \tag{14a}$$

subject to **agent constraints** $- - -$ *for all* $i \in \mathcal{A}$ :

$$x^{[i]}_{k+j+1|k} = f^{[i]}_d(x^{[i]}_{k+j|k}, u^{[i]}_{k+j|k}), \qquad j \in \mathbb{N}_{[0,N-1]} \tag{14b}$$

$$P^{[i]}_x x^{[i]}_{\cdot|k} + P^{[i]}_u u^{[i]}_{\cdot|k} + q^{[i]}_{xu} \leq 0 \tag{14c}$$

$$h_{xu}(x^{[i]}_{\cdot|k}, u^{[i]}_{\cdot|k}) \leq 0 \tag{14d}$$

$$h_N(x^{[i]}_{k+N|k}) \leq 0 \tag{14e}$$

$$x^{[i]}_{k|k} = x^{[i]}_k \tag{14f}$$

**coupling constraints**

$$h_{CA}(x^{[i]}_{k+j|k}, x^{[l]}_{k+j|k}) \leq 0, \qquad \forall i \in \mathcal{A},\ \forall l \in \mathcal{C}(i);\ j \in \mathbb{N}_{[1,N]} \tag{14g}$$

In general, OCP (14) is a (nonconvex) NLP which can, for instance, be solved through sequential quadratic programming (SQP) [6], sequential convex programming [31] or first order proximal methods [54, 55]. Problem (14) constitutes a direct multiple shooting formulation [4] as system dynamics are imposed as equality constraints. Alternatively, by substituting (14b) into the cost (14a) and constraints (14c)–(14e), (14g), a direct single shooting formulation [4] is obtained which requires to solely optimize over control actions $\{u^{[i]}_{\cdot|k}\}_{i=1}^{N_\mathcal{A}}$ (instead of states and control actions), and which is oftentimes easier to solve.

As indicated throughout Sect. 4, Problem (14) exhibits a similar structure for various multi-agent coordination problems. The intersection coordination problem in [25, 42], e.g., can be solved by utilizing the bicycle model (2) instead of the double integrator model (1). Then, for an a priori known route $\mathcal{R}$, the controller only needs to track a constant lateral displacement $\Delta y_{\text{ref}}$ to ensure that every agent stays in its designated lane. All other constraints can be posed in the same way. Along these lines, the dedicated intersection and lane change control schemes, proposed by the author in [22, 25], can even be combined to a single control regime when $h_{CA}$ in (14g) represents a safety region such as a rectangle or a (super)ellipse (see Fig. 4b, c). This observation gives rise to the perspective of generalizing or integrating control schemes instead of solving use cases separately.

### 4.4 Distributed Solution

Solving OCP (14) in a centralized fashion is commonly not the preferred option as the problem does not scale well with an increasing number $N_\mathcal{A}$ of agents. Moreover, the agents' dynamic models and their parameters need to be known to the central optimizer and a central node is prone to a single point of failure. For this reason, there is a rich body of literature which deals with a (semi-)distributed solution of such multi-agent coordination problems [18, 25, 28, 32, 36, 53], where every agent contributes to the solution of OCP (14). In contrast to fully distributed regimes, semi-distributed control schemes apply a (central) coordinator to ensure convergence of distributed computations towards a stationary point.

By virtue of OCP (14), it can be recognized that the problem can be decoupled into $N_{\mathcal{A}}$ independent OCPs with respect to every agent's cost (14a) and its local constraints (14b)–(14f). Only collision avoidance couples the OCPs through coupling constraints (14g). To solve constraint coupled optimization problems in a distributed fashion, there are mature methods in the literature that rely, amongst others, on primal or dual decomposition techniques [45], the alternating direction method of multipliers (ADMM) [1, 45] or ADMM with recent extensions towards nonconvex problems [9, 57]. Distributed numerical optimization algorithms for multi-agent coordination problems are generally challenged by the following three aspects, that is, (i) convergence guarantees for nonconvex problems, (ii) the number of iterations required to converge to a stationary point, which translates into the time required to solve the OCP, and (iii) data privacy related to keeping the agents' dynamic models and their parameters private to the individual agents.

**Convergence** Convergence guarantees for distributed algorithms mostly assume convexity of the cost function and constraints. For instance, ADMM converges to a global optimal solution if OCP (14) is convex [5]. Oftentimes, we can expect multi-agent coordination problems and as such (14) to be nonconvex. If nonconvexity only occurs in the agent's cost, ADMM is still guaranteed to converge to a stationary point under certain regularity assumptions [14]. For a broader class of nonconvex problems, methods such as the extra-layer augmented Lagrangian-based accelerated distributed approximate optimization algorithm (ELLADA) or the augmented Lagrangian alternating direction inexact Newton (ALADIN) algorithm and its fully distributed variant [9] have been proposed. The authors in [20] have demonstrated that ALADIN can be utilized for distributed optimal control of unsignalized intersections. Although convergence can be guaranteed, the algorithm does not yet overcome challenge (ii), that is, it is not able to provide a solution in real-time.

**Solution in Real-Time** In the literature, various methods have been proposed to solve the originally centralized problem in a distributed fashion by introducing suitable relaxations in order to achieve real-time capability along with convergence guarantees. For the purpose of intersection automation, the authors in [17] propose a semi-distributed optimization scheme which assumes the order of agents, crossing the intersection, to be a priori fixed. A proof that this concept even works in a real-world setting is demonstrated in [18]. In our research work, we have shown that OCP (14) can be fully decoupled when collision avoidance constraint (14g) is only imposed on one of the conflicting agents, that is, either on agent $i$ or agent $l \in \mathcal{C}(i)$. Therefore, a bijective prioritization function $\gamma : \mathcal{A} \to \mathcal{A}$ is introduced which assigns a unique priority to every agent. If $\gamma(i) < \gamma(l)$ for $i \in \mathcal{A}$ and $l \in \mathcal{C}(i)$, agent $i$ is said to exhibit a higher priority than agent $l$ and the respective collision avoidance constraint (14g) is only imposed on agent $l$. By leveraging the first order proximal averaged Newton method for optimal control (PANOC) [55], the intersection coordination problem

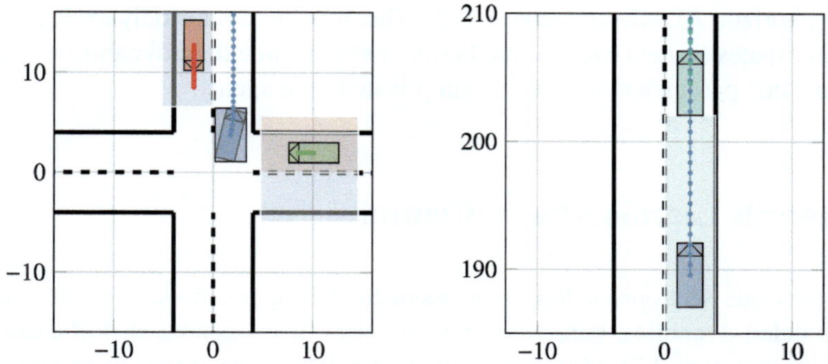


(a) Intersection Collision Avoidance ($t = 9.5$ s)

(b) Rear-end Collision Avoidance ($t = 29.9$ s)

**Fig. 5** Distributed control scheme [25] for a four way intersection with four agents. (Left) Agent 1 (red) and agent 3 (green) yield to agent 2 (blue), having the highest priority. The colored shaded patches indicate the safety areas that must not be violated by the bounding box of the respective agent with the same color. While turning, agent 2 (blue) is approximated by a bounding box that is axis-aligned with agent 1 (red) and agent 3 (green); (Right) Agent 2 (blue) avoids rear-end collisions with agent 4 (cyan). Figure from [25] ©2019 IEEE, reproduced with permission

with four agents and a priori fixed priorities, as illustrated in Fig. 5, has been solved in a fully distributed fashion in less than 50 ms (worst case), using a sample time of 100 ms [25]. Also the author's experiments on the test track [24] have proven the real-time capability of distributed multi-agent coordination algorithms which rely on such priority-based decoupling. An extension towards time-varying priorities, negotiated through consensus-based auction algorithms, has been proposed in [41]. To the author's best knowledge, a fully distributed optimization scheme that solves the originally centralized multi-agent coordination problem (14) with nonconvex constraints in real-time is still part of ongoing research.

**Data Privacy** The third challenge in distributed optimization is to ensure data privacy. For instance, an agent may not be willing to share its objectives, dynamic model or parameters. When the OCP is constrained coupled, it is straightforward to keep the agents' objectives private as those are only associated to the local OCPs. When substituting agent dynamics (14b) into the cost (14a) and constraints (14c)–(14g), and utilizing optimization algorithms like ADMM, the agents exchange their primal variables, i.e. their control trajectories $u^{[i]}_{\cdot|k}$, to solve OCP (14) in a distributed way. As constraints (14g) are constituted in terms of the agents' states, though, the knowledge of the other agents' dynamic model becomes inevitable to evaluate their states as a function of their actions. Conversely, by leveraging a direct multiple shooting formulation, which optimizes over states and actions, agents could simply exchange their state trajectories, thus keeping their dynamic models and their parameters private. As indicated previously, the resulting OCP is however oftentimes more complex to solve. For the automated lane change maneuver, depicted in Fig. 1b, the author has pursued the latter approach, that is, a multiple-shooting OCP has been solved in a receding horizon ADMM framework [22]. That way, the agents only exchange their path coordinates (states) over the prediction horizon while their dynamic model and the associated parameters indeed remain private to the agents.

## 5 Towards Learning-Based Control

In the previous sections, it has been assumed that agent models are not subject to unmodeled dynamics, parametric uncertainties or exogenous disturbances (see Sect. 2, Assumption A7). Moreover, we contemplated every agent to be connected and equally automated (i.e., using the same control strategy), thus being able to reliably share its future intents (see Sect. 2, Assumptions A3–A4). These assumptions have led to the *nominal* control problem outlined in Sect. 4. Relaxing these assumptions gives rise to optimal control schemes which have to deal with model uncertainties or unknown exogenous disturbances. There is a rich body of literature which addresses such kind of problems in a robust or stochastic optimal control framework [38]. While the former considers the worst case realization of the uncertainty or disturbance at the price of higher conservatism and degraded performance, the latter exploits its stochastic nature to be less conservative. In everyday traffic scenarios, human drivers accept a certain risk of collision during their driving task [34]. Therefore, it is more naturalistic to design the controller in a stochastic fashion with sufficiently high probability of constraint satisfaction. In stochastic OCPs, constraints are predominantly imposed as so called *chance constraints*, which have to be satisfied with a minimal user-defined probability [38]. These stochastic optimal control schemes have recently been complemented by machine learning techniques such as to adapt the probabilistic representation of the uncertainty [7, 13, 39] as more data is collected from the environment.

In this section, the discussion is devoted to challenges and perspectives in learning-based model predictive control for multi-agent coordination. As this field of research is very broad and comprehensive, the reader should be aware that only selected topics are contemplated, that is, (i) model uncertainties related to every agent $i$ and (ii) mixed-traffic scenarios with connected and non-connected agents. In the latter case, the main source of uncertainty stems from the presence of non-connected automated or human-driven vehicles whose future motion trajectories are subject to uncertainty.

### 5.1 *Model Uncertainties*

The first perspective of extending the nominal control problem in Sect. 4 is to account for uncertainties in the dynamic model of agent $i$, i.e., to relax Assumption A7 in Sect. 2—while all agents are still connected and equally automated. Section 3 conveys two alternative representations of agent dynamics, that is, the double integrator model and the kinematic bicycle model. Although both models are usually sufficiently accurate for motion planning purposes, they entirely neglect resistance or tire forces and are as such subject to uncertainty which depends on the driving conditions [49]. Moreover, in the scope of intersection automation experimental tests [24] have revealed that the inclusion of drivetrain dynamics (as discussed in Sect. 3.3) into the double integrator model is crucial and has significant influence on the actual speed and position of the agent, and as such on keeping the required distance to other agents. The corresponding dynamic time constant $T_{a_x}$ oftentimes even has a nonlinear dependence on the state $x$ and the input $u$. Eventually, there are also further noise factors that influence the accuracy of the control-oriented model and as such the control performance and the satisfaction of constraints.

Let's assume that such model uncertainty is hard to describe by means of first principle models and may even change over time. Then, learning-based concepts can alleviate the problem of dealing with uncertainty. Such a concept is presented in Hewing et al. [13] where the authors learn the tire force model online while driving on the race track. Adapting this idea to the multi-agent coordination problem, the uncertain dynamics of every agent can be represented as

$$x_{k+1} = f_d(x_k, u_k) + g_d(x_k, u_k) + \nu_k \tag{15}$$

where $f_d(x_k, u_k)$ is the nominal part (see Sect. 3.4) and $g_d(x_k, u_k)$ the unknown part of the dynamics. Moreover, $\nu_k \in \mathbb{R}^{n_x}$ is considered to be independent and identically distributed (i.i.d.) white process noise, i.e., $\nu_k \sim \mathcal{N}(0, \Sigma^\nu)$ with positive semi-definite covariance matrix $\Sigma^\nu$. A prominent approach in the literature is to utilize Gaussian process regression (GPR) models [50]

$$g_d(x_k, u_k) \sim \mathcal{N}(\mu^{g_d}(x_k, u_k), \Sigma^{g_d}(x_k, u_k)) \tag{16}$$

to approximate the unknown dynamics through a vector $g_d : \mathbb{R}^{n_x} \times \mathbb{R}^{n_u} \to \mathbb{R}^{n_x}$ of Gaussian processes with mean $\mu^{g_d}(x_k, u_k)$ and covariance $\Sigma^{g_d}(x_k, u_k)$. The training data $\mathcal{D} = ((x^\iota, u^\iota), y^\iota)_{\iota=1}^{N_\mathcal{D}}$ is defined as a sequence of $N_\mathcal{D}$ visited state-input pairs $(x^\iota, u^\iota)$ and measurements $y^\iota$ where the latter is given by $x(t_{\iota+1}) - f_d(x(t_\iota), u(t_\iota))$, i.e., the difference between the next visited state $x(t_{\iota+1})$ and the nominal mapping $f_d$ of states $x(t_\iota)$ and inputs $u(t_\iota)$ to the next state. While collecting more data, the GPR model's posterior mean and covariance is updated by extending the so called Kernel matrix $K_{\text{GP}}$.

In the context of multi-agent coordination, $g_d(x_k, u_k)$ in (16) could represent the uncertain part of every agent's dynamic model, where the uncertainty originates, e.g., from neglecting resistance and tire forces, or from a time-varying drivetrain time constant. While observing data $\mathcal{D}$ during the maneuver, every agent's model can be adapted online through learning from data. When the learning process is independent from taking decisions for control, it is referred to as *passive learning*. Conversely, if control actions are explicitly chosen such as to gather suitable data to improve the learning objective, such process is referred to as *active learning* [58]. The latter is closely related to the *exploration versus exploitation* dilemma in reinforcement learning [56] as collecting useful data may (temporarily) sacrifice system performance or violate constraints.

To accommodate the uncertainty that resides in the learned model (15), we can integrate GPR models with a (centralized) stochastic OCP formulation. The covariance $\Sigma^{g_d}$ in (16) provides a quantification of uncertainty which can be exploited to impose input and state constraints (14c)–(14e) as well as coupling constraints (14g) as chance constraints. These probabilistic constraints have to be satisfied with a minimal user-defined probability [13, 19]. Compared to the nominal control problem (14), the stochastic OCP commonly minimizes the expected value of the cost (14a) subject to chance constraints.

A distributed formulation of such centralized stochastic OCP is mostly straightforward if the uncertainty does not affect coupling constraints (14g). Then, only the local subproblems can be modified while the rest of the distributed algorithm does not have to be changed. Conversely, if coupling constraints are influenced by the uncertainty, e.g., because uncertain drivetrain dynamics imply an uncertain position of the agent, the entire centralized stochastic OCP has to be transferred into a centralized deterministic counterpart first. Then, an appropriate distributed algorithm can be chosen that is able to accommodate the resulting type of coupling constraints.

Although the potentials are promising, the challenges that arise with GPR models are manifold and we are highlighting only on a few of them. First, GPR models aggregate data points in the kernel matrix $K_{\mathrm{GP}}$, whose dimensionality grows linearly with the number of data points. Moreover, the evaluation of the mean $\mu^{g_d}$ and covariance $\Sigma^{g_d}$ requires the inversion of a (high-dimensional) Kernel matrix [50]. In other words, an embedded implementation is challenged by an increasing demand for memory and computational resources as more data is collected. It is still part of ongoing research to evaluate suitable data-driven models for learning-based optimal control regimes such as to achieve a suitable trade-off between model accuracy, uncertainty quantification, memory demand and computational complexity.

Another important aspect is the (potential) discrepancy between the empirical distribution, being constructed from measured data, and the true underlying distribution. This discrepancy may eventually lead to a lower probability of constraint satisfaction and as such to a loss of rigorous safety guarantees. For that reason, distributionally robust control schemes, which account for the discrepancy between the empirical and true distribution, have recently gained significant attention [12, 48, 52] in the research community.

## 5.2 *Mixed-Traffic Scenarios*

When relaxing the assumption that all agents are connected (Sect. 2, Assumptions A3–A4), we are facing the challenge of so called *mixed-traffic scenarios* [11]. In the scope of a multi-agent coordination problem, the traffic scenario then involves a set of connected and automated agents $\mathcal{A}_{\mathrm{con}}$ (same as set $\mathcal{A}$ in Sect. 2), which apply the same control strategy, and a set of non-connected agents $\mathcal{A}_{\mathrm{ncon}}$ with $\mathcal{A}_{\mathrm{con}} \cap \mathcal{A}_{\mathrm{ncon}} = \emptyset$. The non-connected agents can either be automated (using a control strategy independent from the connected agents) or human-driven.

In contrast to Sect. 5.1, it is now assumed that the dynamic models, the dynamic models of agents $i \in \mathcal{A}_{\text{con}}$ are not subject to uncertainty (see Sect. 2, Assumption A7). That way, only the non-connected agents $l \in \mathcal{A}_{\text{ncon}}$ are introducing uncertainty into the control system due to their uncertain future motion trajectories. In the literature, the intelligent driver model (IDM) [26], e.g., has frequently been utilized in the context of car following problems such as to predict the (nominal) motion trajectory of the frontal vehicle. The IDM is a parametric dynamic model which mimics a human driver tracking a reference speed and keeping a safe distance to the preceding vehicle. Also other predictors like artificial neural networks (ANN) have been applied for that purpose [11]. Following the idea in Sect. 5.1, a meaningful approach would be to complement the nominal model of other road participants, such as the IDM, with unknown dynamics which are successively learned from data. For the sake of simplicity, it is assumed that the dynamics of every non-connected agent $l \in \mathcal{A}_{\text{ncon}}$ depend on the state vector $x_k^{[i]}$ of exactly one other connected agent $i \in \mathcal{A}_{\text{con}}$. Then, the discrete-time state space model of agent $l$ can be written as

$$\xi_{k+1}^{[l]} = f_{d,\xi}^{[l]}(\xi_k^{[l]}, x_k^{[i]}) + g_{d,\xi}^{[l]}(\xi_k^{[l]}, x_k^{[i]}) \tag{17}$$

where $\xi_k^{[l]}$ is the state vector of agent $l$. Moreover, $f_{d,\xi}^{[l]}$ represents the nominal part of the model, which is, e.g., given in terms of the IDM, whereas $g_{d,\xi}^{[l]}$ is learned from data—e.g., by utilizing a GPR model. This approach is meaningful if $f_{d,\xi}^{[l]}$ mimics agent $l$'s behavior generally well, i.e., if agent $l$ is a human-driven vehicle. Conversely, if agent $l$ is an automated vehicle, its actions may be entirely different from the IDM and it may be more reasonable to disregard $f_{d,\xi}^{[l]}$ in (17) and only learn the unknown dynamics $g_{d,\xi}^{[l]}$. Such a concept is pursued by Brüdigam et al. [7] in the context of autonomous racing where the dynamics of the leading vehicle are learned through a GPR model. While [7] follows a passive learning approach, an interesting perspective is to apply active learning, that is, to choose control actions such as to optimize a learning objective, like in reinforcement learning [51]. That way, the autonomous race car could gather relevant data from the leading vehicle that eventually helps to overtake the opponent. Such approach would also be helpful for mixed-traffic multi-agent coordination problems. For instance, consider a lane changing scenario, in which a connected agent intends to change to the adjacent lane, which is densely populated with non-connected agents. The connected agent could actively learn the behavior of the closest agents in the adjacent lane by safely driving towards the lane markings. That way, performance criteria like comfort may be sacrificed, however, meaningful data about agents in the adjacent lane can be collected. The connected agent can as such reason about the probability that the agents in the adjacent lane make sufficient space to allow for a lane change. The problem at hand gets even more complex if the dynamics (17) of agent $l \in \mathcal{A}_{\text{ncon}}$ are a function of two or more other agents, if the agents' actions are mutually dependent, or if the number of interactions with other agents varies over time. The latter can, for instance, be observed in intersection scenarios where the number of conflicting agents usually varies over time, see Sect. 2.

When utilizing a probabilistic model of non-connected agents that provides a measure of uncertainty, such as a GPR model, then this uncertainty measure can be utilized to come up with a centralized stochastic OCP, as outlined in Sect. 5.1. Again, the transition to a distributed stochastic control scheme is mainly challenged by solving the OCP subject to joint chance constraints.

## 6 Conclusion

This chapter has conveyed the challenges and perspectives of modeling and optimization-based control in the context of multi-agent coordination problems. Starting with the nominal control problem, that is, when no uncertainties affect the networked control system, centralized and distributed control schemes have been investigated on the example of intersection automation and lane change scenarios. Furthermore, potentials of generalizing controller design such as to be applicable to multiple scenarios simultaneously have been discussed. In distributed optimal control concepts, especially nonconvexity challenges convergence guarantees and a problem solution in real-time.

In Sect. 5, the potentials and challenges of learning-based control concepts for the safe coordination of agents in road traffic scenarios have been discussed. For the sake of brevity, the section has focused on model uncertainties and mixed-traffic scenarios. The discussion has conveyed that there are many opportunities to leverage machine learning methods for multi-agent coordination problems, and that active learning is a promising research direction to solve challenging scenarios. Along these lines, uncertainty is usually accommodated by solving a chance constrained optimal control problem. As the empirical distribution, derived from collected data, in reality deviates from the true underlying distribution, distributionally robust optimal control schemes are important to provide reliable safety guarantees in learning-based control. Many of these opportunities are still part of ongoing research work and are expected to introduce significant improvements for such kind of multi-agent coordination problems.

## References

1. Bertsekas D, Tsitsiklis J (1989) Parallel and distributed computation: numerical methods. Prentice Hall, Hoboken
2. Bevly D, Cao X, Gordon M, Ozbilgin G, Kari D, Nelson B, Woodruff J, Barth M, Murray C, Kurt A, Redmill K, Ozguner U (2016) Lane change and merge maneuvers for connected and automated vehicles: a survey. IEEE Trans Intell Veh 1(1):105–120
3. Blasi S, Kögel M, Findeisen R (2018) Distributed model predictive control using cooperative contract options. In: IFAC Conf Nonlinear Model Predictive Control 51(20):448–454

4. Bock H, Plitt K (1984) A multiple shooting algorithm for direct solution of optimal control problems. IFAC World Congr 17(2):1603–1608
5. Boyd S, Parikh N, Chu E, Peleato B, Eckstein J (2011) Distributed optimization and statistical learning via the alternating direction method of multipliers. Found Trends Mach Learn 3(1):1–122
6. Boyd S, Vandenberghe L (2004) Convex optimization. Cambridge University Press, Cambridge
7. Brüdigam T, Capone A, Hirche S, Wollherr D, Leibold M (2021) Gaussian process-based stochastic model predictive control for overtaking in autonomous racing. In: International conference on robotics and automation (ICRA)
8. Carvalho A, Lefévre S, Schildbach G, Kong J, Borrelli F (2015) Automated driving: the role of forecasts and uncertainty–a control perspective. Eur J Control 24:14–32
9. Engelmann A, Jiang Y, Houska B, Faulwasser T (2020) Decomposition of nonconvex optimization via bi-level distributed ALADIN. IEEE Trans Control Netw Syst 7(4):1848–1858
10. Febbo H, Liu J, Jayakumar P, Stein JL, Ersal T (2017) Moving obstacle avoidance for large, high-speed autonomous ground vehicles. In: IEEE American control conference, pp 5568–5573
11. Guo L, Jia Y (2021) Anticipative and predictive control of automated vehicles in communication-constrained connected mixed traffic. IEEE Trans Intell Transp Syst, 1–14
12. Hakobyan A, Yang I (2020) Learning-based distributionally robust motion control with Gaussian processes. In: IEEE conference on intelligent robots and systems (IROS), pp 7667–7674
13. Hewing L, Kabzan J, Zeilinger MN (2020) Cautious model predictive control using gaussian process regression. IEEE Trans Control Syst Technol 28(6):2736–2743
14. Hong M, Luo ZQ, Razaviyayn M (2016) Convergence analysis of alternating direction method of multipliers for a family of nonconvex problems. SIAM J Optim 26:337–364
15. Houska B, Ferreau HJ, Diehl M (2011) An auto-generated real-time iteration algorithm for nonlinear MPC in the microsecond range. Automatica 47(10):2279–2285
16. Hu X, Sun J (2019) Trajectory optimization of connected and autonomous vehicles at a multilane freeway merging area. Transp Res Part C Emerg Technol 101:111–125
17. Hult R, Zanon M, Gros S, Falcone P (2016) Primal decomposition of the optimal coordination of vehicles at traffic intersections. In: IEEE conference on decision and control, pp 2567–2573
18. Hult R, Zanon M, Gros S, Falcone P (2019) Optimal coordination of automated vehicles at intersections: theory and experiments. IEEE Trans Control Syst Technol 27(6):2510–2525
19. Janssen NHJ, Kools L, Antunes DJ (2020) Embedded Learning-based model predictive control for mobile robots using Gaussian process regression. In: IEEE American control conference, pp 1124–1130
20. Jiang Y, Zanon M, Hult R, Houska B (2017) Distributed algorithm for optimal vehicle coordination at traffic intersections. In: IFAC world congress, pp 11577–11582
21. Kamal MAS, Imura J, Hayakawa T, Ohata A, Aihara K (2015) A vehicle-intersection coordination scheme for smooth flows of traffic without using traffic lights. IEEE Trans Intell Transp Syst 16(3):1136–1147
22. Katriniok A (2020) Nonconvex consensus ADMM for cooperative lane change maneuvers of connected automated vehicles. IFAC world congress 53:14336–14343
23. Katriniok A, Kleibaum P, Joševski M (2017) Distributed model predictive control for intersection automation using a parallelized optimization approach. IFAC world congress 50:5940–5946
24. Katriniok A, Rosarius B, Mähönen P (2021) Fully distributed model predictive control of connected automated vehicles in intersections: theory and vehicle experiments. In: IEEE transactions on intelligent transportation systems. https://doi.org/10.1109/TITS.2022.3162038
25. Katriniok A, Sopasakis P, Schuurmans M, Patrinos P (2019) Nonlinear model predictive control for distributed motion planning in road intersections using PANOC. In: IEEE conference on decision and control, pp 5272–5278
26. Kesting A, Treiber M, Helbing D (2010) Enhanced intelligent driver model to access the impact of driving strategies on traffic capacity. Philos Trans R Soc A Math Phys Eng Sci 368:4585–4605

27. Khayatian M, Mehrabian M, Andert E, Dedinsky R, Choudhary S, Lou Y, Shirvastava A (2020) A survey on intersection management of connected autonomous vehicles. ACM Trans Cyber-Phys Syst 4(4):1–27
28. Kim KD, Kumar PR (2014) An MPC-based approach to provable system-wide safety and liveness of autonomous ground traffic. IEEE Trans Autom Control 59:3341–3356
29. Kneissl M, Molin A, Esen H, Hirche S (2018) A feasible MPC-based negotiation algorithm for automated intersection crossing. In: European control conference, pp 1282–1288
30. Lim W, Lee S, Sunwoo M, Jo K (2018) Hierarchical trajectory planning of an autonomous car based on the integration of a sampling and an optimization method. IEEE Trans Intell Transp Syst 19(2):613–626
31. Lipp T, Boyd S (2016) Variations and extension of the convex-concave procedure. Optim Eng 17(2):263–287
32. Liu C, Lin C, Shiraishi S, Tomizuka M (2018) Distributed conflict resolution for connected autonomous vehicles. IEEE Trans Intell Veh 3(1):18–29
33. Liu P, Ozguner U, Zhang Y (2017) Distributed MPC for cooperative highway driving and energy-economy validation via microscopic simulations. Transp Res Part C Emerg Technol 77:80–95
34. Liu P, Yang R, Xu Z (2019) How safe is safe enough for self-driving vehicles? Risk Anal 39(2):315–325
35. Maciejowski J (2002) Predictive control with constraints. Prentice Hall, Harlow
36. Malikopoulos AA, Beaver L, Chremos IV (2021) Optimal time trajectory and coordination for connected and automated vehicles. Automatica 125:109469
37. Malikopoulos AA, Cassandras CG, Zhang YJ (2018) A decentralized energy-optimal control framework for connected automated vehicles at signal-free intersections. Automatica 93:244–256
38. Mesbah A (2016) Stochastic model predictive control: an overview and perspectives for future research. IEEE Control Syst Mag 36(6):30–44
39. Mesbah A (2018) Stochastic model predictive control with active uncertainty learning: a survey on dual control. Annu Rev Control 45:107–117
40. Molinari F, Grapentin A, Charalampidis A, Raisch J (2019) Automating lane changes and collision avoidance on highways via distributed agreement. Automatisierungstechnik 67(12):1047–1057
41. Molinari F, Katriniok A, Raisch J (2020) Real-time distributed automation of road intersections. IFAC world congress 52:2606–2613
42. Molinari F, Raisch J (2018) Automation of road intersections using consensus-based auction algorithms. In: IEEE American control conference, pp 5994–6001
43. Namazi E, Li J, Lu C (2019) Intelligent intersection management systems considering autonomous vehicles: a systematic literature review. IEEE Access 7:91946–91965
44. Nocedal J, Wright SJ (2006) Numerical optimization, 2nd edn. Springer, Berlin
45. Notarstefano G, Notarnicola I, Camisa A (2019) Distributed optimization for smart cyber-physical networks. Foundations Trends Syst Control 7(3):253–383
46. Qian X, de La Fortelle A, Moutarde F (2016) A hierarchical model predictive control framework for on-road formation control of autonomous vehicles. In: IEEE intelligent vehicles symposium, pp 376–381
47. Quinlan M, Au TC, Zhu J, Stiurca N, Stone P (2010) Bringing simulation to life: a mixed reality autonomous intersection. In: Proceedings of IROS 2010-IEEE/RSJ international conference on intelligent robots and systems (IROS 2010)
48. Rahimian H, Mehrotra S (2019) Distributionally robust optimization: a review. In: arXiv preprint arXiv:1908.05659
49. Rajamani R (2012) Vehicle dynamics and control, vol 2. Springer, Berlin
50. Rasmussen C, Williams C (2006) Gaussian processes for machine learning. The MIT Press, Cambridge
51. Sadigh D, Dragan AD, Sastry SS, Seshia SA (2017) Active preference-based learning of reward functions. In: Robotics: science and systems

52. Schuurmans M, Katriniok A, Tseng HE, Patrinos P (2020) Learning-based risk-averse model predictive control for adaptive cruise control with stochastic driver models. IFAC world congress 53:15128–15133
53. Shi J, Zheng Y, Jiang Y, Zanon M, Hult R, Houska B (2018) Distributed control algorithm for vehicle coordination at traffic intersections. In: European control conference, pp 1166–1171
54. Sopasakis P, Fresk E, Patrinos P (2020) OpEn: code generation for embedded nonconvex optimization. IFAC world congress 53:6548–6554
55. Stella L, Themelis A, Patrinos P (2017) Forward-backward quasi-Newton methods for nonsmooth optimization problems. Comput Optim Appl 67(3):443–487
56. Sutton RS, Barto AG (2018) Reinforcement learning: an introduction, 2nd edn. The MIT Press, Cambridge
57. Tang W, Daoutidis P (2021) Fast and stable nonconvex constrained distributed optimization: the ELLADA algorithm. Optimization and engineering, Springer, Berlin
58. Taylor AT, Berrueta TA, Murphey TD (2021) Active learning in robotics: a review of control principles. Mechatronics 77:102576
59. Wang D, Hu M, Wang Y, Wang J, Qin H, Bian Y (2016) Model predictive control-based cooperative lane change strategy for improving traffic flow. Adv Mech Eng 8(2):1–17

# Virtual Rings on Highways: Traffic Control by Connected Automated Vehicles

**Tamás G. Molnár, Michael Hopka, Devesh Upadhyay, Michiel Van Nieuwstadt, and Gábor Orosz**

**Abstract** This work gives introduction to traffic control by connected automated vehicles. The influence of vehicle control on vehicular traffic and traffic control strategies are discussed and compared. It is highlighted that vehicle-to-everything connectivity allows connected automated vehicles to access the state of the traffic behind them such that feedback can be utilized to mitigate evolving congestions. Numerical simulations demonstrate that such connectivity-based traffic control is beneficial for smoothness and energy efficiency of highway traffic. The dynamics and stability of traffic flow, under the proposed controllers, are analyzed in detail to construct stability charts that guide the selection of stabilizing control gains.

T. G. Molnár (✉)
Department of Mechanical and Civil Engineering, California Institute of Technology, Pasadena, CA 91125, USA
e-mail: tmolnar@caltech.edu

M. Hopka
Ford Research and Innovation Center, Dearborn, MI 48124, USA
e-mail: mhopka@ford.com

D. Upadhyay
Ford Research and Innovation Center, Dearborn, MI 48124, USA
e-mail: dupadhya@ford.com

M. Van Nieuwstadt
Ford Research and Innovation Center, Dearborn, MI 48124, USA
e-mail: mvanniel@ford.com

G. Orosz
Department of Mechanical Engineering and Department of Civil and Environmental Engineering, University of Michigan, Ann Arbor, MI 48109, USA
e-mail: orosz@umich.edu

 
Y. L. Murphey et al. (eds.), *AI-enabled Technologies for Autonomous and Connected Vehicles*, Lecture Notes in Intelligent Transportation and Infrastructure,
https://doi.org/10.1007/978-3-031-06780-8_16

# 1 Introduction

Traffic congestion is a major factor in reducing the efficiency of road transportation, as it increases travel times, energy consumption of vehicles and air pollution. Human-driven traffic often suffers from the onset of stop-and-go traffic jams on highways. These may occur without any incident and may be triggered by the human driving behavior [1]. This is illustrated in Fig. 1a, where a chain of human-driven vehicles (HVs) is simulated on a single lane (with details given later in the paper). When the lead vehicle (black) decelerates, the following human drivers (gray) tend to overreact and reduce their speed more: they get slower the farther they are behind the lead vehicle. This, so-called string unstable driving behavior is typical in human-driven traffic [2]. It eventually leads to low speeds and stop-and-go motion on highways, giving birth to a traffic congestion [3].

This chapter is dedicated to strategies that allow prevention or mitigation of such driving behavior-induced traffic congestions. Alleviating traffic jams has substantial benefits for each individual vehicle participating in traffic: it improves their travel time and energy consumption. The key factor to obtain these benefits is to ensure that vehicles drive as smoothly as possible without major deceleration. Smoothness is often analyzed via the notion of string stability [4–7], which also has been used recently to evaluate commercial cruise control systems [8–10]. Smooth driving can be achieved either at the level of individual vehicles by *vehicle control* or at large-scale traffic level by *traffic control* [11].

Vehicle control can be realized on automated vehicles (AVs), with various levels of automation. Control strategies include adaptive cruise control (ACC) for individual AVs [12], connected cruise control (CCC) for connected automated vehicles [13], and cooperative adaptive cruise control (CACC) for vehicle platoons [14]. Controlling vehicles to drive smoothly achieves significant benefits in energy efficiency, safety and passenger comfort [15–17]. While vehicle control guarantees benefits for the individual AVs only, a sufficient number of well-designed AVs may positively influence traffic at large and mitigate traffic jams due to their smooth driving behavior [18–21]. The influence of vehicle control on traffic is demonstrated in Fig. 1b, where AVs (green) are mixed with HVs (gray) in the traffic flow. As the AVs drive smoothly and reduce their speed less than the vehicle they follow, the traffic jam is mitigated. This way, vehicle control also controls traffic indirectly, however, there is no direct measurement or feedback of the traffic state.

Traffic control, on the other hand, directly uses the state of the traffic flow as feedback. Traffic state estimation may rely on data from loop detectors or cameras at fixed locations [3, 22, 23] or from vehicles traveling in the traffic flow [24–27]. State estimators may involve Kalman filtering techniques [28–30] and data fusion [31–33]. Via traffic state feedback, traffic control aims to maximize the benefits of large-scale traffic, which eventually leads to benefits for individual vehicles too. Research in traffic control has seen a surge in recent years [34], extending from classical traffic control [11] to reinforcement learning [35, 36]. Traffic control has been approached from so-called Eulerian and Lagrangian perspectives [37]. In Eulerian traffic control,

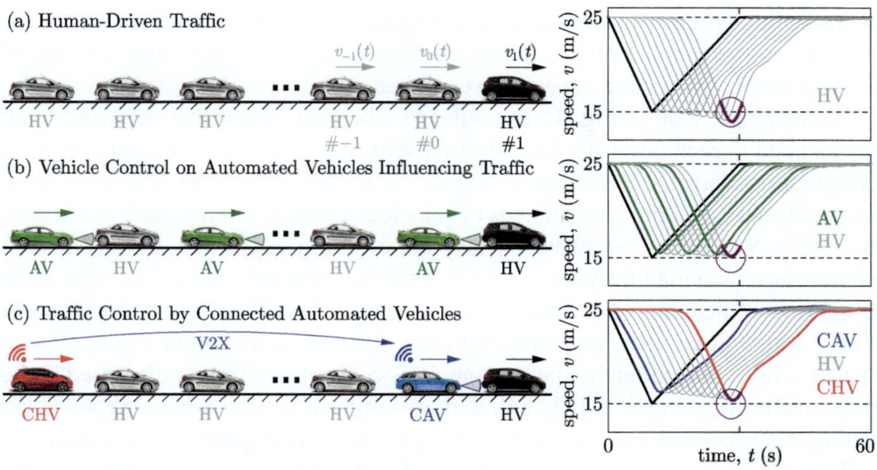


**Fig. 1** Dynamics of traffic flows. **a** A chain of human-driven vehicles (HVs). If the lead vehicle brakes, the vehicle chain amplifies it and the tail vehicle reduces its speed more than the lead. **b** Vehicle control on large penetration of automated vehicles (AVs) which influences traffic. The tail vehicle reduces its speed about as much as the lead. **c** Traffic control with a single connected automated vehicle (CAV) where the tail vehicle is a connected human-driven vehicle (CHV). The tail vehicle reduces its speed less than the lead

traffic flow is regulated at specific locations along the highway based on data about the traffic further down the road [38, 39]. Traffic control measures include speed limits [40] and ramp metering [41–44]. In Lagrangian traffic control, traffic is regulated by the driving behavior of certain vehicles in the flow [45, 46], which directly take into account how they affect traffic, and utilize the traffic state as feedback in their controllers.

Since Lagrangian traffic controllers have been enabled by vehicle automation, they have appeared more recently than Eulerian traffic control, and their literature is still less extensive. Meanwhile, technological advancement in the field of automated vehicles keeps opening new possibilities for traffic control. In particular, the occurrence of connected automated vehicles (CAVs) has significant potential that has not yet been addressed by the literature. Therefore, this chapter is devoted to establishing the notion of Lagrangian traffic control by connected automated vehicles, and highlighting its potential for mitigating traffic congestions.

Namely, automated vehicles must monitor the state of the traffic behind them while traveling on highways in order to achieve Lagrangian control of the traffic behind. Monitoring traffic can be realized by vehicle-to-everything (V2X) connectivity, which becomes the enabling technology of traffic control by connected automated vehicles. A CAV, communicating with vehicles behind it, receives traffic state information that can be used as feedback [47]. This approach is demonstrated in Fig. 1c, where a CAV communicates with and responds to a connected human-driven vehicle (CHV) behind it. Information from the CHV (red) allows the CAV (blue) to mitigate the traffic jam efficiently: the entire vehicle chain reduces its speed less than for the case with a large number of AVs in Fig. 1b. This demonstrates the large potential of CAVs in traffic control that is yet to be exploited.

The feedback loop in Fig. 1c is similar to when vehicles travel on a ring road: they respond to each other in a loop, the first vehicle responding to the last one. Ring configurations have rich and interesting dynamics [3, 17, 48–50] and there have even been successful implementations of vehicle controllers on ring roads to mitigate traffic [51]. While the road geometry may not be a ring in practice, the response of vehicles in Lagrangian traffic control has a similar structure, which we refer to as virtual rings. The concept of virtual rings have been introduced in the experiments of [52] where three vehicles performed car following on a straight highway, with the first vehicle responding to the last one. This concept was further discussed in the context of traffic control in [53, 54], and also utilized in [55] where the so-called leading cruise control was introduced for high-connectivity penetration scenarios.

Here we further elaborate the concept of virtual rings and Lagrangian traffic control by CAVs. As opposed to previous works [54, 55], we put more emphasis on the stability analysis of traffic flow, focus on low-automation and low-connectivity scenarios, and incorporate time delays associated with the response of vehicles, their controllers and the human drivers. We first introduce strategies for vehicle and traffic control, then highlight the benefits of connectivity-enabled feedback loops via simulations, and finally give a comprehensive stability analysis.

## 2 Traffic Control by Connected Automated Vehicles

We consider traffic regulation by means of controlling the motions of automated vehicles (AVs) and connected automated vehicles (CAVs) traveling in the traffic flow. Since the driving behavior of these AVs and CAVs affects the motion of other vehicles behind them, including human-driven vehicles (HVs) and connected human-driven vehicles (CHVs), they regulate the traffic behind.

First, we discuss strategies where AVs and CAVs respond to vehicles ahead of them only. In this case, AVs and CAVs indirectly control the traffic behind them, without feedback of the traffic state. We call this case as *vehicle control influencing traffic*. Then we propose strategies where CAVs respond to vehicles behind them as well. This allows CAVs to directly control traffic by using the traffic state behind as feedback. We refer to this case as *traffic control*.

We consider single lane traffic for simplicity of exposition. The results could be extended to multiple lanes by incorporating the cross-lane dynamics. While lane changes and overtaking requires further investigation and extensive future work, we remark that traffic controllers realized on CAVs consider the interest of the vehicles behind them, which makes it less likely for those vehicles to overtake the CAVs.

## 2.1 *Simplified Models for Longitudinal Vehicle and Traffic Dynamics*

Consider the scenario in Fig. 1, in which vehicles follow each other on a single lane of a straight road. We number the vehicles with indices increasing in the direction of motion. We denote the set of all vehicle indices by $\mathcal{N}$ that comprises of the indices $\mathcal{N} = \{\mathcal{N}_{\mathrm{HV}}, \mathcal{N}_{\mathrm{CHV}}, \mathcal{N}_{\mathrm{AV}}, \mathcal{N}_{\mathrm{CAV}}\}$ representing four vehicle types defined previously. We distinguish a so-called *ego vehicle* of interest by index 0. We denote the length of vehicle $n$ by $l_n$, the position of its rear bumper by $s_n$ and its speed by $v_n$, $n \in \mathcal{N}$.

We model the motion of vehicle $n$ by a delayed double integrator with saturation:

$$\begin{aligned} \dot{s}_n(t) &= v_n(t)\,, \\ \dot{v}_n(t) &= \mathrm{sat}(u_n(t-\tau_n))\,, \end{aligned} \tag{1}$$

$\forall n \in \mathcal{N}$, where $u_n$ is the desired acceleration of vehicle $n$, selected by the human driver for HVs and CHVs ($n \in \{\mathcal{N}_{\mathrm{HV}}, \mathcal{N}_{\mathrm{CHV}}\}$) and prescribed by the longitudinal controller of AVs and CAVs ($n \in \{\mathcal{N}_{\mathrm{AV}}, \mathcal{N}_{\mathrm{CAV}}\}$). We assume that each vehicle realizes the desired acceleration by the help of human action or low-level controllers, respectively, unless this acceleration is above the acceleration capability $a_{\max}$ or below the braking limit $-a_{\min}$ of the vehicle. This is captured by the saturation function:

$$\mathrm{sat}(u) = \min\{\max\{-a_{\min}, u\}, a_{\max}\}\,. \tag{2}$$

Furthermore, we incorporate the time delay $\tau_n$ into the model, that involves the response time of the vehicle, as well as the driver reaction time for HVs and CHVs, and feedback delays for AVs and CAVs. For simplicity, we assume identical delays for HVs and CHVs throughout this study with $\tau_n = \tau$, $n \in \{\mathcal{N}_{\mathrm{HV}}, \mathcal{N}_{\mathrm{CHV}}\}$, and we distinguish the delay of the ego vehicle 0 by the notation $\sigma = \tau_0$.

To capture human driver behavior, we use simple car-following models. Namely, our *human driver model (HDM)* assumes that vehicle $n$ responds to vehicle $n+1$ ahead considering the headway (range) $h_n = s_{n+1} - s_n - l_n$ and the speed difference (range rate) $\dot{h}_n = v_{n+1} - v_n$ between them:

$$u_n = f_{\mathrm{H}}(h_n, \dot{h}_n, v_n)\,, \tag{3}$$

where the specific expression of $f_{\mathrm{H}}$ depends on the choice of the model. Examples for HDM include the optimal velocity model (OVM) [56] and the intelligent driver model (IDM) [57], which were shown to capture human driver behavior observed in experimental data [52, 58]. For simplicity of exposition, throughout this paper we assume that human drivers are identical in their driving behaviors and $f_{\mathrm{H}}$ is the same for each $n \in \{\mathcal{N}_{\mathrm{HV}}, \mathcal{N}_{\mathrm{CHV}}\}$.

***Example 1*** For the numerical examples of this paper, we use the optimal velocity model as human driver model:

$$f_{\rm H}(h_n, \dot{h}_n, v_n) = \alpha_{\rm H}\big(V_{\rm H}(h_n) - v_n\big) + \beta_{\rm H}\dot{h}_n\,. \tag{4}$$

Here $\alpha_{\rm H}$ and $\beta_{\rm H}$ are parameters describing the human driver. The second term with coefficient $\beta_{\rm H}$ captures the response of human drivers to the speed difference $\dot{h}_n = v_{n+1} - v_n$ relative to the vehicle ahead. The first term with parameter $\alpha_{\rm H}$ characterizes the response to the headway $h_n = s_{n+1} - s_n - l_n$ measured from the vehicle ahead. We assume that human drivers track a headway-dependent desired speed given by the range policy:

$$V_{\rm H}(h) = \min\{\max\{0, F_{\rm H}(h)\}, v_{\max}\}\,. \tag{5}$$

This desired speed is nonnegative and it saturates at the speed limit $v_{\max}$. The speed increases strictly monotonically between 0 and $v_{\max}$ at a rate defined by $F_{\rm H}$ that is a function of the headway $h$.

## 2.2 *Vehicle Control Influencing Traffic*

Now we consider vehicle control for (connected) automated vehicles where AVs and CAVs influence the traffic behind them without explicitly responding to its state. Consider the scenario in Fig. 2a–c. We assume that the ego vehicle (vehicle 0) is

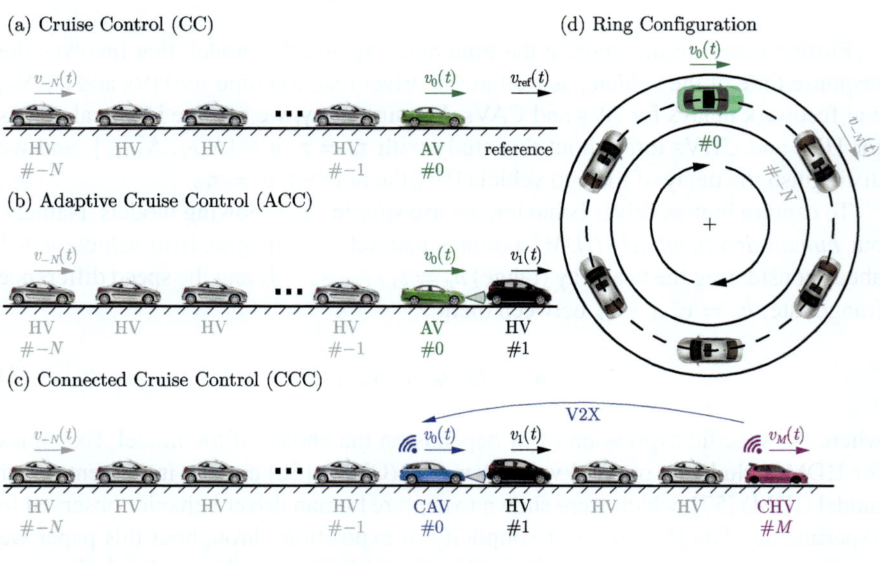


**Fig. 2** Vehicle control strategies where the driving behavior of (connected) automated vehicles influences the traffic behind them, including **a** cruise control, **b** adaptive cruise control, **c** connected cruise control, and **d** these strategies realized on a ring road

automated (an AV in Fig. 2a, b and a CAV in Fig. 2c), and it is followed by $N$ human-driven vehicles (HVs) constituting the traffic influenced by vehicle control.

Our goal is to design a control input $u_0$ for the ego vehicle based on the traffic ahead and observe its effects on the behavior of the traffic behind. If traffic conditions are free-flowing and there are no vehicles ahead of the AV such as in Fig. 2a, then the AV may use *cruise control (CC)* to track a reference speed $v_{\text{ref}}(t)$ such that its speed $v_0(t)$ approaches $v_{\text{ref}}(t)$:

$$u_0 = f_{\text{CC}}(v_0, \underbrace{v_{\text{ref}}}_{\text{reference}}). \tag{6}$$

The specific form of $f_{\text{CC}}$ depends on the controller type. The reference speed $v_{\text{ref}}(t)$ may be constant, or time-varying, or it may even depend on the position $s_0(t)$—an example for the latter case is when vehicles optimize their set speed based on road elevation [59].

If the traffic is more dense, the AV may need to respond to the vehicle ahead; see Fig. 2b. This can be achieved by sensing the position and speed of the vehicle ahead via on-board sensors (radar, lidar, camera or ultrasonics) and using *adaptive cruise control (ACC)* that takes into account the positions and speeds of the AV and the vehicle ahead:

$$u_0 = f_{\text{ACC}}(s_0, v_0, \underbrace{s_1, v_1}_{\text{vehicle ahead}}). \tag{7}$$

Finally, if the ego vehicle is equipped with a communication device, i.e., it is a CAV, then it may respond to connected human-driven vehicles (CHVs) farther ahead, as shown by Fig. 2c. This leads to *connected cruise control (CCC)* that may potentially take into account the positions and speeds of $M$ vehicles ahead:

$$u_0 = f_{\text{CCC}}(s_0, v_0, \underbrace{s_1, v_1, \ldots, s_M, v_M}_{\text{multiple vehicles ahead}}). \tag{8}$$

If a vehicle ahead is not connected to or sensed by the CAV, its state can be omitted from $f_{\text{CCC}}$. An aggregated response to multiple vehicles ahead facilitates smooth driving by making the CAV more clairvoyant about upcoming changes in traffic conditions [16].

***Example 2*** Throughout this paper, we consider the following simple examples as controllers for AVs and CAVs that influence the traffic behind. The simplest CC strategy is a proportional controller for speed tracking with a control gain $\beta$:

$$f_{\text{CC}}(v_0, v_{\text{ref}}) = \beta\big(v_{\text{ref}} - v_0\big). \tag{9}$$

Indeed, this law could be replaced by other more sophisticated controllers.

For ACC, we consider the following control law:

$$f_{\text{ACC}}(s_0, v_0, s_1, v_1) = \alpha\big(V(s_1 - s_0 - l_0) - v_0\big) + \beta\big(W(v_1) - v_0\big). \quad (10)$$

This is an analog to the optimal velocity model (4), with the difference that the control gains $\alpha$, $\beta$ and the range policy $V$ can be selected as control design parameters. For $V$ we keep the form (5) and prescribe the desired speed-headway relationship via $F$. Furthermore, in (10) we include the speed policy $W$ that allows the AV to follow the speed $v_1$ of the vehicle ahead or the speed limit $v_{\text{max}}$, whichever is smaller:

$$W(v_1) = \min\{v_1, v_{\text{max}}\}. \quad (11)$$

Finally, for CCC we extend the controller (10) with response to the speed of multiple vehicles ahead [60]:

$$f_{\text{CCC}}(s_0, v_0, s_1, v_1, \ldots, s_M, v_M) = \alpha\big(V(s_1 - s_0 - l_0) - v_0\big) + \sum_{m=1}^{M} \beta_m\big(W(v_m) - v_0\big), \quad (12)$$

where $\beta_m$ is the control gain associated with the speed $v_m$ of vehicle $m$ ahead. If some vehicle $m$ is not connected and cannot be perceived by the CAV, we omit the response to it by taking $\beta_m = 0$.

## 2.3 *Traffic Control*

Now consider the scenario illustrated in Fig. 3a–c, where the ego vehicle is a CAV and at least one vehicle behind it is connected, e.g., a CHV. Once the CAV syncs with the CHV, it can actively regulate the traffic behind, since it uses traffic state feedback when responding to the CHV. When there are no vehicles ahead of the CAV to respond to, such as in Fig. 3a, the cruise control (6) can be extended to *traffic control (TC)* in response to the positions and speeds of the connected vehicles behind the ego CAV:

$$u_0 = f_{\text{TC}}(\underbrace{s_{-N}, v_{-N}, \ldots, s_{-1}, v_{-1}}_{\text{traffic behind}}, s_0, v_0, \underbrace{v_{\text{ref}}}_{\text{reference}}). \quad (13)$$

The quantities $s_{-N}, v_{-N}, \ldots, s_{-1}, v_{-1}$ represent the state of the traffic behind the CAV. We assume that not all vehicles behind the ego CAV are connected, thus, only some of these states (corresponding to connected vehicles) may be available to the CAV and affect $u_0$. Accordingly, $-N$ denotes the index of the farthest connected vehicle behind communicating with the ego CAV. This scenario is ideal for traffic control in the sense that the CAV does not have to respond to the traffic ahead in order to minimize collision risks. Such large control freedom enables the CAV to stabilize long vehicle chains behind it.

However, when the CAV is traveling in denser traffic environments like the one in Fig. 3b, it has to respond to the vehicle ahead and has less freedom to regulate the traffic behind it. In such a setup, the adaptive cruise control (7) can be extended with feedback from the traffic behind to *adaptive traffic control (ATC)*:

$$u_0 = f_{\mathrm{ATC}}(\underbrace{s_{-N}, v_{-N}, \ldots, s_{-1}, v_{-1}}_{\text{traffic behind}}, s_0, v_0, \underbrace{s_1, v_1}_{\text{vehicle ahead}}). \tag{14}$$

Finally, if vehicles ahead of the CAV are also connected, as illustrated by Fig. 3c, we may establish *connected traffic control (CTC)* that responds to multiple vehicles ahead while also taking into account its effect on the traffic behind:

$$u_0 = f_{\mathrm{CTC}}(\underbrace{s_{-N}, v_{-N}, \ldots, s_{-1}, v_{-1}}_{\text{traffic behind}}, s_0, v_0, \underbrace{s_1, v_1, \ldots, s_M, v_M}_{\text{multiple vehicles ahead}}). \tag{15}$$

***Remark 1*** (Connectivity and automation) We remark that although only CTC includes the word "connected" in its name, TC and ATC also rely on connectivity to obtain information from the traffic behind (unless $N = 1$ and the vehicle behind the AV is detected by on-board sensors). At the same time, automation is not required by any of these strategies for vehicles other than the ego vehicle. Yet, if other connected vehicles happen to possess sufficient levels of automation (they are CAVs), that opens the possibility of coordinating CAVs for traffic control [61–63]. In this paper, we will not discuss details about such coordination.

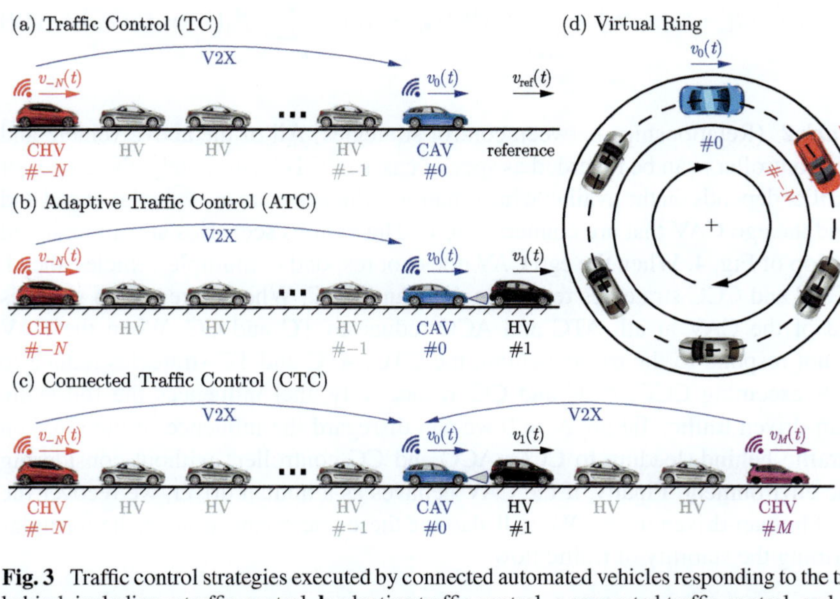


**Fig. 3** Traffic control strategies executed by connected automated vehicles responding to the traffic behind, including **a** traffic control, **b** adaptive traffic control, **c** connected traffic control, and **d** the virtual ring associated with these traffic control strategies

***Example 3*** Analogously to the CCC strategy in (12), response to the traffic behind can be achieved from the feedback of the speed of each connected vehicle behind. This leads to the TC algorithm:

$$u_0 = \beta\big(v_{\rm ref} - v_0\big) + \sum_{n=-N}^{-1} \beta_n\big(W(v_n) - v_0\big)\,, \tag{16}$$

cf. (9). Again, $\beta_n = 0$ is considered if the CAV does not perceive a vehicle due to the lack of sensors or connectivity.

The ACC strategy (10) can also be extended to ATC with a feedback term from the traffic behind:

$$u_0 = \alpha\big(V(s_1 - s_0) - v_0\big) + \beta\big(W(v_1) - v_0\big) + \sum_{n=-N}^{-1} \beta_n\big(W(v_n) - v_0\big)\,. \tag{17}$$

In what follows, we focus on scenarios with lean penetration of connectivity. If only a single vehicle is connected to the CAV, we obtain

$$u_0 = \alpha\big(V(s_1 - s_0) - v_0\big) + \beta\big(W(v_1) - v_0\big) + \beta_{\rm B}(W(v_{-N}) - v_0)\,, \tag{18}$$

where $\beta_{\rm B} = \beta_{-N}$ is the control gain related to the traffic behind.

Finally, the CCC (12) can be extended to CTC according to

$$u_0 = \alpha\big(V(s_1 - s_0) - v_0\big) + \sum_{m=1}^{M} \beta_m\big(W(v_m) - v_0\big) + \sum_{n=-N}^{-1} \beta_n\big(W(v_n) - v_0\big)\,. \tag{19}$$

***Remark 2*** (Relationships between control strategies) All of the above vehicle and traffic controllers can be regarded as special cases of CTC. Ultimately, the choice of controller depends on the available information: whether there are vehicles ahead and behind the ego CAV that are connected to it. The various scenarios are summarized at the top of Fig. 4. When the ego CAV does not respond to multiple vehicles ahead, the CTC and CCC strategies reduce to ATC and ACC. When there are no vehicles ahead of the CAV at all, ATC and ACC reduce to TC and CC. When the CAV does not respond to the traffic behind, the CTC, ATC and TC strategies reduce to a CAV executing CCC, ACC and CC, respectively, that influences the following human-driven traffic. Taking $N = 0$ we can disregard the influence of the CAV on the traffic behind, leading to CCC, ACC and CC controllers without considering traffic environment. Finally, if the CAV behaves as a human driver, we recover the case of human-driven traffic. We will also see these interconnections in the formulas describing the stability of traffic flow.

***Remark 3*** (Rings and virtual rings) Longitudinal controllers of AVs and CAVs are often analyzed on ring roads, as illustrated by Fig. 2d. Mathematically, the ring

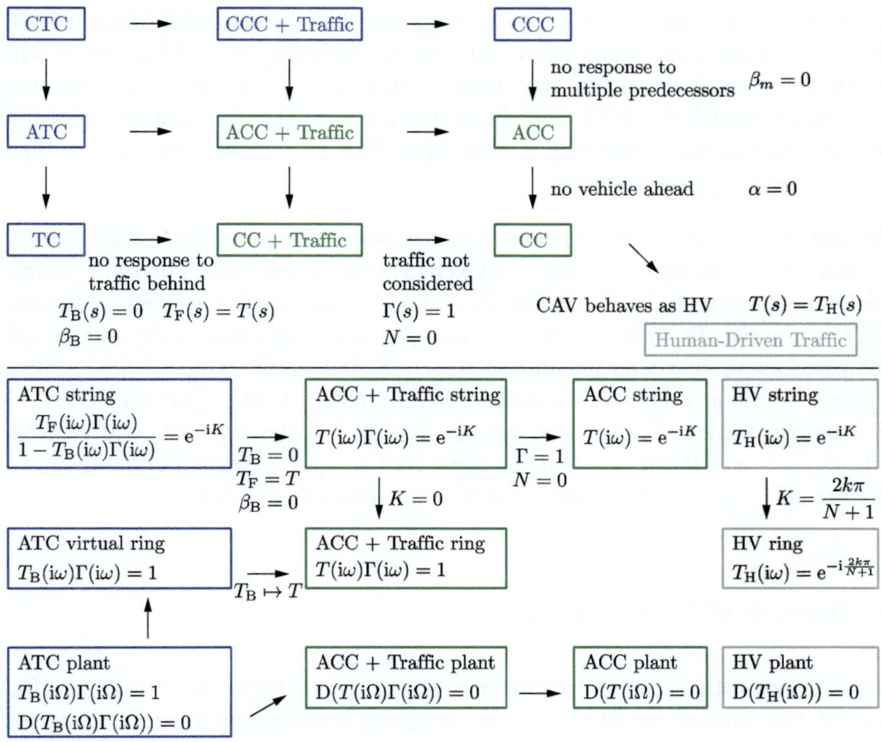


**Fig. 4** Relationships between control strategies (top) and the associated stability conditions (bottom)

configuration has an additional periodic boundary condition. For example, if there are $N + 1$ vehicles (indexed from $-N$ to 0) on a ring, we have:

$$\begin{aligned} s_{-N}(t) &= s_1(t)\,, \\ v_{-N}(t) &= v_1(t)\,. \end{aligned} \tag{20}$$

This is a useful property for the stability analysis of traffic, as discussed below. Apart from analysis [3, 17, 48–50], ring roads have been used in experiments, including successful mitigation of traffic jams [51].

The ring configuration is important conceptually for traffic control that involves feedback of the traffic behind. Namely, the structure of which vehicles respond to which other vehicles includes a ring (a loop), see Fig. 3d, without physically traveling on a ring road. Therefore, we refer to the structure of traffic control as *virtual ring*. The concept of virtual ring appeared in the experiments of [52] and the analysis of [53, 54]. Below we build on these works and give a more detailed analysis.

We also remark that traffic control on single-lane ring roads has a significant advantage compared to single-lane straight roads: the CAV can affect the equilibrium speed of the ring and can decide to travel slower than the set speed of the vehicle ahead. Once the CAV slows down, all vehicles including the vehicle ahead will eventually travel slower on the ring. In contrast, if the CAV decides to travel slower on a straight road, it will fall behind the vehicle ahead and break up the car-following scenario. Hence the virtual ring is more appealing for highway applications than considering a physical ring road.

***Remark 4*** (Scenario in the rest of the paper) In what follows, we analyze the dynamics and stability of traffic influenced by vehicle control and traffic control. For simplicity, we consider ACC, like in Fig. 2b, and ATC with a single CHV behind the ego CAV, like in Fig. 3b. This restricts us to lean penetration of connectivity and we can compare how vehicle control without connectivity influences traffic to traffic control allowed by connectivity. We use human driver model (4), ACC (10) and ATC (18) as example. Vehicle 1 is the lead vehicle, vehicle 0 is the ego CAV, and vehicles $-1, \ldots, -N$ are HVs with $-N$ being connected in the ATC setup. Furthermore, we analyze ACC on a ring road and relate it to the virtual ring of ATC.

## 3 Benefits of Connectivity

Now we demonstrate that information from connectivity is highly beneficial for traffic control. We compare the ACC and ATC setups by investigating the traffic behavior through numerical simulations. We also evaluate the energy consumption of vehicles and show that utilizing connectivity leads to energy savings. The parameters used for numerical case studies throughout this paper are listed in Table 1 in the Appendix.

### *3.1 Simulation Results*

First, we study a baseline scenario without automation. We simulate (1), (3) for human driver model (4) in Example 1 with typical human driver parameters selected from [52]. The simulation results are shown in Fig. 1a. In the simulation, a lead vehicle (black) sequentially decelerates, accelerates and cruises at constant speed, followed by 11 HVs (gray). The following HVs tend to reduce their speed more the farther they are from the lead vehicle as speed perturbations amplify along the vehicle chain. Consequently, the tail vehicle brakes noticeably more than the lead (purple highlight). This is called string unstable behavior and often leads to the formation of stop-and-go traffic jams for large enough number of string unstable drivers.

String instability can be mitigated by vehicle control using AVs or CAVs. To demonstrate how vehicle control influences traffic, we simulate (1), (3), (7) for a heterogeneous chain of HVs and AVs using ACC controller (10) in Example 2. The simulation results shown by Fig. 1b indicate that the AVs can successfully mitigate the onset of a traffic congestion: the tail vehicle brakes as much as the lead vehicle and not more (purple highlight). We remark that congestions are often caused by the reaction time of human drivers: a delayed reaction forces drivers to overreact and brake or accelerate more. As such, automated vehicles must be designed carefully to reduce their response delays, and their controllers must be tuned by considering the presence of delays. There exist strategies for automated vehicles that specifically aim to eliminate the effect of delays [64–68]. Still, even with well-designed automated vehicles it may require a relatively large penetration of AVs in the traffic flow to mitigate congestions if human drivers are string unstable (the penetration is 25% in our example, i.e., every fourth vehicle is automated). Connectivity and a CCC strategy could lead to further benefits: some level of connectivity may reduce the required penetration of automation to stabilize traffic [53]. Yet, with vehicle control it is often hard to achieve stability for CAV penetrations around or below 10%.

The benefits of connectivity are further exploited by traffic control strategies. Simulation of (1), (3), (14) with ATC controller (18) in Example 3 is shown in Fig. 1(c). Here a single CAV responds to the vehicle ahead and to a single CHV $N = 10$ vehicles behind. With ATC strategy, the ego CAV is able to stabilize the traffic flow, and each vehicle including the CHV brakes less than the lead vehicle. This is achieved with about 8% penetration of automation and 17% penetration of connectivity (one automated and two connected vehicles out of 12 vehicles). The required penetration of automation is reduced (compared to when traffic is influenced by vehicle control) thanks to the extra connectivity—which is a significant benefit since automation implies more cost and requirements than communication. Moreover, the CHV has incentive to be connected to the CAV as it can slow down less.

The performance with and without connectivity is further explored in Fig. 5, where the influence of a single ACC vehicle on traffic is compared to traffic control by ATC. These scenarios are illustrated in Figs. 2b and 3b; their only difference is that the vehicle $-N$ is connected in the ATC setup and the ego CAV responds to it. It can be seen that ATC results in less speed reduction and smoother driving for the whole vehicle chain. The price is that the headway of the CAV executing ATC fluctuates more than that for ACC. However, it does not compromise safety, the vehicle is still far from collision. In the meantime, smooth driving leads to less energy consumption, which is discussed next.

### 3.2 Energy Efficiency

We evaluate the energy consumption for each vehicle by the following measure [69]:

$$w_n(t) = \int_{t_0}^{t} v_n(\theta) g\big(\dot{v}_n(\theta) + p(v_n(\theta))\big) \mathrm{d}\theta \,, \tag{21}$$

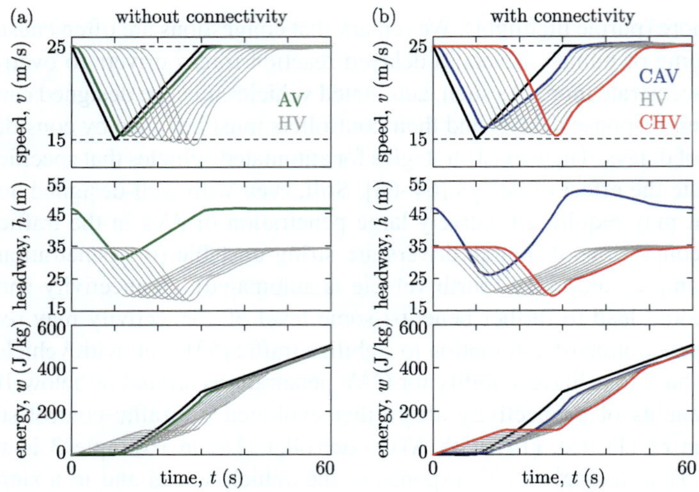


**Fig. 5** Traffic influenced by vehicle control without connectivity using adaptive cruise control (left) and regulated ty traffic control with connectivity using adaptive traffic control (right)

that is, the energy over unit mass consumed during the time interval $[t_0, t]$. It is an integral of the power over unit mass obtained from the product of speed and commanded acceleration. The latter involves the vehicle acceleration and the resistance terms represented by $p$ that vehicles need to overcome. These can be modeled for example by

$$p(v) = a_r + c_r v^2 , \tag{22}$$

where $a_r$ accounts for rolling resistance, while $c_r v^2$ represents air drag. Furthermore, function $g$ in (21) describes how the commanded acceleration is related to energy consumption. For example, if we take

$$g(x) = \max\{0, x\} , \tag{23}$$

then we assume energy consumption during acceleration only and no energy consumption or regeneration during braking.

The energy measure (21), (22), (23) was evaluated for the simulations in Fig. 5; see the bottom of the plot. The energy consumption stays constant during braking, increases linearly during constant speed cruising, and increases at a higher rate during acceleration. Therefore, it is critical for the vehicles to drive smoothly and avoid decelerations after which they need to accelerate to recover the speed.

To highlight the difference between the total energy consumption of ACC and ATC, in Fig. 6 we plot the final value $w_0(t_f)$ of the energy consumed by the CAV over the simulation interval $[t_0, t_f]$. Specifically, the simulations were repeated and the energy consumption was calculated for a grid of $\beta$ and $\beta_B$ control gains representing various control designs. The corresponding energy consumption levels are shown by

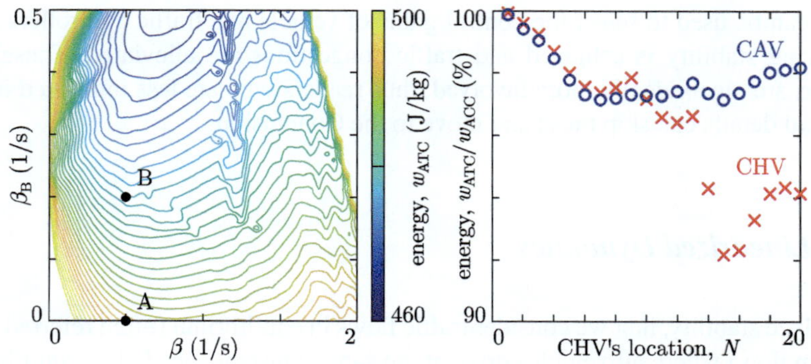


**Fig. 6** Total energy consumption of the ego CAV in the simulation scenario of Fig. 5 with various $\beta$ and $\beta_B$ control gains (left). The energy consumption is reduced in traffic control (ATC, $\beta_B \neq 0$) compared to vehicle control (ACC, $\beta_B = 0$). The relative energy benefits are shown for both the ego CAV and the CHV (right)

the colored contours. It can be seen that increasing $\beta_B$ leads to a lower energy level. One should, however, always keep safety in mind and choose $\beta_B$ small enough to maintain a safe headway similarly to Fig. 5.

The gains corresponding to Fig. 5 are indicated by point A for ACC ($\beta_B = 0$) and point B for ATC ($\beta_B \neq 0$) in Fig. 6. These two cases are compared on the right of Fig. 6 for various values of $N$. It can be seen that ATC consistently saves around 2-3% energy for the ego CAV compared to ACC if $N \geq 5$. The energy consumption $w_{-N}(t_f)$ of the CHV is also indicated. By deciding to stay connected, the CHV saves significant energy, around 6-8% if $N \geq 14$. The trend is that the farther the CHV is from the CAV (i.e., the larger $N$ is), the more energy it saves—this is a consequence of a longer string unstable chain of HVs being more sensitive to the CAV's motion.

We remark that the energy contours in Fig. 6 are specific to the velocity profile of the lead vehicle and to the choice (21) of the energy measure. Besides, extensive numerical simulations were needed to obtain these contours and to identify which controller parameters are energy efficient. Finding energy-optimal parameters through theoretical analysis is a challenging task [69]. On the other hand, energy consumption is associated with the smoothness of driving, thus it is related to the stability of the traffic flow. String stable chains of vehicles attenuate speed fluctuations, drive smoother, and hence tend to consume less energy. Stability analysis is more straightforward than finding energy optima. Hence, below we analyze the dynamics and stability of traffic to drive the controller parameter selection.

## 4 Dynamics of Traffic Flow

In the remaining two sections, we study the dynamics and stability of vehicle and traffic control in detail. The end result is manifested in so-called stability charts, which can be used to select the control gains of vehicle and traffic controllers such that string stability is achieved and traffic congestions are alleviated. These two sections are algebraically more involved, thus readers who are less interested in the technical details can skip these and move to the Conclusions.

## 4.1 Linearized Dynamics

To analyze stability, first we consider traffic flows in equilibrium (often referred to as uniform flow) where all vehicles drive at the same constant speed. The equilibrium speed is dictated by how fast the lead vehicle travels—in practice it can be considered as the average speed of the lead vehicle—and the following vehicles adjust their speed to the lead. Afterwards, we consider deviations from this equilibrium to quantify how much speed perturbations amplify. This will also be used to carry out stability analysis in the next section. We conduct our analysis in the frequency domain by linearization and by constructing transfer functions that allow us to characterize the response of individual vehicles and the overall traffic behavior.

We start with formalizing the equilibrium solution as

$$v_n(t) \equiv v^* , \quad s_n(t) = s_n^*(t) = v^* t + s_n(0) , \tag{24}$$

$n \in \mathcal{N}$, which is associated with constant speed $v^*$ that is identical for each vehicle, and constant headway $h_n^* = s_{n+1}^* - s_n^* - l_n$ that may be different for each vehicle. This equilibrium can be found by substituting (24) into (1), (3) and (7) for ACC or (14) for ATC, and solving for $v^*$ and $h_n^*$.

We consider perturbations around the equilibrium in the form

$$v_n(t) = v^* + \tilde{v}_n(t) , \quad s_n(t) = s_n^*(t) + \tilde{s}_n(t) , \tag{25}$$

and collect position and speed perturbations into the state vector $\mathbf{x}$ given by

$$\mathbf{x}_n(t) = \begin{bmatrix} \tilde{s}_n(t) \\ \tilde{v}_n(t) \end{bmatrix} = \begin{bmatrix} \tilde{v}_n(0) + \int_0^t \tilde{v}_n(\theta) \mathrm{d}\theta \\ \tilde{v}_n(t) \end{bmatrix} , \tag{26}$$

from which the speed fluctuations can be recovered by

$$\tilde{v}_n(t) = \mathbf{c}\mathbf{x}_n(t), \quad \mathbf{c} = \begin{bmatrix} 0 & 1 \end{bmatrix} . \tag{27}$$

Assuming the accelerations of vehicles do not saturate, the linearized dynamics of HVs and CHVs can be derived from (1), (3) in the form

$$\dot{\mathbf{x}}_n(t) = \mathbf{a}\mathbf{x}_n(t) + \mathbf{a}_{\mathrm{H}}\mathbf{x}_n(t - \tau) + \mathbf{b}_{\mathrm{H}}\mathbf{x}_{n+1}(t - \tau) , \tag{28}$$

$n \in \{\mathcal{N}_{\rm HV}, \mathcal{N}_{\rm CHV}\}$. That is, it contains the delay-free state of vehicle $n$ and the delayed states of vehicles $n$ and $n+1$ with coefficient matrices

$$\mathbf{a} = \begin{bmatrix} 0 & 1 \\ 0 & 0 \end{bmatrix}, \quad \mathbf{a}_{\rm H} = \begin{bmatrix} 0 & 0 \\ -\dfrac{\partial f_{\rm H}}{\partial h_n}\Big|_* - \dfrac{\partial f_{\rm H}}{\partial \dot{h}_n}\Big|_* & \dfrac{\partial f_{\rm H}}{\partial v_n}\Big|_* \end{bmatrix}, \quad \mathbf{b}_{\rm H} = \begin{bmatrix} 0 & 0 \\ \dfrac{\partial f_{\rm H}}{\partial h_n}\Big|_* & \dfrac{\partial f_{\rm H}}{\partial \dot{h}_n}\Big|_* \end{bmatrix}, \tag{29}$$

where star indicates that partial derivatives are evaluated at the equilibrium. For AVs performing ACC the linearized dynamics become

$$\dot{\mathbf{x}}_0(t) = \mathbf{a}\mathbf{x}_0(t) + \mathbf{a}_{\rm F}\mathbf{x}_0(t-\sigma) + \mathbf{b}_{\rm F}\mathbf{x}_1(t-\sigma), \tag{30}$$

with

$$\mathbf{a} = \begin{bmatrix} 0 & 1 \\ 0 & 0 \end{bmatrix}, \quad \mathbf{a}_{\rm F} = \begin{bmatrix} 0 & 0 \\ \dfrac{\partial f_{\rm ACC}}{\partial s_0}\Big|_* & \dfrac{\partial f_{\rm ACC}}{\partial v_0}\Big|_* \end{bmatrix}, \quad \mathbf{b}_{\rm F} = \begin{bmatrix} 0 & 0 \\ \dfrac{\partial f_{\rm ACC}}{\partial s_1}\Big|_* & \dfrac{\partial f_{\rm ACC}}{\partial v_1}\Big|_* \end{bmatrix}. \tag{31}$$

whereas CAVs executing ATC are described by

$$\dot{\mathbf{x}}_0(t) = \mathbf{a}\mathbf{x}_0(t) + \mathbf{a}_{\rm FB}\mathbf{x}_0(t-\sigma) + \mathbf{b}_{\rm F}\mathbf{x}_1(t-\sigma) + \mathbf{b}_{\rm B}\mathbf{x}_{-N}(t-\sigma), \tag{32}$$

with

$$\begin{aligned} \mathbf{a} &= \begin{bmatrix} 0 & 1 \\ 0 & 0 \end{bmatrix}, \quad \mathbf{a}_{\rm FB} = \begin{bmatrix} 0 & 0 \\ \dfrac{\partial f_{\rm ATC}}{\partial s_0}\Big|_* & \dfrac{\partial f_{\rm ATC}}{\partial v_0}\Big|_* \end{bmatrix}, \\ \mathbf{b}_{\rm F} &= \begin{bmatrix} 0 & 0 \\ \dfrac{\partial f_{\rm ATC}}{\partial s_1}\Big|_* & \dfrac{\partial f_{\rm ATC}}{\partial v_1}\Big|_* \end{bmatrix}, \quad \mathbf{b}_{\rm B} = \begin{bmatrix} 0 & 0 \\ \dfrac{\partial f_{\rm ATC}}{\partial s_{-N}}\Big|_* & \dfrac{\partial f_{\rm ATC}}{\partial v_{-N}}\Big|_* \end{bmatrix}. \end{aligned} \tag{33}$$

In summary, vehicle control influencing traffic is described by (28), (30) at the linear level, while traffic control is characterized by (28), (32). For the ring configuration, one also has periodicity given by (20) that yields

$$\mathbf{x}_{-N}(t) = \mathbf{x}_1(t). \tag{34}$$

***Example 4*** The coefficient matrices (29) describing human drivers have the following expressions for the case of the optimal velocity model (4):

$$\mathbf{a} = \begin{bmatrix} 0 & 1 \\ 0 & 0 \end{bmatrix}, \quad \mathbf{a}_{\rm H} = \begin{bmatrix} 0 & 0 \\ -\alpha_{\rm H}\kappa_{\rm H} & -(\alpha_{\rm H} + \beta_{\rm H}) \end{bmatrix}, \quad \mathbf{b}_{\rm H} = \begin{bmatrix} 0 & 0 \\ \alpha_{\rm H}\kappa_{\rm H} & \beta_{\rm H} \end{bmatrix}. \tag{35}$$

The ACC controller (10) is associated with the coefficient matrices

$$\mathbf{a} = \begin{bmatrix} 0 & 1 \\ 0 & 0 \end{bmatrix}, \quad \mathbf{a}_{\mathrm{F}} = \begin{bmatrix} 0 & 0 \\ -\alpha\kappa & -(\alpha+\beta) \end{bmatrix}, \quad \mathbf{b}_{\mathrm{F}} = \begin{bmatrix} 0 & 0 \\ \alpha\kappa & \beta \end{bmatrix}, \tag{36}$$

whereas the ATC controller (18) corresponds to

$$\mathbf{a} = \begin{bmatrix} 0 & 1 \\ 0 & 0 \end{bmatrix}, \quad \mathbf{a}_{\mathrm{FB}} = \begin{bmatrix} 0 & 0 \\ -\alpha\kappa & -(\alpha+\beta+\beta_{\mathrm{B}}) \end{bmatrix}, \quad \mathbf{b}_{\mathrm{F}} = \begin{bmatrix} 0 & 0 \\ \alpha\kappa & \beta \end{bmatrix}, \quad \mathbf{b}_{\mathrm{B}} = \begin{bmatrix} 0 & 0 \\ 0 & \beta_{\mathrm{B}} \end{bmatrix}. \tag{37}$$

## 4.2 Transfer Functions

We analyze the linearized dynamics in Laplace domain (assuming zero initial perturbations). Our ultimate goal is to analyze the stability of the system, in particular, whether speed fluctuations amplify or decay as they propagate along the chain of vehicles. This amplification property can be well represented by means of transfer functions describing the response of vehicles to the neighboring traffic.

First, we formulate the so-called *link transfer functions* [60] associated with vehicle pairs. For human drivers, the link transfer function $T_{\mathrm{H}}(s)$ relates the speed fluctuation of an HV to that of the vehicle ahead as

$$V_n(s) = T_{\mathrm{H}}(s) V_{n+1}(s), \tag{38}$$

where $T_{\mathrm{H}}(s)$ can be obtained from (27), (28) in the form

$$T_{\mathrm{H}}(s) = \mathbf{c}\left(s\mathbf{I} - \mathbf{a} - \mathbf{a}_{\mathrm{H}} \mathrm{e}^{-s\tau}\right)^{-1} \mathbf{b}_{\mathrm{H}} \mathrm{e}^{-s\tau} \begin{bmatrix} \frac{1}{s} \\ 1 \end{bmatrix}. \tag{39}$$

Similarly, in ACC where the automated vehicle responds to the vehicle immediately ahead only, the response can be characterized by a single link transfer function $T(s)$:

$$V_0(s) = T(s) V_1(s), \tag{40}$$

whose expression can be derived from (30) as

$$T(s) = \mathbf{c}\left(s\mathbf{I} - \mathbf{a} - \mathbf{a}_{\mathrm{F}} \mathrm{e}^{-s\sigma}\right)^{-1} \mathbf{b}_{\mathrm{F}} \mathrm{e}^{-s\sigma} \begin{bmatrix} \frac{1}{s} \\ 1 \end{bmatrix}. \tag{41}$$

On the other hand, in ATC the ego CAV responds to multiple vehicles, associated with multiple link transfer functions. For example, when responding to a single vehicle ahead and a single vehicle behind, we obtain

$$V_0(s) = T_{\mathrm{F}}(s)V_1(s) + T_{\mathrm{B}}(s)V_{-N}(s)\,, \tag{42}$$

where $T_{\mathrm{F}}(s)$ and $T_{\mathrm{B}}(s)$ are the link transfer functions characterizing the response forward and backward. These are obtained from (32):

$$\begin{aligned} T_{\mathrm{F}}(s) &= \mathbf{c}\left(s\mathbf{I} - \mathbf{a} - \mathbf{a}_{\mathrm{FB}}\mathrm{e}^{-s\sigma}\right)^{-1}\mathbf{b}_{\mathrm{F}}\mathrm{e}^{-s\sigma}\begin{bmatrix}\frac{1}{s}\\ 1\end{bmatrix},\\ T_{\mathrm{B}}(s) &= \mathbf{c}\left(s\mathbf{I} - \mathbf{a} - \mathbf{a}_{\mathrm{FB}}\mathrm{e}^{-s\sigma}\right)^{-1}\mathbf{b}_{\mathrm{B}}\mathrm{e}^{-s\sigma}\begin{bmatrix}\frac{1}{s}\\ 1\end{bmatrix}. \end{aligned} \tag{43}$$

Notice that the denominators of $T_{\mathrm{F}}(s)$ and $T_{\mathrm{B}}(s)$ are the same, $\mathrm{D}(T_{\mathrm{F}}(s)) = \mathrm{D}(T_{\mathrm{B}}(s))$, since the same matrix is inverted.

Using the link transfer functions, we can analyze the overall response of traffic. For a chain of $N$ human drivers, the overall response

$$V_{-N}(s) = \Gamma(s)V_0(s) \tag{44}$$

is characterized by $\Gamma(s)$ given by

$$\Gamma(s) = T_{\mathrm{H}}(s)^N\,. \tag{45}$$

Note that this corresponds to identical human drivers; for the case of a heterogeneous chain of HVs it should be replaced with $\Gamma(s) = \prod_{n=-N}^{-1} T_{\mathrm{H},n}(s)$.

Introducing the ego vehicle (indexed 0) and the vehicle ahead of it (lead vehicle 1) into the human-driven traffic, like in Figs. 2b and 3b, we can describe the overall response of the traffic flow from the head vehicle to the tail vehicle via the *head-to-tail transfer function* [60] given by

$$V_{-N}(s) = G(s)V_1(s)\,. \tag{46}$$

$G(s)$ depends on the human response $\Gamma(s)$ and the control strategy of the ego vehicle. For vehicle control with ACC, we write (40), (44) and obtain

$$G(s) = T(s)\Gamma(s)\,. \tag{47}$$

For the traffic control setup with ATC, we have (42), (44) that lead to

$$G(s) = \frac{T_{\mathrm{F}}(s)\Gamma(s)}{1 - T_{\mathrm{B}}(s)\Gamma(s)}\,. \tag{48}$$

If traffic is considered on a ring road, we have the periodic boundary condition

$$V_{-N}(s) = V_1(s)\,, \tag{49}$$

that leads to

$$G(s) = 1\,, \tag{50}$$

as the characteristic equation of the ring configuration.

***Remark 5*** (Dynamics of rings and virtual rings) If vehicle control (47) is executed on a ring road, (50) leads to

$$1 - T(s)\Gamma(s) = 0\,. \tag{51}$$

Here the left-hand side corresponds to the expression in the denominator of $G(s)$ in (48). Therefore, the dynamics of traffic control, that involves a virtual ring, is closely related to how vehicle control influences traffic when executed on a ring road. As highlighted in the analysis below, the stability properties of the ring configuration give the backbone of the stability analysis of the virtual ring.

***Example 5*** Substituting the coefficient matrices in (35) into the link transfer function (39) describing human drivers leads to

$$T_{\rm H}(s) = \frac{\beta_{\rm H} s + \alpha_{\rm H}\kappa_{\rm H}}{s^2 e^{s\tau} + (\alpha_{\rm H} + \beta_{\rm H})s + \alpha_{\rm H}\kappa_{\rm H}}\,. \tag{52}$$

Similarly, the link transfer function of ACC given by (36) and (41) is

$$T(s) = \frac{\beta s + \alpha\kappa}{s^2 e^{s\sigma} + (\alpha + \beta)s + \alpha\kappa}\,, \tag{53}$$

whereas for ATC (37) and (43) give

$$\begin{aligned} T_{\rm F}(s) &= \frac{\beta s + \alpha\kappa}{s^2 e^{s\sigma} + (\alpha + \beta + \beta_{\rm B})s + \alpha\kappa}\,, \\ T_{\rm B}(s) &= \frac{\beta_{\rm B} s}{s^2 e^{s\sigma} + (\alpha + \beta + \beta_{\rm B})s + \alpha\kappa}\,. \end{aligned} \tag{54}$$

Observe that we have $T(s) = T_{\rm F}(s)/(1 - T_{\rm B}(s))$ that relates ACC and ATC.

***Remark 6*** (Special cases) When the ego CAV does not respond to the traffic behind, ATC reduces to ACC. Mathematically, this means setting $T_{\rm B}(s) = 0$, $T_{\rm F}(s) = T(s)$, e.g. by taking $\beta_{\rm B} = 0$ in Example 5. This reduces $G(s)$ in (48) to $G(s)$ in (47). If no traffic is considered behind the ego CAV, then $N = 0$ and $\Gamma(s) = 1$ due to (45). In such a case, $G(s)$ in (47) and (48) both reduce to $T(s)$ describing the response of the ACC vehicle to the vehicle ahead only (since $T(s) = T_{\rm F}(s)/(1 - T_{\rm B}(s))$ in our example). Consequently, we can analyze the dynamics of ATC and then take these special cases to capture the dynamics of ACC followed by human-driven traffic or ACC on its own; recall Fig. 4.

# 5 Stability of Traffic Flow

Herein we formalize the stability conditions of traffic flows influenced by vehicle control and regulated by traffic control, while linking them to those of the ring configuration.

## 5.1 Stability Conditions

For a chain of vehicles traveling on a straight road, the system is non-autonomous, forced by the speed perturbations of the lead vehicle. Thus, we consider the notions of *plant stability* and *string stability* [5, 6, 60] to analyze the behavior without speed perturbations and the response to the perturbations, respectively. For the ring configuration, the system evolves autonomously, and we apply a single notion of stability; for more details on stability definitions see [48].

String stability is particularly important in traffic control. It has various mathematical definitions; see a comprehensive summary in [5]. Now we consider linear input-to-output string stability with respect to $\mathcal{L}_2$ norm. This string stability notion is directly related to the magnitude ($\mathcal{H}_\infty$ norm) of the transfer function describing system, which makes it is easy to analyze and useful for control design. Other notions are also widely-used, such as $\mathcal{L}_p$ string stability for any value of $p$ [6]. $\mathcal{L}_\infty$ string stability is particularly relevant for traffic safety, since it characterizes the overshoot of the response to perturbations. As this notion is related to the impulse response, it is more challenging to analyze it explicitly, and we rather consider $\mathcal{L}_2$ string stability.

### 5.1.1 Plant Stability of a Vehicle Chain

Plant stability means that each vehicle in the chain can approach a constant (equilibrium) speed in a stable manner. If the vehicle chain is associated with head-to-tail transfer function $G(s)$, the plant stability condition is given by

$$\mathrm{D}(G(s_\ell)) = 0 \quad \Rightarrow \quad \Re(s_\ell) < 0\,, \quad \forall \ell \in \mathbb{N}\,. \tag{55}$$

That is, the roots $s_\ell$ of the characteristic equation $\mathrm{D}(G(s)) = 0$ are located in the left-half of the complex plane, where $\mathrm{D}(G(s))$ is the denominator of the head-to-tail transfer function. The system is at the plant stability boundary if either a real root $s = 0$ is located on the imaginary axis:

$$\mathrm{D}(G(0)) = 0\,, \tag{56}$$

or a complex conjugate pair of roots $s = \pm \mathrm{i}\Omega$, $\Omega > 0$:

$$\mathrm{D}(G(\mathrm{i}\Omega)) = 0\,. \tag{57}$$

### 5.1.2 String Stability of a Vehicle Chain

String stability requires, on one hand, plant stability as a necessary condition, and on the other hand, that the speed perturbations of the head vehicle are not amplified by the vehicle chain. Otherwise large speed fluctuations could arise at the end of the chain, ultimately leading to traffic jams and stop-and-go motion. Specifically, we require that the speed fluctuations $V_1(\mathrm{i}\omega)$ of the head vehicle with any given frequency $\omega > 0$ are smaller in amplitude than those $V_{-N}(\mathrm{i}\omega)$ of the tail vehicle, which, based on (46), can be expressed as

$$|G(\mathrm{i}\omega)| < 1\,, \quad \forall \omega > 0\,. \tag{58}$$

At the string stability boundary we have

$$|G(\mathrm{i}\omega)| = 1 \quad \Longleftrightarrow \quad G(\mathrm{i}\omega) = \mathrm{e}^{-\mathrm{i}K}\,, \tag{59}$$

for some $\omega > 0$ and $K \in [0, 2\pi)$, i.e., the speed fluctuations of the head vehicle are "repeated" by the tail vehicle with a phase lag $K$ (also called as the wave number).

Furthermore, string stability can also be studied when $\omega \to 0$, shortly referred to as $\omega = 0$ string stability. For algebraic convenience, we study this by introducing

$$P(\omega) = \frac{1}{\omega^2}\left(\mathrm{D}\left(|G(\mathrm{i}\omega)|^2\right) - \mathrm{N}\left(|G(\mathrm{i}\omega)|^2\right)\right), \tag{60}$$

where N and D denote numerator and denominator, respectively. With this definition, (58) can be rewritten as

$$P(\omega) > 0\,, \quad \forall \omega > 0\,, \tag{61}$$

and the $\omega = 0$ string stability boundary can be written simply as

$$P(0) = 0\,. \tag{62}$$

### 5.1.3 Stability of the Ring Configuration

Finally, the stability of the ring configuration is determined by the roots of the characteristic equation (50), and stability requires:

$$G(s_\ell) = 1 \quad \Rightarrow \quad \Re(s_\ell) < 0\,, \quad \forall \ell \in \mathbb{N}\,, \tag{63}$$

analogously to (55). Recall that $G(s)$ is the head-to-tail transfer function of the corresponding open chain configuration. Since $G(0) = 1$ automatically holds, we consider the stability boundary where $s = 0$ is a double root of $G(s) - 1$, and where $s = \pm \mathrm{i}\omega$, $\omega > 0$ satisfies the characteristic equation:

$$G(\mathrm{i}\omega) = 1\,. \tag{64}$$

## 5.2 *Relationship of the Stability Conditions*

Now let us take a deeper look into the relationship between the various vehicle configurations and their stability conditions. These relationships are summarized at the bottom of Fig. 4 for the $s = \mathrm{i}\omega$ stability boundaries.

For vehicle control with (47), we have $\mathrm{D}(G(s)) = \mathrm{D}(T(s))\mathrm{D}(\Gamma(s))$, hence the plant stability condition becomes

$$\begin{aligned} \mathrm{D}(T(s_\ell)) = 0 \quad &\Rightarrow \quad \Re(s_\ell) < 0\,, \quad \forall \ell \in \mathbb{N}\,, \quad \text{and} \\ \mathrm{D}(\Gamma(s_\ell)) = 0 \quad &\Rightarrow \quad \Re(s_\ell) < 0\,, \quad \forall \ell \in \mathbb{N}\,, \end{aligned} \tag{65}$$

that is, both the AV and HVs must be plant stable individually. For string stability, we require

$$|T(\mathrm{i}\omega)|\,|\Gamma(\mathrm{i}\omega)| < 1\,, \quad \forall \omega > 0\,, \tag{66}$$

and at the stability boundary we have

$$T(\mathrm{i}\omega)\Gamma(\mathrm{i}\omega) = \mathrm{e}^{-\mathrm{i}K}\,, \tag{67}$$

which depend both on the response of the AV and the chain of HVs. If the human-driven traffic is not considered, i.e., $N = 0$, $\Gamma(\mathrm{i}\omega) = 1$, then we recover the string stability limit $T(\mathrm{i}\omega) = \mathrm{e}^{-\mathrm{i}K}$ for a single ACC vehicle.

Now consider a nonzero (and potentially large) number $N$ of HVs. This leads to one of the most interesting research questions, that several recent works have studied [20, 51, 70]: how many AVs are needed to mitigate traffic congestions? If a human-driven vehicle chain behind an AV is long ($N$ is large), it results in a significant (exponential) increase of speed perturbations. Thus, the AV has to make considerable effort to dampen the perturbation in order to prevent or mitigate a congestion. Mathematically, the magnitude $|T_{\rm H}(\mathrm{i}\omega)|$ of the individual human responses affects the overall response $|\Gamma(\mathrm{i}\omega)| = |T_{\rm H}(\mathrm{i}\omega)|^N$ more significantly for large $N$. Therefore, it may be more challenging to stabilize a long chain of HVs with a single AV.

To highlight this, consider the limit $N \to \infty$. If the human drivers are string stable, i.e., $|T_{\rm H}(\mathrm{i}\omega)| < 1$, then $|\Gamma(\mathrm{i}\omega)| = 0$ for all $\omega > 0$. Unless $|T(\mathrm{i}\omega)| \to \infty$, which can only happen at the plant stability boundary $\mathrm{D}(T(\mathrm{i}\omega)) = 0$ of the AV, (66) is automatically satisfied. Thus string stability is achieved for a long string stable chain of HVs and a plant stable AV. However, when the human drivers are string unstable—which is more likely the case in real traffic [2]—we have $|T_{\rm H}(\mathrm{i}\omega)| > 1$ and $|\Gamma(\mathrm{i}\omega)| \to \infty$ for some $\omega > 0$ and $N \to \infty$. Thus, string stability in (66) cannot be achieved by any AV design (as long as $T(\mathrm{i}\omega) \neq 0$, i.e., the AV responds to what is ahead). This shows the fundamental limitation of influencing traffic by vehicle control: string stability cannot be achieved with arbitrarily small penetration of AVs (i.e., in the limit $N \to \infty$) if the human drivers are string unstable.

If vehicle control is executed on a ring, the stability limit (64) becomes

$$T(\mathrm{i}\omega)\Gamma(\mathrm{i}\omega) = 1\,. \tag{68}$$

Therefore, the stability boundary of the ring can be obtained as the $K = 0$ special case of the string stability boundary (67) of the open chain. Physically, it means that perturbations propagate through the ring and arrive back to the same vehicle in the same phase due to the periodic boundary condition. Another special case to mention is when all vehicles on the ring are identical, achieved by the substitution $T(s) = T_{\rm H}(s)$. Then the characteristic Eq. (68) gives

$$T_{\rm H}(\mathrm{i}\omega)^{N+1} = 1 \quad \Longleftrightarrow \quad T_{\rm H}(\mathrm{i}\omega) = \mathrm{e}^{-\mathrm{i}\frac{2k\pi}{N+1}}\,, \quad k \in \{0, 1, \ldots, N\}\,. \tag{69}$$

When $N \to \infty$, the exponent $K = 2k\pi/(N+1)$ may take any value on $[0, 2\pi)$ and we recover the string stability condition $T_{\rm H}(\mathrm{i}\omega) = \mathrm{e}^{-\mathrm{i}K}$ for individual human drivers. Thus, the stability of an infinitely long homogeneous ring is equivalent to the string stability of a homogeneous chain of vehicles. Conversely, the substitution of $K = 2k\pi/(N+1)$ into the string stability limit $T_{\rm H}(\mathrm{i}\omega) = \mathrm{e}^{-\mathrm{i}K}$ of an individual human driver gives the stability limit (69) of the homogeneous ring.

Additionally, and most importantly, the traffic control setup as a virtual ring is related to both the open chain and ring configurations in terms of stability. Consider formula (48) of $G(s)$. Since $\mathrm{D}(T_{\rm F}(s)) = \mathrm{D}(T_{\rm B}(s))$, we have

$$\mathrm{D}(G(s)) = \mathrm{D}(T_{\rm B}(s)\Gamma(s))\big(1 - T_{\rm B}(s)\Gamma(s)\big)\,. \tag{70}$$

Hence, the plant stability condition (55) leads to

$$\begin{aligned} \mathrm{D}(T_{\mathrm{B}}(s_\ell)\Gamma(s_\ell)) = 0 \quad &\Rightarrow \quad \Re(s_\ell) < 0\,, \quad \forall \ell \in \mathbb{N}\,, \quad \text{and} \\ T_{\mathrm{B}}(s)\Gamma(s) = 1 \quad &\Rightarrow \quad \Re(s_\ell) < 0\,, \quad \forall \ell \in \mathbb{N}\,. \end{aligned} \tag{71}$$

That is, it requires both the plant stability of the associated vehicle control setup and the stability of the associated virtual ring configuration.

The string stability boundary of ATC can be written in the form

$$\frac{T_{\mathrm{F}}(\mathrm{i}\omega)\Gamma(\mathrm{i}\omega)}{1 - T_{\mathrm{B}}(\mathrm{i}\omega)\Gamma(\mathrm{i}\omega)} = \mathrm{e}^{-\mathrm{i}K}\,. \tag{72}$$

This reduces to the string stability boundary of traffic flows influenced by vehicle control when there is no response to the traffic behind, i.e., $T_{\mathrm{B}}(\mathrm{i}\omega) = 0$, $T_{\mathrm{F}}(\mathrm{i}\omega) = T(\mathrm{i}\omega)$, which is achieved by $\beta_{\mathrm{B}} = 0$ in our example. In this sense, the string stability of ATC is related to that of ACC while the plant stability of ATC is related to the stability of its virtual ring.

***Example 6*** The stability analysis can be carried out for the specific choice (4) of human driver model and selection (10), (18) of control laws. The expressions (56), (57) of the plant stability boundaries, (59), (62) of the string stability boundaries and (64) of the stability boundaries for the ring can be expanded by using (52), (53), (54). Then, the stability boundaries can be expressed in terms of the controller parameters such as $\alpha$, $\beta$ or $\beta_{\mathrm{B}}$. This allows one to select control gains such that they ensure stability. The detailed formulas of this analysis are derived in the Appendix.

The resulting stability boundaries can be visualized in the space of control parameters, which we refer to as stability charts. The left panel of Fig. 7 shows the stability chart in the $(\beta, \alpha)$ plane for ACC without considering the traffic behind, i.e., for $N = 0$. This special case can also be found in [60]. Dashed red line indicates the $s = 0$ plant stability boundary, solid red line shows the $s = \pm\mathrm{i}\Omega$ plant stability boundary, and the plant stable region is light gray. Furthermore, dashed blue lines show the $\omega = 0$ string stability boundaries (that overlap with the dashed red lines at $\alpha = 0$), and colorful curves denote the $\omega > 0$ string stability boundaries for various values of $K \in [0, 2\pi)$ indicated by color. The envelope of these curves is presented in the center with a solid blue line, and the string stable region is dark gray. The control gains shall be selected from this region to achieve stability.

Figure 7 also illustrates the characteristics of traffic dynamics under different selections of the control gains $(\beta, \alpha)$. The right panels show the characteristic roots associated with the plant stability condition (55) for a plant unstable point with a complex pair of unstable roots (A), a plant stable point (C), and a plant unstable point with a real unstable root (E). The bottom panels depict the magnitude of the head-to-tail transfer function related to the string stability condition (58) for a point with $\omega = 0$ string instability (B), a string stable point (C) and a point with $\omega > 0$ string instability (D). Out of these points, only point C has desired behavior. This illustrates how stability charts may drive the selection of controller parameters, so that the end result is a stable and smooth traffic flow.

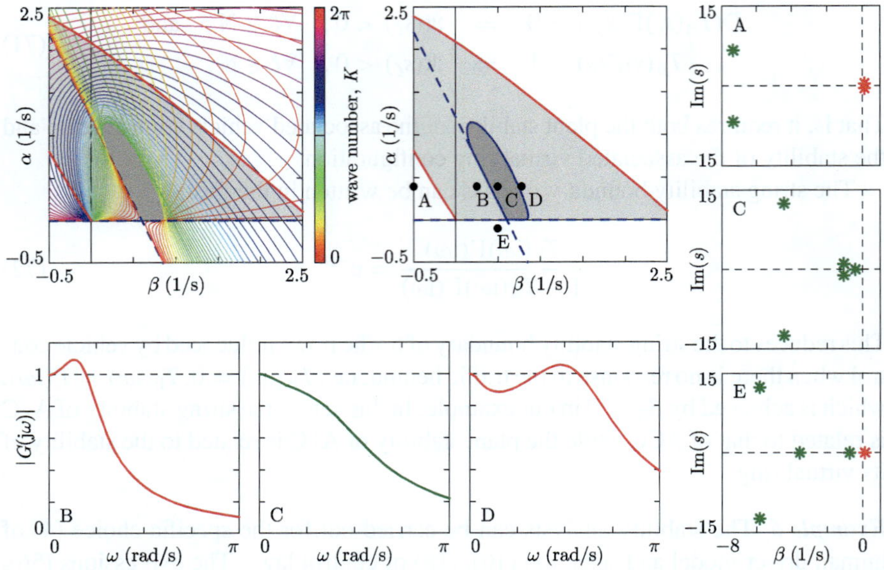


**Fig. 7** Stability chart of adaptive cruise control without considering its influence on the traffic behind ($N = 0$). Detailed (left) and simplified view (center). The light gray region enclosed by red is plant stable, the dark gray region encircled by blue is string stable. Characteristic roots (right) and frequency response (bottom) for specific control gains selected from the stability chart

The stability charts also give another aspect of comparing vehicle and traffic control strategies. The top row of Fig. 8 show the stability charts in the $(\beta, \alpha)$ plane for ACC followed by various numbers of human drivers ($N = 1, 2, 3, 4$). Since the human driver parameters are selected to be string unstable, the string stable region shrinks as the number of vehicles increases. Furthermore, it also shifts towards large gains that may require large control input and acceleration from the ACC vehicle. In comparison, the string stable region of ATC in the second row of Fig. 8 shrinks faster, but is located at lower $\alpha$ gains associated with smaller acceleration and better passenger comfort.

For a fixed, reasonably small gain $\alpha$, the stability charts of ATC are plotted in the $(\beta, \beta_{\mathrm{B}})$ plane in the third row of Fig. 8. Note that the points along $\beta_{\mathrm{B}} = 0$ represent ACC as a special case, and they are located close to or outside the string stable region. Hence for such realistic choice of $\alpha$, traffic control with ATC ($\beta_{\mathrm{B}} \neq 0$) may achieve string stability while vehicle control with ACC ($\beta_{\mathrm{B}} = 0$) may not. For a large number of string unstable human drivers, the string stable region disappears (see $N = 4$). However, ATC may still be beneficial by reducing instability, improving the smoothness of driving behavior, and reducing energy consumption; as these were highlighted in Figs. 1 and 6.

Finally, the last row of Fig. 8 shows the stability charts of TC (as the $\alpha = 0$ special case of ATC). Although the string stable region shrinks as the chain of HVs gets longer ($N$ gets larger), it is significantly larger than that of ATC. In fact, the string

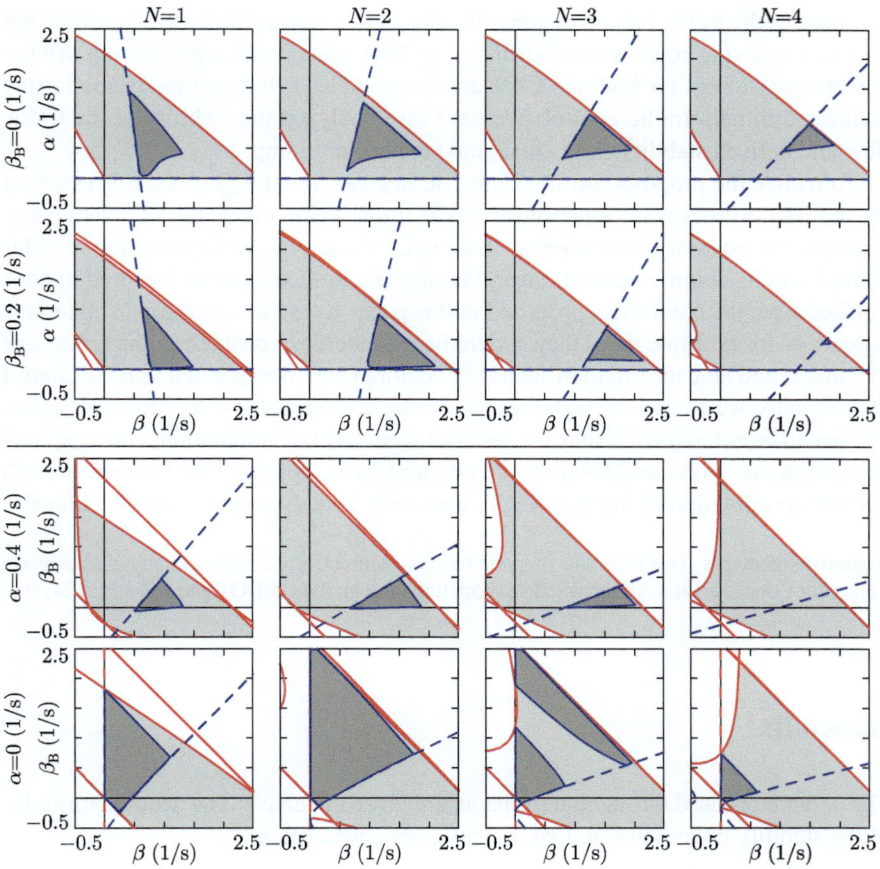


**Fig. 8** Stability charts of vehicle control influencing traffic (ACC, first row) and traffic control (ATC, second and third rows, and TC, fourth row). The color scheme is the same as in Fig. 7

stable region does not vanish even for chains as long as $N = 10$ HVs. This indicates that CAVs can control traffic significantly more efficiently in situations where they do not have to respond to the position of the vehicle ahead.

## 6 Conclusions

We have discussed vehicle control strategies influencing traffic and traffic control strategies regulating traffic by means of injecting connected automated vehicles (CAVs) into the traffic flow. We have shown that vehicle-to-everything (V2X) connectivity allows a CAV to receive information about the traffic behind it, respond to it, and mitigate traffic congestions, while also responding to vehicles ahead. The response to the traffic behind creates a feedback loop for traffic control, in a similar structure to a ring road, termed virtual ring. This strategy achieves smooth driving and energy savings for both the CAV, and the vehicles behind it that decide to stay connected to help traffic control. We have also analyzed the stability of the traffic flow and derived stability charts that guide controller tuning.

To realize the proposed traffic control strategies on real highways, it is essential for the CAV to access information about the traffic behind it. Thus, sufficient penetration of connectivity is required so that the CAV can sync with connected vehicles behind it in its communication range, that may be limited to a few hundred meters. Furthermore, the controllers proposed in this paper were discussed for a single-lane scenario—for multiple lanes they should be extended by considering the cross-lane dynamics, and lane information should be acquired and incorporated into the control laws. Finally, we emphasize that connected vehicles behind the CAV do not necessarily have to be automated. If they possess sufficient level of automation, however, they can coordinate with the CAV for more efficient traffic control. Our future work will explore coordination of multiple CAVs and multi-lane dynamics in traffic control.

**Acknowledgements** This work was supported by Ford Motor Co. and the University of Michigan's Center for Connected and Automated Transportation through the US DOT grant 69A3551747105.

## Appendix

The parameters used throughout the paper are given in Table 1. The detailed formulas of the stability boundaries plotted in Sect. 5 are given below.

### *Stability of ATC*

As highlighted in Fig. 4, the car-following configurations in our examples can all be considered as the special cases of traffic control with ATC. Therefore, we derive the stability charts of ATC and then obtain those of other scenarios as special cases. Throughout the appendix we use (52)–(54) as the expressions of the link transfer functions $T_{\rm H}(s)$, $T(s)$, $T_{\rm F}(s)$ and $T_{\rm B}(s)$.

First, consider the plant stability of ATC. Based on (71), it decomposes into the plant stability of the associated vehicle control setup and virtual ring. Consider the first line of (71) and assume a plant stable chain of HVs described by $\Gamma(s)$ (the conditions of this are given at the end of the Appendix). Then we can take $\mathrm{D}(T_{\rm B}(s)) = 0$, which leads to the characteristic equation

$$s^2 \mathrm{e}^{s\sigma} + (\alpha + \beta + \beta_{\rm B}) s + \alpha\kappa = 0\,. \tag{73}$$

**Table 1** Parameters of the numerical case studies

| Field | Description | Parameter | Value |
|---|---|---|---|
| Vehicle properties | Vehicle length | $l$ | 5 m |
| | Braking limit | $a_{\min}$ | 7 m/s$^2$ |
| | Acceleration limit | $a_{\max}$ | 3 m/s$^2$ |
| Human-driven vehicle | Time delay | $\tau$ | 0.8 s |
| | Gain for headway response | $\alpha_{\mathrm{H}}$ | 0.1 s$^{-1}$ |
| | Gain for speed response | $\beta_{\mathrm{H}}$ | 0.6 s$^{-1}$ |
| | Range policy | $F_{\mathrm{H}}(h)$ | $v_{\max}\left(1-\left(\frac{h_{\mathrm{go}}-h}{h_{\mathrm{go}}-h_{\mathrm{st}}}\right)^2\right)$ |
| Automated vehicle | Time delay | $\sigma$ | 0.6 s |
| | Gain for headway response | $\alpha$ | 0.4 s$^{-1}$ |
| | Gain for speed response | $\beta$ | 0.5 s$^{-1}$ |
| | Gain for traffic control | $\beta_{\mathrm{B}}$ | 0.2 s$^{-1}$ |
| | Range policy | $F(h)$ | $v_{\max}\frac{h-h_{\mathrm{st}}}{h_{\mathrm{go}}-h_{\mathrm{st}}}$ |
| Range policy | Standstill headway | $h_{\mathrm{st}}$ | 5 m |
| | Free flow headway | $h_{\mathrm{go}}$ | 55 m |
| | Speed limit | $v_{\max}$ | 30 m/s |
| Simulations | Time interval | $[t_0, t_{\mathrm{f}}]$ | [0,60] s |
| | Time step | $\Delta t$ | 0.01 s |
| | Lead vehicle braking | $a_1(t)$ | $-1$ m/s$^2$, $t \in [0, 10]$ s |
| | Lead vehicle acceleration | $a_1(t)$ | 0.5 m/s$^2$, $t \in [10, 30]$ s |
| | Initial conditions | $s_n(t), v_n(t)$ $t \in [-\tau_n, 0]$ | Constant $s_n^*, v_n^*$ |
| Energy consumption | Constant resistance coefficient | $a_{\mathrm{r}}$ | 0.0981 m/s$^2$ |
| | Quadratic resistance coefficient | $c_{\mathrm{r}}$ | 0.0003 m$^{-1}$ |
| | Resolution of contour plots | $\Delta\beta \times \Delta\beta_{\mathrm{B}}$ | 0.05 × 0.01 s$^{-1}$ |
| Stability charts | Range policy gradient for HVs | $\kappa_{\mathrm{H}}$ | 0.7 s$^{-1}$ |
| | Range policy gradient for AVs | $\kappa$ | 0.6 s$^{-1}$ |
| | Frequency range for plant stability | $[\Omega_{\min}, \Omega_{\max}]$ | $[0, 2\pi]$ rad/s |
| | Frequency range for string stability | $[\omega_{\min}, \omega_{\max}]$ | $[0, 2\pi]$ rad/s |

The case $s = 0$ gives the plant stability boundary

$$\alpha = 0\,, \tag{74}$$

whereas substitution of $s = \pm i\Omega$, separation into real and imaginary parts and solution for $\alpha$ and $\beta$ lead to the plant stability boundary

$$\begin{aligned} \alpha &= \frac{\Omega^2 \cos(\Omega\sigma)}{\kappa}\,, \\ \beta &= \Omega \sin(\Omega\sigma) - \alpha - \beta_{\mathrm{B}}\,. \end{aligned} \tag{75}$$

This can also be rearranged to

$$\beta_{\mathrm{B}} = \Omega^*(\alpha) \sin\big(\Omega^*(\alpha)\sigma\big) - \alpha - \beta\,, \tag{76}$$

where $\Omega^*(\alpha)$ is the solution of $\alpha = \Omega^2 \cos(\Omega\sigma)/\kappa$ for $\Omega$.

Similarly, the second line of (71) leads to the characteristic equation

$$s^2 e^{s\sigma} + (\alpha + \beta + \beta_{\mathrm{B}})s + \alpha\kappa - \beta_{\mathrm{B}} s \Gamma(s) = 0\,. \tag{77}$$

For $s = 0$, this yields the plant stability boundary

$$\alpha = 0\,, \tag{78}$$

while for $s = \pm i\Omega$ it implies

$$\begin{aligned} \alpha &= \frac{\Omega^2 \cos(\Omega\sigma) - \Omega\Gamma_{\mathrm{I}}\beta_{\mathrm{B}}}{\kappa}\,, \\ \beta &= \Omega \sin(\Omega\sigma) - \alpha - (1 - \Gamma_{\mathrm{R}})\beta_{\mathrm{B}}\,, \end{aligned} \tag{79}$$

or, equivalently,

$$\begin{aligned} \beta_{\mathrm{B}} &= \frac{\Omega^2 \cos(\Omega\sigma) - \alpha\kappa}{\Omega\Gamma_{\mathrm{I}}}\,, \\ \beta &= \Omega \sin(\Omega\sigma) - \alpha - (1 - \Gamma_{\mathrm{R}})\beta_{\mathrm{B}}\,, \end{aligned} \tag{80}$$

where $\Gamma_{\mathrm{R}} = \Re(\Gamma(i\Omega))$ and $\Gamma_{\mathrm{I}} = \Im(\Gamma(i\Omega))$. These equations were used to plot the plant stability boundaries in the $(\beta, \alpha)$ and $(\beta, \beta_{\mathrm{B}})$ planes in Figs. 7 and 8.

String stability can be analyzed by the help of the head-to-tail transfer function (48), which now reads

$$G(i\omega) = \frac{(\beta i\omega + \alpha\kappa)(\Gamma_{\mathrm{R}} + i\Gamma_{\mathrm{I}})}{-\omega^2 e^{i\omega\sigma} + (\alpha + \beta + \beta_{\mathrm{B}})i\omega + \alpha\kappa - \beta_{\mathrm{B}} i\omega(\Gamma_{\mathrm{R}} + i\Gamma_{\mathrm{I}})}\,, \tag{81}$$

for $s = \mathrm{i}\omega$, where $\Gamma_\mathrm{R} = \Re(\Gamma(\mathrm{i}\omega))$ and $\Gamma_\mathrm{I} = \Im(\Gamma(\mathrm{i}\omega))$. Direct substitution into (60) gives the following expression for $P(\omega)$:

$$P(\omega) = \omega^2 - 2\alpha\kappa\cos(\omega\sigma) - 2\beta_\mathrm{B}\omega\Gamma_\mathrm{I}\cos(\omega\sigma) + \alpha^2\kappa^2\frac{1-\Gamma_\mathrm{R}^2-\Gamma_\mathrm{I}^2}{\omega^2} + 2\alpha\kappa\beta_\mathrm{B}\frac{\Gamma_\mathrm{I}}{\omega} + \beta_\mathrm{B}^2\Gamma_\mathrm{I}^2$$
$$-2\big(\alpha+\beta+(1-\Gamma_\mathrm{R})\beta_\mathrm{B}\big)\omega\sin(\omega\sigma) + \big(\alpha+\beta+(1-\Gamma_\mathrm{R})\beta_\mathrm{B}\big)^2 - \beta^2(\Gamma_\mathrm{R}^2+\Gamma_\mathrm{I}^2)\,. \tag{82}$$

We can construct the $\omega = 0$ string stability boundaries via (62). Note that the following hold for $\omega \to 0$:

$$\begin{aligned}\lim_{\omega\to 0}\frac{\Gamma_\mathrm{I}}{\omega} &= -\frac{N}{\kappa_\mathrm{H}}\,,\\ \lim_{\omega\to 0}\frac{\Gamma_\mathrm{R}-1}{\omega^2} &= N\frac{2\kappa_\mathrm{H}-2\beta_\mathrm{H}-(N+1)\alpha_\mathrm{H}}{2\alpha_\mathrm{H}\kappa_\mathrm{H}^2}\,.\end{aligned} \tag{83}$$

This can be shown by substituting (52) into (45), taking $s = \mathrm{i}\omega$, and calculating the limits above by applying L'Hospital's rule (once for $\Gamma_\mathrm{I}/\omega$ and twice for $(\Gamma_\mathrm{R}-1)/\omega^2$). At the limit $\omega \to 0$ we get

$$P(0) = \alpha\left(\alpha + 2\beta - 2\kappa + N\frac{\alpha\kappa^2}{\alpha_\mathrm{H}\kappa_\mathrm{H}^2}(\alpha_\mathrm{H}+2\beta_\mathrm{H}-2\kappa_\mathrm{H}) - 2N\frac{\kappa}{\kappa_\mathrm{H}}\beta_\mathrm{B}\right). \tag{84}$$

Taking $P(0) = 0$ finally gives the $\omega = 0$ string stability boundaries in the form

$$\begin{aligned}\alpha &= 0\,,\\ \alpha &= \frac{2\left(\kappa-\beta+N\frac{\kappa}{\kappa_\mathrm{H}}\beta_\mathrm{B}\right)}{1+\frac{N\kappa^2}{\alpha_\mathrm{H}\kappa_\mathrm{H}^2}(\alpha_\mathrm{H}+2\beta_\mathrm{H}-2\kappa_\mathrm{H})}\,,\end{aligned} \tag{85}$$

which can also be expressed as

$$\beta_\mathrm{B} = \frac{\kappa_\mathrm{H}}{2N\kappa}\left(\alpha+2\beta-2\kappa+N\frac{\alpha\kappa^2}{\alpha_\mathrm{H}\kappa_\mathrm{H}^2}(\alpha_\mathrm{H}+2\beta_\mathrm{H}-2\kappa_\mathrm{H})\right). \tag{86}$$

The $\omega > 0$ string stability boundaries can be found by substitution of (48) into (59). Then, separation into real and imaginary parts leads to a linear system of equations for $\alpha$ and $\beta$, that ultimately gives

$$\alpha = \frac{\omega^2\big(\Gamma_\mathrm{I}\sin(\omega\sigma - K) + \Gamma_\mathrm{R}\cos(\omega\sigma - K) - \cos(\omega\sigma)\big)}{\omega(\Gamma_\mathrm{R}\sin K + \Gamma_\mathrm{I}\cos K) - \kappa(1 + \Gamma_\mathrm{R}^2 + \Gamma_\mathrm{I}^2 - 2\Gamma_\mathrm{R}\cos K + 2\Gamma_\mathrm{I}\sin K)} + \frac{\beta_\mathrm{B}\omega\big((\Gamma_\mathrm{R}^2 + \Gamma_\mathrm{I}^2)\sin K - \Gamma_\mathrm{R}\sin K + \Gamma_\mathrm{I}(1 - \cos K)\big)}{\omega(\Gamma_\mathrm{R}\sin K + \Gamma_\mathrm{I}\cos K) - \kappa(1 + \Gamma_\mathrm{R}^2 + \Gamma_\mathrm{I}^2 - 2\Gamma_\mathrm{R}\cos K + 2\Gamma_\mathrm{I}\sin K)},$$

$$\beta = \frac{\omega^2\cos(\omega\sigma) - \kappa\omega\big(\Gamma_\mathrm{I}\cos(\omega\sigma - K) - \Gamma_\mathrm{R}\sin(\omega\sigma - K) + \sin(\omega\sigma)\big)}{\omega(\Gamma_\mathrm{R}\sin K + \Gamma_\mathrm{I}\cos K) - \kappa(1 + \Gamma_\mathrm{R}^2 + \Gamma_\mathrm{I}^2 - 2\Gamma_\mathrm{R}\cos K + 2\Gamma_\mathrm{I}\sin K)} + \frac{\beta_\mathrm{B}\big(-\omega\Gamma_\mathrm{I} + \kappa\big(1 + (\Gamma_\mathrm{R}^2 + \Gamma_\mathrm{I}^2)\cos K - \Gamma_\mathrm{R}(1 + \cos K) + \Gamma_\mathrm{I}\sin K\big)\big)}{\omega(\Gamma_\mathrm{R}\sin K + \Gamma_\mathrm{I}\cos K) - \kappa(1 + \Gamma_\mathrm{R}^2 + \Gamma_\mathrm{I}^2 - 2\Gamma_\mathrm{R}\cos K + 2\Gamma_\mathrm{I}\sin K)}. \tag{87}$$

This can be re-written also in the form

$$\beta_\mathrm{B} = \frac{\omega^2\big(-\Gamma_\mathrm{I}\sin(\omega\sigma - K) - \Gamma_\mathrm{R}\cos(\omega\sigma - K) + \cos(\omega\sigma))\big)}{\omega\big(\Gamma_\mathrm{I}(1 - \cos K) - \Gamma_\mathrm{R}\sin K + (\Gamma_\mathrm{R}^2 + \Gamma_\mathrm{I}^2)\sin K\big)} + \frac{\alpha\omega(\Gamma_\mathrm{R}\sin K + \Gamma_\mathrm{I}\cos K) - \alpha\kappa\big(1 + \Gamma_\mathrm{R}^2 + \Gamma_\mathrm{I}^2 - 2\Gamma_\mathrm{R}\cos K + 2\Gamma_\mathrm{I}\sin K\big)}{\omega\big(\Gamma_\mathrm{I}(1 - \cos K) - \Gamma_\mathrm{R}\sin K + (\Gamma_\mathrm{R}^2 + \Gamma_\mathrm{I}^2)\sin K\big)},$$

$$\beta = \frac{\omega^2\big(\Gamma_\mathrm{I}\sin(\omega\sigma) - (1 - \Gamma_\mathrm{R})\cos(\omega\sigma)\big)}{\omega\big(\Gamma_\mathrm{I}(1 - \cos K) - \Gamma_\mathrm{R}\sin K + (\Gamma_\mathrm{R}^2 + \Gamma_\mathrm{I}^2)\sin K\big)} + \frac{-\alpha\omega\Gamma_\mathrm{I} + \alpha\kappa\big(1 + (\Gamma_\mathrm{R}^2 + \Gamma_\mathrm{I}^2)\cos K - \Gamma_\mathrm{R}(1 + \cos K) + \Gamma_\mathrm{I}\sin K\big)}{\omega\big(\Gamma_\mathrm{I}(1 - \cos K) - \Gamma_\mathrm{R}\sin K + (\Gamma_\mathrm{R}^2 + \Gamma_\mathrm{I}^2)\sin K\big)}. \tag{88}$$

These formulas allowed us to plot the string stability boundaries in the $(\beta, \alpha)$ and $(\beta, \beta_\mathrm{B})$ planes in Figs. 7 and 8.

## *Stability of ACC*

Now we consider the stability of traffic flows influenced by vehicle control via ACC.

### ACC Followed by Human-Driven Traffic

Since ACC is a special case of ATC, all stability boundaries can be obtained by substituting $\beta_\mathrm{B} = 0$. The plant stability boundaries (74)–(79) reduce to

$$\alpha = 0, \tag{89}$$

and

$$\begin{aligned} \alpha &= \frac{\Omega^2 \cos(\Omega\sigma)}{\kappa}, \\ \beta &= \Omega \sin(\Omega\sigma) - \alpha, \end{aligned} \tag{90}$$

whereas the string stability boundaries become

$$\begin{aligned} \alpha &= 0, \\ \alpha &= \frac{2(\kappa - \beta)}{1 + \frac{N\kappa^2}{\alpha_{\mathrm{H}}\kappa_{\mathrm{H}}^2}(\alpha_{\mathrm{H}} + 2\beta_{\mathrm{H}} - 2\kappa_{\mathrm{H}})}, \end{aligned} \tag{91}$$

and

$$\begin{aligned} \alpha &= \frac{\omega^2\big(\Gamma_{\mathrm{I}} \sin(\omega\sigma - K) + \Gamma_{\mathrm{R}} \cos(\omega\sigma - K) - \cos(\omega\sigma)\big)}{\omega(\Gamma_{\mathrm{R}} \sin K + \Gamma_{\mathrm{I}} \cos K) - \kappa(1 + \Gamma_{\mathrm{R}}^2 + \Gamma_{\mathrm{I}}^2 - 2\Gamma_{\mathrm{R}} \cos K + 2\Gamma_{\mathrm{I}} \sin K)}, \\ \beta &= \frac{\omega^2 \cos(\omega\sigma) - \kappa\omega\big(\Gamma_{\mathrm{I}} \cos(\omega\sigma - K) - \Gamma_{\mathrm{R}} \sin(\omega\sigma - K) + \sin(\omega\sigma)\big)}{\omega(\Gamma_{\mathrm{R}} \sin K + \Gamma_{\mathrm{I}} \cos K) - \kappa(1 + \Gamma_{\mathrm{R}}^2 + \Gamma_{\mathrm{I}}^2 - 2\Gamma_{\mathrm{R}} \cos K + 2\Gamma_{\mathrm{I}} \sin K)}. \end{aligned} \tag{92}$$

### ACC Followed by Human-Driven Traffic on a Ring

When the ACC scenario with subsequent human-driven vehicles is driven on a ring road, the characteristic equation becomes (50). As pointed out in Sect. 5.2, the stability boundary is analogous to the plant stability of the virtual ring in ATC. Thus, for $s = 0$ we obtain

$$\alpha = 0, \tag{93}$$

cf. (74), (78). For $s = \mathrm{i}\omega$, the stability boundary can be obtained as the $K = 0$ special case of the $\omega > 0$ string stability boundary of the associated straight-road configuration. Thus, substituting $K = 0$ into (92) leads to

$$\begin{aligned} \alpha &= \frac{\omega^2\big(\Gamma_{\mathrm{I}} \sin(\omega\sigma) - (1 - \Gamma_{\mathrm{R}}) \cos(\omega\sigma)\big)}{\omega\Gamma_{\mathrm{I}} - \kappa\big((1 - \Gamma_{\mathrm{R}})^2 + \Gamma_{\mathrm{I}}^2\big)}, \\ \beta &= \frac{\omega^2 \cos(\omega\sigma) - \kappa\omega\big(\Gamma_{\mathrm{I}} \cos(\omega\sigma) + (1 - \Gamma_{\mathrm{R}}) \sin(\omega\sigma)\big)}{\omega\Gamma_{\mathrm{I}} - \kappa\big((1 - \Gamma_{\mathrm{R}})^2 + \Gamma_{\mathrm{I}}^2\big)}. \end{aligned} \tag{94}$$

Notice that this gives back the plant stability boundaries (90) if we substitute $\Gamma(\mathrm{i}\omega) = 0$, i.e., $\Gamma_{\mathrm{R}} = 0$, $\Gamma_{\mathrm{I}} = 0$ (which is the special case of $N \to \infty$ string stable human drivers with $|T_{\mathrm{H}}(\mathrm{i}\omega)| < 1$, see Sect. 5.2).

### ACC without Considering the Traffic Behind

The influence of ACC controllers on the traffic flow behind them is often neglected during cruise control design. This can be considered as the $N = 0$ special case of ACC followed by human-driven traffic discussed above. For $N = 0$, we have $\Gamma(s) = 1$ that implies $\Gamma_R = 1$, $\Gamma_I = 0$. The plant stability conditions still yield (89), (90), since we have already considered a plant stable human-driven chain:

$$\alpha = 0\,, \tag{95}$$

and

$$\begin{aligned} \alpha &= \frac{\Omega^2 \cos(\Omega\sigma)}{\kappa}\,, \\ \beta &= \Omega \sin(\Omega\sigma) - \alpha\,. \end{aligned} \tag{96}$$

On the other hand, substitution of $N = 0$ into (91) leads to the $\omega = 0$ string stability boundaries

$$\begin{aligned} \alpha &= 0\,, \\ \alpha &= 2(\kappa - \beta)\,, \end{aligned} \tag{97}$$

while substitution of $\Gamma_R = 1$ and $\Gamma_I = 0$ into (92) results in the $\omega > 0$ string stability boundaries

$$\begin{aligned} \alpha &= \frac{\omega^2\big(\cos(\omega\sigma - K) - \cos(\omega\sigma)\big)}{\omega \sin K - 2\kappa(1 - \cos K)}\,, \\ \beta &= \frac{\omega^2 \cos(\omega\sigma) + \kappa\omega\big(\sin(\omega\sigma - K) - \sin(\omega\sigma)\big)}{\omega \sin K - 2\kappa(1 - \cos K)}\,. \end{aligned} \tag{98}$$

## *Stability of Human-Driven Traffic*

Finally, human-driven traffic can also be analyzed as a special case of ATC or ACC by considering that the ego vehicle that acts like a human driver. In terms of formulas, this case is recovered by replacing $\alpha$, $\beta$, $\kappa$, $\sigma$ with $\alpha_H$, $\beta_H$, $\kappa_H$, $\tau$.

### Human-Driven Traffic on a Straight Road

The straightforward replacement of parameters leads to the plant stability conditions for human drivers in the form

$$\alpha_{\mathrm{H}} = 0\,, \tag{99}$$

and

$$\begin{aligned}\alpha_{\mathrm{H}} &= \frac{\Omega^2 \cos(\Omega\tau)}{\kappa_{\mathrm{H}}}\,,\\ \beta_{\mathrm{H}} &= \Omega \sin(\Omega\tau) - \alpha_{\mathrm{H}}\,,\end{aligned} \tag{100}$$

as well as the string stability conditions

$$\begin{aligned}\alpha_{\mathrm{H}} &= 0\,,\\ \alpha_{\mathrm{H}} &= 2(\kappa_{\mathrm{H}} - \beta_{\mathrm{H}})\,,\end{aligned} \tag{101}$$

and

$$\begin{aligned}\alpha_{\mathrm{H}} &= \frac{\omega^2\big(\cos(\omega\tau - K) - \cos(\omega\tau)\big)}{\omega \sin K - 2\kappa_{\mathrm{H}}(1 - \cos K)}\,,\\ \beta_{\mathrm{H}} &= \frac{\omega^2 \cos(\omega\tau) + \kappa_{\mathrm{H}}\omega\big(\sin(\omega\tau - K) - \sin(\omega\tau)\big)}{\omega \sin K - 2\kappa_{\mathrm{H}}(1 - \cos K)}\,.\end{aligned} \tag{102}$$

### Human-Driven Traffic on a Ring

Human-driven traffic on a ring can be analyzed by the characteristic equation (69). For $s = 0$ the stability boundary is still

$$\alpha_{\mathrm{H}} = 0\,, \tag{103}$$

whereas for $s = \mathrm{i}\omega$ it becomes

$$\begin{aligned}\alpha_{\mathrm{H}} &= \frac{\omega^2\big(\cos\big(\omega\tau - \frac{k2\pi}{N+1}\big) - \cos(\omega\tau)\big)}{\omega \sin\big(\frac{k2\pi}{N+1}\big) - 2\kappa_{\mathrm{H}}\big(1 - \cos\big(\frac{k2\pi}{N+1}\big)\big)}\,,\\ \beta_{\mathrm{H}} &= \frac{\omega^2 \cos(\omega\tau) + \kappa_{\mathrm{H}}\omega\big(\sin\big(\omega\tau - \frac{k2\pi}{N+1}\big) - \sin(\omega\tau)\big)}{\omega \sin\big(\frac{k2\pi}{N+1}\big) - 2\kappa_{\mathrm{H}}\big(1 - \cos\big(\frac{k2\pi}{N+1}\big)\big)}\,.\end{aligned} \tag{104}$$

This latter result can be obtained by substituting $K = 2k\pi/(N+1)$ into the string stability boundaries (102) of the human-driven chain of vehicles, as explained in Sect. 5.2. Alternatively, the same result can be achieved based on (68, 69) by considering $\Gamma(\mathrm{i}\omega) = \mathrm{e}^{\mathrm{i}\frac{2k\pi}{N+1}}$ and substituting $\Gamma_{\mathrm{R}} = \cos\big(\frac{2k\pi}{N+1}\big)$, $\Gamma_{\mathrm{I}} = \sin\big(\frac{2k\pi}{N+1}\big)$ into (92) with $\alpha = \alpha_{\mathrm{H}}$, $\beta = \beta_{\mathrm{H}}$, $\kappa = \kappa_{\mathrm{H}}$ and $\sigma = \tau$.

Finally, we remark that the stability of TC, CC followed by human-driven traffic on a straight road or ring, and CC without considering the traffic behind can all be obtained as the special cases of ATC: when there is no response to the distance from the vehicle ahead, i.e., $\alpha = 0$. For the $s = 0$ stability boundaries, however, one needs to take special care since it is exactly $\alpha = 0$ in ATC.

## References


1. Orosz G, Wilson RE, Szalai R, Stépán G (2009) Exciting traffic jams: Nonlinear phenomena behind traffic jam formation on highways. Phys Rev E 80(4):046205
2. Molnár TG, Upadhyay D, Hopka M, Van Nieuwstadt M, Orosz G (2021) Delayed Lagrangian continuum models for on-board traffic prediction. Transp Res Part C 123:102991
3. Orosz G, Wilson RE, Stépán G (2010) Traffic jams: dynamics and control. Philos Trans Royal Soc A: Math Phys Eng Sci 368(1928):4455–4479
4. Besselink B, Johansson KH (2017) String stability and a delay-based spacing policy for vehicle platoons subject to disturbances. IEEE Trans Autom Control 62(9):4376–4391
5. Feng S, Zhang Y, Li SE, Cao Z, Liu HX, Li L (2019) String stability for vehicular platoon control: definitions and analysis methods. Annu Rev Control 47:81–97
6. Ploeg J, van de Wouw N, Nijmeijer H (2014) $\mathcal{L}_p$ string stability of cascaded systems: application to vehicle platooning. IEEE Trans Control Syst Technol 22(2):786–793
7. Swaroop D, Hedrick JK (1996) String stability of interconnected systems. IEEE Trans Autom Control 41(3):349–357
8. Ciuffo B, Mattas K, Makridis M, Albano G, Anesiadou A, He Y, Josvai S, Komnos D, Pataki M, Vass S, Szalay Z (2021) Requiem on the positive effects of commercial Adaptive Cruise Control on motorway traffic and recommendations for future automated driving systems. Transp Res Part C 130:103305
9. Gunter G, Gloudemans D, Stern RE, McQuade S, Bhadani R, Bunting M, Delle Monache ML, Lysecky R, Seibold B, Sprinkle J, Piccoli B, Work DB (2020) Are commercially implemented adaptive cruise control systems string stable? IEEE Trans Intell Transp Syst 1–12
10. Gunter G, Janssen C, Barbour W, Stern RE, Work DB (2020) Model-based string stability of adaptive cruise control systems using field data. IEEE Trans Intell Veh 5(1):90–99
11. Delle Monache ML, Sprinkle J, Vasudevan R, Work D (2019) Autonomous vehicles: from vehicular control to traffic control. In: 58th IEEE Conference on Decision and Control. Nice, France, pp 4680–4696
12. Xiao L, Gao F (2010) A comprehensive review of the development of adaptive cruise control systems. Veh Syst Dyn 48(10):1167–1192
13. Orosz G (2016) Connected cruise control: modeling, delay effects, and nonlinear behavior. Veh Syst Dyn 54(8):1147–1176
14. Wang Z, Wu G, Barth MJ (2018) A review on cooperative adaptive cruise control (CACC) systems: architectures, controls, and applications. In: 21st International Conference on Intelligent Transportation Systems. Maui, HI, USA, pp 2884–2891
15. Dollar RA, Vahidi A (2018) Efficient and collision-free anticipative cruise control in randomly mixed strings. IEEE Trans Intell Veh 3(4):439–452
16. Ge JI, Avedisov SS, He CR, Qin WB, Sadeghpour M, Orosz G (2018) Experimental validation of connected automated vehicle design among human-driven vehicles. Transp Res Part C 91:335–352
17. Zheng Y, Wang J, Li K (2020) Smoothing traffic flow via control of autonomous vehicles. IEEE Internet of Things J 7(5):3882–3896
18. Ard T, Dollar RA, Vahidi A, Zhang Y, Karbowski D (2020) Microsimulation of energy and flow effects from optimal automated driving in mixed traffic. Transp Res Part C 120:102806
19. van Arem B, van Driel C, Visser R (2006) The impact of cooperative adaptive cruise control on traffic-flow characteristics. IEEE Trans Intell Transp Syst 7(4):429–436

20. Cui S, Seibold B, Stern R, Work DB (2017) Stabilizing traffic flow via a single autonomous vehicle: possibilities and limitations. In: 2017 IEEE Intelligent Vehicles Symposium. Los Angeles, CA, USA, pp 1336–1341
21. Spiliopoulou A, Manolis D, Vandorou F, Papageorgiou M (2018) Adaptive cruise control operation for improved motorway traffic flow. Transp Res Rec 2672(22):24–35
22. Mehran B, Kuwahara M, Naznin F (2011) Implementing kinematic wave theory to reconstruct vehicle trajectories from fixed and probe sensor data. In: 19th International Symposium on Transportation and Traffic Theory, vol 17. Berkeley, CA, USA, pp 247–268
23. Yu H, Gan Q, Bayen A, Krstic M (2020) PDE traffic observer validated on freeway data. IEEE Trans Control Syst Technol 1–13
24. Bekiaris-Liberis N, Roncoli C, Papageorgiou M (2017) Highway traffic state estimation per lane in the presence of connected vehicles. Transp Res Part B: Methodol 106:1–28
25. Chu K, Saigal R, Saitou K (2016) Stochastic Lagrangian traffic flow modeling and real-time traffic prediction. In: IEEE International Conference on Automation Science and Engineering. Fort Worth TX, USA, pp 213–218
26. Chu K, Saigal R, Saitou K (2019) Real-time traffic prediction and probing strategy for Lagrangian traffic data. IEEE Trans Intell Transp Syst 20(2):497–506
27. Delle Monache ML, Liard T, Piccoli B, Stern R, Work D (2019) Traffic reconstruction using autonomous vehicles. SIAM J Appl Math 79(5):1748–1767
28. Herrera JC, Bayen AM (2010) Incorporation of Lagrangian measurements in freeway traffic state estimation. Transp Res Part B: Methodol 44(4):460–481
29. Work DB, Tossavainen OP, Blandin S, Bayen AM, Iwuchukwu T, Tracton K (2008) An ensemble Kalman filtering approach to highway traffic estimation using GPS enabled mobile devices. In: 47th IEEE Conference on Decision and Control. Cancun, Mexico, pp 5062–5068
30. Work DB, Tossavainen OP, Jacobson Q, Bayen AM (2009) Lagrangian sensing: traffic estimation with mobile devices. In: American Control Conference. St. Louis, MO, USA, pp 1536–1543
31. Duret A, Yuan Y (2017) Traffic state estimation based on Eulerian and Lagrangian observations in a mesoscopic modeling framework. Transp Res Part B: Methodol 101:51–71
32. Yuan Y, Van Lint H, Van Wageningen-Kessels F, Hoogendoorn S (2014) Network-wide traffic state estimation using loop detector and floating car data. J Intell Transp Syst 18(1):41–50
33. Yuan Y, van Lint JWC, Wilson RE, van Wageningen-Kessels F, Hoogendoorn SP (2012) Real-time Lagrangian traffic state estimator for freeways. IEEE Trans Intell Transp Syst 13(1):59–70
34. Siri S, Pasquale C, Sacone S, Ferrara A (2021) Freeway traffic control: A survey. Automatica 130:109655
35. Kreidieh AR, Wu C, Bayen AM (2018) Dissipating stop-and-go waves in closed and open networks via deep reinforcement learning. In: 21st International Conference on Intelligent Transportation Systems. Maui, HI, USA, pp 1475–1480
36. Wu C, Bayen AM, Mehta A (2018) Stabilizing traffic with autonomous vehicles. In: 2018 IEEE International Conference on Robotics and Automation. Brisbane, Australia, pp 1–7
37. Laval JA, Leclercq L (2013) The Hamilton-Jacobi partial differential equation and the three representations of traffic flow. Transp Res Part B 52:17–30
38. Bekiaris-Liberis N, Delis AI (2021) PDE-based feedback control of freeway traffic flow via time-gap manipulation of ACC-equipped vehicles. IEEE Trans Control Syst Technol 29(1):461–469
39. Yu H, Koga S, Krstic M (2018) Stabilization of traffic flow with a leading autonomous vehicle. In: ASME Dynamic Systems and Control Conference, DSCC2018–9239. Atlanta, GA, USA
40. Karafyllis I, Papageorgiou M (2018) Feedback control of scalar conservation laws with application to density control in freeways by means of variable speed limits. Automatica 105:228–236
41. Koehler S, Mehr N, Horowitz R, Borrelli F (2016) Stable hybrid model predictive control for ramp metering. In: 19th IEEE International Conference on Intelligent Transportation Systems. Rio de Janeiro, Brazil, pp 1083–1088
42. Pasquale C, Sacone S, Siri S (2018) Closed-loop stability of freeway traffic systems with ramp metering control. In: 57th IEEE Conference on Decision and Control. Miami, FL, USA, pp 223–228

43. Yu H, Krstic M (2019) Traffic congestion control for Aw-Rascle-Zhang model. Automatica 100:38–51
44. Yu H, Park S, Bayen A, Moura S, Krstic M (2021) Reinforcement learning versus PDE backstepping and PI control for congested freeway traffic. arXiv:1904.12957
45. Čičić M, Johansson KH (2018) Traffic regulation via individually controlled automated vehicles: a cell transmission model approach. In: 21st International Conference on Intelligent Transportation Systems. Maui, HI, USA, pp 766–771
46. Piacentini G, Čičić M, Ferrara A, Johansson KH (2019) VACS equipped vehicles for congestion dissipation in multi-class CTM framework. In: 18th European Control Conference . Naples, Italy, pp 2203–2208
47. Chen C, Wang J, Xu Q, Wang J, Li K (2021) Mixed platoon control of automated and human-driven vehicles at a signalized intersection: dynamical analysis and optimal control. arXiv:2010.16105
48. Giammarino V, Lv, M, Baldi S, Frasca P, Delle Monache ML (2019) On a weaker notion of ring stability for mixed traffic with human-driven and autonomous vehicles. In: 58th IEEE conference on decision and control. Nice, France, pp 335–340
49. von Allwörden H, Gasser I (2021) On a general class of solutions for an optimal velocity model on an infinite lane. Transportmetrica A 17(3):258–277
50. Wang J, Zheng Y, Xu Q, Wang J, Li K (2020) Controllability analysis and optimal control of mixed traffic flow with human-driven and autonomous vehicles. arXiv:2002.02099
51. Stern RE, Cui S, Delle Monache ML, Bhadani R, Bunting M, Churchill M, Hamilton N, Haulcy R, Pohlmann H, Wu F, Piccoli B, Seibold B, Sprinkle J, Work DB (2018) Dissipation of stop-and-go waves via control of autonomous vehicles: field experiments. Transp Res Part C 89:205–221
52. Avedisov SS, Bansal G, Kiss AK, Orosz G (2018) Experimental verification platform for connected vehicle networks. In: 21st IEEE International Conference on Intelligent Transportation Systems. Maui, HI, USA, pp 818–823
53. Avedisov SS, Bansal G, Orosz G (2021) Impacts of connected automated vehicles on freeway traffic patterns at different penetration levels. IEEE Trans Intell Transp Syst 1–14
54. Molnár TG, Upadhyay D, Hopka M, Van Nieuwstadt M, Orosz G (2020) Open and closed loop traffic control by connected automated vehicles. In: 59th IEEE Conference on Decision and Control. Online Jeju Island, Republic of Korea, pp 239–244
55. Wang J, Zheng Y, Chen C, Xu Q, Li K (2020) Leading cruise control in mixed traffic flow. In: 59th IEEE Conference on Decision and Control. Online, Jeju Island, Republic of Korea, pp 226–232
56. Bando M, Hasebe K, Nakanishi K, Nakayama A (1998) Analysis of optimal velocity model with explicit delay. Phys Rev E 58(5):5429–5435
57. Treiber M, Hennecke A, Helbing D (2000) Congested traffic states in empirical observations and microscopic simulations. Phys Rev E 62(2):1805
58. Dollar RA. Molnár TG, Vahidi A, Orosz G (2021) MPC-based connected cruise control with multiple human predecessors. In: American Control Conference. Online, New Orleans, LA, USA, pp 404–410
59. He CR, Maurer H, Orosz G (2016) Fuel consumption optimization of heavy-duty vehicles with grade, wind, and traffic information. J Comput Nonlinear Dyn 11(6):061011
60. Zhang L, Orosz G (2016) Motif-based design for connected vehicle systems in presence of heterogeneous connectivity structures and time delays. IEEE Trans Intell Transp Syst 17(6):1638–1651
61. Mahbub AMI, Malikopoulos AA (2021) Conditions to provable system-wide optimal coordination of connected and automated vehicles. arXiv:2005.14551
62. Mahbub AMI, Malikopoulos AA (2021) A platoon formation framework in a mixed traffic environment. arXiv:2103.03393
63. Piacentini G, Goatin P, Ferrara A (2018) Traffic control via moving bottleneck of coordinated vehicles. IFAC-PapersOnLine 51(9):13–18

64. Bekiaris-Liberis N, Roncoli C, Papageorgiou M (2018) Predictor-based adaptive cruise control design. IEEE Trans Intell Transp Syst 19(10):3181–3195
65. Molnár TG, Qin WB, Insperger T, Orosz G (2018) Application of predictor feedback to compensate time delays in connected cruise control. IEEE Trans Intell Transp Syst 19(2):545–559
66. Qi J, Mo S, Krstic M (2021) Delay-compensated distributed PDE control of traffic with connected/automated vehicles. arXiv:2107.08651
67. Wang M, Hoogendoorn SP, Daamen W, van Arem B, Shyrokau B, Happee R (2018) Delay-compensating strategy to enhance string stability of adaptive cruise controlled vehicles. Transportmetrica B: Transp Dyn 6(3):211–229
68. Zhang Y, Bai Y, Hu J, Wang M (2020) Control design, stability analysis, and traffic flow implications for cooperative adaptive cruise control systems with compensation of communication delay. Transp Res Rec 2674(8):638–652
69. He CR, Ge JI, Orosz G (2020) Fuel efficient connected cruise control for heavy-duty trucks in real traffic. IEEE Trans Control Syst Technol 28(6):2474–2481
70. Qin WB, Orosz G (2020) Experimental validation of string stability for connected vehicles subject to information delay. IEEE Trans Control Syst Technol 28(4):1203–1217

# Socioeconomic Impact of Emerging Mobility Markets and Implementation Strategies

**Ioannis Vasileios Chremos and Andreas A. Malikopoulos**

**Abstract** Emerging mobility systems such as connected and automated vehicles (CAVs) provide the most intriguing opportunity for more accessible, safe, and efficient transportation. CAVs are expected to significantly improve safety by eliminating the human factor and ensure transportation efficiency by allowing users to monitor transportation network conditions and make better operating decisions. However, CAVs could alter the users' tendency-to-travel, leading to a higher traffic demand than expected, thus causing rebound effects (e.g., increased vehicle-miles-traveled). In this chapter, we focus on tackling the social factors that could drive an emerging mobility system to unsustainable congestion levels. We propose a mobility market that models the economic in-nature interactions of the travelers in a smart city network with roads and public transit infrastructure. Using techniques from mechanism design, we introduce appropriate monetary incentives (e.g., tolls, fares, fees) and show how a mobility system consisting of selfish travelers that seek to travel either with a CAV or use public transit can be socially efficient. Furthermore, the proposed mobility market ensures that travelers always report their true travel preferences and always benefit from participating in the market; lastly, we also show that the market generates enough revenue to potentially cover its operating costs.

## 1 Introduction

Nowadays, it is nearly impossible to commute in a major urban area without the frustration of congestion and traffic jams. Moreover, congestion is one of the leading factors behind road accidents and altercations, negatively impacting the economic success of cities and the quality of life of their citizens. For these reasons, congestion has been broadly recognized as one of the major challenges to address for next-generation cities. One incoming highly transformative innovation that promises to

I. V. Chremos (✉) · A. A. Malikopoulos
University of Delaware, Newark, DE 19716, USA
e-mail: ichremos@udel.edu

A. A. Malikopoulos
e-mail: andreas@udel.edu

 
Y. L. Murphey et al. (eds.), *AI-enabled Technologies for Autonomous and Connected Vehicles*, Lecture Notes in Intelligent Transportation and Infrastructure,
https://doi.org/10.1007/978-3-031-06780-8_17

address congestion though is autonomous driving, and in particular, connected and automated vehicles (CAVs). Recent advancements in emerging mobility systems with CAVs are highly expected to eliminate congestion and increase mobility efficiency in terms of energy consumption and travel time [1]. In addition, CAVs are expected to have vast technological, commercial, and regulatory dimensions [2].

There has been a significant amount of work on the technological impact of CAVs, mostly focusing on congestion, emissions, energy consumption, and safety [3, 4]. It is apparent that CAVs will transform today's urban transportation system and revolutionize mobility [5]. However, one of the most novel and defining characteristics of an emerging mobility system is its socioeconomic complexity. Mobility is an indispensable prerequisite for social, cultural, and economic development as well as social participation. Thanks to the unprecedented improvements in mobility, we expect a significant alteration in human behavior and, most importantly, on tendency-to-travel. This may lead to unintended consequences, i.e., rebound effects, in the sense of additional energy use and greenhouse gas emissions, as well as leading to decreases in the density of urban areas and negatively impacting congestion. In addition, future mobility systems will enable human-vehicle interactions between people of any age and abilities, thus allowing enhanced and universal accessibility. One key reason why connectivity (e.g., Internet of Things) and automation in mobility may lead to rebound effects is because of the high levels of comfort and convenience—factors that urge drivers, passengers, and travelers to change their commute and travel tendencies, and thus use their vehicles quite more frequently and more unexpectedly. As urban social life has been greatly associated with the technological impact of the car, this compels us to reassess the relationship between automobility and social life [6, 7]. To add to our argument, evident from similar technological revolutions, for example, the impact of elevators on building design and social class hierarchies [8], human social perspective and view can have a tremendous effect on how technological innovations are utilized and implemented. For all these reasons, it is vital to study the impact of CAVs in a sociotechnical context focusing on the social implications and attempt to provide optimal solutions for the efficient CAV-utilization in society.

There is a solid body of research now available for optimizing the efficiency of emerging mobility systems with CAVs. Over the last decade, several research efforts reported in the literature [9–13] have aimed at addressing questions regarding the CAVs' impact on transportation efficiency. For example, can we consider the problem of optimizing fuel economy and emissions by coordinating a mobility system consisting of CAVs? What would be the appropriate conceptual approaches for modeling and optimizing emerging mobility systems? Recent technological developments can answer the above questions, indicating that CAVs will most likely help us eliminate congestion, significantly decrease fuel consumption, and minimize road accidents. Analytical frameworks have been proposed to quantify and evaluate the impacts of CAVs from the technological perspective [14, 15]. Furthermore, coordination of CAVs at different traffic scenarios (e.g., intersections, vehicle-following) have been extensively evaluated in the literature [16–20]. Moreover, the impact of CAVs has been identified as one that will enable traffic administrators to monitor transportation network conditions efficiently and effectively, thus improving the operating decisions that are required daily [1, 21]. However, they are challenges. The cyber-physical nature of emerging mobility systems is associated with significant control challenges and gives rise to a new level of complexity in modeling and control [22]. These challenges tend to focus on the technological dimension, and what is mostly missing is a complementary study to the broader social implications of CAVs. For example, the impact of selfish social behavior in routing networks of regular and autonomous vehicles has been studied [23–25], as well as "how people learn" in mobility systems with behavioral dynamics [26–29]. However, it seems that the problem of how CAVs will affect human tendency-to-travel and decision-making has not been adequately approached yet. Understanding this "social" aspect of CAVs is critical in our effort to design efficient mobility systems.

One of the standard approaches to alleviate congestion in a transportation system has been the management of demand size due to the shortage of space availability and scarce economic resources in the form of congestion pricing (alternatively called "tolling mechanisms" [30, 31]). Such an approach focuses primarily on intelligent and scalable traffic routing, in which the objective is to guide and coordinate users in path-choice decision-making. For example, one computes the shortest path from a source to a destination regardless of the changing traffic conditions [32]. Interestingly, by adopting a game-theoretic approach, advanced systems have been proposed to assign users concrete routes or minimize travel time and studying the Nash equilibria under different tolling mechanisms [33–38]. This motivates us to ask: "How can we design an emerging mobility system that ensures that all travelers reach their destination safely, efficiently, and in a timely manner?" This question is quite important as it is widely accepted that CAVs will revolutionize the way people travel. We aim to provide a first-attempt answer to this question in this chapter and argue that a sociotechnical approach focusing on the social dimension of a mobility problem can help us design the next-generation mobility systems. To achieve this, we consider a mobility system with decentralized information (alternatively called "asymmetric information") and multiple selfish and intelligent decision-makers (e.g., travelers), who, in turn, may misreport their true travel preferences for better individual benefits. Hence, based on their background and unique behavioral tendencies, travelers make decisions that generally do not lead to system-wide optimal performance. We tackle this *discrepancy between individual and collective interests* [39] by reverse-engineering the mobility system from its optimal solution (e.g., efficiency, congestion-free) to what should each traveler do via the implementation of monetary incentives. This method in economics is known as "mechanism design," in which by treating systems as economic institutions, we can control and coordinate the selfish agents' "economic activity" (e.g., which mode of transportation to use).

The theory of mechanism design was developed as an objective-first approach to efficiently align the individuals' and system's interests in problems of asymmetric information, where the individual agents have private preferences [40, 41]. It can be viewed as the art of designing the rules of a game to achieve a specific desired outcome. A well-established and broadly-used mechanism that has been successful in widely different applications (e.g., auctions, public projects, and cost-minimization problems) is the Vickrey-Clarke-Groves (VCG) mechanism [42–44].

The VCG mechanism ensures the existence and implementation of a dominant strategy equilibrium, which is an efficient solution and allows selfish agents to make a decision (alternatively choose a strategy) that is best no matter what other agents may decide. Agents are also incentivized to report their private preferences truthfully and have no reason (e.g., chance of receiving negative utility) not to participate in the mechanism. However, the VCG mechanism is known to be an extravagant mechanism, i.e., it can generate big surpluses. Overall, mechanism design has broad applications spanning surprisingly many different fields, including microeconomics, social choice theory, and control engineering. Applications in engineering include communication networks [45], social media [46], transportation routing [47], online advertising [48], smart grid [49], multi-agent systems [50], and in general resource allocation problems [51]. We provide a formal overview of mechanism design in Sect. 2.

The application of mechanism design is not new in transportation and mobility problems [52–56]. For example, it has been used to provide solutions to individual route selection under different congestion traffic scenarios (e.g., first-mile ridesharing, selfish routing, tradable driving permits). In particular, auction-based mechanisms treat traffic congestion as an economic problem of supply and demand, focusing on travel time allocation or routing. So, on the one hand, auctions have been proposed to design pricing schemes with tolls in a network of roads leading to a spark of studies in auctioning techniques. On the other hand, this approach has important limitations: (i) the implementability of auction-based tolling on highways is not straightforward due to the dynamic and fast-changing nature of transportation systems; (ii) it is also uncertain how the public (e.g., drivers, passengers, travelers) will respond concerning toll roads in an auction setting. Therefore, understanding the travelers' interests (willingness-to-pay, value of time) and the impacts on different sociodemographic groups become imperative for a socially-efficient design of an emerging mobility system. For these reasons, it is essential to design an emerging mobility system whose focal point is the social aspect and societal impact of CAVs. In conjunction, it is the authors' belief that the emerging mobility systems—CAVs, shared mobility, electric vehicles—will be characterized by their socioeconomic complexity: (1) improved productivity and energy efficiency, (2) widespread accessibility, and (3) drastic urban redesign and evolved urban culture. This characteristic can naturally be modeled and analyzed using game theory/mechanism design and behavioral economics alongside control and optimization techniques. One of the main arguments in this chapter is that the social interactions of human travelers with CAVs, and other modes of transportation can be modeled as an economically-inspired mobility market, where monetary incentives (tolls) are used to induce the desired socially-efficient outcome.

Our aim is to develop a holistic and rigorous framework to capture the societal impact of connectivity and automation in emerging mobility systems and provide solutions that prevent any potential rebound effects (e.g., increased vehicle-miles-traveled, increased travel demand, empty trips). To achieve this aim, as a first attempt, we study an emerging mobility system consisting of a finite group of travelers who seek to travel in a "smart city," where a central authority (alternatively called social planner) seeks to ensure the efficient distribution and operation of the different modes of transportation offered by the city. We call these different modes of transportation "mobility services." A few examples of mobility services are CAVs, shared vehicles, and public transit (e.g., train, bus, light rail, subway). The travelers request to use at most one service to satisfy their mobility needs, i.e., to reach their destination, via a smartphone app easily accessible to all travelers. The social planner (e.g., a central computer) compiles all travelers' origin-destination requests and other information (e.g., preferred travel time, value of time, and maximum willingness-to-pay) in order to provide a travel recommendation to each traveler. The social planner's goal is to ensure that the aggregate travel recommendations are *socially-efficient*. Informally, by socially-efficient, we mean that the endmost collective travel recommendation must achieve two objectives: (i) respect and satisfy the travelers' preferences regarding mobility, and (ii) ensure the alleviation of congestion in the system. Since our focus is to provide socially-efficient solutions, we consider a city that supports connected and automated mobility technologies on its roads and public transit infrastructure. Subsequently, the social planner is fully aware of the system's capabilities and network's capacity. In other words, the social planner is fully capable of computing the maximum capacity of each mobility service and the associated costs aimed at providing travel recommendations to all travelers.

Our objective in this chapter is to design a mobility market of an emerging mobility system and provide a socially-efficient solution consisting of well-designed and appropriate monetary incentives (e.g., tolls, fares, fees) for a social planner to guarantee the realization of the desired outcome, i.e., maximize the social welfare of all travelers. At the same time, our solution will ensure to provide such incentives to travelers so that the usage of any mobility service will not lead to congestion in the mobility system. In other words, we design a mobility market that efficiently assigns each traveler to the "right" mode of transportation.

Our contributions are the following: we design a socially-efficient mobility market that assigns mobility services to a finite group of travelers by taking into consideration their travel preferences. We achieve that by implementing a special case of the VCG mechanism after modifying it accordingly for a mobility problem. We show that the proposed mobility market is incentive compatible and individually rational, two properties that ensure all selfish travelers are truthful in their communication with the social planner and voluntarily participate in the mobility market. We also show that the proposed market is economically sustainable, i.e., it generates revenue from each traveler and ensures that the operating costs of each mobility service are covered. It is through the appropriate design of monetary incentives that we successfully incentivize all travelers to truthfully report their travel preferences and voluntarily participate in the market. Thus, we are guaranteed a socially-efficient mobility solution. The proposed mobility market also provides an incentive to central authorities to implement it, since as we show, the market ensures that there are minimum acceptable payments to cover the operating costs of the mobility services.

The chapter is structured as follows. In Sect. 2, we review the main concepts of mechanism design and briefly discuss the VCG mechanism. In Sect. 3, we present the mathematical formulation of the emerging mobility market, which forms the basis for the rest of the chapter. In Sect. 3.2, we present the imposed optimization problem. In Sect. 4, we present the methodology used to design the monetary incentives for each traveler. In Sect. 5, we study the properties of the mobility market, and finally, in Sect. 6, we draw conclusions and offer a discussion for future research.

## 2 Theoretical Preliminaries

In this section, we provide the theoretical preliminary material related to this chapter's proposed modeling framework, and we formally introduce all important concepts needed to prove our principal results.

### *2.1 An Introduction to Mechanism Design*

Most generic control systems can be viewed as a specification of how decisions (e.g., how to utilize a number of resources) are determined as a function of the information that is known by the agents in the system. What interests us in most cases is *efficiency*, i.e., realizing the best possible allocation of resources with the best use of information to achieve an outcome where collectively agents are satisfied, and there is no overutilization of the system's resources [57]. One key challenge in ensuring efficiency in a control system is the fact that different agents may have conflicting interests and act selfishly. In other words, systems that incorporate human decision-making, if remained uninfluenced, are not guaranteed to exhibit optimal performance. This is well-known to be the case in control theory, and economics [58, 59]. There are various different theories and approaches that attempt to guarantee efficiency in such systems and can provide solutions of varying degrees of success. One such theory is mechanism design, in which we are concerned with how to implement system-wide optimal solutions to problems involving multiple selfish agents, each with private information about their preferences [60, 61]. Within the context of mobility, agents are the travelers, and their private information can be either tolerance to traffic delays, value of time, preferred travel time, or any disposition to a specific mode of transportation. Our goal in mechanism design is to design appropriate incentives in order to align the interests of agents with the interests of the system [51]. For example, in mobility, given that each traveler/driver/passenger "competes" with everyone else to reach their destination first, we want to ensure that given this inherent conflict of interest, we can still guarantee uncongested roads, no traffic accidents, and no travel time delays. Mechanism design can help us design the rules of systems where information is decentralized (different agents know different things), and agents do not necessarily have an immediate incentive to cooperate [62]. In particular, mechanism design helps us design rules that align all agents' decision-making by providing the right incentives to achieve a well-defined objective for the system (e.g., aggregate optimal performance, system-level efficiency). Thus, mechanism design entails solving an optimization problem with sometimes unverifiable and always incomplete information structure [63]. We call such a problem *an incentive design and preference elicitation problem*.

We start by supposing that there is a system consisted of a finite group of agents, each competing with each other for a limited, fixed allocation of resources. Each agent evaluates different allocations based on some private information that is known only to them. We consider a *social planner*, playing the role of a centralized entity, whose task is to align the selfish and conflicting interests of the agents with the overall system's objective (e.g., an efficient allocation of resources or the maximization of social welfare). As it can be seen in Fig. 1, there are four components: (1) There is a group of decision-makers, (2) who make a decision based on their personal information, and (3) that decision is reported as a message to the social planner who is tasked to design the rules of which (4) it can be determined what each agent gets. What follows next is a mathematically formal presentation of the social planner's task.

Consider a set of selfish agents $\mathcal{I}$, $|\mathcal{I}| = n \in \mathbb{N}$ with preferences over different outcomes in a set $\mathcal{O}$. Each agent $i \in \mathcal{I}$ is assumed to possess private information, denoted by $\theta_i \in \Theta_i$. Since an agent $i$'s $\theta_i$ can characterize and influence their decision-making in a significant way, we call $\theta_i$ the *type* of agent $i$. We write $(\theta_i)_{i \in \mathcal{I}} = \theta \in \Theta = \prod_{i \in \mathcal{I}} \Theta_i$ to represent the type profile of all agents. Next, an agent $i$'s preferences over different outcomes can be represented by a utility function $u_i : \mathcal{O} \times \Theta_i \to \mathbb{R}$. Although the exact form of $u_i$ can vary depending on the application of the problem [64–67], what is common in the literature [41, 50, 62] is a quasilinear function of the form

$$u_i(o, \theta_i) = v_i(o, \theta_i) - p_i, \tag{1}$$

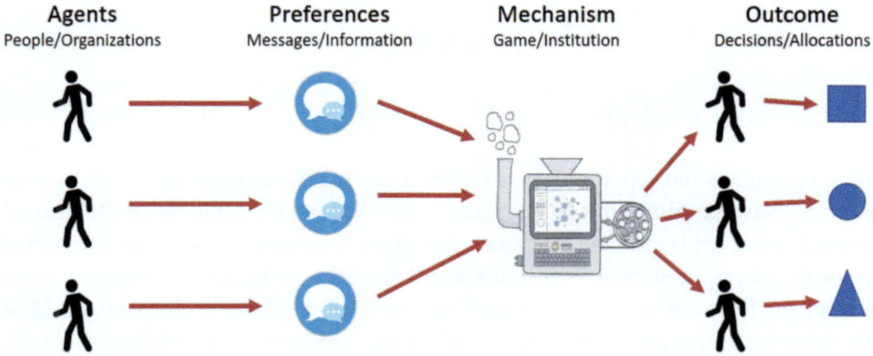


**Fig. 1** A visualization of how an arbitrary control system (agents, preferences, allocations) can be viewed under a mechanism design framework. Agents hold private information, of which they send reports to the social planner who is responsible for designing a mechanism. How efficient the mechanism is can depend on whether the agents' messages are truthful or not

where $v_i : \mathcal{O} \times \Theta_i \to \mathbb{R}_{\geq 0}$ represents an arbitrary valuation function, and $p_i \mapsto \mathbb{R}$ is a monotonically increasing function. If outcome $o \in \mathcal{O}$ represents an allocation of a resource, then $p_i$ can be thought of as a transfer of agent $i$'s wealth or a cost imposed to agent $i$ for that particular allocation $o$. Intuitively, a quasilinear function defined as in (1) ensures that the marginal value of $v_i$ does not depend on how large $p_i$ becomes, and vice-versa. Furthermore, (1) assumes $u_i$ is linear with respect to $p_i$. We can now naturally define the *social welfare* as the collective summation of all agents' valuations, i.e.,

$$w(o, \theta) = \sum_{i \in \mathcal{I}} v_i(o, \theta_i). \tag{2}$$

If our system objective is to maximize $w$, then immediately we observe that there is an important obstacle, i.e., any agent $i$ may misreport their type $\theta_i$ in the hopes to increase their own utility. So, the question is now: How can we incentivize agents to truthfully report their type? The answer is through the appropriate design of $p_i$. Next, we outline the building blocks that can help us design $p_i$. Formally, we can define a *mechanism* as the tuple $\langle f, p\rangle$ composed of a *social choice function* (SCF) $f : \Theta \to \mathcal{O}$ and a vector of *payment functions* $p = (p_i)_{i \in \mathcal{I}}$, with $p_i : \Theta \to \mathbb{R}$. In words, a mechanism $\langle f, p\rangle$ defines the rules of which we can implement a system objective by mapping the agents' types to an outcome while using the payments to ensure the optimality or efficiency of that outcome (see Fig. 2 for an illustration of the mechanism design framework). We can now state the social planner's problem as follows

$$\max_{o \in \mathcal{O}} w(o, \theta) \tag{3}$$

$$\text{subject to: } \hat{\theta}_i = \theta_i, \quad \forall i \in \mathcal{I}, \tag{4}$$

$$\sum_{i \in \mathcal{I}} v_i(o, \theta_i) \geq \sum_{i \in \mathcal{I}} v_i(o', \theta_i), \quad \forall o' \in \mathcal{O}, \tag{5}$$

$$\sum_{i \in \mathcal{I}} p_i(s(\theta)) \geq 0, \quad \forall \theta \in \Theta, \tag{6}$$

$$v_i(f(s(\theta))) - p_i(s(\theta)) \geq 0, \quad \forall i \in \mathcal{I}, \forall \theta \in \Theta, \tag{7}$$

where $\hat{\theta}_i$ denotes the reported type of agent $i$, $s(\cdot)$ is the equilibrium strategy profile (e.g., Nash equilibrium). Constraints (4) ensure the truthfulness in the agents' reported types, (5) impose an efficiency condition, (6) make certain that no external payments are required, and (7) incentivize all agents to voluntarily participate in the mechanism. If we could know for certain the true types of all agents, then we would be able solve the optimization problem (3)–(7) using standard optimization techniques. However, as this is unreasonable to expect from selfish decision-makers, the social planner needs to elicit $\theta = (\theta_i)_{i \in \mathcal{I}}$ by designing the appropriate $p = (p_i)_{i \in \mathcal{I}}$. We discuss in the next subsection one such mechanism that elicits the private information of agents truthfully.

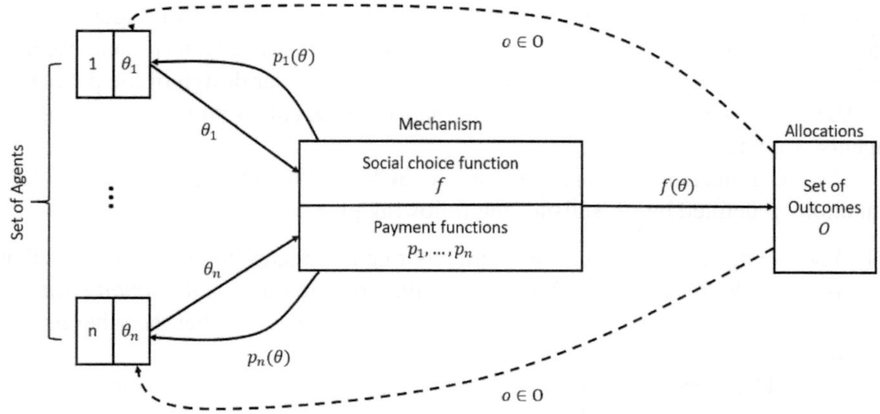


**Fig. 2** A theoretical representation of the mechanism design framework

## 2.2 The Vickrey-Clarke-Groves Mechanism

In the previous subsection, we reviewed the main concepts of mechanism design and formulated the incentive design and preference elicitation problem. In words, we asked "How can we design the payments $p = (p_i)_{i \in \mathcal{I}}$ so that every agent makes the decision that agrees with what *we* have chosen as the system's objective (e.g., efficiency)? To answer this question, in this subsection, we review the Vickrey-Clarke-Groves (VCG) mechanism [42–44], one of the most successful mechanisms as it incentivizes agents to be truthful and guarantees efficiency.

As we discussed earlier, a mechanism is a tuple $\langle f, p \rangle$. In a VCG mechanism, the SCF $f$ is defined as an allocation rule (who gets what) based on the optimization problem (3)–(7), i.e.,

$$f(\hat{\theta}) = \arg\max_{o \in \mathcal{O}} W(o, \hat{\theta}_i), \tag{8}$$

where $\hat{\theta} = (\hat{\theta}_i)_{i \in \mathcal{I}}$. In words, assuming that the agents disclose their true information, (8) provides to the social planner who attempts to maximize the social welfare a formal way to compute the allocations of each agent. At the same time, the VCG mechanism charges each agent for their allocation as follows

$$p_i(\hat{\theta}) = \sum_{j \neq i} v_j(f(\hat{\theta}_{-i})) - \sum_{j \neq i} v_j(f(\hat{\theta})), \tag{9}$$

where $\hat{\theta}_{-i}$ denotes the type profile of all agents except agent $i$. Note that the payments defined in (9) do not depend on an agent $i$'s own declaration $\hat{\theta}_i$. Let us assume for a moment that all agents declare their types truthfully. Then, the first sum in (9) computes the value of the social welfare with agent $i$ not participating in the mechanism. The second sum in (9) computes the value of the social welfare of all other agents $j \neq i$ with agent $i$ participating in the mechanism. Thus, agent $i$ when they report $\hat{\theta}_i$ are made to pay the *marginal effect* of their decision (in our case that is agent $i$'s reported type $\hat{\theta}_i$). In other words, this particular design of the payments in (9) internalizes an agent $i$'s social externality, i.e., agent $i$'s impact on every other agents' welfare.

The VCG mechanism represented by the SCF $f$ defined by (8) and the payment functions $p$ defined by (9) satisfies the following properties:

1. For any agent, truth-telling is a strategy that dominates any other strategy that is available for that agent. We say then that truth-telling is a *dominant strategy*. Note that such strategies are "always optimal" no matter what the other agents decide.
2. The VCG mechanism successfully aligns the agents' individual interests with the system's objective. In our case, that objective was to maximize the social welfare of all agents. We call this property, *economic efficiency*.
3. For any agent, the VCG mechanism incentivizes them to voluntarily participate in the mechanism as no agent loses by participation (in terms of utility).
4. The VCG mechanism ensures no positive transfers are made from the social planner to the agents. Thus, the mechanism does not incur a loss. We call this *weakly budget balanced*.

The VCG mechanism essentially ensures the realization of a *socially-efficient outcome*, i.e., satisfying properties 1–3, in a system of selfish agents, where each possesses private information. It is noteworthy to note how powerful the VCG mechanism is as it induces a dominant strategy equilibrium maximizing the social welfare while also making sure no agent is hurt by participating.

We conclude Sect. 2 with the following remark: although the main motivation of mechanism design is the microeconomic study of institutions and relies heavily on game-theoretic techniques, it can prove a powerful theory providing a systematic methodology in the design of systems of asymmetric information, consisted of strategic decision-makers, and whose performance must attain a specified system objective. The rest of the chapter shall present how we can use this theory to design a socially-efficient mobility system consisting of travelers who compete with each other for the utilization of a limited number of mobility services.

## 3 The Emerging Mobility Market

We consider an emerging mobility system consisting of a transportation city network managed by a social planner and a finite group of travelers who seek to travel in the network. Informally, this network represents the high-level mobility connections of multiple and different city neighborhoods. In other words, we move away from the concept of "personally-owned" modes of transportation and focus our modeling towards mobility provided as a service. This means that a social planner (e.g., a central computer) offers travelers a unified gateway of public and private transportation providers capable of providing mobility solutions to manage and realize their trip. For example, travelers can plan their journey via a smartphone app by specifying their preferences (e.g., cost, time, and convenience) and their desired destination. The social planner then is tasked to offer a travel recommendation to each traveler, i.e., which mode of transportation to take. In addition, we consider that multiple and different travel options can be offered to each traveler focusing on urban modes of transportation (e.g., CAVs, bus, train). We call these options "mobility services" or "services" for short. Within this framework, we propose a mobility market for a socially-efficient implementation of connectivity and automation in an emerging mobility system. The goal of the mobility market is twofold: (i) ensure that all travelers voluntarily participate and truthfully report their travel preferences, and (ii) be economically sustainable by generating revenue from each traveler and setting a minimum acceptable mobility payment for each traveler to potentially cover the operating costs.

## 3.1 Mathematical Formulation of the Emerging Mobility Market

The proposed mobility market is managed by a social planner who aims to allocate $m \in \mathbb{N}$ mobility services to $n \in \mathbb{N}$ travelers, where $n \geq m \neq 0$. We denote the set of travelers by $\mathcal{I}$, $|\mathcal{I}| = n$ and the set of mobility services by $\mathcal{J}$, $|\mathcal{J}| = m$. For example, each service $j \in \mathcal{J}$ can either represent a shared CAV, a train, or a bus. Both sets $\mathcal{I}$ and $\mathcal{J}$ are nonempty, disjoint, and finite. The set of all mobility services $\mathcal{J}$ can be partitioned to a finite number of disjoint subsets, each representing a specific "type" of a mobility service, i.e., $\mathcal{J} = \bigcup_{h=1}^{H} \mathcal{J}_h$, where $H \in \mathbb{N}$ is the number of subsets of $\mathcal{J}$. For example, $\mathcal{J} = \mathcal{J}_1 \cup \mathcal{J}_2$, where $|\mathcal{J}_1|$ represents the number of all available CAVs, and $|\mathcal{J}_2|$ represents the number of all available busses. Next, travelers seek to travel using these mobility services in a transportation network represented by an undirected multigraph $\mathcal{G} = (\mathcal{V}, \mathcal{E})$, where each node in $\mathcal{V}$ represents a different city area or neighborhood, and each link $e \in \mathcal{E}$ represents a sequence of city roads or a public transit connection. For our purposes, we think of $\mathcal{G} = (\mathcal{V}, \mathcal{E})$ as a representation of a smart city network with a road and public transit infrastructure. In $\mathcal{G}$, a traveler $i \in \mathcal{I}$ seeks to travel from their current location $o_i \in \mathcal{V}$ to their desired destination $d_i \in \mathcal{V}$. So, on one hand, each traveler $i \in \mathcal{I}$ is associated with a origin-destination pair $(o_i, d_i)$. On the other hand, each type of mobility services (e.g., one type is shared CAVs, another is trains) is associated with a unique link that connects any two nodes. At the same time, we do not limit the number of different mobility services that connect any origin $o_i$ to any destination $d_i$ of any traveler $i \in \mathcal{I}$. We suppose that any traveler $i \in \mathcal{I}$ has at least two travel options for their origin-destination pair $(o_i, d_i)$. Furthermore, each traveler $i \in \mathcal{I}$ can travel in $\mathcal{G}$ with any mobility service $j \in \mathcal{J}$ that satisfies their origin-destination pair $(o_i, d_i)$ and each service $j \in \mathcal{J}$ can be used by multiple travelers.

***Remark 1*** Network $\mathcal{G}$ represents the upper-level connections of different city neighborhoods. By connections, we mean either roads or public transit routes. Instead of modeling each node to represent travelers' exact location, we consider dividing a city into zones. By grouping travelers' exact locations into such zones, we can use network $\mathcal{G}$ to model the mobility connections between the different city zones.

Next, we partition the set of travelers $\mathcal{I}$ into different smaller subsets characterized by a common origin-destination pair.

**Definition 1** The set of travelers with the exact same origin-destination pair is $\mathcal{I}_k = \{i \in \mathcal{I} \mid (o_i, d_i) = (o_k, d_k)\}$, $k = 1, 2, \ldots, K$, where $K \in \mathbb{N}$ is the number of subclasses over the complete set of travelers, i.e., $\mathcal{I} = \bigcup_{k=1}^{K} \mathcal{I}_k$.

The justification of Definition 1 is that in an emerging mobility system, we can acquire verifiable location data of travelers either by using a global positioning system or estimating the average number of travelers using public transit [68, 69].

Mathematically, the allocation of the finite number of mobility services to travelers can be described by a vector of binary variables.

**Definition 2** The *traveler-service assignment* is a vector $\mathbf{a} = (a_{ij})_{i \in \mathcal{I}, j \in \mathcal{J}}$, where $a_{ij}$ is a binary variable of the form:

$$a_{ij} = \begin{cases} 1, & \text{if } i \in \mathcal{I} \text{ is assigned to } j \in \mathcal{J}, \\ 0, & \text{otherwise.} \end{cases} \tag{10}$$

Note that we have $(a_{ij})_{i \in \mathcal{I}, j \in \mathcal{J}} = (a_{11}, \ldots, a_{ij}, \ldots, a_{nm})$. By partitioning the set of travelers in $K \in \mathbb{N}$ subclasses, the traveler-service assignment of subclass $\mathcal{I}_k$ is given by $\mathbf{a}_k = (a_{ij})_{i \in \mathcal{I}_k, j \in \mathcal{J}}$.

Naturally, we need to impose a physical limit on the use of each mobility service $j \in \mathcal{J}$ in network $\mathcal{G}$ as well as a connection capacity of a mobility service for each link in the network. Note that each link in $\mathcal{G}$ represents a road or a public transit connection, which means that multiple mobility services of one type use that one link. For example, one link can be a bus lane with stops between two different city areas; another can be a train route between two stations.

**Definition 3** The *usage capacity* of any mobility service $j \in \mathcal{J}$ is given by $\varepsilon_j \in \mathbb{N}$. The *link capacity* in network $\mathcal{G}$ is given by $\gamma_e \in \mathbb{R}_{\geq 0}$.

For example, $\varepsilon_j$ can represent the maximum number of travelers (or passengers) in a shared vehicle or the maximum number of travelers in a train vehicle (seated and standing). Similarly, $\gamma_e$ can represent a critical traffic density of mobility services, which means that any additional input of vehicles or trains can lead to a reduced traffic flow and eventually to traffic congestion. For example, we can use the GreenShields model to define explicitly the critical traffic density [70].

As in any mobility problem that involves travelers, we need to consider the travelers' preferences (e.g., preferred travel time, value of time, willingness-to-pay for service). Hence, we formally define the notion of "personal travel requirements" by introducing three important parameters (our selection of those three parameters is justified by recent transportation studies [71, 72].)

**Definition 4** For any traveler $i \in \mathcal{I}_k$, $k = 1, \ldots, K$, let $\alpha_i \in (0, 1)$ be the *value of time*, $\theta_i \in \mathbb{R}_{\geq 0}$ the *preferred travel time*, and $\bar{v}_i \in \mathbb{R}_{\geq 0}$ the *maximum willingness-to-pay*. Then, the *personal travel requirements* of traveler $i$ is a tuple of the form $\pi_i = (\alpha_i, \theta_i, \bar{v}_i)$.

We offer the intuition behind each parameter: traveler $i$'s value of time $\alpha_i$ transforms the traveler's time urgency in monetary units as it can model, for example, the acceptable amount of compensation for lost time. Similarly, a traveler $i$'s preferred travel time $\theta_i$ is a non-negative real value representing how fast traveler $i$ wishes to reach their destination. The last term in $\pi_i$ captures how much traveler $i$ appraises a direct and completely convenient mobility service. For example, $\bar{v}_i$ can measure the maximum willingness-to-pay of traveler $i$ traveling with the fastest and most convenient service (e.g., taking a taxicab with no co-travelers) to their destination.

For each traveler $i \in \mathcal{I}_k$, the tuple $\pi_i$ is considered private information, known only to traveler $i$. Hence, as the social planner does not know $(\pi_i)_{i \in \mathcal{I}}$, each traveler $i$ must report their $\pi_i$. This is one of the key challenges in the proposed mobility market: *How can we incentivize the travelers to be truthful and elicit the private information needed to provide a socially-efficient solution to the emerging mobility market?* The answer to this question will be given in Sect. 4.

Next, we introduce an "inconvenience" metric for any traveler $i \in \mathcal{I}_k$ using any mobility service $j \in \mathcal{J}$. Quantitatively, the inconvenience metric can represent the extra monetary value of travel disutility from any costs, travel delays, or violation of personal travel requirements caused by the use of a mobility service.

**Definition 5** The *mobility inconvenience metric* for traveler $i \in \mathcal{I}_k$, $k = 1, \ldots, K$, assigned to service $j \in \mathcal{J}$ is a continuous, increasing, and convex function $\phi_i \left(\alpha_i, \theta_i, \tilde{\theta}_i(\mathbf{a}_k)\right) \mapsto \mathbb{R}_{\geq 0}$, where $\tilde{\theta}_i(\mathbf{a}_k) \in \mathbb{R}_{\geq 0}$ is the experienced travel time.

Note that the mobility inconvenience metric $\phi_i$ increases when $\tilde{\theta}_i(\mathbf{a}_k)$ increases. From a modeling perspective, traveling with time delays or during peak times can cause significant inconveniences to any traveler $i \in \mathcal{I}_k$. Although, an exact form of $\phi$ is beyond the scope of this chapter, our definition of $\phi$ is consistent with general inconvenience functions in the literature [73, 74].

Next, a traveler $i$'s satisfaction is captured by a valuation function $v_i$, which can reflect the traveler's *willingness-to-pay* for their travel, i.e.,

$$v_i(\mathbf{a}_k) = \bar{v}_i - \phi_i \left(\alpha_i, \theta_i, \tilde{\theta}_i(\mathbf{a}_k)\right), \tag{11}$$

where $\bar{v}_i \in \mathbb{R}_{\geq 0}$ is the value gained by traveler $i \in \mathcal{I}_k$ when their origin-destination pair $(o_i, d_i)$ is satisfied using service $j \in \mathcal{J}$ without any travel delays, i.e., $\theta_i = \tilde{\theta}_i(\mathbf{a}_k)$. Naturally, for any traveler $i$ and any service $j$, we have $v_i(\mathbf{a}_k) \in [0, \bar{v}_i]$, where $v_i(\mathbf{a}_k) = 0$ means that traveler $i$ is unwilling to use service $j$. Below we summarize the two extreme cases and their interpretation:

$$v_i(\mathbf{a}_k) = \begin{cases} \bar{v}_i, & \text{if } \phi_i = 0, \\ 0, & \text{if } \phi_i = \bar{v}_i. \end{cases} \tag{12}$$

When $\phi_i = 0$, we say that traveler $i$ travels to their destination in the fastest and most convenient mobility service offered by the mobility market (e.g., a taxicab with no co-travelers). On the other hand, when $\phi_i = \bar{v}_i$, we say that traveler $i$'s personal travel requirements are not satisfied, and the traveler is most inconvenienced with regards to mobility.

Although our analysis can treat the valuation function $v_i$ in its most general form, given by (11), we explicitly define the second component of (11) in our mathematical exposition. Thus, the explicit form for the inconvenience mobility metric $\phi_i$ is

$$\phi_i\left(\alpha_i, \theta_i, \tilde{\theta}_i(\mathbf{a}_k)\right) = \alpha_i \cdot (\tilde{\theta}_i(\mathbf{a}_k) - \theta_i), \tag{13}$$

Basically, (13) gives the monetary value of the difference between the travel times (experienced vs preferred), and can be interpreted as the travel time tolerance that the traveler can accept (in monetary units).

In our modeling framework, the total utility $u_i(\mathbf{a}_k)$ of traveler $i \in \mathcal{I}_k$, $k = 1, \ldots, K$, is given by

$$u_i(\mathbf{a}_k) = v_i(\mathbf{a}_k) - p_i(\mathbf{a}_k), \tag{14}$$

where $v_i(\mathbf{a}_k)$ is the willingness-to-pay and $p_i(\mathbf{a}_k) \in \mathbb{R}_{\geq 0}$ is the mobility payment that traveler $i$ is required to make to use service $j \in \mathcal{J}$ (e.g., pay road tolls or buy a public transit fare). Hence, (14) establishes a "quasi-linear" relationship between a traveler's satisfaction and payment, both measured in monetary units [40].

In contrast to the traveler's satisfaction, we also introduce an "operating cost" to capture the needed investment that public and private mobility providers and operators make to ensure the proper function of their mobility services.

**Definition 6** The operating cost of service $j \in \mathcal{J}$ can be computed by

$$c_j(\mathbf{a}_k) = \sum_{i \in \mathcal{I}_k} c_{ij}(a_{ij}), \tag{15}$$

where $c_{ij}(a_{ij}) \in \mathbb{R}_{\geq 0}$ is traveler $i$'s corresponding share of the operating cost of vehicle $j \in \mathcal{J}$. In the case of $a_{ij} = 0$, we have $c_{ij} = 0$.

Intuitively, the operating cost $c_{ij}$ captures traveler $i$'s fair share of the costs of service $j \in \mathcal{J}$. These costs can be associated with fuel/energy consumption, drivers' labor reimbursement, maintenance, and environmental impact.

**Definition 7** Given the traveler-service assignment $\mathbf{a}_k = (a_{ij})_{i \in \mathcal{I}_k, j \in \mathcal{J}}$, the travelers' payments are given by the vector $\mathbf{p}_k = (p_i(a_{ij}))_{i \in \mathcal{I}_k, j \in \mathcal{J}}$. Then, for a subclass $\mathcal{I}_k$, $k = 1, \ldots, K$, the proposed mobility market can be fully described by the tuple

$$\left\langle \mathcal{I}_k, \mathcal{J}, (\pi_i)_{i \in \mathcal{I}_k}, (u_i)_{i \in \mathcal{I}_k}, \mathbf{a}_k, \mathbf{p}_k \right\rangle, \tag{16}$$

where $(\pi_i)_{i \in \mathcal{I}_k}$ is considered private information (unknown to the social planner), and the experienced travel time $\tilde{\theta}_i$ and operation costs $c_j$ of all mobility services are considered known to the social planner.

Note that in Definition 7, we have also defined the informational structure of the proposed market. The operation costs $(c_j)_{j \in \mathcal{J}}$ are considered public information as well as the minimum acceptable mobility payments $(\sigma_i)_{i \in \mathcal{I}}$. In general, though, any VCG-based mechanism requires agents to report their entire valuation function [75]. In our case, we can take advantage of more advanced and sophisticated data gathering techniques so that we may infer the form and shape of a traveler's valuation (and utility) function using, for example, historical and empirical data [76, 77]. Hence, the functional form of $v_i$ can be considered known, but the realization of $v_i(\cdot)$ is agent $i$'s private information. It is important to note that the evaluation of any traveler $i$'s valuation function can be learned using the three-parameter tuple $\pi_i$, which provides the personal travel requirements of any traveler $i \in \mathcal{I}_k$. In addition, we expect any social planner of a generic transportation system to have the ability (e.g., using regression analysis [78]) to approximate the experienced travel time of any mobility service and its operating costs. Hence, the only private information that we are required to elicit from the travelers is their personal travel requirements $(\pi_i)_{i \in \mathcal{I}_k}, k = 1, \ldots, K$. At the same time, receiving communication in the form of messages from all travelers regarding the $(\pi_i)_{i \in \mathcal{I}_k}, k = 1, \ldots, K$ can be an unrealistic burden. That is why, in our framework, any traveler $i \in \mathcal{I}$ is expected to report the evaluation of their valuation function $v_i$, which depends on their $\pi_i$. Essentially, we parameterize the private information of travelers into a one-dimensional number. In future research, we plan to address a multi-dimensional mechanism to ensure there is no loss of information of the traveler's preferences.

On a different note, a natural question to ask here is whether there is any guarantee that the travelers' mobility payments will meet the providers' operating costs. As we saw in Sect. 2, the VCG mechanism can only charge travelers their social cost or impact into the mobility system. Thus, this might lead to very low mobility payments for a significant number of travelers, leading to deficits to cover operating costs for the providers. Since, in reality, we cannot expect any providers to serve travelers indefinitely when their costs have not been met, we introduce a "pricing base" for the mobility payments. Essentially, these bases can be chosen by the providers to ensure that no payment by any traveler is below a set value (e.g., minimum acceptable payment), which can be determined approximately by the traveler's location and destination, supply and demand, and operator's reimbursement fee [79].

**Definition 8** The *minimum acceptable mobility payment* of any service $j \in \mathcal{J}$ is given by $\sigma_i(\mathbf{a}_k) \in \mathbb{R}_{\geq 0}$, for any traveler $i \in \mathcal{I}_k$, $k = 1, \ldots, K$. If for an arbitrary traveler $i$, we have $p_i(\mathbf{a}_k) \geq \sigma_i(\mathbf{a}_k)$, then we say that the mobility market, defined in (16), is *economically sustainable*.

The minimum acceptable mobility payments $\sigma = (\sigma_i)_{i \in \mathcal{I}}$ are considered public information set by the providers and may be different for each traveler $i \in \mathcal{I}_k$, $k = 1, \ldots, K$.

In the modeling framework described above, we impose the following assumption:

**Assumption 1** For all subclasses $\mathcal{I}_k$, $k = 1, \ldots, K$, $K \in \mathbb{N}$, any traveler $i \in \mathcal{I}_k$ is modeled as a selfish decision-maker with private information $\pi_i = (\alpha_i, \theta_i, \bar{v}_i)$. Traveler $i$'s objective is to maximize their total utility $u_i(\mathbf{a}_k) = v_i(\mathbf{a}_k) - p_i(\mathbf{a}_k)$ in a non-cooperative game-theoretic setting.

Assumption 1 essentially says that each traveler is selfish in the sense that they are only interested in their own well-being. In economics, such behavior is called "strategic" since agents attempt to misreport their private information to the social planner if that means higher individual benefits.

**Assumption 2** The aggregate usage capacities of all mobility services can adequately serve all travel requests of travelers. Mathematically, we have $\sum_{j \in \mathcal{J}} \varepsilon_j = n = |\mathcal{I}|$.

Intuitively, Assumption 2 ensures that no traveler will remain unassigned. We can justify this assumption as follows: our focus is on efficiently allocating the different mobility services to travelers in a mobility market, a multimodal mobility system that incorporates public transit services with high traveler capacity capabilities. A relaxation of this assumption must consider scenarios where the existing mobility services cannot meet the travelers' demand, thus transforming our problem into a "mobility and equity" problem (giving priority to a subset of travelers in a fair way).

## 3.2 *The Optimization Problem Statement of the Emerging Mobility Market*

In the proposed mobility market, travelers request (via a smartphone app), in advance, a travel recommendation from the social planner that satisfies their origin-destination. Given the travelers' origin-destination pairs, the social planner partitions all travelers to different subclasses, as described in Definition 1. Thus, travelers from the same neighborhood have the same origin. Similarly, travelers going to the same neighborhood have the same destination. The social planner's task is to elicit the travelers' preferences, attempt to satisfy all travel requests, and provide recommendations to the travelers (e.g., which mobility service to use) by considering the social optimum subject to the city network's physical constraints. Hence, we are interested in minimizing the travel inconvenience of all travelers and the operating costs.

***Remark 2*** Without loss of generality and to simplify the mathematical analysis in our exposition, we consider that both the mobility inconvenience metrics $(\phi_i)_{i \in \mathcal{I}_k}$,

$k = 1, \ldots, K$, the minimum mobility payments $(\sigma_i)_{i \in \mathcal{I}_k}, k = 1, \ldots, K$, and the operating costs $(c_j)_{j \in \mathcal{J}}$ are normalized. This ensures that $\phi_i$, $\sigma_i$, and $c_j$ do not dominate each other in Problem 1 next, while all three are measured in the same monetary units.

**Problem 1** For each subclass $\mathcal{I}_k, k = 1, \ldots, K$, the optimization problem is

$$\min_{\mathbf{a}_k} \sum_{i \in \mathcal{I}_k} \left[ \phi_i \left( \alpha_i, \theta_i, \tilde{\theta}_i(\mathbf{a}_k) \right) + \sigma_i(\mathbf{a}_k) \right] + \sum_{j \in \mathcal{J}} c_j(\mathbf{a}_k), \tag{17}$$

subject to:

$$\sum_{j \in \mathcal{J}} a_{ij} \leq 1, \quad \forall i \in \mathcal{I}_k, \tag{18}$$

$$\sum_{i \in \mathcal{I}_k} a_{ij} \leq \varepsilon_j, \quad \forall j \in \mathcal{J}, \tag{19}$$

$$\sum_{j \in \mathcal{J}_h} \sum_{i \in \mathcal{I}_k} a_{ij} \leq \gamma_e, \quad \forall h \in \{1, 2, \ldots, H\}, \quad \forall e \in \mathcal{E}, \tag{20}$$

where (18) assures that each traveler $i \in \mathcal{I}_k$ will be assigned at most one mobility service, and (19) stipulates that service $j$'s maximum usage capacity $\varepsilon_j$ must not be exceeded. Lastly, (20) ensures that there will be no congestion on the links that represent roads or public transit connections. Note also that even though in Problem 1 we focus only on the $k$th partition of the set of travelers $\mathcal{I}$, we do not need to do the same for the mobility services. In other words, since each type of mobility services is associated with a unique link that connects any two nodes, any services that do not satisfy $(o_k, d_k)$ will not be considered in the optimization.

Problem 1 is similar to the many-to-one assignment problem, and standard algorithmic approaches (e.g., Jonker-Volgenant algorithm [80]) exist to find its global optimal solution or, in worst-case scenarios, a second-best optimal approximation of a solution. We can also reformulate Problem 1 to a linear program by relaxing to a non-negativity constraint the binary optimization variable $a_{ij}$ for all $i \in \mathcal{I}$ and $j \in \mathcal{J}$. We can then guarantee that an optimal solution of zeros and ones exists by noting that the constraint matrix formed by (18)–(20) satisfies the total unimodularity property [81]. Note, though, that these approaches assume complete information of all parameters and variables in the model. Such an assumption is unreasonable to expect from strategic decision-makers, so, in our framework, travelers are not expected to report their private information truthfully. This turns our problems to a *preference elicitation problem*. Our task in Sect. 4 is to provide a theoretical approach that elicits the necessary private information of all travelers using monetary incentives in the form of mobility payments (e.g., tolls, fares, fees).

## 4 Methodology for the Design of Mobility Incentives

We can reformulate Problem 1 as a standard social welfare maximization problem. First, recall that $\phi_i\left(\alpha_i, \theta_i, \tilde{\theta}_i(\mathbf{a}_k)\right) = \bar{v}_i - v_i(\mathbf{a}_k)$, so the objective function (17) becomes

$$\max_{\mathbf{a}_k} \sum_{i \in \mathcal{I}_k} [v_i(\mathbf{a}_k) - \sigma_i(\mathbf{a}_k)] - \sum_{j \in \mathcal{J}} c_j(\mathbf{a}_k). \tag{21}$$

This reformulation will prove useful as the design of the monetary payments relies on the social welfare impact (or mobility externality) caused by one traveler to the rest of the travelers in the proposed mobility market.

**Problem 2** We rewrite Problem 1 as follows:

$$\max_{\mathbf{a}_k} \sum_{i \in \mathcal{I}_k} [v_i(\mathbf{a}_k) - \sigma_i(\mathbf{a}_k)] - \sum_{j \in \mathcal{J}} c_j(\mathbf{a}_k), \tag{22}$$

subject to:

$$\sum_{j \in \mathcal{J}} a_{ij} \leq 1, \quad \forall i \in \mathcal{I}_k, \tag{23}$$

$$\sum_{i \in \mathcal{I}_k} a_{ij} \leq \varepsilon_j, \quad \forall j \in \mathcal{J}, \tag{24}$$

$$\sum_{j \in \mathcal{J}_h} \sum_{i \in \mathcal{I}_k} a_{ij} \leq \gamma_e, \quad \forall h \in \{1, 2, \ldots, H\}, \quad \forall e \in \mathcal{E}, \tag{25}$$

where $\mathbf{a}_k = (a_{ij})_{i \in \mathcal{I}_k, j \in \mathcal{J}}$ denotes the solution of Problem 2.

In order for the solution of Problem 2 to be socially-efficient, we would need a control input in utility function (14) to incentivize all travelers to report their personal travel requirements truthfully. In our case, this control input is the payments $\mathbf{p}_k$, $k = 1, \ldots, K$, which can be designed to be the difference between the *maximum social welfare with traveler* $\ell \in \mathcal{I}_k$ *not participating* and the *maximum social welfare of other travelers with traveler* $\ell$ *participating*. Thus, to capture the first term, we revise Problem 2 by adding constraint (30) to help us capture the "mobility externality" of traveler $\ell$ rejecting any travel recommendations from the social planner. For example, traveler $\ell$ may use a taxicab with no other co-travelers. Thus, Problem 2 takes the following form.

**Problem 3** For each traveler $i \in \mathcal{I}_k$, $k = 1, \ldots, K$, we fix traveler $\ell \in \mathcal{I}_k$ and solve the following optimization problem:

$$\max_{\mathbf{b}_k} \sum_{i \in \mathcal{I}_k} [v_i(\mathbf{b}_k) - \sigma_i(\mathbf{b}_k)] - \sum_{j \in \mathcal{J}} c_j(\mathbf{b}_k), \tag{26}$$

subject to:

$$\sum_{j \in \mathcal{J}} b_{ij} \leq 1, \quad \forall i \in \mathcal{I}_k, \tag{27}$$

$$\sum_{i \in \mathcal{I}_k} b_{ij} \leq \varepsilon_j, \quad \forall j \in \mathcal{J}, \tag{28}$$

$$\sum_{j \in \mathcal{J}_h} \sum_{i \in \mathcal{I}_k} b_{ij} \leq \gamma_e, \quad \forall h \in \{1, 2, \ldots, H\}, \quad \forall e \in \mathcal{E}, \tag{29}$$

$$b_{\ell j} = 0, \quad \forall j \in \mathcal{J}, \tag{30}$$

where $\mathbf{b}_k = (b_{ij})_{i \in \mathcal{I}_k, j \in \mathcal{J}}$ defined similarly as in (10) denotes the solution of Problem 3, and (30) states that traveler $\ell \in \mathcal{I}_k$ is not considered in the optimization problem.

***Remark 3*** In what follows, to simplify the mathematical exposition, we introduce the following notation:

$$w_2(\mathbf{a}_k) = \sum_{i \in \mathcal{I}_k} [v_i(\mathbf{a}_k) - \sigma_i(\mathbf{a}_k)] - \sum_{j \in \mathcal{J}} c_j(\mathbf{a}_k), \tag{31}$$

$$w_3(\mathbf{b}_k) = \sum_{i \in \mathcal{I}_k} [v_i(\mathbf{b}_k) - \sigma_i(\mathbf{b}_k)] - \sum_{j \in \mathcal{J}} c_j(\mathbf{b}_k), \tag{32}$$

where $w_2$ and $w_3$ denote the objective functions of Problems 2 and 3, respectively.

We can now propose the exact form of the mobility payment $p_\ell$ for an arbitrary traveler $\ell \in \mathcal{I}_k$, $k = 1, \ldots, K$, of the proposed mobility market. For any subclass $\mathcal{I}_k$, $k = 1, \ldots, K$, traveler $\ell \in \mathcal{I}_k$ makes the following payment:

$$p_\ell(\mathbf{a}_k, \mathbf{b}_k) = w_3(\mathbf{b}_k) - [w_2(\mathbf{a}_k) - v_\ell(\mathbf{a}_k)]. \tag{33}$$

Since $w_3(\mathbf{b}_k)$ yields the maximum social welfare from the traveler-service assignment $\mathbf{b}_k$ when traveler $\ell \in \mathcal{I}_k$ does not participate in the mobility market, it can be viewed by traveler $\ell \in \mathcal{I}_k$ in (33) as a constant, regardless of what traveler $\ell$ reports to the social planner about their own personal travel requirements $\pi_\ell$. The term $[w_2(\mathbf{a}_k) - v_\ell(\mathbf{a}_k)]$ in (33) represents the maximum social welfare of all travelers other than traveler $\ell \in \mathcal{I}_k$, when traveler $\ell \in \mathcal{I}_k$ partakes in the mobility market. As a consequence, $p_\ell$ can be interpreted as the externality caused by traveler $\ell \in \mathcal{I}_k$ to all other travelers. In addition, the computation of the mobility payments (33) requires solving Problem 3 repeatedly for each traveler. As shown in Algorithm 1, first we derive the optimal solution of Problem 2, and then we use the optimal solution of Problem 3 to compute the monetary payment of each traveler $\ell \in \mathcal{I}_k$.

**Algorithm 1:** Solution of Problem 2 with Problems 3

**Data**: $\mathcal{I}_k$, $\mathcal{J}$, $(\pi_i)_{i\in\mathcal{I}_k}$, $(u_i)_{i\in\mathcal{I}_k}$
**Result**: $\mathbf{a}_k^*$ and $\mathbf{p}_k$
Solve for the optimal solution $\mathbf{a}_k^*$ of Problem 2;
**for** $\ell \in \mathcal{I}_k$ **do**
  Solve for the optimal solution $\mathbf{b}_k^*$ of Problem 3;
  Next, compute
$$p_\ell(\mathbf{a}_k^*, \mathbf{b}_k^*) = w_3(\mathbf{b}_k^*) - \left[w_2(\mathbf{a}_k^*) - v_\ell(\mathbf{a}_k^*)\right].$$
**end**

Before we move on to the next section, we note that informally we talked about a traveler not participating in the mobility market in solving Problem 3. This idea helps us design the mobility payments in (33) by identifying the mobility externalities in the welfare of all travelers. Thus, we introduce the notion of "mobility exclusion," which will help us capture the socioeconomic impact of any traveler on the rest of the mobility market.

**Definition 9** For any subclass $\mathcal{I}_k, k = 1, \ldots, K$, given a traveler-service assignment $\mathbf{a}_k$ of Problem 2, a traveler $\ell \in \mathcal{I}_k$ is said to be *mobility excluded* if they are not assigned to any mobility service in the traveler-service assignment $\mathbf{b}_k$ of Problem 3.

Problem 3 is used to compute the mobility payments for each traveler in the mobility market by identifying the mobility externality caused by the decision-making of the traveler to the rest of the market. In addition, however, we are also interested in identifying the traveler's impact on (i) operating costs and (ii) overall welfare. We shall see in the next section how we can achieve this.

## 5 Properties of the Mobility Market

Our first result is an immediate and straightforward consequence of Definition 9. Recall that the operating cost $c_{ij}(a_{ij})$ captures traveler $i$'s fair share of the mobility service $j$'s costs that they use under the traveler-service assignment $\mathbf{a}_k$.

**Corollary 1** *Let $\mathbf{b}_k^\ell$ be a feasible traveler-service assignment of Problem 3. Given that traveler $\ell \in \mathcal{I}_k$ is mobility excluded, the operating cost that is associated with the traveler-service assignment $\mathbf{b}_k^\ell$ is smaller than or equal than the operating cost associated with the optimal assignment $\mathbf{a}_k^*$ of Problem 2, i.e., we have*

$$\sum_{i\in\mathcal{I}_k} c_{ij}(a_{ij}^*) \geq \sum_{i\in\mathcal{I}_k\setminus\{\ell\}} c_{ij}(b_{ij}^\ell). \tag{34}$$

Similarly, using Definition 9, we show that the sum of valuations (or welfare) of all travelers other than the traveler, who is mobility excluded specifically in Problem

3, is greater or equal than the sum of valuations evaluated at the traveler-service assignment of Problem 2.

**Lemma 1** *Let $\mathbf{b}_k^\ell$ be a feasible traveler-service assignment of Problem 3, in which traveler $\ell \in \mathcal{I}_k$ is mobility excluded. Then, we have*

$$\sum_{i \in \mathcal{I}_k \setminus \{\ell\}} v_i(\mathbf{a}_k) \leq \sum_{i \in \mathcal{I}_k} v_i(\mathbf{b}_k^\ell). \tag{35}$$

***Proof*** Given that traveler $\ell \in \mathcal{I}_k$ is mobility excluded in the traveler-service assignment $\mathbf{b}_k^\ell$ of Problem 3, we know that there is one less traveler required to be served by any mobility service in the market. Naturally, this affects the experienced travel times of any other traveler $i \in \mathcal{I}_k$, i.e., we have either a decreased or constant $\tilde{\theta}_i(\mathbf{b}_k^\ell)$. So, mathematically this means that with traveler-service assignment $\mathbf{a}_k$ of Problem 2, we have

$$\tilde{\theta}_i(\mathbf{b}_k^\ell) \leq \tilde{\theta}_i(\mathbf{a}_k), \tag{36}$$

where $\tilde{\theta}_i(\mathbf{b}_k^\ell)$ is the experienced travel time of traveler $i \in \mathcal{I}_k$ evaluated at $\mathbf{b}_k^\ell$ and $\tilde{\theta}_i(\mathbf{a}_k)$ is the experienced travel time of traveler $i$ evaluated at $\mathbf{a}_k$. Intuitively, (36) means there is one less traveler leading to better travel times for other travelers (better here means less). Hence, since the explicit form of traveler $i$'s valuation is given by

$$v_i(\mathbf{a}_k) = \bar{v}_i - \phi_i\left(\alpha_i, \theta_i, \tilde{\theta}_i(\mathbf{a}_k)\right) = \bar{v}_i - \alpha_i \cdot (\tilde{\theta}_i(\mathbf{a}_k) - \theta_i), \tag{37}$$

if we compare the two valuations $v_i(\mathbf{a}_k)$ and $v_i(\mathbf{b}_k^\ell)$, we get $v_i(\mathbf{a}_k) \leq v_i(\mathbf{b}_k^\ell)$. This completes the proof. □

Next, we show that for any traveler, their valuation will always be greater or equal than the minimum mobility payment. This will be instrumental in our attempt to show individual rationality later on.

**Lemma 2** *Let $\mathbf{a}_k^*$ denote the optimal solution of Problem 2. Then the minimum mobility payment $\sigma_\ell$ in the objective function* (22) *of Problem 2 ensures that, for any $\ell \in \mathcal{I}_k$, $k = 1, \ldots, K$, $v_\ell(\mathbf{a}_k^*) \geq \sigma_\ell(\mathbf{a}_k^*)$.*

***Proof*** Let $\mathbf{a}_k^*$ denote the optimal solution of Problem 2 and $\mathbf{b}_k^{\ell *}$ the corresponding solution of Problem 3. Hence, traveler $\ell$ has been assigned a mobility service in the optimal traveler-service assignment $\mathbf{a}_k^*$, but they are mobility excluded in $\mathbf{b}_k^{\ell *}$. Thus, we have

$$\begin{aligned} w_3(\mathbf{b}_k^{\ell *}) &= \sum_{i \in \mathcal{I}_k} \left[v_i(\mathbf{b}_k^{\ell *}) - \sigma_i(\mathbf{b}_k^{\ell *})\right] - \sum_{j \in \mathcal{J}} c_j(\mathbf{b}_k^{\ell *}) \\ &\geq \sum_{i \in \mathcal{I}_k \setminus \{\ell\}} v_i(\mathbf{a}_k^*) - \sum_{i \in \mathcal{I}_k} \sigma_i(\mathbf{b}_k^{\ell *}) - \sum_{j \in \mathcal{J}} c_j(\mathbf{a}_k^*), \end{aligned} \tag{38}$$

where (38) follows from Corollary 1 and Lemma 1. Next, we look at the welfare of an arbitrary traveler $i \in \mathcal{I}_k$ under $\mathbf{a}_k^*$, i.e.,

$$\begin{aligned} w_2(\mathbf{a}_k^*) &= \sum_{i\in\mathcal{I}_k} \left[v_i(\mathbf{a}_k^*) - \sigma_i(\mathbf{a}_k^*)\right] - \sum_{j\in\mathcal{J}} c_j(\mathbf{a}_k^*) \\ &= \sum_{i\in\mathcal{I}_k} v_i(\mathbf{a}_k^*) - \sum_{i\in\mathcal{I}_k} \sigma_i(\mathbf{a}_k^*) - \sum_{j\in\mathcal{J}} c_j(\mathbf{a}_k^*), \end{aligned} \tag{39}$$

where it also follows that $w_2(\mathbf{a}_k^*) \geq w_3(\mathbf{b}_k^{\ell*})$ from the fact that $\mathbf{b}_k^{\ell*}$ is not an optimal solution of Problem 2. Thus, if we compare (38) and (39), we get

$$\begin{aligned} &\sum_{i\in\mathcal{I}_k} v_i(\mathbf{a}_k^*) - \sum_{i\in\mathcal{I}_k} \sigma_i(\mathbf{a}_k^*) - \sum_{j\in\mathcal{J}} c_j(\mathbf{a}_k^*) \\ &\qquad \geq \sum_{i\in\mathcal{I}_k\setminus\{\ell\}} v_i(\mathbf{a}_k^*) - \sum_{i\in\mathcal{I}_k} \sigma_i(\mathbf{b}_k^{\ell*}) - \sum_{j\in\mathcal{J}} c_j(\mathbf{a}_k^*). \end{aligned} \tag{40}$$

So, by simplifying and rearranging (40), we have

$$\begin{aligned} \sum_{i\in\mathcal{I}_k} v_i(\mathbf{a}_k^*) - \sum_{i\in\mathcal{I}_k\setminus\{\ell\}} v_i(\mathbf{a}_k^*) &\geq \sum_{i\in\mathcal{I}_k} \sigma_i(\mathbf{a}_k^*) - \sum_{i\in\mathcal{I}_k} \sigma_i(\mathbf{b}_k^{\ell*}), \\ &= \sigma_\ell(\mathbf{a}_k^*) - \sigma_\ell(\mathbf{b}_k^{\ell*}) = \sigma_\ell(\mathbf{a}_k^*), \end{aligned} \tag{41}$$

since $\sigma_\ell(\mathbf{b}_k^{\ell*}) = 0$ as traveler $\ell$ is not assigned any mobility service under the traveler-service assignment $\mathbf{b}_k^{\ell*}$. Therefore, (41) simplifies to $v_\ell(\mathbf{a}_k^*) \geq \sigma_\ell(\mathbf{a}_k^*)$. □

Our first main result is incentive compatibility, which means that all travelers are incentivized to report their private information truthfully. Formally, for an arbitrary traveler $i \in \mathcal{I}_k$, $k = 1, \ldots, K$, given that $u_i'$ is the utility gained with misreported $\pi_i$ and $u_i$ is the "actual" utility, showing that $u_i' \leq u_i$ guarantees truthfulness.

**Theorem 1** *The mobility market defined in* (16) *provides the appropriate monetary incentives to each traveler* $i \in \mathcal{I}_k$, $k = 1, \ldots, K$ *to report their personal travel requirements* $\pi_i = (\alpha_i, \theta_i, \bar{v}_i)$ *truthfully regardless of what other travelers report.*

***Proof*** It is sufficient to show incentive compatibility only for an arbitrary mobility market for some arbitrary $k \in \{1, \ldots, K\}$. Suppose some traveler $\ell \in \mathcal{I}_k$ misreports their personal travel requirements denoted by $\pi_\ell = (\alpha_\ell', \theta_\ell', \bar{v}_\ell')$ to the social planner. Thus, we have

$$v_\ell'(\mathbf{a}_k) = \bar{v}_\ell' - \phi_\ell\left(\alpha_\ell', \theta_\ell', \tilde{\theta}_\ell(\mathbf{a}_k)\right). \tag{42}$$

The objective function of Problem 2 becomes

$$w_2'(\mathbf{a}_k) = \sum_{i\in\mathcal{I}_k\setminus\{\ell\}} [v_i(\mathbf{a}_k) - \sigma_i(\mathbf{a}_k)] - \sum_{j\in\mathcal{J}} c_j(\mathbf{a}_k) + v_\ell'(\mathbf{a}_k), \tag{43}$$

where the feasible solution of (43) is subject to the same constraints as in Problem 2. We denote the optimal solution of the optimization problem that traveler $\ell$ has misreported their personal travel requirements $\pi_\ell$ with (43) as the objective function by $\tilde{\mathbf{a}}_k^*$. Then, for traveler $\ell \in \mathcal{I}_k$ their mobility payment can be computed as follows:

$$p'_\ell(\tilde{\mathbf{a}}_k^*, \tilde{\mathbf{b}}_k^*) = w_3(\tilde{\mathbf{b}}_k^*) - \left[w_2^\ell(\tilde{\mathbf{a}}_k^*) - v'_\ell(\tilde{\mathbf{a}}_k^*)\right] = w_3(\mathbf{b}_k^*) - \left[w_2^\ell(\tilde{\mathbf{a}}_k^*) - v'_\ell(\tilde{\mathbf{a}}_k^*)\right], \quad (44)$$

where $\tilde{\mathbf{b}}_k^*$ denotes the optimal solution of Problem 3 with traveler $\ell \in \mathcal{I}_k$ misreporting. However, $w_3(\tilde{\mathbf{b}}_k^*) = w_3(\mathbf{b}_k^*)$ since, in Problem 3, it does not matter what traveler $\ell \in \mathcal{I}_k$ reports. Thus, the total utility of traveler $\ell \in \mathcal{I}_k$ is

$$u_\ell(\tilde{\mathbf{a}}_k^*) = v_\ell(\tilde{\mathbf{a}}_k^*) - p'_\ell(\tilde{\mathbf{a}}_k^*, \mathbf{b}_k^*), \quad (45)$$

where for traveler $\ell \in \mathcal{I}_k$ the term $v_\ell(\tilde{\mathbf{a}}_k^*)$ is the actual satisfaction gained by misreporting their private information. Substituting (44) into (45) yields

$$u_\ell(\tilde{\mathbf{a}}_k^*) = v_\ell(\tilde{\mathbf{a}}_k^*) - \left[w_3(\mathbf{b}_k^*) - \left(w_2^\ell(\tilde{\mathbf{a}}_k^*) - v'_\ell(\tilde{\mathbf{a}}_k^*)\right)\right], \quad (46)$$

which after a few simplifications gives

$$\begin{aligned} u_\ell(\tilde{\mathbf{a}}_k^*) &= v_\ell(\tilde{\mathbf{a}}_k^*) - w_3(\mathbf{b}_k^*) \\ &- \left[\left(\sum_{i\in\mathcal{I}_k\setminus\{\ell\}} \left[v_i(\tilde{\mathbf{a}}_k^*) - \sigma_i(\tilde{\mathbf{a}}_k^*)\right] - \sum_{j\in\mathcal{J}} c_j(\tilde{\mathbf{a}}_k^*) + v'_\ell(\tilde{\mathbf{a}}_k^*)\right) - v'_\ell(\tilde{\mathbf{a}}_k^*)\right]. \end{aligned} \quad (47)$$

Hence, as the term $v'_\ell(\tilde{\mathbf{a}}_k^*)$ appears in opposite signs in (47), we have

$$\begin{aligned} u_\ell(\tilde{\mathbf{a}}_k^*) &= \left[\sum_{i\in\mathcal{I}_k} \left[v_i(\tilde{\mathbf{a}}_k^*) - \sigma_i(\tilde{\mathbf{a}}_k^*)\right] - \sum_{j\in\mathcal{J}} c_j(\tilde{\mathbf{a}}_k^*)\right] - w_3(\mathbf{b}_k^*) \\ &= w_2(\tilde{\mathbf{a}}_k^*) - w_3(\mathbf{b}_k^*). \end{aligned} \quad (48)$$

Note that $\tilde{\mathbf{a}}_k^*$ is not necessarily the optimal solution of Problem 2. Thus, we have $w_2(\tilde{\mathbf{a}}_k^*) \leq w_2(\mathbf{a}_k^*)$. So, we observe that

$$u_\ell(\tilde{\mathbf{a}}_k^*) = w_2(\tilde{\mathbf{a}}_k^*) - w_3(\mathbf{b}_k^*) \leq w_2(\mathbf{a}_k^*) - w_3(\mathbf{b}_k^*) = u_\ell(\mathbf{a}_k^*). \quad (49)$$

Therefore, from (49), it follows immediately that the proposed mobility market is incentive compatible. □

Our next result is individual rationality, which implies that all travelers voluntarily participate in the proposed mobility market. Formally, for any traveler $i \in \mathcal{I}_k$, $k = 1, \ldots, K$, if traveler $i$'s utility $u_i$ is non-negative, i.e., $u_i \geq 0$, then we say traveler $i$

voluntarily participates in the mobility market. This is important as we can guarantee for any traveler $i$ that what they are willing to pay, $v_i$, will never be less than what they actually pay, $p_i$.

**Theorem 2** *The mobility market is individually rational. For any subclass $\mathcal{I}_k$, $k = 1, \ldots, K$, and for any traveler $i \in \mathcal{I}_k$, the utility of any traveler is non-negative, i.e., we have for all $i \in \mathcal{I}_k$, $u_i(\mathbf{a}_k) \geq 0$. Equivalently, $v_i(\mathbf{a}_k) \geq p_i(\mathbf{a}_k)$.*

***Proof*** It is sufficient to show the result only for one instance of a mobility market for some $k = \{1, \ldots, K\}$. There are two cases to consider. First, let us suppose that traveler $\ell \in \mathcal{I}_k$ rejects any travel recommendations from the social planner; denote such an assignment by $\hat{\mathbf{a}}_k$. From (33), traveler $\ell$ would be required to make a monetary payment equal to their maximum willingness-to-pay, i.e., $p_\ell = \bar{v}_\ell$. This implies that $u_\ell(\hat{\mathbf{a}}_k) = 0$. This is justifiable as traveler $\ell$ seeks to travel and the only alternative travel option to our mobility market is a taxicab service.

For the second case, let us consider the utility of an arbitrary traveler $i \in \mathcal{I}_k$ evaluated at the optimal solution $\mathbf{a}_k^*$ is given by

$$u_i(\mathbf{a}_k^*) = v_i(\mathbf{a}_k^*) - p_i(\mathbf{a}_k^*, \mathbf{b}_k^*). \tag{50}$$

Note that by Theorem 1 all travelers report their true private information at equilibrium. So, substituting (33) into (50) yields

$$u_i(\mathbf{a}_k^*) = v_i(\mathbf{a}_k^*) - \left[w_3(\mathbf{b}_k^*) - \left[w_2(\mathbf{a}_k^*) - v_i(\mathbf{a}_k^*)\right]\right] = w_2(\mathbf{a}_k^*) - w_3(\mathbf{b}_k^*). \tag{51}$$

Note that for each $k = 1, \ldots, K$, the feasible regions of Problems 2 and 3, say $\mathcal{F}_2$ and $\mathcal{F}_3$, respectively, satisfy the relation $\mathcal{F}_3 \subset \mathcal{F}_2$. This is because Problem 3 has the exact same constraints plus an additional one, i.e., (30), thus the maximization of $w_3$ (which is almost similar to the one in Problem 2) will always be less or equal than the maximization of $w_2$. Hence, it follows that $u_i(\mathbf{a}_k^*) = w_2(\mathbf{a}_k^*) - w_3(\mathbf{b}_k^*) \geq 0$. Therefore, the result follows. □

Next, we establish that the proposed mobility market is economically sustainable (see Definition 8).

**Theorem 3** *The mobility market is economically sustainable, i.e., it is guaranteed to generate revenue from each traveler and always meet the minimum acceptable mobility payments. In other words, for each subclass $\mathcal{I}_k$, $k = 1, \ldots, K$, and for an arbitrary $\ell \in \mathcal{I}_k$, we have*

$$p_\ell(\mathbf{a}_k^*, \mathbf{b}_k^*) = w_3(\mathbf{b}_k^*) - \left[w_2(\mathbf{a}_k^*) - v_\ell(\mathbf{a}_k^*)\right] \geq \sigma_\ell(\mathbf{a}_k^*). \tag{52}$$

***Proof*** Let $\mathbf{b}_k^*$ be an optimal solution of Problem 3 and $\mathbf{b}_k^{\ell *}$ be the corresponding feasible solution of Problem 3 with $\mathbf{a}_k^*$ an optimal solution of Problem 1. Since $\mathbf{b}_k^{\ell *}$ is only a feasible solution, we have

$$w_3(\mathbf{b}_k^*) \geq w_3(\mathbf{b}_k^{\ell *}). \tag{53}$$

Given the mobility payments (33), if we subtract the term $\left[w_2(\mathbf{a}_k^*) - v_\ell(\mathbf{a}_k^*)\right]$ from both sides of (53), we have

$$p_\ell(\mathbf{a}_k^*, \mathbf{b}_k^*) = w_3(\mathbf{b}_k^*) - \left[w_2(\mathbf{a}_k^*) - v_\ell(\mathbf{a}_k^*)\right] \geq w_3(\mathbf{b}_k^{\ell*}) - \left[w_2(\mathbf{a}_k^*) - v_\ell(\mathbf{a}_k^*)\right]. \quad (54)$$

The RHS of (54) can be expanded as follows:

$$\begin{aligned} w_3(\mathbf{b}_k^{\ell*}) - \left[w_2(\mathbf{a}_k^*) - v_\ell(\mathbf{a}_k^*)\right] &= \sum_{i \in \mathcal{I}_k} \left[v_i(\mathbf{b}_k^{\ell*}) - \sigma_i(\mathbf{b}_k^{\ell*})\right] - \sum_{j \in \mathcal{J}} c_j(\mathbf{b}_k^{\ell*}) \\ &- \left[\left(\sum_{i \in \mathcal{I}_k} \left[v_i(\mathbf{a}_k^*) - \sigma_i(\mathbf{a}_k^*)\right] - \sum_{j \in \mathcal{J}} c_j(\mathbf{a}_k^*)\right) - v_\ell(\mathbf{a}_k^*)\right]. \end{aligned} \quad (55)$$

After a few simplifications and rearranging of (55), we have

$$\begin{aligned} p_\ell(\mathbf{a}_k^*, \mathbf{b}_k^*) \geq &\left[\sum_{i \in \mathcal{I}_k} v_i(\mathbf{b}_k^{\ell*}) - \sum_{i \in \mathcal{I}_k \setminus \{\ell\}} v_i(\mathbf{a}_k^*)\right] \\ &+ \sum_{i \in \mathcal{I}_k} \left[\sigma_i(\mathbf{a}_k^*) - \sigma_i(\mathbf{b}_k^{\ell*})\right] + \left[\sum_{j \in \mathcal{J}} c_j(\mathbf{a}_k^*) - \sum_{j \in \mathcal{J}} c_j(\mathbf{b}_k^{\ell*})\right]. \end{aligned} \quad (56)$$

So, by Corollary 1, the last term in (56) is non-negative. Similarly, by Lemma 1, the first term in (56) is non-negative. So, we get

$$p_\ell(\mathbf{a}_k^*, \mathbf{b}_k^*) \geq \sigma_\ell(\mathbf{a}_k^*) - \sigma_\ell(\mathbf{b}_k^{\ell*}) = \sigma_\ell(\mathbf{a}_k^*), \quad (57)$$

since under $\mathbf{b}_k^{\ell*}$ traveler $\ell$ has not been assigned any mobility service, thus $\sigma_\ell(\mathbf{b}_k^{\ell*}) = 0$, and so the result follows immediately. □

# 6 Conclusion

This chapter demonstrates how we can model and study the mobility decision-making of selfish travelers who are faced with the dilemma of "which mode of transportation to use" as an economically-inspired mobility market. First, the proposed market provides a *socially-efficient solution*, i.e., the endmost collective travel recommendation respects and satisfies the travelers' preferences regarding mobility and ensures that, implicitly, there will be an alleviation of congestion in the system. We achieve the latter by introducing appropriate constraints in the optimization problem; thus, our solution efficiently allocates all the available mobility services to the travelers.

Furthermore, we showed that the proposed mobility market attains the properties of *incentive compatibility* and *individual rationality*. In other words, all travelers are incentivized to participate in the market while also truthfully reporting their personal travel requirements. Last, we introduced the notion of *minimum acceptable mobility payments* to ensure that the tolls and fares collected by the social planner will meet the mobility services' operating costs. Hence, the proposed market satisfies a status of *economic sustainability*.

One particular limitation of the proposed mobility market is that we require all travelers to book in advance, so the traveler-service assignment is static. This implies that the social planner would have to recompute all optimization problems in the mobility market to get an updated traveler-service assignment if the travelers' information changes. However, the static aspect of the proposed model is quite fitting in our case as our aim was to design a mobility market that considers the travelers' personal travel requirements to provide a socially-efficient assignment, i.e., "who should use which mode of transportation." Future work will focus on translating our model and results in a real-time environment. Furthermore, we have implicitly assumed that the travelers' utilities are not interdependent, i.e., a traveler's utility does not depend on the other travelers' private information. It remains an open problem the design of dynamic mechanisms with interdependent utility functions for mobility systems.

Ongoing work includes extending and enhancing the traveler-behavioral model, motivated by a social-mobility survey. Our objective is to observe any correlations between behavioral tendencies or attitudes of travelers and their mode of transportation preference (including CAVs). For example, how likely are people to use CAVs instead of public transit? Will CAVs impact travelers' tendencies and behavior; if yes, then in what way? Answers can help us refine the proposed mobility market and improve our understanding of the socioeconomic impact of CAVs. Our future research efforts will also focus on using methods, techniques, and insights from behavioral economics and mixed integer optimization theory to develop a holistic framework of the societal impact of connectivity and automation in mobility and provide socially-efficient, real-time solutions while tackling any potential rebound effects.

## References

1. Zhao L, Malikopoulos AA (2021) Enhanced mobility with connectivity and automation: a review of shared autonomous vehicle systems. IEEE Intell Transp Syst Mag (2021 in press)
2. Marletto G (2019) Who will drive the transition to self-driving? A socio-technical analysis of the future impact of automated vehicles. Technol Forecast Soc Change 139:221–234
3. Taiebat M, Brown AL, Safford HR, Qu S, Xu M (2018) A review on energy, environmental, and sustainability implications of connected and automated vehicles. Environ Sci Technol 52(20):449–465
4. Zmud JP, Ipek NS (2017) Towards an understanding of the travel behavior impact of autonomous vehicles. Transp Res Procedia 25:2500–2519

5. Barnes P, Turkel E (2017) Autonomous vehicles in delaware: analyzing the impact and readiness for the first state. Available via http://udspace.udel.edu/handle/19716/21596. Cited 1 October 2021
6. Bissell D, Birtchnell T, Elliott A, Hsu EL (2020) Autonomous automobilities: the social impacts of driverless vehicles. Curr Sociol 68(1):116–134
7. Sheller M, Urry J (2000) The city and the car. Int J Urban Reg Res 24(4):737–757
8. Bernard A (2014) Lifted: a cultural history of the elevator. NYU Press, NY
9. Malikopoulos AA, Cassandras CG, Zhang YJ (2018) A decentralized energy-optimal control framework for connected automated vehicles at signal-free intersections. Automatica 93:244–256
10. Malikopoulos AA, Beaver LE, Chremos IV (2021) Optimal time trajectory and coordination for connected and automated vehicles. Automatica 125:109469
11. Zardini G, Lanzetti N, Salazar M, Censi A, Frazzoli E, Pavone M (2020) Towards a co-design framework for future mobility systems. Ann Meeting Transp Res Board
12. Zhang J, Pourazarm S, Cassandras CG, Paschalidis IC (2018) The price of anarchy in transportation networks: data-driven evaluation and reduction strategies. Proc IEEE 106(4):538–553
13. Zhang Y, Cassandras CG (2019) An impact study of integrating connected automated vehicles with conventional traffic. Ann Rev Control 48:347–356
14. Rios-Torres J, Malikopoulos AA (2016) An overview of driver feedback systems for efficiency and safety. In 2016 IEEE 19th international conference on intelligent transportation systems (ITSC), 667–674
15. Rios-Torres J, Malikopoulos AA (2016) Energy impact of different penetrations of connected and automated vehicles: a preliminary assessment. In Proceedings of the 9th ACM SIGSPATIAL international workshop on computational transportation science
16. Beaver LE, Chalaki B, Mahbub AMI, Zhao L, Zayas R, Malikopoulos AA (2020) Demonstration of a time-efficient mobility system using a scaled smart city. Veh Syst Dyn 58(5):787–804
17. Rios-Torres J, Malikopoulos AA (2017) A survey on the coordination of connected and automated vehicles at intersections and merging at highway on-ramps. IEEE Trans Intell Transp Syst 18(5):1066–1077
18. Jang K, Vinitsky E, Chalaki B, Remer B, Beaver LE, Malikopoulos AA, Bayen AM (2019) Simulation to scaled city: zero-shot policy transfer for traffic control via autonomous vehicles. In 2019 10th ACM/IEEE international conference on cyber-physical systems (ICCPS), 291–300
19. Mahbub AMI, Malikopoulos AA, Zhao L (2020) Decentralized optimal coordination of connected and automated vehicles for multiple traffic scenarios. Automatica 117:108958
20. Xiao W, Cassandras CG (2021) Decentralized optimal merging control for connected and automated vehicles with safety constraint guarantees. Automatica 123:109333
21. Sarkar R, Ward J (2016) DOE SMART mobility: systems and modeling for accelerated research in transportation. In: Meyer G, Beiker S (eds) Road vehicle automation. Springer, Cham, pp 39–52
22. Ferrara A, Sacone S, Siri S (2018) Freeway traffic modeling and control. Springer Nature
23. Biyik E, Lazar D, Pedarsani R, Sadigh D (2021) Incentivizing efficient equilibria in traffic networks with mixed autonomy. IEEE Trans Netw Syst
24. Lazar DA, Coogan S, Pedarsani R (2020) Routing for traffic networks with mixed autonomy. IEEE Trans Autom Control 66(6):2664–2676
25. Mehr N, Horowitz R (2019) How will the presence of autonomous vehicles affect the equilibrium state of traffic networks? IEEE Trans Netw Syst
26. Cabannes T, Shyu F, Porter E, Yao S, Wang Y, Vincentelli MAS, Hinardi S, Zhao M, Bayen AM (2018) Measuring regret in routing: assessing the impact of increased app usage. In 2018 IEEE 21st international conference on intelligent transportation systems (ITSC), 2589–2594
27. Krichene W, Drighès B, Bayen AM (2015) Online learning of Nash equilibria in congestion games. SIAM J Control Optim 53(2):1056–1081
28. Krichene W, Castillo MS, Bayen AM (2016) On social optimal routing under selfish learning. IEEE Trans Netw Syst 5(1):479–488

29. Lam K, Krichene W, Bayen AM (2016) On learning how players learn: estimation of learning dynamics in the routing game. In: 2016 ACM/IEEE 7th international conference on cyber-physical systems (ICCPS), 1–10
30. Pigou AC (2013) The economics of welfare. Palgrave Macmillan, London
31. Vickrey WS (1969) Congestion theory and transport investment. Am Econ Rev 59(2):251–260
32. Schmitt E, Jula H (2006) Vehicle route guidance systems: classification and comparison. In 2006 IEEE Intell Transp Syst Conf 242–247
33. Chen O, Ben-Akiva M (1998) Game-theoretic formulations of the interaction between dynamic traffic control and dynamic traffic assignment. Transp Res Record 1617:178–188
34. Chremos IV, Malikopoulos AA (2020) Social resource allocation in a mobility system with connected and automated vehicles: a mechanism design problem. In: 2020 59th IEEE conference on decision and control (CDC), 2642–2647
35. Chremos IV, Malikopoulos AA (2021) Design and stability analysis of a shared mobility market. In 2021 European control conference (ECC), 374–379
36. Joksimovic D, Bliemer MCJ, Bovy PHL (2004) Road pricing problem including route choice and elastic demand—a game theory approach. In: The fifth triennial symposium on transportation analysis. Le Gosier, Guadeloupe, TRISTAN V
37. Salazar M, Lanzetti N, Rossi F, Schiffer M, Pavone M (2019) Intermodal autonomous mobility-on-demand. IEEE Trans Intell Transp Syst 21(9):3946–3960
38. Wada K, Akamatsu T (2011) Auction mechanisms for implementing tradable network permit markets. J Jpn Soc Civil Eng 67(3):376–389
39. Chremos IV, Beaver LE, Malikopoulos AA (2020) A game-theoretic analysis of the social impact of connected and automated vehicles. In 2020 IEEE 23rd international conference on intelligent transportation systems (ITSC), 2214–2219
40. Mas-Colell A, Whinston MD, Green JR (1995) Microeconomic theory. OUP, NY
41. Nisan N, Roughgarden T, Tardos E, Vazirani V (2007) Algorithmic game theory. CUP, Cambridge
42. Clarke EH (1971) Multipart pricing of public goods. Public Choice 11(1):17–33
43. Groves T (1973) Incentives in teams. J Econ Soc 617–631
44. Vickrey WS (1961) Counterspeculation, auctions, and competitive sealed tenders. J Fin 16(1):8–37
45. Renou L, Tomala T (2012) Mechanism design and communication networks. Theor Econ 7(3):489–533
46. Dave A, Chremos IV, Malikopoulos AA (2022) Social media and misleading information in a democracy: a mechanism design approach. IEEE Trans Autom Control (in press)
47. Bian Z, Liu X, Bai Y (2020) Mechanism design for on-demand first-mile ridesharing. Transp Res Part B Methodol 138:77–117
48. Kakade SM, Lobel I, Nazerzadeh H (2013) Optimal dynamic mechanism design and the virtual-pivot mechanism. Oper Res 61(4):837–854
49. Samadi P, Mohsenian-Rad H, Schober R, Wong VW (2012) Advanced demand side management for the future smart grid using mechanism design. IEEE Trans Smart Grid 3(3):1170–1180
50. Shoham Y, Leyton-Brown K (2008) Multiagent systems: algorithmic, game-theoretic, and logical foundations. CUP, Cambridge
51. Hurwicz L, Reiter S (2006) Designing economic mechanisms. CUP, Cambridge
52. Iwanowski S, Spering W, Coughlin WJ (2003) Road traffic coordination by electronic trading. Transp Res Part C Emerg Technol 11(5):405–422
53. Olarte R, Haghani A (2018) Introducing and testing a game-theoretic model for a lottery-based metering system in minneapolis. Transp Policy 62:63–78
54. Raphael J, Maskell S, Sklar E (2015) From goods to traffic: first steps toward an auction-based traffic signal controller. In: International conference on practical applications of agents and multi-agent systems, 187–198
55. Teodorovic D, Triantis K, Edara P, Zhao Y, Mladenovic S (2008) Auction-based congestion pricing. Transp Plann Technol 31(4):399–416

56. Vasirani M, Ossowski S (2011) A computational market for distributed control of urban road traffic systems. IEEE Trans Intell Transp Syst 12(2):313–321
57. Maskin E, Sjöström T (2002) Implementation theory. Handb Soc Choice Welf 1:237–288
58. Brown PN, Marden JR (2017) Optimal mechanisms for robust coordination in congestion games. IEEE Trans Autom Control 63(8):2437–2448
59. Myerson RB (2008) Perspectives on mechanism design in economic theory. Am Econ Rev 98(3):586–603
60. Diamantaras D, Cardamone E, Campbell KAC, Deacle S, Delgado LA (2009) A toolbox for economic design. Palgrave Macmillan, NY
61. Nisan N, Ronen A (2001) Algorithmic mechanism design. Games Econ Behav 35(1–2):166–196
62. Börgers T (2015) An introduction to the theory of mechanism design. OUP, NY
63. Maskin E (2008) Mechanism design: how to implement social goals. Am Econ Rev 98(3):567–576
64. Bitar E, Xu Y (2016) Deadline differentiated pricing of deferrable electric loads. IEEE Trans Smart Grid 8(1):13–25
65. Kazumura T, Mishra D, Serizawa S (2020) Strategy-proof multi-object mechanism design: Ex-post revenue maximization with non-quasilinear preferences. J Econ Theory 188:105036
66. Tang W, Jain R (2011) Stochastic resource auctions for renewable energy integration. In 2011 49th annual Allerton conference on communication, control, and computing (Allerton), 345–352
67. Tavafoghi H, Teneketzis D (2016) Multidimensional forward contracts under uncertainty for electricity markets. IEEE Trans Netw Syst 4(3):511–522
68. Coleri S, Cheung SY, Varaiya P (2004) Sensor networks for monitoring traffic. In: Allerton conference on communication, control and computing, 32–40
69. Tubaishat M, Zhuang P, Qi Q, Shang Y (2009) Wireless sensor networks in intelligent transportation systems. Wirel Commun Mobile Comput 9(3):287–302
70. Garber NJ, Hoel LA (2014) Traffic and highway engineering. Cengage Learning
71. Ho CQ, Hensher DA, Mulley C, Wong YZ (2018) Potential uptake and willingness-to-pay for mobility as a service (MaaS): a stated choice study. Transp Res Part A Policy Pract 117:302–318
72. Polydoropoulou A, Tsouros I, Pagoni I, Tsirimpa A (2020) Exploring individual preferences and willingness to pay for mobility as a service. Transp Res Record 2674(11):152–164
73. Cordeau JF, Laporte G (2007) The dial-a-ride problem: models and algorithms. Ann Oper Res 153(1):29–46
74. Dumas Y, Soumis F, Desrosiers J (1990) Optimizing the schedule for a fixed vehicle path with convex inconvenience costs. Transp Sci 24(2):145–152
75. Sanghavi S, Hajek B (2008) A new mechanism for the free-rider problem. IEEE Trans Autom Control 53(5):1176–1183
76. Abou-Zeid M, Ben-Akiva M (2011) The effect of social comparisons on commute well-being. Transp Res Part A Policy Pract 45(4):345–361
77. Carrasco JA, Hogan B, Wellman B, Miller EJ (2008) Collecting social network data to study social activity-travel behavior: an egocentric approach. Environ Plann B Plann Des 35(6):961–980
78. Ben-Akiva M, Lerman SR (2018) Discrete choice analysis: theory and application to travel demand. MIT Press, Cambridge
79. Hall J, Kendrick C, Nosko C (2015) The effects of Uber's surge pricing: a case study. The University of Chicago Booth School of Business
80. Jonker R, Volgenant A (1987) A shortest augmenting path algorithm for dense and sparse linear assignment problems. Computing 38(4):325–340
81. Schrijver A (1998) Theory of linear and integer programming. Wiley, NY
82. Damberg O, Storøy S, Sørevik T (1996) A data parallel augmenting path algorithm for the dense linear many-to-one assignment problem. Comput Optim Appl 6(3):251–272

83. Kakhbod A, Nayyar A, Sharma S, Teneketzis D (2014) Power allocation and spectrum sharing in wireless networks: an implementation theory approach. In: Alpcan T, Boche H, Honig ML, Poor HV (eds) Mechanisms and games for dynamic spectrum allocation. CUP, Cambridge, pp 108–144
84. Stoenescu TM, Teneketzis D (2005) Decentralized resource allocation mechanisms in networks: realization and implementation. In: Abed EH (eds) Advances in control, communication networks, and transportation systems. Birkhäuser Boston, pp 225–263

# A Real-Time Seq2Seq Beamforming Prediction Model for C-V2X Links

**Weidong Xiang, Vivekanandh Elangovan, and Sridhar Lakshmanan**

**Abstract** The chapter presents the research on a real-time deep-learning beamforming prediction model for C-V2X systems. Machine Learning (ML) for modeling and predicting wireless channels of vehicular communications has attracted increasing interest recently. In C-V2X systems, long-haul communication is critically needed, sometimes, without sacrificing congestion factor. Beamforming emerges as such an enabling approach to enlarge coverage by choosing right beams, instantaneously. In both the latest Wi-Fi and C-V2X standards, beamforming selections are to scan all the beams and pick up optimum beams during antenna training phase within a beacon interval (BI). Simulations and experiments conducted by the authors identified such scheme will lead to medium to significant degradation in performances sometimes during the whole BI under certain situations. To respond to this vital situation, this paper presented and studied a deep-learning beamforming prediction model to forecast optimum beams within each BI. In this chapter, a real-time sequence-to-sequence (Seq2Seq) beamforming prediction model is presented and implemented. Experiment data validated the effectiveness of the proposed prediction model under the Dearborn Campus of the University of Michigan, resulting in an enhancement of prediction accuracy of 50–75%.

## 1 Backgrounds

Cellular vehicle to everything (C-V2X) is an emerging technology which encompasses Vehicle to Vehicle (V2V) connectivity, Vehicle to Infrastructure (V2I), Vehicle to Pedestrians (V2P) and Vehicle to Network (V2N). C-V2X communication

W. Xiang (✉) · V. Elangovan · S. Lakshmanan
University of Michigan, Dearborn, MI, USA
e-mail: xwd@umich.edu

V. Elangovan
e-mail: velango@umich.edu

S. Lakshmanan
e-mail: lakshman@umich.edu



Y. L. Murphey et al. (eds.), *AI-enabled Technologies for Autonomous and Connected Vehicles*, Lecture Notes in Intelligent Transportation and Infrastructure,
https://doi.org/10.1007/978-3-031-06780-8_18

is envisioned to enhance the safety of drivers, passengers and pedestrians. C-V2X is considered as an upgrade to Dedicated Short-Range Communications (DSRC). C-V2X is more often referred to as PC5 interfaces in Long-Term Evolution (LTE) Release 14 standard released by the 3rd Generation Partnership Project (3GPP) [1]. C-V2X communication has been explained in detail in [2, 3].

C-V2X system is governed by the National Highway Traffic Safety Administration (NHTSA) and Department of Transportation (DOT). In 2017, the NHTSA and DOT issues Notice of Proposed Rulemaking (NPRM) [4] for the V2V communication. At that times, V2V communication was based on the DSRC defined in SAE J2735 [5]. The technology behind V2V communication expects an implementation of 360°C "awareness" and a range of 300 m where omnidirectional antennas are adopted.

Omnidirectional antenna gives complete coverage of 300 m but increases congestion factor, which is regulated in SAE J2945/1 [6]. In a highly congested vehicular location, a network experiences high data loads, which requires reduced radiation powers. On the other hand, reducing power reduces the coverage. An effective way to communicate in longer range without increasing the congestion is to adopt directional antennas. Beamforming is a technique in which an antenna array can be steered in predefined directions. The input RF signal is fed to the antenna array in parallel and signals are added constructively and destructively, depending on the phases, in such a way that they concentrate the energy into a narrow beam.

In both Wi-Fi and 5G standards, during the antenna training phase of each beacon interval (BI) scanning is performed across all the beams and the optimum one is chosen and adopted during the whole BI. If the same method is performed in the C-V2X system, it will lead to medium or significant non-optimum selection of beam due to rapid variation of direction of arrival (DoA) of multipath signals. To this end, a real-time beam prediction model is presented and researched in this paper through adopting the so-called msequence to sequence (Seq2Seq) prediction model. The receiving signal strength indicator (RSSI) is used as the metric to select the optimum beam.

Wireless Channel prediction is a well-researched topic using methods from Markov Chain to machine learning [7–10]. There has been numerous neural network research being performed in wireless communication. In [11, 12], the authors designed deep learning models for addressing hybrid precoding and beamforming issues in MIMO systems. Interference alignment for wireless communications were studied in [13]. A convolutional neural network used for distance estimation between vehicles for DSRC system is presented in [14]. Similarly, the authors in [15] adopted a reinforcement learning (RL) to study on a heterogeneous 5G architecture including both DSRC and V2X communications. The research in [16] presented machine learning for analyzing the Doppler profile to improve the road safety in V2V network. The work of [17] performs reinforcement learning for intelligent traffic signal control systems, the authors of [18] performed a smoothed LSTM on Vehicular GPS prediction. A comprehensive survey on the use of deep learning in wireless network has been presented at [19, 20]. Chen et al. presented a tutorial elaborating on various types of machine learning used for corresponding wireless networks.

The rest of the chapter is organized as follows. Section 2 briefs system model. The implementation and filed test are detailed in Sect. 3. The real-time Seq2Seq beamforming prediction model are presented and discussed in Sect. 4. Finally, conclusions are given in Sect. 5.

## 2 System Model

A 4-element uniform linear array (ULA) beamforming is developed and implemented to evaluate the enhancement of performances for C-V2X systems when compared to adopting omni-directional antennas. The beamforming antenna array is illustrated in Fig. 1, of which the spacing between the antennas is half a wavelength, $\lambda$, at $5.9GHz$. The arriving signal with an incident angel of $\theta$ creates a phase difference of $\phi = kdcos\theta$ between two adjacent antennas within the array, where $k = 2/\lambda$ is the wave number and $d = /2$ the antenna spacing. The array factor (AF) of the array is given by,

$$AF(\theta) = \left| \frac{1}{N} \sum_{n=0}^{N-1} e^{j(nkdcos\theta + \alpha_n)} \right|^2 \tag{1}$$

where $N$ is the total number of antennas and $\alpha_n$ is additional phase shift of $n$th antenna element.

A $4 \times 4$ Butler matrix is used to offer addition phase shift for each antenna element in order to generate predefined beams. Moreover, an omnidirectional antenna is added as a benchmark during the project, which is selected by a single pole four

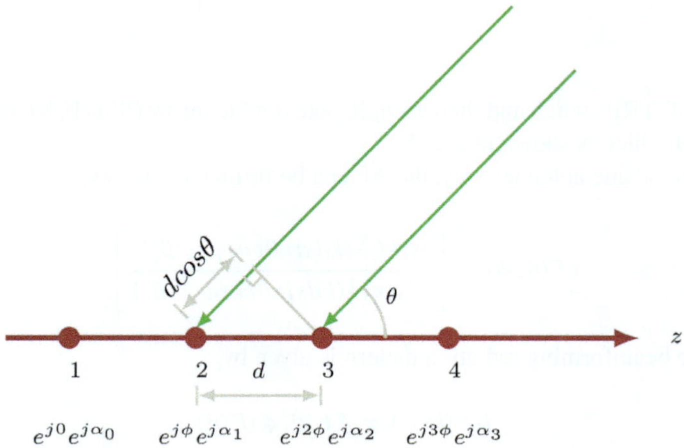


**Fig. 1** The configuration of a 4-element ULA

**Fig. 2** The diagram of beamforming antenna array

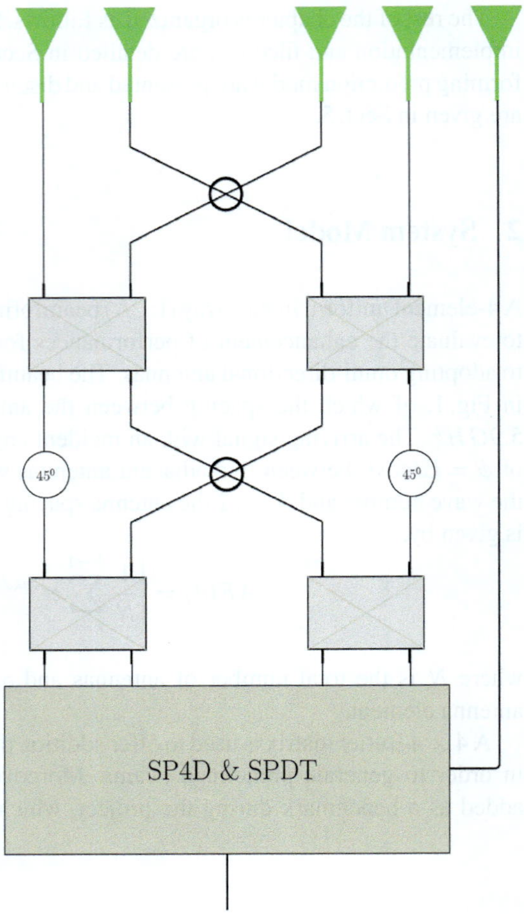


throw (SP4T) RF switch and then a single pole double throw (SPDT) RF switch, the diagram of which is shown in Fig. 2.

For a broadside antenna array, the AF can be further written as,

$$AF(\theta, \phi) = \left[ \frac{sin(\frac{N}{2}(kdsin\theta cos\phi + \beta))}{Nsin(\frac{1}{2}(kdsin\theta cos\phi + \beta))} \right]^2 \tag{2}$$

finally, the beamforming radiation pattern is given by,

$$BF(\theta, \phi) = AF(\theta, \phi)P(\theta) \tag{3}$$

where $P(\theta)$ is the radio pattern of an element antenna. The generated four beams are shown in Fig. 3.

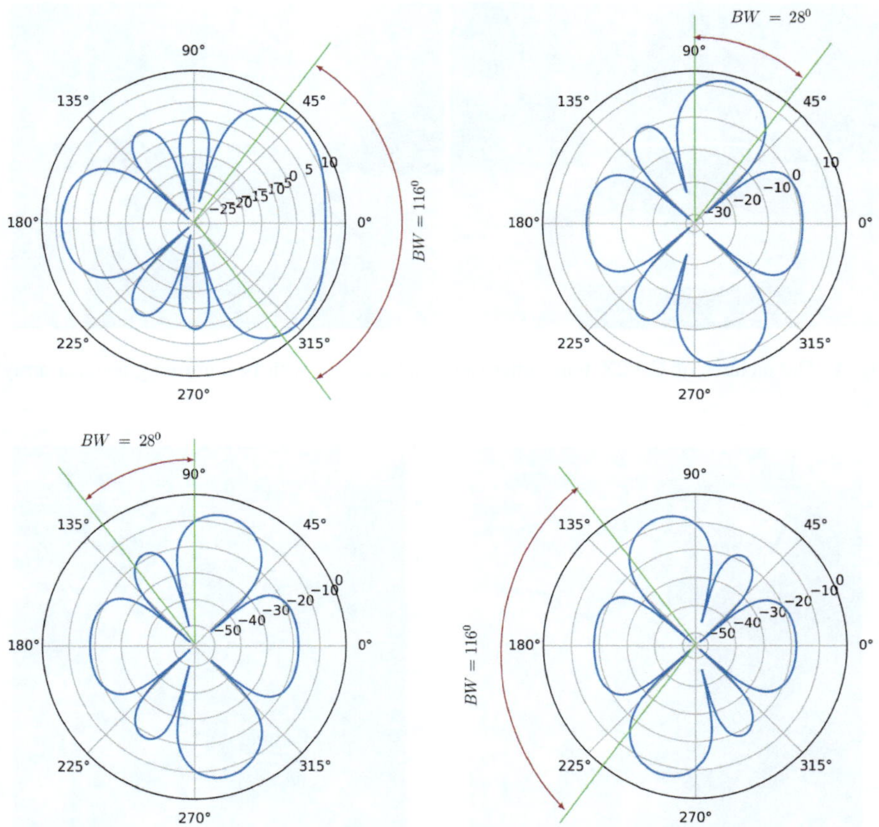


**Fig. 3** The radiation pattern of 4-element ULA at Port 0, 1, 2 and 3 where $BW = 116°$, $28°$, $28°$ and $116°$, respectively

# 3 Implementation and Field Test

The beamforming antenna array is adopted at the receiver of C-V2X road side unit (RSU) while the transmitter adopts omnidirectional antenna, the pictures of which are shown in Fig. 4. During the experiments, the vehicle mounting the transmitter was parked at the corner of a parking lot in the University of Michigan, Dearborn campus while the vehicle with the receiver driving around the parking lot.

The left figure of Fig. 5 illustrates the best beam, varying with the positioning of the receiver. The red dot shows the beam direction aligns with the line-of-sight (LOS) between transmitter and receiver, while the blue dot does not, due to in lack of LOS. The right figure of Fig. 5 illustrates the gain of beamforming when comparing to omnidirectional antenna. It validates the effectiveness of adopting beamforming can enlarge coverage, enhance signal strength as well as anti-interference. As shown in the figure, the best beam varying with positions.

**Fig. 4** The pictures of C-V2X transmitter (left) and receiver with beamforming antenna array (right)

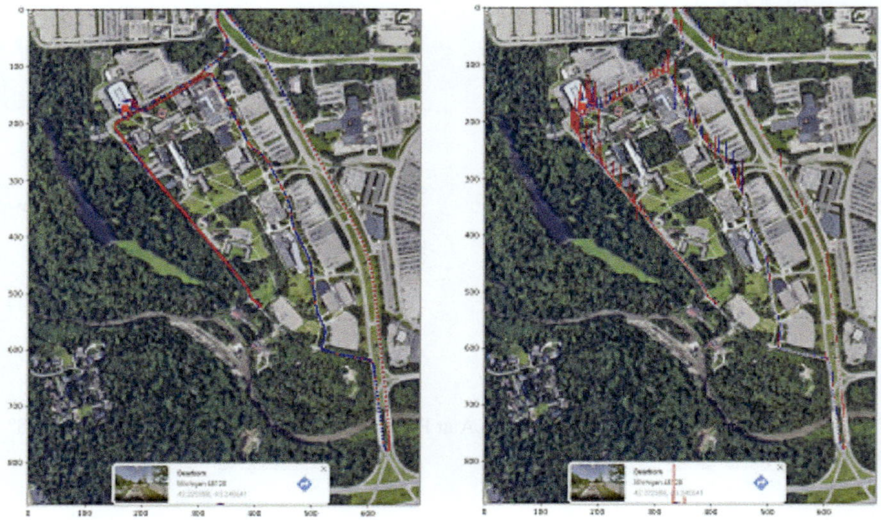

**Fig. 5** The illustration of the best beam varying with positions (left) and gain of beamforming when comparing omnidirectional antenna

# 4 Real-Time Seq2Seq Beamforming Prediction Model

Current WiFi systems have not adopted the beamforming antenna array, but it has been included in the IEEE 802.11ad standard. In the IEEE 802.11ad standard, a beacon interval (BI), defined as the duration between two beacon frames, consists of a beacon header interval (BHI) for information exchange of management and network information, and a data transmission interval (DTI) for data delivery. Furthermore, BHI is divided into beacon a transmission interval (BTI), for network announcement and beamforming training, an association beamforming training (A-BFT), for antenna training and pairing by the aid of access points (APs) and an announce-

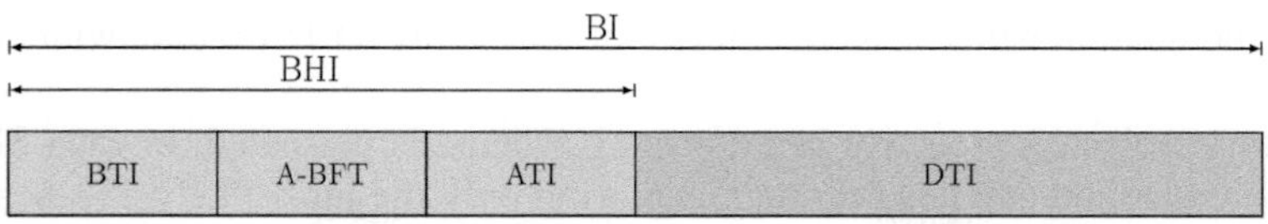


**Fig. 6** The configuration of a beacon interval

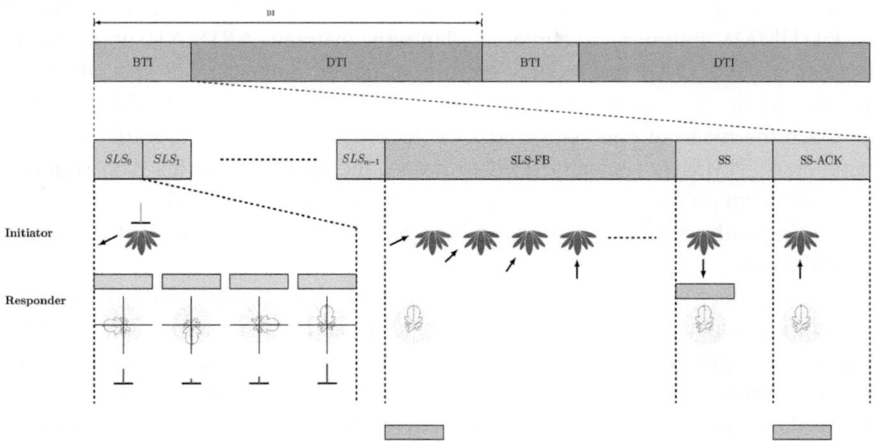


**Fig. 7** The processing of SLS defined by the IEEE 802.11ad standard

ment transmission interval (ATI), for management information exchange with the associated stations. Figure 6 shows the configuration.

Th access point (AP) and stations may adopt a different number of beams; say 8 and 4 beams, as an example. Beamforming will be performed in two phases of sector-level sweep (SLS) for the exchange of bidirectional frame sequence and beam refinement phase (BRP) for the stations to select the right beam accordingly.

However, the main issue that existed in the IEEE 802.11ad beamforming process is the overhead of SLS. For example, if there is single AP and multiple stations, a lot amount of time is spent in the training frames delivery. In fact, SLS is performed in the following four steps: (1) the initiator broadcast a training frame per beam, and responders listen through all the beams. (2) A responder replies with a frame in all its beams. (3) The imitator then sends the Sector Sweep Feedback (SSW-FB) frames, and (4) The responder sends back with the sector sweep acknowledges (SSW-ACKs). The above process is illustrated in Fig. 7. SLS and BRP are two essential beamforming training processing, the former is used to establish a link between two users, and the latter is to refine the beam over existing links.

In general, an AP broadcasts beacon frames through all beams alternatively, while the station tries to receive all of them. The station reads the corresponding headers and data and then acknowledges the received packets. The payload helps the station to associate with the AP during A-BFT. However, if the station is far from the AP, it will enter control mode. During control mode, it will adopt a single-carrier differential

BPSK modulation with a spreading factor of 32, ending up 15 dB gain with broader coverage.

In this project, a proactive beamforming scheme is presented, in each SLS, to largely reduce the overhead through exploring only the most likely 1–3 beams in both AP and stations, which is predicted by a novel real-time reinforcement learning beam prediction model.

Traditional prediction models are statistically based, including hidden Markov model (HMM), autoregressive integrated moving average (ARIMA), support vector regression (SVR). All the models require sufficient data and not specialized for specific applications and scenarios. It is important to note that ARIMA tends to approach the mean of past data, and SVR lacks the structure to identify some key parameters. HMM is typically used to model the user time-series data sequence and to perform prediction through Bayesian interference. Its main disadvantages are over-fitting with specific conditions and the generation of the unsatisfied outcome to general cases.

Recently, deep learning based on neural networks has attracted intensive attention due to their superior performance in data fitting and classification. The proposed beamforming prediction model can comprehensively depict, in a non-mathematical way yet statistically accurately, the time-varying rssi values from all the beams as well as the omnidirectional antenna, used as a reference, under various environments.

In this project, a novel Seq2Seq is adopted to predict the best beam within the next several transmission slots. The training and prediction process is illustrated in Fig. 8, shown below. The $(M + N)$ data set of RSSI vectors, each containing four RSSI values from four beams in addition to one from omnidirectional antenna, are normalized between 0 and 1 and fed into the predictor at the current time slot $t_0$. The first $M$ data are set as input data, and the latest N data as targets. After training is accomplished, the predictor generates predicted RSSI vectors in the next several transmission slots, from which the best beam is chosen. For example, for the settings of $M = 30$ and $N = 10$, the data set of the previous 3 seconds will be adopted to train the predictor in a real-time mode, and the predicator predicts the next 10 RSSI vectors in next one second where the RSSI vectors update rate is set as $10Hz$.

Figure 8 is the state diagram of the Seq2Seq predictor consisting of three phases, namely, training, prediction, and monitoring. The predictor will enter training mode, periodically or triggered by an event. Every $t = nt_p$ the predictor will preform training based on the previous $(M + N)$ data set and update corresponding parameters, where $n$ is an integer and $t_p$ is the training interval, both can be predefined or adaptively set. The predictor can also be triggered by an event, which is defined as the case that there are several consecutive estimate errors. At $t_q + t_p$, predictor finishes training and enters the prediction mode generating prediction results within the interest time duration, where $t_q$ is the time that predictor preform a prediction. During the prediction phase and at every $nt_f$, the prediction monitor will calculate the estimate accuracy where $t_f$ can be set as the beacon update rate, that is, $t_f = 100ms$ in this work. If the prediction is reconsidered workable, it will return acknowledge information to prediction at $nt_f + t_r$ where $t_r$ is the processing time; otherwise, an event will be released to force the predictor to initiate a new training. Such a Seq2Seq

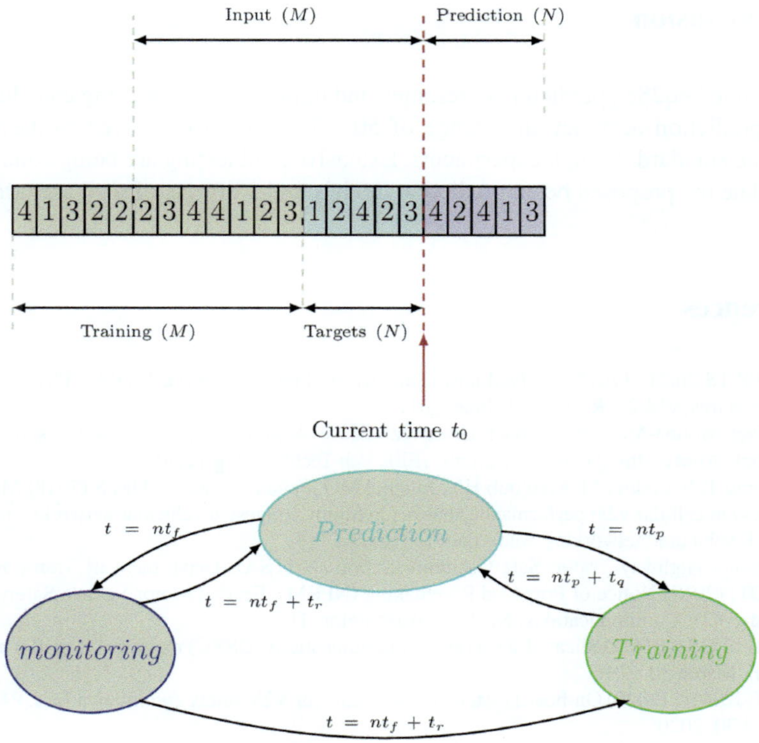


**Fig. 8** The training and prediction of the proposed real-time Seq2Seq beam predictor (left) and its state diagram (right)

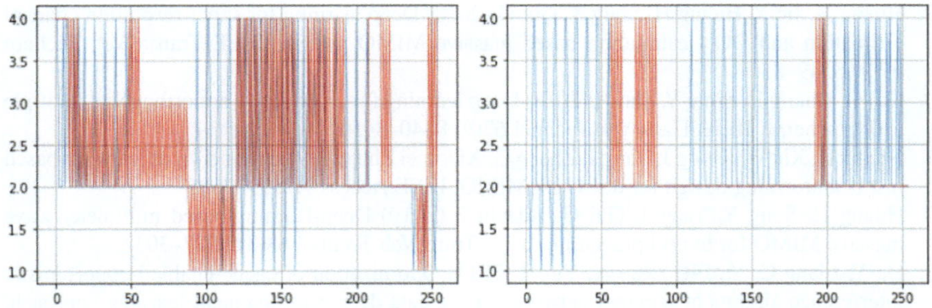

**Fig. 9** The comparison of the predicted and real best beams at two locations during the experiments

predictor is developed and implemented and the enhancement of prediction accuracy ranging from $50\% - 75\%$ for various road conditions. Figure 9 illustrate two typical such results where the gray line shows the actual best beam measured and red line the selected best beam.

## 5 Conclusion

A real-time Seq2Seq predictor is presented and implemented achieving an enhanced beam prediction accuracy in a range of $50 - 75\%$ when compared to the IEEE 802.11ad standard through experiments. Extensive road testing are being conducted to validate the proposed beam prediction models under diverse of road scenarios.

## References

1. 3GPP TS 36.213 (2017) Evolved universal terrestrial radio access (E-UTRA); Physical layer procedures. v14.2.2 Release 14, June, 2017
2. Rafael Molina-Masegosa, Javier Gozalvez (2017) A new 5G technology for short-range vehicle-to-everything communications. IEEE Veh Technol Mag 12:30–39
3. Behrad T, Saifuddin M, Mahjoub HN, Mughal MO, Fallah YP, Rao J, Das S (2018) Multiple access in cellular v2x: performance analysis in highly congested vehicular networks. In: 2018 IEEE vehicular networking conference (VNC), pp 1–8
4. National Highway Traffic Safety Administration (NHTSA), Department of Transportation (DOT) (2017) Notice of Proposed Rulemaking (NPRM): Federal Motor Vehicle Safety Standards; V2V Communications. NHTSA, Washington, DC
5. SAE J2735 (2016) Dedicated short range communications (DSRC) message set dictionary. PA SAE, March 30, 2016
6. SAE J2945/1 (2020) On-board system requirements for V2V safety communications. PA SEA, April 30, 2020
7. Wang X, Gao L, Mao S, Pandey S (2017) CSI-based fingerprinting for indoor localization: a deep learning approach. IEEE Trans Veh Technol 66(1):763–776
8. Tan LT, Hu RQ (2018) Mobility-aware edge caching and computing in vehicle networks: a deep reinforcement learning. IEEE Trans Veh Technol 67(11):10190–10203
9. Huang H, Jie Y, Huang H, Song Y, Gui G (2018) Deep learning for super-resolution channel estimation and DOA estimation based massive MIMO system. IEEE Transa Veh Technol 67(9):8549–8560
10. Gui G, Huang H, Song Y, Sari H (2018) Deep learning for an effective nonorthogonal multiple access scheme. IEEE Trans Veh Technol 67(9):8440–8450
11. Huang H, Xia W, Xiong J, Yang J, Zheng G, Xiaomei Zhu (2019) Unsupervised learning-based fast beamforming design for downlink MIMO. IEEE Access 7:7599–7605
12. Huang H, Song Y, Yang J, Gui G, Adachi F (2019) Deep-learning-based millimeter-wave massive MIMO for hybrid precoding. IEEE Trans Veh Technol 68(3):3027–3032
13. He Y, Liang C, Yu FR, Zhao N, Yin H (2017) Optimization of cache-enabled opportunistic interference alignment wireless networks: a big data deep reinforcement learning approach. In: 2017 IEEE international conference on communications (ICC), pp 1–6
14. Tuzi G, Medenica Z, Miucic R (2018) Using convolutional neural networks for distance estimation between dedicated short-range communications equipped vehicles. In: 2018 IEEE 87th vehicular technology conference (VTC Spring), pp 1–6
15. Sheng Z, Pressas A, Ocheri V, Ali F, Rudd R, Nekovee M (2018) Intelligent 5g vehicular networks: an integration of DSRC and mmWave communications. In: 2018 international conference on information and communication technology convergence (ICTC), pp 571–576
16. Kihei B, Copeland JA, Chang Y (2017) Automotive doppler sensing: the doppler profile with machine learning in vehicle-to-vehicle networks for road safety. In: 2017 IEEE 18th international workshop on signal processing advances in wireless communications (SPAWC)

17. Zhang R, Ishikawa A, Wang W, Striner B, Tonguz OK (2021) Using reinforcement learning with partial vehicle detection for intelligent traffic signal control. IEEE Trans Intell Transp Syst 22(1):404–415
18. Liu S, Elangovan V, Xiang W (2019) A vehicular GPS error prediction model based on data smoothing preprocessed LSTM. In: IEEE VTC Fall
19. Mao Q, Hu F, Hao Q (2018) Deep learning for intelligent wireless networks: a comprehensive survey. IEEE Commun Surv Tutorials 20(4):2595–2621
20. Chen M, Challita U, Saad W, Yin C, Debbah M (2019) Artificial neural networks-based machine learning for wireless networks: a tutorial. IEEE Commun Surv Tutorials 21(4):3039–3071

# Big Data in Road Transport and Mobility Research

**Carol A. Flannagan**

**Abstract** This chapter covers topics related to big data sources, methods and applications in transportation and mobility research. Big data sources covered include data from vehicle-based and infrastructure-based sensors. Methods from traditional regression to machine-learning and AI are discussed in terms of prediction and inference goals. Finally, example use cases of Big Data and AI are discussed in the context of safety, travel and micromobility.

## 1 Introduction

In the early days of the previous transportation revolution around the turn of the twentieth century, it became clear that traffic crashes involving the new automotive technology were a significant problem. From that time to now, crash data have been recorded by police officers at the scene of crashes, though the data contents and systems have improved dramatically over the last century. Notably, in 1975, the National Highway Traffic Safety Administration (NHTSA) established the Fatality Analysis Reporting System (FARS), a census of crashes in the U.S. that occur on public roads and in which someone dies of injuries sustained in the crash within 30 days. Every state has a FARS office that gathers additional data about fatal crashes and submits those data electronically to NHTSA. The dataset is publicly available and provides accurate annual counts of fatalities in the U.S., along with extensive detail that has substantially improved our understanding of causes of fatal crashes.

In 1979, NHTSA introduced the National Automotive Sampling System (NASS), which collects two key *samples* of crashes in the U.S.:

1. Police-reported crashes of all types and severities called the General Estimates System (GES) until 2016 when it was replaced by the Crash Report Sampling System (CRSS)

C. A. Flannagan (✉)
Ann Arbor, USA
e-mail: cacf@umich.edu

 
Y. L. Murphey et al. (eds.), *AI-enabled Technologies for Autonomous and Connected Vehicles*, Lecture Notes in Intelligent Transportation and Infrastructure,
https://doi.org/10.1007/978-3-031-06780-8_19

2. Towaway crashes involving light vehicles called the Crashworthiness Data System (CDS) until 2016 when it was replaced by the Crash Investigation Sampling System (CISS).

These additional datasets are carefully selected, representative samples of their respective crash types. Both involve significant work by individuals to code police reports to a common standard (in the case of GES/CRSS) and to do in-depth on-site investigation of crashes (in the case of CDS/CISS). Both have led to significant insights and changes in vehicle design, state law, and infrastructure design that have improved safety over nearly half a century.

While these datasets have led to significant benefits in fatality-rate reduction, the vehicle-based countermeasures they support are primarily related to injury mitigation rather than crash prevention. According to NHTSA, airbags, seat belts, crush zones and other improvements related to crash-data-supported federal motor vehicle safety standards (FMVSS) have saved over 500,000 lives since 1975 [1].[1] These achievements are extremely important, but in the last two decades, sensor and computing technology advancements have supported a shift to development of safety systems that *prevent* crashes altogether. These systems, which include Advanced Driver Assistance Systems (ADAS) as well as Automated Driving Systems (ADS) hold the promise of preventing not just fatalities or injuries, but the broad cost of crashes, including costs of congestion, property damage, emergency response, medical bills, and so on.

The advances in sensor and computing technology not only support development of these safety systems, but they also support large-scale collection of data related to mobility in general, i.e., *Big Data in transportation*. These new data sources have opened up whole new areas of research into mobility topics and enabled new insights. However, Big Data comes with new challenges as well.

In this chapter, I discuss Big Data and Big Data methods as they relate to transportation research. The first section covers common large transportation-data sources and Big Data analytical methods. Detailed description of methods is out of scope for this chapter. Instead, the goal of this section is to introduce the reader to a wide variety of methods used in transportation research.

Following this introduction to methods, the next section describes examples of the use of Big Data analysis to understand certain transportation problems. The section focuses on applications to transportation safety and travel behavior. These cover large subsets of research in the mobility field and provide key use cases for the methods and methodological issues describe in this chapter.

Finally, the last section addresses challenges, pros, and cons of Big Data, Artificial Intelligence (AI), and Machine Learning (ML). The transition from carefully constructed, representative samples to less expensive, larger, but non-representative samples introduces potential benefits and new cautions for the analyst. The chapter ends with recommendations for the future in this area.

---

[1] This number was obtained from Table 2–3 by summing lives saved from 1975 forward.

# 2 Big Data and Methods in Transportation Research

## *2.1 Big Data Sources*

### 2.1.1 Vehicle-Based Data Sources

A key source of information about driver behavior, safety, route choice, and travel is data captured from vehicles while in motion. "Naturalistic" Driving Study (NDS) is a term used to refer to studies in which participants drive instrumented vehicles in the course of their daily travel and data about their driving is captured. These studies emphasize driver behavior because the studies are meant to capture typical driving without interventions.

Field Operational Tests (FOTs) are similar to NDSs in that study participants drive vehicles with instrumentation capable of capturing driving data. However, these studies include evaluation of some type of technology, such as Forward Collision Warning (FCW; [2]) or multiple integrated ADAS [3]. A common, though not universal, study design includes a baseline period when the system is turned off, followed by a test period when the system is enabled. This allows comparison of an individual driver's behavior with and without the system.

Classic NDSs and FOTs, which involve extensive vehicle instrumentation and relatively frequent data downloads from the vehicle, are very expensive to run. These studies tend to have extensive instrumentation, including video cameras capturing interior and exterior views, multi-axis accelerometers capturing vehicle motions in detail, Global Positioning Systems (GPSs) capturing location and route information, and in some cases, radar(s) capturing proximity to other road users. Some studies include data from the vehicle's Controller Area Network (CAN) bus, which provides access to the vehicle's own sensors and internal states.

While this level of detail is very useful for many research topics, sample sizes and geographic diversity are very limited. For example, the largest NDS in the world was the second Strategic Research Program (SHRP2) NDS, which was conducted from 2012–2014 [4]. That study captured data from over 3000 drivers in six U.S. locations. However, even the SHRP2 sample is limited with respect to rare events such as crashes, especially serious crashes. That dataset contains approximately 100 serious crashes, with an additional ~150 crashes considered to be police-reportable level. These ~250 crashes can be compared to the GES/CRSS data sample, which includes 50,000 raw cases per year representing approximately 6 million crashes annually. While SHRP2 and other NDS provide data critical for insights into driver behavior, they are far too expensive to conduct regularly as a means of tracking how behavior changes, or in the form of FOTs, as a means of testing new technologies.

In Fig. 1, the gray rectangle represents all possible information that could be captured from a vehicle while driving. Rows represent cases, events, drivers, or even moments in time, while columns represent variables, or types of information (e.g., location, speed, distance to lead vehicle, etc.). In principle, a dataset covering all of the gray rectangle would provide complete information for research purposes.

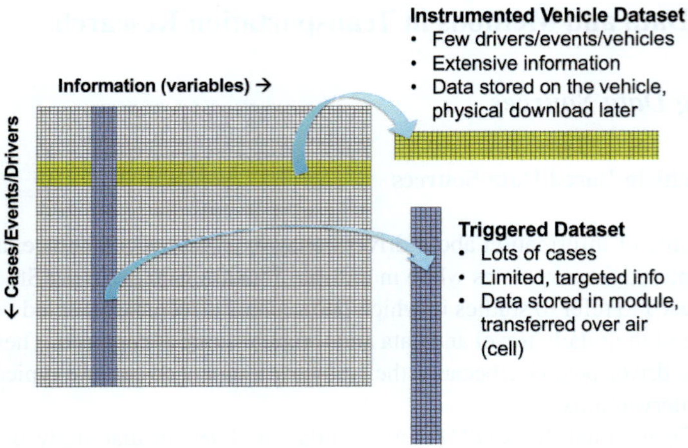


**Fig. 1** Illustration of two strategies for vehicle-based data collection

However, with three trillion vehicle miles traveled per year in the U.S., this approach is prohibitively expensive. Instead, there are two basic approaches to capturing data that have different advantages and disadvantages. The first, which is used by classic NDSs and FOTs, is to capture extensive information through on-board instrumentation, but to study only a small subset of driver, events, locations, etc. This can be called a "horizontal" dataset with high detail and low sample size. The alternative is to use on-vehicle telematic capability to capture a small amount of information about a large number of vehicles, miles, etc. This information is often triggered, meaning that data are only captured when certain conditions (e.g., hard braking) apply. This can be called a "vertical" dataset with low detail and high sample size.

Triggered datasets have been used for a number of observational FOTs. In these studies, drivers opted in to having their data captured through General Motors (GM) Corporation's OnStar® system. Data included information triggered by alerts from the systems under study (e.g., Forward Collision Warning (FCW) and Lane-Departure Warning (FCW); [5]; or Automatic Emergency Braking (AEB); [6]) as well as background information captured in the form of histograms of speeds or in some cases, 1-Hz speed and location.

The datasets used in those FOTs include thousands of alert events from thousands of drivers at relatively low cost, but the data capture systems must be purpose-built for the study goals. That is, the on-vehicle production modules must be capable of capturing the desired data in the first place, to be transferred for storage and analysis over the air later.

In contrast, another form of triggered data collection is often used by insurance companies, which provide drivers with a dongle or a smartphone app that can capture simple triggered data about vehicle motions and send it to a data repository telematically. These dongles are relatively inexpensive and have been used for both commercial [7] and research [8] purposes.

Finally, another form of "vertical" moving-base (in this case, person-based) data capture is the use of location traces from smart phones. Because of the ubiquitous use of smart phones and their GPS sensors, cell phone companies and researchers can capture location data as individuals move around on a large scale. Data aggregators purchase such data from companies and can use algorithms to estimate whether an individual is traveling by car, bus, bicycle or on foot. Researchers (and practitioners) can purchase the aggregated data or can collect their own data with a smartphone app and participants who opt in. Location traces can be mined for a variety of types of information including route choices, travel speeds, and trip origin–destination pairs and estimated travel mode(s).

### 2.1.2 Infrastructure-Based Data Sources

The other primary source of Big Data for transportation research is data collected from fixed infrastructure locations. Here, sensors are mounted in a fixed location and collect information about the movements of people and vehicles past that location. These data are crucial to understanding congestion, effects of signal phase and timing, and volume of various road users, among other topics.

Traffic volume has been measured using sensors for decades. Sensors include both permanent installations such as inductive loops embedded in the road surface and temporary sensors such as pneumatic tubes stretched across the road. Newer sensing options include cameras, which can be used with computer vision algorithms to count and identify types of vehicles, as well as pedestrians and bicyclists.

A key future source of infrastructure-based data will be Connected Vehicle to Anything (CV2X) messages captured by roadside units. This data source has been studied in a number of pilot demonstration projects, of which the largest is the Safety Pilot Model Deployment (SPMD), conducted in Southeast Michigan from 2012–2013 [9]. This study included installation of roadside units along with over 2800 passenger cars, buses and trucks equipped with devices capable of sending and receiving Basic Safety Messages (BSMs). The SPMD data have supported analysis of effects of distance and orientation on message transmission [10], potential benefits of safety countermeasures [11], and traffic volume estimation [12], to name just a few topics.

In general, infrastructure-based sensing can be used inexpensively to capture comprehensive data at a single location (a "horizontal" dataset). The challenge for research is that sensing must be set up at a large number of (ideally representative) locations to ensure that results of analyses are generalizable. The promise of communication-based data is that roadside units will be deployed widely for practical purposes, but that the data will also serve broad research purposes.

## 2.2 Methods

Two key goals of research data analysis are inference and prediction. *Prediction* models and methods are aimed at accurately estimating values of a variable for instances outside of the data sample being analyzed. For prediction models, the means of generating predictions and the variables used are unimportant and only the accuracy of the out-of-sample prediction are of concern. *Inference* refers to the goal of understanding relationships between variables that will generalize to the real world. Most such analysis is aimed at *causal inference* in which the causes of an outcome of interest (e.g., crashes) are identified and their relationship measured.

### 2.2.1 Prediction

A prediction model is one that is intended to simply estimate a value for observations not in the original dataset. For example, prediction models are used to estimate annual daily traffic volume on segments of roads where traffic volume is not directly measured; to estimate how many crashes will occur at certain intersections; or to predict how long a trip will take based on the time of day.

Prediction models can be developed using a wide variety of methods. Classical methods are typically parametric–that is, they assume a distribution for the error in the model and often assume a shape of the relationship between predictors and the outcome. Multple linear regression is the most well-known parametric model, but in transportation research, binary-outcome models (e.g., logistic or probit models) and count-outcome models (Poisson or negative binomial regression) are used extensively. Binary outcomes occur frequently in analyses of injury outcome or crash outcome (i.e., what is the probability that an occupant will be injured given the speed of a crash and the direction of impact?). Count outcomes occur frequently in analyses of crash counts or other event counts (e.g., near misses, pedestrians crossing the street, etc.) as a function of characteristics of a road or intersection.

In contrast, non-parametric models do not make distribution assumptions and thus can be more flexible. Machine-learning approaches, such as random forest, support vector machines or neural networks, are generally non-parametric. These approaches, enabled by large training datasets and fast computers, have revolutionized prediction and are the engine behind AI applications such as ADS. They are able to represent very complex interactions between predictors and outcomes. Importantly, unlike classical models, where the model form and predictor variables are selected by the human analyst, machine learning models essentially can, within limits, select their own structures and the form of the relationships between variables that they model. That is, they can flexibly select cutpoints in continuous variables and/or choose combinations of variables and categories that are used together to predict an outcome.

In the context of prediction, machine-learning approaches generally outperform classical models on out-of-sample cross-validation accuracy. However, machine learning models are black boxes, where the relationships between predictors and the outcome are not apparent to the analyst. While extrapolation of any prediction model to cases outside of its training sample's distribution can produce errors, the behavior of machine learning models can be unexpected in extrapolation. Validation of any prediction model should include more detailed examination of its performance in subsets of the population rather than assuming that average performance (e.g., 90% accuracy on a decision task) will apply across the board.

### 2.2.2 Inference

In contrast to causal inference, *inference* is much more complex. A detailed discussion of causal inference is out of scope for this chapter, but a few key elements are worth introducing. First, there are two major approaches to causal inference: (1) counterfactuals [13] and (2) potential outcomes [14]. Both approaches aim to estimate the relationship between a causal factor and an outcome to answer questions such as "If a pedestrian crossing had been installed at this intersection, how many pedestrian crashes would have occurred?" The counterfactual approach identifies of the state of a causal factor that would be changed in a hypothetical version of reality (in the example, "intersection has pedestrian crossing" is the counterfactual). The potential outcomes approach focuses on the difference in outcomes between the two conditions. The existing condition (no pedestrian crossing) is in the dataset and the outcome under the alternative case (with a pedestrian crossing) is treated as missing data.

A key aspect of both approaches is that *observational data alone cannot prove a causal relationship.* Instead, certain assumptions must be brought to the analysis, often in the form of a causal diagram or model that defines how the researcher envisions the relationships between variables. This diagram is, in essence, an argument for a particular causal structure among variables, and that causal structure can be discussed and modified through the research process. The corresponding data analysis is then a means of estimating the magnitude of the relationships and thus the potential effect of implementing changes in practice.

Another key aspect of all causal inference from observational data is that the data themselves must support the inferences in the model. A key reason that the Randomized Controlled Trial (RCT) is considered the gold standard of causal inference is that the experimental design itself produces qualities in the data that are needed for causal inference. For example, random assignment of individuals to groups means that people in each group are *exchangeable* (i.e., that information on outcomes of members of Group A are informative for what would have happened to members of Group B if they had been in Group A (the counterfactual condition). Moreoever, all individuals have a non-zero probability of having been assigned to either group, a condition known as *positivity,* or *common support.* Thus, it is possible to look at the difference in outcomes between the two groups and make the inference that the difference in treatment between Group A and Group B has *caused* the difference in outcomes.

In observational data, such as nearly all Big Data, these assumptions are not automatically met by the sampling design. Instead, people typically self-sort into groups; for example, people who live in urban areas are more likely to use public transportation than people who live in rural areas. If we are interested in the causal effect of using public transportation as opposed to driving on individual's risk of injury, we cannot assume that people in cities are completely exchangeable with people in rural areas, even if we could imagine public transportation being in rural areas. Residents or rural areas are older, on average, than residents of urban areas, and as a result, they have a higher risk of injury in general if they are in a crash. Thus, age and transportation mode are confounded. Confounding in itself does not prevent us from making causal inferences, and is actually expected in observational datasets. However, in this example, if all rural residents always drive, then the data do not support inferences about public transportation risks for rural residents because the rural group has failed to meet the positivity requirement (i.e., there is no empirical information about their outcomes on public transportation). The only recourse, then, is to ignore the effect of rural/urban residence and assume that older adults in cities are exchangeable with older adults in rural areas in terms of their injury risk. Because older adults in the dataset can be in either treatment condition, we can estimate the difference in injury risk, conditioned on age and any other available variables with appropriate common support.

The challenge in ensuring positivity is especially significant when ML models are used for causal inference. Because ML models tend to represent finely-detailed interactions among variables, often as categories, some of those categories will not meet the positivity requirement. Using machine learning for causal inference is a new field and will be discussed more later in this chapter.

It is important to note that a great deal of transportation research involves causal inference. Many research questions revolve around how to make changes to meet some goal such as improving safety, reducing congestion, or shortening commute times. These are all questions that seek to identify countermeasures or factors that will cause a change in an outcome of interest.

### 2.2.3 Choosing Methods

A new area of methods work has been in combining machine learning and causal inference (e.g., [15]). In this approach, a machine learning model is developed to predict the outcome of interest. Then, counterfactual inputs are used to predict the outcome for each case as though it were in a different condition. For example, variables describing the characteristics of intersections might be used to predict pedestrian crash count at each intersection, such as the presence/absence of crosswalks. The training sample contains actual observations, but after the model is developed, the counts for all intersections without crosswalks can be predicted by changing their true input (crosswalk = 0) to the counterfactual input (crosswalk = 1) while keeping all other values the same. In principle, this approach can estimate the effect of changing something such as implementing a countermeasure. However, given the opacity of the ML model and the lack of an associated causal diagram, it is easy to create a counterfactual condition that would not exist. For example, the presence of a crosswalk may lead pedestrians to choose to cross at that intersection rather than elsewhere. If so, then the crosswalk will cause both an increase in pedestrians and potentially a reduction in their individual risk when crossing. On balance, the total number of pedestrian crashes could go up or down, even if crosswalks are shown to improve safety (causally). Changing the input to the ML model simulates the case where the crosswalk does not change pedestrian volumes, which is an intermediate effect that is not represented anywhere in the ML model but could (and generally would) be represented in a different type of model created by an analyst. However, failing to account for intermediate effects will lead to failure to correctly predict the effect of changing the crosswalk.

In general, the real strength of ML models is in their prediction capability. In choosing a modeling approach, the analyst should determine whether their research question is aimed at prediction or inference. If the goal of analysis is to learn something about countermeasures or the relationship between variables, then (causal) inference techniques are required and care should be taken to build a causal model, check data for support, and use methods that are appropriate. However, if the goal is to predict or label (e.g., a model that labels whether a driver is attentive based on their lane-keeping and speed-keeping), then the flexibility of ML methods is likely to produce better results.

## 3 Transportation Research Using Big Data

This section contains a set of general topics and examples of transportation research using Big Data and/or ML/AI approaches. The examples fall primarily into the categories of safety and travel behavior. The goal is to familiarize the reader with the kind of things that are being done with these cutting-edge datasets and methods.

With respect to safety, research in transportation is about understanding crashes and how to prevent them, prevent injury from them and/or respond to them. Historically, crash data have been the key data source that safety research was based on. Other datasets that provide context such as population, vehicle-miles traveled, travel behavior, may be used to understand the exposure of different road users, but crash datasets remain the key data sources for this research.

In general, crash data do not qualify as “Big Data” even though some crash databases are very large (e.g., state police-reported crash data over years). Information about each crash was hand-collected by someone (a police officer or crash investigator) and the data are not truly crowd-sourced or derived from sensors. However, crash data can serve as training data for a variety of ML models and ML approaches can enhance the interpretation of data present in crash datasets.

Because this chapter is intended to go beyond traditional crash data analysis and look at Big Data sources and the use of AI/ML methods, I focus on a small set of key safety use cases:

1. Use of AI/ML to add to information in transportation safety data
2. Use of sensors and AI/ML to obtain safety-relevant data at large scales
3. Use of AI/ML as a safety-research tool.

In contrast with safety data research, mobility, or travel-related research depends primarily on data that can shed light on where, when, why, and how people travel. The gold standard nationally representative travel dataset is the National Household Travel Survey (NHTS; [16]), which is collected approximately every seven years. The infrequent collection of these data is due to the high cost, so many researchers have been looking at other data sources for understanding travel. These sources and associated research are discussed in presentation of three travel-related use cases for ML and Big Data:

1. Travel demand estimation
2. Route choice
3. Understanding micromobility.

### *3.1 Use of AI/ML to Add to Information in Large Transportation Datasets*

As described earlier, a limitation of traditional crash datasets is that they require the crash to have occurred before data are collected. Vehicle-based sensor data through NDSs and FOTs provide a new opportunity to learn about the precursors to crashes by observing kinematics and driver behavior while driving and just prior to crashing. In addition, police reports include diagrams and narratives that typically include the officer's assessment of how the crash arose (albeit inferred after the fact). These data sources can substantially improve our understanding of crash indicators and causal factors, but they both require substantial work to annotate and interpret.

One use of AI in safety research has been to add information to crash or driving datasets in a scalable way by using prediction models to replace some human labor in the data-enhancement process. For example, text mining has been used in a few applications to interpret police-report narratives. For example, [17] used natural language processing (NLP) to interpret crash narratives. Their vocabulary and association rules were based on a training dataset of 10,000 crashes in the UK. The development of the domain-specific grammar rules and vocabulary required some substantial human work, but the resulting knowledge-representation model could be used to parse narratives into a common machine-interpretable structure from which information could be extracted per the needs of a research question.

Text mining approaches have also been used for more specific goals that do not require building a complete grammar and association rules. For example, NLP-based classifiers were used to identify work-zone crashes [18]; random forest, support vector machines, and logistic regression have been used on Kentucky police-report narratives to identify secondary crashes that were not coded on the forms [19] and to identify causes of highway-grade rail crossing crashes [20]; finally, topic models were used with California AV crash reports to identify five themes, or clusters of AV crash characteristics [21].

Another area where ML approaches are being used to extract additional information from transportation safety datasets has been in annotating and identifying events in NDS data. While horizontal (i.e., highly detailed) NDS data provide a unique view of driver behavior and crash precursors, key information about driver involvement in secondary tasks and eye-gaze location or fatigue is based on in-cabin camera views that must be annotated. The Federal Highway Administration (FHWA) Exploratory Advanced Research (EAR) program has funded a series of studies developing computer vision algorithms for use with SHRP2 data [22]. Approaches include CNNs designed to extract information about features or objects present in forward and face views as well as automated masking of identity to better enable sharing of face views without violating privacy requirements.

Similarly, neural networks have been used with street scene photographs (e.g., from Google® Street View or other available sources) to identify elements of the roadway infrastructure that are not available in datasets. For example, Campbell et al. [23] use deep learning to extract street signs from street images, and Ning et al. [24] used a similar method to identify sidewalks from images. These roadway features can be used for many research questions related to safety (e.g., linking roadway features to crash data) and travel (e.g., predicting pedestrian routing as a function of sidewalk locations).

Computer vision algorithms have the potential to substantially increase the usability of NDS and roadway asset datasets, but their accuracy needs to be much higher than is currently being achieved to ensure that errors in labels don't affect research findings. That said, the same algorithms can be used effectively to find potential events of interest for review by human video coders. For rare events such as crashes or conflicts with pedestrians, this can save substantial amounts of time that would be spent viewing irrelevant video or looking at thousands of street view images. This type of machine-assisted labeling has significant potential to improve the use of large transportation datasets.

### *3.2 Use of Sensors and Algorithms to Obtain Safety-Relevant Data at Large Scales*

The primary use of ML techniques to obtain safety-relevant data on a large scale is to define triggers for data collection from sensors that are widespread. As described earlier, triggered (vertical) datasets are used when data must be sent over the air to be stored and data volumes are too large for it to be practical to obtain all data that way.

Newer vehicles, in particular, are equipped with production sensing systems that can be used for data capture as well. However, onboard storage and bus capacity are limited, so stored data must be very small in size, and the data triggering system must be designed to capture only the most important information. For example, Event Data Recorders (EDRs) capture a variety of data elements from a few seconds before a crash to just after (e.g., 500 ms). EDR trigger criteria and minimum data elements are regulated [25], but many manufacturers capture additional information, such as pre-crash steering, passenger belt status, and even occupant size beyond that required (e.g., [26]). EDR data capture is triggered by a sustained deceleration pulse and/or airbag deployment. The acceleration pulse trigger is designed to exclude false positives such as those caused by potholes and other roadway-based acceleration signals, while still capturing crashes with appreciable injury risk. The same type of data capture approach used for EDRs can also be used to gather large-scale data on advanced driver assistance systems in the field (e.g., [6]).

While EDR contents and triggers are pre-defined in regulation, other data are captured by a variety of companies using triggers determined by ML algorithms. The most well-known such use is dongle-based and smartphone-based data capture systems used by insurance companies to adjust rates. Tselentis and colleagues [27] reviews a variety of such systems and the data behind them. The relevant data elements can be divided into usage-based data and driver-behavior data. The former includes total miles or time, number of trips, time of day, and trip location. The latter includes acceleration, braking, and speeding, among others. Tselentis and colleagues [27] summarized the approaches to determining triggers as well as models determining insurance rates. Most of the approaches to date use linear models, but some have explored non-linear models including neural networks and decision trees e.g., [28].

Finally, driver monitoring systems with in-vehicle and forward cameras plus limited kinematic sensing also trigger on crash-related kinematics, capturing and saving only a small video clip that include a few seconds prior to the event. Notably, Lytx, for example, uses ML models to refine its triggering algorithm for its driver monitoring system [29]. Lytx's commercial purpose is to identify crashes and limit the cost of review time by their staff. For research purposes, most trigger algorithms to date have been constructed by hand (e.g., [5, 6]). However, ML models have the potential to influence future research-level triggering in addition to commercial purposes.

For all of these high-volume data-collection approaches, a key element is that the triggering algorithm must be pre-determined and designed for the purpose that the system serves. This creates a sort of chicken-and-egg problem where complete data are needed to develop the triggering algorithm so that large-scale data can be obtained. NDS data can be useful for such algorithm development, especially to provide training data for ML algorithms.

## 3.3 *Machine Learning and Artificial Intelligence in Safety Research*

### 3.3.1 Naturalistic Driving Study Data

NDS data is useful for understanding an array of aspects human driver behavior, including behaviors related to safety. To investigate safety-related questions, unsafe events must be identified in the data. Many crashes are very minor and do not result in the abrupt end of a trip or even very significant accelerations, so these crashes are found using an algorithm followed by human video review. That algorithm is generally a set of kinematic thresholds such as hard deceleration or swerving, but many events that exceed the threshold are not crashes. Thus, video review by human coders is essential to finding these events for later use.

In addition, crashes are rare events and even in a very large NDS such as SHRP2, there are relatively few of them to support analyses. One approach to addressing this limitation is to use crash *surrogates*, which are events that are similar to crashes (e.g., near-crashes) but are not actually crashes. A good surrogate is influenced by the same factors as crashes are and can predict crash counts in an area. This notion, which was inspired by Heinrich's Triangle [30] and developed into the Traffic Conflict Technique for traffic safety by Hyden [31], assumes that there is a constant ratio between more common, less severe surrogates and crashes (or more severe crashes).

Work by Wu and Jovanis ([32–34]) and Tarko [35], among others, shows that surrogates do not work equally well for all crash types and that there is not a constant ratio between near crashes and crashes in all locations and situations, for example. Instead, they recommend focusing on a particular crash mechanism when using surrogates. For example, Tarko [35] proposes measuring maxima of longitudinal time to collision (TTC) over periods of driving and employing extreme value analysis (EVA) to estimate the rate of rear-end striking crashes. This approach uses TTC as a surrogate to enable estimation of one particular type of crash.

While surrogates have been shown to be influenced by variables that also influence crash rates, their use results in effect estimates that are biased towards the null [36]. That is, surrogates appear to partially reflect crash mechanisms, but dampen the size of effects, presumably because they add noise to analyses that include them.

Finally, because NDS are essentially observations of driver behavior, much of the research using NDS has focused on behavioral risk factors. Attention and secondary tasks have received a lot of attention in the literature ([37, 38]). However, NDS have also been used to develop driver state prediction models based on kinematics. These models use machine learning techniques and/or computer vision techniques (with face video) to predict when a driver is distracted (e.g., [39]) or fatigued (e.g., [16]) based on driving performance data such as lane-keeping or speed-keeping.

### 3.3.2 Triggered Data Modeling

In contrast with instrumented NDS data, most triggered (vertical) large-scale data does not include video. The exception to this is data from commercial driver-monitoring systems that captures clips from in-cab cameras when certain kinematic triggers are met (e.g., used in [40]). Crucially, without video, algorithms and corresponding inferences need to be made with limited or non-existent ground truth. Moreover, triggering itself imposes a filter on what enters the research dataset, so care must be taken to understand the context from which the triggered data comes.

The large-scale FCW/LDW study [5] described earlier in this chapter obtained data from almost 2000 drivers across the U.S. who produced over 250,000 FCW events and over 10 million LDW events. Data collection also included contextual information such as total driving time and miles, as well as histograms of speeds and headways to lead vehicles. The limited context data is crucial because ADAS often work only at certain speeds and, for example, FCW is irrelevant if there is no lead vehicle.

One of the key results of that study categorized FCW events into scenarios that were defined on the basis of very limited kinematic data. Kinematic variables were captured at three timepoints around FCW warning events—3–6 s before, at the event, and four seconds after. Braking onset (if any) was also captured between the at-event and post-event timepoints. Scenarios are essentially clusters of events from which inferences can be made, but without video, validation is not possible. The notable finding of this analysis was that over 80% of FCW events occur in situations that can resolve themselves without braking, whereas 19% occur in lead-vehicle decelerating scenarios and <1% in lead-vehicle stopped scenarios. In this study, scenarios were defined by the researchers, but in principle, clustering methods could also be used on such data. The key is that methodologically, useful machine-learning approaches would generally be unsupervised clustering methods rather than supervising learning approaches (because there is no ground truth to learn). This is true for most datasets without video, which would be a typical source of ground-truth data for NDS.

As described in the previous section, another key source of triggered data comes from EDRs. Smaller-scale EDR data with context available (e.g. in the Crashworthiness Data System (CDS) database) has been used to validate and/or supplement detailed crash investigation data [41], estimate pre-crash time-to-collision in intersection crashes [42], and understand pre-crash braking behavior in rear-end crashes [43]. However, these studies benefit from the associated (expensive) detailed in-depth crash investigation. EDR supplements the information with key measurements but is not being used in the absence of knowledge of, for example, the crash configuration. In contrast, use of EDR alone could be done on a large scale, but so far, few published studies have done so. Once exception is [26] which describes distributions of crash severity as a function of speed prior to the crash and crash direction.

Triggered data from driver-monitoring systems with video provide an interesting intermediate approach between detailed instrumented-vehicle NDS and the triggered data collection approach described above. The challenge with data from such sources is that they seldom include baseline data because all of the events meet the triggering conditions. In general, this type of source has been used primarily to identify crashes and review pre-crash conditions in the video and kinematic data around those crashes (e.g., [44, 45]). That said, a challenge of using commercial data in a research setting is that because both the triggering method and the context are unknown, it is difficult to assess representativeness. As will be discussed later in this chapter, this is an important issue for research using Big Data.

### 3.4 Travel Demand Estimation

As described earlier in the chapter, travel-data sources include Big Data such as cell phone location traces, and in the future, BSMs that include location traces from CV2X data. Over-the-air vehicle-based data collection and instrumented-vehicle-based data can also include location traces. While location traces are in some ways very simple, they can be used to infer a variety of additional pieces of information, such as travel mode, travel speeds, and origins and destinations.

A number of studies have used cell-phone-based data to estimate travel demand in localized areas (e.g., Senegal [46]; Israel [47]). Travel demand has also been estimated by mode, by using patterns in each cell-phone record to infer travel mode. Mode inference is most commonly based on GPS traces (e.g., [48, 49]), but has also been inferred based on coarse information from simple call data that provides only total travel time and original and destination [50]. ML approaches are useful when trip-mode ground truth is present, because they can represent complex interactions in predicting mode.

Travel demand for specific locations is crucial for planning applications. However, it is also the observed outcome in activity-based models of travel choices. For example, Clifton and colleagues have developed models to estimate pedestrian activity and travel choices ([51, 52]). Their work is based on a number of data sources including travel surveys that help identify choices people make to walk and characteristics of locations that make them more or less attractive as trip destinations and/or routes.

### 3.5 Route Choice Models

Routing is another research topic that benefits from Big Data on travel choices. Routing has been investigated both as an observed behavior and to develop algorithms to aid travelers in choosing routes that conform to certain criteria. Chen et al. [53] reviewed the literature on travel behavior with particular emphasis on the differences between big and small data sources. They describe the use of Big Data for a number of uses including inferring activity locations (i.e., endpoints of travel), inferring activity purpose (based on the nature of the location), inferring mode and route choices, and deriving origin–destination matrices that summarize travel patterns in an area. The use of Big Data for these purposes is still being developed, with areas of research needed in validating the inferences made and in handling data that are not representative of the population. However, these data sources show great promise for travel behavior research.

Advisory routing algorithms often seek to optimize certain route characteristics such as travel time and route preferences (e.g., avoiding tolls). However, the quality of travel time estimates depends on the ability of the algorithm to accurately predict speeds on road segments at particular times of day and under particular conditions. Park et al. [54] used a machine learning model trained on traffic sensor data from California to predict travel speeds as a function of time of day, day of the week, and speeds on neighboring road segments. Using NDS data from two drivers, Dai et al. [55] developed a machine-learning algorithm to estimate those specific drivers' route preference at each link as a function of past route choices between common origin–destination pairs. This produced a personalized route choice model that could, in principle, learn patterns of any individual driver over time.

### 3.5.1 Micromobility

Sensors and smartphone apps are enabling technology that have led to the explosion of a variety of shared small-vehicle transportation options, collectively known as "micromobility." These include shared bicycles, e-bikes, e-scooters, and mopeds, among others.

In 2021, Abduljabbar and colleagues [56] published a systematic literature review of research related to micromobility and its effect on city transportation, especially as related to sustainability. They identified four clusters of research that they called Benefits, Policy, Technology, and Determinants of Usage. This robust research area relied primarily on publicly available data provided by micromobility companies and simulation studies. Typically, publicly available data was in the form of origin–destination pairs rather than complete trajectories.

One of the key findings of work on micromobility was that these modes can create a net *increase* in energy use because of the resources required for manufacturing, maintenance and redistribution [57]. Moreover, the modes tend to replace public transportation rather than use of private vehicles. However, the potential to decrease energy use is there, but requires certain effort on the part of policy-makers and mobility companies to reduce resource use and encourage travelers to replace private-vehicle use.

## 4 Pros and Cons of Big Data and ML/AI

While Big Data in combination with Machine Learning and Artificial Intelligence show great potential for transportation research, they are tools that have advantages and disadvantages. The advantages, demonstrated above, include the ability to model complex interactions in very large samples of travel. In addition, prediction performance of ML methods nearly always outstrips that of traditional parametric models. That said, two major challenges to using these tools are (1) non-representativeness of Big Data, and (2) opaqueness, or "black box" qualities of ML methods.

## *4.1 Big Data Challenges*

### 4.1.1 Sample Characteristics

A significant challenge for using Big Data is to understand sample representation. In 2016, [58] reviewed a number of issues related to Big Data use and pointed out two key factors. First, Big Data are convenience samples, which have an unknown relationship to the population of interest. Second, Big Data is often provided by private companies whose interest is not in producing a carefully constructed sample for public use. Even when data are made available, they reflect the behavior of the companies' customers and the algorithm used to produce the data.

These issues lead to what is known as the Big Data paradox [59] where increasing the size of a poor-quality sample can lead to results that are *increasingly* biased relative to the true population statistics. This has also been phrased as estimates that are "precisely inaccurate" [60].

In transportation, key Big Data sources include smartphone-based travel data, which reflect the sample of smartphone users, typically those with contracts with large companies; instrumented vehicle data in NDS, which uses convenience samples of drivers in localized areas; over-the-air vehicle data from a specific manufacturer, reflecting that company's sales demographics; and infrastructure-based sensors typically installed at high-traffic-volume locations that emphasize vehicle traffic, such that motorized-vehicle travelers who live nearby tend to be overrepresented.

A related challenge in the area of travel behavior is that only people who travel are included in sensor-based datasets. This means that unmet demand is never found in such datasets, even if people in areas with low vehicle ownership and fewer transportation options have smartphones. Travel that is needed but not undertaken because of lack of resources can only be measured using surveys (though notably, the NHTS also only measures travel actually taken as of this writing). The combination of the failure of Big Data sources to represent those less likely to have smartphones (e.g., elderly and low-income travelers) and the assumption that *observed* travel represents *needed* travel doubly disadvantages certain subgroups of travelers in research and in planning applications that use the same data sources.

Addressing issues of representation in Big Data requires there to be carefully constructed probability samples or complete administrative databases (e.g., U.S. Census or state police crash databases) that allow for adjustment of Big Data via weights or other methods. A significant body of work in survey methods is aimed at developing methods to achieve the goal of addressing the representation issue in Big Data in transportation and other fields (e.g., [60–62]).

What this means, however, is that surveys like NHTS are more important than ever. Such surveys can be designed to maximize their value in combining with Big Data, and they can even include sensor-based data collection (e.g., travel surveys where participants wear location-data collection devices to track their travel in addition to answering questions about those trips). However, Big Data cannot replace such surveys without leaving researchers in the dark about how estimates may be biased for the population as a whole.

### 4.1.2 Algorithm Testing

As discussed above, developing a good prediction model on an unrepresentative dataset, and then applying it to subgroups who were not represented in the original data (e.g., people without smartphones) potentially further exacerbates the failures of representation in the original data source. It is unlikely that people who are not represented in Big Data are missing at random. Instead, these are typically groups of people who do not have the same resources as those who are counted. Because Big Data are less expensive to collect, it can be tempting to ignore these issues and hope that the large sample size and prediction performance will overcome these disadvantages. However, a good algorithm evaluation strategy can help to mitigate issues introduced into research (or application) by the algorithm itself.

The most common approach to evaluating algorithms is to hold out a randomly sampled subset of the original dataset and test the algorithm's performance on that. Accuracy is measured as the number of correct predictions divided by the total number of samples. However, because the validation sample is a random subset of the original sample, it will have the same characteristics as the training sample. Thus, it will tend to overrepresent the same groups as in the training sample and thus overrepresent performance on these groups in accuracy assessment. To address this, algorithm performance testing needs to go beyond the standard accuracy measure.

In recent years, there has been substantially more attention paid to "fairness" in AI, including definitions of fairness as well as solutions to improve fairness in algorithms (e.g., [63, 64]). One element of this is that not all definitions of fairness can necessarily be met simultaneously. Thus, for each AI algorithm, its use needs to be considered in determining appropriate evaluation metrics.

Notably, it is important to assess whether the output of an algorithm is used to provide benefits or punishments. In the transportation case, most AI is likely to identify where to provide benefit—e.g., repaving a road, adding crosswalks, adding transit stops. Thus, evaluating algorithms to be sure that the needed benefits are equally likely to be identified (under the same conditions) for different subgroups of travelers might be critical. Further details can be found in [64] and implementation code can be found in [65].

### 4.1.3 Opaque Algorithms

Another challenge in using ML approaches is that their "black box" nature makes inference, especially causal inference, very difficult. As described earlier, causal inference relies on external assumptions, including a model of how variables are related to each other causally. In standard practice (e.g., [66]), a causal diagram informs the nature of the analysis, such as which variables are included or excluded. Moreover, the form of the relationship between two variables is pre-specified and transparent in traditional approaches.

That said, causal inference using ML is a rising field, led by Judea Pearl (accessible summary here: [67]). Pearl argues that by appropriately tying ML techniques to the causal inference framework that emphasizes these external causal models and assessment of assumptions, the challenges brought by opaqueness of ML approaches can be overcome.

### 4.1.4 Protecting Privacy

A significant challenge in the use of Big Data in transportation is privacy protection. NDS and other traditional studies make use of consent forms and Institutional Review Board (IRB) requirements for ethical treatment of participants' data. However, sensor-based Big Data is often collected without participants' awareness and generally contains personally identifying information (PII).

In NDS, PII is generally in the form of face video. Measures to protect PII often make it very difficult to for researchers to access face video, which is key to understanding driver behavior. One possible approach to serving both privacy goals and access goals is to use computer vision (CV) algorithms to label driver face video in place of human annotators. FHWA's support of work in this field is aimed at maintaining privacy protection while enabling the use of more data. To the extent that CV can accurately mirror human annotations, this approach shows promise.

A more common form of PII in Big Data in transportation is travel traces. Smartphone-based and dongle-based data collect a location "bread-crumb trail" in the form of a series of GPS points following a driver's path. Since drivers generally repeat certain trips and end their travel day at home, it is relatively easy to re-identify them based on their most common final location. A number of methods have been developed to mask geographical data to preserve privacy [68]. For travel data, these might include removal of the starts and ends of all trips (ideally with a random component on the length of the removed portion) or sharing only summary statistics, such as trip length, trip distance, hard-braking events, and other key elements of travel (excluding the specific location). ML approaches, such as Generalized Adversarial Networks (GANs), have also been used for this purpose [69].

A key aspect of geo-privacy protection measures is that their use trades off with the accuracy of analytic results. For example, [68] evaluates a number of approaches on their relative ability to prevent re-identification and their effect on accuracy of analysis. This is true for the use of CV algorithms to replace human annotators as well. In general, the tolerance of different research topics to inaccuracy in data due to privacy protection varies. The more options can preserved, the better the overall value for research. That is, approaching privacy protection and data sharing with a tiered approach will produce better research results, such that some well-de-identified data can be shared easily but re-identifiable data is shared using more secure methods such as in data enclaves.

### 4.2 Going Forward

Improvements in sensor and computing technology have enabled an explosion of Big Data sources in transportation. Moreover, developments in Machine Learning and Artificial Intelligence methods have made prediction models substantially more flexible and accurate. The combination of the two has great potential to enable researchers to answer questions that could not previously be answered.

However, with great power comes great responsibility. Big Data sources are often convenience samples that will fail to represent subgroups of people who are not measured. Proper sample surveys (possibly using sensor-based measurement techniques) are even more crucial than ever to ensure that Big Data insights apply to everyone.

Similarly, ML models applied without care can produce impressive prediction performance that will fail on cases it hasn't seen (e.g., subgroups not included a in Big Data sample) and that are opaque to researchers. In particular, it is easy to rely on measures such as variable "importance" to make unjustified causal inferences (i.e., if a variable is a key predictor of an outcome, it must therefore be causally related). Instead, causal inference requires careful external construction of models of how variables are related to each other (causally) in conjunction with ML or traditional parametric models to estimate causal effects.

This chapter was aimed at introducing the reader to some key sources of Big Data in transportation, as well as some key methods and issues in the use of ML/AI and Big Data in transportation. There are extensive resources available for more detail on any particular topic and the reader is encouraged to take a deeper dive.

## References

1. Kahane CJ (2015) Lives saved by vehicle safety technologies and associated Federal Motor Vehicle Safety Standards, 1960 to 2012—passenger cars and LTVs—with reviews of 26 FMVSS and the effectiveness of their associated safety technologies in reducing fatalities, injuries, and crashes. (Report No. DOT HS 812 069). National Highway Traffic Safety Administration, Washington, DC
2. Ervin R, Sayer J, LeBlanc D, Bogard S, Mefford M, Hagan M, Winkler C (2005) Automotive collision avoidance system field operational test report: methodology and results (No. HS-809 900)

3. Nodine E, Lam A, Stevens S, Razo M, Najm W (2011) Integrated vehicle-based safety systems (IVBSS) light vehicle field operational test independent evaluation (No. DOT-VNTSC-NHTSA-11-02). National Highway Traffic Safety Administration, United States
4. Antin JF, Lee S, Perez MA, Dingus TA, Hankey JM, Brach A (2019) Second strategic highway research program naturalistic driving study methods. Saf Sci 119:2–10
5. Flannagan CA, LeBlanc DJ, Kiefer RJ, Bogard SE, Leslie A, Zagorski CT, Beck CS (2018) Field study of light-vehicle crash avoidance systems: automatic emergency braking and dynamic brake support (No. DOT HS 812 615). Department of Transportation. National Highway Traffic Safety Administration, United States
6. Flannagan C, LeBlanc D, Bogard S, Nobukawa K, Narayanaswamy P, Leslie A, Kiefer R, Marchione M, Beck C, Lobes K (2016) Large-scale field test of forward collision alert and lane departure warning systems. Office of Advanced Safety Research, Washington, D.C. Report No. DOT HS 812 247
7. Bergasa LM, Almería D, Almazán J, Yebes JJ, Arroyo R (2014) Drivesafe: an app for alerting inattentive drivers and scoring driving behaviors. In: 2014 IEEE intelligent vehicles symposium proceedings. IEEE, pp 240–245
8. Li G, Eby DW, Santos R, Mielenz TJ, Molnar LJ, Strogatz D, Andrews HF (2017) Longitudinal research on aging drivers (LongROAD): study design and methods. Inj epidemiol 4(1):1–16
9. Nodine E, Stevens S, Lam A, Jackson C, Najm WG (2015) Independent evaluation of light-vehicle safety applications based on vehicle-to-vehicle communications used in the 2012–2013 safety pilot model deployment (No. DOT HS 812 222). National Highway Traffic Safety Administration, United States
10. Bogard SE, Bao S, LeBlanc D, Li J, Qiu S, Liu B (2017) Performance of DSRC during safety pilot model deployment. SAE Int J Passeng Cars-Electron Electr Syst 10:165–172, (2017-01-0077)
11. He Z, Qin X, Liu P, Sayed MA (2018) Assessing surrogate safety measures using a safety pilot model deployment dataset. Transp Res Rec 2672(38):1–11
12. Zheng J, Liu HX (2017) Estimating traffic volumes for signalized intersections using connected vehicle data. Transp Res Part C: Emerg Technol 79:347–362
13. Pearl J (2000) Causality: models, reasoning, and inference, 2nd edn. Cambridge University Press, New York, 2009
14. Rubin D (2005) Causal inference using potential outcomes: design, modeling, decisions. J Am Stat Assoc 100:322–331
15. Blakely T, Lynch J, Simons K, Bentley R, Rose S (2020) Reflection on modern methods: when worlds collide—prediction, machine learning and causal inference. Int J Epidemiol 49(6):2058–2064
16. Rani TP (2021) Smart surveillance of driver using machine learning. In: 2021 3rd International Conference on Signal Processing and Communication (ICPSC). IEEE, pp 85–88
17. Wu J, Heydecker BG (1998) Natural language understanding in road accident data analysis. Adv Eng Softw 29(7–9):599–610
18. Sayed MA, Qin X, Kate RJ, Anisuzzaman DM, Yu Z (2021) Identification and analysis of misclassified work-zone crashes using text mining techniques. Accid Anal Prev 159:106211
19. Zhang X, Green E, Chen M, Souleyrette RR (2020) Identifying secondary crashes using text mining techniques. J Transp Saf Secur 12(10):1338–1358
20. Soleimani S, Mohammadi A, Chen J, Leitner M (2019) Mining the highway-rail grade crossing crash data: a text mining approach. In: 2019 18th IEEE International conference on machine learning and applications (ICMLA). IEEE, pp 1063–1068
21. Alambeigi H, McDonald AD, Tankasala SR (2020) Crash themes in automated vehicles: a topic modeling analysis of the California Department of Motor Vehicles automated vehicle crash database. arXiv preprint arXiv:2001.11087
22. Federal Highway Administration (2015) Exploratory advanced research program video analytics research projects. FHWA-HRT-15-025
23. Campbell A, Both A, Sun QC (2019) Detecting and mapping traffic signs from Google Street View images using deep learning and GIS. Comput Environ Urban Syst 77:101350

24. Ning H, Ye X, Chen Z, Liu T, Cao T (2021) Sidewalk extraction using aerial and street view images. Environ Plann B: Urban Anal City Sci, 2399808321995817
25. United States Code of Federal Regulations, 49 CFR Part563 "Event Data Recorders"
26. Iyoda M, Trisdale T, Sherony R, Mikat D, Rose W (2016) Event data recorder (EDR) developed by Toyota Motor Corporation. SAE Int J Transp Saf 4(1):187–201
27. Tselentis DI, Yannis G, Vlahogianni EI (2017) Innovative motor insurance schemes: a review of current practices and emerging challenges. Accid Anal Prev 98:139–148
28. Paefgen J, Staake T, Thiesse F (2013) Evaluation and aggregation of pay-as-you-drive insurance rate factors: a classification analysis approach. Decis Support Syst 56:192–201
29. https://www.lytx.com/en-us/about-us/our-technology. Accessed 22 July 2021
30. Heinrich HW (1931) Industrial accident prevention: a scientific approach. McGraw-Hill, New York, NY
31. Hyden C, Linderholm L (1984) The Swedish traffic-conflicts technique. In: International calibration study of traffic conflict techniques. Springer, Berlin, Heidelberg, pp 133–139
32. Wu KF, Jovanis PP (2012) Crashes and crash-surrogate events: exploratory modeling with naturalistic driving data. Accid Anal Prev 45:507–516
33. Wu KF, Jovanis PP (2013) Defining and screening crash surrogate events using naturalistic driving data. Accid Anal Prev 61:10–22
34. Wu KF, Aguero-Valverde J, Jovanis PP (2014) Using naturalistic driving data to explore the association between traffic safety-related events and crash risk at driver level. Accid Anal Prev 72:210–218
35. Tarko AP (2018) Surrogate measures of safety. In: Safe mobility: challenges, methodology and solutions. Emerald Publishing Limited
36. Guo F, Klauer SG, Hankey JM, Dingus TA (2010) Near crashes as crash surrogate for naturalistic driving studies. Transp Res Rec 2147(1):66–74
37. Klauer SG, Guo F, Simons-Morton BG, Ouimet MC, Lee SE, Dingus TA (2014) Distracted driving and risk of road crashes among novice and experienced drivers. N Engl J Med 370(1):54–59
38. Dozza M, Flannagan CA, Sayer JR (2015) Real-world effects of using a phone while driving on lateral and longitudinal control of vehicles. J Saf Res 55:81–87
39. McDonald AD, Ferris TK, Wiener TA (2020) Classification of driver distraction: a comprehensive analysis of feature generation, machine learning, and input measures. Hum Factors 62(6):1019–1035
40. Pipkorn L, Piccinini GB (2020) The role of off-path glances: a quantitative analysis of rear-end conflicts involving Chinese professional truck drivers as the striking partners. J Saf Res 72:259–266
41. Gabauer DJ, Gabler HC (2006) Comparison of delta-v and occupant impact velocity crash severity metrics using event data recorders. In: Annual proceedings/association for the advancement of automotive medicine, vol 50. Association for the Advancement of Automotive Medicine, p 57
42. Kusano KD, Gabler H (2011) Method for estimating time to collision at braking in real-world, lead vehicle stopped rear-end crashes for use in pre-crash system design. SAE Int J Passeng Cars-Mech Syst 4:435–443, (2011-01-0576)
43. Gabler HC, Hinch J (2009) Feasibility of using event data recorders to characterize the pre-crash behavior of drivers in rear-end collisions. In: Proceedings of the 21st international conference on the enhanced safety of vehicles, Stuttgart, Germany
44. Bärgman J (2016) Methods for analysis of naturalistic driving data in driver behavior research. Chalmers Tekniska Hogskola (Sweden)
45. Carney C, Harland KK, McGehee DV (2018) Examining teen driver crashes and the prevalence of distraction: recent trends, 2007–2015. J Saf Res 64:21–27
46. Chajka-Cadin L, Petrella M, Timmel C, Futcher E, Mittleman J (2017) Federal highway administration research and technology national household travel survey program (No. FHWA-HRT-16-082). Federal Highway Administration. Office of Research, Development, and Technology

47. Demissie MG, Phithakkitnukoon S, Sukhvibul T, Antunes F, Gomes R, Bento C (2016) Inferring passenger travel demand to improve urban mobility in developing countries using cell phone data: a case study of Senegal. IEEE Trans Intell Transp Syst 17(9):2466–2478. https://doi.org/10.1109/TITS.2016.2521830
48. Bekhor S, Shem-Tov IB (2015) Investigation of travel patterns using passive cellular phone data. J Location Based Serv 9(2):93–112
49. Zheng Y, Liu L, Wang L et al (2008) Learning transportation mode from raw GPS data for geographic applications on the web. In: Proceedings of the 17th international conference on World Wide Web, pp 247–256
50. Zheng Y, Chen Y, Li Q et al (2010) Understanding transportation modes based on GPS data for web applications. ACM Trans Web 4(1)
51. Clifton KJ, Singleton PA, Muhs CD, Schneider RJ (2016) Representing pedestrian activity in travel demand models: framework and application. J Transp Geogr 52:111–122
52. Clifton KJ, Singleton PA, Muhs CD, Schneider RJ (2016) Development of destination choice models for pedestrian travel. Transp Res Part A: Policy Pract 94:255–265
53. Chen C, Ma J, Susilo Y, Liu Y, Wang M (2016) The promises of big data and small data for travel behavior (aka human mobility) analysis. Transp Res Part C: Emerg Technol 68:285–299
54. Park J, Murphey YL, McGee R, Kristinsson JG, Kuang ML, Phillips AM (2014) Intelligent trip modeling for the prediction of an origin–destination traveling speed profile. IEEE Trans Intell Transp Syst 15(3):1039–1053
55. Dai Y, Ma Y, Wang Q, Murphey YL, Qiu S, Kristinsson J, Feldkamp T (2016) Dynamic prediction of drivers' personal routes through machine learning. In: 2016 IEEE symposium series on computational intelligence (SSCI). IEEE, pp 1–8
56. Abduljabbar RL, Liyanage S, Dia H (2021) The role of micro-mobility in shaping sustainable cities: a systematic literature review. Transp Res Part D: Transp Environ 92:102734
57. Liu J, Li J, Li W, Wu J (2016) Rethinking big data: a review on the data quality and usage issues. ISPRS J Photogramm Remote Sens 115:134–142
58. Meng XL (2018) Statistical paradises and paradoxes in big data (I): law of large populations, big data paradox, and the 2016 US presidential election. Ann Appl Stat 12(2):685–726
59. McFarland DA, McFarland HR (2015) Big Data and the danger of being precisely inaccurate. Big Data & Soc 2015:1–4. https://doi.org/10.1177/2053951715602495 bds.sagepub.com
60. Rafei A, Flannagan CA, Elliott MR (2020) Big Data for finite population inference: applying Quasi-random approaches to naturalistic driving data using bayesian additive regression trees. J Surv Stat Methodol 8(1):148–180. https://doi.org/10.1093/jssam/smz060
61. Rafei* A, Flannagan C, Elliott MR (2019) Big Data for finite population inference: calibrating pseudo-weights. In: ITACOSM 2019-survey and data science
62. Elliott MR, Alexa Resler AJ, Flannagan CA, Rupp JD (2009) Appropriate analysis of CIREN data: using NASS-CDS to reduce Bias in estimation of injury risk factors in passenger vehicle crashes. Accid Anal Prev 42(2):530–539
63. Saleiro P, Rodolfa KT, Ghani R (2020) Dealing with bias and fairness in data science systems: a practical hands-on tutorial. In: Proceedings of the 26th ACM SIGKDD international conference on knowledge discovery & data mining, pp 3513–3514
64. Saleiro P, Kuester B, Hinkson L, London J, Stevens A, Anisfeld A, Ghani R (2018) Aequitas: a bias and fairness audit toolkit. arXiv preprint arXiv:1811.05577
65. Bellamy RK, Dey K, Hind M, Hoffman SC, Houde S, Kannan K, Zhang Y (2019) AI fairness 360: an extensible toolkit for detecting and mitigating algorithmic bias. IBM J Res Dev 63(4/5):4-1
66. Hernán MA, Robins JM (2010) Causal inference
67. Pearl J (2019) The seven tools of causal inference, with reflections on machine learning. Commun ACM 62(3):54–60

68. Kwan MP, Casas I, Schmitz B (2004) Protection of geoprivacy and accuracy of spatial information: how effective are geographical masks? Cartographica: The Int J Geogr Inf Geovisualization 39(2):15–28
69. Liu X, Chen H, Andris C (2018) trajGANs: using generative adversarial networks for geo-privacy protection of trajectory data (vision paper). In: Location privacy and security workshop, pp 1–7

# Machine Learning for Automotive Cybersecurity: Challenges, Opportunities and Future Directions

**Rafi Ud Daula Refat, Abdulrahman Abu Elkhail, and Hafiz Malik**

**Abstract** Connected autonomous vehicles (CAVs) hold the promise of not only improving functional safety but also improving mobility and the efficiency of transportation systems. CAVs can be viewed as a cyber-physical system that contains a large number of minicomputers called electronic control units (ECUs). In order for ECU subsystems to share information and operate efficiently, they are typically networked via various in-vehicle networks (IVNs). Such IVNs include the controller area network (CAN), local interconnected network (LIN), media-oriented system transport (MOST), FlexRay and automotive Ethernet. These IVNs are used to connect safety-critical and non-critical components of the vehicle, including brakes, airbags, engine control, active safety devices, the electronic stability program and adaptive cruise control. Although these IVNs provide some luxury functions and improve the functional safety of the vehicles, the use of in-vehicle communication networks can pose serious security threats to CAVs. Several incidents have been reported showing that intruders are able to access vehicle information, even for safety critical tasks. As the IVNs architecturally are not designed to defend against these attacks, additional methods are needed for security. In recent years researchers are taking advantage of advances in more powerful computing hardware, as well the availability of huge amounts of network data and proposing machine learning-based frameworks to secure these IVNs. To the best of our knowledge, these frameworks lack details such as how to apply machine learning for IVN security. Most of them are focused on the selection of machine learning algorithms to improve attack detection rates. As a result, these frameworks become uninterpreted since they took a lot of time in order to reproduce their result. An efficient successful machine learning system depends not only on the selected machine learning algorithm but also on the quality of data. This chapter aims to bridge this research gap by developing a gener-

R. U. D. Refat · A. A. Elkhail · H. Malik (✉)
University of Michigan, Dearborn, USA
e-mail: hafiz@umich.edu

R. U. D. Refat
e-mail: rerafi@umich.edu

A. A. Elkhail
e-mail: abdkhail@umich.edu

 
Y. L. Murphey et al. (eds.), *AI-enabled Technologies for Autonomous and Connected Vehicles*, Lecture Notes in Intelligent Transportation and Infrastructure,
https://doi.org/10.1007/978-3-031-06780-8_20

alized machine learning pipeline designed to defend against existing and emerging cyberattacks on IVNs. The chapter starts with an overview of IVNs, threat modeling of IVNs followed by machine learning-based defense mechanisms against existing and emerging cyberattacks targeted at these IVNs. The last section of the chapter outlines future directions of using the proposed machine learning approach as a solution against vehicle-based cyberattacks for the next generation of vehicles.

## 1 Introduction

Connectivity and automation are foundational attributes of autonomous vehicles. Connectivity is the backbone of connected vehicle systems by which vehicles communicate. On the other hand, automation is the foundation of self-driving cars. These two technologies can improve the luxury features, functional safety and traffic mobility in vehicles. As a demonstration, recent work shows that these technologies can reduce traffic accidents and improve transport system efficiency [1]. According to the author in [2], connectivity and automation can be used to lower the rate of traffic accidents by lowering the rate of human errors. They can also be used to improve the lifestyle of citizens; for example by providing transportation for people who are not able to drive. Another important example is trucking and product delivery services. If human involvement can be reduced, there will potentially be less injury to humans associated with these risky tasks.

These two technologies are merged by the automotive industry to evolve the term *connected autonomous vehicle* (CAV). The connectivity gathers information from the vehicle's surroundings and passes it to the intelligent decision-making unit of the vehicle. The vehicle can take the right decision using this valuable information. Although we have not reached the highest level of automation, there are some vehicles in the market that are at level 3 (controlled automation). Fully automated vehicles i.e. level 5 automation is not possible unless the decision-making unit of the vehicle is trustworthy and capable of performing driving tasks like an actual human without fear of cyberattack. In order to do so, it is necessary to provide security to the vehicle in different layers, including (1) sensing layer or data gathering layer, (2) data processing layer and (3) onboard data-sharing layer. The scope of this chapter is focused on the security of the data-sharing layer inside a vehicle. This is called in-vehicle network security.

Recently, machine learning-based approaches are becoming popular for providing network security [3–5]. Machine learning algorithms are a good fit for this purpose since they analyze the data to generalize system behavior. These algorithms try to mimic the human learning system by finding patterns from past incidents and taking decisions according to the knowledge. Based on this, unwanted system behavior can be detected. That's why a machine learning algorithm is a perfect approach for attack detection of any kind of system. However, to build a machine learning system, sequential generalized steps should be considered. The overall workflow is called the machine learning pipeline, which is difficult to manage and time-consuming.

The main goal of this chapter is to present a description of the machine learning framework for in-vehicle network security. This chapter begins with a section that discusses the security analysis of CAVs. This is followed by a description of the in-vehicle networks that are commonly employed today. The chapter further discusses the architectural loopholes of the in-vehicle networks in terms of security. Furthermore, the security mechanism of an in-vehicle network is modeled with the generalized machine learning pipeline. Finally, this chapter includes a discussion of the challenges and future opportunities of using machine learning for in-vehicle network security. Our existing work on automotive cybersecurity [6–8] employs machine learning tools and techniques to secure in-vehicle networks. These methods have demonstrated good accuracy in detecting attacks. This chapter provides a detailed discussion of the most promising technologies that can provide in-vehicle network (IVN) security.

## 2 Security Analysis of CAVs

Current vehicular systems are considered intelligent when they perceive and respond to their surroundings in an efficient and safe manner, equalling or even surpassing human performance on the same task. These advanced vehicles gather and exchange information with other vehicles or infrastructures by establishing v2x communication. They use the gathered information with or without the captured data from sensors and provide advanced features like lane detection, cruise control, object detection, etc. To achieve these complex driving tasks, modern vehicles utilize electronic control units (ECUs) or mini embedded computers, known as the brain of the vehicle. State-of-the-art vehicles [9] currently utilize 70–100 ECUs. Depending on the data they receive as input, these ECUs control one or more sub-system functionality of the vehicle. Depending on the types of functionalities they perform, the ECUs can be divided into safety critical ECUs and non-safety critical ECUs. Routine control of the window or windshield wiper blades is not related to passenger safety, thus the ECUs controlling them fall into the category of non-safety critical ECUs. On the other hand, the ECUs that control the brakes, gas padel, etc. and can affect the life of passengers and pedestrians are obviously safety critical. Regardless of these differences. ECUs communicate among each other to control different vehicular systems efficiently through an automotive network (CAN bus) which has some known security vulnerabilities.

Until recently, the unguarded automotive network was considered safe, as the network was accessible only through the OBD-II port. Nowadays, vehicles can be connected with a variety of technologies, such as WiFi, Bluetooth, 3G, etc. which has caught the attention of hackers. Intruders can use these connectivity ports as a backdoor to access the automotive network. Recent work has shown that intruders can take control of the car remotely and can misguide the vehicle. From 2010 to 2018 there were several high profile reports where hackers gained access to the vehicle using wireless or wired connectivity ports. In 2010, Karl Koscher and his team performed an experiment by taking control of a car using the OBD port [10]. A recent report in 2018 also indicates that hackers were able to inject malware into a BMW vehicle via the OBD2 unit. In 2014 Stephen and his group showed that someone can remotely take control over a vehicle [11]. Miller and Valasek proved Stephen's conjecture and were able to misguide a Jeep Cherokee vehicle by establishing remote access in 2015 [12]. The Keen Security Lab of Tencent was able to gain access to a Tesla through the wifi/cellular network of the vehicle. They even injected malicious messages into the vehicle and were able to misguide not only a parked vehicle but also a driving one [13].

These attacks can be divided by two different categories.

Message injection attack: As the name suggests, a message injection attack happens when the attacker injects a suspicious message into the CAN bus system and pretends to be an authorized entity in the vehicular system. For example, in a CAN bus system, by default, the CAN message does not contain sender or receiver information embedded in its packer, so the CAN bus will not be able to distinguish between the CAN message sent by an authorized or unauthorized ECU. By injecting the message, the attacker can misguide the vehicle and cause unintended behaviors like turning on the light indicators, unlocking the door, or decelerating the vehicle. This type of attack is also called a spoofing attack or fuzzy attack.

Denial of service (DoS) attack: The motivation of this attack is to prevent authorized ECUs from using the IVN to communicate with other ECUs. This type of attack is easy to implement as the goal is to make the bandwidth of the bus unavailable so that data can not transmit through the channel. For example. in a CAN bus system, the attacker can inject a high priority CAN message. Subsequently, it causes delays of other messages and causes threats in regards to availability with no reaction to the driver's commands since all ECUs share a single bus.

## 3 In Vehicular Networks (IVN)

Current automotive vehicles use multiple communication protocols to exchange information between ECUs, vehicle-to-vehicle (V2V), vehicle-to-infrastructure (V2I), and vehicle-to-pedestrians (V2P). The IVN protocols have their own characteristics, such as speed, specification, and use cases. They help to reduce the complex wiring system of a car, making it easier to maintain, thereby reducing cost. This section presents the conventional automotive networking protocols such as LIN, MOST, CAN, FlexRay, and Ethernet.

### *3.1 LIN Protocol*

Local Interconnect Network (LIN) was introduced in the late 90's as a low-cost alternative of CAN bus protocol to connect components in a car [14]. It allows serial communication in a master-slave architecture and has a data rate of 20–25 kbit/s. However, due to the low bandwidth of data communication, it is impractical to use in a high-speed communication system. But it became popular to use for connecting non-critical subsystems where the speed of communication is not considered as an issue because the protocol itself is low cost. It has been used in subsystems, such as adjusting seats, mirrors, or controlling car windows, etc. [15].

### 3.2 MOST Protocol

Media Oriented System Transport (MOST) is a high-speed multimedia network protocol and was developed in 1988 by MOST corporation [9]. By architecture, it supports up to 64 ECUs to connect in a ring topology structure. It has a maximum data rate of 24.8 Mbit/s [15]. Due to the support of high-speed communication, it became popular for use in car infotainment systems. MOST has been used by renowned car manufacturers, such as BMW, Mercedes-Benz, Porsche, Audi, Volkswagen, Jaguar, Hyundai, Toyota, Land Rover and many others [16].

### 3.3 FlexRay Protocol

FlexRay is another in-vehicle communication protocol and was first introduced in vehicles by BMW in 2006 [9]. It has a data rate of 10 Mbit/s. This data rate is high enough to support high-speed communication between ECUs and is used for applications like active suspensions, adaptive cruise control, etc. [15]. BMW 7 Series is the first car that used FlexRay fully in the car [9]. Other interesting characteristics of FlexRay are, it can fit any kind of topological network structure.

### 3.4 CAN Bus Protocol

Controller area network (CAN) is the most widely used in-vehicle communication protocol. It was developed by the German company Robert Bosch GmbH in 1980 [17] and was first made public in 1986 [9]. The introduction of the CAN protocol dramatically solved the complex wiring issue in a vehicle and it became popular in a short time. By architecture, it is a broadcasting system and any ECU can access the bus at any time. For their access control, Carrier Sense Multiple Access with Collision Detection (CSMA/CD) is used with the help of a few bits in the CAN message frame called arbitration ID [18]. In terms of speed, it has a data rate of 1 Mbit/s. Due to its simplicity and easy plug and play feature it is used in communication between safety-critical ECUs and is used by most of the leading car manufacturers in the world.

### 3.5 *Ethernet Protocol*

Another popular IVN used nowadays is the Ethernet. It is a physical network that connects different components of a vehicle. The popularity of Ethernet is demonstrated by its recent use by Hyundai, BMW and Volkswagen [19]. As a communication protocol, Ethernet has been used in computer networks for decades, but was not considered for the automotive industry because of OEM's automotive requirement. But extensive research to develop driverless vehicles and camera-based ADAS systems has convinced the automotive industry that its components need high speed and high bandwidth buses. As a result, IEEE 802.3 Ethernet was introduced by automotive makers. Modern software and electronics in vehicles can introduce exciting new functionalities because of the speed and bandwidth offered by Ethernet. Unlike the CAN protocol, the Ethernet protocol has sender and receiver information in the message packet and is considered more secure.

## 4 Security Analysis of IVN Architecture

The section security analysis of CAV shows that vehicle connectivity can be leveraged by intruders and they can misguide the vehicle. It is considered a serious issue because the driver's and/or passenger's life can be at risk if malicious hackers can remotely take control of the vehicle. In this section, we discuss the architectural loopholes in each of the IVNs outlined above, save for the MOST protocol, which, to best to our knowledge, has no history of demonstrated attacks.

### 4.1 *LIN*

LIN is known as a low-cost IVN that connects non safety critical sensors and ECUs in a vehicle. By architecture, it is a single wire system and has a UART serial interface [20]. Modern vehicle systems connect a master and several slaves to form a LIN network and they send two types of messages in the bus: (1) unconditional message frame and (2) event triggered message frame. In an unconditional message frame, the master specifies a slave to respond with a message and the slave follows the command. The event-triggered message frame is sent to the bus by the master when information is requested from the slave. The main difference between the unconditional message frame and the event-triggered message frame is the response of the slaves. Unlike the unconditional message frame, the slave did not respond in an unchanged state upon receiving the event trigger message frame. From an intruder point of view, this feature can be used to perform attacks. As the master initiates action for all the slaves, gaining access to one of those gives the attacker access to the bus. Apart from this, another type of attack has been reported by researchers: sending the sleep command [21].

Although considered as a unique advantage of the LIN communication protocol, the ability to enter sleep mode to save power can be exploited by the attacker to disable the LIN bus system in a vehicle. Lastly, the normal state of a LIN bus system can be hampered by an attacker using electromagnetic interference as it is a single wire system [20].

### 4.2 FlexRay

FlexRay is another IVN protocol that provides high-speed communication between ECUs. It has been used recently in several applications such as adaptive cruise control and active suspensions [15]. It is tailored to the demands of today's automotive industry, with features such as flexible data transfers, support for any kind of topological network structure, fault-tolerant operation, and greater data throughput than prior standard protocols. It is also a dual-channel system that supports both asynchronous and real-time data transfer modes [22]. Although it provides some protection features of data availability and data integrity since it uses CRC as a kind of data protection against transmission mistakes, these features do not imply any guarantee of data confidentiality, authenticity, or freshness. A FlexRay network, in reality, does not have directly accessible interfaces; instead, it connects to other network protocols through gateways. Access to the FlexRay bus can be gained through such gateways. This makes the FlexRay protocol insufficiently protected against attacks and makes the FlexRay bus a likely target for attackers since the ECUs attached to it is utilized to provide control and mobility in the vehicles. These cyberattacks can target the in-vehicle network's control and maneuverability ECUs, causing significant harm to the driver. A variety of attacks have been easily created and developed on the FlexRay standard protocol including the Nilsson-Larson attacker model [23], in which an adversary has access to the in-vehicle network via the wireless gateway and can read, modify, flood, steal, drop, monitor, record, broadcast, spoof and replay messages. Also, the Man-in-the-middle attacks, or intercepting and dropping messages, are possible with such an adversary [24].

### 4.3 CAN (Controller Area Network)

Controller area network (CAN) is known as the de facto standard for IVN communication. Architecturally it is a broadcasting system that means any CAN messages sent to the bus can be accessed by all the components connected to the network. This makes the vehicular networking system more simple and solves the complex wiring problem. But the design of the protocol lacks authentication features because it does not have any field containing sender or receiver information in their message frame. A typical standard CAN data frame has 111 bits at most. Out of them, it has a unique arbitration ID (11 bits), that is used to control the accessibility of the CAN bus by establishing a priority scheme. According to the scheme, a lower arbitration ID has a higher priority to send messages to the bus. That means if two ECUs try to send CAN message to the bus at the same time, the protocol lets the sender with a lower arbitration ID send messages first before the other one. Theoretically, if an ECU sends a CAN message with an arbitration ID of 0000 continuously, then the other ECUs will not get a chance to transmit messages to the bus. There is also a 15 bit CRC field in the CAN data frame that is calculated from the data fields (0-8 bytes), but that only protects the data loads. By default, there are no other ways to provide security to the CAN message frame. The attacker can take advantage of the CAN bus protocol characteristics discussed above and manipulate the bus. In particular, there are three important properties of the protocol that an intruder can utilize to perform state-of-the-art attacks [25]. First, the broadcasting nature provides a way for the attackers to read all the CAN messages in the network without getting noticed. Second, the priority scheme of the protocol can be misused by the intruder and a denial of service attack can be performed by sending the highest priority CAN message continuously. Finally, due to the lack of sender identification, an attacker can act as an authorized ECU and can send CAN messages to the bus to perform a spoofing attack.

## *4.4 Ethernet*

The Ethernet has recently been employed in camera-based ADAS systems and self-driving vehicles [26]. It is a physical network that allows various components of a vehicle to be connected to each other. The Ethernet protocol offers greater features compared to the standard CAN protocol such as speed, flexibility, bandwidth, cost-effectiveness, and interoperability. The Ethernet protocol is also considered more secure compared to the standard CAN protocol since it relies on the TCP/IP protocol, which includes sender and receiver information in the message packet, which eliminates the need to broadcast each message. However, because Ethernet has long been the actual standard protocol for linking computers, years of expertise in hacking computers may be applied to hacking vehicles. Furthermore, the Ethernet protocol might be vulnerable to attacks since it relies on TCP/IP protocol which focuses on communication for resource sharing. A variety of attacks have been implemented on TCP/IP protocol, including source address attacks, sequence number spoofing attacks and mad authentication attacks [27]. Another possible issue is that open-source protocol is not highly appreciated in the automotive sector due to common copyleft rules that compel companies to publicly reveal code updates and enhancements. Furthermore, the lack of validation of the open-source protocol implementations may cause further implementation issues, which potentially expose serious vulnerabilities.

# 5 Machine Learning as Defence Mechanism

Although the IVNs are not security proven at least in terms of their architecture, nevertheless they have been used in automotive systems for decades and the production of such vehicles is increasing due to high market demand. For providing a better passenger experience, these vehicles are equipped with intelligent software and are treated as cyber-physical systems. As the safety of passengers is directly related to the security IVN, there are two popular approaches that can increase the security of safety-critical IVNs (i.e. CAN bus). The first is implementing an encryption algorithm to secure the CAN bus channel. The second is to monitor the network traffic data and/or analog CAN signal and report unusual behavior. One process of providing solutions by using CAN data is machine learning technology, which is now widely used in the domain of network security. In this section, a brief description of the intersection point between machine learning and IVN security will be discussed. The section will present an overview of the machine learning pipeline when it is used in IVN security. The overall diagram of the machine learning pipeline is displayed in Fig. 1.

As the CAN bus in modern vehicles is exposed to a large number of threats and becomes an attractive target for attackers, the need for an Intrusion Detection System (IDS) for the CAN bus is becoming one of the most important security components in modern vehicles. The existing IDSs for the CAN bus can be divided into behavior-based IDS and fingerprinting-based IDS. Behavior-based IDS operate on the data link layer and fingerprinting-based IDS operate on the physical layer. Behavior-based IDS are used to monitor network traffic data in the data link layer and report unusual behavior while fingerprinting-based IDS is used to utilize the physical characteristics of the analog CAN signal in the physical layer and report any anomalies. One process of using the network traffic data and the physical characteristics of the analog CAN signal in order to build an automated solution is machine learning technology, which is currently becoming one of the most popular technologies in the domain of security.

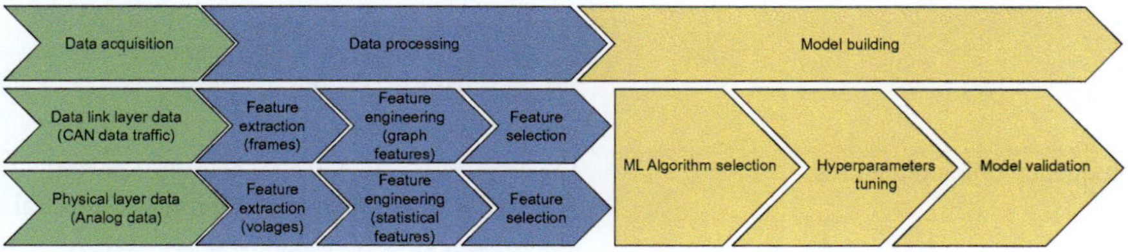


**Fig. 1** Machine learning pipeline in CAN security

## 5.1 CAN Data Acquisition

The first step for implementing a machine learning system to defend the CAN bus against cyberattacks is to acquire CAN data. This data acquisition is a complex process where the researcher needs to access the CAN bus of the vehicle first. To access the CAN bus, a standardized connector to a vehicle is needed: OBD-II. Since 2006, the OBD-II port was required for every consumer vehicle and was mandated by law in the USA. The OBD-II port has 16 pins and they are summarized in Table 1. The CAN—H and CAN—L pins can be wired to a microcontroller unit like an Arduino with the help of a CAN bus shield (CAN bus transceiver) to capture raw CAN bus data traffic. There are also a few commercial dongle interfaces available that can be used to capture raw CAN bus data like panda OBD-II OBD2 interface, OpenXC, OBDLink SX, etc. The streamed CAN bus raw data can be captured in the data link layer of Open Systems Interconnection (OSI) model as network traffic and is presented in Fig. 2. As the data acquisition from an actual vehicle needs to be accessed through the OBD-II port, preparation should be taken for safety according to [28]. To avoid this complex task people have also used car simulators with a virtual CAN like the popular ICSim. Researchers in [28–30] have gathered CAN bus traffic data using the ICSim simulator. The simulator simulates a car's instrument cluster, e.g. speedometer, its controls, e.g. throttle and steering, and the corresponding CAN traffic. It is built based on the SocketCAN—a Linux-based library for the CAN network. According to [30], the CAN traffic generated from ICSim seems like real vehicular traffic. On the other hand, the CAN bus data can be collected in the physical layer of the OSI model shown in Fig. 2 that is the analog data (voltages). To collect the analog CAN signals data TivaC, TM4C123GXL, MCP, TriCore, NXP MPC, STM32, and oscilloscope instruments can be used [6, 7, 31–36].

**Table 1** Physical layer features

| Feature type | Feature name |
|---|---|
| Time-domain features | Maximum, Minimum, Mean, Variance, Std.-Dev., Average deviation, Non-negative count, Zero Crossing Rate (ZCR), Root Mean Square (RMS), Amplitude, Energy, Power, Skewness, Kurtosis |
| Frequency-domain features | Spectral std.-dev., Spectral entropy, Spectral spread, Spectral flux, Spectral roll off, Spectral skewness, Spectral brightness, Spectral kurtosis, Spectral flatness, Spectral centroid, Irregularity K |
| Signal descriptive features | Ratio max plateau, Plateau, Overshoot height, Maximum |
| Deep features | Deep features were extracted using Recurrent Neural Network (RNN) |

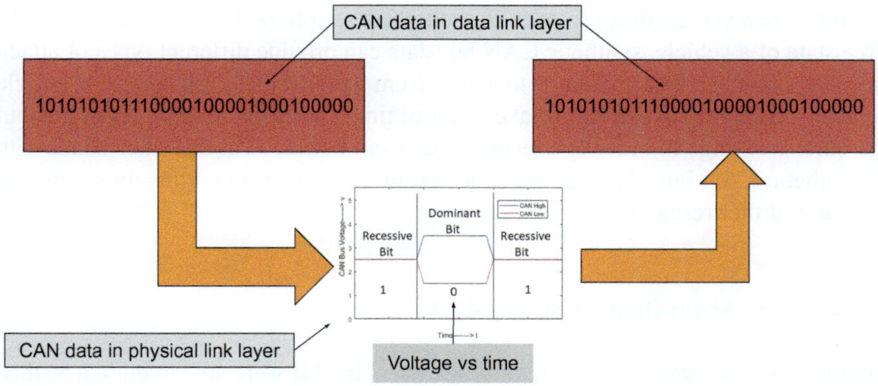


**Fig. 2** CAN bus data acquisition in data link layer and physical link layer

### 5.1.1 Available Public Dataset

While capturing CAN bus traffic in the data link layer is pretty straightforward, there are few variations when analog CAN signals are captured. The researchers in [7, 31, 36] have used an oscilloscope with 2 GS/s in order to collect 144k samples, 4.147 M samples, and 3.15 M samples of the CAN signals where the CAN signals are captured with different CAN cable types with various lengths, as well as different number of ECUs with the same input of the CAN bus message. Similarly, work in [32, 33] have used an oscilloscope with 2.5 and 1 GS/s, in order to record the CAN signals frames. Other work in [35] have used an MCP2515 controller and an MCP2551 transceiver and by using a 20 MS/s, they were able to capture 56.56k frames of the CAN signals where the CAN signals were recorded at the output of ten ECUs. Similarly, work in [6] also used the MCP2515 controller and the MCP2551 transceiver to capture the CAN signal and they were able to capture 48.128k frames of the CAN signals of five ECUs by using only 2 MS/s. Additional work in [34] have used a TM4C123GXL microcontroller integrated with TivaC instrument and by using 50 MS/s, they were able to collect 10k frames of the CAN signal of ten ECUs. Unfortunately, none of the aforementioned datasets are publicly available for researchers. On the other hand, for the data link layer, the CAN traffic is captured and is publicly available for researchers. The CAN datasets are presented in Table 2.

**Table 2** Available public datasets

| Dataset reference | CAN data source type |
|---|---|
| [37] | KIA soul |
| [38] | Ford escape |
| [39] | Synthetic CAN bus data |

While real vehicle data provides an accurate description of the attacked or attack-free state of a vehicle, synthetic CAN bus data can provide different types of attack scenarios that can be a challenge to gather from a real vehicle. By design, a vehicle is a very complex system and it takes a lot of time and effort to learn enough about its inner workings in order to execute a successful attack. On the other hand, with a synthetic CAN bus, data can be collected in a cost and time-efficient manner to conduct further research.

#### 5.1.2 Data Acquisition Using Test-Bed

According to the state-of-the-art, the researchers in [34] were able to build a testbed consisting of ten Arduino Unos, each with two identical CAN shields, and use a TM4C123GXL microcontroller combined with a TivaC instrument with using 50 MS/s in order to capture CAN signal frames. Other works in [6, 35] were able to create testbeds in order to collect the CAN signals frames from the Fiat 500 and the Porsche Panamera vehicles. The testbeds in [6, 35] consist of ten Arduino Unos and five Arduino Unos, respectively. Each Arduino Uno is equipped with two identical CAN shields and both work in [6, 35] used an MCP2515 controller and an MCP2551 transceiver using 20 and 2 MS/s, respectively in order to record CAN signal frames from both the Fiat 500 and the Porsche Panamera vehicles. Additionally, two Raspberry Pis where each one is equipped with a CAN Shield were connected to increase the number of ECUs. One of the Raspberry Pis was connected to the OBD-II port and the second one was connected to the CAN bus in order to capture CAN signal frames.

### *5.2 Data Processing*

Data processing is a step of using domain knowledge to extract information (characteristics, properties, attributes) from raw data. The success of machine learning depends on this information which makes the learning easier if the extracted characteristics correlate with the target class. On the other hand, if the class is a very complex function of the extracted information, one may not be able to learn it properly. Often, the raw data is not in a form that is amenable to learning, but suitable features can be constructed. In machine learning, the extraction of salient features is what accounts for most of the development effort. It is also one of the most interesting parts of development, where intuition, creativity, and "black art" are as important as the technical requirements. To implement machine learning, it is important to choose attributes that are different between benign CAN bus traffic and malicious CAN bus traffic. This subsection will provide an overview of the feature extraction, engineering and selection process when applying machine learning in the domain of CAN bus security.

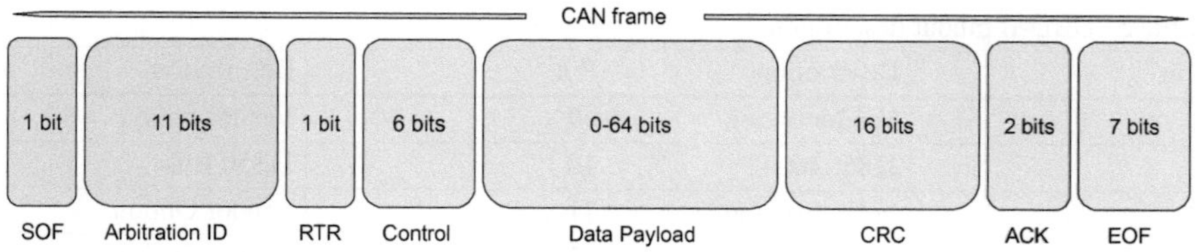


**Fig. 3** CAN bus message frame

### 5.2.1 Feature Extraction

To extract important information from the CAN data, it is important to first understand CAN bus traffic. For the standard CAN bus protocol, each ECU sends a CAN message of at most 111 bits. Out of the 111 bits, 11 bits are known as the arbitration ID, 0-8 bytes contain the data payload, and the remaining bits consist of a start bit, RTR bit, CRC bit and end bits, etc. Figure 3 shows a packet of CAN messages. These portions of CAN bus messages have been used as machine learning features for years. In particular, the CAN arbitration ID and the data fields have been used in [8, 40, 41] for detecting benign and malicious CAN bus messages. Apart from these traditional features, authors in [42] used the timestamp of each message and used it as a feature for classification.

### 5.2.2 Feature Engineering

The above subsection shows that there are few features in the CAN bus data. Obviously, too few features can not show the complexity of the data, which will affect intrusion detection performance. So, to increase the performance of machine learning models, the idea is to generate features that represent hidden differentiable characteristics of the CAN bus. To do this, one popular approach is to construct entropy-based features [43, 44], where two or more features are used to generate a new feature that improves the attack detection rate of the model. Another popular technique is to extract graph theory-based features from CAN message [25, 45]. This interesting approach considers a window of CAN messages to construct a graph and then extracts graph-based attributes like a number of edges, nodes, degree of nodes, etc. to use as machine learning features. For example, in [45], the author took batches of 200 CAN messages into consideration and built a graph using them. He further used these graphs to gather features i.e. number of edges, number of nodes, radius, diameter, density, reciprocity, average cluster coefficient, and assortativity coefficient. The result section shows the effectiveness of feature engineering in CAN security. The graph-based features approach shows better performance than traditional CAN bus message features by at most 1.44% when using the same machine learning algorithm as the attack detector.

**Table 3** OBD-II pinout description

| Pin | Description | Pin | Description |
|---|---|---|---|
| 1 | Vendor option | 9 | Vendor option |
| 2 | J1850 bus+ | 10 | J1850 Bus |
| 3 | Vendor option | 11 | Vendor Option |
| 4 | Chassis ground | 12 | Vendor Option |
| 5 | Signal ground | 13 | Vendor Option |
| 6 | CAN (J-2234) high | 14 | CAN (J-2234) low |
| 7 | ISO 9141-2 K-line | 15 | ISO 9141-2 low |
| 8 | Vendor option | 16 | Battery power |

When working with analog CAN signals, feature engineering is a mandatory step. According to the state-of-the-art, statistical features were extracted from CAN analog voltages and used in intrusion detection. Fingerprinting-based IDSs [6, 7, 31–36] utilize this approach in order to detect anomalies, where these statistical features represent the signal fingerprint in addition to the associated ECU. These statistical features can be classified into four main types: time-domain features, frequency-domain features, signal descriptive features, and deep features. These statistical features are presented in Table 3.

### 5.2.3 Feature Selection

The last important task before training the model is to select appropriate features. This process can be done manually or automatically. The main goal of this process is to select the features that correlate with the target variable. This process is sometimes considered of lesser importance than the feature extraction or engineering steps. In fact, by using irrelevant features the model memorizes rather than generalizes the problem from the training data. Thus the inclusion of non-salient features decrease the accuracy of the model. In CAN bus security research, the two most common approaches are selecting features based on feature differences [45] or based on a feature rank score on the model prediction [41]. For example, in [45], the author plotted the graph features as box plots for attack-free and attacked graphs and selected 7 features out of the 8 features which have high feature differences in terms of data distribution. The feature involving the number of nodes in the graph is the same for both the attacked and attack-free graphs, hence does not have a significant effect on the model's prediction. To determine a ranking score between the features using a machine learning algorithm on the training data is another way for selecting the features. Tree-based algorithms calculate the importance of each feature based on every single tree and then average the output of the trees to make the result more reliable. Additionally, different traditional feature selection methods such as information gain, entropy, and the Gini coefficient can be utilized also for this purpose.

## 5.3 *Algorithms to Detect a CAN Bus Attack*

In a machine learning project, the next task after feature selection is modeling. It is the process of constructing a mathematical equation by iterating a set of data to find a perfect line that can either categorize the data or predict future values. This process can be broken down into three parts: algorithm selection, hyperparameter tuning, and model validation. In the following subsections, the description of these three procedures will be explained in terms of CAN bus data.

### 5.3.1 Algorithm Selection

The selection of machine learning algorithms in CAN data security is dependent on the quality of data and available differential features. In a modern vehicle, there are 70–100 ECUs and they communicate with each other very frequently, hence we can assume there is a large amount of data available for detecting CAN bus attacks. KNN, decision trees, or kernel SVM algorithms can be used. Refat et al. [45], Alshammari et al. [46] used KNN and kernel SVM, [8, 41] used decision trees to detect CAN bus attacks and achieved high accuracy in detection. The authors in [39, 47, 48] used deep learning-based algorithms to detect CAN bus intrusion which is computationally expensive as shown in Table 4.

When working with physical CAN signals, according to the state-of-the-art, fingerprinting-based IDSs [6, 7, 31–36], work in [34, 35] used an Artificial Neural Network (ANN) algorithm to detect anomalies in the CAN bus. Similarly, work in [7, 31] also used an ANN algorithm to detect anomalies in the CAN bus, where 70 and 65% of the collected samples were used for training the model and the 30 and 35% of the collected samples were used for testing. Other work in [36] used a recurrent neural network with long short-term memory (RNN-LSTM) to detect anomalies in the CAN bus. Additional work in [32, 33] used a Support Vector Machine (SVM), ANN, and Bagged Decision Tree (BDT) to detect any anomalies in the CAN bus. Similarly, work in [6] used the same algorithms as [32, 33] in addition to Logistic Regression (LR) and Naive Bayes to detect anomalies in the CAN bus.

**Table 4** Experimental environment for deep learning-based algorithms

| Paper | CPU | GPU (GHz) | RAM (GB) |
|---|---|---|---|
| [39] | 3.50 | N/A | 32 |
| [47] | 3.60 | Yes | 32 |
| [48] | 2.20 | Yes | 16 |

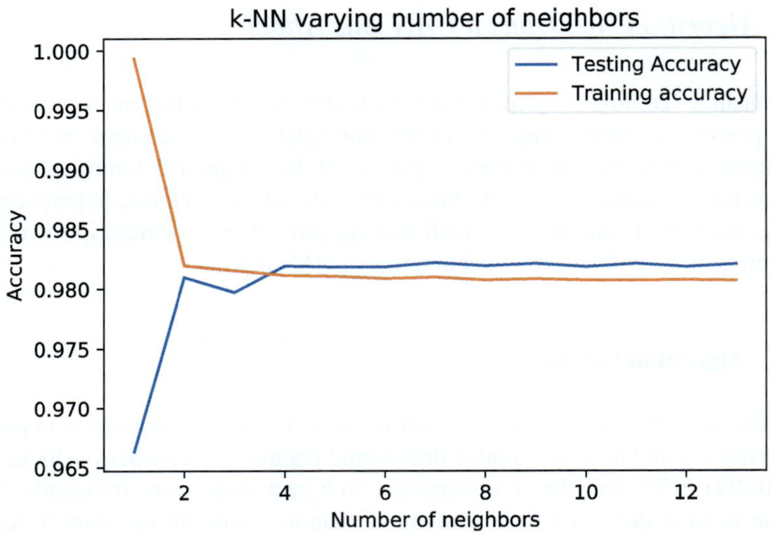


**Fig. 4** Tuning number of nearest neighbors (k) in KNN algorithm

### 5.3.2 Hyperparameter Tuning

To train an unbiased, highly efficient machine learning model, it is important to tune the hyperparameters of the selected algorithm and select the parameters before moving to model validation. In CAN bus security, state-of-the-art papers do not explain the hyperparameter tuning process, hence an example is given how the hyperparameter can be selected using the KNN algorithm. This will help to interpolate the model and provide an idea of how tuning can be done. In the KNN algorithm, $k$ is a parameter that needs to be tuned, but unfortunately, there is no conventional way to tune the parameter. The go-to approach is to try different values of $k$, monitor the training and test error rates and finally choose the $k$ that yields the smallest amount of test and training errors. Figure 4 shows the results of applying the KNN algorithm with different values of $k$ with the selected features in paper [45] while using the same public dataset [37]. According to the plot the value of $k = 4$ is chosen on the training data as it provides the smaller training and test errors. The entire hyperparameter selection process can be automated as shown by the authors in [49, 50].

### 5.3.3 Model Validation

Once the hyperparameters are set, the model is finally trained and ready to be validated. It is always advisable to validate the model with an unknown set of data. This will help the researcher to understand whether the model has successfully generalized from the training data or it has only been memorized. To execute this step, the total data set can be divided into three chunks into a ratio of 60%:20%:20%: (1)

training data, (2) test data and (3) validate data. The training data and test data are used till the hyperparameter tuning and the validation dataset will be used to measure the final performance of the model. For example, with $k = 4$, the KNN achieved an accuracy of 98.00% when working with 8 graph-based CAN features from [45].

## *5.4 Challenges of ML in IVN Research*

The first challenge of applying machine learning to secure IVN is to supply quality network data to the model. CAN data acquisition is not a straightforward task because access to the network of the vehicle is required to capture such data. Acquiring network data from an actual vehicle is a complex process, hence people usually build vehicular networking prototypes which is also challenging. Because this process requires analog data acquisition devices, such as an oscilloscope, it can be expensive and not very portable. The oscilloscope can be replaced by low-cost microcontrollers, but they require additional programming. To capture quality physical layer data, a higher sampling rate should be used to read analog voltages, which is also a challenging task [6]. On the other hand, while using the automotive network's data link layer and data in the machine learning model, feature engineering needs particular attention because of the lack of direct features. Another challenge is monitoring and detecting any anomalies in the CAN bus in real-time, where the majority of research so far has been done to detect intrusions using datasets and is not suitable for real-time detection. Moreover, there is no well-known and widely recognized dataset that can be used to assess the effectiveness of intrusion detection systems for vehicles and unfortunately, no datasets for the physical layer are publicly available for researchers. The limited computing power of ECUs to process complex machine learning to detect anomalies in IVN is also challenging.

## *5.5 Future Opportunities in This Domain*

One critical threat to modern vehicles is malware, which is malicious software designed to gain unauthorized access to data or to disrupt computer operations. Malware can infect vehicles via a variety of interface vulnerabilities, including wireless connections with roadside networks, Wi-Fi hotspots, Internet connectivity, Bluetooth, and cellular networks like 5G. It can also infect vehicles through cell phones, removable media, iPods and laptops that are connected directly with the vehicle's network [11]. Malware can cause a wide range of disturbances and harm to the vehicle system once it is inside the vehicle [51]. Some examples of how malware affects the vehicle's regular operations are: tampering with the in-car radio so the driver can't turn it on, locking automotive functions such as locking the car doors, occupying the memory and CPU cycles of the ECUs, and disabling the vehicle's safety features [51]. The above-mentioned examples are considered high priority that must be appropriately handled in order to effectively safeguard them. According to the state-of-the-art [52], machine learning approaches are successful in production level applications for defending malware, and we expect machine learning techniques can be used in the future to address such cases and effectively protect the vehicle systems from malware.

## 6 Conclusion

The advances in the technologies related to CAV are the main catalyst behind the research works in the intelligent transportation system. This chapter shows that researchers are achieving significant results in securing the in-vehicle network. The introduction of CAV will reduce human-related errors and misjudgements in decision-making while driving by providing more control to the vehicle on the road. The transition to autonomous driving has the potential to prevent thousands of lives lost to road crashes, save millions of work hours lost to road congestion, improve the environment by reducing carbon emission, and make our lives better by providing the flexibility of mobility without the worries of the driving and parking tasks. At the same time, the errors made by the self-driven vehicle need to be minimized. To eliminate the external bias of an unauthorized entity on the decision of a self-driven vehicle, a strong trustworthy intrusion detection system is needed. Data-centric machine learning algorithms can be a good choice for building such systems. The success of machine learning-based systems depends not only on the selection of algorithms but also on the quality of the underlying data. This book chapter discusses the generalized steps for developing a machine learning system when working with in-vehicular network security. The factors involved in the development of a high quality machine learning-based solution were indicated. The challenges such as the difficulty of data acquisition (in-vehicle network), scarcity of differential features, and real-time attack detection are also specified. Finally, the future direction of machine learning in an in-vehicular network was pointed to the area of defending malware in CAVs.

**Acknowledgements** The authors extend their appreciation to the Deputyship for Research and Innovation, Ministry of Education in Saudi Arabia for supporting this work through the project # DRI-KSU-934. This research is also partly supported by National Science Foundation (NSF) under the award #2035770.

## References

1. Bajpai J (2016) Emerging vehicle technologies & the search for urban mobility solutions. Urban Plan Transport Res 4:83–100
2. Elliott D, Keen W, Miao L (2019) Recent advances in connected and automated vehicles. J Traffic Transport Eng (English Edn) 6:109–131

3. Mulinka P, Casas P (2018) Stream-based machine learning for network security and anomaly detection. In: Proceedings of the 2018 workshop on big data analytics and machine learning for data communication networks. pp 1–7
4. Sommer R, Paxson V (2010) Outside the closed world: on using machine learning for network intrusion detection. In: 2010 IEEE symposium on security and privacy, pp 305–316
5. Furdek M, Natalino C, Lipp F, Hock D, Di Giglio A, Schiano M (2020) Machine learning for optical network security monitoring: a practical perspective. J Lightwave Technol 38:2860–2871
6. Kneib M, Schell O, Huth C (2020) EASI: edge-based sender identification on resource-constrained platforms for automotive networks. In: Network and distributed system security symposium (NDSS), pp 1–16
7. Hafeez A, Ponnapali S, Malik H (2020) Exploiting channel distortion for transmitter identification for in-vehicle network security. SAE Int J Transport Cybersecur Privacy 3:5–17
8. Minawi O, Whelan J, Almehmadi A, El-khatib K (202) Machine learning-based intrusion detection system for controller area networks
9. Jadhav S, Kshirsagar D (2018) A survey on security in automotive networks. In: 2018 Fourth international conference on computing communication control and automation (ICCUBEA), pp 1–6
10. Koscher K, Czeskis A, Roesner F, Patel S, Kohno T, Checkoway S, McCoy D, Kantor B, Anderson D, Shacham H, Savage S (2010) Experimental security analysis of a modern automobile. In: 2010 IEEE symposium on security and privacy. pp 447–462
11. Checkoway S, McCoy D, Kantor B, Anderson D, Shacham H, Savage S, Koscher K, Czeskis A, Roesner F, Kohno T et al (2011) Comprehensive experimental analyses of automotive attack surfaces. USENIX security symposium, vol 4, pp 447–462
12. Miller C, Valasek C (2015) Remote exploitation of an unaltered passenger vehicle. In: Black Hat USA, vol 2015
13. Nie S, Liu L, Du Y (2017) Free-fall: Hacking tesla from wireless to can bus. Briefing, Black Hat USA 25:1–16
14. Hafeez A, Topolovec K, Zolo C, Sarwar W (2020) State of the art survey on comparison of CAN, FlexRay, LIN protocol and simulation of LIN protocol. In: SAE technical paper
15. Talic A (2017) Security analysis of ethernet in cars
16. Muyshondt H, Sanchez C (2009) MOST—The audio/video backbone of the car. In: Audio engineering society conference: 36th international conference: automotive audio
17. Johansson K, Törngren M, Nielsen L (2005) Vehicle applications of controller area network. In: Handbook of networked and embedded control systems, pp 741–765
18. Islam R, Refat R (2020) Improving CAN bus security by assigning dynamic arbitration IDs. J Transport Secur 13:19–31
19. IXIA automotive ethernet: an overview. (IXIAcom.com), https://support.ixiacom.com/sites/default/files/resources/whitepaper/ixia-automotive-ethernet-primer-whitepaper_1.pdf
20. Ernst J, Michaels A (2018) LIN bus security analysis. In: IECON 2018—44th annual conference of the IEEE industrial electronics society, pp 2085–2090
21. Kleberger P, Olovsson T, Jonsson E (2011) Security aspects of the in-vehicle network in the connected car. In: 2011 IEEE intelligent vehicles symposium (IV), pp 528–533
22. Consortium F et al (2005) FlexRay communications system protocol specification. Version 2:198–207
23. Nilsson D, Larson U (2008) Simulated attacks on can buses: vehicle virus. IASTED International conference on communication systems and networks (AsiaCSN), pp 66–72
24. Püllen D, Anagnostopoulos N, Arul T, Katzenbeisser S (2019) Security and safety co-engineering of the FlexRay bus in vehicular networks. In: Proceedings of the international conference on omni-layer intelligent systems, pp 31–37
25. Islam R, Refat R, Yerram S, Malik H (2020) Graph-based intrusion detection system for controller area networks. IEEE Trans Intell Transport Syst
26. Huo Y, Tu W, Sheng Z, Leung VA (2015) Survey of in-vehicle communications: requirements, solutions and opportunities in IoT. In: 2015 IEEE 2nd world forum on internet of things (WF-IoT), pp 132–137

27. Bellovin S (1989) Security problems in the TCP/IP protocol suite. ACM SIGCOMM Comput Commun Rev 19:32–48
28. Payne B (2019) Car hacking: accessing and exploiting the CAN bus protocol. J Cybersecur Educ Res Practice 2019:5
29. Saber A, Di Troia F, Stamp M (2021) Intrusion detection and can vehicle networks. In: Digital forensic investigation of Internet of Things (IoT) devices. pp 125–154
30. Schappin C (2017) Intrusion Detection on the Automotive CAN bus. Technische Universiteit Eindhoven
31. Avatefipour O, Hafeez A, Tayyab M Malik H (2017) Linking received packet to the transmitter through physical-fingerprinting of controller area network. In: 2017 IEEE workshop on information forensics and security (WIFS). pp 1–6
32. Choi W, Jo H, Woo S, Chun J, Park J, Lee D (2018) Identifying ECUs using inimitable characteristics of signals in controller area networks. IEEE Trans Veh Technol 67:4757–4770
33. Choi W, Joo K, Jo H, Park M, Lee D (2018) VoltageIDS: low-level communication characteristics for automotive intrusion detection system. IEEE Trans Inf Forensic Secur 13:2114–2129
34. Foruhandeh M, Man Y, Gerdes R, Li M, Chantem T (2019) SIMPLE: single-frame based physical layer identification for intrusion detection and prevention on in-vehicle networks. In: Proceedings of the 35th annual computer security applications conference, pp 229–244
35. Kneib M, Huth C (2018) Scission: signal characteristic-based sender identification and intrusion detection in automotive networks. In: Proceedings of the 2018 ACM SIGSAC conference on computer and communications security, pp 787–800
36. Yang Y, Duan Z, Tehranipoor M (2020) Identify a spoofing attack on an in-vehicle CAN bus based on the deep features of an ECU fingerprint signal. Smart Cities 3:17–30
37. Lee H, Jeong S, Kim H (2017) OTIDS: a novel intrusion detection system for in-vehicle network by using remote frame. In: 2017 15th annual conference on privacy, security and trust (PST). pp 57–5709
38. Miller C, Valasek C (2013) Adventures in automotive networks and control units. Def Con 21:15–31
39. Hanselmann M, Strauss T, Dormann K, Ulmer H (2020) CANet: an unsupervised intrusion detection system for high dimensional CAN bus data. IEEE Access 8:58194–58205
40. Avatefipour O, Al-sumaiti A, El-sherbeeny A, Awwad E, Elmeligy M, Mohamed M, Malik H (2019) An intelligent secured framework for cyberattack detection in electric vehicles' CAN bus using machine learning. IEEE Access 7:127580–127592
41. Yang L, Moubayed A, Hamieh I, Shami A (2019) Tree-based intelligent intrusion detection system in internet of vehicles
42. Theissler A (2017) Detecting known and unknown faults in automotive systems using ensemble-based anomaly detection. Knowl-Based Syst 123:163–173
43. Tian D, Li Y, Wang Y, Duan X, Wang C, Wang W, Hui R, Guo P (2018) An intrusion detection system based on machine learning for CAN-bus. In: Industrial networks and intelligent systems, pp 285–294
44. Tian D, Li Y, Wang Y, Duan X, Wang C, Wang W, Hui R, Guo P (2017) An intrusion detection system based on machine learning for CAN-bus. In: International conference on industrial networks and intelligent systems, pp 285–294
45. Refat R, Elkhail A, Hafeez A, Malik H (2021) Detecting CAN bus intrusion by applying machine learning method to graph based features. In: Proceedings Of SAI intelligent systems conference
46. Alshammari A, Zohdy M, Debnath D, Corser G (2018) Classification approach for intrusion detection in vehicle systems. Wirel Eng Technol 9:79–94
47. Seo E, Song H, Kim H (2018) GIDS: GAN based intrusion detection system for in-vehicle network
48. Hossain M, Inoue H, Ochiai H, Fall D, Kadobayashi Y (2020) LSTM-based intrusion detection system for in-vehicle can bus communications. IEEE Access 8:185489–185502
49. Yogatama D, Mann G (2014) Efficient transfer learning method for automatic hyperparameter tuning. In: Artificial intelligence and statistics, pp 1077–1085

50. Feurer M, Hutter F Hyperparameter optimization. In: Automated machine learning, pp 3–33
51. Zhang T, Antunes H, Aggarwal S (2014) Defending connected vehicles against malware: challenges and a solution framework. IEEE Internet of Things J 1:10–21
52. Gardiner J, Nagaraja S (2016) On the security of machine learning in malware c&c detection: a survey. ACM Comput Surveys (CSUR) 49:1–39

